Extreme depositional environments: Mega end members in geologic time

Edited by

Marjorie A. Chan
Department of Geology & Geophysics
135 S. 1460 E.
University of Utah
Salt Lake City, Utah 84112
USA

and

Allen W. Archer
Department of Geology
Kansas State University
Manhattan, Kansas 66506
USA

THE
GEOLOGICAL
SOCIETY
OF AMERICA

Special Paper 370

3300 Penrose Place, P.O. Box 9140 ▪ Boulder, Colorado 80301-9140 USA

2003

Published by The Geological Society of America, Inc.
3300 Penrose Place, P.O. Box 9140, Boulder, Colorado 80301
www.geosociety.org

Printed in U.S.A.

GSA Books Science Editor: Abhijit Basu

Library of Congress Cataloging-in-Publication Data

Extreme depositional environments : mega end members in geologic time / edited by Marjorie A. Chan and Allen W. Archer
p. cm. — (Special paper ; 370)
Includes bibliographic references and index.
ISBN 0-8137-2370-1 (pbk)
1. Sedimentation and deposition. 2. Geology, stratigraphic. I. Chan, Marjorie A. II. Archer, Allen W. (Allen William), 1952- III. Special Papers (Geological Society of America) ; 370.

QE571.E97 2003
551.3'03—dc21

2003048532

Cover: This image depicts the Jurassic Navajo Sandstone, the largest erg of North America. Photograph is from Coyote Buttes of the Paria Wilderness area near the Utah-Arizona border. Photo by Marjorie A. Chan.

10 9 8 7 6 5 4 3 2 1

Contents

1. ***Introduction: A look at extreme depositional systems*** 1
   Marjorie A. Chan and Allen W. Archer

2. ***Timing of late Paleozoic glaciation in Gondwana: Was glaciation responsible for the development of northern hemisphere cyclothems?*** 5
   John L. Isbell, Molly F. Miller, Keri L. Wolfe, and Paul A. Lenaker

3. ***Subglacial outburst floods and extreme sedimentary events in the Labrador Sea***
   John Shaw and Jerome-Etienne Lesemann 25

4. ***Limits on extreme eolian systems: Sahara of Mauritania and Jurassic Navajo Sandstone examples*** 43
   Gary Kocurek

5. ***Quaternary loess-paleosol sequences as examples of climate-driven sedimentary extremes*** 53
   Daniel R. Muhs and E. Arthur Bettis III

6. ***Lessons from large lake systems—Thresholds, nonlinearity, and strange attractors*** 75
   Kevin M. Bohacs, Alan R. Carroll, and Jack E. Neal

7. ***Organic carbon burial by large Permian lakes, northwest China*** 91
   Alan R. Carroll and Marwan A. Wartes

8. ***Features and origin of the giant Cucomungo Canyon alluvial fan, Eureka Valley, California*** 105
   Terence C. Blair

9. ***Desmoinesian coal beds of the Eastern Interior and surrounding basins: The largest tropical peat mires in Earth history*** 127
   Stephen F. Greb, William M. Andrews, Cortland F. Eble, William DiMichele, C. Blaine Cecil, and James C. Hower

10. ***Highest tides of the world*** 151
    Allen W. Archer and Mary S. Hubbard

11. ***Giant submarine canyons: Is size any clue to their importance in the rock record?*** 175
    William R. Normark and Paul R. Carlson

12. ***Remnant-ocean submarine fans: Largest sedimentary systems on Earth*** 191
    Raymond V. Ingersoll, William R. Dickinson, and Stephan A. Graham

13. ***Megareefs in Middle Devonian supergreenhouse climates*** 209
Paul Copper and Christopher R. Scotese

14. ***Origin and evolution of large Precambrian iron formations*** 231
Bruce M. Simonson

15. ***Extreme paleoceanographic conditions in a Paleozoic oceanic upwelling system: Organic productivity and widespread phosphogenesis in the Permian Phosphoria Sea*** 245
Eric E. Hiatt and David A. Budd

Geological Society of America
Special Paper 370
2003

Introduction: A look at extreme depositional systems

Marjorie A. Chan
Department of Geology & Geophysics, 135 S. 1460 E., University of Utah, Salt Lake City, Utah 84112, USA

Allen W. Archer
Department of Geology, 108 Thompson Hall, Kansas State University, Manhattan, Kansas 66506, USA

WHAT IS EXTREME?

Although we live in a world of norms, we are commonly fascinated by the unusual and the extreme. Attempting to understand the extreme includes the desire to stretch the envelope, to reach the outer limits, to extend our imagination, or to step into a world beyond. In the process, humans push to the edge and attempt to grasp the unusual and the unexplained. Sometimes we fall short and fail.

An extreme sedimentary system is one that can be described with such adjectives as unique, rare, distinctive, intense, radical, of great severity, drastic, giant, mega, long-lived, aerially extensive, unusually thick, unparalleled, unexplained, or perhaps all of the above. Within an extreme sedimentary environment, a depositional setting must attain the greatest, highest, and/or utmost degree that extends far beyond the norm.

Although microenvironments may sometimes be extreme environments, they are biological in origin, and we will not attempt to address extreme conditions for life in this volume. The study of biological microenvironments is a whole different endeavor, and one that others address in the arenas of biology, geomicrobiology, astrogeology, cryptozoology, and other areas of specialization.

In terms of modern and ancient sedimentary environments, an extreme depositional system is one that stands out from all others and is unique in terms of size, scale, or other attributes. In examining sedimentary extremes, we wish to push the envelope, particularly in terms of depositional systems. We try to stretch our thinking beyond our normal methodical descriptions and limited calculations. We strive to examine sedimentary rocks from a different perspective and beyond the traditional facies models in an attempt to look for new or unique analogs, whether they be modern or ancient.

Why narrowly focus on extreme depositional systems? First, recognition of extreme depositional systems allows us to better understand the range, scales, and variability of the geologic record. Secondly, it will allow us to better isolate what the most extreme conditions or controls might be. Obviously, the basic controls, such as tectonics, climate, sea level, and even biology, are fairly well known and understood. But do we know which controls might pull a system out of line, out of the norm, or out of the ordinary? This kind of study can help clarify what the boundary conditions and limits are and to what extent they dominate or affect a depositional system. Third, an understanding of the controls on a depositional system will enable us to better define the role and magnitude of processes—wind, waves, currents, etc.—in an environment. Fourth, there may be limits on a depositional system. Are there any limits, either theoretical and/or real? Part of the wonder of earth science is the fact that much of what we see astounds us. It is not uncommon for systems to exceed what we can imagine, as we are limited by our biased knowledge of the ongoing modern world and present-day processes. Here we can attempt to address the issue of exactly what our earth is capable of producing. Fifth and finally, the recognition of the extreme systems and understanding their controls will provide insights that can be used to better model geologic systems whether for understanding geologic history or for predictions and practical applications of resource exploration.

This book is very different from the topical theme books that focus on regional areas, particular processes, or single depositional environments. Instead, this volume is aimed at taking a broad look at depositional systems in a relevant and stimulating manner. Herein, we concentrate on what controls the extreme states. This volume covers a gamut of diverse environments and tectonic settings, yet still within the context of a coherent theme. The settings and controlling parameters described herein offer much potential for modeling. This kind of synthesis can appeal to students who are investigating depositional environments. A student may casually wonder, "How big are meandering river channels?" Similarly, a professional geologist may grapple with such questions as, "What are the biggest meandering channels, and why do they evolve to that state?"

Chan, M.A., and Archer, A.W., 2003, Introduction: A look at extreme depositional systems, *in* Chan, M.A., and Archer, A.W., eds., Extreme depositional environments: Mega end members in geologic time: Boulder, Colorado, Geological Society of America Special Paper 370, p. 1–4.

IDENTIFICATION OF EXTREME DEPOSITIONAL SETTINGS

To identify the extremes, we might look at the depositional systems that have one of the following characteristics:

1. Great lateral and aerial extent
2. Exceptional volume
3. Great thickness
4. Unusual geometries
5. Long temporal span
6. Magnitude of processes

Some of these characteristics may exceed other examples from the same depositional environments by an order of magnitude or more.

Difference between Uniformitarianism and the "Norm"

Uniformitarianism and the related concepts of actualism and gradualism are inherently useful, but they may constrain our understanding of sedimentary extremes because some extremes may exist only in the ancient record with no modern analog. To understand an extreme depositional setting, we must first recognize what would make the extreme setting different, and how these extremes compare to the norms and the commonplace. Derek Ager's (1973) book, *The Nature of the Stratigraphic Record,* helped us review and reevaluate uniformitarianism, which is commonly and simply defined as "the present is the key to the past." Our view, which is based upon our limited human existence, is certainly biased by the present. As we examine modern processes, we are struck by the repetitive nature, cyclicity, and seasonality of winds, waves, and sea-level change. Yet there are some geological deposits that do not seem to be repeated either in size, scale, magnitude, composition, or extent.

Here we attempt to look at these anomalous deposits with the hope that a new look at the geologic extremes will truly help us understand the processes. Does uniformitarianistic thinking prevent us from seeing the forest because we are only able to comprehend the individual trees? Our facies models tend to focus on the "norms," yet it might be the extremes that can actually tell us more about important earth processes. We recognize that the stratigraphic record is punctuated, episodic, and perhaps slanted by the rare and unusual events that have better preservation potential. However, the extreme examples may hold the key to isolating the most important boundary conditions that can yield important clues to geologic limits.

Boundary Conditions

What are the important boundary conditions that control the geologic record? Our standard answers invariably invoke factors such as climate, tectonics, sediment supply, sea level, and biological activity, where biology affects initiation, affects the rates, or provides feedback. Yet not all extreme systems are affected by the same simple control or complex, interrelated controls. Each type of depositional setting may be more susceptible or responsive to specific controls as compared to others. Here we may get a better idea of the importance, on one hand, and the degree of influence of the controlling factors. Boundary conditions will provide constraints and some limits. For example, we might think that climate would greatly affect lacustrine systems, since continental environments are highly susceptible to climate change. Conversely, marine systems are more buffered by the large oceanic body. However, studies here (Bohacs and others, this volume) indicate that the role of tectonics greatly overshadows that of climate. The extent to which controls affect both the formation and development of depositional systems gives us boundary conditions for what causes the initiation of systems, what sustains them over time, and what—in combination with other factors—makes them develop. Forward modelers have long sought help in defining the boundary conditions and their effect. Here we can help sort the magnitude of the effects for certain parameters.

Input Parameters to Modeling

Input parameters to modeling may include basin tectonics, subsidence, time, climate conditions, and sediment delivery. These input parameters may not "pull" or exert the same control as the boundary conditions, but they still may affect the internal characteristics of the environment and the deposits. Some of the typical parameters of sediment input may vary from the direction, magnitude, and strength to velocity. Our thinking even on directional parameters may require reexamination. Traditional ideas of proximal and distal equated to grain size may not be a simple solution. Recent studies on deep-water clastics and the understanding of turbidite processes now suggests that the long-held idea that coarsest material is closest to the source is not as simple or as straightforward as we have thought in the past. These deep-water studies suggest, in fact, that more sand may be pushed farther into the basin (e.g., Beaubouef et al., 1999).

Extreme Depositional Systems Versus Convulsive Events

How do extreme depositional systems differ from extreme catastrophic or convulsive events? Extreme depositional environments might represent conditions that existed over a relatively long period of geologic time, long enough to be preserved in the geologic record. The processes in the environment are also likely to be relatively continuous in order to produce temporally widespread and/or mega-sized environments. In some instances, this might even be at the scale of sequences or longer, spanning even several higher frequency orders of global climate or sea-level change. A depositional system involves multiple processes even if one process is the stronger or the most dominant. There is likely a convergence of processes and/or events—either continually or episodically—that, within unusual conditions, reinforce each other and result in giant-scale systems.

In contrast, convulsive events are likely to be short, episodic, and reflect discontinuous events in geologic time. Thus, these convulsive events are likely to be temporarily limited, representing only seconds to minutes to perhaps days or weeks. The control on a convulsive event is likely to be a single process such as a catastrophic flood, an earthquake, or a meteorite impact. Single events might be likely to happen more than once in geologic time where it could be uncommon, but still repeated in some forms in geologic history, even if at different scales. A number of professional meeting sessions and other books have already addressed catastrophic, convulsive events in the stratigraphic record, which have a high likelihood of preservation (e.g., Clifton, 1988).

Conversely, extreme environments have received little discussion in the literature. In fact, it really requires workers who are very familiar with the literature and have observed a number of similar depositional systems to synthesize the magnitude of a system and be able to compare it with other depositional systems.

RATIONALE

Is it possible that mere humans can comprehend the largest depositional phenomena of all geologic time? Given the immensity of time that has passed, even low probability events become near certainties. Were the biggest rivers only on the biggest supercontinents? Are there theoretical limits regarding the maximum extent and thickness of river, reef, erg, evaporite, lacustrine, tillite, or flood deposits? How wide or deep can a river channel cut and why? How big can a reef grow? How much sand can the wind pile up? What was the world's biggest flood? Even the experts can't answer some of these questions, although we may be able to make educated guesses. In some instances, we have to take the most extreme case study and try to answer some of the basic questions before we can truly determine if it is, in fact, the one-and-only ultimate. How do we identify the greatest environmental and/or facies extremes? What are their dimensions, controls, and, more specifically, how did they evolve to this extreme state? Does the rock record help provide boundary conditions for such immense-scale events? Are they reaching theoretical limits, and/or are they telling us something very important about dynamic processes that have been unmatched by a convergence of the right factors?

Questions to Address

For each paper in this volume, we asked authors to address the following questions:

1. What makes your particular deposit unique in the stratigraphic record? Why are there no other examples either modern or ancient of the type and magnitude you discuss?

2. What are the conditions required to create the extremely large, in area or extent, depositional system? Is it a convergence of one or more factors?

3. What are the most important controls: tectonics, eustasy, climate, biology, and/or others? Qualify and/or quantify the contribution of those controls.

Papers here focus on depositional systems and phenomenon that perhaps were not as catastrophic, but simply represent some extreme conditions that may produce large depositional systems for which we have few or no analogs. We have asked experts and synthesizers in their fields to write, in each paper, about one of the most extreme examples they can think of and to try to give a perspective on why or how this example is extreme. Not every depositional environment is extreme in the same way; some may be extreme in size, some may be long-lived, still others may be unusually thick.

Within this volume, the papers are arranged to first present the clastic systems, ranging from non-marine to marine (ever in the seaward direction). These are subsequently followed by carbonate and chemical compositional systems. The non-marine extreme systems all show a consistent tectonic control, with climate making noticeable overprints on the larger tectonic framework. In the marine realm, active tectonics is again an important factor. Along the transition to the marine system—for example, shorelines and tidal systems—the depositional record is affected in large part by tectonics and the role that it plays on the geometry of shorelines and embayments. In the deep marine setting, climate and sea level impart weaker signals that are superimposed on tectonically controlled sediment packages. Lastly, in the carbonate and compositional systems, tectonics is necessarily subdued for the non-clastic environments to exist. There, other controls combine to affect the water chemistry and mineralogical growth.

The papers on continental clastic depositional systems begin with the climatically driven examples of late Paleozoic glaciation in Gondwana (Isbell, Miller, Wolfe, and Lenaker) and huge Laurentide glacial meltwaters (Shaw, Piper, Hesse, and Rashid), followed by large eolian records comparing the Jurassic of the western interior United States and the Sahara (Kocurek) and the loess deposits that indicate strong wind conditions (Muhs and Bettis). Lacustrine systems are examined from a quantitative perspective (Bohacs, Carroll, and Neal) as well as from a mega-lake example from the Permian of northwest China (Carroll and Wartes). Giant alluvial fans in active tectonic settings (Blair) and thick Pennsylvanian coals of the Eastern Interior Basin (Greb, Andrews, Eble, DiMichele, Cecil, and Hower) round out the continental systems.

In the marine realm, a paper on giant tidal systems (Archer and Hubbard) is followed by a discussion of giant submarine canyons and their deposits (Normark and Carlson) and perhaps the largest sedimentary system on Earth, remnant-ocean turbidite fans (Ingersoll, Dickinson, and Graham).

Carbonate and chemical compositional systems represent their own special conditions that include Siluro-Devonian mega-reefs in a super greenhouse (Copper and Scotese), Precambrian iron formations (Simonson), and phosphogenesis of the Permian Phosphoria Formation (Hiatt and Budd).

SUMMARY

This compilation of papers is a synthesis of some of the largest depositional systems, designed to stretch our thinking beyond our sometimes limited uniformitarian views. Herein, we attempt to explore the hows and whys of sedimentary events that exceed the present norms by as much as several orders of magnitude. The papers explore a range of sedimentary processes and deposits, from the present to the past, the normal to the unusual, and the rare to extreme. The focus on extremes necessitates examination of the forcing parameters that cause systems to evolve into an extreme state. Sometimes the extreme state may develop out of one main control. Other times it is a unique combination and coincidence of events that may never be repeated in such a fashion again for all of geologic time.

ACKNOWLEDGMENTS

We thank the many contributors to this volume, and some who also presented oral papers at Pardee Symposium (cosponsored by the Sedimentary Division of the Geological Society of America [GSA]) titled "Sedimentary Extremes: Modern and Ancient," which was held during the GSA Annual Meeting in Reno, Nevada, in November 2000. That symposium was the original impetus for this Special Paper. We also acknowledge the many reviewers who carefully read the manuscripts and offered constructive advice: J. Wellner, D. Sharpe, R. Dalrymple, T. Demko, L. Eisenberg, R. Lanford, N. James, J. Werne, P. Heckel, E. Kvale, S. Coleman, P. Jewell, W. Dean, J. Oviatt, M. Elrick, G. Stanley, N. Beukes, P. Link, O. Catuneanu, J. Collinson, J. O'Connor, H. Miller, G. Kukla, J. Beget, D. Lowe, S. Shanmugam, M. Hendrix, and A. Miall. This volume would not be possible without the time and effort the reviewers provided. We thank Robert Dott Jr., Eric Roberts, and Cari Johnson for reviewing this introduction. Although this volume could not include every extreme depositional system, we hope it can form a springboard to better defining boundary conditions and input parameters of unique sedimentary systems.

REFERENCES

Ager, D.V., 1973, The nature of the stratigraphic record: New York, John Wiley & Sons, 62 p.

Beaubouef, R.T., Rossen, C., Zelt, F.B., Sullivan, M.D., Mohrig, D.C., and Jennette, D.C., 1999, Field Guide for AAPG Hedberg Field Research Conference: Deep-water sandstones, Brushy Canyon Formation, West Texas: Tulsa, Oklahoma, American Association of Petroleum Geologists.

Clifton, E., editor, 1988, Sedimentologic consequences of convulsive geologic events: Boulder, Colorado, Geological Society of America Special Paper 229, 157 p.

Manuscript Accepted by the Society January 22, 2003

Printed in the USA

Geological Society of America
Special Paper 370
2003

Timing of late Paleozoic glaciation in Gondwana: Was glaciation responsible for the development of northern hemisphere cyclothems?

John L. Isbell
Department of Geosciences, University of Wisconsin, Milwaukee, Wisconsin 53201, USA

Molly F. Miller
Department of Geology, Vanderbilt University, Nashville, Tennessee 37235, USA

Keri L. Wolfe
Paul A. Lenaker
Department of Geosciences, University of Wisconsin, Milwaukee, Wisconsin 53201, USA

ABSTRACT

The formation of upper Paleozoic (Viséan to Sakmarian-Artinskian) Euramerican cyclothems, which resulted from base-level fluctuations of up to 100 m, commonly are attributed to large-scale waxing and waning of Gondwanan glaciers. However, evaluation of the geographic and chronostratigraphic distribution of Gondwana deposits reveals that glaciation was not the primary cause of base-level changes of that magnitude.

Gondwana strata contain three non-overlapping glacial successions. Glacial I (Frasnian to possibly Tournaisian) and Glacial II (Namurian to lowermost Westphalian) rocks were deposited by alpine glaciers. Water sequestered by these glaciers was insufficient to account for the base-level changes. In contrast, Upper Carboniferous (Stephanian) to Lower Permian (Sakmarian-Artinskian) Glacial III rocks were widespread and indicate deposition by ice sheets that may have covered a total area of between 17.9 and 22.6 × 10^6 km^2. Complete ablation of a single ice sheet of this size would produce eustatic changes of ~100 m. However, multiple ice sheets were likely present, which would have resulted in considerably smaller fluctuations in sea level during Glacial III deposition.

The argument that Glacial I and II deposits were originally comparable in extent to those of Glacial III, but were eroded during the advance of Glacial III ice-sheets, is untenable. Weathered granite profiles on the pre-Glacial III unconformity occur scattered over a 1200-km length of the central Transantarctic Mountains. The profiles indicate prolonged subaerial exposure and, thus, an absence of ice cover. These and non-glacial successions in Gondwana constrain the size of ice sheets before Glacial III deposition and imply that glaciation prior to Glacial Episode III was not the primary cause of base-level changes linked to upper Paleozoic Euramerican cyclothems.

Keywords: cyclothems, glacioeustasy, Gondwana glaciation, Devonian, Carboniferous, Permian.

Isbell, J.L., Miller, M.F., Wolfe, K.L., and Lenaker P.A., 2003, Timing of late Paleozoic glaciation in Gondwana: Was glaciation responsible for the development of northern hemisphere cyclothems? *in* Chan, M.A., and Archer, A.W., eds., Extreme depositional environments: Mega end members in geologic time: Boulder, Colorado, Geological Society of America Special Paper 370, p. 5–24.

INTRODUCTION

Upper Paleozoic strata of both the northern and southern hemispheres contain distinctive rock suites. In North America and Europe, Carboniferous to Lower Permian rocks are characterized by cyclothems deposited in marine and nonmarine settings during multiple transgressive and regressive events. Southern hemisphere rocks of equivalent age are contained within basins scattered across Gondwana (Fig. 1) and consist of diverse facies, including strata deposited by ice sheets and alpine glaciers. Previous reconstructions of Gondwana glaciation imply waxing and waning of ice sheets during the late Paleozoic from the Late Devonian (Frasnian) to the Late Permian (Kazanian-Tatarian; Fig. 2; Veevers and Powell, 1987; Frakes and Francis, 1988; Veevers, 1994; Veevers et al., 1994c; Crowell, 1999) with the main glaciation extending from the Early Carboniferous (Viséan) to the Late Permian (Kazanian; Frakes and Francis, 1988; Crowell, 1999). Maximum glaciation was hypothesized to have occurred from the Middle Carboniferous (earliest Namurian) to the Early Permian (Fig. 2; either Sakmarian-Artinskian or Artinskian-Kungurian boundary; Veevers and Powell, 1987; Frakes and Francis, 1988; Crowell, 1999). Because of the apparent temporal synchroneity between cyclothems and glaciation, it is widely accepted that late Paleozoic sea-level fluctuations were caused by changes in ice sheet volume (e.g., Wanless and Shepard, 1936; Crowell, 1978; Veevers and Powell, 1987). The simplicity

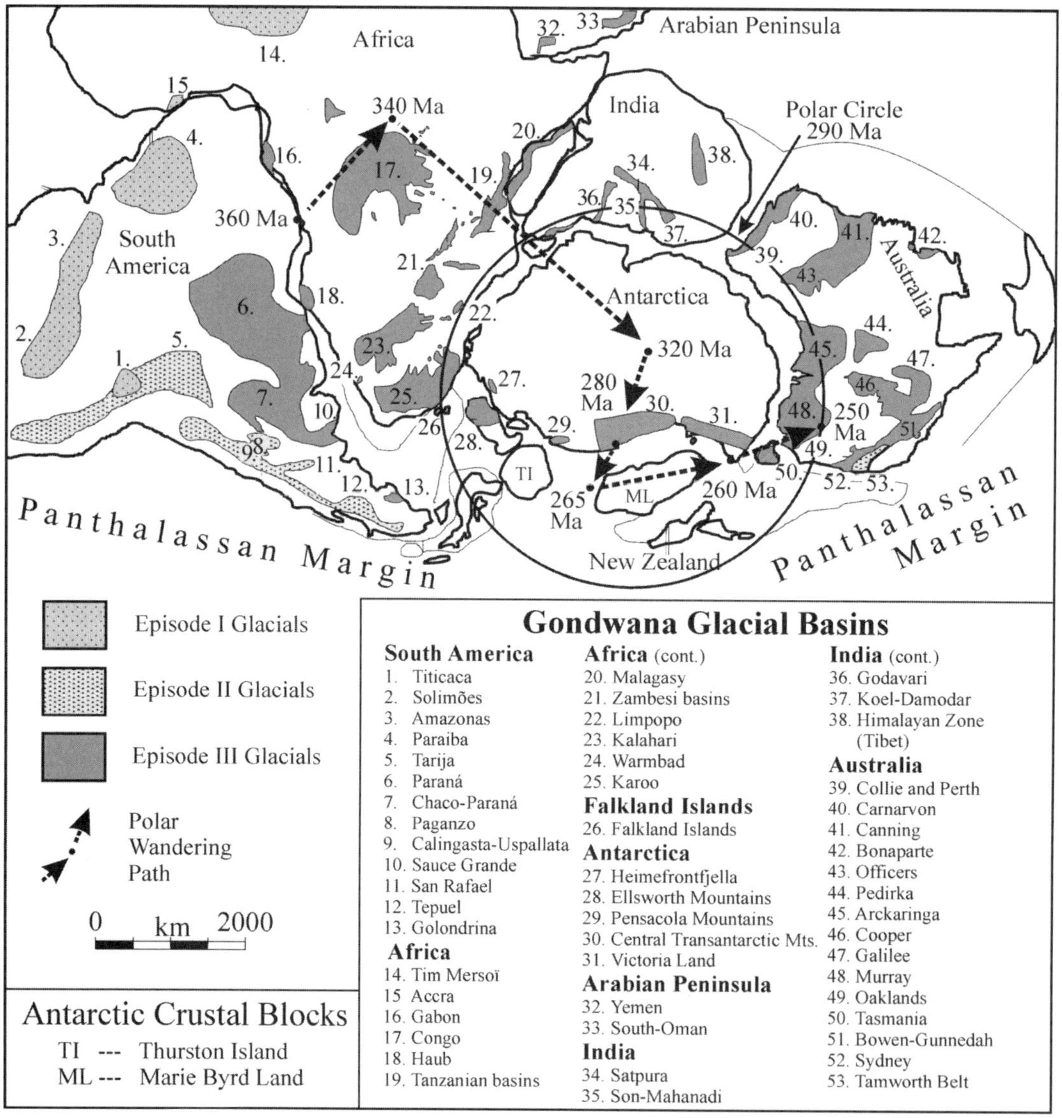

Figure 1. Reconstruction of Gondwana showing basins with Upper Paleozoic glacigenic successions deposited during three discrete glacial episodes. Reconstruction and polar wander path are from Powell and Li (1994).

of this explanation is so powerful that it never has been rigorously tested. However, before the cause and effect relationship between Gondwana glaciation and cyclothems can be accepted, at least two questions must be addressed: (1) was glaciation temporally continuous throughout the period of cyclothem deposition, and (2) were the late Paleozoic ice sheets large enough to account for the inferred changes in sea level?

In this paper, we summarize cyclothem deposition and examine the sedimentary record of late Paleozoic glaciation in the southern hemisphere to: (1) determine the timing of glacial deposition and its temporal relationships to northern hemisphere cyclothems; (2) constrain the type of glaciation that occurred and the size of the glaciers present to determine the influence of glacial events on sea level; and (3) assess the fidelity of the Gondwana glacigenic record and the extent to which that record either was removed by erosion during glacial advance, or accurately reflects the history of late Paleozoic glaciation.

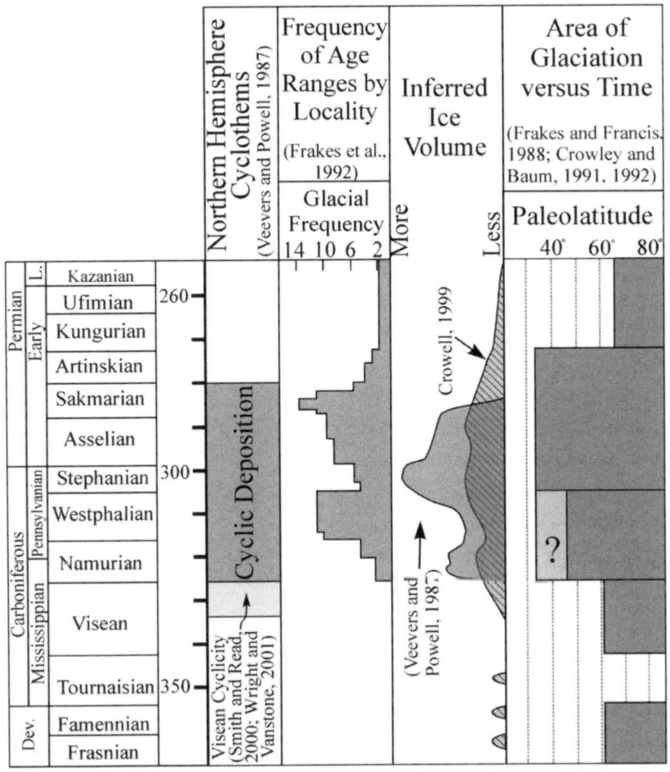

Figure 2. Previous temporal relationships between late Paleozoic cyclic deposition and ice sheet distribution/ice volume as inferred by Veevers and Powell (1987), Frakes and Francis (1988), Crowley and Baum (1991, 1992), Frakes et al. (1992), Crowell (1999), Smith and Read (2000), and Wright and Vanstone (2001). Carboniferous time scale is from Menning et al. (Time Scale B; 2000) and the Permian time scale is from Roberts et al. (1996).

CYCLOTHEMS AND BASE LEVEL

Cyclic sedimentation recorded in extensively exposed and well-studied Carboniferous cyclothems in the northern hemisphere reflect repeated fluctuations in base level (e.g., Wanless and Shepard, 1936; Ramsbottom, 1979; Ross and Ross, 1985; Chesnut, 1994; Heckel, 1994). These rocks were deposited in the equatorial regions of Pangea (cf., Scotese and McKerrow, 1990), and are known primarily from North America and Europe, where they form vertically stacked transgressive-regressive successions of nonmarine, nearshore, and offshore clastic and carbonate deposits (e.g., Moore, 1964; Ramsbottom, 1979; Maynard and Leeder, 1992; Heckel, 1994). Middle to Upper Carboniferous coal-bearing rocks, the cyclothems of Weller (*in* Wanless and Weller, 1932), are well-developed in central and eastern North America; marine dominated shales and carbonate successions in Kansas give way to mixed clastic-carbonate cycles in the Illinois Basin, which, in turn, give way to clastic-dominated, nonmarine and shallow marine deposits in the Appalachian Basin. Individual cyclothems are a few to tens of meters thick, and although individual units are not traceable over large distances, marine intervals and paleosols can often be correlated between the various basins (Heckel, 1994). In western North America (central Utah, Arizona, New Mexico, and Nevada) stacked successions of shoaling-upward carbonate and mixed carbonate/clastic cycles also suggest high-frequency eustatic changes (Dickinson et al., 1994; Langenheim, 1994; Soreghan, 1994). Cyclic deposits that display apparent transgressive-regressive synchroneity with the North American cyclothems occur in northwestern Europe, the Russian Platform, the Moscow Basin, and in the Ural Mountains (Ross and Ross, 1987, 1988).

Veevers and Powell (1987) reported that upper Paleozoic cyclic deposits equivalent to fifth-order sea-level fluctuations occur from the Early Carboniferous (Viséan-Namurian boundary) to the Early Permian (Sakmarian-Artinskian boundary). Recent work by Smith and Read (2000) on mixed carbonate/clastic rocks in the Illinois Basin (Brigantian) and by Wright and Vanstone (2001) on shallow marine platform carbonates in the United Kingdom (Early Asbian) extend that record farther back into the Early Carboniferous (Viséan).

Periodicities of eustatic cycles have been estimated to be between 44,000 and 4.3 million yr (Table 1). Except for cycles hypothesized to have periodicities > 412,000 yr, eustatic signals from Carboniferous rocks are thought to approximate the duration of Milankovitch cycles (cf., Heckel, 1994), and hence fluctuations in mass balance of glacial ice.

Eustatic amplitudes during the Carboniferous and Permian can be estimated using sedimentologic/stratigraphic, oceanographic, paleoecologic, and isotopic criteria. The depth of Lower Carboniferous incised valleys suggests changes in accommodation space of up to 95 m (Smith and Read, 2000), which is similar to the 100-m depth estimated by Heckel (1977, 1994) for sub-pycnoclinal accumulation of Middle to Upper Carboniferous phosphatic black shales in Kansas, Nebraska, Missouri, and Iowa, and amplitudes of 100+ m estimated from the preserved relief on Upper Carboniferous bioherms in the southwestern

TABLE 1. EUSTATIC PERIODICITIES ESTIMATED FROM UPPER PALEOZOIC CYCLOTHEMS

Location, rock type	Age	Cycle duration	References
United Kingdom, carbonates	Viséan	100 ka	Wright and Vanstone (2001)
Northern England, Pennine Basin	Namurian to Westphalian	120 ka	Maynard and Leeder (1992)
Eastern North America, Appalachian Basin, coal-bearing siliciclastic cyclothems	Lower to Middle Pennsylvanian (upper Bashkirian to Moscovian)	(a) 60–109 ka (b) 200–700 ka (avg. 400 ka) (c) 1.1–4.3 m.y. (avg. 2.5 m.y.)	Maynard and Leeder (1992) Chesnut (1994)
North American midcontinent, mixed clastic and carbonate cyclothems	Westphalian to mid-Stephanian	44–400 ka	Heckel (1986, 1994) Maynard and Leeder (1992)
Western United States, carbonate and mixed carbonates/clastic cycles	Stephanian	353–471 (avg. 412 ka)	Dickinson et al. (1994)
Northern Hemisphere, coastal onlap analysis	Carboniferous and Permian	2 m.y.	Ross and Ross (1987, 1988)

United States (Soreghan and Giles, 1999). Smaller eustatic amplitudes of ~60–70 m are suggested by the occurrence of benthic fusulinids within limestones (67 m; Moore, 1964), oxygen isotopes from Upper Carboniferous brachiopods within north Texas cyclothems (70 m; Adlis et al., 1988), and by area, volume, and water equivalence relationships of estimated late Paleozoic ice sheets (60 ± 15 m; Crowley and Baum, 1991). Larger eustatic fluctuations of between 100 and 200 m are estimated from studies of coastal onlap of Carboniferous and Permian strata (Ross and Ross, 1987).

Bruckschen and Veizer (1997), Mii et al. (1999), and Saltzman et al. (2000) reported large positive excursions in $\delta^{13}C$ and $\delta^{18}O$ (–0.4‰ to –3‰) values from Lower Carboniferous (Tournaisian) rocks in North America and Europe. High $\delta^{18}O$ values (–1‰ to –3‰) were also recovered from Middle to Upper Carboniferous (Namurian to Stephanian) rocks in North America (Mii et al., 1999). The $\delta^{13}C$ and $\delta^{18}O$ values suggest cool global temperatures and possible glaciation during much of the Carboniferous (Bruckschen and Veizer, 1997; Mii et al., 1999; Saltzman et al., 2000).

Summary: Cyclothems and Base Level

In summary, upper Paleozoic cyclothemic deposits are widely interpreted to have resulted from glacioeustasy (e.g., Wanless and Shepard, 1936; Heckel, 1986, 1990, 1994, 1995; Smith and Read, 2000; Wright and Vanstone, 2001). The Carboniferous data set suggests that most deposits display cyclicity between 44,000 to 412,000 yr, which is within the range of Milankovitch band orbital parameters (cf., Imbrie and Imbrie, 1980; Imbrie, 1985), and eustatic amplitudes of 60 to 100 m, which are within the range of sea-level changes during the Pleistocene (Fairbanks, 1989; Crowley and Baum, 1991). Therefore, many studies on upper Paleozoic rocks assume continuous waxing and waning of large ice sheets in Gondwana from at least the Early Carboniferous (Viséan) to the Late Permian (Kazanian), with maximum glaciation having occurred between the Middle Carboniferous (earliest Namurian) and the Early Permian (Fig. 2; either Sakmarian-Artinskian or Artinskian-Kungurian boundary; Crowell, 1978, 1999; Veevers and Powell, 1987; Frakes and Francis, 1988; Crowley and Baum, 1991, 1992; Frakes et al., 1992; Wright and Vanstone, 2001).

GLACIAL EUSTASY

The relationship between glacial mass balance and glacial eustasy is important in determining a link between upper Paleozoic cyclothems and Gondwana glaciation. During the Carboniferous and Permian, complete melting of an ice sheet covering an area of between 13.4 and 20.3×10^6 km^2 would have corresponded to a change in sea level of between 60 and 100 m (Fig. 3; cf., Crowley and Baum, 1991). However, because a single ice sheet contains more ice by volume than multiple ice sheets covering an equivalent area, the potential change in sea level for melting of a given area of ice cover decreases as the number of ice sheets increases (Fig. 3). Between one to 10 ice-spreading centers (ice domes, ice caps, and ice sheets) may have occurred in Gondwana during the late Paleozoic (cf., Crowell and Frakes, 1970; Veevers and Powell, 1987; Zeigler et al., 1997; Crowell, 1999). Therefore, depending on the volume of ice in each sheet, a change of 100 m in sea level required an area of ice cover of between 20.3 (single ice sheet) and 31.1×10^6 km^2 (10 equally sized ice sheets; Fig. 3).

Could alpine glaciation have caused the late Paleozoic changes in sea level? Assuming that the average alpine glacier covers an area of 1000 km^2 (100 km long × 10 km wide; a gross overestimate of ice volume for a single alpine glacier), the total ice cover required for a 100-m change in sea level is 198×10^6 km^2,

A Glacial area, ice volume, and sea level conversions

$$\log(V) = 1.23[\log(S) - 1]$$

S = Area of ice cover in km^2

$$WE = 0.917(V)$$

V = Volume

WE = Water equivalent

$$SLE = WE\ 374.5 \times 10^6\ km^2$$

SLE = Sea level equivalent

$$SLEI = SLE - 0.284(SLE)$$

SLEI = Sea level equivalent with isostasy

* 374.5 x10^6 km^2 is the estimated size of the Westphalian ocean

Procedures from Crowley and Baum (1991) and Paterson (1994)

B Single ice sheet -- area, volume, and sea level

Ice sheet area	Ice volume	Sea level equivalent	Sea level after isostasy
6.55 x 10^6 km^2	14.3 x 10^6 km^3	34.9 m	25 m
11.6 x 10^6 km^2	28.8 x 10^6 km^3	70.5 m	50.5 m
13.5 x 10^6 km^2	34.7 x 10^6 km^3	85 m	60.8 m
17.9 x 10^6 km^2	49.1 x 10^6 km^3	120.2 m	86 m
20.3 x 10^6 km^2	57.3 x10^6 km^3	140.3 m	100.4 m
22.6 x 10^6 km^2	65.4 x 10^6 km^3	160 m	114.6 m
30 x 10^6 km^2	92.6 x10^6 km^3	226.9 m	162.4 m
35.6 x10^6 km^2	114.3 x 10^6 km^3	280 m	200.5 m

C Comparison of area, volume, and sea level for a single vs. multiple ice sheets

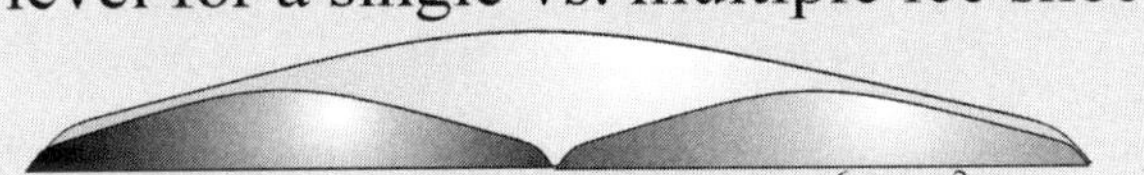

Ice cover area = 20.3 x 10^6 km^2

	1 Ice sheet	2 Ice sheets	10 Ice sheets
Volume	57.3 x 10^6 km^3	48.9 x 10^6 km^3	33.7 x 10^6 km^3
Sea level with isostasy	100.4 m	85.6 m	59.1 m

D Estimated areas of Gondwana ice cover and potential eustatic change (complete ablation of single or multiple ice sheets)

Area \ # of Ice sheets	1 Ice sheet	5 Ice sheets	10 Ice sheets
17.6 x 10^6 km^2	86 m	59.4 m	50.6 m
22.6 x 10^6 km^2	114.6 m	79.2 m	67.5 m

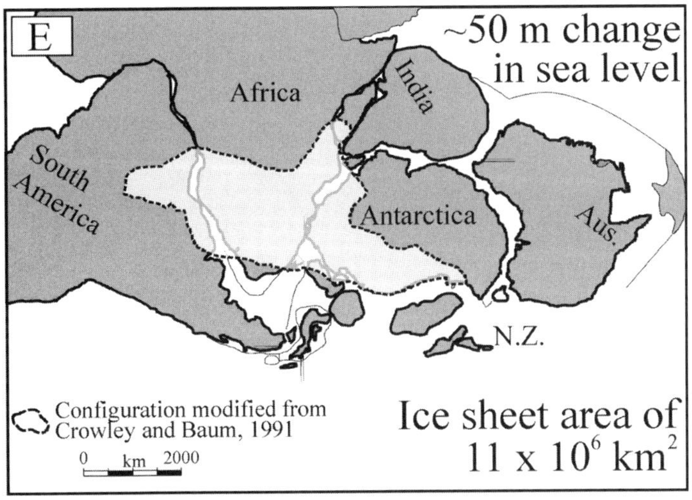

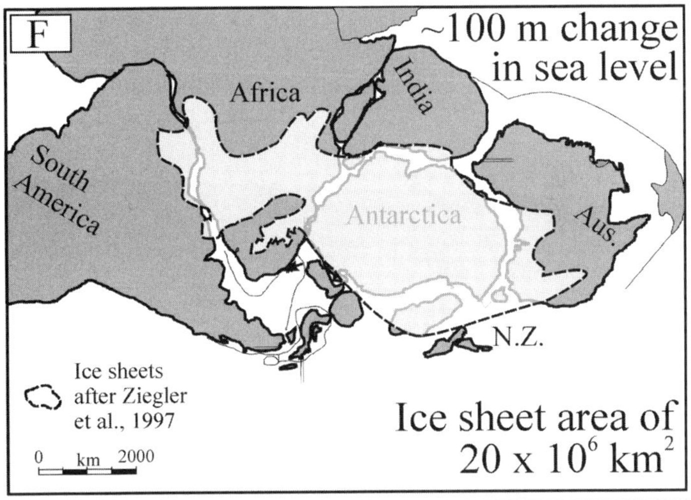

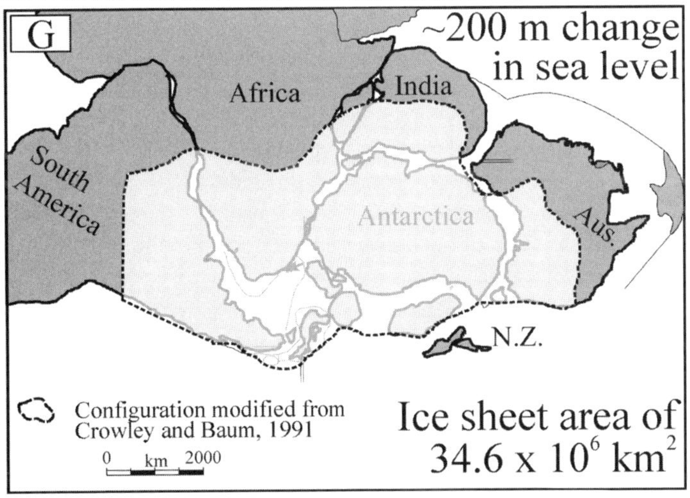

Figure 3. Relationships between areal extent of ice cover, ice volume, and glacioeustasy. A: Procedures for calculating ice volume and potential change in sea level for a known ice cover area. Procedures are those of Crowley and Baum (1991) and Paterson (1994). B: Calculated ice volumes and sea level equivalences for different areas of late Paleozoic ice cover following procedures given in Figure 3A. C: Effects of multiple ice sheets on ice volume and glacioeustasy for given area of ice cover. D: Relationship between estimated ice cover in Gondwana (e.g., Crowley et al., 1991; Crowley and Baum, 1992; Veevers, 1994; Ziegler et al., 1997) and eustatic change for single versus multiple ice sheets. E, F, and G: Hypothetical Gondwanan ice sheets of various sizes and their potential effect on sea level assuming entire ice sheet melted. No attempt is made here to suggest that these configurations were those of the ice sheets that covered Gondwana during the late Paleozoic.

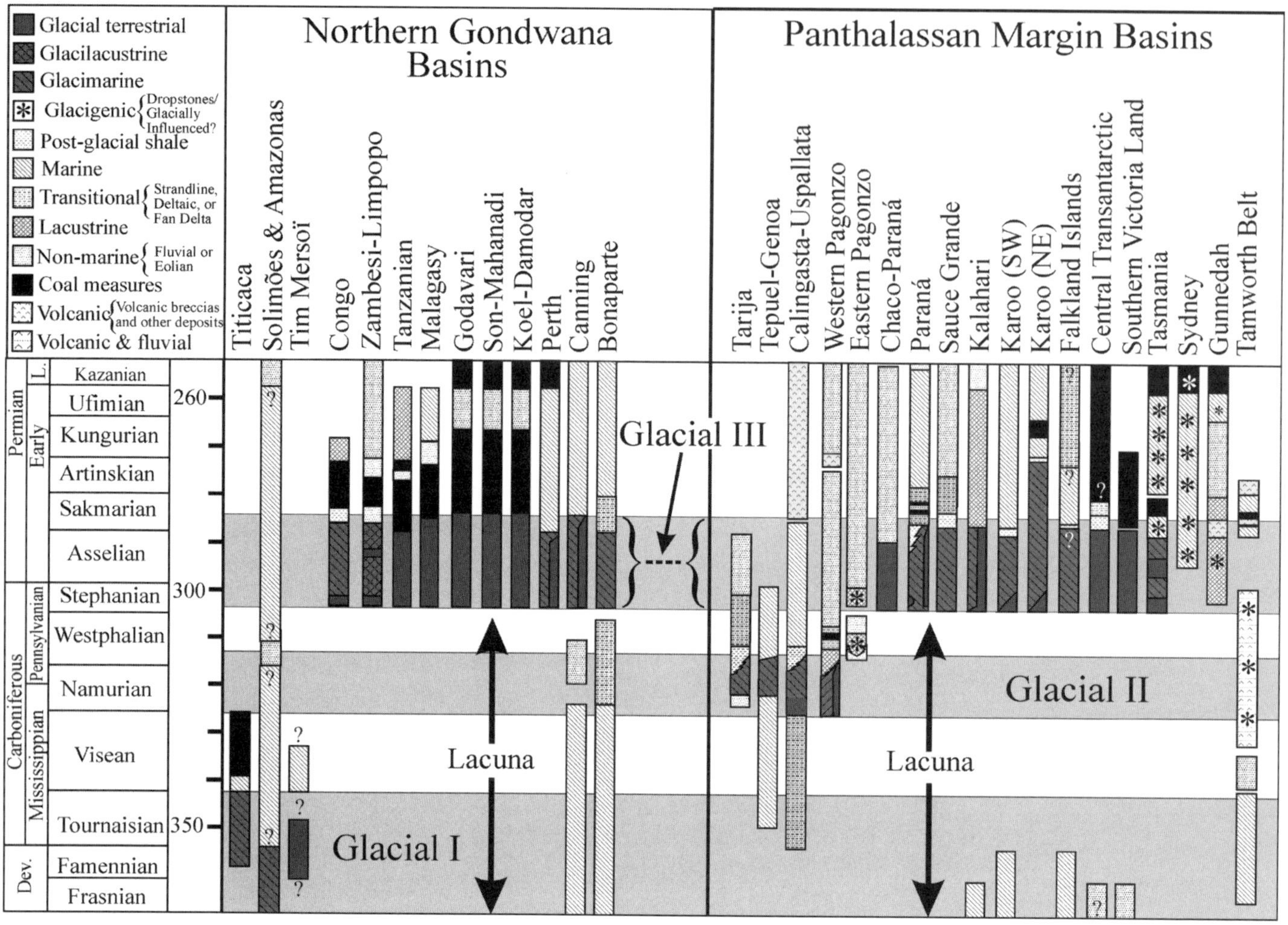

Figure 4. Temporal and spacial distribution of Gondwana glacigenic deposits and associated facies. Note three packages of glacial rocks deposited during three distinct intervals. Carboniferous time scale is from Menning et al. (Time Scale B; 2000) and Permian time scale is from Roberts et al. (1996). Question marks denote poorly dated units. Data compiled from references listed in Appendix.

or roughly 40% of the Earth's surface area. This calculation clearly demonstrates that alpine glaciation is not a major contributor to sea-level fluctuations, but rather changes in the mass balance of ice sheets drive glacioeustasy.

THE GONDWANA GLACIAL RECORD

How does the Gondwana glacial record compare to that of the glacial record inferred from cyclothems? Is there sufficient temporal overlap in the two records? Can the style of glaciation that occurred and the volume of glacial ice that was present adequately explain the observed changes in late Paleozoic base level?

Upper Paleozoic glacigenic deposits of the southern hemisphere are as widespread and well studied as the northern hemisphere cyclothems. Glacigenic rocks occur in South America, the Falkland Islands, Africa, the Arabian Peninsula, the Indian subcontinent, Antarctica, and Australia, and were deposited within cratonic, rift, foreland, successor, pull-apart, and fore-arc basins scattered across Gondwana (Fig. 1). The Gondwana strata contain three distinct and separate packages of upper Paleozoic glacial deposits (e.g., Veevers and Powell, 1987; López-Gamundí, 1997); Glacial I (Late Devonian to earliest Carboniferous; Frasnian to possibly Tournaisian), Glacial II (Early to Late Carboniferous; Namurian to earliest Westphalian), and Glacial III (latest Carboniferous to Permian; Stephanian to Sakmarian-Artinskian; Fig. 4). These rocks are correlated using pollen and spores (e.g., Kyle and Schopf, 1982; Foster and Waterhouse, 1988; Lindström, 1995), invertebrate faunas (e.g., Amos and López-Gamundí, 1981; Archbold, 1999), flooding surfaces (e.g., López-Gamundí, 1989; Isbell et al., 1997a), and radiometric dates (SHRIMP) obtained from zircon grains within volcanic tuffs (e.g., Claqué-Long et al.; 1995; Roberts et al., 1996; Bangert et al., 1999). The glacial rocks consist of massive diamictites resting on grooved and striated surfaces deposited as sub-glacial lodgment tills, sheared diamictites reworked into sub-glacial deformation tills, massive diamictites with gradational upper and lower contacts and stratified diamictites deposited by ice-proximal glacimarine processes such as rain-out and subaqueous sediment gravity flows, and laminated fine-grained rocks containing dropstones deposited as ice-rafted debris within lacustrine and marine settings.

Continental reconstructions of southern Pangea for the late Paleozoic are loosely constrained (Grunow, 1999), but indicate polar latitudes, with Gondwana drifting across the South Pole during the Carboniferous and Permian. Powell and Li's (1994) reconstructions place Africa over the pole in the Early Carboniferous (Mississippian), Antarctica over the pole from the Middle Carboniferous (Pennsylvanian) until the end of the Early Permian, and southeastern Australia over the pole in the Late Permian (Fig. 1). Glacial depocenters are inferred to have shifted across Gondwana as the continent drifted across the South Pole (e.g., Du Toit, 1921; Crowell, 1983, 1999; Caputo and Crowell, 1985).

Glacial I

Basins

The oldest upper Paleozoic glacial deposits, Glacial I (Table 2, Fig. 4), occur in Upper Devonian to possibly lowermost Carboniferous rocks only in Peru and Bolivia (the proto-Andean basins; e.g., Titicaca Basin), northern Brazil (the Acre, Solimões, Amazonas, and Parnaíba Basins); and in central Africa (the Tim Mersoï Basin; Figs. 1 and 4; Hambrey and Kluyver, 1981; Caputo and Crowell, 1985; Veevers and Powell, 1987; Díaz Martínez and Isaacson, 1994; Isaacson and Díaz Martínez, 1995; López-Gamundí, 1997; Crowell, 1999). Although the tectonic setting of some of these basins is ambiguous, the basins are interpreted to have formed adjacent to uplands (Hambrey and Kluyver, 1981; Sablock, 1993; López-Gamundí, 1997). Isaacson and Díaz Martínez (1995) interpreted the Titicaca Basin as a backarc basin.

Facies

Although Glacial I rocks reflect both terrestrial and marine processes, the association of abundant massive and deformed diamictites with dropstone-bearing laminated mudstones generally suggests glacimarine deposition near the tidewater terminus of alpine glaciers (Table 2; Díaz Martínez and Isaacson, 1994). Soft sediment deformation is abundant and is interpreted to have occurred during submarine slumping associated with the advance of glacimarine grounding lines. Elsewhere, the presence of striated surfaces confirms a glacial origin for the Glacial I rocks (Rocha-Campos, 1981a; Melo, 1988).

TABLE 2. GLACIAL I DEPOSITS

Basin, location	Formation and age	Glacial facies	Environmental interpretation	References
Titicaca Basin, Bolivia	Cumaná Formation—Base contains Upper Fammenian (*pusillites-lepidophyta* palynozone) palynomorphs; unit overlain by Tournaisian to lowermost Viséan Kasa Formation	Thick, laminated mudstones with striated dropstones Massive diamictites and deformed sandstones	Glacially influenced marine setting with ice-rafted debris Glaciomarine tidewater glaciers with ice-proximal sediment gravity flows and possible subglacial tills	Díaz Martínez and Isaacson (1994) Isaacson and Díaz Martínez (1995)
Amazonas Basin, Brazil	Curuá Formation B contains Frasnian and Famennian palynomorphs and an Upper Devonian brachiopod and molluscan fauna	Deformed diamictites interbedded with deformed graded and cross-bedded sandstones, siltstones, and shales	Submarine slumps and sediment gravity flows associated with a glaciomarine grounding line	Rocha-Campos (1981a) Caputo and Crowell (1985)
Parnaíba Basin, Brazil	Middle to Upper Devonian	Deformed diamictites interbedded with deformed graded and cross-bedded sandstones, siltstones, and shales Striated surfaces Laminated shales containing dropstones	Submarine slumps and sediment gravity flows associated with a glaciomarine grounding line Subglacial erosion Glacially influenced marine setting with ice-rafted debris	Rocha-Campos (1981b)
Tim Mersoï Basin, north central Africa	Téragh Formation—Late Devonian (?)	Diamictites resting on polished Precambrian granite Beds overlying diamictites consist of laminated mudstones with dropstones Top of succession is cross-bedded sandstones	Subglacial lodgement till Glaciomarine Glaciofluvial	Caputo and Crowell (1985) Hambrey and Kluyver (1981)

Upper Devonian non-glacial rocks across the rest of Gondwana indicate that Glacial I deposits are geographically restricted to Peru, Bolivia, northern Brazil, and central Africa. Upper Devonian marine shales occur in southern Brazil (Frasnian of the Paraná Basin; Melo, 1988); mixed clastic and carbonate successions occur in Morocco (Frasnian and Famennian; Wendt, 1988); thick shallow marine and deltaic quartzarenites and mudstones containing bivalves, bryozoans, brachiopods, and cephalopods occur in South Africa (Frasnian and Famennian; Theron and Loock, 1988); nonmarine quartzarenites and redbeds containing miospores and fossil fish occur in Antarctica's southern Victoria Land (Frasnian; Bradshaw and Webers, 1988); and reef complexes occur in Western Australia (Frasnian and/or Famennian of the Canning and Bonaparte Basins; Cockbain and Playford, 1988; Mory, 1990).

Age

Within the various basins, palynomorphs, molluscs, brachiopods, and bryozoans constrain the age of the glacigenic rocks to the Late Devonian (Frasnian and Famennian; Rocha-Campos, 1981a, 1981b; Caputo and Crowell, 1985; Melo, 1988; Isaacson and Díaz Martínez, 1995). In the Titicaca Basin, Lower Carboniferous (Tournaisian to lowermost Viséan) sandstones and scattered deformed diamictites are interpreted by some as reflecting glacially influenced deposition associated with retreating glaciers; however, the evidence is inconclusive (Isaacson and Díaz Martínez, 1995).

Glacial II

Basins

Glacial II rocks occur in mid Carboniferous (Namurian to lowermost Westphalian) strata, and were deposited in the convergent margin (the Calingasta-Uspallata, Pagonzo, San Rafael, and Tepuel), and intracratonic (Tarija Basins) basins of South America, in the convergent margin setting (New England Fold Belt) of eastern Australia (Tamworth block forearc basin), and along the Himalayas in southern Tibet (Figs. 1 and 4, Table 3; Crowell and Frakes, 1971; Veevers and Powell, 1987; López-Gamundí et al., 1994; Veevers et al., 1994b; Dickins, 1996; Garzanti and Sciunnach, 1997; López-Gamundí, 1997). Glacial II rocks are only known from western South America, eastern Australia, and southern Tibet.

During the Late Devonian and Early Carboniferous, pre-Carboniferous strata in western South America were deformed by compressional, transtension, and oblique-slip processes during the Chañic Orogeny (Tankard et al., 1995). The Calingasta–Uspallata, Pagonzo, and San Rafael Basins are interpreted to have formed as backarc to foreland basins (Eyles et al., 1995; López-Gamundí, 1997); however, formation as pull apart basins during wrench tectonics is at least possible for the Pagonzo Basin (cf., Tankard et al., 1995). The Tepuel Basin developed in an intra- to forearc setting, while the Tarija (Chaco-Tarija) Basin was an elongate intracratonic basin (López-Gamundí, 1997). The basins are floored by folded and faulted pre-Carboniferous rocks and were bounded along basin margins by either tectonic highlands or by crystalline basement rocks (e.g., Sierras Pampeana and the Northern Patagonian Massif). Within the basins, basal sedimentary rocks onlap and overstep topographic relief on the underlying rocks (Eyles et al., 1995; Tankard et al., 1995; López-Gamundí, 1997).

In Australia, Glacial II rocks occur within the New England Fold Belt of New South Wales. The New England Fold Belt developed as a volcanic arc along the east coast of Australia during the middle Carboniferous (McPhie, 1987).

In southern Tibet, glacigenic rocks are associated with uplifted blocks that formed during initial rifting between the Indian subcontinent and the Peri-Gondwanian blocks (Garzanti and Sciunnach, 1997). In South America, Australia, and Tibet, the basins containing glacigenic deposits were relatively small and generally covered an area of less than 0.2×10^6 km^2.

Facies

In each of the South American basins containing Glacial II deposits, there is compelling evidence for a glacial origin. Evidence includes one or more of the following: striated boulder pavements, diamictites overlying polished and striated surfaces, diamictites with striated pebbles, and dropstone-bearing laminated mudstones (Table 3). However, in each basin the predominant facies includes diamictite, sandstone, and/or shale that lack unequivocal signatures of glacial deposits. In South America, for example, common massive and weakly stratified sandstones, deformed sandstones, and large deformational features underscore the prevalence of syn-sedimentary deformational processes and sediment gravity flows. Lonestones in mudstones reflect deposition of ice-rafted debris. Glacial II rocks interfinger with rocks containing marine fossils including brachiopods, bryozoans, and molluscs, indicating that deposition occurred in glacimarine or glacially influenced marine settings. Glacial II deposits in South America are interpreted to have been associated with small, discontinuous ice centers in uplifted areas along the continental margin (López-Gamundí, 1997). Radial paleocurrent directions away from basement highs support the hypothesis that Glacial II rocks were deposited by mountain glaciers (cf., López-Gamundí et al., 1994).

In eastern Australia, a few diamictites containing striated stones and lonestone-bearing fine-grained laminated "varved" units occur within thick clast-supported conglomerate and ignimbrite/tuffaceous successions. Much of the conglomeratic detritus is of volcanic and pyroclastic origin; however, a few plutonic, metamorphic, and striated pebbles are present (McKelvey, 1981). Rare striated pavements have also been reported (Herbert, 1980). Although most of the eastern Australian sedimentary succession appears to have been deposited in alluvial fan, fluvial, and lacustrine settings (cf., McPhie, 1987), White (1968) and McKelvey (1981) interpreted the diamictites as morainal deposits. The occurrence of striated and faceted clasts, rare striated pavements, and lonestone-bearing "varves" also suggests a glacial and/or

TABLE 3. GLACIAL II DEPOSITS

Basin, location	Formation and age	Glacial facies	Environmental interpretation	References
Calingasta-Uspallata Basin, western Argentina	Hoyada-Verde Formation B Namurian to earliest Westphalian (*Levipusala* zone)	Massive and stratified diamictites Dropstone-bearing shales Thin graded sandstone beds Locally massive diamictites rest on striated boulder pavements	Glaciomarine ice-proximal rainout and sediment gravity flow to possible ice contact deposits Ice-rafted debris in glacially influenced marine settings Turbidity currents Subglacial	González (1990) López-Gamundí (1987, 1989, 1991) López-Gamundí, et al. (1994)
Pagonzo Basin, western Argentina	Guandacol Formation—Namurian to early Westphalian (unit is overlain by rocks containing Westphalian to Stephanian fossils)	Massive to weakly stratified thin-bedded diamictites Thin sandstones and shales containing dropstones Massive diamictites with striated clasts in northeast corner of basin	Glaciomarine and glacially influenced marine settings with rainout and sediment gravity flows (locally glaciolacustrine settings occurred) Turbidity currents Ice-rafted debris in glacially influenced marine settings Subglacial	Azcuy et al. (1987, 1991) Limarino and Cesari (1988) López-Gamundí et al. (1994) López-Gamundí (1997)
San Rafael and Tepuel Basins, western and southwestern Argentina	Namurian to earliest Westphalian rocks (*Levipusala* zone)	Massive and stratified diamictites Dropstone-bearing laminated mudstones Deformed sandstones	Glaciomarine mass flow and proximal rain-out deposits Ice-rafted debris in glacially influenced marine settings Mass-flow deposits	Andreis et al. (1987a, 1991) González (2001) López-Gamundí (1989, 1997)
Tarija Basin (Chaco-Tarija), northern Argentina, Bolivia, southern Peru	Tarija, Itacuami, and Chorro Formations—Namurian to earliest Westphalian (*Levipusala* zone)	Local striated pavements occur in Argentina Massive and stratified diamictites with lenses of deformed sandstone Rocks grade northward into highly deformed successions of muddy diamictites cut by thick sandstone and mudstone channels	Local grounded ice conditions in northern Argentina Subaqueous sediment gravity flows Amalgamated channel-filling turbidite deposits and associated overbank fines Deformation structures resulted from mass movements and slope failures	Rocha Campos et al. (1977) López-Gamundí (1987, 1989) Salfity et al. (1987) Eyles et al. (1995) Sempere (1995) Starck (1995) Tankard et al. (1995)
New England Fold Belt, Australia	Currabubula Formation Spion Kop Conglomerate Seaham Formation Earliest Namurian to earliest Westphalian	Currabubula Formation—2600-m-thick conglomerate, tuff, sandstone, and mudstone succession containing rare dropstone-bearing units up to 18 m thick; few striated clasts occur Spion Kop Conglomerate—orthoconglomerates, cross-bedded sandstones and tuffs with a few thin (up to 3 m thick) massive and stratified diamictites containing rare striated clasts, and dropstone-bearing shales Seaham Formation—conglomerates, diamictites, dropstone-bearing fine-grained units, tuffs, and volcanic lithic sandstones. Some striated clasts occur; unit overlies a polished and striated surface cut on the underlying Paterson Volcanics	Currabubula Formation—fluvial, alluvial fan, volcanic debris flow, and pyroclastic deposition associated with erosion of nearby volcanic arc; isolated lakes with dropstones deposited as ice-rafted debris Spion Kop Conglomerate—braided fluvial deposits with either rare ablation moraines and/or mass flow deposits Seaham Formation—deposited in fluvial and lacustrine setting with associated influx of pyroclastic debris; striated clasts and underlying striated surface suggest a glacial influence. Deposition by alpine glaciers in glacioterrestrial, glaciomarine, and glaciolacustrine settings adjacent to a volcanic arc	Coombs (1958) Whetten (1965) White (1968) Crowell and Frakes (1971) Herbert (1980) Benson (1981) McKelvey (1981) McPhie (1987) Claqué-Long et al. (1995) Roberts et al. (1995a, 1995b)

glacilacustrine origin for at least a small portion of the rocks (cf., Benson, 1981; Claqué-Long et al., 1995). However, evidence for a glacial origin is not unequivocal (cf., Coombs, 1958; Lindsay, 1966; Crowell and Frakes, 1971; Herbert, 1980; Benson, 1981; McKelvey, 1981; Dickins, 1985, 1996; Eyles, 1993; Eyles and Young, 1994). Lindsay (1969) pointed out that, except for the occurrence of striated and faceted clasts, the diamictites could just as easily have been deposited by mass flow processes, and Coombs (1958) indicated that at least some of the "varves" are of pyroclastic origin. In spite of possible multiple interpretations, most workers agree that limited alpine glaciation occurred, and that it was associated with convergent margin volcanoes (cf., McPhie, 1987; Dickins, 1996, 1997).

In southern Tibet, subglacial and ice-contact deposits have not been identified. However, the occurrence of faceted pebble and trapezoidal cobble/boulder-bearing diamictites, dropstone-bearing laminated mudstones, and quartz sand grains displaying grooves and arc-shaped surface textures suggest that alpine glaciers, which drained uplifted rift shoulders, locally supplied sediment to a glacially influenced marine setting (Garzanti and Sciunnach, 1997).

Age

In South America, Glacial II rocks are underlain by rocks containing a Lower Carboniferous (Viséan) paleoflora (*Archeosigillaria-Lepidodendropsis* Zone) and marine fossils (Viséan or possibly the Tournaisian-Viséan) of the *Protocanites* Zone (Amos, 1964; Sessarego and Cesari, 1986; Archangelsky et al., 1987a, 1987b, 1987c). The occurrence of brachiopods, clams, bryozoans, and crinoids of the *Levipustula* Zone within the Glacial II units and within overlying shales places an upper limit on the age of the glacial deposits in western South America (Rocha Campos et al., 1977; Andreis et al., 1987a; Salfity et al., 1987; López-Gamundí, 1989; Starck, 1995; Gonzalez, 2001). These rocks are thought to be mid Carboniferous (Namurian to Early Westphalian) based on correlations with fossils of the *Levipustula* Zone in Australia, and due to the position of the glacigenic strata below rocks containing marine fossils of the Middle and Upper Carboniferous (Middle Westphalian to Stephanian) "Zona de Intervalo" and fossil plants of the Middle and Upper Carboniferous (Westphalian-Stephanian) *Nothorhacopteris-Botrychiopsis-Ginkgophyllum* Zone (González, *in* López-Gamundí et al., 1994; Archangelsky et al., 1987a, 1987b, 1987c, 1991; Veevers, *in* López-Gamundí et al., 1994).

In Australia, Roberts et al. (1991, 1993) originally reported the range of the *Levipustula* Zone as mid Carboniferous (Namurian) to the end of the Carboniferous based on zircon dates from rhyolite beds. Subsequent field observations indicated that the dated rocks were sills rather than erupted units (Roberts et al., 1995a, 1995b), which allowed for revision of the Australian *Levipustula* Zone to entirely mid Carboniferous (Namurian). SHRIMP zircon dates and hornblende K-Ar dates obtained from tuff beds within the glacigenic strata and from underlying volcanic rocks indicate that the Glacial II rocks in Australia are mid Carboniferous (earliest Namurian to Early Westphalian) in age (Claqué-Long et al., 1995; Roberts et al., 1995a, 1995b).

In Tibet, thin diamictites occur near the middle third of an approximately 1-km-thick shale and thin sandstone succession. The succession overlies limestones containing a Tournaisian conodont fauna and underlies black shales containing a brachiopod fauna, which includes *Levipustula sp.* of Early Bashkirian (Late Namurian) age (Garzanti and Sciunnach, 1997).

Glacial III

Basins

Upper Carboniferous (Stephanian) to Lower Permian (Asselian/Sakmarian) Glacial III deposits are extensive across Gondwana (Figs. 1 and 4, Table 4; Veevers et al., 1994a, 1994b, 1994c; Veevers and Tewari, 1995; López-Gamundí, 1997; Wopfner and Casshyap, 1997), and are far greater in extent than is the distribution of Glacial I or II strata (Fig. 4). Glacial III rocks occur in upper Paleozoic cratonic (e.g., Kalahari and Paraná Basins; dos Santos et al., 1996; Veevers et al., 1994a), rift (e.g., Koel, Damodar, Son, and Mahanadi Basins; Veevers and Tewari, 1995; Wopfner and Casshyap, 1997), foreland (e.g., Karoo and Sydney Basins; Veevers et al., 1994b; Catuneanu et al., 1998), and successor basins (e.g., central Transantarctic Mountains; Isbell et al., 1997a, 1997b; Isbell, 1999) now exposed in South America, the Falkland Islands, Africa, the Arabian Peninsula, Madagascar, Antarctica, India, and Australia (Fig. 1). These basins were large features, with the largest, the Paraná Basin, covering an area $> 1.6 \times 10^6$ km^2.

Facies

In general, Glacial III rocks can be grouped into two different facies associations (Table 4). The first consists of massive diamictite resting on striated surfaces, sheared diamictite, sandstone, and shale. These rocks were deposited sub-glacially as lodgement and deformation till, and in glacifluvial and glacilacustrine settings. The second association consists of massive diamictite overlying gradational and sharp contacts, stratified diamictite, lonestone-bearing shales, and shales without lonestones. These rocks represent diverse styles of glacimarine deposition, including deposition at or near a grounding line or ice front, rain out of debris from ice shelves or ice tongues, deposition as ice-rafted debris, and as open marine sedimentation. Large-scale facies patterns suggest that, over distances of several hundred kilometers, environments changed across basin margins and down the basins from dominantly glaciterrestrial to dominantly glacimarine. Paleocurrent orientations within the same rocks indicate that ice flowed transversely across basin margins and then longitudinally down the basin axis (e.g., Kalahari and Karoo Basins, Visser, 1983, 1997a, 1997b; Transantarctic Basin, Isbell et al., 1997a; Isbell, 1999). The size of the basins, the nature of the facies and paleocurrents all suggest that Glacial III rocks may have been deposited by ice sheets rather than by alpine glaciers.

TABLE 4. GLACIAL III DEPOSITS

Basin, location	Age of glacial deposits	Glacial facies	Environmental interpretation	References
Paraná Basin, Brazil, Paraguay, Uruguay	Late Carboniferous to Early Permian (Stephanian or possibly latest Westphalian to Asselian-Sakmarian)	Stratified and massive diamictites Massive and graded sandstone beds Massive and laminated mudstones Dropstone-bearing shales Diamictites resting on grooved or glacial tectonized basement along basin margins	Glacially influenced marine facies within the basin and glaciomarine rainout, resedimentation, and grounded ice facies along basin margins	Eyles et al. (1993) França (1994) Santos et al. (1996) López Gamundí (1997)
Karoo Basin, South Africa	Late Carboniferous to Early Permian, (Stephanian to Asselian; Stephanian to Artinskian-Kungurian along basin margins)	Stratified and massive diamictites with gradational bases Dropstone-bearing shales Deep paleo-valleys occur along basin margins	Deposition in glaciomarine environments, which included ice shelf, tidewater, and grounded ice-front conditions; ice streams and outlet glaciers fed ice into the basin from the present north, east, and possibly from the south	Visser (1983, 1990, 1996, 1997a, 1997b) Catuneanu et al. (1998) Bangert et al. (1999)
Kalahari Basin, Namibia Botswana, and South Africa	Late Carboniferous to Early Permian, (Stephanian to Early Sakmarian)	Stratified and massive diamictites with gradational bases Dropstone-bearing shales Deep paleo-valleys occur along basin margins	Glacioterrestrial deposition occurred along basin margins, while glaciomarine and glacially influenced marine conditions occurred within the basin; ice streams and outlet glaciers fed ice into the basin from the present north, east, and south	Veevers et al. (1994c) Visser (1997a, 1997b) Key et al. (1998)
Transantarctic Basin, central Transantarctic Mountains, Antarctica	Late Carboniferous to Early Permian, (Stephanian to Asselian-Sakmarian)	Head of basin Diamictites and sheared diamictites resting on striated surfaces Cross-bedded sandstone Down Basin Stratified and massive diamictites with gradational bases Dropstone-bearing shales	Glacioterrestrial deposition with subglacial lodgment and deforming bed conditions; environments grade down the basin to glaciomarine ice shelf and glacially influenced marine conditions. Deposits likely fed by ice streams	Aitchison et al. (1988) Miller (1989) Collinson et al. (1994) Isbell et al. (1997a, 2001) Isbell (1999) Askin (1998)
Officer and Canning Basins, Western Australia	Late Carboniferous to Early Permian, (Stephanian to Early Sakmarian)	Officer Basin Diamictites with striated and faceted clasts, cross-bedded sandstones and shales Canning Basin Breccias, massive and graded sandstones, and conglomerates Diamictites Swaley cross-bedded sandstones Dropstone-bearing shales Glacially grooved surface	Officer Basin—Glacioterrestrial Canning Basin—several interpretations occur Mass flow and wave reworked sediment deposits (Eyles and Eyles, 2000) Environments include glaciofluvial within the basin with subglacial along basin margins during relative sea-level lowstands, and glaciomarine to glacially influenced marine facies during relative sea-level highstands (O'Brien and Christie-Blick, 1992; Lindsay, 1997; O'Brien et al., 1998)	Veevers (1984) Eyles and Eyles (2000) O'Brien and Christie-Blick (1992) Lindsay (1997) O'Brien et al. (1998)
Gondwana Master Basins, Peninsular India	Late Carboniferous to Early Permian, (Stephanian to Early Sakmarian)	Diamictites contained striated and faceted clasts Thin diamictites interstratified with thin sandstones Conglomerates Cross-bedded sandstone	Deposition in glacioterrestrial environments as subglacial tills, glaciolacustrine, and sediment gravity flow facies	Veevers and Tewari (1995) Wopfner and Casshyap (1997)

Age

The age of Glacial III deposits are constrained by palynomorphs, brachiopods, and SHRIMP zircon dates obtained from tuff beds. Basal beds within Glacial III deposits contain either Upper Carboniferous (Stephanian) or Lower Permian (Asselian-Tastubian) palynomorphs (e.g., Stapleton, 1977; Visser, 1990, Lindström, 1995; dos Santos et al., 1996; Key et al., 1998; R.A. Askin, 2002, personal commun.). SHRIMP zircon dates from tuff beds in the lower portion of the glacigenic rocks in the Karoo Basin also return a Late Carboniferous age (Stephanian age, 302.0 Å 3.0 Ma and 299.2 Å 3.2 Ma, latest Kasimovian; Bangert et al., 1999).

Within Lower Permian (Asselian-Sakmarian) strata, an abrupt change from glacial to post-glacial deposits records the rapid withdrawal of ice from the depositional basins. In nonmarine basins, a sharp Gondwana-wide contact between glacial (below) and fluvial and/or lacustrine (above) deposits occurs within the ~2 million-year-long *Pseudoreticulatispora confluens* palynomorph zone (Miller, 1989; Collinson et al., 1994; Lindström, 1995; Seegers, 1996; Isbell et al., 1997a, 1997b; Wopfner and Casshyap, 1997; Askin, 1998). At approximately the same stratigraphic level (Lindström, 1995), a flooding surface at the base of the post-glacial *Eurydesma* (Lower Sakmarian) transgression abruptly overlies glacimarine deposits (Veevers and Powell, 1987; Dickins, 1996; López-Gamundí, 1997). Although evidence for continuation of glaciation beyond the Early Permian (Asselian-Sakmarian) is sparse, limited data suggest that ice continued along basin margins and in upland areas until the Late Permian (Fig. 4). The evidence includes: (1) interstratification of shales containing the *Eurydesma* fauna with diamictites along the margins of the Karoo and Kalahari Basins (Visser, 1982, 1990, 1991, 1997a, 1997b; von Brunn, 1987; López-Gamundí, 1997); (2) rare, striated lonestones within flood-plain deposits of Lower Permian coal measures in southern Victoria Land, Antarctica (Francis et al., 1994; Smith et al., 1998); (3) ice-rafted debris within mid to Upper Permian (Sakmarian to Kungurian) strata in the Sydney Basin, Australia (Eyles et al., 1998); (4) diamictites with striated clasts in the subsurface of the western Sydney Basin (Eyles et al., 1998); and (5) ice-rafted debris within Upper Permian rocks in Tasmania (Banks and Clarke, 1987).

Summary: Gondwana Glacial Deposits

In summary, the Gondwana glacial deposits record three distinct glacial successions. Glacial I and II represent predominantly glacimarine/glacilacustrine sedimentation near the termini of relatively small-scale alpine glaciers flowing into small basins (Figs. 1 and 4). In contrast, Glacial III deposits were widespread throughout Gondwana and were deposited as glaciterrestrial and glacimarine/glacilacustrine deposits associated with large basins and ice sheets. Although the glacial record was not temporally continuous (Fig. 4), it does suggest that ice sheets were widespread only during Glacial III deposition.

PRE-GLACIAL III LACUNA

Does the record accurately represent the history of late Paleozoic glaciation in Gondwana, or is the record misleading because of massive removal of older glacial deposits by subglacial erosion?

Across most of Gondwana, a major unconformity separates Glacial III deposits above from Devonian and older sedimentary and crystalline basement rocks below (Fig. 4; Collinson et al., 1994; López-Gamundí et al., 1994; Veevers et al., 1994a, 1994b, 1994c; Veevers and Tewari, 1995; López-Gamundí, 1997; Wopfner and Casshyap, 1997). The lacuna below Glacial III deposits can be interpreted in two ways. First, because ice sheets leave a meager record due to glacial erosion, the lacuna may reflect widespread sub-glacial erosion and, hence, glaciation (Veevers and Powell, 1987; González-Bonorino and Eyles, 1995). Alternatively, we interpret the rocks to faithfully reflect the paucity of glaciation across much of Gondwana during the period represented by the lacuna.

The fidelity of the upper Paleozoic glacial record can be evaluated by examining Gondwana sites that contain continuous Middle Carboniferous to Permian successions, and by examining weathering profiles that developed on the pre-Glacial III unconformity. Only in western and northern South America and in western and eastern Australia have nearly complete Carboniferous to Permian deposits been reported in Gondwana (Fig. 4). The Ellsworth Mountains, which contain a thick glacigenic succession, may also be a key area. However, the age of rocks at the base of the Ellsworth succession is not known, and the nature of the contact with the underlying unit is poorly constrained (cf., Matsch and Ojakangas, 1991, 1992; Spörli, 1992). The South American and Australian sites are summarized in Table 5, and their locations are shown in Figure 1. Of the complete successions, none contain Westphalian glacial deposits. Instead, rocks within these basins were deposited in fluvial, deltaic, and shallow marine settings (e.g., Bonaparte Basin, Mory, 1990; Calingasta-Uspallata and western Pagonzo Basins, López-Gamundí et al., 1994; López-Gamundí, 1997; Solimões Basin, Tsubone et al., 1991; Tarija Basin, López-Gamundí et al., 1994). Therefore, available evidence indicates that ice was not present in basins located along the margins of Gondwana at that time.

For mid Carboniferous (Westphalian) rocks, the strongest evidence for glaciation in South America is the presence of lonestones within lacustrine turbidites contained within narrow, fault-bounded paleovalleys in Argentina (e.g., Malanzán sub-basin in the eastern Pagonzo Basin). However, a direct glacial link for these strata cannot be established, as the lacustrine deposits grade laterally into clast-supported, basin-margin, alluvial fan conglomerates rather than into glacial deposits (Azcuy et al., 1987, 1991; López-Gamundí et al., 1994).

In eastern Australia, it has been suggested that glaciation was continuous from the time of Glacial II into Glacial III deposition and that during that interval, glaciation expanded westward into central Australia (Veevers and Powell, 1987; Veevers et al.,

1994b, 1994c; Veevers and Tewari, 1995). These conclusions have been challenged because supporting data are sparse (Dickins, 1985, 1996), and because the available record suggests that only limited mountain glaciation occurred in eastern Australia (cf., Crowell and Frakes, 1971; Herbert, 1980; Lindsay, 1997).

In examining the pre-Glacial III unconformity, a weathered granite profile occurs in the central portion of the central Transantarctic Mountains, Antarctica (Isbell et al., 2001). The profile indicates that an extended period of exposure, and therefore, an absence of glacial ice, occurred near the paleo South Pole (cf., Powell and Li, 1994) prior to Glacial III deposition. Three other localities, scattered over a distance of 1200 km along the central Transantarctic Mountains, also have weathered granite profiles. These occurrences underscore the absence of ice prior to Glacial III deposition across a large area near the paleo South Pole and document that the stratigraphic record has retained sufficient information to indicate that glaciation was not widespread prior to Glacial III deposition.

In summary, there is no evidence for the occurrence of large ice sheets prior to Glacial III deposition. However, evidence supporting the hypothesis that large ice sheets were absent during the mid Carboniferous (Westphalian) includes: (1) absence of glacial deposits of this age (Dickins, 1996); (2) presence of non-glacial, fluvio-deltaic, and shallow-marine successions of this age within Gondwana basins that were actively subsiding; and (3) absence of glacial deposits in the central Transantarctic Mountains, which were at or near the South Pole during the interval between the deposition of Glacial II and III rocks (Figs. 1 and 4).

DISCUSSION

Although it has been suggested that glaciation was continuous throughout the late Paleozoic (Veevers and Powell, 1987; Frakes and Francis, 1988; Crowley and Baum, 1991, 1992; Frakes et al., 1992; Crowell, 1999), available stratigraphic and sedimentologic data indicate that glaciation occurred in three discrete, non-overlapping episodes: Glacial Episode I, Glacial Episode II, and Glacial Episode III (Fig. 4, Tables 2–4). Of the three glacial episodes, the first two were relatively small in scale. Their deposits contain reliable indicators of glaciation, including striated pebbles and striated pavements overlain by diamictites. However, these unequivocal glacial deposits are overshadowed by glacimarine, turbidite, and slump deposits (Tables 2 and 3). Because of their occurrence at convergent plate boundaries, their intimate association with glacimarine deposition and syn-sedimentary deformation features, and their limited geographical extent, Glacial I and II deposits are interpreted to have accumulated at the termini of alpine glaciers. There is no evidence of glaciterrestrial deposition by large ice sheets. The implication from these deposits is that there was not enough water tied up in ice during Glacial Episodes I and II to cause a major sea-level rise upon melting. In contrast, glaciterrestrial deposition was widespread, as was glacimarine deposition, during Glacial Episode III, which reflects ice sheet development. Although the max-

TABLE 5. ROCKS DEPOSITED FOLLOWING GLACIAL II AND PRIOR TO GLACIAL III DEPOSITION

Basin, location	Age	Rock type	Environment	Overlain by Glacial III rocks	References
Amazonas and Solimões basins, northern Brazil	Carboniferous to Permian deposits	Sandstones, shales, evaporites, and limestones	Fluvial, lacustrine, eolian, shallow marine, and restricted marine environments	No	Tsubone et al. (1991)
Calingasta-Uspallata and western Pagonzo Basins, western Argentina	Westphalian to Late Permian deposits	Sandstones, shales, coals, tuffs, and silicic volcanic rocks	Post-glacial marine, peat forming fluvial-deltaic, eolian and volcanic environments	No	López-Gamundí et al. (1994) Pazos (2002)
Tarija Basin, Bolivia, Paraguay, and northern Argentina	Westphalian to Early Permian deposits	Sandstones and shales	Post-glacial marine and fluvial-deltaic environments	No	López-Gamundí et al. (1994) López-Gamundí (1997)
Bonaparte Basin, Western Australia	Frasnian to Westphalian rocks overlain by Stephanian to Artinskian glacial deposits	Limestones, sandstones, shales, and conglomerates	Shallow marine, reef, deltaic, and fluvial environments	Yes	Mory (1990)
Central Transantarctic Mountains, Antarctica	Stephanian to Lower Permian Glacial III deposits	Weathering profile developed on granite overlain by shale and sandstone beds containing abundant trace fossils and conchostracans; succession is cut by an erosion surface at the base of the first Glacial III diamictite	Physical weathering of an exposure surface followed by glacilacustrine deposition	Yes	Isbell et al. (2001)

imum areal extent of Gondwana glaciation is unknown (Dickins, 1985, 1997), estimates range from 17.9 to 22.6 × 10^6 km^2 (Crowley et al., 1991; Crowley and Baum, 1992; Veevers, 1994; Ziegler et al., 1997). During Glacial Episode III, ice was likely divided among multiple spreading centers (ice sheets, ice domes, and/or ice caps) and perhaps divided among as many as 10 different ice masses. Therefore, complete ablation of an ice cover of this size would have produced eustatic changes of no more than 50 (10 equally sized ice sheets) to 115 m (a single ice sheet; Fig. 3).

Overlap Between Cyclothems and Glaciation

On a gross scale, cyclothems and episodes of late Paleozoic glaciation overlap temporally, but they do not coincide on a finer scale. Cyclothems were abundant prior to the Stephanian and the advent of Glacial Episode III, yet there was not enough water tied up in ice during Episodes I and/or II for glacial waxing and waning to have caused the observed base-level fluctuations.

It could be argued that erosion by Glacial Episode III glaciers removed the glacial sediments deposited between Glacial Episodes II and III. Assuming that to be the case, any glaciers that occurred between Glacial Episodes II and III could not have been continent-wide. Ice sheets extant during that interval would have had to meet the following requirements: (1) a location far from any depocenter, so that no glacial signature was recorded by rocks within the basins; (2) a location within the interior of Gondwana, far away from major sources of moisture along the continental margins; and (3) a non-polar location (Figs. 1 and 4, Table 5). With these constraints, any extant ice sheets would have been too small to account for the sea-level fluctuations of 60–100 m suggested from cyclothem and incised valley studies.

Why no Gondwana Cyclothems?

If upper Paleozoic cyclothems in North America and Europe were caused by the waxing and waning of Gondwana glaciers, the associated sea-level fluctuations must have been global in scale. Given this, contemporaneous shallow and marginal marine deposits in Gondwana would also have been affected. Although non-glacial successions in Gondwana have been described (e.g., Table 5), no cyclic deposits associated with Milankovitch-scale (< 412,000 yr) changes in base level (4th or 5th order sequences/cycles) are identified (cf., Pazos, 2002). An absence of Gondwana cyclothems is consistent with data presented herein indicating that the disparate timing of cyclothems and major glaciation during Glacial Episode III precludes waxing and waning of Gondwana ice sheets as a sole cause of cyclothems.

CONCLUSIONS

Upper Paleozoic (Carboniferous to Permian) cyclothems in North America and Europe have long been attributed to fluctuations of glacial ice volume in Gondwana. Reevaluation of the Gondwana glacial record does not substantiate this relationship. Three glacial episodes occurred in Gondwana during the late Paleozoic. Episodes I and II were characterized by alpine glaciers of limited extent. Sea-level fluctuations produced by changes in mass balance of alpine glaciation would be substantially smaller than that inferred from cyclothems. Thus, cyclothems during Glacial I (Late Devonian to earliest Carboniferous; Frasnian, Famennian, and possibly earliest Tournaisian) and II (Carboniferous; Namurian to earliest Westphalian) do not reflect fluctuations in ice sheet volume. Only during Glacial Episode III (latest Carboniferous to Permian; Stephanian to Sakmarian, possibly Kungurian) were ice sheets possibly present in Gondwana, and only during this interval could waxing and waning of these massive ice sheets have produced the changes in base level recorded by cyclothems.

Evidence for nondeposition within the central Transantarctic Mountains prior to the onset of Glacial Episode III supports the fidelity of the Gondwana stratigraphic record and indicates that deposits of large ice sheets prior to Glacial Episode III were never present. The record was not completely removed during the advance of ice sheets in that episode.

ACKNOWLEDGMENTS

Discussions with Allen Archer, Rosemary Askin, Loren Babcock, Octavian Catuneanu, Douglas Cole, James Collinson, Rubén Cúneo, Stephen Greb, Phillip Heckel, Cristal Jansen, and Gina Seegers-Szablewski are greatly appreciated, especially the views expressed that differed from those of the authors. We also thank Gerardo Bossi, Joel Carneiro de Castro, Octavian Catuneanu, Doug Cole, Rubén Cúneo, John Hancox, Noel Kemp, Bruce Rubidge, Roger Smith, Jurie Viljoen, and De Ville Wickens for field introductions to rocks in the Karoo Basin of South Africa, the Paraná Basin of Brazil, the Paganzo, Chaco-Paraná, Malanzán, and Calingasta-Uspallata basins of Argentina, and in Tasmania. Octavian Catuneanu, Nicholas Christie-Blick, James Collinson, and William Mode made helpful comments on earlier versions of the manuscript. Antarctic Support Associates, Raytheon Polar Services, the U.S. Navy Squadron VXE-6, New York Air National Guard, Kenn Borek Air, Helicopters New Zealand, Petroleum Helicopters (PHI), and the National Science Foundation provided logistic support for fieldwork in Antarctica. This work was supported by National Science Foundation grants OPP-9615045, OPP-9909637, and OPP-0126086 to John Isbell, and grants OPP-9417978, OPP-9614709, and OPP-9614989 to Molly Miller, and by a University of Wisconsin–Milwaukee Graduate School Research Committee award to John Isbell.

APPENDIX: REFERENCES DESCRIBING GONDWANA BASINS CONTAINING GLACIGENIC DEPOSITS SHOWN IN FIGURE 4

(1) Titicaca Basin (López-Gamundí, 1997); (2) Solimões Basin (Rocha-Campos, 1981a; Tsubone et al., 1991); (3) Tim

Mersoï Basin (Hambrey and Kluyver, 1981); (4) Congo and Zambesi–Limpopo basins (Rust, 1975; Anderson, 1981; Bond, 1981a, 1981b; Rocha-Campos, 1981c; Veevers et al., 1994a; Visser, 1997b); (5) Tanzanian Basins (Veevers et al., 1994a; Wopfner and Casshyap, 1997); (6) Malagasy Basin (Veevers et al., 1994a; Visser, 1997b; Wopfner and Casshyap, 1997); (7) Godavari, Son–Mahanadi and Koel–Damodar Basins (Veevers and Tewari, 1995; Wopfner and Casshyap, 1997); (8) Perth Basin (Kemp et al., 1977; Van de Graaff, 1981; Veevers, 1984; Veevers and Tewari, 1995); (9) Canning Basin (Veevers and Powell, 1987; Cockbain and Playford, 1988; Foster and Waterhouse, 1988; Middleton, 1990; O'Brien and Christie-Blick, 1992; Lindsay, 1997; O'Brien et al., 1998; Eyles and Eyles, 2000); (10) Bonaparte Basin (Mory, 1990); (11) Tarija Basin (Helwig, 1972; Salfity et al., 1987; López-Gamundí, 1989, 1997); (12) Tepuel–Genoa Basin (Andreis et al., 1987a, 1991; López-Gamundí, 1989; Cúneo et al., 1991; López-Gamundí, 1997); (13) Calingasta–Uspallata Basin (López-Gamundí, 1989, 1997; López-Gamundí et al., 1992, 1994); (14) western Pagonzo Basin (López-Gamundí et al., 1992, 1994; López-Gamundí, 1997; Pazos, 2002); (15) eastern Pagonzo Basin (López-Gamundí et al., 1994); (16) Chaco–Paraná Basin (Vergel, 1991; López-Gamundí, 1997; López-Gamundí et al., 1994); (17) Paraná Basin (Castro, 1988; Daemon and Marques-Toigo, 1991; França, 1994; dos Santos et al., 1996; López-Gamundí, 1997; López-Gamundí and Rossello, 1998); (18) Sauce Grande Basin (Coates, 1969; Andreis et al., 1987b; López-Gamundí, 1997; López-Gamundí et al., 1994; López-Gamundí and Rossello, 1998); (19) Kalahari Basin (Veevers et al., 1994a; Visser and Praekelt, 1996; Visser, 1997a, 1997b; Key et al., 1998); (20) Karoo Basin (Devonian—Veevers et al., 1994a; glacial deposits Visser, 1990, 1997a, 1997b; Bangert et al., 1999; overlying beds—Catuneanu et al., 1998); (21) Falkland Islands (Marshall, 1994; López-Gamundí and Rossello, 1998; Visser and Praekelt, 1996); (22) Central Transantarctic Mts. (Kyle and Schopf, 1982; Farabee et al., 1991; Masood et al., 1994; Collinson et al., 1994); (23) Southern Victoria Land (Kyle, 1977; Kyle and Schopf, 1982; Collinson et al., 1994; Askin, 1995; Isbell and Cúneo, 1996); (24) Tasmania (Banks and Clarke, 1987; Domack et al., 1993; Veevers et al., 1994b), (25) Sydney Basin (Herbert, 1980; Veevers et al., 1994b; Eyles et al., 1998); and (26) Gunnedah Basin and Tamworth Belt (Veevers et al., 1994b).

REFERENCES CITED

Adlis, D.S., Grossma, E.L., Yancey, T.E., and McLerran, R.D., 1988, Isotope stratigraphy and paleodepth changes of Pennsylvanian cyclical sedimentary deposits: Palaios, v. 3, p. 487–506.

Aitchison, J.C., Bradshaw, M.A., and Newmann, L., 1988, Lithofacies and origin of the Buckeye Formation: Late Paleozoic glacial and glaciomarine sediments, Ohio Range, Transantarctic Mountains, Antarctica: Palaeogeography, Palaeoclimatology, Palaeoecology, v. 64, p. 93–104.

Amos, A.J., 1964, A review of the marine Carboniferous stratigraphy of Argentina, *in* 12th International Geological Congress: Calcutta, India, Calcutta International Geological Congress, p. 53–72.

Amos, A.J., and López Gamundi, O., 1981, Late Paleozoic diamictites of the Central Patagonian Basin, Argentina, *in* Hambrey, M.J., and Harland, W.B., eds., Earth's pre-Pleistocene glacial record: Cambridge, Cambridge University Press, p. 878–881.

Anderson, J.M., 1981, World Permo-Triassic correlations: Their biostratigraphic basis, *in* Cresswell, M.M., and Vella, P., eds., Gondwana Five: Selected papers and abstracts of papers presented at the Fifth International Gondwana Symposium: Rotterdam, A.A. Balkema, p. 3–10.

Andreis, R.R., Amos, A.J., Archangelsky, S., and Gonzalez, C.R., 1987a, Cuenca Sauce Grande (Sierras Australes y Colorado), *in* Archangelsky, S., ed., El Sistema Carbonífero en la República Argentina: Córdoba, Academia Nacional de Ciencias de Córdoba, p. 213–223.

Andreis, R.R., Archangelsky, S., Gonzalez, C.R., López-Gamundí, O., Sabattini, N., Aceñolaza, G., Azcuy, C.L., Cortiñas, J., Cuerda, A., and Cúneo, R., 1987b, Cuenca Tepuel-Genoa, *in* Archangelsky, S., ed., El Sistema Carbonífero en la República Argentina: Córdoba, Academia Nacional de Ciencias de Córdoba, p. 169–196.

Andreis, R.R., Cúneo, R., López-Gamundí, O., Sabattini, N., and Gonzalez, C.R., 1991, Cuenca Tepuel-Genoa, El Sistema Pérmico en la República Argentina y en la Republica Oriental del Uruguay: Buenos Aires, XII Congreso Internacional de la Estatigrafia y Geología del Carbonífero y Pérmico: Córdoba, Academia Nacional de Ciencias de Córdoba, p. 69–95.

Archangelsky, S., Azcuy, C.L., Césari, S., Gonzalez, C.R., Hünicken, M., Mazzoni, A., and Sabattini, N., 1991, Correlacion esad de las biozonas, El Sistema Pérmico en la República Argentina y en la Republica Oriental del Uruguay: Buenos Aires, XII Congreso Internacional de la Estatigrafia y Geología del Carbonífero y Pérmico: Córdoba, Academia Nacional de Ciencias de Córdoba, p. 209–232.

Archangelsky, S., Azcuy, C.L., Gonzalez, C.R., and Sabattini, N., 1987a, Correlacion general de biozonaso, *in* Archangelsky, S., ed., El Sistema Carbonífero en la República Argentina: Córdoba, Academia Nacional de Ciencias de Córdoba, p. 281–292.

Archangelsky, S., Azcuy, C.L., Gonzalez, C.R., and Sabattini, N., 1987b, Esad de las biozonas, *in* Archangelsky, S., ed., El Sistema Carbonífero en la República Argentina: Córdoba, Academia Nacional de Ciencias de Córdoba, p. 293–301.

Archangelsky, S., Azcuy, C.L., Gonzalez, C.R., Sabattini, N., and Aceñolaza, G., 1987c, Paleontologia, bioestratigrafia, y paleoecologia de las Cuencas Paganzo, Calingasta-Uspallata y Rio Blanco, *in* Archangelsky, S., ed., El Sistema Carbonífero en la República Argentina: Córdoba, Academia Nacional de Ciencias de Córdoba, p. 133–151.

Archbold, N.W., 1999, Permian Gondwanan correlations: The significance of the western Australian marine Permian: Journal of African Earth Sciences, v. 29, p. 63–75.

Askin, R.A., 1995, Permian palynomorphs from southern Victoria Land, Antarctica: Antarctic Journal of the United States, v. 30, p. 47–48.

Askin, R.A., 1998, Floral trends in the Gondwana high latitudes: Palynological evidence from the Transantarctic Mountains: Journal of African Earth Sciences, v. 27, p. 12-13.

Azcuy, C.L., Andreis, R.R., Cuerda, A., Hünicken, M.A., Pensa, M.V., Valencio, D.A., Vilasy, J.F., Archangelsky, S., Berkowski, D.A., and Leguizamón, R., 1987, Cuenca Paganzo, *in* Archangelsky, S., ed., El sistema Carbonifero en la Republica Argentina: Córdoba, Academia Nacional de Ciencias de Córdoba, p. 41–100.

Azcuy, C.L., Ottone, E.G., Bossi, G.E., and Andreis, R., 1991, Pagonzo Basin, International Congress of Carboniferous and Permian Stratigraphy and Geology Excursion Guide 3: Buenos Aires, CONICET, 51 p.

Bangert, B., Armstrong, R., Stollhofen, H., and Lorenz, V., 1999, The geochronology and significance of ash-fall tuffs in the glaciogenic Carboniferous-Permian Dwyka Group of Namibia and South Africa: Journal of African Earth Sciences, v. 29, p. 33–49.

Banks, M.R., and Clarke, M. J., 1987, Changes in the geography of the Tasmania Basin in the late Paleozoic, *in* McKenzie, G.D., ed., Gondwana Six: Stratigraphy, sedimentology, and paleontology: Washington, D.C., American Geophysical Union, p. 1–14.

Benson, J.M., 1981, Late Carboniferous and Early Permian tillites north of Newcastle, New South Wales, *in* Hambrey, M.J., and Harland, W.B., eds.,

Earth's pre-Pleistocene glacial record: Cambridge, Cambridge University Press, p. 480–484.

Bond, G., 1981a, Late Palaeozoic (Dwyka) glaciation in the middle Zamezi region, *in* Hambrey, M.J., and Harland, W.B., eds., Earth's pre-Pleistocene glacial record: Cambridge, Cambridge University Press, p. 55–57.

Bond, G., 1981b, Late Palaeozoic (Dwyka) glaciation in the Sabi-Limpop region, Zimbabwe, *in* Hambrey, M.J., and Harland, W.B., eds., Earth's pre-Pleistocene glacial record: Cambridge, Cambridge University Press, p. 58–60.

Bradshaw, M.A., and Webers, G.F., 1988, The Devonian rocks of Antarctica, *in* McMillan, N.J., Embry, A.F., and Glass, D.J., eds., Devonian of the world: Calgary, Canadian Society of Petroleum Geologist Memoir 14, v.1, p. 783–795.

Bruckschen, P., and Veizer, J., 1997, Oxygen and carbon isotopic composition of Dinantian brachiopods: Paleoenvironmental implications for the Lower Carboniferous of western Europe: Palaeogeography, Palaeoclimatology, Palaeoecology, v. 132, no. 1–4, p. 243–264.

Caputo, M.V., and Crowell, J.C., 1985, Migration of glacial centers across Gondwana during Paleozoic Era: Geological Society of America Bulletin, v. 96, p. 1020–1036.

Castro, J.C., 1988, Sedimentology, stratigraphy and paleontology of the Gondwana sequence of the Paraná Basin, Seventh International Gondwana Symposium, São Paulo, Brazil, Guidebook, Excursion A2: Rio de Janeiro, Petróleo Brasilerio S.A., p. 100.

Catuneanu, O., Hancox, P.J., and Rubidge, B.S., 1998, Reciprocal flexural behavior and contrasting stratigraphies: A new basin development model for the Karoo retroarc foreland system, South Africa: Basin Research, v. 10, p. 417–439.

Chesnut, D.R. J., 1994, Eustatic and tectonic control of depositional of the Lower and Middle Pennsylvanian strata of the central Appalachian Basin, *in* Dennison, J.M., and Ettensohn, F.R., eds., Tectonic and eustatic controls on sedimentary cycles: Tulsa, Oklahoma, SEPM (Society for Sedimentary Geology) Concepts in Sedimentology and Paleontology, v. 4, p. 51–64.

Claqué-Long, J.C., Compston, W., Roberts, J., and Fanning, C.M., 1995, Two Carboniferous ages: A comparison of SHRIMP zircon dating with conventional zircon ages and $^{40}Ar/^{39}Ar$ analysis, *in* Berggren, W.A., Kent, D.V., Aubry, M.-P., and Hardenbol, J., eds., Geochronology, time scales and global stratigraphic correlation: Tulsa, Oklahoma, SEPM (Society of Sedimentary Geology) Special Publication 54, p. 173–222.

Coates, D.A., 1969, Stratigraphy and sedimentation of the Sauce Grande Formation, Sierra de la Ventana, Southern Buenos Aries Province, Argentina, *in* Amos, A.J., ed., Gondwana Stratigraphy: Paris, UNESCO, p. 799–816.

Cockbain, A.E., and Playford, P.E., 1988, The Devonian of Western Australia: a review, *in* McMillan, N.J., Embry, A.F., and Glass, D.J., eds., Devonian of the world: Calgary, Canadian Society of Petroleum Geologist Memoir 14, v. 1, p. 743–754.

Collinson, J.W., Isbell, J.L., Elliot, D.H., Miller, M.F., and Miller, J.M. G., 1994, Permian-Triassic Transantarctic basin, *in* Veevers, J.J., and Powell, C.M., eds., Permian-Triassic Pangean basins and foldbelts along the Panthalassan Margin of Gondwanaland: Boulder, Colorado, Geological Society of America Memoir 184, p. 173–222.

Coombs, D.S., 1958, Zeolitized tuffs from the Kuttung glacial beds near Seaham, New South Wales: Australian Journal of Science, v. 21, p. 18–19.

Crowell, J.C., 1978, Gondwana glaciation, cyclothems, continental positioning, and climate change: American Journal of Science, v. 278, p. 1345–1372.

Crowell, J. C, 1983, Ice ages recorded on Gondwanan continents: Geological Society of South Africa Transactions, v. 86, p. 238–261.

Crowell, J. C, 1999, Pre-Mesozoic ice ages: Their bearing on understanding the climate system: Boulder, Colorado, Geological Society of America Memoir 192, p. 1–106.

Crowell, J.C., and Frakes, L.A., 1970, Phanerozoic glaciation and the causes of ice ages: American Journal of Science, v. 268, p. 193–224.

Crowell, J.C., and Frakes, L.A., 1971, Late Paleozoic glaciation: Part IV, Australia: Geological Society of America Bulletin, v. 82, p. 2515–2540.

Crowley, T.J., and Baum, S.K., 1991, Estimating Carboniferous sea-level fluctuations from Gondwana ice extent: Geology, v. 19, p. 975–977.

Crowley, T.J., and Baum, S.K., 1992, Modeling late Paleozoic glaciation: Geology, v. 20, p. 507–510.

Crowley, T.J., Baum, S.K., and Hyde, W.T., 1991, Climate model comparison of Gondwanan and Laurentide glaciations: Journal of Geophysical Research, v. 96, p. 9217–9226.

Cúneo, R., Andreis, R., and Cúneo, A., 1991, Field Guide: Patagonia Excursion, *in* XII International Congress on Carboniferous and Permian Geology and Stratigraphy, Buenos Aires, Museo Argentino de Ciencias Naturales, p. 55.

Daemon, R.F., and Marques-Toigo, M., 1991, An integrated biostratigraphic column for the Parana Basin, Brazil, *in* XII International Congress on Carboniferous and Permian Geology and Stratigraphy, Buenos Aires, Museo Argentino de Ciencias Naturales, p. 25.

Díaz Martínez, E., and Isaacson, P.E., 1994, Late Devonian glacially influenced marine sedimentation in western Gondwana: The Cumana Formation, Altiplano, Bolivia, *in* Embry, A.F., Beauchamp, B., and Glass, D.J., eds., Pangea: Global environments and resources: Calgary, Canadian Society of Petroleum Geologists Memoir 17, p. 511–522.

Dickins, J.M., 1985, Late Palaeozoic glaciation: Journal of Australian Geology and Geophysics, v. 9, p. 163–169.

Dickins, J.M., 1996, Problems of a late Palaeozoic glaciation in Australia and subsequent climate in the Permian: Palaeogeography, Palaeoclimatology, Palaeoecology, v. 125, p. 185–197.

Dickins, J.M., 1997, Some problems of the Permian (Asselian) glaciation and the subsequent climate in the Permian, *in* Martini, I.P., ed., Late glacial and postglacial environmental changes: Quaternary, Carboniferous-Permian, and Proterozoic: Oxford, UK, Oxford University Press, p. 243–245.

Dickinson, W.R., Soreghan, G.S., and Giles, K.A., 1994, Glacio-eustatic origin of Permo-Carboniferous stratigraphic cycles: Evidence from the southern Cordilleran foreland region, *in* Dennison, J.M., and Ettensohn, F.R., eds., Tectonic and eustatic controls on sedimentary cycles: Tulsa, Oklahoma, SEPM (Society for Sedimentary Geology) Concepts in Sedimentology and Paleontology, v. 4, p. 25–34.

Domack, E.W., Burkley, L.A., Domack, C.R., and Banks, M.R., 1993, Facies analysis of glacial marine pebbly mudstones in the Tasmania Basin: Implications for regional paleoclimates during the late Paleozoic, *in* Findlay, R.H., Unrug, R., Banks, M.R., and Veevers, J.J., eds., Gondwana Eight: Assembly, evolution and dispersal: Rotterdam, A.A. Balkema, p. 471–484.

dos Santos, P.R., Rocha-Campos, A.C., and Canuto, J.R., 1996, Patterns of late Palaeozoic deglaciation in the Paraná Basin, Brazil: Palaeogeography, Palaeoclimatology, Palaeoecology, v. 125, p. 165–184.

Du Toit, A.L., 1921, The Carboniferous glaciation of South Africa: Geological Society of South Africa Transactions, v. 24, p. 188–277.

Eyles, N., 1993, Earth's glacial record and its tectonic setting: Earth-Science Reviews, v. 35, p. 1–248.

Eyles, C.H., and Eyles, N., 2000, Subaqueous mass flow origin for Lower Permian diamictites and associated facies of the Grant Group, Barbwire Terrace, Canning Basin, Western Australia: Sedimentology, v. 47, p. 343–356.

Eyles, N., and Young, G.M., 1994, Geodynamic controls on glaciation in Earth history, *in* Deynoux, M., Miller, J.M. G., Domack, E.W., Eyles, N., Fairchild, I.J., and Young, G.M., eds., Earth's glacial record: Cambridge, UK, Cambridge University Press, p. 1–28.

Eyles, C.H., Eyles, N., and Franca, A.B., 1993, Glaciation and tectonics in an active intracratonic basin: The Late Palaeozoic Itararé Group, Paraná Basin, Brazil: Sedimentology, v. 40, p. 1–25.

Eyles, C.H., Eyles, N., and Gostin, V.A., 1998, Facies and allostratigraphy of high-latitude, glacially influenced marine strata of the Early Permian southern Sydney Basin, Australia: Sedimentology, v. 45, p. 121–161.

Eyles, N., González Bonorino, França, A.B., Eyles, C.H., and López Paulsen, O., 1995, Hydrocarbon-bearing late Paleozoic glaciated basins of southern and central South America, *in* Tankard, A.J., Suárez, S.R., and Welsink, H.J., eds., Petroleum basins of South America: Tulsa, Oklahoma, American Association of Petroleum Geologists Memoir 62, p. 165–183.

Fairbanks, R.G., 1989, A 17,000-year glacio-eustatic sea level record: Influence of glacial melting rates on Younger Dryas event and deep-ocean circulation: Nature, v. 342, p. 637–642.

Farabee, M.J., Taylor, T.N., and Taylor, E.L., 1991, Late Permian palynomorphs from the Buckley Formation, central Transantarctic Mountains, Antarctica: Review of Palaeobotany and Palynology, v. 69, p. 353–368.

Foster, C.B., and Waterhouse, J.B., 1988, The *Granulatisporites confluens* Oppel-zone and Early Permian marine faunas from the Grant Formation on the Barbwire Terrace, Canning Basin Western Australia: Australian Journal of Earth Sciences, v. 35, p. 135–157.

Frakes, L.A., and Francis, J.E., 1988, A guide to Phanerozoic cold polar climates from high-latitude ice-rafting in the Cretaceous: Nature, v. 333, p. 547–549.

Frakes, L.A., Francis, J.E., and Syktus, J.I., 1992, Climate modes of the Phanerozoic: Cambridge, Cambridge University Press, 274 p.

França, A.B., 1994, Itararé Group: Gondwanan Carboniferous-Permian of the Paraná Basin, Brazil, *in* Deynoux, M., Miller, J.M. G., Domack, E.W., Eyles, N., Fairchild, I.J., and Young, G.M., eds., Earth's glacial record: Cambridge, UK, Cambridge University Press, p. 70–82.

Francis, J.E., Woolfe, K.J., Arnott, M.J., and Barrett, P.J., 1994, Permian climates of the southern margins of Pangea: Evidence from fossil wood in Antarctica, *in* Embry, A.F., Beauchamp, B., and Glass, D.J., eds., Pangea: Global environments and resources: Calgary, Canadian Society of Petroleum Geologists Memoir 17, p. 275–282.

Garzanti, E., and Sciunnach , D., 1997, Early Carboniferous onset of Gondwanian glaciation and Neo-tethyan rifting in South Tibet: Earth and Planetary Science Letters, v. 148, no. 1-2, p. 359–365.

González, C.R., 1990, Development of the late Paleozoic glaciations of the South American Gondwana in western Argentina: Palaeogeography, Palaeoclimatology, Palaeoecology, v. 79, p. 275–287.

González, C.R., 2001, New data on the late Palaeozoic glaciations in Argentina: Newsletter on Carboniferous Stratigraphy, v. 19, p. 44–45.

González-Bonorino, G., and Eyles, N., 1995, Inverse relationship between ice extent and the late Paleozoic glacial record of Gondwana: Geology, v. 23, p. 1015–1018.

Grunow, A.M., 1999, Gondwana events and palaeogeography: A palaeomagnetic review: Journal of African Earth Sciences, v. 28, p. 53–69.

Hambrey, M.J., and Kluyver, H.M., 1981, Evidence of Devonian or Early Carboniferous glaciation in the Agades region of Niger, *in* Hambrey, M.J., and Harland, W.B., eds., Earth's pre-Pleistocene glacial record: Cambridge, Cambridge University Press, p. 188–189.

Heckel, P.H., 1977, Origin of phosphatic black shale facies in Pennsylvanian cyclothems of mid-continent North America: American Association of Petroleum Geologists Bulletin, v. 61, no. 7, p. 1045–1068.

Heckel, P.H., 1986, Sea-level curve for Pennsylvanian eustatic marine transgressive-regressive depositional cycles along midcontinent outcrop belt, North America: Geology, v. 14, p. 330–334.

Heckel, P.H., 1990, Evidence for global (glacio-eustatic) control over Upper Carboniferous (Pennsylvanian) cyclothems in midcontinent North America, *in* Hardman, R.F. P., and Brooks, J., eds., Tectonic events responsible for Britain's oil and gas reserves: London, Geological Society Special Publication 55, p. 35–47.

Heckel, P.H., 1994, Evaluation of evidence for glacio-eustatic control over marine Pennsylvanian cyclothems in North America and consideration of possible tectonic effects, *in* Dennison, J.M., and Ettensohn, F.R., eds., Tectonic and eustatic controls on sedimentary cycles: Tulsa, Oklahoma, SEPM (Society for Sedimentary Geology) Concepts in Sedimentology and Paleontology, v. 4, p. 65–87.

Heckel, P.H., 1995, Glacial-eustatic base-level—climatic model for Late Middle to Late Pennsylvanian coal-bed formation in the Appalachian Basin: Journal of Sedimentary Research, v. B65, p. 348–356.

Helwig, J., 1972, Stratigraphy, sedimentation, paleogeography, and paleoclimates of Carboniferous ("Gondwana") and Permian of Bolivia: American Association of Petroleum Geologists Bulletin, v. 56, p. 1008–1033.

Herbert, C., 1980, Evidence of glaciation in the Sydney Basin and Tamworth synclinorial zone: New South Wales Geological Survey Bulletin, v. 26, p. 275–293.

Imbrie, J., 1985, A theoretical framework for the Pleistocene ice ages: Journal Geological Society , v. 142, p. 417–432

Imbrie, J., and Imbrie, J.Z., 1980, Modeling the climate response to orbital variations: Science, v. 207, p. 943–953.

Isaacson, P.E., and Díaz Martínez, E., 1995, Evidence for a middle–late Paleozoic foreland basin and significant paleolatitudinal shift, central Andes, *in* Tankard, A.J., Suárez, S.R., and Welsink, H., eds., Petroleum basins of South America: Tulsa, Oklahoma, American Association of Petroleum Geologists Memoir 62, p. 231–249.

Isbell, J.L., 1999, The Kukri Erosion Surface; a reassessment of its relationship to rocks of the Beacon Supergroup in the central Transantarctic Mountains, Antarctica: Antarctic Science, v. 11, p. 228–238.

Isbell, J.L., and Cúneo, N.R., 1996, Depositional framework of Permian coal-bearing strata, southern Victoria Land, Antarctica: Palaeogeography, Palaeoclimatology, Palaeoecology, v. 125, p. 217–238.

Isbell, J.L., Gelhar, G.A., and Seegers, G.M., 1997a, Reconstruction of pre-glacial topography using a post-glacial flooding surface: upper Paleozoic deposits, central Transantarctic Mountains, Antarctica: Journal of Sedimentary Research, v. 67, p. 264–272.

Isbell, J.L., Miller, M.F., Babcock, L.E., and Hasiotis, S.T., 2001, Ice-marginal environment and ecosystem prior to initial advance of the late Paleozoic ice sheet in the Mount Butters area of the central Transantarctic Mountains, Antarctica: Sedimentology, v. 48, p. 953–970.

Isbell, J.L., Seegers, G.M., and Gelhar, G.A., 1997b, Upper Paleozoic glacial and postglacial deposits, central Transantarctic Mountains, Antarctica, *in* Martini, I.P., ed., Late glacial and postglacial environmental changes: Quaternary, Carboniferous-Permian, and Proterozoic: Oxford, UK, Oxford University Press, p. 230–242.

Kemp, E.M., Balme, B.E., Helby, R.J., Kyle, R.A., Playford, G., and Price, P.L., 1977, Carboniferous and Permian palynostratigraphy in Australia and Antarctica: A review: Australian Bureau of Mineral Resources Journal, v. 2, p. 177–208.

Key, R.M., Tidi, J., McGeorge, I., Aitken, G., Cadman, A., and Anscombe, J., 1998, The Lower Karoo Supergroup geology of the south-eastern part of the Gemsbok Sub-Basin of the Kalahari Basin, Botswana: South African Journal of Geology, v. 101, no. 3, p. 225–236.

Kyle, R.A., 1977, Palynostratigraphy of the Victoria Group of south Victoria Land, Antarctica: New Zealand Journal of Geology and Geophysics, v. 20, p. 1081–1102.

Kyle, R.A., and Schopf, J.M., 1982, Permian and Triassic Palynostratigraphy of the Victoria Group, Transantarctic Mountains, *in* Craddock, C., ed., Antarctic Geosciences: Madison, University of Wisconsin Press, International Union of Geological Sciences, p. 649–659.

Langenheim, R.L., 1994, Comparison of Pennsylvanian carbonate cycles in the Cordilleran region with mid-continental cyclothems suggest a common eustatic origin, *in* Dennison, J.M., and Ettensohn, F.R., eds., Tectonic and eustatic controls on sedimentary cycles: Tulsa, Oklahoma, SEPM (Society for Sedimentary Geology), Concepts in Sedimentology and Paleontology v. 4, p. 248.

Limarino, C.O., and Cesari, S.N., 1988, Paleoclimatic significance of the lacustrine Carboniferous deposits in northwest Argentina: Palaeogeography, Palaeoclimatology, Palaeoecology, v. 65, p. 115–131.

Lindsay, J.F., 1966, Carboniferous subaqueous mass-movement in the Manning-Macleay Basin, Kempsey, New South Wales: Journal of Sedimentary Petrology, v. 36, p. 719–732.

Lindsay, J.F., 1969, The glacial origin of Carboniferous conglomerates west of Barraba, New South Wales: Discussion: Geological Society of America Bulletin, v. 80, p. 911–914.

Lindsay, J.F., 1997, Permian postglacial environments of the Australian Plate, *in* Martini, I.P., ed., Late glacial and postglacial environmental changes: Quaternary, Carboniferous-Permian, and Proterozoic: Oxford, UK, Oxford University Press, p. 213–229.

Lindström, S., 1995, Early Permian palynostratigraphy of the northern Heimefrontfjella Mountain-Range, Dronning Maud Land, Antarctica: Review of Palaeobotany and Palynology, v. 89, p. 359–415.

López-Gamundí, O.R., 1987, Depositional models for the glaciomarine sequences of Andean late Paleozoic basins of Argentina: Sedimentary Geology, v. 52, p. 109–126.

López-Gamundí, O.R., 1989, Postglacial transgressions in late Paleozoic basins of western Argentina: A record of glacioeustatic sea level rise: Palaeogeography, Palaeoclimatology, Palaeoecology, v. 71, p. 257–270.

López-Gamundí, O.R., 1991, Thin-bedded diamictites in the glaciomarine Hoyada Verde Formation (Carboniferous), Calingasta-Uspallata Basin, western Argentina: A discussion on the emplacement conditions of subaqueous cohesive debris flows: Sedimentary Geology, v. 73, p. 247–256.

López-Gamundí, O.R., 1997, Glacial-postglacial transition in the late Paleozoic basins of southern South America, *in* Martini, I.P., ed., Late glacial and postglacial environmental changes: Quaternary, Carboniferous-Permian, and Proterozoic: Oxford, UK, Oxford University Press, p. 147–168.

López-Gamundí, O.R., and Rossello, E.A., 1998, Basin fill evolution and paleotectonic patters along the Samfrau geosyncline: The Sauce Grande Basin-Ventana Foldbelt (Argentina) and Karoo Basin-Cape Foldbelt (South Africa) revisited: Geologische Rundschau, v. 86, p. 819–834.

López-Gamundí, O.R., Limarino, C.O., and Cesari, S.N., 1992, Late Paleozoic paleoclimatology of central west Argentina: Palaeogeography, Palaeoclimatology, Palaeoecology, v. 91, p. 305–329.

López-Gamundí, O.R., Espejo, I.S., Conaghan, P.J., and Powell, C.M., 1994, Southern South America, *in* Veevers, J.J., and Powell, C.M., eds., Permian-Triassic Pangean basins and foldbelts along the Panthalassan Margin of Gondwanaland: Boulder, Colorado, Geological Society of America Memoir 184, p. 281–329.

Marshall, J.E. A., 1994, The Falkland Islands: A key element in Gondwana paleogeography: Tectonics, v. 13, p. 499–514.

Masood, K.R., Taylor, T.N., Horner, T., and Taylor, E.L., 1994, Palynology of the Mackellar Formation (Beacon Supergroup) of East Antarctica: Review of Palaeobotany and Palynology, v. 83, p. 329–337.

Matsch, C.L., and Ojakangas, R.W., 1991, Comparison in depositional style of "polar" and "temperate" glacial ice; late Paleozoic Whiteout Conglomerate (West Antarctica) and Late Proterozoic Mineral Fork Formation (Utah), *in* Anderson, J.B., and Ashley, G.M., eds., Glacial marine sedimentation; paleoclimatic significance: Boulder, Colorado, Geological Society of America Special Publication 261, p. 191–206.

Matsch, C.L., and Ojakangas, R.W., 1992, Stratigraphy and sedimentology of the Whiteout Conglomerate—A late Paleozoic glacigenic sequence in the Ellsworth Mountains, West Antarctica, *in* Webers, G.F., Craddock, G.F., and Splettstoesser, J.F., eds., Geology of the Ellsworth Mountains, Antarctica: Boulder, Colorado, Geological Society of America Memoir 170, p. 37–62.

Maynard, J.R., and Leeder, M.R., 1992, On periodicity and magnitude of Late Carboniferous glacio-eustatic sea-level changes: Journal of the Geological Society, v. 149, p. 303–311.

McKelvey, B.C., 1981, Carboniferous tillites in the New England area of New South Wales, *in* Hambrey, M.J., and Harland, W.B., eds., Earth's pre-Pleistocene glacial record: Cambridge, Cambridge University Press, p. 476–479.

McPhie, J., 1987, Andean analogue for Late Carboniferous volcanic arc and arc flank environments of the western New England Orogen, New South Wales, Australia: Tectonophysics, v. 138, p. 269–288.

Melo, J.H.G., 1988, The Malvinokaffric realm in the Devonian of Brazil, *in* McMillan, N.J., Embry, A.F., and Glass, D.J., eds., Devonian of the World: Calgary, Canadian Society of Petroleum Geologist Memoir 14, p. 669–703.

Menning, M., Weyer, D., Drozdzewski, G., and van Amerom, H.W.J., 2000, A carboniferous time scale 2000: Discussion and use of geological parameters as time indicators from central and western Europe: Geologische Jahrbuch, v. A 156, p. 3–44.

Middleton, M.F., 1990, Canning Basin, Geology and Mineral Resources of Western Australia: Perth, Geological Survey of Western Australia Memoir 3, p. 425–457.

Mii, H.-S., Grossman, E.L., and Yancey, T.E., 1999, Carboniferous isotope stratigraphies of North America: Implications for Carboniferous paleoceanography and Mississippian glaciation: Geological Society of America Bulletin, v. 111, no. 7, p. 960–973.

Miller, J.M.G., 1989, Glacial advance and retreat sequences in a Permo-Carboniferous section, central Transantarctic Mountains: Sedimentology, v. 36, p. 419–430.

Moore, R.C., 1964, Paleoecological aspects of Kansas Pennsylvanian and Permian cyclothems, *in* Merrian, D.F., ed., Symposium on cyclic sedimentation: Kansas Geological Survey Bulletin 169, p. 287–380.

Mory, A.J., 1990, Bonaparte Basin, Geology and Mineral Resources of Western Australia: Geological Survey of Western Australia Memoir 3, p. 380–415.

O'Brien, P.E., and Christie-Blick, N., 1992, Glacially grooved surfaces in the Grant Group, Grant Range, Canning Basin and the extent of late Paleozoic Pilbara ice sheets: BMR Journal of Australian Geology and Geophysics, v. 13, p. 87–92.

O'Brien, P.E., Lindsay, J.F., Knauer, K., and Sexton, J., 1998, Sequence stratigraphy of a sandstone-rich Permian glacial succession, Fitzroy Trough, Canning Basin, Western Australia: Australian Journal of Earth Sciences, v. 45, p. 533–545.

Paterson, W.S.B., 1994, The physics of glaciers: Oxford, UK, Pergamon, 480 p.

Pazos, P.J., 2002, The Late Carboniferous glacial to postglacial transition: Facies and sequence stratigraphy, western Paganzo Basin, Argentina: Gondwana Research, v. 5, p. 467–487.

Powell, C. McA., and Li, Z.X., 1994, Reconstruction of the Panthalassan margin of Gondwanaland, *in* Veevers, J.J., and Powell, C.M., eds., Permian-Triassic Pangean basins and foldbelts along the Panthalassan margin of Gondwanaland: Boulder, Colorado, Geological Society of America Memoir 184, p. 5–9.

Ramsbottom, 1979, Rates of transgression and regression in the Carboniferous of NW Europe: Journal of the Geological Society (London), v. 136, p. 147–153.

Roberts, J., Claqué-Long, J.C., and Jones, P.J., 1991, Calibration of the Carboniferous and Early Permian of the Southern New England Orogen by SHRIMP ion microprobe zircon analyses: Newsletter on Carboniferous Stratigraphy, v. 9, p. 15–17.

Roberts, J., Claqué-Long, J.C., and Jones, P.J., 1993, SHRIMP zircon dating and Australian Carboniferous time, *in* 12 Congrès International de Stratigraphie et de Géologie du Carbonifére, Compte Rendu, v. 2, Buenos Aires, p. 319–338.

Roberts, J., Claqué-Long, J.C., and Foster, C.B., 1996, SHRIMP zircon dating of the Permian System of eastern Australia: Australian Journal of Earth Sciences, v. 43, p. 401–421.

Roberts, J., Claqué-Long, J.C., and Jones, P.J., 1995a, Australian Early Carboniferous time, *in* Berggren, W.A., Kent, D.V., Aubry, M.-P., and Hardenbol, J., eds., Geochronology, time scales and global stratigraphic correlation: Tulsa, Oklahoma, SEPM (Society of Sedimentary Geology) Special Publication 54, p. 23–40.

Roberts, J., Claqué-Long, J.C., Jones, P.J., and Foster, C.B., 1995b, SHRIMP zircon age control of Gondwana sequences in Late Carboniferous and Early Permian Australia, *in* Dunay, R.E., and Hailwood, E.A., eds., Non-biostratigraphical methods of dating and correlation: London, Geological Society Special Publication 89, p. 145–174.

Rocha-Campos, A.C., 1981a, Late Devonian Curuá Formation, Amazon Basin, Brazil, *in* Hambrey, M.J., and Harland, W.B., eds., Earth's pre-Pleistocene glacial record: Cambridge, Cambridge University Press, p. 888–891.

Rocha-Campos, A.C., 1981b, Middle-Late Devonian Cabeças Formation, Parnaíba Basin, Brazil, *in* Hambrey, M.J., and Harland, W.B., eds., Earth's pre-Pleistocene glacial record: Cambridge, Cambridge University Press, p. 892–895.

Rocha-Campos, A.C., 1981c, Late Palaeozoic 'Série Tilítica' of Mozambique, *in* Hambrey, M.J., and Harland, W.B., eds., Earth's pre-Pleistocene glacial record: Cambridge, Cambridge University Press, p. 52–54.

Rocha-Campos, A.C., de Carvalho, R., and Amos, A.J., 1977, A Carboniferous (Gondwana) fauna from Subandean Bolivia: Boletim Instituto de Geociencias, v. 6, p. 181–191.

Ross, C.A., and Ross, J.R.P., 1985, Carboniferous and Early Permian biogeography: Geology, v. 13, p. 27–30.

Ross, C.A., and Ross, J.R.P., 1987, Late Paleozoic sea levels and depositional sequences: Cushman Foundation for Formaniferal Research Special Publication 24, p. 137–149.

Ross, C.A., and Ross, J.R.P., 1988, Late Paleozoic transgressive-regressive deposition, *in* Wilgus, C.K., Hastings, B.S., Kendall, C.G. St.C., Posamentier, H.W., Ross, C.A., and Van Wagoner, J.C., eds., Sea-level changes: An integrated approach: Tulsa, Oklahoma, Society of Economic Paleontologists and Mineralogists Special Publication 42, p. 227–247.

Rust, I.C., 1975, Tectonic and sedimentary framework of Gondwana basins in southern Africa, *in* Campbell, K.S.W., ed., Gondwana Geology: Canberra, Australian National University Press, p. 537–564.

Sablock, P.E., 1993, A warm to cold paleoclimate change in Gondwana (Devonian through Carboniferous): A consequence of coupled continental shift and onlap, *in* Compte Rends XII Congrès International de la Stratigraphie et Géologie du Carbonifere et Permien, Buenos Aries, p. 355–367.

Salfity, J., Azcuy, C.L., López-Gamundí, O., Valencio, D.A., and Vilas, J.F., 1987, Cuenca Tarija, *in* Archangelsky, S., ed., El Sistema Carbonífero en la República Argentina: Córdoba, Academia Nacional de Ciencias de Córdoba, p. 15–39.

Saltzman, M.R., González, L.A., and Lohmann, K.C., 2000, Earliest Carboniferous cooling step triggered by the Antler orogeny?: Geology, v. 28, no. 4, p. 347–350.

Scotese, C.R., and McKerrow, W.S., 1990, Revised world maps and introduction, *in* McKerrow, W.S., and Scotese, C.R., eds., Palaeozoic palaeogeography and biogeography: London, Geological Society Memoir 12, p. 1–21.

Seegers, G.M., 1996, Sedimentology of the Permian Mackellar Formation, Central Transantarctic Mountains, Antarctica [M.S. Thesis]: Milwaukee, University of Wisconsin, 106 p.

Sempere, T., 1995, Phanerozoic evolution of Bolivia and adjacent regions, *in* Tankard, A.J., Suárez, S.R., and Welsink, H., eds., Petroleum basins of South America: Tulsa, American Association of Petroleum Geologists Memoir 62, p. 207–230.

Sessarego, H., and Cesari, S.N., 1986, La Zona (de conjunto) Archeosigillaria-Lepidodendropsis del Carbonífero temprano de Argentina: Annual Meeting IUGS-IGCP Project 211, Late Paleozoic of South America Abstracts, p. 69–70.

Smith, L.B., and Read, J.F., 2000, Rapid onset of late Paleozoic glaciation on Gondwana: Evidence from Upper Mississippian strata of the Midcontinent, United States: Geology, v. 28, p. 279–282.

Smith, N.D., Barrett, P.J., and Woolfe, K.J., 1998, Glacier-fed(?) sandstone sheets in the Weller Coal Measures (Permian), Allan Hills, Antarctica: Palaeogeography, Palaeoclimatology, Palaeoecology, v. 141, p. 25–51.

Soreghan, G.S., 1994, Stratigraphic responses to geologic processes: Late Pennsylvanian eustasy and tectonics in the Pedregosa and Orogrande basins, Ancestral Rocky Mountains: Geological Society of America Bulletin, v. 106, p. 1195–1211.

Soreghan, G.S., and Giles, K.A., 1999, Amplitudes of Late Pennsylvanian glacioeustasy: Geology, v. 27, p. 255-258.

Spörli, K.B., 1992, Stratigraphy of the Crashsite Group, Ellsworth Mountains, West Antarctica, *in* Webers, G.F., Craddock, G.F., and Splettstoesser, J.F., eds., Geology of the Ellsworth Mountains, Antarctica: Boulder, Colorado, Geological Society of America Memoir 170, p. 21–35.

Stapleton, R.P., 1977, Carboniferous unconformity in southern Africa: Nature, v. 268, p. 222–223.

Starck, D., 1995, Silurian-Jurassic stratigraphy and basin evolution of northwestern Argentina, *in* Tankard, A.J., Suárez, S., R., and Welsink, H., eds., Petroleum basins of South America: Tulsa, Oklahoma, American Association of Petroleum Geologists Memoir 62, p. 251–267.

Tankard, A.J., Uliana, M.A., Welsink, H., Ramos, V.A., Turic, M., França, A.B., Milani, E.J., de Brito Neves, B.B., Eyles, N., Skarmeta, J., Santa Ana, H., Wiens, F., Cribian, M., López Paulsen, O., Germs, G.J. B., De Wit, M.J., Machacha, T., and Miller, R.M., 1995, Structural and tectonic controls of basin evolution in southwestern Gondwana during the Phanerozoic, *in* Tankard, A.J., Suárez-Soruco, S., R., and Welsink, H., eds., Petroleum basins of South America: Tulsa, American Association of Petroleum Geologists Memoir 62, p. 5–52.

Theron, J.N., and Loock, J.C., 1988, Devonian deltas of the Cape Supergroup, South Africa, *in* McMillan, N.J., Embry, A.F., and Glass, D.J., eds., Devonian of the world: Calgary, Canadian Society of Petroleum Geologist Memoir 14, v. 1, p. 729–740.

Tsubone, K., Neto, A.F. A., and Brazil, I.R., 1991, Tectonics and sedimentation of the Solimões Basin, northern Brazil, *in* Ulbrich, H., and Rocha Campos, A.C., eds., Gondwana Seven Proceedings: Papers presented at the Seventh International Gondwana Symposium: São Paulo, Instituto de Geociências, Universidade de São Paulo, p. 119–132.

Van de Graaff, W.J. E., 1981, Early Permian Nangetty Formation, Perth Basin, Western Australia, *in* Hambrey, M.J., and Harland, W.B., eds., Earth's pre-Pleistocene glacial record: Cambridge, Cambridge University Press, p. 459–462.

Veevers, J.J., 1984, Phanerozoic Earth history of Australia: Oxford, Clarendon Press, 418 p.

Veevers, J.J., 1994, Case for the Gamburtsev Subglacial Mountains of East Antarctica originating by mid-Carboniferous shortening of an intracratonic basin: Geology, v. 22, p. 593–596.

Veevers, J.J., and Powell, C.M., 1987, Late Paleozoic glacial episodes in Gondwanaland reflected in transgressive-regressive depositional sequences in Euramerica: Geological Society of America Bulletin, v. 98, p. 475–487.

Veevers, J.J., and Tewari, R.C., 1995, Gondwana master basin of Peninsular India between Tethys and the interior of the Gondwanaland province of Pangea: Boulder, Colorado, Geological Society of America Memoir 187, 72 p.

Veevers, J.J., Cole, D.I., and Cowan, E.J., 1994a, Southern Africa: Karoo Basin and Cape Fold Belt, *in* Veevers, J.J., and Powell, C.M., eds., Permian-Triassic Pangean basins and foldbelts along the Panthalassan Margin of Gondwanaland: Boulder, Colorado, Geological Society of America Memoir 184, p. 223–279.

Veevers, J.J., Conaghan, P.J., Powell, C.M., Cowan, E.J., McDonnell, K.L., and Shaw, S.E., 1994b, Eastern Australia, *in* Veevers, J.J., and Powell, C.M., eds., Permian-Triassic Pangean basins and foldbelts along the Panthalassan margin of Gondwanaland: Boulder, Colorado, Geological Society of America Memoir 184, p. 11–171.

Veevers, J.J., Powell, C.M., Collinson, J.W., and López-Gamundí, O.R., 1994c, Synthesis, *in* Veevers, J.J., and Powell, C.M., eds., Permian-Triassic Pangean basins and foldbelts along the Panthalassan Margin of Gondwanaland: Boulder, Colorado, Geological Society of America Memoir 184, p. 331–353.

Vergel, M., d.M., 1991, Palynostratigraphy of the Neopaleozoic sequence in the Chacoparanense Basin, Argentina, *in* XII International Congress on Carboniferous and Permian Geology and Stratigraphy, Buenos Aires, Museo Argentino de Ciencias Naturales, p. 83.

Visser, J.N.J., 1982, Implications of a diachronous contact between the Dwyka Formation and Ecca Group in the Karoo Basin: South African Journal of Science, v. 78, p. 249–251.

Visser, J.N.J., 1983, An analysis of the Permi-Carboniferous glaciation in the marine Kalahari Basin, southern Africa: Palaeogeography, Palaeoclimatology, Palaeoecology, v. 44, p. 295–315.

Visser, J.N.J., 1990, The age of the late Palaeozoic glacigene deposits in southern Africa: South African Journal of Geology, v. 93, p. 366–375.

Visser, J.N.J., 1991, Self-destructive collapse of the Permo-Carboniferous marine ice sheet in the Karoo Basin: Evidence from the southern Karoo: South African Journal of Geology, v. 94, p. 255–262.

Visser, J.N.J., 1996, A late Carboniferous subaqueous glacial valley fill complex: Fluctuations in meltwater output and sediment flux: South African Journal of Geology, v. 99, p. 285–291.

Visser, J.N.J., 1997a, Deglaciation sequences in the Permo-Carboniferous Karoo and Kalahari basins of southern Africa: a tool in the analysis of cyclic glaciomarine basin fills: Sedimentology, v. 44, p. 507–521.

Visser, J.N.J., 1997b, A review of the Permo-Carboniferous glaciation in Africa, *in* Martini, I.P., ed., Late glacial and postglacial environmental changes: Quaternary, Carboniferous-Permian, and Proterozoic: Oxford, UK, Oxford University Press, p. 169–191.

Visser, J.N.J., and Praekelt, H.E., 1996, Subduction, mega-shear systems and late Palaeozoic basin development in the African segment of Gondwana: Geologische Rundschau, v. 85, p. 632–646.

von Brunn, V., 1987, A facies analysis of Permo-Carboniferous glacigenic deposits along a paleoscarp in northern Natal, South Africa, *in* McKenzie, G.D., ed., Gondwana Six: Stratigraphy, sedimentology, and paleontology: Washington, D.C., American Geophysical Union, p. 113–122.

Wanless, H.R., and Shepard, F.P., 1936, Sea level and climate changes related to late Paleozoic cycles: Geological Society of America Bulletin, v. 47, p. 1177–1206.

Wanless, H.R., and Weller, J.M., 1932, Correlation and extent of Pennsylvanian cyclothems: Geological Society of America Bulletin, v. 43, p. 1003–1016.

Wendt, J., 1988, Facies pattern and paleogeography of the middle and late Devonian in eastern Anti-Atlas (Morocco), *in* McMillan, N.J., Embry, A.F., and Glass, D.J., eds., Devonian of the world: Calgary, Canadian Society of Petroleum Geologist Memoir 14, v. 1, p. 467–480.

Whetten, J.T., 1965, Carboniferous glacial rocks from the Werrie Basin, New South Wales, Australia: Geological Society of America Bulletin, v. 76, p. 43–56.

White, A.H., 1968, The glacial origin of Carboniferous conglomerates west of Barraba, New South Wales: Geological Society of America Bulletin, v. 79, p. 675–686.

Wopfner, H., and Casshyap, S.M., 1997, Transition from freezing to subtropical climates in the Permo-Carboniferous of Afro-Arabia and India, *in* Martini, I.P., ed., Late glacial and postglacial environmental changes: Quaternary, Carboniferous-Permian, and Proterozoic: Oxford, UK, Oxford University Press, p. 192–212.

Wright, V.P., and Vanstone, S.D., 2001, Onset of late Paleozoic glacio-eustasy and the evolving climates of low latitude areas: A synthesis of current understanding: Journal of the Geological Society (London), v. 158, p. 579–582.

Ziegler, A.M., Hulver, M.L., and Rowley, D.B., 1997, Permian world topography and climate, *in* Martini, I.P., ed., Late glacial and postglacial environmental changes: Quaternary, Carboniferous-Permian, and Proterozoic: Oxford, UK, Oxford University Press, p. 111–146.

Manuscript Accepted by the Society January 22, 2003

Printed in the USA

Geological Society of America
Special Paper 370
2003

Subglacial outburst floods and extreme sedimentary events in the Labrador Sea

John Shaw*
Department of Earth and Atmospheric Sciences, University of Alberta, Edmonton, Alberta T6G 2E3, Canada

Jerome-Etienne Lesemann
Department of Geography, Simon Fraser University, Burnaby, British Columbia V5A 1S6, Canada

ABSTRACT

The meltwater hypothesis for the formation of subglacial landforms is described and discussed in terms of its acceptance. The reluctance by some to accept this hypothesis is outlined in light of a pressing need to treat evidence critically. A call is made for more rigorous evaluation of the meaning of the evidence used to test theories for subglacial landform genesis. One of the main objections to the meltwater hypothesis is its apparent failure to account for the fate of the eroded sediment. This sediment is shown not to be on land, and it is not evident on continental shelves, although there is growing evidence for flood sediments in the deep ocean. Extraordinary gravel deposits around the margins of the Laurentide Ice Sheet are best explained by outburst floods and represent extreme sedimentary events. The magnitude of subglacial outbursts for selected floods is presented, and some simple sediment budget estimates establish that meltwater drainage, with magnitudes of 10^{6-7} m^3/s and total discharge of 8.4×10^4 km^3, transported sediment loads equivalent to about 4.2×10^3 km^3 of surficial sediment and rock. Such extreme erosional events must have sedimentary counterparts. The sediments of the Labrador Sea in the vicinity of the North Atlantic Mid-Ocean Channel are targeted as promising candidates for outburst flood deposits. The morphology of the North Atlantic Mid-Ocean Channel system, with dissected levees, an erosional braid plain, and giant linguoid bedforms marks the passage of immense, hyperpycnal flows. These underflows were more than 100 m deep, 100 km wide, and extended for 4000 km, from the Labrador Sea to the North Atlantic. Detrital carbonate beds within the North Atlantic Mid-Ocean Channel levees reach thickness in excess of 14 m and conform to the levee topography. Powerful currents spilling from the channel evidently deposited them. These beds are separated by bioturbated hemipelagic sediment including ice-rafted deposits. Lithofacies, the virtual absence of biogenic sediment, the lack of bioturbation, magnetic susceptibility, density and colour indicate that the detrital carbonate beds were deposited quickly, as single events. Their lithology and grain shapes and sizes seen in scanning electron microscope scans are explained by subglacial meltwater erosion of carbonate rocks in Hudson Bay, suggesting a direct link between the deposits and meltwater processes. Extreme sedimentary events best explain these carbonate megabeds. Giant meltwater outburst floods such as those inferred from terrestrial evidence are the most likely causes of the events. Much thinner carbonate beds alternate with bioturbated muds. These beds also formed during glaciation and probably represent smaller outbursts than the megabeds. This possibility complicates the interpretation of carbonate events in the Labrador Sea.

Keywords: subglacial floods, meltwater landforms, erosion volumes, submarine channel, levee sedimentation, detrital carbonate flood beds, Heinrich layers, braid plain, climate change.

*John.Shaw@ualberta.ca

Shaw, J. and Lesemann, J.-E., 2003, Subglacial outburst floods and extreme sedimentary events in the Labrador Sea, *in* Chan, M.A., and Archer, A.W., eds., Extreme depositional environments: Mega end members in geologic time: Boulder, Colorado, Geological Society of America Special Paper 370, p. 25–41.

INTRODUCTION

Glaciers and ice sheets store immense amounts of the Earth's freshwater, even in the present day. Great Ice Ages of the past saw Earth in the frozen grip of continental-scale ice sheets, and there is sound evidence suggesting a "snowball Earth" sheathed in ice (Hoffman et al., 1998). We are well aware of the work of ice in sculpting the land surface and transporting and depositing sediment. Indeed, some of the most spectacular landscapes on Earth bear witness to glacial action in high mountains, and vast areas of Earth's surface show subtle changes brought about by continental ice sheets.

There has been less appreciation of the potential geomorphological work by glacial meltwater. It is true that landforms such as the vast arrays of eskers and the major glaciofluvial moraines of the mid-latitude, Pleistocene ice sheets of North America and Scandinavia are attributed to meltwater action. As well, thick and extensive outwash deposits in major valleys such as the Mississippi (Saucier, 1991) and Rhine (Schirmer, 1981) speak of melting ice sheets and immense, braided streams. This geomorphological work is seen as gradual, the product of annual melting related to weather and climate. Yet, some modern ice caps and glaciers discharge stored meltwater in cataclysmic, short-lived discharges capable of eroding and transporting enormous volumes of sediment (Maizels, 1997). These jökulhlaups are the greatest floods on Earth. J Harlen Bretz reconstructed similar, but larger, floods from the landforms of the Washington Scabland, created by discharges measured on the magnitude 10^6–10^7 m^3/s (Bretz, 1923; Baker and Bunker, 1985). Obviously, such spectacular erosional terrain must have a depositional counterpart, yet it was only recently that these sediments were identified on the floor of the Pacific Ocean (Brunner et al., 1999; Zuffa et al., 2000). Zuffa et al. (2000) described 60-m-thick sand beds and suggested that they are the results of single flood events. Baker and Bunker (1985) suggested that these floods endured for a matter of days. Thus, these were truly extreme sedimentation events.

Similarly, large jökulhlaups or outburst floods are inferred from landforms such as drumlins and fields of erosional marks in bedrock extending over thousands of square kilometers in areas formerly occupied by the Laurentide Ice Sheet (Fig. 1; e.g., Shaw, 1996). These floods are considered to be probable causes of abrupt climate change (Shaw, 1989) and were probable reasons for step-like rises in sea level (Blanchon and Shaw, 1994). These outburst floods are estimated to have been of much greater magnitude than the Scabland floods, yet there has been little research on their deposits.

The outburst flood or meltwater hypothesis is not widely accepted, and it is probably for this reason that the question of outburst deposits has not been pursued. Here, we review this hypothesis, drawing attention to questions raised on its validity. The magnitude of the floods and estimates of the volume of sediment they eroded indicate that, if the meltwater hypothesis is true, the sedimentary effects must have been dramatic. Finally, Labrador Sea deposits are investigated in search of evidence for such dramatic sedimentation. Clearly, if there is no evidence for extreme sedimentation events on the ocean floor, the meltwater hypothesis is probably wrong. Thus, the Labrador Sea investigation offers a demanding test of this controversial hypothesis.

THE MELTWATER HYPOTHESIS OF SUBGLACIAL FLOODS

Introduction

Over the past 20 years, the view of subglacial hydrology has expanded to include the likelihood of immense storage reservoirs and regional-scale floods coursing beneath the mid-latitude, Late Pleistocene ice sheets. This view began with the idea that suites of subglacial landforms comprise depositional and erosional landscapes resulting from subglacial floods (Fig. 1; e.g., Shaw, 1996). Following this interpretation, Shoemaker (1992) presented the theory that now underpins our understanding of broad channels, sheet flows and linked-channel-reservoir networks beneath ice sheets. Although both theoretical and observational evidence supports the view that extensive landscapes originated through subglacial meltwater processes, it is not popular among some geologists, oceanographers, and ice-sheet and climate modelers. For example, Aylsworth and Shilts (1989) mapped glacial landforms of the Keewatin sector of the Laurentide Ice Sheet and concluded that there was no evidence of large-scale meltwater activity as advocated by Shaw (1983) and Shaw and Kvill (1984). But drumlins, Rogen moraines, and erosional marks in bedrock (s-forms), all of which are found in great numbers over extensive fields in Keewatin, are the very evidence on which the outburst flood theory is based. There is a pressing need for critical reasoning regarding the significance of evidence used in the interpretation of landforms that have not been observed under formation. Before we can simply assert that there is no evidence, additional reasoning must be undertaken.

In a study of the Coppermine area, Northwest Territories, St.-Onge and McMartin (1995) suggested that drumlins there were formed by subglacial deformation, not meltwater. They arrived at this conclusion despite observations that drumlin axes and clast fabrics within the drumlins are aligned differently, the drumlins carry a stone lag, and they show classical erosional forms corresponding to those produced by turbulent flows (Potschin, 1989). Evidence of this sort has been used elsewhere to argue that drumlins and flutings are a product of meltwater erosion (Shaw et al., 2000). Again, the evidence itself is not the issue; the observations are straightforward and not open to question. It is the way the evidence is used in argument that requires careful evaluation. We believe that St.-Onge and McMartin (1995) err in their argument against the meltwater hypothesis. On the one hand, they state that it is flawed because there are striations, aligned with the drumlins, on bedrock between the drumlins. They correctly point out that the striations are unlikely to have survived the intense abrasion of an outburst flood. Yet, such aligned striations were reported from the Livingstone Lake drumlin field (Shaw and Kvill, 1984) and

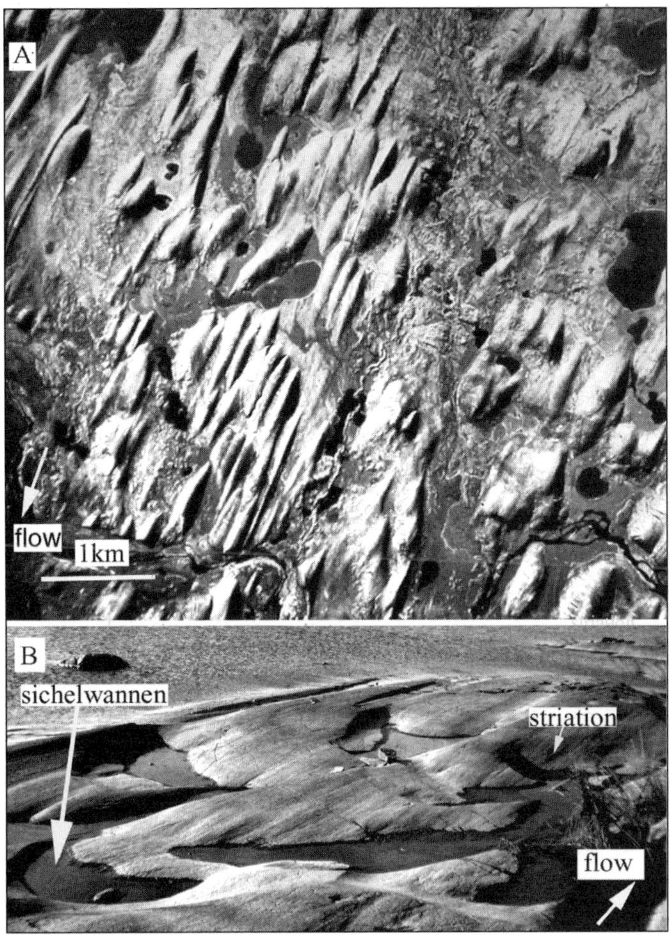

Figure 1. Landforms as evidence for outburst floods. A: Drumlins of Livingstone Lake drumlin field, Saskatchewan. These features are unlike classical drumlins, which are broad upstream and taper downflow. Many drumlins shown here are of the parabolic type with sharply defined, pointed upstream ends and broad, distal slopes that merge with surrounding land surface. They appear like positive counterparts of flutes cut by turbidity currents and are explained as infills of cavities cut upward by meltwater into bed of ice sheets. Eskers are common in the figure, and in places it appears that subglacial streams that formed eskers "scavenged" the drumlins for sediment. B: Sichelwannen eroded in granite, French River, Ontario. These bed forms are part of a field extending over 100 km across the flow. They are identical to classical flutes, and several common form elements are annotated. Rims are sharp and smoothly curved, and rock faces immediately below rims are smooth and do not carry striations. Rock surfaces sloping into flow are heavily striated. Since deeply eroded furrows are not striated and higher rock slopes are, it is most unlikely that glacial abrasion eroded these forms. The striations clearly postdate the formation of the sichelwannen. The remarkable similarity between these forms and flutes produced by turbulent flows supports a meltwater origin. The geographical extent of the field of these erosional bedforms and the coherent paleoflow field require a regional-scale outburst flood for their formation (Kor et al., 1991).

are common on other meltwater forms (Shaw, 1988; Kor et al., 1991). Obviously, these striations were not formed prior to the meltwater forms on which they are carved. Nor could they have formed at the same time as the meltwater marks. But, in each case, the striations could easily have formed *after* the drumlin-forming or bed-scouring flood when the ice sheet settled back to its bed. Seen in this way, what is presented as insurmountable objection to the meltwater hypothesis in fact presents no difficulty. On the other hand, St.-Onge and McMartin (1995) used considerations of sediment transport and deposition as an argument against the meltwater hypothesis. But, rather than refuting the meltwater hypothesis, their concerns argue strongly against their favored hypothesis, the formation of drumlins by subglacial deformation. They point to the absence of major deposits at the former ice margin and argue that it means that the drumlins could not have been eroded by meltwater. But Potschin (1989) showed that the drumlins are *erosional,* and (if subglacial deformation caused this erosion) there should be a major deposit of deformation till. Since there is no such deposit, deformation is an unlikely cause of the drumlins. By contrast, meltwater, unlike deformation processes, is quite capable of carrying sediment well beyond an ice margin. After all, it is the sparseness of deposits around former ice-sheet margins that prompted this study. Of course, the meltwater hypothesis requires that there be sediment associated with erosional tracts of land, but the bulk of this sediment may well lie in deltas and submarine fans fed by glacial meltwater or on the abyssal plains beyond. This paper is concerned with the latter possibility for the Labrador Sea.

Some authors of recently published textbooks dismiss the meltwater hypothesis without clear justification (Hambrey, 1994; Bennet and Glasser, 1996; Benn and Evans, 1998). Bennet and Glasser (1996) describe the meltwater process for drumlin formation briefly and skeptically. They express a preference for the subglacial deformation theory and present it in great detail. The reader is not told *why* one theory is preferred over another, though the extended discussion of the deformation process suggests that it is considered most worthy of attention. Hambrey (1994) also expresses a preference for the subglacial deformation theory, and, without any sound evidence, ties drumlin formation to the work of fast-flowing ice. Benn and Evans (1998) present the meltwater theory with extreme prejudice; it is dismissed as unscientific and non-testable because it states that sorted material in landforms may reflect the landforming process or may simply be sediment of an earlier time making up part of a residual landform. The reality is that sorted material may originate in many landforms as a result of different histories. In deserts, sediment in dunes accumulates as part of the dune-forming processes, and sediment in yardangs is residual.

Flood Implications and Magnitude

Subglacial floods are extreme events even in the modern day, as witnessed by the dramatic 1996 jökulhlaup at Skeidarársandur,

Iceland (Russell and Knudsen, 1990; Worsley, 1997). Reinterpretation of glaciated landscapes suggests the possibility of enormous outburst floods from beneath the mid-latitude Pleistocene Ice Sheets (Shaw, 1996). The diagnosis of such floods in the landscape record has implications for the marine record. It stands to reason that extreme flows from the land must be represented in the marine record: large volumes of rapidly deposited marine sediment are expected to match the immensity of the terrestrial erosion.

Emphasis is placed here on the marine record because the volume of sediment indicated by the extent of erosion does not appear in terrestrial environments. Nevertheless, recent accounts of thick and widespread boulder gravel in marginal zones of the Great Lakes ice lobes call for a serious review of glacial sedimentation associated with these classical lobate moraines (Fisher and Taylor, 2002). A major, subglacial flood path extending from the inner parts of the ice sheet is mapped to the north of the boulder gravel deposits (Kor et al., 1991: Kor and Cowell, 1998), and it is likely that these gravels are a product of regional-scale flood. As well, James Wilde (2001, personal commun.) finds thick sheets of coarse gravel around the margins of the Laurentide Ice Sheet in Montana. These gravels are located at the downstream ends of eroded tracts extending hundreds of kilometers to the north (Rains et al., 1993; Shaw et al., 1996).

An initial search for flood deposits in marine environments was conducted at the Scotian Shelf, where submersible survey, marine seismic traverses, and swath bathymetry revealed mainly eroded sea bed with exposed bedrock, drumlins, and tunnel channels in many areas (Boyd et al., 1988; Loncarevic et al., 1992; Sankeralli, 1998). Thick, Late Wisconsin deposits are absent except in deep tunnel channels (Boyd et al., 1988). Uchupi et al. (2001) report bed forms and sediment lobes on the New Jersey-southern New England continental shelf and slope, which they attribute to catastrophic drainage of lakes, perhaps augmented by subglacial outbursts. At the same time, they present evidence for erosion on the shelf and coast in conjunction with records of gravel deposition in deep water, which they attribute to strong meltwater currents. In retrospect, it is not surprising that the situation on the shelf and land are similar; the shelf was glaciated by grounded ice. Thus, subglacial drainage behaved similarly on what is now shelf and what is now land; it eroded the ground and carried away sediment. Consequently, the bulk of the sediment must be on continental slopes, or, more likely, the abyssal plains. Fluting and drumlins in sediment and bedrock on the Antarctic continental shelf in the vicinity of tunnel channels are interpreted as products of subglacial deformation (Shipp et al., 1999; Anderson et al., 2001). Yet, their formation on till and bedrock and the morphological similarity with hairpin scours generated by horseshoe vortices make it much more probable that they are meltwater forms (Shaw, 1994).

Ó Cofaigh et al. (2002) and Wellner et al. (2001) go part way toward a meltwater interpretation by acknowledging that crescentic scours around the up-flow margins of bedforms indicate meltwater erosion. Yet, these crescentic scours are part of hairpin scours generated by horseshoe vortices in water currents (Shaw, 1994). The associated furrows or lineations are also part of the hairpin scours, and they, too, must be meltwater forms. Lowe and Anderson (2002) recognize abundant p- or s-forms on the Antarctic shelf near Pine Island and interpret them as evidence for extensive, catastrophic meltwater events. At the same time, they consider large-scale lineations to be products of glacial action and deformation. Yet, some similar scale lineations are produced by the wind and others are inferred to be products of subglacial meltwater flow (Shaw, 1994, 1996; Shaw et al., 2000). These fluvial and eolian lineations are erosional, and—if the crescentic scours on the Antarctic Shelf were produced by meltwater—it is again probable that the lineations were formed by erosion, too. Thus, the lineations are as well, if not better, interpreted by meltwater erosion as by subglacial deformation, even if they are in till. This interpretation overcomes the difficulties in the Wellner et al. (2001) explanation of incorporating meltwater into till and in the Ó Cofaigh et al. (2000) explanation of producing sufficient water to erode these bedforms by steady-state melting. Robert Gilbert (2001, personal commun.) also reports suites of bedforms on the shelf around the Antarctic Peninsula with all the attributes of Laurentide subglacial bedforms interpreted as fluvial (Shaw, 1996). Similar bedforms, including flutes and transverse ridges, are reported from the Barents shelf (Solheim et al., 1990), indicating the probability of broad, subglacial meltwater flows across this shelf. The indication, then, for the continental shelf is that sediment transport and erosion dominated over deposition, and a more complete sedimentary record is to be expected on the slope and abyssal plain.

There is some hint of such sediment in the major submarine fans and submarine channel systems connected to drainage pathways of the Pleistocene ice sheets (Shaw et al., 1989; Stelting et al., 1985; Brunner et al., 1999; Zuffa et al., 2000). Gravel in the Mississippi Fan sequence (Stelting et al., 1985) is best explained by high discharges carried along the Mississippi Valley. The floods that converged on the Mississippi Valley, were they from Lake Agassiz drainage (Kehew and Teller, 1994) or outburst floods, would have transported gravels to the Mississippi fan.

Flood Magnitude and Sediment Transport

Before discussing the sediment itself, it is important to grasp the magnitude of the outburst floods and their rapidity. Flood discharges have been calculated in a number of ways for floods in western and eastern Canada and for outbursts from beneath the Pleistocene, Antarctic Ice Sheet (Shaw, 1983; Shaw and Kvill, 1984; Sharpe and Shaw, 1989; Shaw, 1989; Shaw et al., 1989; Shoemaker, 1992, 1995; Rains et al., 1993; Brennand and Shaw, 1994; Sawagaki and Hirakawa, 1997; Beaney and Shaw, 1999; Beaney and Hicks, 2000). Instantaneous discharges are in the range 10^6–10^7 m^3/s. Shaw et al. (1989) calculated the volume of flow for a filament of the Livingstone Lake event to have been about 8.4×10^4 km^3, sufficient to raise sea level by 0.23 m. Assuming a suspended sediment concentration of 100 g/L for this floodwater (see Maizels, 1997) and a density of 2000 kg/m^3 for the sediment and rock removed by erosion, the total volume of sediment transported by the flood would have been 4.2×10^3 km^3. Taking the flood path to be about 1200 km long and 150 km wide, the thickness of surficial sediment and rock removed along this track would have been about 23 m. This estimate, which is based on plausible values, implies that the average height of erosional bedforms is less than the average depth of erosion. Residual landforms within the inferred flood tracts are generally about 10–30 m high (Munro-Stasiuk and Shaw, 1997; Beaney and Shaw, 1999; Shaw et al., 2000). The areas between the landforms were eroded to a depth at least equal to the height of the form. Consequently, the estimate is of the right order.

Another way of making this calculation is to assume that, if such landforms as drumlins and hummocky terrain are erosional and remnant, a volume of sediment greater than the space between the landforms themselves must have been removed. We can estimate this volume conservatively by taking the surface just touching landform crests and calculating the volume between it and the present land surface. This volume is a minimum estimate of the eroded material because it is probable that the landform crests were also eroded and are now below the antecedent land surface. As well, late-stage deposition may blanket the land surface, partially filling depressions and channels and masking erosion that occurred in these low areas.

A 100 km^2 area of the Peterborough, Ontario, drumlin field was digitized to estimate the volume of sediment removed during drumlin formation. A first digital terrain model (DTM) was obtained for the present land surface. A second DTM of the former land surface was generated using the high points of prominent drumlins to form an elevation grid. Two-dimensional multiquadratic functions were then used to fit surfaces to the elevation values (Hardy, 1971). The volume removed was calculated by subtracting the volume under the modern surface from that beneath the initial surface, where volume is calculated as $\sum$area $\times$ elevation for the DTM grids. The total volume removed is estimated to be 2.3 km^3, equivalent to 23 m of stripping averaged over the 100 km^2 area.

The two estimates for the average depth of erosion during drumlin and flood tract formation turn out to be the same. Given the crude measurements and very general assumptions used in the calculations, it would be misleading to place too much significance on this coincidence. Nevertheless, that the results are plausible and similar justifies extending this order of magnitude approach to broader considerations of the sediment removed from the area of the Laurentide Ice Sheet draining to Hudson Strait. This area was about 1.7×10^6 km^2 (Dowdeswell et al., 1995); with a 20 m depth and assuming flood scouring over 50% of the area, the total volume of sediment removed would have been about 1.7×10^3 km^3

Equally significant is the total amount of meltwater released from the Laurentide Ice Sheet during flood events (Blanchon and Shaw, 1994). An estimate equivalent to several meters rise in sea level is conservative if floods around the Laurentide Ice Sheet were simultaneous (Brennand et al., 1995; Shaw, 1996).

LABRADOR SEA: MORPHOLOGY AND DEPOSITS

As already discussed, it is clear that a sedimentary counterpart of the volume of material inferred to have been transported by floods is not to be found on the continent or on the continental shelf. Attention is thus directed toward more distal marine environments. The 1999 cruise of the Marion Dufresne was partly dedicated to a search for potential flood deposits on the abyssal floor of the Labrador Sea in the vicinity of the North Atlantic Mid-Ocean Channel and the remarkable braid plain that skirts this channel (Hesse and Rakofsky, 1992; Klaucke and Hesse, 1996; Klaucke et al., 1998; Hesse et al., 2001).

The Labrador Sea Abyssal Plain

The Labrador Sea lies between Greenland and Labrador, Canada, and opens southward into the North Atlantic Ocean. Canyons extending from channels or saddles on the Labrador continental shelf lace the slope and seafloor on the Canadian side (Fig. 2). These canyons dissect the continental slope and act like Yazoo streams on the abyssal plain, where they course alongside the North Atlantic Mid-Ocean Channel, prevented from joining this trunk channel by a high levee along its western bank (Fig. 3; Hesse, 1989; Klaucke et al., 1998). The North Atlantic Mid-Ocean Channel, which extends the length of the Labrador Sea, is of low sinuosity and is joined by major tributaries. Levees on the right and left banks of the channel differ in size, because the underflows that formed them fall under the influence of the Coriolis effect and deposit more on the right levee (Fig. 3).

A sandy braid plain runs alongside the North Atlantic Mid-Ocean Channel and, in places, the flows that deposited the braid plain have breached and highly dissected the levees and deposited giant linguoid bars across the North Atlantic Mid-Ocean Channel (Figs. 4 and 5; Hesse et al., 2001). Multibeam bathymetry illustrates that the bars completely obliterate the North Atlantic Mid-Ocean Channel in places (Fig. 5), indicating that the channel has not been reestablished since the braid plain was formed.

The braid plain shows sinuous flutes or grooves, covering immense tracts of the ocean floor. Individual flutes are several tens of kilometers long, and the flutes appear in clusters tens of kilometers wide (Fig. 5). These flutes have the form of fluting in glaciated areas (Smith, 1948; Evans, 2000; Shaw et al., 2000; Munro-Stasiuk and Shaw, 2002; Clark et al., 2000). They also have morphological counterparts in yardangs, where the flutes are formed between the yardangs (McCauley et al., 1977). Similar large-scale lineation is observed on Mars (Lucchitta, 1982). Recently, similar fluting has been reported in till on glaciated continental shelves (Shipp et al., 1999). Fluting is also noted in areas of meltwater flow beyond ice sheet margins (Baker and Bunker, 1985; Kehew and Lord, 1986). Although there is debate about the origin of such fluting—was it of glacial or fluvial origin?—there can be no doubt that the braid plain fluting is not of glacial origin; the water depth is between 3 and 4 km. Rather, fluting on this regional scale and under such deep water lends credibility to

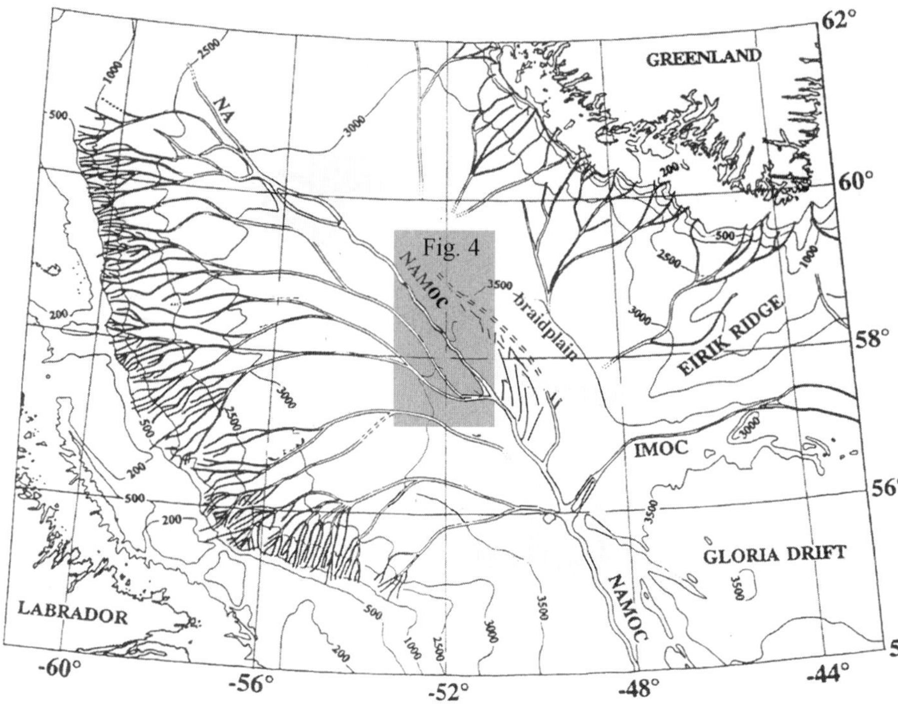

Figure 2. Labrador Sea drainage system (adapted from Klaucke et al., 1998).

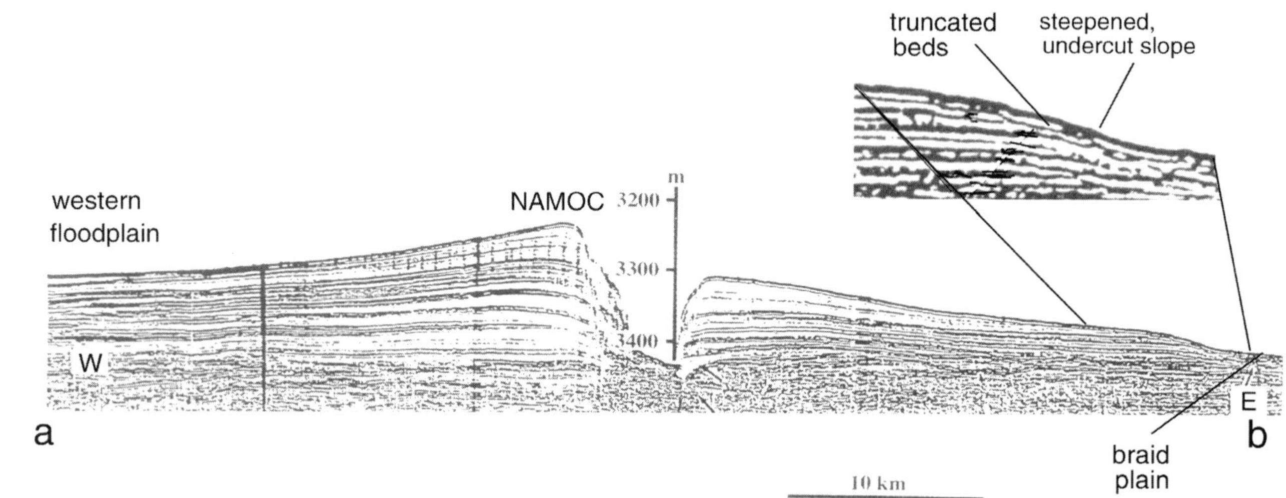

Figure 3. Sleeve gun seismic profile across North Atlantic Mid-Ocean Channel, levees, and braid plain (after Klaucke et al., 1998). See Figure 4 for location. The steepening of the levee profile and truncation of strata at levee toe indicate that the formation of the braid plain is, in part, an erosional process. Dissection of the levees also points to the erosive power of the braid plain flows.

fluvial interpretations of fluting formed subglacially elsewhere. This is the same point as was made earlier regarding large-scale lineations on the Antarctic Shelf. The flutes on the braid plain surface clearly record broad flows of enormous magnitude that descended from the continental shelf, off Hudson Strait (Hesse et al., 2001).

Hesse et al. (2001) described the braid plain in detail and recorded thick, sandy units beneath it. As well, the braid plain sediments commonly rest on erosional surfaces cut into levee deposits, and numerous unconformities are seen in seismic sections in a zone of transition between the braid plain and the levee. In other seismic sections, interfingering between the coarse-grained braid plain deposits and the fine-grained levee deposits that resulted from spill-over from the North Atlantic Mid-Ocean Channel is clearly visible (Hesse et al., 2001).

Labrador Sea Sediments

The Calypso cores obtained from three locations on the North Atlantic Mid-Ocean Channel levees are all over 30 m long, affording access to a much longer sedimentary record than any

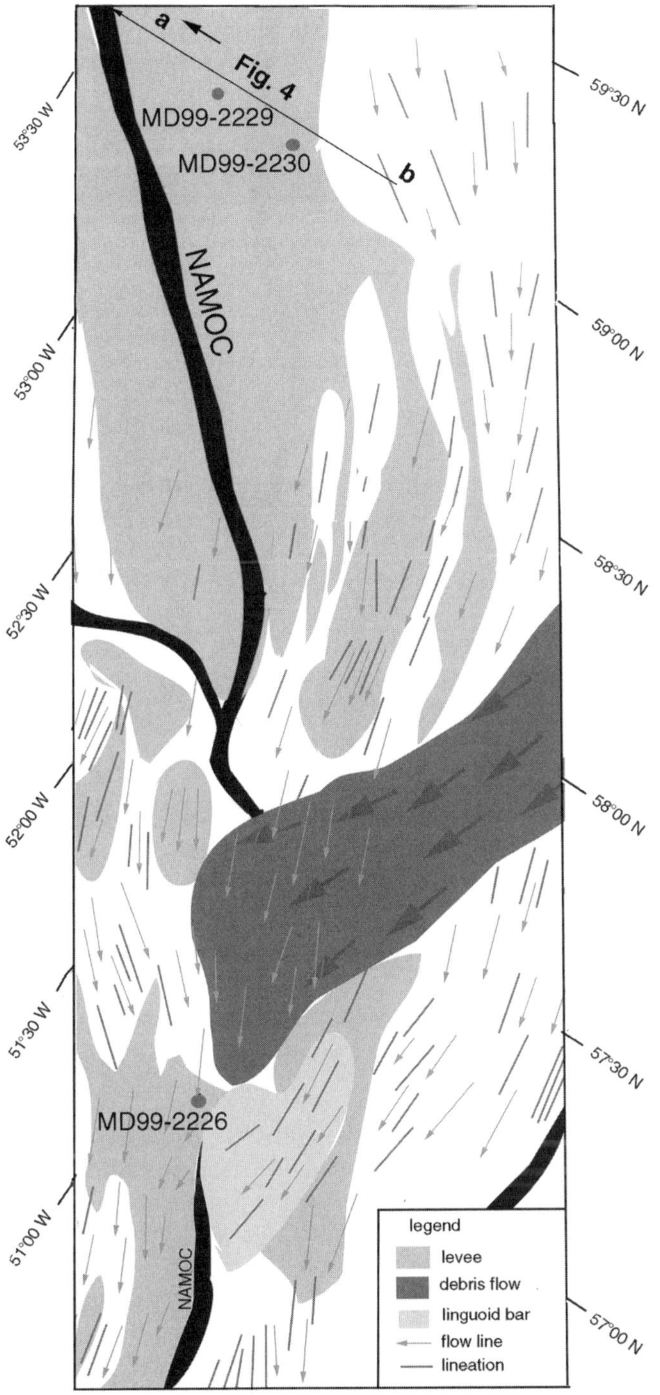

Figure 4. Northwest Atlantic Mid-Ocean Channel (NAMOC) system based on HAWAII-MR-1 side-scan sonar imagery. Flowlines for braid-plain complex are plotted from flutings (lineations) and indicate broadly sinuous, coherent flow. Flow lines indicate that these flows swept over levee remnants and must have been sufficiently deep to submerge morphological elements of the seabed in excess of 150 m high. Note also the linguoid bar deposited across the North Atlantic Mid-Ocean Channel by flow entering the channel zone from the braid plain. Thus, the levees were dissected from the "outside in."

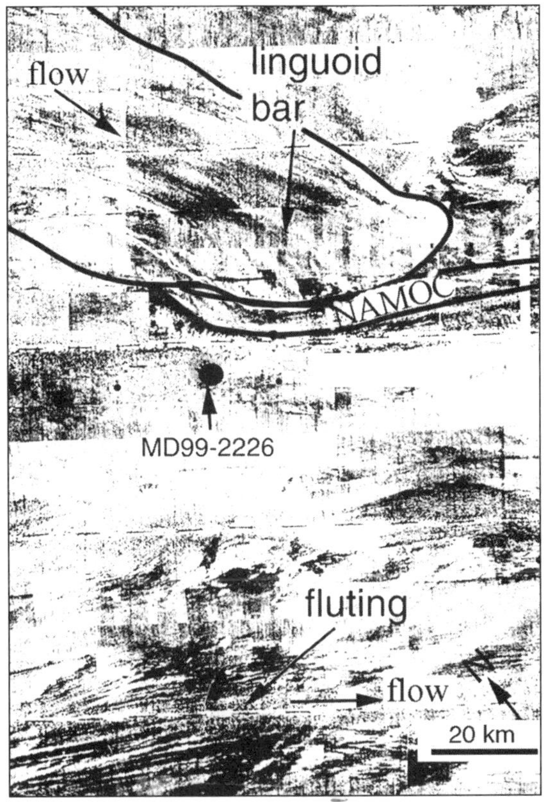

Figure 5. Large-scale fluting and linguoid bar in the vicinity of site MD99-2226. The linguoid bar shows superimposed lineations and streamlined residual ridges, indicating flow from the north obliquely across the North Atlantic Mid-Ocean Channel. Fluting on the scale shown is common in formerly glaciated areas and is usually interpreted as evidence of direct glacial action, fast glacier flow, and a deforming glacier bed. In the meltwater hypothesis, it is interpreted as the product of meltwater sheet flows. Obviously, direct glacial action is ruled out for depths of water at ~3,000 m and a broad, hyperpycnal, underflow is the only reasonable way to explain these flutings. The fluting and linguoid bar indicate the immense scale of braid plain flows.

previous cores. Three cores are described and interpreted in this paper (Figs. 6, 7, and 8). The locations of the cores are also shown in Figures 6, 7, and 8. For two of the cores, MD99-2226 and MD99-2230, lithologies are described in detail with Geotek Multisensor Track (MST) and spectrophotometer data. For core MD99-2629, only MST physical properties are presented, with an X-radiograph of a short interval of a detrital carbonate bed.

MST took high-resolution measurements of physical properties of the core sediment. Magnetic susceptibility was measured using a Barington loop sensor (MS2B) in which a low intensity, non-saturating, alternating magnetic field was produced by an oscillator circuit. Changes in the oscillator frequency caused by the core sediment were converted into magnetic susceptibility values (SI units).

Bulk density was measured using gamma ray attenuation. A 10-millicurie Caesium-137 capsule and a sodium iodide scintillation detector are mounted diametrically across the core. Photon

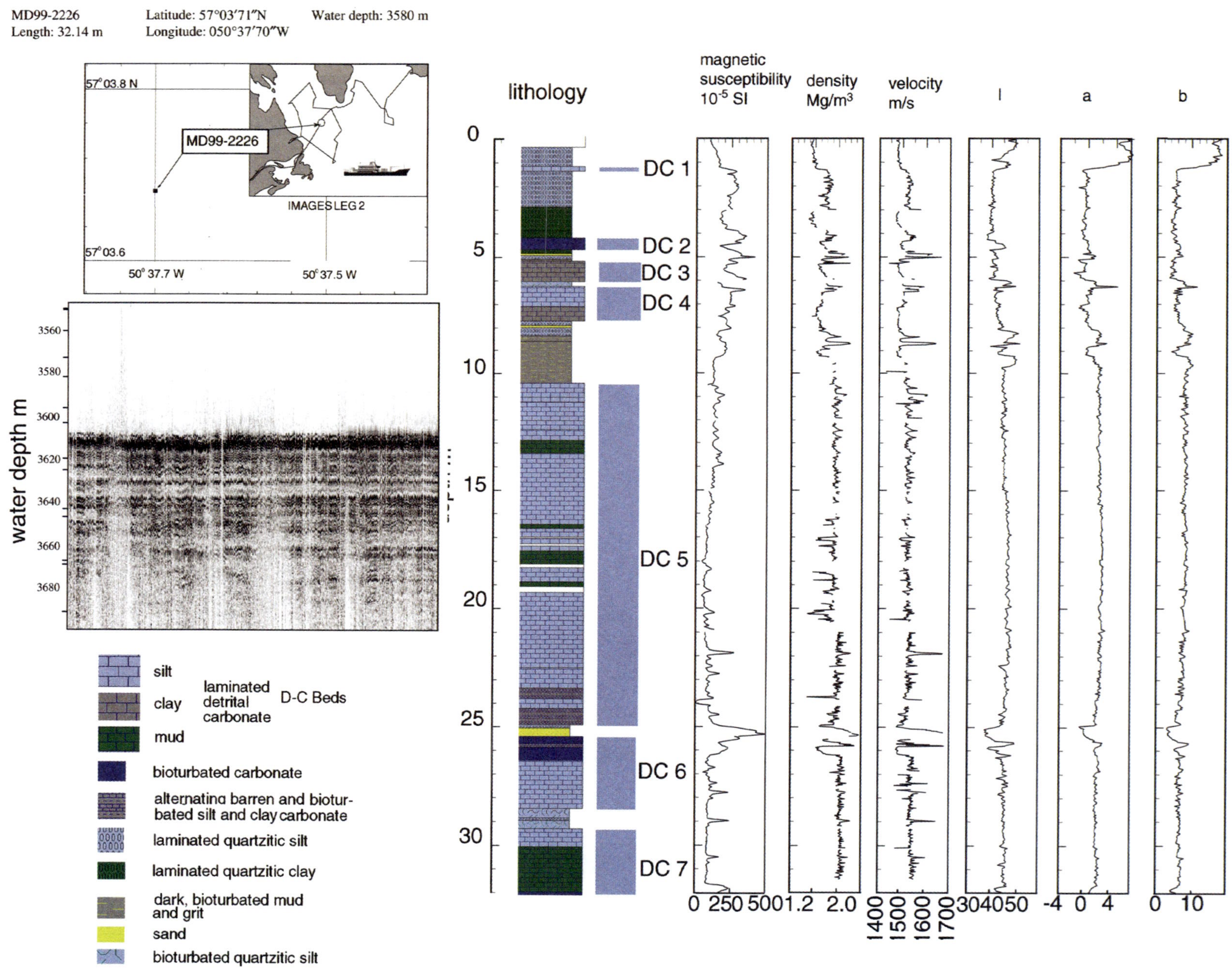

Figure 6. Lithological and physical properties of core MD99-2226 taken at the crest of the western (right) levee of the North Atlantic Mid-Ocean Channel. The 3.5-kHz seismic profile is from the coring site.

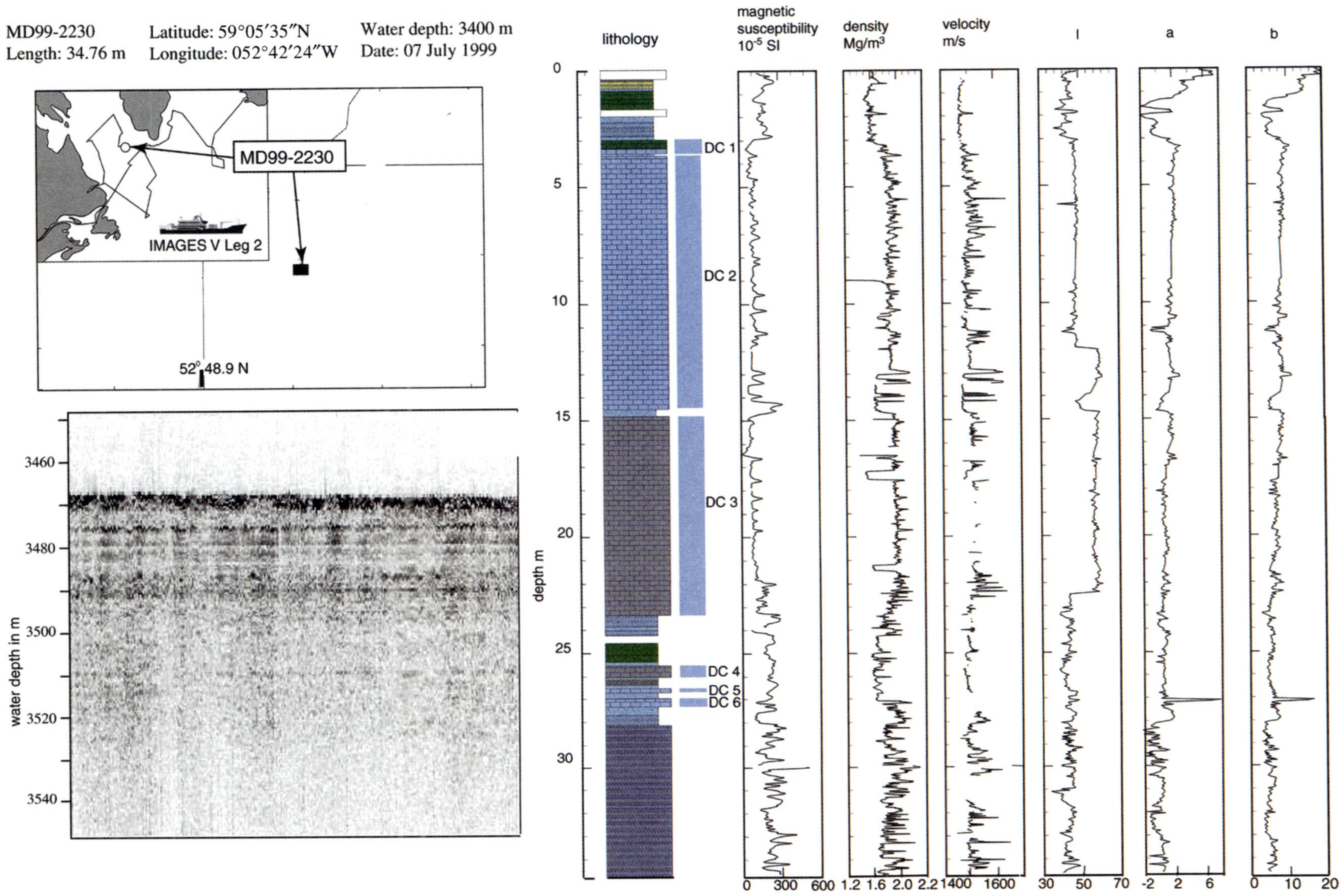

Figure 7. Lithological and physical characteristics of core MD99-2230 from the eastern (left) levee of the North Atlantic Mid-Ocean Channel. The 3.5-kHz seismic profile is from the coring site. See Figure 6 for lithological legend.

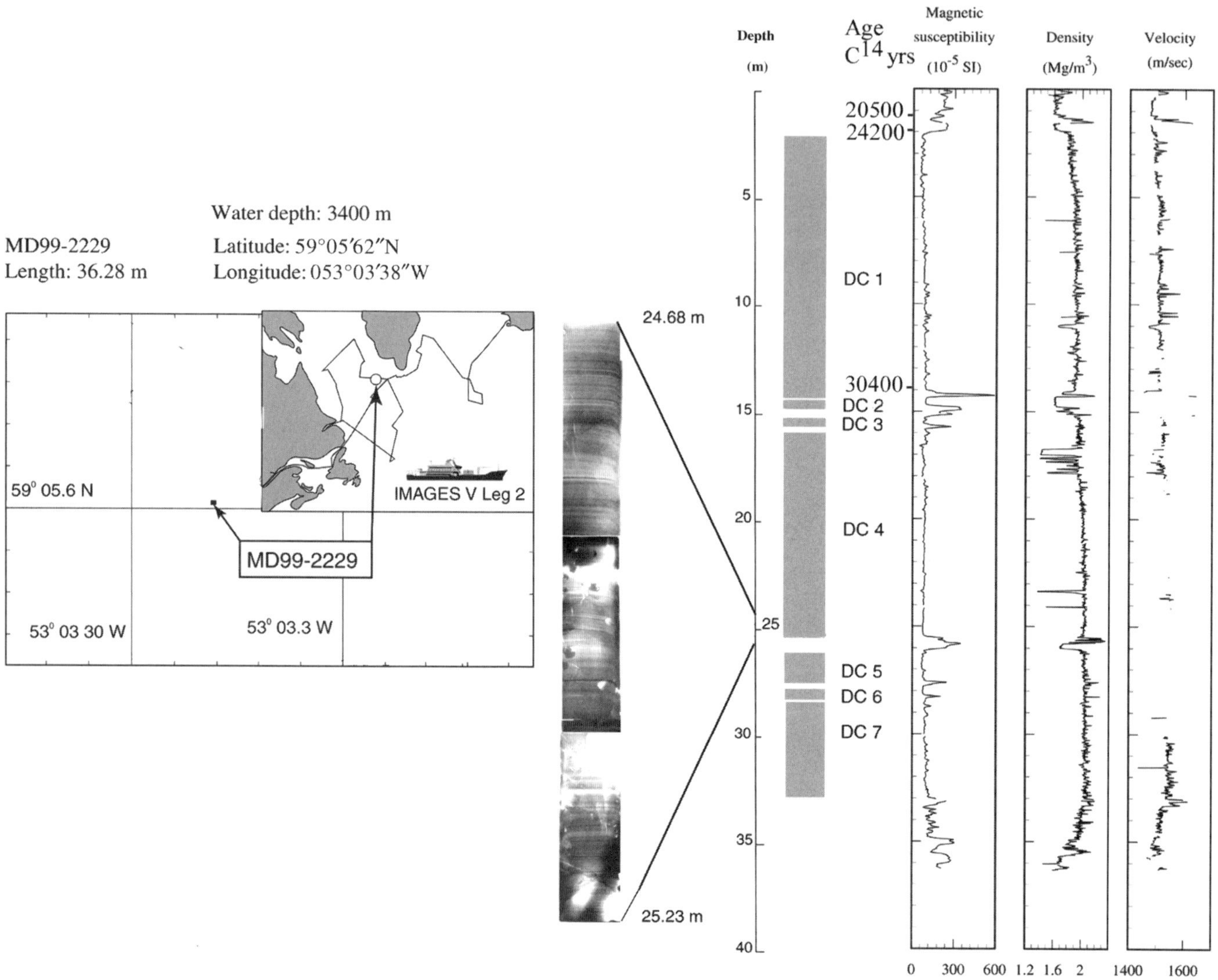

Figure 8. Physical properties of core MD99-2229 taken near junction of braid plain and toe of braid plain's eastern (left) levee. Detrital carbonate layers are recognized by their low magnetic susceptibility. Step-like changes in density from one detrital carbonate layer to another and the relatively constant density within a layer mark consolidation, showing that layers were deposited in a relatively short period with relatively long periods between them. The X-radiograph shows the style of lamination in detrital carbonate beds.

scattering causes attenuation loss that is measured by the detector. The bulk density of the core is calculated in comparison with the attenuation of the gamma rays through aluminum with a correction for pore-water hydrogen.

Wave velocity was measured using spring-loaded, compressional-wave transducers and rectilinear-displacement transducers. Velocities were calculated for 500-kHz compressional wave pulses produced by the transmitting transducer at a repetition rate of 1 kHz. The P-wave travel time is corrected for the delay caused by the core liner and the electronics of the system.

Spectrophotometry on split cores was taken with a Minolta CM-2002 handheld spectrophotometer soon after core splitting. The cores were cleaned, and, for the sections that were soupy, excess water was removed before the surface to be imaged was dried. Measurements were made every 5 cm along the core, and a white calibration was performed at the end of each section. Spectral reflectance was measured in the range 400–700 nm, divided into 31 channels. Measurements are presented in the **L, a, b** model of the Commission Internationale d'Eclaraige (CIE). In this model, the letters express color difference: **L** is the lightness component, whereby +**L** lightness to –**L** darkness; **a** represents the difference in redness and greenness with +**a** redness to –**a** greenness; and **b** represents the difference in yellowness and blueness with +**b** yellowness to –**b** blueness. The lightness component is particularly effective for distinguishing detrital carbonates, which—being light in color—show high values of **L**.

The seismic records and bathymetry indicate thinning sedimentary wedges away from the levee crest (Fig. 3), and, in the vicinity of the core sites, the levees are about 180 m and 110 m above the channel bed for the western and eastern banks, respectively.

From their geometry and architecture, it is clear that the levee sediments were deposited as spillover sedimentation from turbidity currents in the North Atlantic Mid-Ocean Channel. The coarse nature of the braidplain deposits suggests that these are more in keeping with bed-load transport. Consequently, they probably originated from direct meltwater input from the ice-sheet to the slope proximal to the braid plain. This view is supported by the more westerly extension of the braid plain relative to the North Atlantic Mid-Ocean Channel (R. Hesse, 2002, personal commun.). Previous observations on the levee sediment indicate that deposition rates were highest on the western levee as a result of Coriolis effects (Hesse et al., 1987; Klaucke and Hesse, 1996).

Cores mainly reveal laminated or massive silt and clay, which Hesse and Chough (1980) and Klaucke et al. (1998) interpret to have been deposited from suspension in spill-over flows. Thick, meter-scale detrital carbonate units correlate with Heinrich layers and are considered to register meltwater events (Hesse and Khodabakhsh, 1998). Similar, detrital carbonate layers, described from the Labrador Sea and Baffin Island shelf, are explained by meltwater and iceberg events (Hillaire-Marcel et al., 1994; Andrews et al., 1998). Thus, the modern view is that these beds in ice-proximal regions of the Labrador Sea are considered to reflect more complex processes than the simple ice rafting of the Heinrich layers in the North Atlantic (Heinrich, 1988; Bond et al., 1992). However, the nature of the meltwater events in terms of sources, mechanisms of release, and magnitudes is not clearly defined. Understanding these aspects of the meltwater events, which are probably key to explaining the climatic changes that accompanied the deposition of the detrital carbonate beds, is the main reason for the sedimentological examination of the Marion Dufresne cores.

AMS Dating

Three accelerator mass spectrometry (AMS) dates were obtained from picks of *N. pachyderma* from core MD99-2229 (courtesy of Harunur Rashid). The dateable units in the sediments are limited, due to the prevalence of barren, detrital carbonate beds in which the only organic material is broken fragments of foraminifera and spicules. The dates obtained with sample depths are: 24,200 ± 110 ^{14}C yr at 1.95 m, and 30,400 ± 340 ^{14}C at 14.81 m. These dates are uncorrected for reservoir effect, and a correction of 450 yr may be applied.

Detrital Carbonate Facies

Description

The thickness of these beds is measured in meters, with a maximum thickness of about 15 m. Detrital carbonate accounts for >30% of their volume, and the grain size varies from sandy mud to silty clay to clayey silt. The sand fraction may be introduced by core disturbance and should, perhaps, not be considered as a primary characteristic. Occasional sand or coarse silt grains within silty clay and clayey silt are usually quartz. The carbonate grains are angular to subangular. The detrital carbonate beds appear massive to the eye, but X-radiographs show graded and usually laminated layers. The graded laminae are irregular in thickness, grading is continuous in that it is not interrupted by other styles of bedding, and there are few sharp internal boundaries (Fig. 8). The detrital carbonate beds are light and invariable in color (5Y 4/1) and show a systematically low magnetic susceptibility because of their high carbonate content (Figs. 6, 7, and 8). Their bulk density is generally low as a result of high water content. While the bulk density varies about a relatively constant value for an individual detrital carbonate bed, this value decreases upward from one bed to another in a step-like fashion. This step-wise decrease in density marks bed consolidation and indicates that the beds were deposited over a short time, and a much longer time elapsed between detrital carbonate events. Since the beds between the detrital carbonate beds are very thin, the deposition rates during detrital carbonate events must have been extremely high. This conclusion follows from the low rate of change of density within carbonate beds compared to the abrupt change within the intervening beds. Evidently there is little difference in consolidation within the carbonates, which indicates rapid deposition.

Infilled cavities caused when sections of the core were pulled apart by stretching during coring are common in this facies. Consequently, corrections are required for measured bed thickness. Detrital carbonate 2 in Core MD 99-2230 (Fig. 7) is 10.45 m thick if injected cavity fills are included. The actual bed thickness, after subtraction of 1.62 m (16%) of cavity fill, is 8.53 m. The thickness of detrital carbonate 3 is 8.30 m prior to adjustment for cavities; subtraction of a cavity thickness of 1.49 m (18%) results in a corrected thickness of 6.81 m. Similar adjustments were not made to the detrital carbonate beds of MD99-2239, though a correction of ~17%—the average of the values for core MD99-2230—to the measured thicknesses is probably in order.

Outsized clasts attributable to ice rafting are rare in the detrital carbonate facies. Only two such clasts were found in MD99-2226. There was one outsized, granitic clast at 7.30 m in the core and a basaltic clast at 8.90 m, both within the same carbonate bed. No outsized clasts were noted in the carbonate facies of MD99-2230.

The two uppermost carbonate layers in MD99-2226, detrital carbonate 1 and 2 (Fig. 6) coarsen upward to sandy mud, and the coarser layers include soft-sediment clasts of detrital carbonate in silty clay and clayey silt. The soft-sediment clasts indicate erosion of previously deposited carbonate beds and the coarsening upward points to increased current strength. Lithic granules are found within detrital carbonate 4 at 808–820 cm. Detrital carbonate 3 is thicker than detrital carbonates 1 and 2, and it is made up of relatively thick, alternating units of clayey silt and silty clay. Two quartzitic sand units (at 21.85 m and 22.95 m) within the detrital carbonate 5 mark sudden increase in flow power. Detrital carbonate 7 is dominated by clayey silt, although the full thickness of this unit, which is lowermost in the core, was not sampled.

Megabeds of great relative thickness stand out among the detrital carbonate beds. In core MD99-2226, beds detrital carbonate 5, detrital carbonate 6, and detrital carbonate 7 have uncorrected thicknesses of 14.7 m, 3 m, and 2.5 m, respectively, and are considerably thicker than the other detrital carbonate beds in the core (Fig. 6). Of the beds in core MD99-2230, detrital carbonate 2 at 10.45 m and detrital carbonate 3 at 8.3 m stand out (Fig. 7). Three major carbonate beds are recognized from the magnetic susceptibility measurements on core MD99-2229. Detrital carbonate 1, detrital carbonate 4, and detrital carbonate 7 are 12.2 m, 9.5 m, and 4.5 m thick, respectively (Fig. 8). The proximity of cores MD99-2229 and MD99-2230 suggests that they should correlate well. This is the case, and there is little doubt that detrital carbonate 1 in MD99-2229 correlates with detrital carbonate 2 in MD99-2230 (Figs. 7 and 8). Their thicknesses of 12.2 m and 10.45 m, respectively, are considerably greater than for other beds. They show similar position relative to the floor of the ocean and relative to the next thickest bed. The next thickest beds in cores MD99-2229 and MD99-2230 are detrital carbonate 4 and detrital carbonate 3, which are 9.5 m and 8.4 m, respectively. In both cases, the correlated beds show the expected thinning toward the toe of the levee (Fig. 3). Correlation between these two proximal cores and the distal MD99-2226 is less straightforward. Detrital carbonate 5, in MD99-2226, has a thickness of 14.7 m and correlates well with detrital carbonate 1 (12.2 m) and detrital carbonate 2 (10.45 m) of cores MD99-2229 and MD99-2230, respectively. Since core MD99-2226 is from close to the levee crest, the expected sequence of thinning toward the toe is also observed, though the effects of thinning downflow cannot be take into consideration here. However, correlation for beds below detrital carbonate 5 in core MD99-2226 is problematical, since the next bed is only 3 m thick as compared with ~9 m in the other two cores. It is possible that detrital carbonate 6 and detrital carbonate 7 in MD99-2226 should be combined, in which case the carbonate events are well matched.

The detrital carbonate beds denoted as detrital carbonate in Figures 6, 7, and 8 are selected on the basis of their carbonate content, the absence of biogenic sediment within the carbonate beds, and their lack of bioturbation. There is also a very low incidence of ice-rafted debris. The absence of bioturbated intervals and distinct intervals of ice-rafted debris may indicate rapid and continuous sedimentation of each bed. Alternatively, the detrital carbonate units may record many turbidites from numerous events, the time between events being too short for benthic fauna to become established. The intervening time must also have been too short for the accumulation of horizons of ice-rafted debris. The carbonate grains are mainly angular and are commonly in the form of flakes or tabular particles of calcium carbonate, even down to clay sizes (Fig. 9). Larger grains have smaller grains cemented to them; these cemented grains remained attached despite vigorous ultrasound treatment prior to SEM imaging (Fig. 9). An SEM image of the surface of fluvially abraded Paleozoic carbonate rock from Wilton Creek, Ontario, shows that similar angular particles were

Figure 9. Scanning electron microscope (SEM) images of detrital carbonate grains and their potential source. A: SEM image of highly polished Paleozoic carbonate rock surface. Sample is from a fluvially abraded surface covered by scallops, sinuous grooves, and spindle-form erosional marks, Wilton Creek, Ontario (Shaw, 1988). Note angular "scars," where scale-shaped particles were removed by abrasion. B: Angular carbonate grains from core MD99-2230. Grains are well sorted, mostly clay sized, and grain shape corresponds closely to that of grains removed from rock surface in A. C: Angular carbonate grain with finer grains attached (center). Many smaller grains are platy.

removed in the erosion process (Fig. 9). Particles attached to the surface, which were probably close to being removed, are also angular like those in the detrital beds.

Interpretation

The thick detrital carbonate beds are clearly integral to the internal architecture of the North Atlantic Mid-Ocean Channel levees and, consequently, the deposits probably originated in flows spilling from the channel itself. In keeping with this view, where detrital carbonate beds are correlated with each other, the thickest is in MD99-2226, which was located near the crest of the levee; the thinnest is in MD99-2230, located at the junction of the levee and the braid plain (Fig. 4). In the absence of bioturbation, the beds represent continuous sedimentation of detrital carbonate or sedimentation events, which prevented continuous occupancy by benthic fauna, and between which any pauses were too short for faunal establishment. The graded laminae in the carbonates, which are not interrupted by other styles of deposition, show few sharp boundaries and are of irregular thickness. This suggests deposition by a gradually varying flow, perhaps related to large-scale turbulent eddies rather than a series of discrete events. The sorting suggests transport by currents rather than by icebergs, and the graded lamination indicates deposition from suspension, as would be expected for spillover sedimentation. The angularity of the particles suggests that they were never abraded during bed-load transport.

Bioturbated and coarser-grained beds in the cores show a high proportion of quartz particles. The dominance of carbonate in the detrital carbonate beds suggests that they are of a distinct provenance (Andrews and Tedesco, 1992; Hesse et al., 1997) and that either the quartz source was swamped by carbonate or that the quartz source was not providing sediment at the time of detrital carbonate sedimentation. The former is the most likely, and in conjunction with the evidence of continuous sedimentation, it suggests that the thick carbonate beds represent huge sedimentary events. The virtual absence of drop-stones from icebergs, in an area where they are commonplace on the seabed, also points to a very high rate of sedimentation or a dearth of icebergs when the carbonate beds were deposited.

These findings follow those of others who recognized the detrital carbonate beds as marking significant events correlated with the Heinrich events of the north Atlantic (Andrews and Tedesco, 1992; Hillaire-Marcel et al., 1994; Hesse et al., 1997). The findings also raise some questions about the conventional interpretation of these beds and the origin of the carbonate. In the case of the North Atlantic Mid-Ocean Channel levee deposits, their architectural integrity with the levee makes a strong case for their origin in turbidity currents. The bedding structures also indicate deposition from turbidity currents, but probably not currents originating in failure of the ocean bed that would be expected to produce normally graded units. Continuous, but unsteady input of meltwater with high sediment concentration, as is the case of Icelandic jökulhlaups (Maizels, 1997; Russell and Knudsen, 1999), would account for this style of sedimentation and the thickness of the detrital carbonate beds. The outburst floods inferred from terrestrial landforms (Shaw, 1996) are the most probable sources of the meltwater events responsible for depositing the carbonate megabeds. Short-lived depositional events are also indicated by the upward, step-like decrease in bulk density of detrital carbonate beds (Fig. 8). Rashid et al. (2000) made a similar suggestion based on the short-lived spikes of isotopically light meltwater associated with the carbonate beds. They suggested that the short duration of the spikes indicates that the meltwater events that caused them were also short-lived but of high magnitude, the exact characteristics of outburst floods. Furthermore, it is difficult to envisage another meltwater source when ice extends onto the continental shelf. Storage in and release of meltwater from proglacial lakes (Johnson and Lauritzen, 1995) are not likely in the absence of suitable sites or space for such lakes.

The origin of the detrital carbonate has been a matter of simple assumption: it is "glacial flour" produced by ice abrasion and transported by glacier ice and in icebergs. This assumption is evident in interpretations of the carbonates as depositional events (Andrews and Tedesco, 1992; Hillaire-Marcel et al., 1994) and in explanations of debris accretion prior to transportation in ice sheets (MacAyeal, 1993). In the meltwater hypothesis, detrital carbonate may originate as glacial flour, but it may also be the product of fluvial abrasion (Sharpe and Shaw, 1989). The tills of the Peterborough drumlin field also contain finely abraded carbonate. Kor et al. (1991) suggested that subglacial, fluvial erosion of resistant, crystalline bedrock of the Canadian Shield amounted to several meters. Shaw (1988), Tinkler and Stenson (1992), and Kor and Cowell (1998) inferred deep fluvial erosion of carbonate bedrock in Ontario. Erosion of similar rocks in Hudson Bay produces identical fluting to that described by Josenhans and Zevenhuisen (1990). The surface of the eroded carbonate rock is smooth and polished, and the scale and angularity of the residual surface-roughness show that erosion by scaling of flat particles produced silt and clay grains (Fig. 9).

A primary source of carbonate in fluvial erosion explains the monolithological composition of the carbonates more successfully than a glacial source, especially if the glacial source was deforming bed sediment. It also unifies erosion, transport, and deposition of the detrital carbonate under the rubric of meltwater processes. This view introduces the likelihood that the cooling accompanying Heinrich events may be partly a result of the carbonate production. Carbonate grains of silt and clay size present a large surface area for weathering, which would deplete CO_2 in the meltwater in which they were transported as suspended sediment. Eventually, with sediment deposition and concomitant reduction in density, the meltwater depleted in CO_2 would have risen to the surface in buoyant plumes and taken up CO_2 from the atmosphere. Reduced atmospheric CO_2 is expected to cause reduced temperature. This effect is likely to have been subordinate to the direct atmospheric cooling by the cold, brackish plumes.

The timing of detrital carbonate bed 1 in core MD99-2229 is well constrained by the two oldest dates from this core, 24,200 ± 110 ^{14}C yr at 1.95 m and 30,400 ± 340 ^{14}C at 14.81 m (Fig. 8). This timing raises questions about some fundamental assumptions of ours at the beginning of this study. We believed that meltwater flood events and detrital events would be correlated one to one and that they would be the same as Heinrich events. This is clearly not the case. MD99-2229 shows no record of major carbonate events at the times of H_0, H_1, and H_2. Yet, the timing is right for correlation between detrital carbonate-1 (core MD99-2229) and H_3.

Alternating Barren and Bioturbated Silt/Clay

Description

Sedimentary facies states in core MD99-2230 include gray, barren beds alternating with dark grayish yellow, heavily bioturbated and mottled beds; both the barren and bioturbated beds are in silty clay or clayey silt. The barren beds are colored 5Y 4/1 on the Munsell scale, and the bioturbated beds 2.5Y 5/2. Barren detrital carbonate units are also colored 5Y 4/1. The boundary between the barren beds and underlying bioturbated units is mainly sharp, although in one case the barren bed was loaded into the bioturbated unit. The couplets of barren and bioturbated beds range in thickness from ~1 m to ~15 cm. Toward the top of the core, the massive, barren unit may contain a lower bed of sand-sized rip-up clasts of the underlying material with intercalated laminae. These couplets show variable density and low magnetic susceptibility. Although they are almost certainly products of rainout from overspill currents deposited as turbidites, they are termed alternating, barren and bioturbated silt/clay. The silt/clay designation signifies that silt and clay are present in about equal proportions.

In the lower part of core MD99-2230, this facies is more complex with the addition of pronounced erosion cross-cutting a considerable thickness of the underlying bed and a basal sand unit. The sand is plane-bedded or cross-laminated. Rip-up clasts are common in the sands and in the barren unit. The sands at the base of the units contain a high proportion of non-carbonate sediments, which accounts for the highly variable magnetic susceptibility data in this part of the sequence. A similar variability is seen at the bottom of core MD99-2229, and it is probable that this signifies a sequence of alternating barren and bioturbated beds in silt and clay with sandy basal units.

Interpretation

These alternating beds evidently represent periodic rapid deposition with intervening periods of slower sedimentation with faunal colonization and bioturbation. The lower part of this couplet is clearly a turbidite given the erosion, stratification indicating bedload transport, and rip-up clasts. The absence of bioturbation in this unit points to rapid deposition, as would be expected of turbidity current deposition. However, the bedding is thicker and the grain size coarser than the typical spillover turbidites described by Wang and Hesse (1996).

The bioturbation generally affects a thickness of about 10–20 cm, presumably the depth of contemporary burrowing. Thus, strong currents that deposited detrital carbonate beds, which were bioturbated between turbidity events, periodically swept the ocean floor. The period between currents must have been long compared to the duration of the depositional events, an expected characteristic of spillover turbidity currents crossing the North Atlantic Mid-Ocean Channel levee (Klaucke et al., 1998). Jaeger and Nitrouer (1999) described similar turbidite beds associated with outbursts and surges of the Bering Glacier, Alaska. Since similar turbidites are not seen in the modern sedimentary sequences of the cores, the depositional conditions during the accumulation of these beds must have been quite different from the present day interglacial environment in which nanofossil ooze is accumulating. As well, the detrital carbonate composition of these beds requires glacial or glaciofluvial transport of carbonates from the Hudson Bay-Hudson Strait-Cumberland Sound areas (Andrews and Tedesco, 1992). In essence, these beds mark a second type of detrital carbonate event during which powerful, erosive currents and high sedimentation rates deposited beds of decimeter thickness. These currents were more powerful than those that deposited the thick detrital carbonate beds, though thinner beds suggest that the currents were of shorter duration and much smaller magnitude. The occurrence of two distinct detrital carbonate events—*Type A*, high-magnitude (total discharge), long-duration events that produce thick beds (meter scale) and *Type B*, low-magnitude, high-power, short-duration events that produce thinner beds (decimeter scale)—calls for caution in the interpretation of Labrador Sea detrital carbonate beds from Labrador Sea sediments.

Dark Bioturbated Mud and Grit Facies

Description

This facies marks a noticeable change in color from the 5Y 4/1 of the detrital carbonate to 5Y 5/1. This change is accompanied by bioturbation and mottling with crudely bedded mud and gritty beds; the grittiness arises from a mixture of sand and mud. Sand grains include carbonate, quartz, pyroxene, and hornblende minerals. This facies is closely associated with *Type A* detrital carbonate beds, which commonly lie between these thick carbonate units. Bioturbation extends downward into the underlying carbonate bed, and escape burrows are even seen in the lowest few centimeters of the overlying carbonate. The dark, bioturbated mud and grit beds are usually only a few centimeters thick.

Interpretation

This facies represents hemipelagic sedimentation with low sedimentation rates and abundant benthic fauna. The grit was most likely ice rafted, and the high proportion of this material also attests to the low sedimentation rates for this facies. Escape burrows in the bottom of the overlying detrital carbonate beds is further confirmation of their rapid emplacement.

DISCUSSION

The evidence on land for meltwater outbursts is strong, and a case has been made in this paper that the meltwater hypothesis cannot stand unless there is a depositional counterpart in the deep ocean of the erosion inferred on land. The volumes of meltwater and sediment transported in a few days are measured in thousands of cubic kilometers, and the rate of denudation is estimated on the order of 1 m per day over thousands of square kilometers. This sediment is not on land, nor is it on the continental shelves around the margins of the Laurentide Ice Sheet. If it is anywhere, it has to be in the deep ocean.

The Labrador Sea is an obvious place to look, and the unique sedimentological and morphological system around the North Atlantic Mid-Ocean Channel (Hesse et al., 2001) requires explanation by special events and processes. The braid plain, ornamented by flutings tens of kilometers long and recording coherent flows over a width of tens of kilometers and a length of hundreds of kilometers, must represent colossal flow events (Hesse et al., 2001) taking place outside the confines of the North Atlantic Mid-Ocean Channel. These flows somehow by-passed or avulsed from the channel. They seem to have deposited just coarse grain sizes, mainly sand, on the braid plain, while the levee deposits described here are almost entirely in silt and clay.

The sorting of the levee and braid plain deposits requires that the inflowing turbidity current splits proximal to the channel/levee/braid plain system. Hesse et al. (1997) follow Syvitsky et al. (1987) and assume that the separation occurs as a relatively low density plume carrying fine-grained sediment rises to the surface, and a sand-rich hyperpycnal flow continues along the seabed. This may be the case, but it does not explain the characteristics of the thick detrital carbonate beds on the levees. Currents that were in places erosive, as is indicated by the frequency of soft-sediment rip-up clasts, evidently deposited these beds. The beds are also integral parts of the levee architecture, indicating deposition from turbidity currents spilling over from the North Atlantic Mid-Ocean Channel. It is unlikely that sandy currents, depleted of silt and clay would have followed the braid plain and not continued in part along the North Atlantic Mid-Ocean Channel. Consequently, another possibility might be considered whereby avulsion-involved deflection of the lower and slower moving, sand-rich flow out of the channel. This lower part of the flow would follow the side slope of the levee and continue along the braid plain. Its momentum along the channel would carry the upper, faster moving part of the current as a hyperpycnal flow carrying mainly fine-grained sediment. Spillover from this current would explain the thick detrital carbonate beds on the levees, and they would have been contemporaneous with the braid plain events. In this way, there is no need to invoke separate extreme events for the braid plain deposits and the thick detrital carbonate beds on the levees—they are both related to the same inputs. Evidently, these inputs themselves must have been extreme events, and the likelihood that they originated in subglacial outburst events remains to be considered.

The braid plain and its deposits record hyperpycnal flows on the order of 100 km wide. The depth of the braid plain flow and the absence of topographic constraints suggest the full width of the plain was occupied during its formation. Sand beds several meters in thickness were deposited by individual flow events. As well, the flows dissected levees over 100 m high, and lineations on the levee remnants indicate that they were submerged in these flows. The lineations on the braid plain itself show that the current was capable of scouring sands. Assuming that this erosion requires a flow velocity of about 1 m/s, a flow width of 100 km, and a flow depth of 150 m, an instantaneous discharge of 1.5×10^6 m^3/s is obtained. Only outburst floods are capable of generating flows of this magnitude, for example, non-catastrophic meltwater discharge estimates to Hudson Strait are about 2.0×10^5 m^3/s (Marshall and Clarke, 1999), almost an order of magnitude less than the discharge rate for braid plain floods.

Considering the detrital carbonate beds in the levees, deposition of several meters of sediment in short-lived events is truly extreme. The characteristics of the graded laminated beds with angular silt and clay sized particles of carbonate illustrate a subglacial source and transport by meltwater in hyperpycnal flows. The virtual absence of ice-rafted debris at the time of accumulation of the detrital carbonate beds eliminates the possibility that the detrital carbonate was transported by icebergs and emphasizes the view that meltwater events were the probable cause of these beds (Andrews et al., 1998; Hesse et al., 1997; Hesse and Khodabakhsh, 1998; Rashid et al., 2000; Hesse et al., 2001). More specifically, these events are hard to explain other than by outburst floods from beneath the Laurentide Ice Sheet. This conclusion adds to the evidence in support of the meltwater hypothesis for terrestrial landforms and sediments. With this added support, there is a pressing need to include the effects of such enormous outpouring of meltwater in ocean and climate modeling of glacial times.

Nevertheless, the timing of depositional events is surprising, and the correlation with Heinrich events is not as straightforward as expected. Clearly, with only 1.25 m of sedimentation in the last 20,000 years or so, with much of that sediment rich in organics and lacking high proportions of detrital carbonate and also lacking obvious ice-rafted clasts, Heinrich events H_0, H_1, and H_2 did not contribute much sediment to the levees. However, the timing is right for correlation of detrital carbonate 1 in core MD99-2229 and H_3. Event H_3, dated at 27 k.y. B.P., was at a time when the Laurentide Ice Sheet was expanding to a maximum at about 20 k.y. This suggests the possibility of meltwater floods from the Hudson Bay-Ungava sector of the Laurentide Ice Sheet during ice sheet build up toward the Late Wisconsin maximum. However, the hypothesis presented here suggests that would call for the Labrador-Ungava drumlin fields to have formed during the advancing stage, too. It also limits the sedimentary effects on the North Atlantic Mid-Ocean Channel levees of deglacial floods through Hudson Strait.

ACKNOWLEDGMENTS

We thank the Natural Sciences and Engineering Research Council of Canada for research support to John Shaw and for a special ship time grant to Reinhard Hesse, David Piper, and John Shaw. This work would not have been possible without the professionalism of the officers and crew, especially the coring team, of the Marion Dufresne. John Shaw particularly thanks Commandant Gilles Foubert for his kindness. David Piper has offered advice and encouragement throughout this work. Reinhard Hesse made many constructive, critical comments on an earlier version of this paper. We are also indebted to him for his expertise in the planning stages of the voyage. Harunur Rashid generously provided the X-radiographs of Figure 8 and the dates from MD99-2229. George Braybrook processed the SEM images, and Xavier Morin provided seismic and bathymetric images from the Marion Dufresne 99 cruise. We are grateful to them for their expertise and kindness.

REFERENCES CITED

Anderson, J.B., Wellner, J.S., Lowe, A., Mosola, A., and Shipp, S., 2001, Footprint of the expanded West Antarctic Ice Sheet: Ice stream history and behavior: GSA Today, v. 11, no. 10, p. 4–9.

Andrews, J.T., and Tedesco, K., 1992, Detrital carbonate-rich sediments, northwestern Labrador Sea: Implications for ice-sheet dynamics and iceberg rafting (Heinrich) events in the North Atlantic: Geology, v. 20, p. 1087–1090.

Andrews, J.T., Kirby, M., Jennings, A.E., and Barber, D.C., 1998, Late Quaternary stratigraphy, chronology, and depositional processes on the slope of S.E. Baffin Island, detrital carbonate and Heinrich events: Implications for onshore glacial history: Géographie physique et Quaternaire, v. 52, p. 91–105.

Aylsworth, J.M., and Shilts, W.W., 1989, Glacial features of the west-central Canadian Shield: Geological Survey of Canada Paper 85-1B, p. 375–381.

Baker, V.R., and Bunker, R.C., 1985, Cataclysmic Late Pleistocene flooding from Glacial Lake Missoula: A review: Quaternary Science Reviews, v. 4, p.1–41.

Beaney, C.L., and Hicks, F.L., 2000, Hydraulic modelling of subglacial tunnel channels, south-east Alberta: Canada. Hydrologic Processes, v. 14, p. 2545–2547.

Beaney, C.L., and Shaw, J., 1999, The subglacial morphology of southeast Alberta: Evidence for subglacial meltwater erosion: Canadian Journal of Earth Science, v. 37, p. 511–561.

Benn, D.I., and Evans, D.J.A., 1998, Glaciers and Glaciation: London, Arnold, 734 p.

Bennet, M.R., and Glasser, N.F., 1996, Glacial Geology: Chichester, Wiley, 364 p.

Blanchon, P., and Shaw, J., 1994, Reef drowning events during the last deglaciation: evidence for catastrophic sea level rise and ice sheet collapse: Geology, v. 23, p. 4–8.

Bond, G., Heinrich, H., Broecker, W.S., Labyrie, L., McManus, J., Andrews, J.T., Huon, S. Jantschick, R., Clasen, S., Simet, C., Tedesco, K., Klas, M., Bonani, G., and Ivy, S., 1992, Evidence for massive discharges of icebergs into the glacial Northern Atlantic: Nature, v. 360, p. 245–249.

Boyd, R., Scott, D.B., and Douma, M., 1988, Glacial tunnel valleys and Quaternary history of the Scotian Shelf: Nature, v. 333, p. 61–64.

Brennand, T.A., and Shaw, J., 1994, Tunnel channels and associated landforms south central Ontario: Their implications for ice sheet hydrology: Canadian Journal of Earth Sciences, v. 31, p. 505–522.

Brennand, T.A., Shaw, J., and Sharpe, D.R., 1995, Regional scale meltwater erosion and deposition patterns, northern Quebec, Canada: Annals of Glaciology, v. 22, p. 928–944.

Bretz, J H., 1923, The channeled scabland of the Columbia Plateau: Journal of Geology, v. 31, p. 617–649.

Brunner, C.A, Normark, W.R., Zuffa, G.G., and Serra, F., 1999, Deep-sea sedimentary record of the Late Wisconsin cataclysmic floods from the Columbia River: Geology, v. 27, p. 463–466.

Clark, C.D. Knight, J.K., and Gray, J.T., 2000, Geomorphological reconstruction of the Labrador Sector of the Laurentide Ice Sheet: Quaternary Science Reviews, v. 19, p. 1343–1366.

Dowdeswell, J.A., Maslin, M.A., Andrews, J.T., and McCave, I.N., 1995, Iceberg production, debris rafting, and the extent and thickness of Heinrich layers (H-1, H-2) in North Atlantic sediments: Geology, v. 23, p. 301–304.

Evans, D.J.A., 2000, Quaternary geology and geomorphology of the Dinosaur Park area and surrounding plains, Alberta, Canada: The identification of former glacial lobes, drainage diversions and former flood tracks: Quaternary Science Reviews, v. 19, 931–958.

Fisher, T.G., and Taylor, L.D., 2002, Sedimentary and stratigraphic evidence for subglacial flooding, south central Michigan, U.S.A: Quaternary International, v. 90, p. 87–115.

Hambrey, M.J., 1994 Glacial environments: Vancouver, UBC Press, 296 p.

Hardy, R.L., 1971, Multiquadratic equations of topography and other irregular surfaces: Journal of Geographic Research, v. 78, p. 1905–1916.

Heinrich, H., 1988, Origin and consequences of cyclic ice rafting in the Northeast Atlantic Ocean during the past 130,000 years: Quaternary Research, v. 29, p. 143–152.

Hesse, R., 1989, "Drainage systems" associated with mid-ocean channels and submarine yazoos: Alternative to submarine fan depositional systems: Geology, v. 17, p. 1148–1151.

Hesse, R., and Chough, S.K., 1980, The Northwest Atlantic Mid-Ocean Channel of the Labrador Sea. II. Deposition of parallel laminated levee-mud from the viscous sublayer of low-density turbidity currents: Sedimentology, v. 27, p. 679–711.

Hesse, R., and Khodabakhsh, S., 1998, Depositional facies of Late Pleistocene Heinrich events in the Labrador Sea: Geology, v. 26, p. 103–106.

Hesse, R., and Rakofsky, A., 1992, Deep-sea channel/submarine-yazoo system of the Labrador Sea: American Association of Petroleum Geology Bulletin, v. 76, p. 680–707.

Hesse, R., Chough, S.K., and Rakofsky, A., 1987, The Northwest Atlantic Mid-Ocean Channel of the Labrador Sea; V, Sedimentology of a giant deep-sea channel: Canadian Journal of Earth Sciences, v. 24, p. 1594–1624.

Hesse, R., Klaucke, I., Ryan, W.B.F., and Piper, D.J.W., 1997, Ice-sheet sourced juxtaposed turbidite systems in Labrador Sea: Geoscience Canada, v. 24; p. 3–12.

Hesse, R., Klaucke, I., Khodabakhsh, S., Piper, D.J.W, Ryan, B.F., and North Atlantic Mid-Ocean Channel Study Group, 2001, Sandy submarine braid plains: Potential deep water reservoirs: American Association of Petroleum Geologists, v. 85, p. 1499–1521.

Hillaire-Marcel, C., de Vernal, A., Bilodeau, G., and Wu, G., 1994, Isotope stratigraphy, sedimentation rates, deep circulation, and carbonate events in the Labrador Sea during the last ~200 kyr: Canadian Journal of Earth Science, v. 31, p. 63–68.

Hoffman, P.F., Kaufman, A.J., Halverson, G.P., and Schragg, D.P., 1998, A Neoproterozoic Snowball Earth: Science, v. 281, p. 1342–1346.

Jaeger, J.M., and Nittrouer, C.A., 1999, Marine record of surge induced outburst floods from the Bering Glacier, Alaska: Geology, v. 27, p. 847–850.

Johnson, R.G., and Lauritzen, S.-E., 1995, Hudson Bay-Hudson Strait jökulhlaups and Heinrich events: A hypothesis: Paleogeography, Paleoclimatology, Paleoecology, v. 117, p. 123–137.

Josenhans, H., and Zevenhuizen, J., 1990, Dynamics of the Laurentide Ice Sheet in Hudson Bay, Canada: Marine Geology, v. 92, p.1–26.

Kehew, A.E., and Lord, M.L., 1986, Origin and large-scale erosional features of glacial-lake spillways in the northern Great Plains: Geological Society of America Bulletin, v. 97, p. 162–177.

Kehew, A.E., and Teller, J.T., 1994, History of late glacial runoff along the southwestern margin of the Laurentide Ice Sheet: Quaternary Science Reviews, v.13, p. 859–877.

Klaucke, I., and Hesse, R., 1996, Fluvial features in the deep-sea: New insights from the glacigenic submarine drainage system of the Northwest Atlantic Mid-Ocean Channel in the Labrador Sea: Sedimentary Geology, v. 106, p. 223–234.

Klaucke, I, Hesse, R., and Ryan, W.B.F., 1998, Seismic stratigraphy of the Northwest Atlantic Mid-Ocean Channel; growth pattern of a mid-ocean channel levee complex: Marine and Petroleum Geology, v. 15, p. 575–585.

Kor, P.S.G., and Cowell, D.W., 1998, Evidence for catastrophic subglacial sheetflood events on the Bruce Peninsula: Canadian Journal of Earth Sciences, v. 35, p. 1180–1202.

Kor, P.S.G., Shaw, J., and Sharpe, D.R., 1991, Erosion of bedrock by subglacial meltwater, Georgian Bay, Ontario: A regional view: Canadian Journal of Earth Sciences, v. 28, p. 623–642.

Loncarevic, B.D., Piper, D.J.W., and Fader, G.B.J., 1992, Application of high quality bathymetry to geological interpretation on the Scotian Shelf: Geoscience Canada, v. 19, p. 5–13.

Lowe, A.L., and Anderson, J.B., 2002, Reconstruction of the West Antarctic ice sheet in Pine Island Bay during the Last Glacial Maximum and its subsequent retreat history: Quaternary Science Reviews, v. 21, p. 1879–1897.

Lucchitta, B.K., 1982, Ice sculpture in the Martian outflow channels: Journal of Geophysical Research, v. 87, p. 9951–9973.

MacAyeal, D.R., 1993 Binge/purge oscillations of the Laurentide Ice Sheet as a cause of the North Atlantic's Heinrich events: Paleoceanography, v. 8, p. 775–784.

Maizels, J.A., 1997, Jökulhlaup deposits in proglacial areas: Quaternary Science Reviews, v. 16, p. 793–819.

Marshall, S., and Clarke, G.K.C., 1999, Modeling North American freshwater runoff through the last glacial cycle: Quaternary Research, v. 52, p. 300–315.

McCauley, J.F., Grolier, M.J., and Breed, C.S., 1977, Yardangs of Peru and other desert regions: United States Geological Survey Interagency Report, Astrogeology 81, 177 p.

Munro-Stasiuk, M.J., and Shaw, J., 2002, The Blackspring Ridge Flute Field, south-central Alberta, Canada: evidence for subglacial sheetflow erosion: Quaternary International v. 90, p. 75–86.

Ó Cafaigh, C., Pudsey, C.J., Dowdeswell, J., Morris, P., 2002, Evolution of subglacial bedforms along a paleo-ice stream, Antarctic Peninsula continental shelf: Geophysical Research Letters, v. 29, p. 41-1–41-4.

Potschin, M.B., 1989, The origin of drumlins at Bernard Harbour, District of Mackenzie, N.W.T., Canada [M.A. thesis]: Ottawa, Ontario, Carleton University, 203 p.

Rains, B., Shaw, J., Skoye, R., Sjogren, D., and Kvill, D., 1993, Late Wisconsin subglacial megaflood paths in Alberta: Geology, v. 21, p. 323–326.

Rashid, H., Hesse, R., and Piper, D.J., 2000, Was it melt-water outflow or ice-sheet discharge? A high-resolution study of melt-water discharge from the high-latitude sector of the LIS during the LGM [abs.]: Eos (Transactions, American Geophysical Union), v. 81, p. S1

Russell, A.J., and Knudsen, Ó., 1999, Controls on the sedimentology of the November 1996 jökulhlaup deposits, Skeidarársandur, Iceland: Special Publications International Association of Sedimentology, v. 28, p. 315–329.

Saucier, R.T., 1991, Geomorphology, stratigraphy and chronology, lower Mississippi Valley, *in* Morrisson, R.B., ed., Quaternary nonglacial geology: Coterminous U.S.: The Geology of North America: Boulder, Geological Society of America, v. K-2, p. 550–563.

Sawagaki, T., and Hirakawa, K., 1997, Erosion of bedrock by subglacial meltwater, Soya Coast, East Antarctica: Geografiska Annaler, v. 79A, p. 223–238.

Sankeralli, L.M., 1998, The origin of the morphology of the Eastern Scotian Shelf, Atlantic Canada [M.Sc. thesis]: Alberta, University of Alberta, 207 p.

Schirmer, W., 1981. Abflüssverhalten des Mains im Junquartær, *in* Boenigk, W., and Tillmanns, W., eds. Festschrift zur Vollendung des 60; Lenbensjahres von Prof. Dr. Karl Brunnacker: Sonderveroeffentlichungen des Geologischen instituts der Universität Koeln, Cologne, Germany, v. 41, p. 197–208.

Sharpe, D.R., and Shaw, J., 1989, Erosion of bedrock by subglacial meltwater, Cantley, Quebec: Geological Association of America Bulletin, v. 101, p. 1011–1020.

Shaw, J., 1983, Drumlin formation related to inverted meltwater erosional marks: Journal of Glaciology, v. 29, p. 461–479.

Shaw, J. 1988, Subglacial erosional marks, Wilton Creek, Ontario: Canadian Journal of Earth Sciences, v. 25, p. 1256–1267.

Shaw, J., 1989, Drumlins, subglacial meltwater floods, and ocean responses: Geology, v. 17, p. 853–856.

Shaw, J., 1994, Hairpin erosional marks, horseshoe vortices and subglacial erosion: Sedumentary Geology, v. 91, p. 269–284.

Shaw, J., 1996, A meltwater model for Laurentide subglacial landscapes, *in* McCann, B., and Ford, D.C., eds., Geomorphologie sans frontiers: Chichester, Wiley, p.181–236.

Shaw, J., and Kvill, D., 1984, A glaciofluvial origin for drumlins of the Livingstone Lake area, Saskatchewan: Canadian Journal of Earth Sciences, v. 21, p.1442–1459.

Shaw, J., Kvill, D., and Rains, R., 1989, Drumlins and catastrophic subglacial floods: Sedimentary Geology, v. 62, p. 177–202.

Shaw, J., Faragini, D.M., Kvill, D.R., and Rains, R.B., 2000, The Athabasca fluting field, Alberta, Canada: Implications for the formation of large-scale fluting (erosional lineations): Quaternary Science Reviews, v. 19, p. 959–980.

Shaw, J., Rains, B., Eyton, R., Weissling, L., 1996, Laurentide subglacial outburst floods; landform evidence from digital elevations models: Canadian Journal of Earth Sciences, v. 33, p. 1154–1168.

Shipp, S., Anderson, J., and Domack, E., 1999, Late Pleistocene-Holocene retreat of the West Antarctic Ice Sheet system in the Ross Sea: Part 1—Geophysical results: Geological Society of America Bulletin, v. 111, p. 1486–1516.

Shoemaker, E.G., 1992, Water sheet outburst floods from the Laurentide Ice Sheet: Canadian Journal of Earth Sciences, v. 29, p.1250–1260.

Shoemaker, E.G., 1995, On the meltwater genesis of drumlins: Boreas, v. 24, p. 3–10.

Smith, H.T.U., 1948, Giant glacial grooves in northwest Canada: American Journal of Science, v. 246, p. 503–514.

Solheim, A., Russwurm, L., Elverhøi, A., and Berg, M.N., 1990, Glacial geomorphic features in the northern Barents Sea: Direct evidence for grounded ice and implications for the pattern of glaciation and late glacial sedimentation, *in* Dowdeswell, J.A., and Scourse, J.D., eds., Glaciomarine environments: Processes and sediments: Geological Society [London] Special Publication 53, p. 253–268.

Stelting, C.E., Pickering, K.T., and Bouma, A.H., and DSDP Leg 96 Shipboard Scientists, 1985, Drilling results in the Middle Mississippi Fan, *in* Bouma, A., Normark, W.R., and Barnes, N.E., eds., Submarine fans and related turbidite systems: New York, Springer-Verlag, p. 275–282.

St.-Onge, D.-A., and McMartin, I., 1995, Quaternary geology of the Inman River area, Northwest Territories: Geological Survey of Canada Bulletin 446, 59 p.

Syvitski, J.P.M., Burell, D.C., and Skej, J.M., 1987, Fjords: Processes and Products: New York, Springer-Verlag, 375 p.

Tinkler, R.J., and Stenson, R.E., 1992, Sculpted bedrock forms along the Niagara escarpment, Niagara Peninsula, Ontario: Géographie physique et Quaternaire, v. 46, p. 195–207.

Uchupi, E., Driscoll, N., Ballard, R.D., and Bolmer, S.T., 2001, Drainage of Late Wisconsin glacial lakes and the morphology and Late Quaternary stratigraphy of the New Jersey-southern New England continental shelf and slope: Marine Geology, v. 172, p. 117–145.

Wang, D., and Hesse, R., 1996, Continental slope sedimentation adjacent to an ice-margin. II. Glaciomarine depositional facies on Labrador Slope and glacial cycles: Marine Geology, v. 135, p. 65–96.

Wellner, J.S., Lowe, A.S., Shipp, S.S., Anderson, J.B., 2001, Distribution of glacial geomorphic features on the Antarctic continental shelf and correlation with substrate: Implications for ice behavior: Journal of Glaciology, v. 47, p. 397–411.

Worsley, P.W., 1997, The 1996 volcanically induced glacial mega flood in Iceland—cause and consequences: Geology Today, November-December, p. 211–216.

Zuffa, G.G., Normark, W.R., Serra, F., and Brunner, C.A., 2000, Turbidity megabeds in an oceanic rift valley recording jökulhlaups of Late Pleistocene glacial lakes of the Western United States: Journal of Geology, v. 108, p. 253–274.

Manuscript Accepted by the Society January 22, 2003

Printed in the USA

Geological Society of America
Special Paper 370
2003

Limits on extreme eolian systems: Sahara of Mauritania and Jurassic Navajo Sandstone examples

Gary Kocurek
Department of Geological Sciences, University of Texas, Austin, Texas 78712, USA

ABSTRACT

The construction, accumulation, and preservation of eolian systems are distinct phases with largely independent controls. The limits on these controls indicate that extreme eolian systems require the coincidence of: (1) construction in an arid, sufficiently windy climate that follows an antecedent condition of a climatic-tectonic-eustatic template conducive for the generation and storage of a large volume of sandy sediment; (2) accumulation (positive angle of climb) within a dry eolian system as a result of dunes decelerating as they migrate into a topographic basin; and (3) preservation of the accumulations by continuous basin subsidence and sediment influx, and rising sea level. The Sahara, in spite of construction during a windy, arid period, has a limited sand supply, conditions broadly unfavorable for accumulation, and preservation potential confined to the flooded shelf. In contrast, the Jurassic Navajo Sandstone of the Western Interior of the United States had basin-wide favorable conditions for construction, accumulation, and preservation, with the exception of episodic tectonic stripping of sediment.

Keywords: eolian, sand seas, accumulation, preservation, Sahara, Jurassic.

INTRODUCTION

In a quest for the "extreme" eolian depositional environment on Earth, one approach is to look for representatives, both in the rock record and on the planet today. The Jurassic Navajo Sandstone of the Colorado Plateau of the United States may well be a candidate for such an extreme, judging by the scale of the preserved sets of cross-strata and the regional extent of the unit. Another candidate may be the Cretaceous of Brazil (Scherer, 2000). There is a high probability, however, that greater eolian systems have existed on this planet but left no rock record, or the record itself has been consumed through tectonic processes. The modern Sahara Desert is an impressive eolian depositional environment, but for most of the Saharan craton, the rock record will be a surface (Kocurek, 1998).

The difficulty in identifying the "extreme" eolian environment is that it requires an understanding of the limits imposed upon the system. This realization suggests an alternate approach: identification of the controls on eolian systems, from inception to incorporation into the rock record, and the limits imposed upon these controls. The underlying premise is that (1) system construction, (2) accumulation of strata, and (3) preservation of these strata are distinct phases and are governed by controls that are largely independent of one another (Kocurek, 1999). Therefore, the definition of "extreme" in this paper is inclusive and is defined in all three phases: construction of an eolian system of regional coverage with immense dunes, maximum rates of accumulation, and a high degree of preservation.

In this paper, the phases of construction, accumulation, and preservation are first outlined in theory, which is more completely presented in Kocurek (1999). Second, using this theory, what is the recipe for construction, accumulation, and preservation of an extreme eolian environment? Finally, by way of examples, the Sahara of Mauritania and the Jurassic Navajo Sandstone are considered in terms of how closely they approach and fall short of the "ideal." The focus here is entirely on dune eolian systems; other potentially extreme eolian systems, such as loess environments, are not addressed.

Kocurek, G., 2003, Limits on extreme eolian systems: Sahara of Mauritania and Jurassic Navajo Sandstone examples, *in* Chan, M.A., and Archer, A.W., eds., Extreme depositional environments: Mega end members in geologic time: Boulder, Colorado, Geological Society of America Special Paper 370, p. 43–52.

EOLIAN SYSTEM CONSTRUCTION

Theory

Eolian system construction is a function of three separate controls: (1) sediment supply, (2) sediment availability, and (3) transport capacity of the wind. Together, these controls define the sediment state of the system (Kocurek and Lancaster, 1999). Sediment supply is defined as the volume of sediment of a suitable grain size generated per time that contemporaneously or at some later point in time serves as the source material for system construction. On Earth, the suitable grain size for dunes with slipfaces is typically 0.1–0.3 mm. Rates of weathering and wind deflation of rock nearly always dictate that the sediment supply for eolian systems will be secondary in the sense that it is derived from fluvial, alluvial, coastal, or lacustrine deposits. Any of these deposits are potential eolian sources essentially, but whether or not they will be utilized in dune construction is a function of the availability of the sediment to eolian deflation. Sediment availability is affected by numerous factors, including vegetation, moisture content, surface binding and cementation, and grain parameters such as sorting. Because of the range of factors involved, Kocurek and Lancaster (1999) have suggested that the actual eolian sediment transport rate, given as a volume per time, is a functional definition of sediment availability. In contrast, every wind has a potential transport rate, which is a solely a function of wind power and can be given also as a volume per time.

There are only nine possible sediment states in a plot of sediment supply, availability, and transport capacity against time (Fig. 1). At any given time, sediment that is not utilized by the wind is stored sediment, which must be such because it is (1) availability-limited (S_{AL}), (2) transport-limited (S_{TL}), or (3) some combination of the two (S_{TAL}). For example, coastal sediments below the water table are availability-limited, whereas glacial outwash may produce sediment at a rate beyond which it can be transported by the wind, also resulting in stored sediment. Sediment that is not stored is transported and utilized in dune construction. Sediment transported from a contemporary sediment supply is contemporary influx (CI) and must either be (4) limited only by the capacity of the wind (CI_{TL}), or (5) limited by availability (CI_{AL}). Alternatively, dune construction may occur from lagged influx (LI), which represents deflation of previously stored sediment, and be (6) transport-limited (LI_{TL}) or (7) availability-limited (LI_{AL}). The only other possibilities are that dune construction proceeds from a combination of contemporaneous and lagged influx (CLI), which must again be (8) transport-limited (CLI_{TL}) or (9) availability-limited (CLI_{AL}).

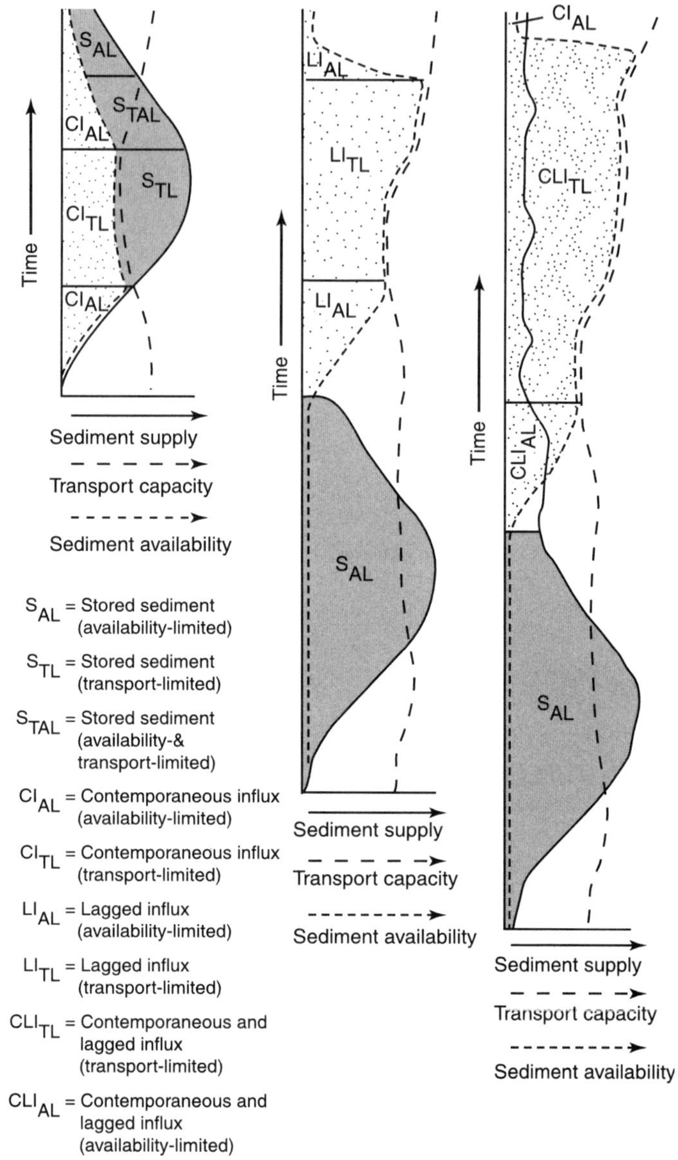

Figure 1. Sediment state diagrams in which sediment supply, transport capacity, and sediment availability (all volumetric rates) are plotted against time, defining nine possible sediment states for eolian systems. From Kocurek and Lancaster (1999).

Eolian Extreme

The extreme of a dune environment, manifested by immense dune size and regional coverage, requires an enormous quantity of sand. For creation of the maximum sediment supply, optimum or long-lasting conditions for maximum erosion and transport of sediment must occur. Sediment yield by streams increases with stream power, rates of weathering and erosion, and a decrease in vegetative cover. Langbein and Schumm (1958) first recognized that the scenario that yields maximum erosion and transport is during the transition from subhumid to semiarid conditions, during which vegetation is reduced, and flood magnitude increases as the maximum/mean precipitation ratio increases. Tectonic uplift promotes enhanced erosion by providing the source for the sediment and increasing stream gradient. A relative fall in sea level promotes enhanced stream downcutting along coastal plains and also exposes shelf sands. Conversely, transgressive conditions promote stockpiling of sediment along the coast and inland.

The availability of the sediment is enhanced by aridity that reduces or eliminates vegetation, and by a falling water table, as with regional drying or a marine regression. To a large extent, sediment availability varies inversely with sediment supply (e.g., profound aridity and a low water table enhance availability, but production of sediment supply is suppressed by aridity). For this reason, for maximum eolian construction, creation of the sediment supply should predate the eolian constructional event. Although local geomorphic conditions can foster high wind energy (i.e., high transport capacity) at a global scale, wind energy increases with pressure gradient, which, in turn, increases with the global equator-to-pole temperature gradient. Because sand transport increases roughly as a cubic function of wind shear stress (e.g., Bagnold, 1941), immense amounts of sand can be transported under conditions of high availability and high wind speed, such that construction of a vast sand sea is geologically rapid.

From the above, the recipe for creation of the extreme eolian depositional environment is one in which extremely arid, profoundly windy conditions occur during a marine regression and follow an intense humid–subhumid period of weathering during a marine highstand and adjacent tectonic uplift of suitable rocks that yield fine- to medium-grained sand upon weathering. Without the antecedent humid period during which enormous sediment supplies of sand are stockpiled on the continent, the extreme eolian constructional event is not possible. Moreover, the arid phase of high availability and transport cannot be so prolonged as to exhaust the sediment supply; if it did, the sand sea would shift into a destructional phase. Excluding the special case of the pre-vegetated Earth, this extreme sand sea is most likely to occur within the subtropical desert belt, where aridity can be sufficient to reduce or eliminate vegetation at a regional scale. The high temperature and pressure gradients needed to produce strong regional winds most likely occur during an Icehouse Earth. Because depth of flow is essentially unlimited, dune height may be limited only by sand volume. Star dunes commonly achieve great heights because migration is slight, yet given the supposed enormous sand influx, both crescentic and linear dunes are also possible candidates. Because of the dune size, these extreme bedforms will almost certainly be compound or complex features with superimposed dunes.

EOLIAN SYSTEM ACCUMULATION

Theory

Accumulation is the buildup of a body of sediment; this causes the accumulation surface upon which the bedforms rest to rise over time (Fig. 2A). The space generated for the accumulations by the rise of the accumulation surface is termed accumulation space by Kocurek and Havholm (1993). The alternatives to accumulation are bypass and erosion. All three possibilities are straightforwardly addressed by the first-order principle of continuity and transformed into the sediment conservation equation by Middleton and Southard (1984),

$$\frac{\partial h}{\partial t} = -\left(\frac{\partial q}{\partial x} + \frac{\partial c}{\partial t}\right)$$

where h is the height of the accumulation surface, t is time, q is the transport rate in the x direction and assuming that all transport occurs as dunes, and c is the concentration of sediment in transport, taken as average dune height (see Kocurek, 1999). Solutions to the sediment conservation equation by sign alone allow only five conditions under which accumulation occurs, three in which bypass occurs, and five in which erosion occurs (Fig. 2B). These basic possibilities can be subdivided and realistically portrayed for the three basic types of eolian systems (dry, wet, and stabilizing) as defined by Kocurek and Havholm (1993). Dry systems are those in which the wind is the sole control on the behavior of the accumulation surface over time. Wet systems

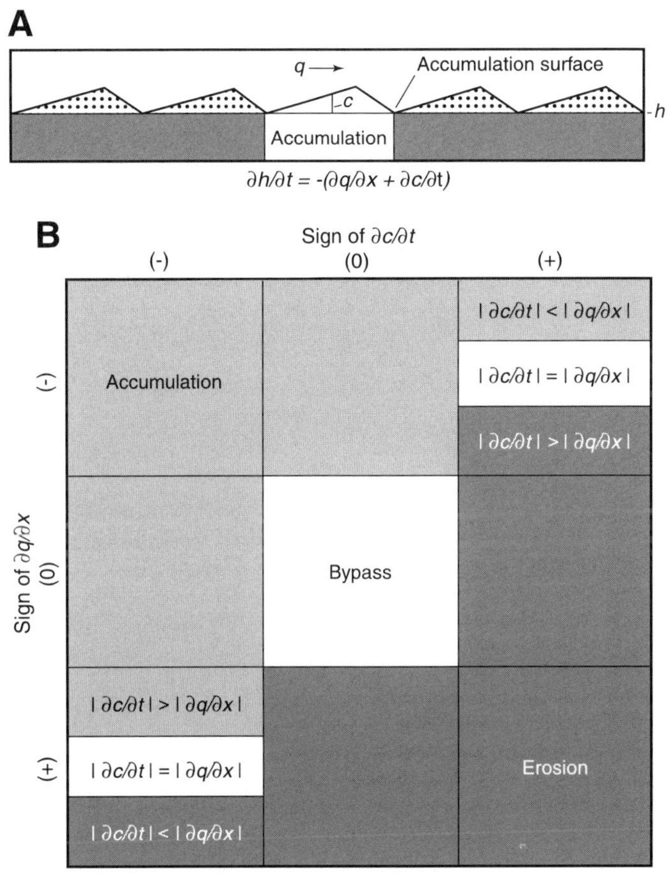

Figure 2. A: Definition diagram in which accumulation surface separates accumulation from sediment in transport. The elevation of surface (h) over time (t) is a function of transport rate (q) in the transport direction (x) and concentration of sediment in transport (c). B: Possible solutions to sediment conservation equation by sign, defining fields of accumulation, bypass, and erosion. From Kocurek (1999).

are those in which the water table is at or near the accumulation surface and governs its behavior over time. Stabilizing systems are those in which surface-stabilizing factors such as vegetation control the behavior of the accumulation surface over time.

Eolian Extreme

The extreme eolian system is one in which accumulation is at a maximum, which, for bedforms, is the angle of climb θ, defined as $\tan \theta = V_y / V_x$), where V_y is the vertical accumulation rate and V_x is the downwind migration rate. Scenarios in Figure 2B that yield bypass or erosion are, therefore, automatically eliminated. Bypass systems are marked only by migration of dunes over the surface, which remains static over time (angle of climb is zero). Erosional systems are those characterized by dunes that cannibalize the substrate over time (i.e., angle of climb is negative). From the sediment conservation equation, accumulation (positive angle of climb) will occur only where (1) the transport rate decreases downwind, (2) the concentration of sediment in transport decreases over time, or (3) a combination of both conditions occurs.

It is unlikely that the extreme eolian system will occur with a stabilizing system because the most common stabilizing agent, vegetation, will restrict sediment availability and is unlikely to pace extreme depositional rates. Similarly, for wet systems, even rapid rates of water-table rise associated with climatic change and eustatic sea-level rise during deglaciation will not result in extreme values in the angle of climb because the rate of water-table rise (V_y) will be small in comparison to the rate of bedform migration (V_x). Hence, the extreme eolian accumulation rate is most likely to be found among dry systems.

For dry eolian systems, satisfying the sediment conservation equation in natural settings occurs with (1) a downwind decrease in wind energy, thus yielding a decrease in the transport rate (e.g., Rubin and Hunter, 1982); (2) a decrease in wind energy over time, thus yielding a decrease in concentration, assuming the yet unproven, but likely, positive correlation between dune size and wind energy (Kocurek, 1999); or (3) a combined temporal and downwind decrease in wind strength. Decreasing transport rates in the migration direction occur under regional flow paths of decreasing pressure gradients and with flow into geomorphic basins where the flow expands vertically. The first case is not the best candidate for the extreme system because semi-permanent pressure gradients tend to be more gradual. Flow expansion into a pronounced topographic basin, however, can yield dramatic decreases in wind energy. A temporal decrease in wind strength occurs, for example, with a shift from glacial to interglacial conditions in the subtropical desert belt, but this rate of change would be small in comparison to likely bedform migration rates. With respect to dune type, accumulation in dry eolian systems does not occur until after interdune areas have been restricted by dune growth to mere depressions between dunes (Kocurek and Havholm, 1993). Given the common large spacing of linear and star dunes, the tendency for sand to be swept from interdune areas, and the slow lateral migration rates of these dunes, it is likely that the extreme of bedform climb will occur with a crescentic dune.

Thus, theory argues that extreme eolian rates of dune accumulation will occur within a dry eolian system that consists of cresentic dunes migrating into a pronounced topographic basin.

EOLIAN SYSTEM PRESERVATION

Theory

Preservation is the incorporation into the rock record of a body of accumulated strata. This space for the preserved accumulation is termed *preservation space* by Kocurek and Havholm (1993), and for eolian systems it can differ from accumulation space. For eolian accumulations, preservation occurs with (1) subsidence and burial, and/or (2) a rise in the water table through the accumulation (Fig. 3) (Kocurek, 1999). Subsidence with burial is necessary because an eolian accumulation, even one housed within a topographic basin characterized by wind deceleration, can experience deflation when the upwind sand supply is exhausted, thereby causing an erosional front that begins at the upwind margin of the accumulation and progresses downwind. Hence, a continued saturated eolian influx into the system or deposition within a new environment is necessary to prevent removal of the eolian accumulations. Likewise, raising

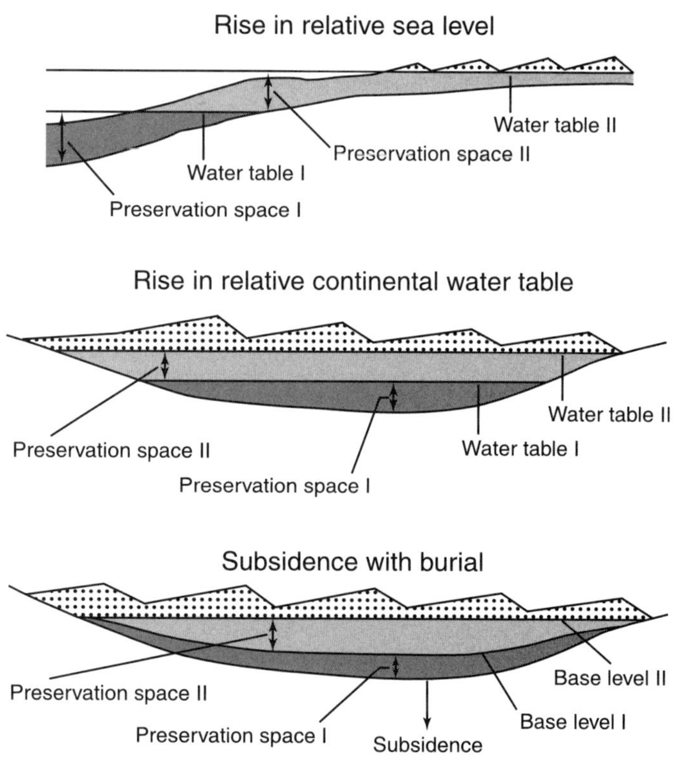

Figure 3. Modes of preservation of eolian accumulations. From Kocurek (1999).

the water table through the accumulation effectively shields the accumulation from deflation. The water table can be raised either in an absolute sense, as with a marine transgression; or through a shift to a more humid climate; or in a relative sense, as with subsidence of the accumulation through a static water table. Other factors, such as vegetation or surface armoring, may stabilize an accumulation, but they are not means of incorporating the accumulation into the rock record.

Eolian Extreme

From the above theory, preservation of the accumulations of the eolian extreme system is largely restricted to: (1) tectonic basins with pronounced sediment influx and high rates of subsidence, (2) coastal regions that are marine transgressed with an associated inland rise of the water table (see Kocurek et al., 2001), and (3) interior continental basins that experience a rise in the water table because of subsidence or climate (Kocurek, 1999). Continental tectonic basins with high rates of subsidence and prominent source areas to provide sediment influx are rift and cratonic basins, passive margins of oceanic basins, and foreland basins; the latter two afford the greatest regional extent. Preservation is enhanced with a marine system occupying the basin interior, in which a net progressive transgression occurs over time. Hence, extreme preservation is most likely to occur within a foreland basin or along a passive margin with a high subsidence rate, and in which the eolian system lies adjacent to a more basinal marine system that transgresses progressively inland through time.

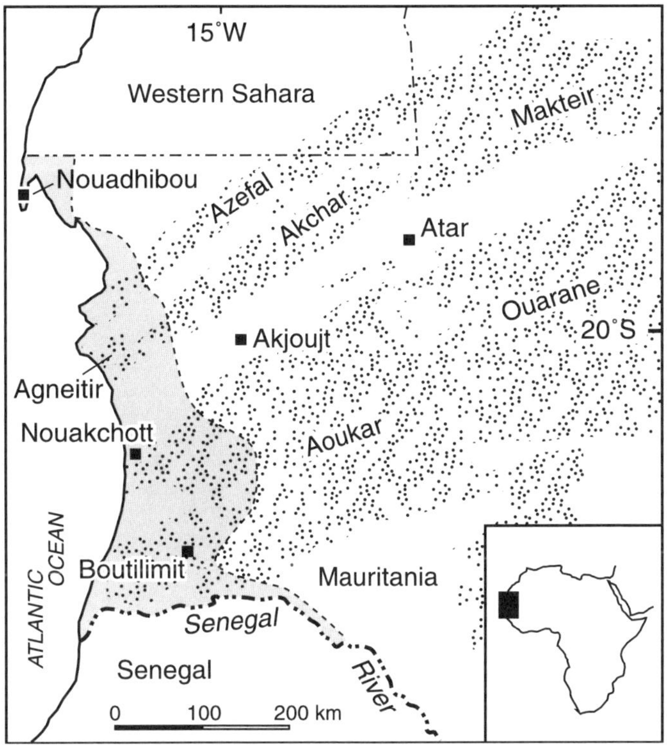

Figure 4. Western Sahara Desert in Mauritania. Stippled areas are eolian sand seas. Unmarked areas between sand seas are Precambrian basement. Mauritanian Basin (shaded) is rimmed by Miocene-Pliocene continental deposits, which yield to progressively younger strata toward the Atlantic Ocean.

SAHARA OF MAURITANIA

Mauritania is dominated by extensive sand seas that form the westernmost Sahara. These sand seas rest upon Precambrian basement until the vicinity of the Mauritanian Basin, which is rimmed by Miocene-Pliocene continental sediments with progressively younger deposits toward the coast (Fig. 4). The Sahara is the largest warm-climate desert on Earth today; it is an icehouse desert with an origin in the Pliocene-Pleistocene that coincides with the onset of glacial conditions (see review in Kocurek, 1998). Sand seas of Mauritania consist largely of compound/complex linear dunes that can rise to 75 m above the interdune floors. Do the sand seas of Mauritania, their accumulations, and their ultimate rock record approach an extreme for eolian systems?

Eolian System Construction

It has long been recognized (e.g., Glennie, 1970; Wilson, 1973; Kocurek, 1998) that the Sahara follows a Milankovitch and sub-Milankovitch scale climatic cycle (Fig. 5). During humid periods, fluvial/lacustrine systems are active and yield a sediment supply that is largely stored (S_{AL}) because sediment availability is low, owing to vegetation and a relatively high water table. Dunes are largely stabilized and undergo pedogenesis. Eolian construction begins with the onset of aridity, during which sediment availability increases as vegetation decreases and the water table falls (LI_{AL}). Transport capacity increases with strengthening of the Hadley Cells during glaciation (e.g., Parkin and Shackleton, 1973; Talbot, 1984). Maximum eolian construction as dry systems (LI_{TL}) proceeds under full glacial-hyperarid conditions to the point where the sediment supply is exhausted, at which time an eolian destructional phase begins.

Recent work in the western portions of the Azefal, Agneteir, and Akchar Sand Seas of Mauritania shows that the sand seas consist of three superimposed generations of construction: 15–25 ka (Last Glacial Maximum), 12–10 ka (Younger Dryas return to near glacial-maximum conditions), and after 5 ka (onset of recent arid conditions) (Lancaster et al., 2002). Humid conditions (Middle Holocene African Humid Period), characterized by dune stabilization and pedogenesis, fluvial activity, and widespread lakes, occurred between constructional events (Kocurek et al., 1991).

In terms of eolian system construction, foremost, the Sahara has to be considered a relatively sand-poor desert. Sand covers no more than 30% of the surface, which otherwise consists of deflated bedrock and reg, forming, in the terminology of Mainguet and Chemin (1983), a negative sediment budget desert system. Even in Mauritania, which forms the downwind terminus of the general east-to-west sediment transport (Wilson, 1973),

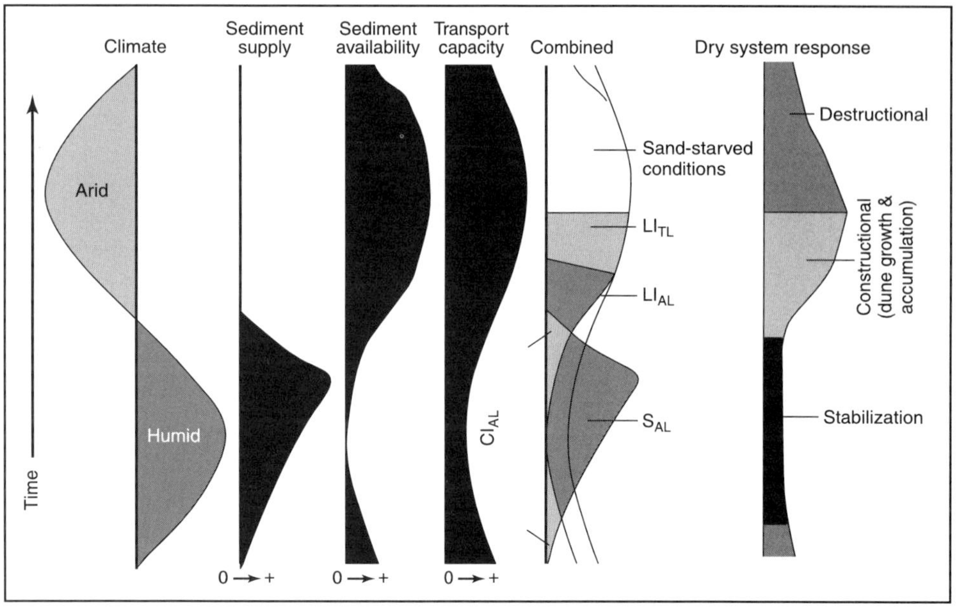

Figure 5. Sediment state diagram for Saharan arid-humid climatic cycles, as discussed in text. Modified from Kocurek (1998).

appreciable sediment cover between sand seas does not occur until near the coast. Presently, the upwind portions of the Akchar and, especially, the Agneteir are strongly deflationary and deposits are being cannibalized, with formation of coarse-grained lags that are followed upwind by bedrock. Fluvial and lacustrine deposits from the Middle Holocene African Humid Period consist of thin, muddy sands, and organic carbonate and evaporite deposits, respectively. These deposits do not show a major influx of sand suitable for later dune construction (Kocurek et al., 1991).

At the larger time scale, it has been argued that the last major influx of terrigenous sediment to the Sahara craton occurred during the Miocene (Kocurek, 1998, 1999). The Miocene was a humid period during which large fluvial systems drained from the Neogene-uplifted Central African Highlands into the foreland basin along the northern African margin. Uplift of the Atlas Mountains effectively stockpiled large quantities of stored sediment within the basin and upon the Saharan craton, providing the sediment supply during the subsequent Pliocene-Pleistocene onset of glacial arid conditions and sand-sea construction. If this hypothesis is correct, then through time the Sahara has undergone a net loss of sediment because of the westward regional transport of sand into the Atlantic Ocean. Marine regression during the Last Glacial Maximum and previous glacial periods did expose large portions of the shelf to eolian deflation, but given the regional east-to-west transport, this sand supply was irrelevant for the more interior Sahara.

In terms of sediment availability and transport capacity, the Icehouse, hyperarid Sahara is ideal for eolian construction. Dating of the composite dunes of Mauritania (Lancaster et al., 2002) shows that under ideal conditions, major sand-sea construction can be very rapid, such as during the relatively short Younger Dryas.

Eolian System Accumulation

Accumulation of Mauritanian and Saharan sand-sea deposits in general is lacking. Results from Lancaster et al. (2002) in Mauritania show that the sand seas there are composite bedforms that initially formed during the Last Glacial Maximum and were significantly reworked during the Younger Dryas, then more lightly imprinted during recent arid conditions. This degree of reworking is precisely the type of process to be expected when accumulation does not occur, but rather the accumulation surface and bedforms resting upon it are exposed and subject to repeated reworking. Older accumulations, representing what must be numerous glacial-interglacial cycles since the origin of the Sahara, are present only as isolated, erosional remnants (Kocurek, 1998).

The case for accumulation, however, is potentially higher along the Mauritanian coast and offshore. Dunes prograded to the shelf edge during the last glacial lowstand (Sarnthein, 1978), but it is not known if dry-system accumulations formed during this eolian constructional period. It is more likely that wet-system accumulation occurred with the subsequent rise in sea level, in which coastal sabkha and lagoonal interdune deposits formed adjacent to and flanked linear dune accumulations, as these do today (see Kocurek et al., 1991). Minor (<1 m) Pleistocene eolian accumulations beneath marine deposits have been documented inland within the Mauritanian Basin (Giresse et al., 2000).

Eolian System Preservation

Coastal and the offshore shelf of Mauritania have a preservation potential associated with subsidence of the shelf and within the Mauritanian Basin (Fig. 4) and rising sea level of the adjacent Atlantic Ocean. For example, whatever accumulations formed during the Last Glacial Maximum lowstand and during the subsequent transgression had a preservation potential owing to the rise of sea level of about 120 m and subsequent burial by marine deposits. However, the occurrence of eolian-sand-derived turbidites in the deep Atlantic (Sarnthein and Diester-Haas, 1977), as well as the relatively high energy of the Atlantic shelf, indicates that very significant volumes of eolian sand were reworked during the transgression. Interestingly, plotting of the generation of preservation space because of the rise of sea level during the last 120 ka (Fig. 6) shows that until the most recent rise, generation of preservation space was countered by an equal or greater loss during regressions.

For the inland portions of Mauritania, as well as much of the Sahara, processes of preservation, such as subsidence with burial and/or a rise in the water table, are not occurring (Kocurek, 1998). For example, the Chott Rhasa Basin of Tunisia lies within the active foreland basin inboard of the Atlas Mountains. During portions of the Quaternary, initial preservation of multiple generations of eolian construction and lacustrine accumulations occurred with a rising water table and progressive burial (e.g., eolian accumulations overlain by lacustrine deposits), but these are presently being deflated because of a loss of preservation space owing to a fall in the water table and a greatly diminished sediment influx (Blum et al., 1998; Swezey et al., 1999). Data are scant for elsewhere within the Sahara, but preservation potential may exist along the foreland basin in Algeria, where fluvial sediment influx is higher; in the Lake Chad area, where the water table is maintained by fluvial systems fed from the more humid south; and perhaps in some of the basins on the craton.

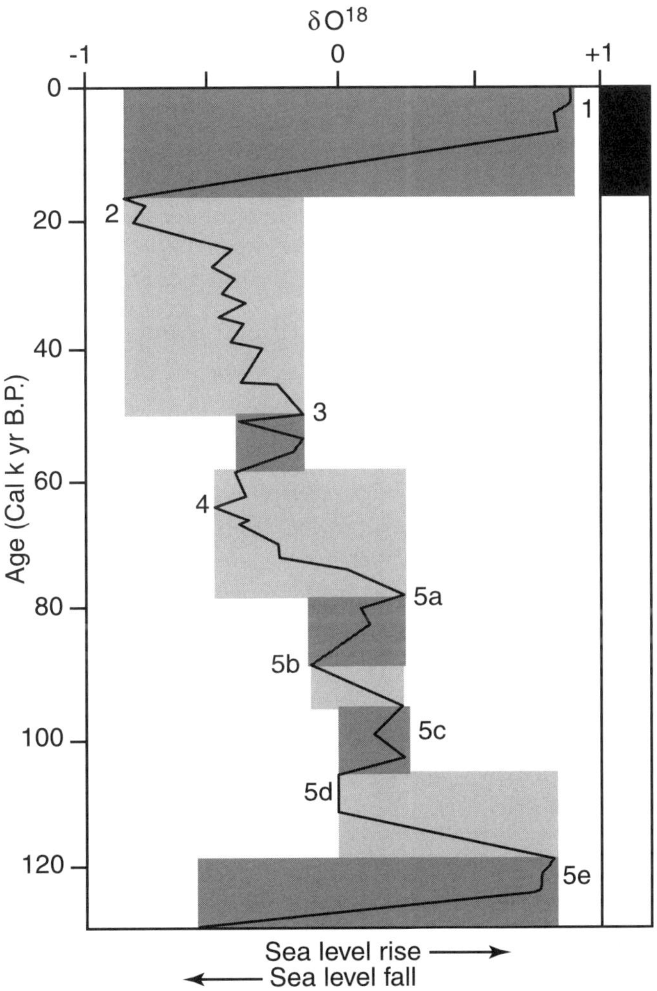

Figure 6. Plot of eustatic sea level (from Martinson et al., 1987) for last 120 ka with isotope stages indicated. Note gain (dark shading) and loss (light shading) of preservation space with changing sea level. Net preservation (black, last column) occurs only with rise in sea level since Last Glacial Maximum. From Kocurek (1998).

JURASSIC NAVAJO SANDSTONE

The Early Jurassic (Pliensbachian-Toarcian) Navajo Sandstone and equivalent Nugget and Glen Canyon Sandstones are among the thickest, most widespread and best exposed eolian units on the planet. The Navajo Sandstone reaches nearly 700 m in thickness in the Utah-Idaho trough, and the Navajo-Nugget-Glen Canyon complex extends over 265,000 km^2 in five states of the Western Interior of the United States (see Fig. 18 in Blakey et al., 1988). Judging by outlying equivalent formations (e.g., Aztec Sandstone and other units) in Nevada and California, and the eastern erosional contact of the formation by the J-2 regional unconformity, the original extent of the Navajo Sand Sea may have been 2.5 times larger than the remaining outcrop (Marzolf, 1988). There is considerable variety in the unit, including wet-system accumulations, fluvial intertongues (Kayenta Formation and Navajo), and distinct zones of biogenic activity (e.g., Verlander, 1995). In a broad sense, the Navajo Sandstone consists of a lower part in which eolian accumulations intertongue to the south and west with fluvial facies of the Kayenta Formation, and an eolian-dominated upper part (Marzolf, 1983; Blakey, 1994). The apex of the Navajo dry eolian system occurs in the Zion area, where sets over 20 m in thickness are vertically stacked. The Navajo and correlated units occur within the Late Paleozoic-Jurassic sequence of the Colorado Plateau, which contains the most extensively preserved eolian accumulations globally.

Eolian System Construction

By any consideration, the collective Navajo sand sea was a constructive event of enormous proportions. Approximations of sand volume by Marzolf (1988) indicate a range of 57–143,000 km^3 of deposited sand, the equivalent of 6–14 m.y. of sand delivery by the modern Mississippi River to the Gulf of Mexico. The bulk of the sand is a quartzarenite, reflecting a mature source area and/or intense weathering and transport.

In terms of sand supply, the origin of the Navajo sands remains speculative, but the tectonic configuration, antecedent conditions, and climate are conducive for the creation of enormous sand supplies within the Western Interior. During the Jurassic, uplands (Appalachian-Ouachita systems) bordered the eastern rim of North America. A foreland basin lay inboard of a magmatic arc along the western edge of the continent (Burchfiel and Davis, 1972; Riggs and Blakey, 1993; Allen et al., 2000). In a scenario similar to that envisioned by Johansen (1988) for the Late Paleozoic, essentially continent-wide drainage for North America would have been directed toward the west. The foreland basin, which housed the Navajo sand sea, provided a sediment sink, and the magmatic arc provided an orographic barrier to sediment loss from the continent. This configuration is broadly similar to that described earlier for the Miocene-Pliocene Sahara, in which north-draining fluvial systems emptied into a foreland basin, and the emergent Atlas Mountains provided an orographic barrier.

Evidence for the magnitude of fluvial activity occurs in the antecedent conditions that, to an extent, continue into Navajo times. The Triassic was a period of widespread fluvial activity in western North America, as is evident from the Chinle, Chugwater, and Dockam Formations, and many other units. It is not unreasonable to believe that fluvial deposits from much of the remaining craton of North America drained into the length of the foreland basin; however, since the area lacks outcrops, this theory is speculative. Triassic fluvial systems were driven by a monsoon climate (Parrish and Peterson, 1988), and although the influx of monsoon moisture was diminished by the Jurassic as the region drifted north in paleolatitude, fluvial systems of the Moenave Formation intertongue with the Lower Jurassic eolian Wingate Sandstone (Clemmensen et al., 1989), the fluvial Kayenta Formation intertongues with the Navajo Sandstone (e.g., Blakey et al., 1988; Blakey 1994; Peterson 1994), and fluvial channels occur in the eolian portion of the Navajo (Verlander, 1995). Blakey (1994) emphasized sand sources of the Ouachita Uplands to the south and east, and a model of fluvial-eolian recycling within the basin. Monsoon wind directional shifts have long been recognized within Jurassic cycles of sets of eolian cross-strata (Hunter and Rubin, 1983; Kocurek et al., 1991), and, more recently, evidence for monsoon rains have been cited in Navajo cross-strata (Loope et al., 2001). It is also possible that streams emptying into the Navajo sand sea along its southern and southeastern margins were fed by streams with headwaters sufficiently south to be within the region still receiving relatively abundant monsoon rains (Riggs and Blakey, 1993).

The Triassic-Jurassic transition to a more arid climate was optimum for enhanced production of a sediment supply and subsequent sediment availability. As with other Jurassic units of the Colorado Plateau, the Navajo system occurred during a lowstand of sea level (Kocurek and Chan, 1988). Moreover, tectonic readjustment of western North America, evident by the lowermost Jurassic J-0 unconformity, must have liberated considerable volumes of sediment for subsequent sand sea construction. The widespread arid conditions of Navajo and other Jurassic sand seas has long been recognized to have been the result of the paleolatitudinal position of the region within the subtropical desert belt (e.g., Kocurek and Dott, 1983; Parrish and Peterson, 1988; Chandler et al., 1992).

Both the cross-strata and paleoclimate models for Navajo sand sea support a dominant winter wind from the northwest and a dune-modifying summer monsoon wind from the northeast (Parrish and Peterson, 1988). The dominant dune-building winds were oriented to transport sediment to the sand sea from more basinal fluvial deposits, and they were oriented onshore for transport of recycled marine sands to the sand sea during periods when marine bodies occupied the basin to the west and north (see Peterson, 1994; Blakey, 1994). During Jurassic Greenhouse time, general circulation model simulations indicate a diminished intensity of the trade-wind-generating Hadley Cell in comparison to Icehouse conditions (Chandler et al., 1992), thereby suggesting that the transport capacity of the Navajo winds was less than that of the Last Glacial Maximum Sahara.

Eolian System Accumulation

Clearly there was basin-wide accumulation within the collective Navajo sand sea. Determining the cause of this accumulation is more speculative. Although wet-system accumulations are present in the formation, primarily along its eastern margins and within the lower part of the formation, most of the accumulations, especially the very large sets at Zion, are those of a dry system in which interdune flats have been reduced to interdune depressions between dunes. Basin-wide accumulation of a dry system argues for a regional cause. Deceleration of the dominant northwest trade winds owing to a regional decrease in pressure gradients, and/or flow expansion into the topographic basin formed by subsidence within the foreland basin that housed the sand sea are the most likely possibilities. Because set thickness generally increases from east to west into the foreland basin, it is most likely that basin geomorphic expression increased from east to west, parallel with basin subsidence, and was the primary cause of accumulation. Indeed, the great set thickness within the Zion area argues for a significant height of unfilled accumulation space (i.e., an unfilled, subsided basin) prior to the onset of the Navajo sand sea. The argument against a regional deceleration of the wind is that the east-to-west direction of set thickening is obliquely upwind, and, hence, is not consistent with a deceleration of the winds down-flow within the sand sea.

Mechanically, accumulation requires a positive angle of climb for migrating dunes (e.g., positive θ where $\tan \theta = V_y / V_x$). Although angle of climb nor the types of dunes that populated the Navajo sand sea have been determined systematically, the few published examples indicate large features with superimposed dunes (e.g., Rubin, 1987), and overall set width, preserved thickness, and grain-flow thickness within the cross-strata argue that Navajo dunes were very large, perhaps even the largest dunes for which evidence has been found on Earth. Large dunes are the manifestation of the ample sand supply (see above), and large dunes tend to have slow migration rates (V_x). When combined, these factors favor a strong positive angle of climb. Ironically, if low wind speeds were characteristic of Jurassic Greenhouse conditions (in comparison to Icehouse conditions), then slower dune migration rates (V_x) would have occurred and contributed to a higher angle of climb. The monsoon nature of the winds, with lee reworking due to summer northeast winds, would also have contributed to a slower migration rate.

Eolian System Preservation

Preservation of Navajo sand sea accumulations is accredited to regional subsidence within the retroarc foreland basin, and a relative and absolute rise of the water table through the accumulations. Although the literature shows a disagreement about the timing of the onset of foreland basin loading and asymmetric basin subsidence (Bjerrum and Dorsey, 1996; Allen et al., 2000), the pronounced east-to-west thickening of the Navajo and related units supports the subsidence-curve models of Allen et al. (2000), in which a Lower Jurassic (Navajo) initiation of loading events in the Cordilleran orogen is given. Until regional unconformities are documented within the Navajo, it must be assumed that sediment influx, accumulation, and burial of the accumulations continued uninterrupted. Loss of preservation space within the basin occurs with regional uplift and creation of the J-1 and J-2 unconformities that truncate the Navajo and are interpreted as the response to the migration of the forebulge as a result of episodic thrusting in the Cordilleran Belt (Bjerrum and Dorsey, 1996). The forebulge, located east of the Capitol Reef area in the Allen et al. (2000) model, appears to divide the Navajo sand sea into a well-developed dry system to the west and a more wet system with fluvial influx to the east (Verlander, 1995).

Subsequent to the J-1 unconformity, western portions of Navajo accumulations were marine transgressed (marine portions of Temple Cap Sandstone, Gypsum Springs Formation). The eastward extent of the transgression is not well known because the transgressive units are truncated by the J-2 unconformity (see Fig. 15 in Peterson, 1994). The absolute rise in the water table because of this marine transgression and the corresponding inland rise in the water table (i.e., Kocurek et al., 2001), as well as relative rise of the water table as a result of subsidence, ultimately defined the limits of preservation of the Navajo accumulations. The J-2 unconformable surface that bounds the Navajo and correlated units is widely marked by water table features (e.g., polygonal fractures, chert-replaced evaporite nodules; Kocurek and Hunter, 1986), indicating deflation to the water table. Preservation of eolian accumulation by a rise of the water table associated with a marine transgression is a theme repeated throughout the Jurassic, in which each eolian unit is capped by a marine unit with a progressive eastward shoreline migration (Fig. 7). This eastward shift of the shoreline within this tectonically active basin seems best accredited to an eastward migration of the forebulge (c.f., Bjerrum and Dorsey, 1996).

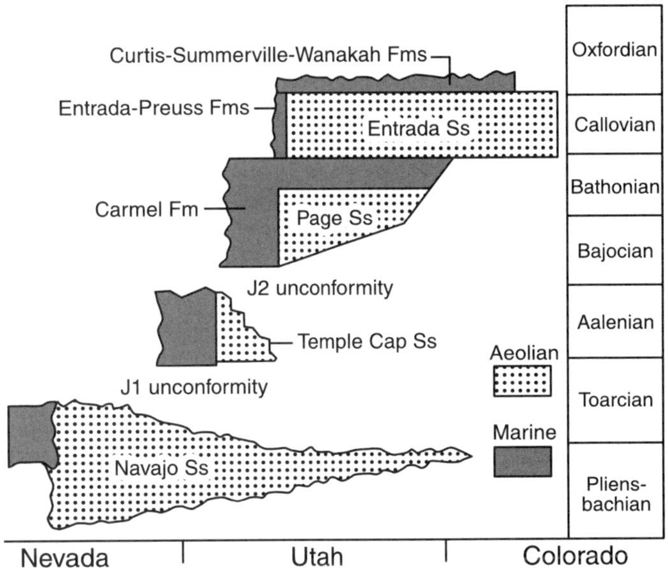

Figure 7. Wheeler-type diagram for portion of Jurassic of the U.S. Western Interior, showing preserved eolian accumulations and marine strata. Note progressive eastward migration of marine strata. From Kocurek (1999).

CONCLUSIONS

The underlying theme of this paper is that the controls on the dynamics of eolian sand seas are understood to the point that the conditions favorable for an "extreme" eolian example in the rock record can be given. The theory underlying this theme is that the construction, accumulation, and preservation of eolian systems are distinct phases and largely independent of each other. The construction, accumulation, and preservation of an eolian extreme, therefore, call for the coincidence of favorable events and conditions.

For the Mauritanian Saharan example given, although it was constructed during an Icehouse period of high sediment availability and though transport potential occurred, sediment supply was not great, and conditions for accumulation are largely absent at the regional scale. Although preservation potential exists for shelf eolian accumulations owing to a eustatic rise in sea level and subsidence of the shelf and the Mauritanian Basin, sediment reworking during the transgression has been significant. In contrast, the Jurassic Navajo Sandstone represents a unit that experienced significantly greater sediment supply and favorable conditions for accumulation and preservation. Loss of preservation space and stripping of accumulations was largely confined to the post-Navajo tectonic readjustment of the Western Interior by migration of the forebulge and formation of the J-1 and J-2 unconformities. If comparisons are to be drawn, the now offshore Sahara represents, in a diminished way, the eolian record of the Jurassic of the Western Interior, whereas the present Sahara Desert represents the Jurassic east of the outcrop belt where the sand sea is represented by a surface.

ACKNOWLEDGMENTS

I appreciate the constructive reviews of this paper by Rip Langford and Len Eisenberg, and the infinite patience of the volume editor, Margie Chan.

REFERENCES CITED

Allen, P.A., Verlander, J.E., Burgess, P.M., and Audet, D.M., 2000, Jurassic giant erg deposits, flexure of the United States continental interior, and timing of the onset of Cordilleran shortening: Geology, v. 28, p. 159–162.

Bagnold, R.A., 1941, The physics of blown sand and desert dunes: London, Chapman and Hall, 265 p.

Bjerrum, C.J., and Dorsey, R.J., 1996, Tectonic controls on deposition of Middle Jurassic strata in a retroarc foreland basin, Utah-Idaho trough, western interior, United States: Tectonics, v. 14, p. 962–978.

Blakey, R.C., 1994, Paleogeographic and tectonic controls on some Lower and Middle Jurassic erg deposits, Colorado Plateau, *in* Caputo, M.V., Peterson, J.A., and Franczyk, K.J., eds., Mesozoic Systems of the Rocky Mountain Region, USA: Tulsa, Oklahoma, Rocky Mountain Section Society for Sedimentary Geology (SEPM), p. 273–298.

Blakey, R.C., Peterson, F., and Kocurek, G., 1988, Synthesis of Late Paleozoic and Mesozoic eolian deposits of the Western Interior of the United States: Sedimentary Geology, v. 56, p. 3–125.

Blum, M., Kocurek, G., Swezey, C.S., Deynoux, M., Lancaster, N., Price, D.M., and Pion, J.-C., 1998, Quaternary wadi, lacustrine, aeolian depositional cycles and sequences, Chott Rharsa Basin, southern Tunisia, *in* Alsharhan, A., Glennie, K., Whittle., G., and Kendall, C., eds., Quaternary deserts and climatic change: Rotterdam, Balkema, p. 539–551.

Burchfiel, B.C., and Davis, G.A., 1972, Structural framework and evolution of the southern part of the Cordilleran Orogen, western United States: American Journal of Science, v. 272, p. 97–118.

Chandler, M.A., Rind, D., and Ruedy, R., 1992, Pangaean climate during the Early Jurassic: GCM simulations and the sedimentary record of paleoclimate: Geological Society of America Bulletin, v. 104, p. 543–559.

Clemmensen, L.B., Olsen, H., and Blakey, R.C., 1989, Erg-margin deposits in the Lower Jurassic Moenave Formation and Wingate Sandstone, southern Utah: Geological Society of America Bulletin, v. 101, p. 759–773.

Giresse, P., Barusseau, J.-P., Causse, C., and Diouf, B., 2000, Successions of sea-level changes during the Pleistocene in Mauritania and Senegal distinguished by sedimentary facies study and U/Th dating: Marine Geology, v. 170, p. 123–139.

Glennie, K.W., 1970, Desert sedimentary environments: Amsterdam, Elsevier, 210 p.

Hunter, R.E., and Rubin, D.M., 1983, Interpreting cyclic crossbedding, with an example from the Navajo Sandstone, *in* Brookfield, M.E., and Ahlbrandt, T.S., eds., Aeolian sediments and processes: Amsterdam, Elsevier, p. 429–454.

Johansen, S.J., 1988, Origins of Upper Paleozoic quartzose sandstones, American southwest: Sedimentary Geology, v. 56, p. 153–166.

Kocurek, G., 1998, Aeolian system response to external forcing factors–a sequence stratigraphic view of the Saharan region, *in* Alsharhan, A., Glennie, K., Whittle, G., and Kendall, C., eds., Quaternary deserts and climatic change: Rotterdam, Balkema, p. 327–337.

Kocurek, G., 1999, The aeolian rock record (Yes, Virginia, it exists, but it really is rather special to create one), *in* Goudie, A.S., Livingstone, I., and Stokes, S., eds., Aeolian environments, sediments and landforms: New York, John Wiley and Sons, p. 239–259.

Kocurek, G., and Chan, M.A., 1988, Complexities in eolian and marine interactions: Processes and eustatic controls on erg development: Sedimentary Geology, v. 56, p. 283–300.

Kocurek, G., and Dott, R.H., 1983, Jurassic paleogeography and paleoclimate of the central and southern Rocky Mountains region, *in* Reynolds, M.W., and Dolly, E.D., eds., Mesozoic paleogeography of west-central United States: Tulsa, Oklahoma, Rocky Mountain Section Society for Sedimentary Geology (SEPM), p. 101–116.

Kocurek, G., and Havholm, K., 1993, Eolian sequence stratigraphy–a conceptual framework, *in* Weimer, P., and Posamentier, H., eds., Siliciclastic sequence stratigraphy: American Association of Petroleum Geologists Memoir 58, p. 393–409.

Kocurek, G., and Hunter, R.E., 1986, Origin of polygonal fractures in sand, uppermost Navajo and Page Sandstones, Page, Arizona: Journal of Sedimentary Petrology, v. 56, p. 895–904.

Kocurek, G., and Lancaster, N., 1999, Aeolian system sediment state: Theory and Mojave Desert Kelso dune field example: Sedimentology, v. 46, p. 505–515.

Kocurek, G., Havholm, K.G., Deynoux, M., and Blakey, R.C., 1991, Amalgamated accumulations resulting from climatic and eustatic changes, Akchar Erg, Mauritania: Sedimentology, v. 38, p. 751–772.

Kocurek, G., Robinson, N.I., and Sharp, J.M., 2001. The response of the water table in coastal aeolian systems to changes in sea level: Sedimentary Geology, v. 139, p. 1–13.

Lancaster, N., Kocurek, G., Singhvi, A., and Deynoux, M., 2002, Late Pleistocene and Holocene dune activity in the western Sahara of Mauritania: Geology, v. 30, p. 991–994.

Langbein, W., and Schumm, S., 1958, Yield of sediment in relation to mean annual precipitation: Eos (Transactions, American Geophysical Union), v. 39, p. 1076–1084.

Loope, D.B., Rowe, C.M., and Joeckel, R.M., 2001, Annual monsoon rains recorded by Jurassic dunes: Nature, v. 412, p. 64–66.

Mainguet, M., and Chemin, M., 1983, Sand seas of the Sahara and Sahel: An explanation of their thickness and sand dune type by the sand budget principle, *in* Brookfield, M.E., and Ahlbrandt, T.S., eds., Aeolian sediments and processes: Amsterdam, Elsevier, p. 353–363.

Martinson, D., Pisias, N., Hays, J., Imbrie, J., Moore, T., and Shackleton, N., 1987, Age dating and the orbital theory of the ice ages: Development of a high-resolution 0 to 300,000-year chronostratigraphy: Quaternary Research, v. 27, p. 1–29.

Marzolf, J.E., 1983, Changing wind and hydrologic regimes during deposition of the Navajo and Aztec Sandstone, Jurassic (?), southwestern United States, *in* Brookfield, M.E., and Ahlbrandt, T.S., eds., Aeolian sediments and processes: Amsterdam, Elsevier, p. 635–660.

Marzolf, J.E., 1988, Controls on Late Paleozoic and Early Mesozoic eolian deposition of the western United States: Sedimentary Geology, v. 56, p. 167–192.

Middleton, G., and Southard, J., 1984, Mechanics of sediment movement: Society for Sedimentary Geology (SEPM) Short Course 3.

Parkin, D.W., and Shackleton, N.J., 1973, Trade wind and temperature correlations down a deep-sea core off the Sahara coast: Nature, v. 245, p. 455–457.

Parrish, J.T., and Peterson, F., 1988, Wind directions predicted from global circulation models and wind directions determined from eolian sandstones of the western United States—a comparison: Sedimentary Geology, v. 56, p. 261–282.

Peterson, F., 1994, Sand dunes, sabkhas, streams, and shallow seas: Jurassic paleogeography in the southern part of the Western Interior Basin, *in* Caputo, M.V., Peterson, J.A., and Franczyk, K.J., eds., Mesozoic systems of the Rocky Mountain Region, USA: Tulsa, Oklahoma, Rocky Mountain Section, Society for Sedimentary Geology (SEPM), p. 233–272.

Riggs, N.R., and Blakey, R.C., 1993, Early and Middle Jurassic paleogeography and volcanology of Arizona and adjacent areas, *in* Dunn, G., and McDougall, K., eds., Mesozoic paleogeography of the western United States II: Pacific Section Society for Sedimentary Geology (SEPM) Book 71, p. 347–375.

Rubin, D.M., 1987, Cross-bedding and paleocurrents: Society for Sedimentary Geology (SEPM) Concepts in Sedimentology, no. 1, 187 p.

Rubin, D.M., and Hunter, R.E., 1982, Bedform climbing in theory and nature: Sedimentology, v. 29, p. 121–138.

Sarnthein, M., 1978, Sand deserts during glacial maximum and climatic optimum: Nature, v. 272, p. 43–46.

Sarnthein, M., and Diester-Haas, L., 1977, Eolian sand turbidites: Journal of Sedimentary Petrology, v. 47, p. 868–890.

Scherer, C.M.S., 2000, Eolian dunes of the Botucatu Formation (Cretaceous) in southernmost Brazil: Morphology and origin: Sedimentary Geology, v. 137, p. 63–84.

Swezey, C., Lancaster, N., Kocurek, G., Deynoux, M., Blum, M., Price, D., and Pion, J.-C., 1999, Response of aeolian systems to Holocene climatic and hydrologic changes on the northern margin of the Sahara: A high-resolution record from the Chott Rharsa Basin, Tunisia: The Holocene, v. 9, p. 141–147.

Talbot, M.R., 1984, Late Pleistocene rainfall and dune building in the Sahel, *in* Coetzee, J.A., and van Zinderen, E.M., eds., Palaeoecology of Africa: Rotterdam, Balkema, p. 203–221.

Verlander, J.E., 1995, Basin-scale stratigraphy of the Navajo Sandstone: Southern Utah, USA [Ph.D. dissertation]: Oxford, UK, Oxford University, 159 p.

Wilson, I., 1973, Ergs: Sedimentary Geology, v. 10, p. 77–106.

Manuscript Accepted by the Society January 22, 2003

Printed in the USA

Geological Society of America
Special Paper 370
2003

Quaternary loess-paleosol sequences as examples of climate-driven sedimentary extremes

Daniel R. Muhs
U.S. Geological Survey, MS 980, Box 25046, Federal Center, Denver, Colorado 80225, USA

E. Arthur Bettis III
Department of Geoscience, University of Iowa, Iowa City, Iowa 52242, USA

ABSTRACT

Loess is a widespread, wind-transported, silt-dominated deposit that contains geologic archives of atmospheric circulation and paleoclimate on continents. Loess may cover as much as 10% of the Earth's land surface. It is composed mainly of quartz, feldspars, and clay minerals, with varying amounts of carbonate minerals. The geochemistry of loess differs from region to region, depending on source materials, but all loess is very high in SiO_2 with lesser amounts of other major elements. Trends in loess downwind from source areas include systematic decreases in thickness and amounts of sand and coarse silt, and increases in amounts of fine silt and clay. Loess particle size also varies at a given locality over time within individual depositional packages. This variability may be a function of changing wind strengths, different source sediments, or some combination of the two.

The classical concept of loess is that it is a product of glacial grinding, with subsequent entrainment by wind from outwash deposits. However, it is now known that other processes contribute to silt particle formation, including frost shattering, salt weathering, fluvial and colluvial comminution, eolian abrasion, and ballistic impact. Much debate has taken place over the concept of "desert" (nonglaciogenic) loess, which is widespread in some regions but of limited distribution elsewhere. Nevertheless, glacial silt production probably exceeds the amount of silt generated by all other processes. Much of the loess in or adjacent to deserts may be inherited silt-sized particles from siltstone, mudstone, shale, and volcanic ash.

In many regions, loess is near dune fields or eolian sand sheets. A question that arises from this geographic assocation is whether or not eolian sand and loess should be considered facies of the same depositional unit. There are regions such as China where these deposits are interbedded, which supports the facies concept. In other regions, such as North America, detailed geochemical and isotopic analyses show that the majority of loess particles were derived from a different, and more distant source than eolian sand.

Key to understanding loess stratigraphy and interpreting environments of the past is the recognition of buried soils (paleosols). Ancient soils can be recognized by their distinctive morphological features and by vertical changes in particle size, chemistry, and mineralogy. Paleosols represent past periods when loess sedimentation rates decreased to zero or slowed significantly. Thus, loess and their interstratified soils represent end members of a continuum of sedimentary extremes: high rates of sedimentation yield relatively unaltered loess in the stratigraphic record, whereas low or

Muhs, D.R., and Bettis, E.A., III, 2003, Quaternary loess-paleosol sequences as examples of climate-driven sedimentary extremes, *in* Chan, M.A., and Archer, A.W., eds., Extreme depositional environments: Mega end members in geologic time: Boulder, Colorado, Geological Society of America Special Paper 370, p. 53–74.

episodic rates of sedimentation commonly leave a record of buried soils. The shift between these sedimentary extremes is preserved in the long-term glacial-interglacial record of the Quaternary.

Although it is now known that not all eolian silt is glaciogenic, in almost all loess regions, eolian sedimentation rates were much higher during glacial periods than during interglacial periods. Drier, colder climates, a decreased intensity of the hydrologic cycle, stronger or more-persistent winds, increased sediment supply, decreased vegetation cover, and increased sediment availability all probably contributed to the sedimentary "extreme" of rapid loess accumulation during the last glacial period. The present interglacial period represents an opposite sedimentary "extreme" of minimal loess sedimentation and is characterized primarily by soil formation within loess deposits of last glacial age.

Keywords: loess, eolian, Quaternary, paleosols, paleoclimate, climate extremes, glacial-interglacial cycles.

INTRODUCTION

Loess is an important archive of Quaternary climate changes, and may have one of the most complete terrestrial records of interglacial-glacial cycles (e.g., Kukla et al.,1988; Hovan et al., 1989; Porter and An, 1995; Porter, 2001; Palmer and Pillans, 1996; Muhs et al., 1999a). Because loess is deposited from the atmosphere, it is also one of the few geologic deposits that directly records atmospheric circulation and can be used to test atmospheric general circulation models (e.g., Muhs and Bettis, 2000). Furthermore, airborne dust itself may be significant in bringing about climate change through its radiative transfer properties or through its role in the fertilization of primary producers in the world's oceans. Loess records hold promise for evaluating the role of dust in climate change (Tegan et al., 1996; Mahowald et al., 1999; Kohfeld and Harrison, 2000, 2001; Harrison et al., 2001). Finally, there has been an increasing recognition that loess, in the form of loessite, may be a more common part of the sedimentary rock record than previously supposed (Johnson, 1989; Soreghan, 1992; Chan, 1999).

In this paper, we review recent studies that report the composition of loess, its origins, and its paleoclimatic significance. Because loess covers a significant portion of the Earth's surface, it is an important part of Quaternary geologic and climatic records. Loess is also the surficial deposit on which millions of humans live, and it forms the parent material for some of the world's most productive agricultural soils. The history of loess studies was recently reviewed by Smalley et al. (2001). In this paper, we concentrate on issues regarding loess origins and its paleoclimatic importance. We present data to show that loess-paleosol sequences in the Quaternary geologic record represent good examples of sedimentary extremes. By sedimentary "extremes" we mean periods of unusually high or low rates of eolian silt sedimentation. We attempt to show that in the Quaternary geologic record, the stratigraphic sequences of loesses and paleosols represent sedimentary extremes that are largely climate-controlled, either directly or indirectly.

DEFINITION OF LOESS

Loess is often perceived as a rather homogenous deposit that has considerable similarity from continent to continent. It is surprising, therefore, how much variation there is in various workers' definitions of loess. Smalley and Smalley (1983) defined loess in terms of process and considered four mechanisms to be a part of its formation: (1) loess-sized material (20–60 μm) forms, (2) the material is transported by the wind, (3) the sediment is deposited, and (4) the sediment experiences post-depositional changes. Pésci (1990) listed 10 characteristics that, in his view, define loess. These include the following: loess is wind-deposited, homogenous, porous, permeable, pale-yellow, predominantly coarse silt (10–50 μm) with small amounts of fine silt and clay, dominantly quartz with some feldspars and carbonate, usually unstratified (but contains paleosols), slightly cemented, stable when dry, easily eroded when wet, and contains fossil flora and fauna. Pésci (1990) and other investigators have referred to post-depositional compaction and minor diagenetic cementation of loess particles by carbonate as an important process of "loessification." Many European workers feel that this is a key process that defines loess as a sedimentary body. We think that the concept of "loessification" is unnecessarily restrictive. Although minor cementation does occur in many calcareous loess bodies, other loesses (such as those in Alaska and New Zealand) do not contain carbonates and are not cemented. However, few workers would disagree that the eolian silts in these two regions should be called loess. Pye (1987) defines loess simply as a terrestrial, windblown silt consisting chiefly of quartz, feldspar, mica, clay minerals, and carbonate grains in varying proportions.

Despite the differences apparent in these workers' views, common to all definitions are two factors: loess is wind-deposited and is dominated by silt-sized particles. Pye's (1987) definition is the one we adopt here, with the modification that many loesses, such as most of those in Alaska and New Zealand, do not contain carbonates. We consider loess to be any terrestrial, wind-deposited sediment dominated by silt-sized particles, whether calcareous or not. Thus, eolian sediments in the deep-sea record are excluded in this definition, although they may have an origin common with some loess deposits (e.g., Hovan et al., 1989). A corollary definition we present here is that loess is an eolian silt of sufficient thickness that it is recognizable as a sedimentary body in the field.

GEOGRAPHY OF LOESS

Compilation of loess maps worldwide shows that wind-blown silt covers a significant amount of the Earth's land surface, perhaps as much as 10% (Figs. 1–4). In the western hemisphere, loess is present in both North and South America (Fig. 1). In South America, there are two major loess belts, the Pampas loess in central Argentina and the Chaco loess in northern Argentina and Paraguay. Although the distribution of loess in South America appears at small scale to be continuous, the compositions of these two named loess bodies are distinct. The different origins for South American loesses were reviewed recently by Muhs and Zárate (2001). In North America, there are five distinct areas of loess: (1) extensive but discontinuous bodies of loess in Alaska and adjacent Yukon Territory, (2) the Palouse loess of eastern Washington and adjacent Oregon, (3) the Snake River Plain and adjacent uplands of Idaho, (4) the Great Plains region of the midcontinent, and (5) the central lowlands region of the midcontinent. As with the Pampas and Chaco region loess bodies of South America, the Great Plains (Nebraska, Kansas, and Colorado) and central lowlands (Missouri, Iowa, Illinois, Indiana, and areas to the north and south) loesses of the North American midcontinent appear to be continuous at small scale (Fig. 1). At a larger scale (Fig. 2), it is apparent that these loess bodies have very different thickness trends that are not part of a larger regional trend, and as will be shown, Great Plains and central lowlands loesses have different origins. Loess is not present in Canada except in small, isolated patches because most areas were covered by the Laurentide ice sheet at the most recent times of eolian silt deposition in North America. As the Laurentide ice sheet receded, glaciogenic silt became available, but in much of eastern and central Canada, this silt was deposited in proglacial lakes rather than entrained by the wind and deposited as loess (Flint, 1971). In fact, the termination of loess deposition farther south in the midcontinent may be, in part, the result of this silt deposition in proglacial lakes. Thick (>1 m) loess has not been reported in Mexico or most parts of the southwestern United States, and its absence in those regions has significance for the origin of loess.

In the eastern hemisphere, loess is abundant over much of Eurasia (Fig. 3). Most loess in Eurasia is distributed in a latitudinal belt between about 40° and 60°N, covering areas south of the limits of continental or mountain glaciers of Quaternary age. An important exception is China (Fig. 4), where loess covers large areas at lower latitudes that were not close to either continental or mountain glaciers. Loess is largely absent from the subtropical and tropical latitudes of Eurasia.

Loess is not extensive over Africa, nor is it widespread in adjacent subtropical parts of the Middle East. There are, however, well-

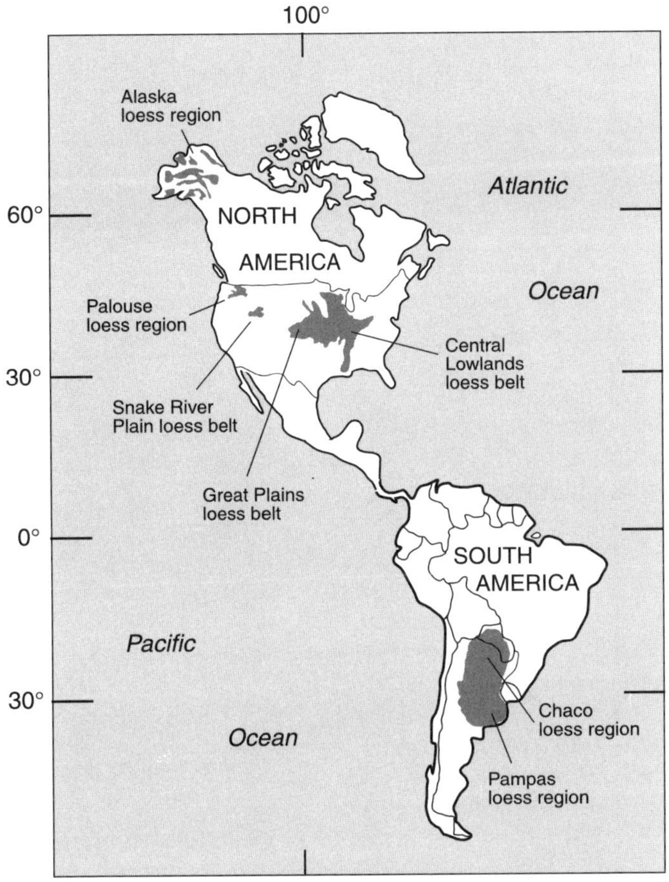

Figure 1. Map showing the distribution of loess in North America and South America and names of loess belts used in this paper. Loess distribution from compilation in Muhs and Zárate (2001) and sources therein.

documented but geographically limited loess deposits in Tunisia, Libya, Nigeria, Namibia, and Israel (e.g., Bruins and Yaalon, 1979; McTainsh, 1987). Loess is also largely absent in Australia, although there are limited occurrences of clay-rich eolian deposits termed *parna* that some workers interpret as essentially a clay-rich loess (Butler, 1974). However, loess is widespread over much of New Zealand, where it has been studied in considerable detail (Palmer and Pillans, 1996; Graham et al., 2001).

SEDIMENTOLOGY OF LOESS

Although loess is silt-dominated by all definitions, there is a surprising range of particle size distributions reported, even within the same sedimentary body. Mean particle sizes for loess vary from coarse silt to fine silt, and individual loess bodies span this entire range (Fig. 5). Variation in loess particle size can occur either spatially or temporally. In China, for example, loess in the northwesternmost part of the Loess Plateau is described as "sandy" and has a mean particle size between about 4.75–5.0 phi (37–31 microns). At the southeastern por-

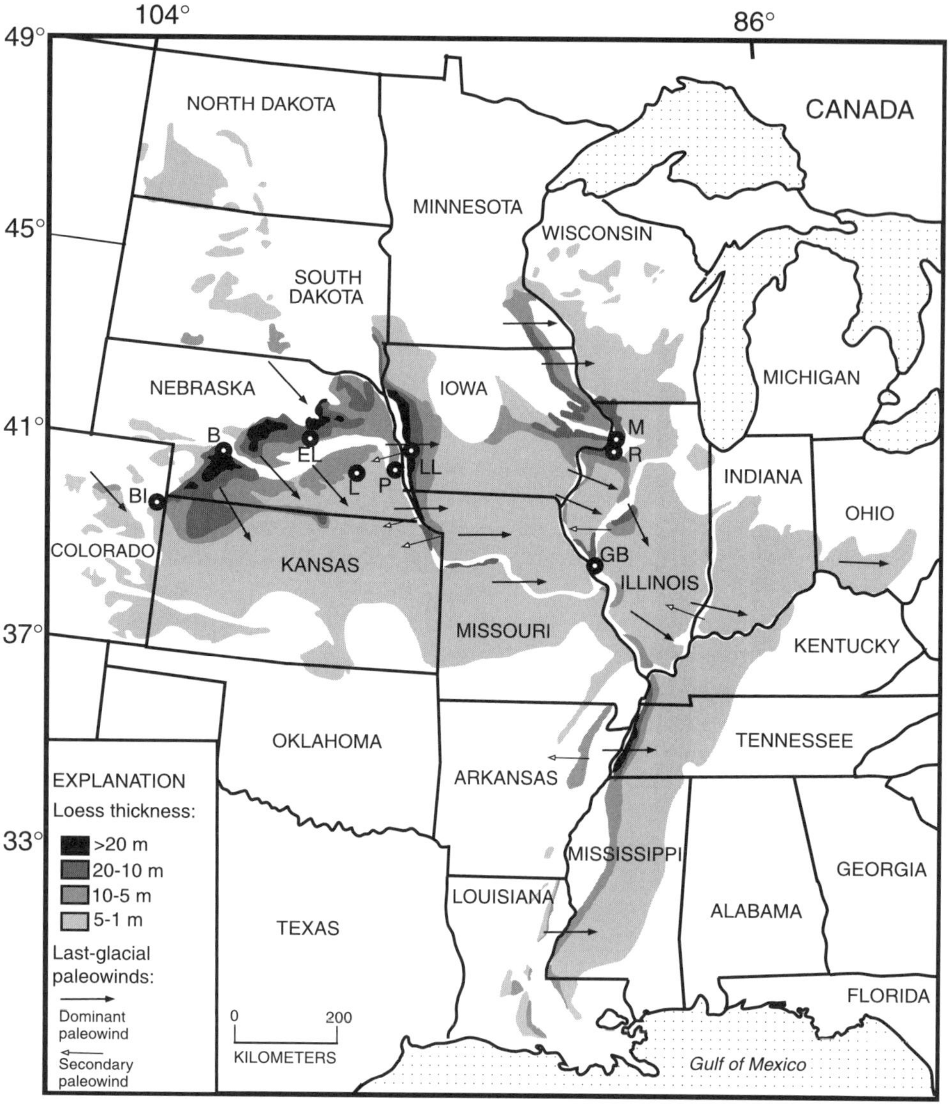

Figure 2. Map showing distribution and thickness of loess in central North America and stratigraphic sections referred to in text. Loess distribution from compilation by Bettis et al. (2003). Arrows indicate inferred paleowinds based on loess thickness, particle size trends, and other data (see compilation by Muhs and Bettis, 2000). Stratigraphic sections referred to in text: BI—Beecher Island; B—Bignell Hill; EL—Elba; L—Lincoln; P—Plattsmouth; LL—Loveland; M—Morrison; R—Rapid City; GB—Greenbay Hollow.

tion of this loess body, what is referred to as "clayey" loess has a mean particle size of about 6 phi (15–16 microns). Standard deviations of loess can have a range of several phi units within a loess body, as illustrated by the range for loess in the Yakutia region of Russia and to a lesser degree by the loess of eastern Colorado in central North America. The wide range of mean particle size and relatively poor sorting in a loess body compared to eolian sand can be the result of (1) multiple sources, (2) clay-sized particles being transported as silt-sized aggregates, (3) loess bodies extending considerable distances from their sources, or (4) varying wind strengths over time. It is apparent from the data in Figure 5 that eolian sands and loesses are distinctively different in terms of their sedimentological properties, as there is neither overlap in mean particle size nor much in degree of sorting. This distinction is important for understanding the origin of loess, which we discuss later.

Numerous workers have shown that a wide variety of loess sedimentological parameters show strong distance-decay functions away from probable sources. Smith (1942) and Ruhe (1983) summarize many of the trends for North American loess bodies, and Porter (2001) shows similar trends for Chinese loess. Loess thickness, mean particle size, sand content, and coarse silt content all decrease away from a source while fine silt and clay contents increase away from a source (Figs. 2 and 6). The decrease in overall loess thickness reflects a net decrease in the sediment load away from the source, at least when vegetation cover is sufficient to trap particles (Tsoar and Pye, 1987). The decrease in sand and coarse silt contents and increase in fine silt and clay contents reflect a winnowing of the coarse load away from the source.

Loess also shows particle size variations at individual localities, even within the same depositional package. In China, Porter and An (1995) showed that mean diameters of the quartz fraction

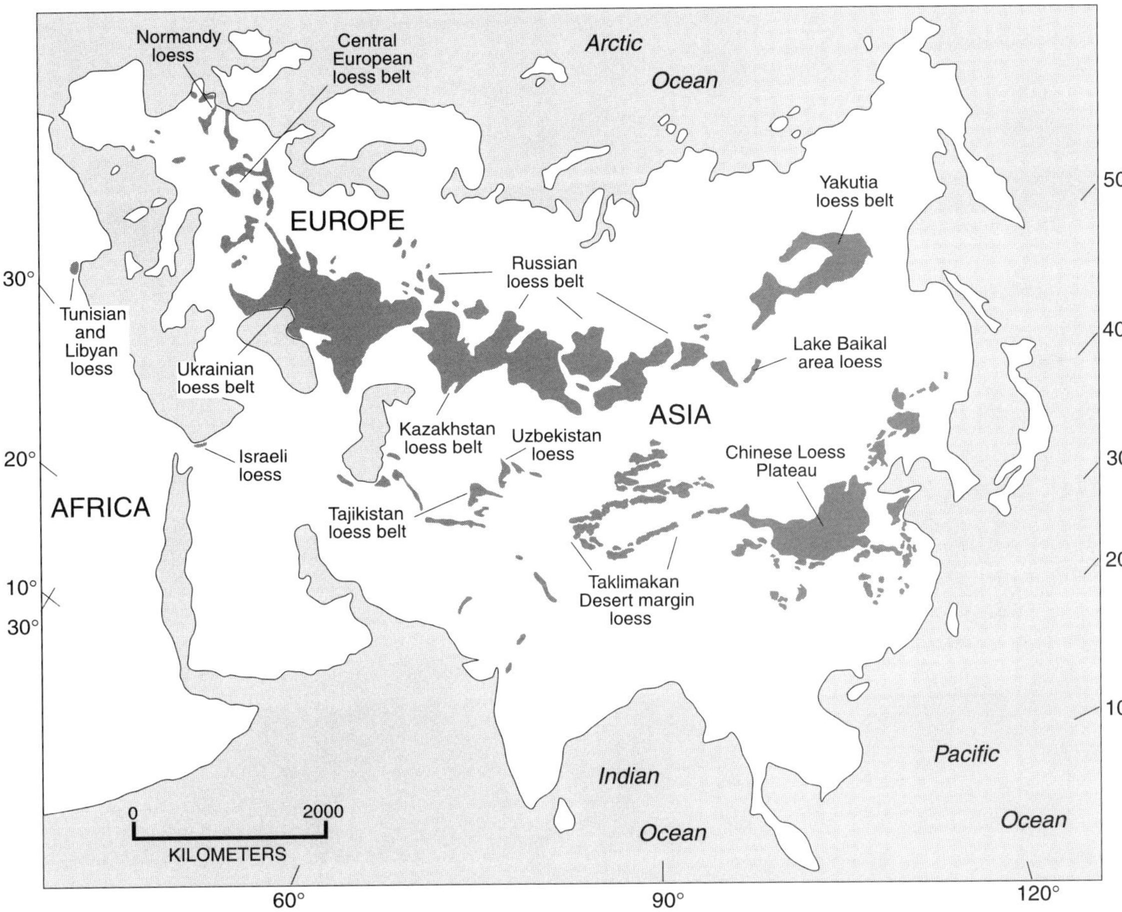

Figure 3. Map showing distribution of loess in Eurasia and localities or regions referred to in text. Compiled by the authors from Rozycki (1991) and Liu (1985).

in loess varied significantly over the last glacial period (Fig. 7). They interpret these data to indicate varying wind strengths over the period of loess deposition. In other regions, loess particle size variability has been interpreted to indicate changing loess sources over time. For example, at a locality in North America (Loveland, Iowa; see Fig. 2), Muhs and Bettis (2000) demonstrated that last-glacial-age (Peoria) loess, which is ~40 m thick, has three distinct zones based on particle size distribution. The differentiation of the upper two zones at Loveland, Iowa, is interpreted to be a function of changing dominance by two sources, the nearby Missouri River and distant Great Plains.

GEOCHEMISTRY OF LOESS

There are now many data available on loess geochemistry that yield important information about loess mineralogy and origins. In all loesses, the dominant constituent is SiO_2, which ranges from ~45% to 75%, but is typically 55–65%. Plots of SiO_2 versus Al_2O_3 (Fig. 8) show that most loess has a composition that falls close to the range of average upper crustal rock (Taylor and McLennan, 1985). Mineralogical studies show that the high SiO_2 contents of loess reflect a dominance of quartz, but smaller amounts of feldspars and clay minerals also contribute to this value. Most loesses also fall between fields spanning the average composition of shales and quartz-dominated sandstones. Sandstones, particularly quartz arenites, are very high in SiO_2 whereas shales are clay-dominated and are therefore high in Al_2O_3. An exception to these generalizations is shown by the highly variable composition of loesses from the North Island of New Zealand, near Wanganui (Graham et al., 2001). Amounts of SiO_2 in these loesses range from 40% to 50% to greater than 70%, a range that spans compositions from basalt to granite and shows little relation to a continuum from shale to sandstone. It seems likely that the loesses in this region had sources that ranged in composition from highly mafic to highly felsic and that the relative amount of sediment from these sources changed over time.

Although most loesses have a composition between that of shale and sandstone, loesses from different localities nevertheless show considerable variation in composition. Plots of complimentary element pairs that represent the non-quartz mineral fractions

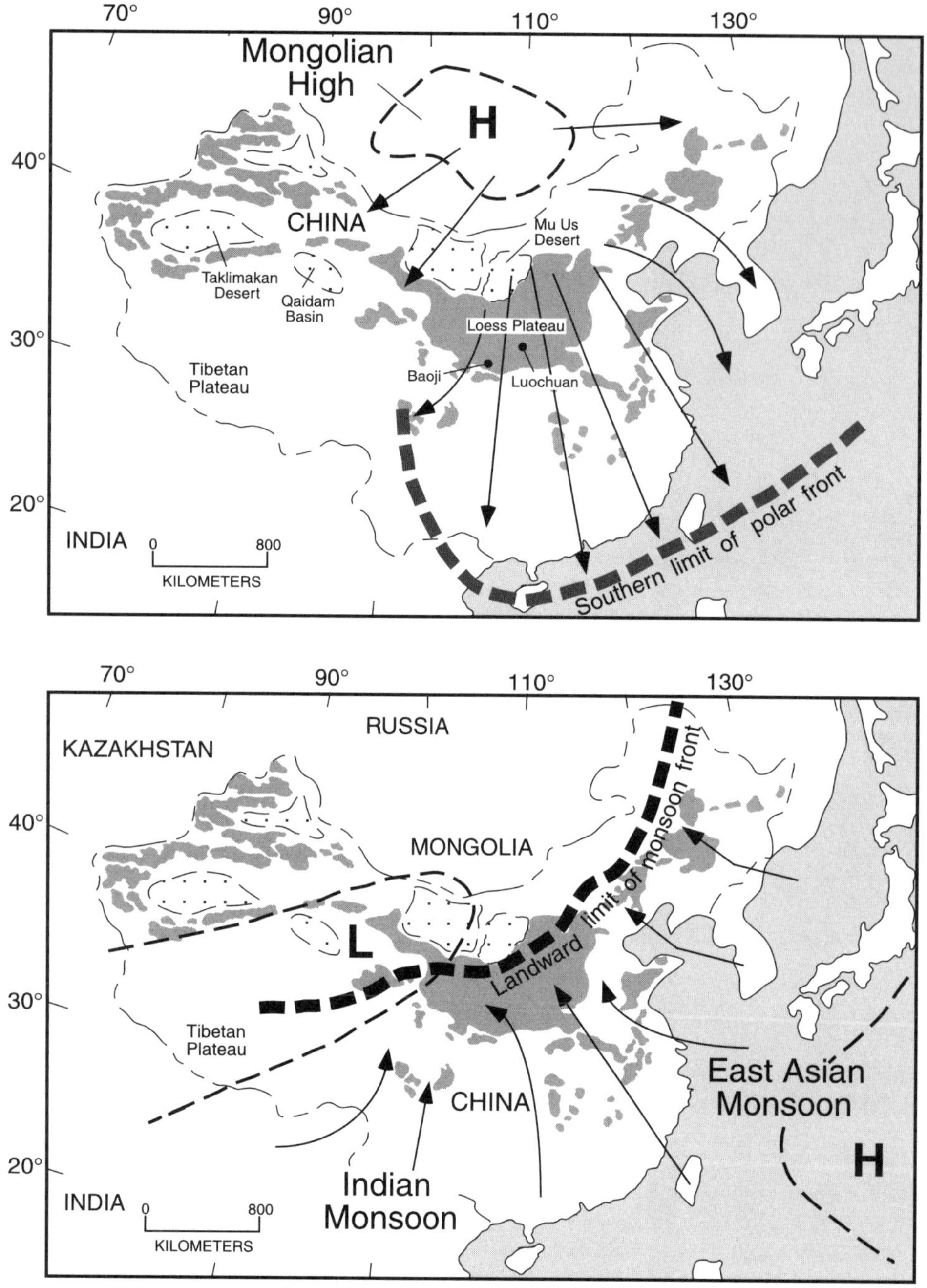

Figure 4. Distribution of loess (shaded areas), sandy deserts (dotted areas), localities referred to in text, and synoptic climatology of China during winter (upper diagram) and summer (lower diagram) showing pressure systems and dominant surface winds (arrows). Climatic data from Porter and An (1995); loess distribution from Liu (1985).

are particularly revealing in this regard because they define geochemical fields that are distinctive for each loess body (Fig. 9). North American (Illinois) loess derived from outwash of the Laurentide ice sheet has abundant carbonate minerals, particularly dolomite (McKay, 1979; Grimley et al., 1998). Thus, compared to Russian or Chinese loesses, MgO contents in Illinois loesses are high. In contrast, Russian and Chinese loesses have greater amounts of silt-sized feldspars and micas, represented by Na_2O and K_2O, and clay minerals, represented by Al_2O_3 and Fe_2O_3, compared to loess from Illinois.

LOESS ORIGINS: PROCESSES OF SILT PARTICLE PRODUCTION

"Glacial" versus "Desert" Loess

A common and probably oversimplified view is that silt-sized particles in loess are produced almost exclusively by glacial grinding, deposited in till, reworked by fluvial processes as outwash, and finally entrained and deposited by wind (Fig. 10). This classical model of loess formation has led to the view that

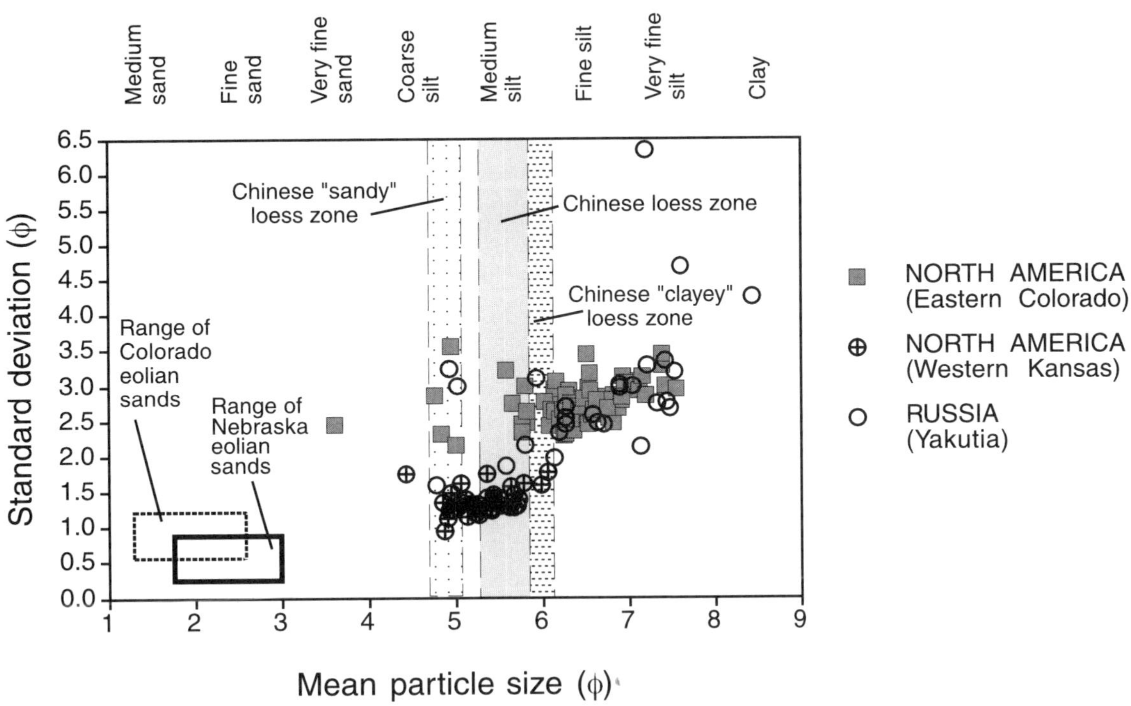

Figure 5. Mean particle size and standard deviation (sorting) of North American, Chinese, and Russian loesses. Shown for comparison are ranges of these parameters for eolian sands in North America (Nebraska and Colorado). Chinese loess data from Liu (1985); Russian loess data from Péwé and Journaux (1983); Kansas loess data from Swineford and Frye (1951); Colorado loess data from Muhs et al. (1999b). Ranges of Nebraska eolian sand data from Ahlbrandt and Fryberger (1980); Colorado eolian sand data from Muhs et al. (1999a).

loess deposits are primarily markers of glacial periods, because no significant mechanism of silt particle formation existed during interglacial periods. The model is reinforced by observations of the geographic proximity of loess bodies to the southern limits of the Laurentide Ice Sheet in North America and the Fennoscandian Ice Sheet in Europe, as well as smaller glaciers in Asia and South America. In the 1950s and 1960s, widespread application of radiocarbon dating showed that the youngest loess deposits in North America coincided with the ages of the last major expansion of the Laurentide Ice Sheet (see summaries in Willman and Frye, 1970, and Ruhe, 1983).

There is little question that silt is produced by glacial grinding. In North America, tills in Canada and the United States that were deposited by the Laurentide and Cordilleran Ice Sheets have abundant silt, based on hundreds of careful and detailed particle size analyses. In North America, for example, tills of last-glacial age have, on average, silt contents of ~40% (Willman and Frye, 1970; Kemmis et al., 1981; Clague, 1989). Outwash deposits derived from modern glaciers also contain much silt. In areas of active glaciers, rivers draining glacierized valleys have abundant silt-sized particles in suspension that give the waters a distinctive milky appearance; we have observed this in the Alaska Range and Chugach Mountains; Canadian Rockies; French and Swiss Alps; and Vatnajökull and Myrdalsjökull, Iceland. Detailed studies in Alaska by Hallet et al. (1996) show that sediment yields in rivers are up to an order of magnitude higher in glacierized basins than in those that are not. In glacierized areas of Alaska and Iceland, we have observed spectacular dust storms derived from silt-rich outwash plains and valleys.

Despite the abundance of geological, geographical, and geochronological support for the classical "glacial" concept of loess formation, there have been challenges to this model for at least 50 years (see Thorp, 1945, *in* Bryan, 1945) and perhaps longer (Smalley et al., 2001). The debate has continued to this day and centers on the issue of "glacial" loess versus "desert" loess. "Desert" loess is a term used loosely to describe eolian silt generated in and derived from arid or semiarid regions that were not glaciated. We feel that "glacial loess" and "desert loess" are terms that are inappropriate for what is probably a complex of processes, some of which are common to both environments. Nevertheless, we have retained their usage in this review simply because the terms have been used in loess origin debates for more than half a century and it is convenient for reference to the abundant literature on the issue (Bryan, 1945; Smalley and Krinsley, 1978; Whalley et al., 1982; Tsoar and Pye, 1987; Wright, 2001a, 2001b).

The debate on desert loess versus glacial loess centers on whether silt-sized particles can be produced by mechanisms other than glacial grinding and whether they can be produced in hot

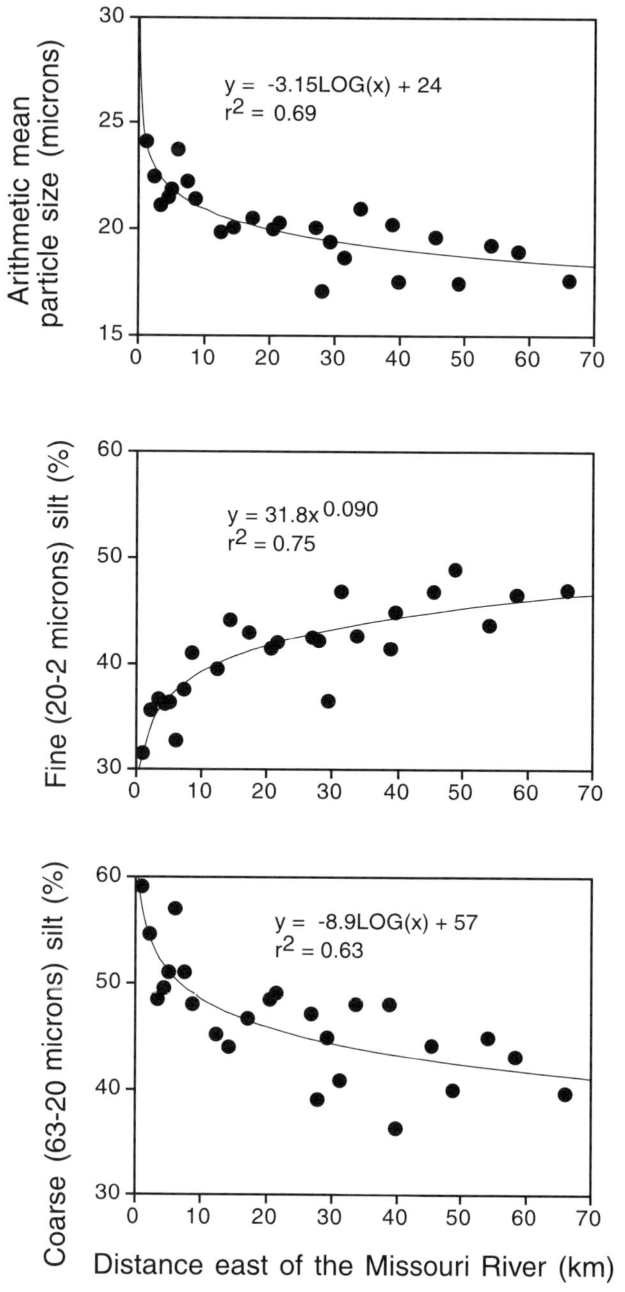

Figure 6. Mean particle diameter, coarse-silt (63–20 µm) content, and fine silt (20–2 µm) content in last-glacial-age loess as a function of distance east of Missouri River bluffline in western Iowa. Particle size data are derived from previously unpublished sedigraph analyses of the authors; sample localities are identical to those in Muhs and Bettis (2000).

deserts. A variety of mechanisms can, in principle, produce silt-sized particles in arid regions; we have summarized these in a highly simplified model (Fig. 11). Processes of silt production that have been proposed for arid regions (or the mountainous areas adjacent to them) include frost shattering (Wright et al., 1998), colluvial and fluvial comminution (Wright and Smith, 1993; Wright, 1995; Derbyshire et al., 1998), salt weathering (Goudie et al., 1979; Wright et al., 1998), eolian abrasion (Whalley et al., 1982; Wright et al., 1998), and ballistic impacts (Dutta et al., 1993). Experimental studies, many of them conducted in laboratory settings by J.S. Wright and summarized by her (Wright, 2001b), show that frost weathering, salt weathering, fluvial comminution, and eolian abrasion can all produce silt-sized particles. An important follow-up question, then, is: if silt *can* be produced in deserts, *has* sufficient silt been generated in deserts to produce loess deposits? A first step toward answering the question of desert loess formation is to determine simply whether there are loess deposits adjacent to deserts.

Desert Loess in China

The region that has been cited most often with regard to desert loess is China. Thorp (1945, in Bryan, 1945) was one of the first North American scientists to point out that nearby deserts could be the sources of thick loess deposits in China. Studies of loess in China have accelerated in the past couple of decades, particularly since the publication of Liu's (1985) excellent synthesis of loess studies in this region. On the basis of modern dust storm observations, geochemical and isotopic provenance studies, and loess thickness and particle size trends, most workers now agree that the desert basins of China and Mongolia (Fig. 4) are the immediate sources of loess in China (Liu, 1985; Liu et al., 1994; Derbyshire et al., 1998; Porter, 2001). Nevertheless, a question that has been debated intensively is whether the silt-sized sediments in the desert basins owe their origin to processes operating within the basin itself or whether they were formed by glacial grinding in nearby mountain ranges. Smalley and Krinsley (1978) proposed that much of the silt in Chinese loess was produced by glacial grinding in mountain ranges rimming the desert basins. Derbyshire (1983) challenged this interpretation by pointing out that glaciation in the mountains of northern and western China was of limited extent and that tills derived from these mountains are not particularly silt-rich. He suggested instead that much of the silt in Chinese loess was produced by salt weathering and frost shattering, either in the desert basins themselves or in the nearby mountains. Wright (2001a) reiterated Derbyshire's (1983) arguments against a glacial origin for Chinese loess and agreed that salt weathering and frost shattering in the desert basins are important processes of silt particle formation. Furthermore, she suggested that chemical weathering, fluvial comminution, and eolian abrasion have all been important processes in Chinese silt particle formation.

It is important to point out that Smalley and Krinsley (1978), Derbyshire (1983), and Wright (2001a) do not provide much in the way of quantitative field data to support their arguments. For example, although Derbyshire (1983) describes the tills of the Tian Shan as silt-poor, he gives no particle size data to support this statement. Smalley and Krinsley (1978), on the other hand, provide no maps of the extent of the glaciers that they propose

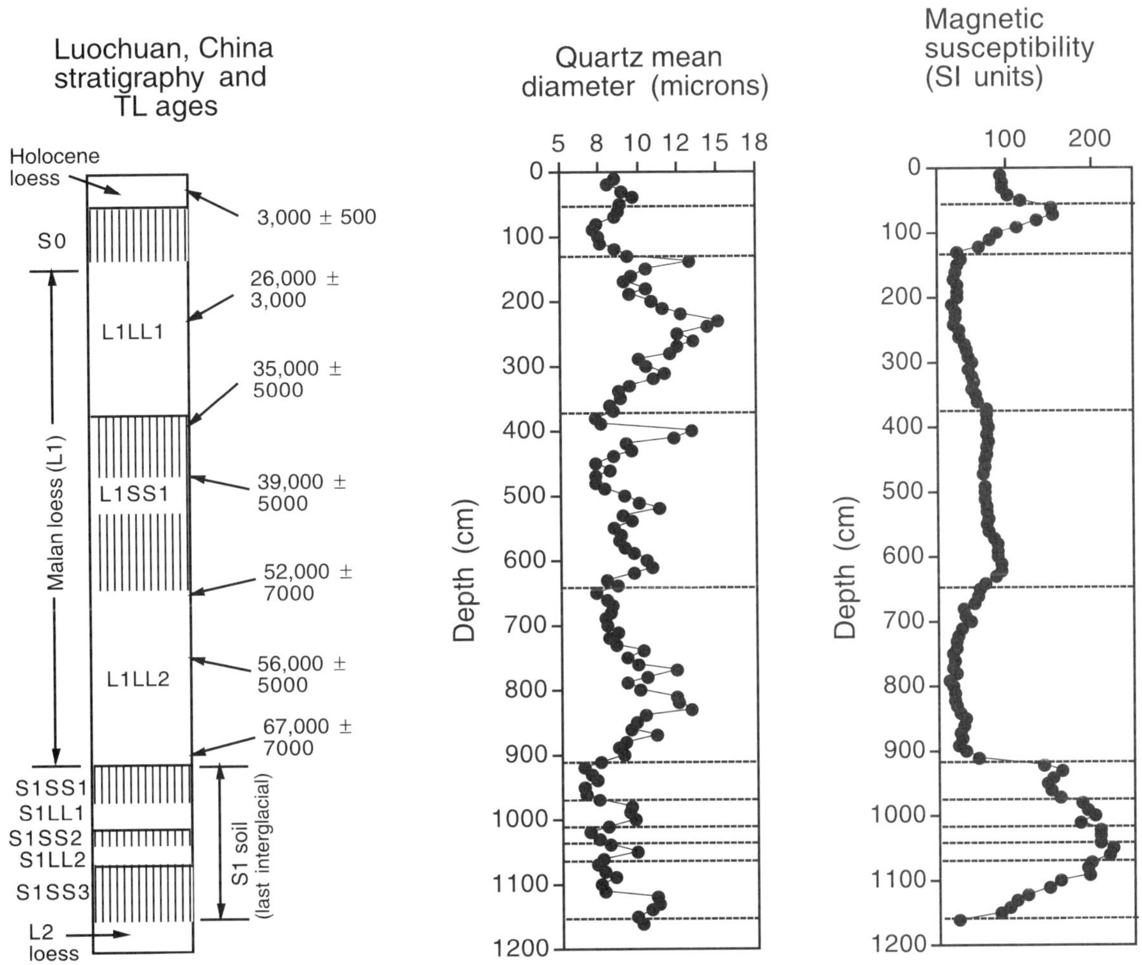

Figure 7. Stratigraphy, thermoluminescence (TL) ages, quartz mean diameter, and magnetic susceptibility of loess and paleosols over the last interglacial-glacial cycle at Luochuan, China. Loess units indicated by "L" prefix; paleosols indicated by "S" prefix. Stratigraphy, quartz mean diameter, and magnetic susceptibility data from An et al. (1991), Porter and An (1995), and Xiao et al. (1995, 1999); TL data from Forman (1991).

generated the silt. More research needs to be conducted on Chinese loess and possible source sediments that might help answer the question of silt particle formation in this important region. We feel it is premature to debate the efficacy of silt particle formation in the Chinese desert basins as a source of loess and to make paleoclimatic interpretations when the sediments in those basins and in the mountains surrounding them have not been adequately characterized.

Desert Loess in North America, Africa, and Australia

The largest area of dominantly non-glacial loess in North America is the semiarid Great Plains region of Nebraska, Kansas, and Colorado (Fig. 2). During the last glacial period, most of this region was upwind and upstream of the Laurentide Ice Sheet and the rivers that drained it. Detailed isotopic analyses indicate that loess in Colorado and Nebraska is probably derived mostly from volcaniclastic siltstones of the Tertiary White River Group, with small contributions from Rocky Mountain glaciers (Aleinikoff et al., 1998, 1999). Despite the dominantly non-glacial source sediment, loess of last-glacial age in the Great Plains is as much as 48 m thick (Maat and Johnson, 1996).

Other arid and semiarid regions in the world show evidence of only modest silt production and little evidence of loess formation along their margins. Elsewhere in North America, thick loess deposits have not been reported in desert regions of the southwestern United States and Mexico, although silt-dominated eolian mantles less than one meter thick have been described (Muhs, 1983; McFadden et al, 1986; Reheis et al., 1995). A detailed study with good stratigraphic control has shown that dust fluxes were greater during the last glacial period than during the Holocene in the Mojave Desert of the southwestern United States

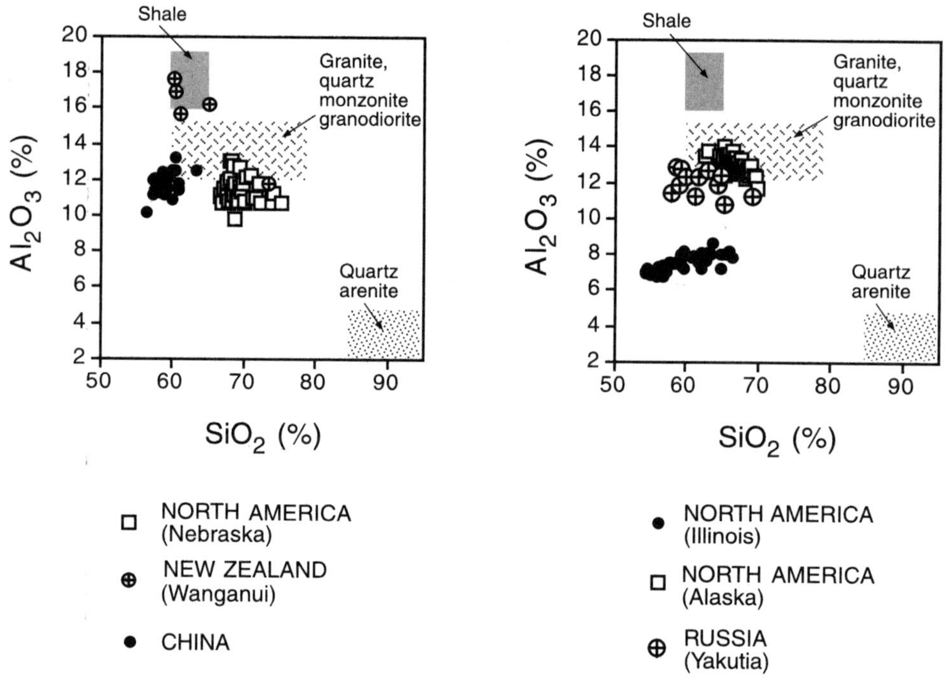

Figure 8. Plots of SiO_2 and Al_2O_3 concentrations in North American (Alaska, Illinois, and Nebraska) loess compared to similar data for loess in New Zealand (Wanganui), Russia (Yakutia), and China (Luochuan area). Illinois localities are Morrison and Rapid City; Nebraska localities are Elba, Lincoln, and Plattsmouth (Fig. 2). North American loess data from Muhs and Bettis (2000) and Muhs et al. (2001, 2003); New Zealand data from Graham et al. (2001); Russian loess data from Péwé and Journaux (1983); Chinese loess data from Gallet et al. (1996) and Jahn et al. (2001). Ranges of values in shale from Condie (1993); ranges of values in quartz arenite from Pettijohn et al. (1972).

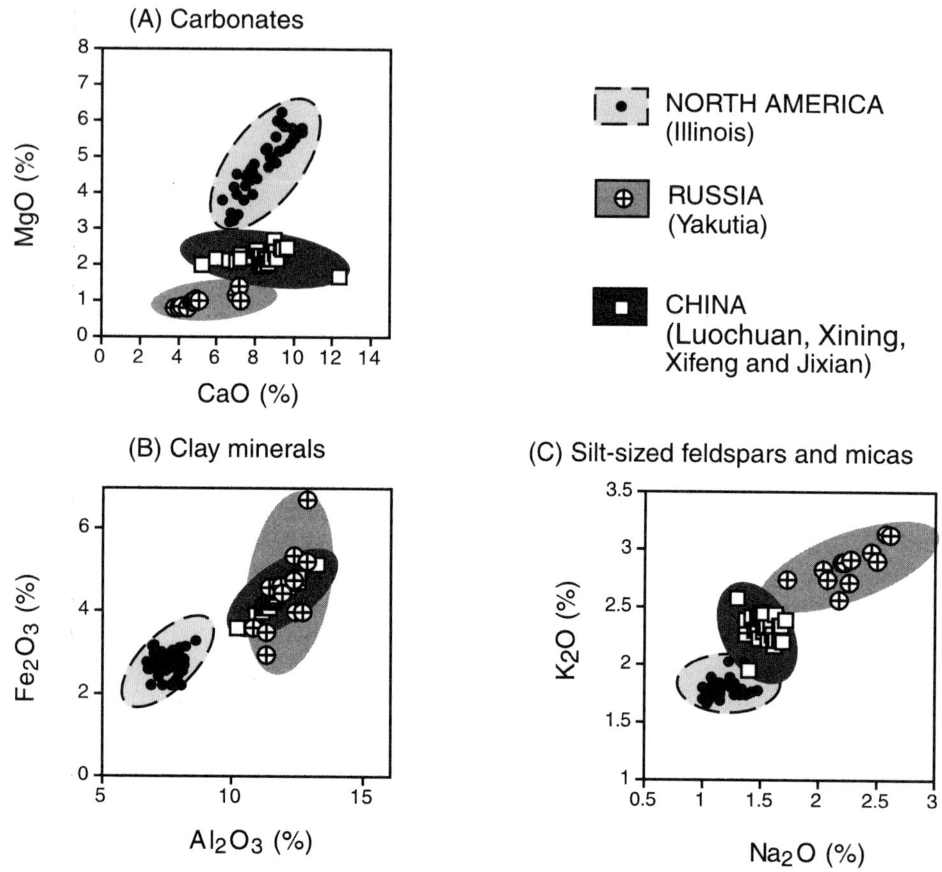

Figure 9. Plots of element pairs (A) CaO-MgO, (B) Al_2O_3-Fe_2O_3, and (C) K_2O-Na_2O that reflect the relative abundances of (A) carbonates, (B) clay minerals, and (C) silt-sized feldspars and micas in loess from various regions. Data sources as in Figure 8.

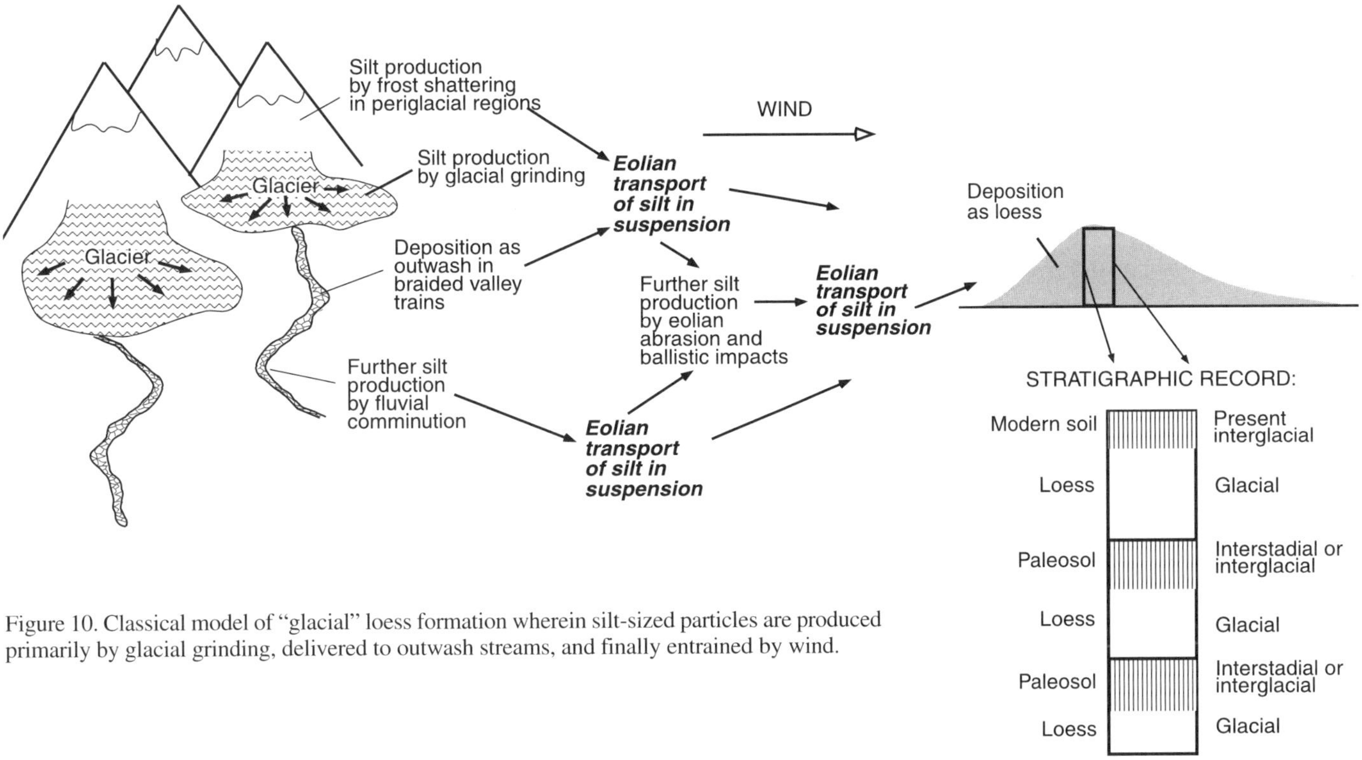

Figure 10. Classical model of "glacial" loess formation wherein silt-sized particles are produced primarily by glacial grinding, delivered to outwash streams, and finally entrained by wind.

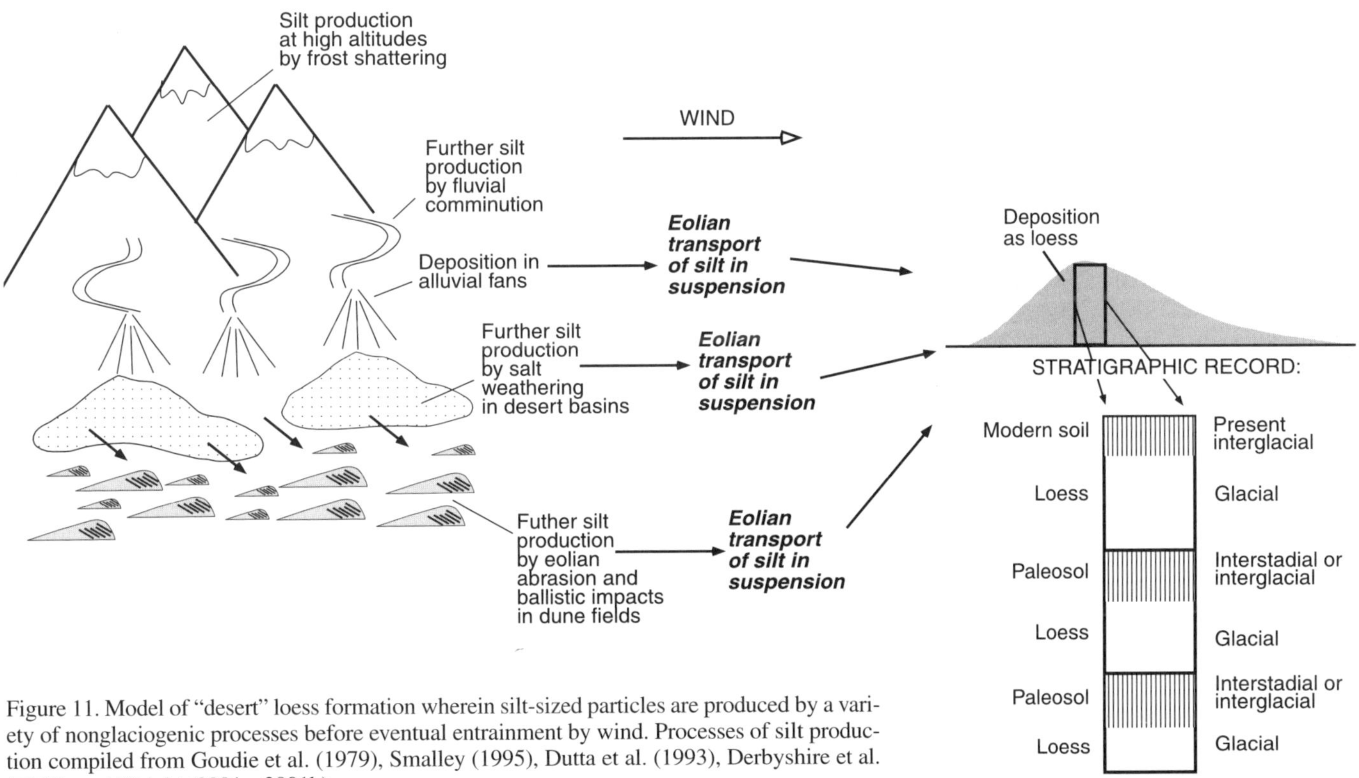

Figure 11. Model of "desert" loess formation wherein silt-sized particles are produced by a variety of nonglaciogenic processes before eventual entrainment by wind. Processes of silt production compiled from Goudie et al. (1979), Smalley (1995), Dutta et al. (1993), Derbyshire et al. (1998), and Wright (2001a, 2001b).

(Reheis et al., 1995). Nevertheless, the magnitude of last-glacial dust flux in this region is quite modest, generally less than 50 $g/m^2/yr$. These rates are much lower than those for loess in the North American midcontinent, which range from 400 to 4,000 $g/m^2/yr$ (Bettis et al., 2003).

A region that has been cited extensively as evidence for silt production in deserts is the Sahara. Dust, virtually all of it less than 20 μm in diameter, is removed by wind from the Sahara and adjacent semiarid Sahel region and carried west across the Atlantic Ocean on the Trade Winds as far as Barbados and Florida (Prospero et al., 1970). Observations of dramatic dust storms in the Sahara and distant transport westward suggest that loess occurrence along the margins of the desert should be common. Despite the impressive evidence for dust production and transport from the Sahara, there is actually evidence of only modest tracts of loess around the margins of this enormous desert. Although loess deposits on the margins of the Sahara sometimes have impressive thicknesses (e.g., Israel, see Bruins and Yaalon, 1979), they are of very limited areal extent and do not form continuous loess bodies over large areas such as those in North America, South America, Europe, or Asia. Based on records from cores across parts of the Atlantic (Ruddiman, 1997), it is apparent that Saharan mass accumulation rates are actually relatively low (2–15 $g/m^2/yr$) compared to loess fluxes in the midcontinent of North America or China.

Dust storms are common in Australia and show distinctive tracks toward offshore regions in the Indian and Pacific Oceans and the Tasman Sea (McTainsh, 1989). Some of the best records of dust derived from the deserts of Australia are from the Tasman Sea (Hesse and McTainsh, 1999), but fluxes are relatively low (McTainsh, 1989). As a result, silt-dominated loess deposits have not been reported in Australia (McTainsh, 1989). Eolian deposits that are finer-grained than sand are limited to small areas of eolian clay, or "parna" deposits (Butler, 1974).

Thus, the long-term record of particle flux from the Sahara and Australia (whether in desert margins or in the oceans) is not nearly as high as that in areas that were adjacent to large glaciers in North America, South America, Europe, and Asia. We conclude, therefore, that the magnitude of production of "desert" loess, as proposed by Wright (2001b), may be somewhat overstated. It is possible that much of the silt in Saharan dust and other deserts is derived not from silt-sized particles newly formed from sand, but from sediments eroded chiefly from siltstones. For example, geologic mapping in Libya has shown that Paleozoic and Mesozoic siltstones or silt-rich shales are extensive over much of the Sahara (Conant and Goudarzi, 1964), including siltstone facies of the Cretaceous Nubian Sandstone, which is widespread in Libya and Egypt. These rocks could be the sources of some of the loess found in Tunisia and Libya. In at least some regions of Australia, such as the Lake Eyre Basin, the sources of silt in dust storms are siltstones and mudstones of the Rolling Downs Group of Cretaceous age (McTainsh, 1989) rather than newly formed silt-sized particles produced in the desert basin. As discussed above, recent studies in eastern Colorado and Nebraska show that the same is true for much of the loess in the semiarid Great Plains of North America, where silt-sized particles are inherited from volcaniclastic siltstone (Aleinikoff et al., 1998, 1999). In South America, Zárate and Blasi (1993) interpreted many of the loess particles of the Pampas region to be inherited from a volcanic source. This conclusion is supported by recent isotopic analyses of loess from the Pampas region (Gallet et al., 1998). Inheritance of silt-sized particles is not limited to desert regions. Palmer and Pillans (1996) point out that some of the most important loess sources in New Zealand are volcanogenic silts and Pliocene-Pleistocene siltstones and mudstones. Thus, much desert silt actually appears to be inherited from silt-rich protoliths rather than particles newly produced in the desert.

LOESS AND ITS RELATION TO EOLIAN SAND

In many regions, loess belts are proximal to dune fields or eolian sand sheets. The Great Wall of China separates a region of eolian sand in the Mu Us Desert (Fig. 4) from the Loess Plateau to the southeast (Baosheng et al., 2000). At the boundary between the Mu Us Desert and the Loess Plateau, the stratigraphic record and thermoluminescence ages of sediments from the last interglacial-glacial cycle show that sediment deposition and paleosol formation took place at times similar to those on the Loess Plateau (Sun et al., 1998). However, along the desert margin (as opposed to the central part of the Loess Plateau) loess is interbedded with eolian sand. Within the Loess Plateau itself, areas immediately southeast (downwind) of the deserts are referred to as "sandy loess," whereas silt-dominated loess occurs farther to the southeast and clayey loess occurs still farther downwind (Liu, 1985; see also Fig. 5).

In periglacial regions, a facies relation between eolian sand and silt has also been recognized. For example, in Greenland, eolian sand and silt are described as facies of an eolian sediment continuum where active fine-particle production and eolian deflation are occurring at present (Dijkmans and Tornqvist, 1991). In southwestern Alaska, loess bodies occur downwind of areas of eolian sheet sand, which are, in turn, downwind of dune fields (Lea, 1990). Eolian sheet sands in this region have a greater component of silt in a downwind direction. In the Pampas loess belt of Argentina, there is a west-to-east transition from dunes to sheet sands to loess (Zárate and Blasi, 1993). In the Great Plains of North America, loess occurs immediately downwind of eolian sand in Nebraska and Colorado (Muhs et al., 1999a).

All of these observations raise the question of whether loess should be considered as a distinct sediment body or whether it is essentially a fine-grained facies of eolian sand with a common source. Pye (1995) presented several models of eolian sediment changes downwind from a source (Fig. 12). His simplest case (Fig. 12A) is typical of many loess bodies in areas adjacent to continental-scale ice sheets, such as North America, Europe, and parts of Asia; these sediments are referred to as "periglacial" loess. In two other models, loess is shown as a finer, downwind facies of eolian sand, either with an intervening zone of sediment

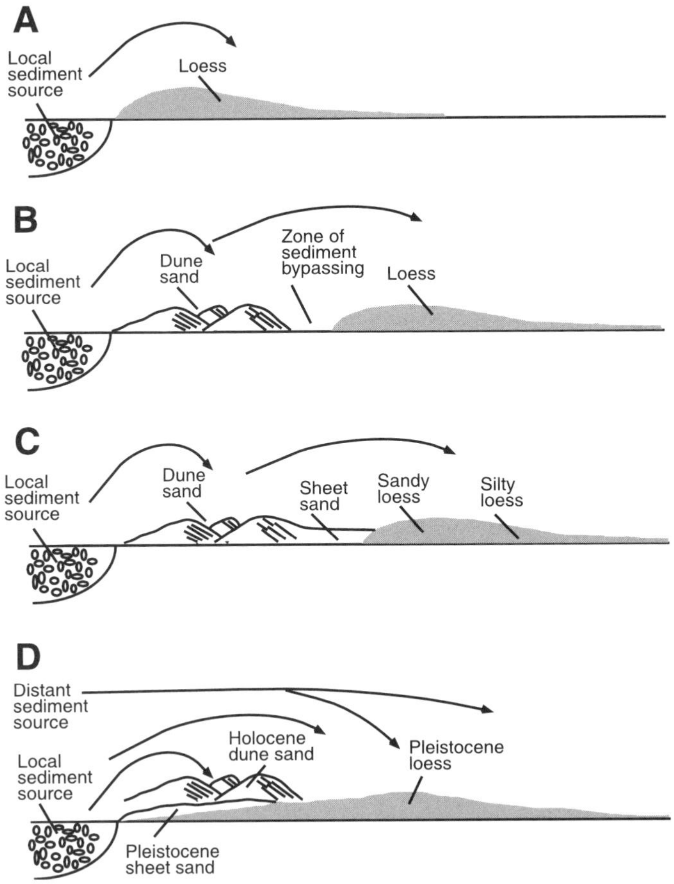

Figure 12. Models of eolian sand-loess facies changes downwind from a source. Models (A), (B), and (C) are redrawn from Pye (1995); model (D) is from the present study.

bypassing (Fig. 12B) or as part of a zone of continuous deposition with a gradual fining downwind (dune sand to sheet sand to sandy loess to silty loess). The models shown in Figure 12B and C call upon saltation-dominated transport downwind from a source to explain the origin of the dunes and sand sheets and suspension-dominated transport to explain the origin of the loess body. In the case of the model shown in Figure 12C, vegetation cover is sufficient over much of the region to trap particles continuously in a downwind direction, with a gradual downwind fining as large particles fall out. This situation contrasts strongly with that shown in the model in Figure 12B, where there is a zone of sediment bypassing. Pye's sediment-bypassing scenario is applicable to geomorphic settings where the source sediment for eolian sand and loess is in an arid basin. Because sand-sized particles are coarse, they may be deposited close to the source; however, if there is an insufficient vegetation cover farther downwind, finer-grained, silt-sized particles remain in suspension. Ultimately, much farther downwind, these finer particles are deposited when a more humid climate, with a greater degree of vegetation cover, is reached.

In places, loess occurs downwind of eolian sand, but the two sediments have different sources and possibly even different times of deposition. This variation on the models of Pye (1995), described above, is shown in Figure 12D, and an example can be found in eastern Colorado. Mineralogical, geochemical, and geochronological data show that although loess is found downwind of eolian sand, it has a different source and was deposited mainly in Pleistocene time, whereas much of the eolian sand was deposited in Holocene time (Muhs et al., 1999b; Muhs and Zárate, 2001; Aleinikoff et al., 1999).

PALEOSOLS IN QUATERNARY LOESS SEQUENCES

Soil-forming Processes

Some of the most important components of Quaternary loess sequences are buried soils or paleosols. Paleosols formed in loess have been studied at many localities and are significant for both stratigraphic interpretations and paleoclimatic reconstructions. Buried soils represent former land surfaces. Soils form in a deposit when the rate of sedimentation has slowed so that the soil-forming processes can take place and leave an imprint on the deposit. For these pedologic processes to operate, it is also essential that the land surface has enough geomorphic stability such that little or no erosion takes place.

A loess-paleosol sequence we studied at Greenbay Hollow, Illinois, a short distance from the Mississippi River Valley source (Fig. 2), illustrates some of the changes in properties that reflect humid-climate processes of soil additions, removals, translocations, and transformations (Fig. 13). Carbonate minerals, calcite and dolomite, are depleted in the modern soil, the Farmdale soil (interstadial, about 30–55 ka), the Sangamon soil (last interglacial, about 55-130 ka), and older buried soils. These depletions are apparent in the low CaO and MgO contents compared to unaltered Peoria (last glacial) Loess. Translocation of fine particles to form clay-rich, B-horizons is also apparent in the field in all soils. This is shown by the presence of clay coatings on soil structural unit ("ped") faces, but also in the higher amounts of clay in the soil B-horizons of the modern soil and the paleosols at depths of ~1200 and ~1600 cm (Fig. 13). Transformations in this loess section include the possible alteration of Na-plagioclase by hydrolysis to clay minerals within the paleosols. This is shown in the Greenbay Hollow section as relatively low Na_2O/TiO_2 values in the modern soil, the Sangamon soil, and the oldest pre-Sangamon buried soil, found at at the bottom of the section, compared to the unaltered loesses. Note that clay content is highest where Na_2O/TiO_2 values are lowest, suggesting that the clay increases, in addition to translocation, are due to the transformation of plagioclase to clay.

The Greenbay Hollow section illustrates how paleosols in loess could be used to interpret past climates. Muhs et al. (2001) showed that in a transect of modern loess-derived soils, Na_2O/TiO_2 values—as well as other indicators of chemical weathering—decrease from north to south, which parallels a

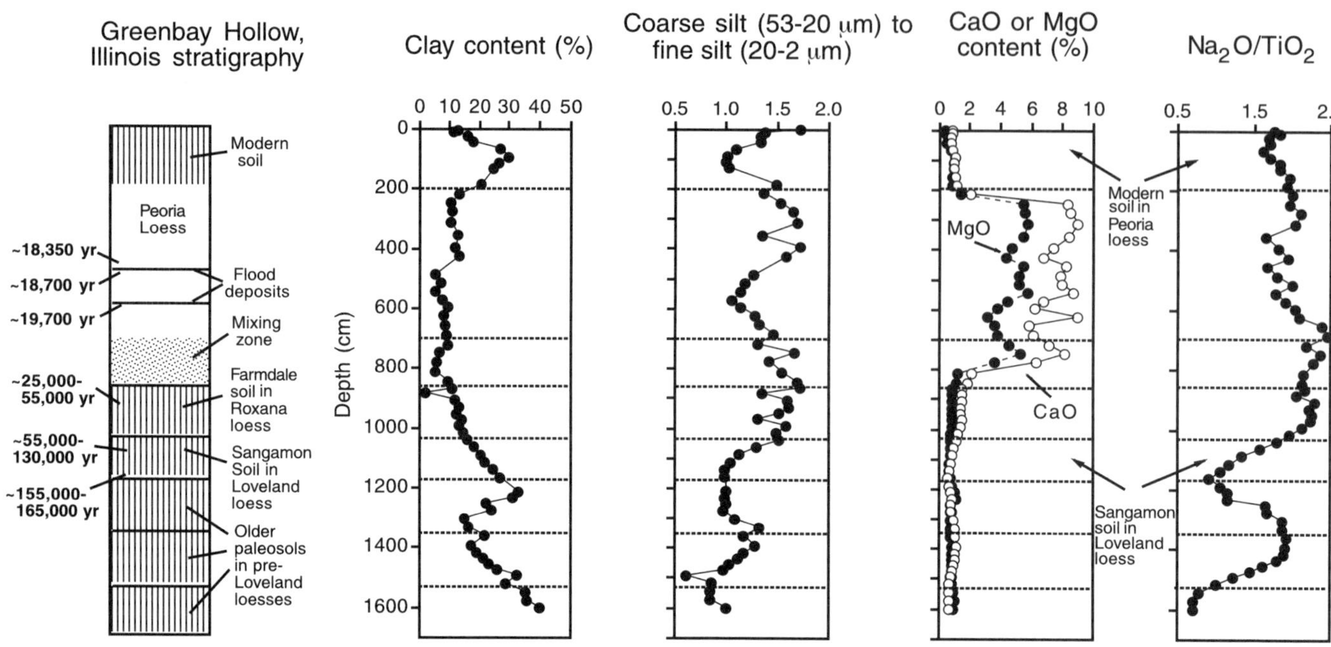

Figure 13. Stratigraphy, ages, clay content, silt ratios, CaO content and MgO content, and Na_2O/TiO_2 ratios of Greenbay Hollow loess section, western Illinois (see Fig. 2 for location; see Hajic (1990) and Grimley et al. (1998) for additional data on this section). Stratigraphy, particle size data, and chemical data from present study and Muhs et al. (2001). Peoria loess ages are correlated from a nearby section reported by Grimley et al. (1998); other ages given are approximate and based on radiocarbon, thermoluminescence (TL), and ^{10}Be age estimates for these units at other localities in the Mississippi River Valley reported by Leigh (1994), Curry and Pavich (1996), and Markewich et al. (1998).

southward-increasing mean annual temperature and precipitation gradient. Thus, the lower Na_2O/TiO_2 values in the Sangamon soil compared to the modern soil at Greenbay Hollow could be interpreted to mean that conditions during the Sangamon interglacial period were warmer and more humid than at present. However, such an interpretation is complicated by the fact that the Sangamon interglacial period could have lasted several tens of thousands of years, compared to only 10,000 years for the present interglacial. Thus, the duration of pedogenesis may also explain the lower-than-modern Na_2O/TiO_2 values.

Stratigraphic Significance of Paleosols in Glacial-Interglacial Cycles

The stratigraphic record of loess with intercalated paleosols shows sedimentary extremes in glacial-interglacial cycles. The Greenbay Hollow loess section contains examples of pedogenesis that occurred when there was little or no loess sedimentation. If glaciogenic silt is the major source of loess in this region (see earlier discussion on loess origins), then little or no eolian sediment deposition took place during periods when the Laurentide Ice Sheet had retreated from the headwaters of the Mississippi River. In essence, therefore, loess sedimentation in this region is a "turn-on, turn-off" process that is a function of glacial sediment sources. Thus, the loess record in midcontinental North America is a good example of sedimentary extremes: abundant loess deposition occurred during glacial periods, whereas very little or no loess deposition took place during interglacial or interstadial periods, which are periods of soil formation. The loess stratigraphic record in much of Europe is similar to that of regions that were near glaciers in North America. Stratigraphic and geochronologic studies in western Europe show that the last major period of loess deposition was during the last glacial period (e.g., Antoine et al., 2001; Rousseau et al., 2002).

In other regions, there is a less distinct record of loess sedimentation versus soil formation. In China, for example, the loci of modern dust storms match closely the distribution of Quaternary loess (Derbyshire et al., 1998). Modern dust storms in China are a function of dry, strong, northwesterly winds that are generated by the Mongolian high pressure cell that develops over Asia in late fall, winter, and early spring (Fig. 4). In contrast, a thermal low develops over this region in summer, and high pressure offshore generates weak, moist, southeasterly winds associated with the summer monsoon. During the summer period, rainfall is abundant and winds are not passing over dust source regions; thus, little or no eolian sediment transport takes place. In addition to observations of modern dust storms, the stratigraphic record indicates abundant loess deposition throughout the Holocene, both due to natural and anthropogenic causes (e.g., Roberts et al., 2001). Thus, many of the surface soils in China are receiving at least small increments of dust. The fact that soils mantle the surface of the landscape merely indicates that the rate of soil development exceeds the rate of dust accretion. Similar observations have been made in Alaska (Begét, 1996; Muhs et al., 2003). Alaskan paleosols contain higher amounts of fine silt than the loess units in which they are developed, indicating that sedimentation (albeit with a decreased wind competence and reduced sediment supply) is continuing simultaneously with pedogenesis. All of these observations show that loess deposition in China and Alaska is not an exclusively glacial-period process. In fact, recent stratigraphic studies in central Alaska indicate that there is only a modest record of last-glacial loess deposition, although this may be due largely to a lack of loess *preservation* rather than a lack of last-glacial loess *deposition* (Muhs et al., 2003).

In China, it appears, as proposed by Verosub et al. (1993), that loess deposition and soil formation are essentially competing processes: loess deposition is greater during glacial periods and soil formation is greater during interglacial periods, but both processes proceed simultaneously. Porter et al. (1992) formulated these concepts into a simple model that explains much of the stratigraphic record of loess in China (Fig. 14). During glacial periods, the Mongolian high-pressure cell is stronger over Asia and is dominant during a greater part of the year while the summer monsoon is weaker during such periods (Fig. 4). As a result, dust accumulation rates are high and soil development cannot keep pace with sedimentation. In contrast, during interglacial periods, although dust deposition still occurs—primarily during the late fall, winter and spring—the strength and residence time of the Mongolian high pressure cell are diminished whereas the summer monsoon is strengthened. Thus, dust deposition rates are lower and soil development can keep pace with or exceed sedimentation. The stronger summer monsoon, with its increased rainfall, enhances pedogenesis. During interglacial periods, the mean diameter of loess particles is smaller and the ratio of clay to silt increases (Fig. 7). The greater abundance of clay-sized particles is due not only to pedogenesis (i.e., greater clay production through alteration of silt-sized particles via chemical weathering under a strong summer monsoon) but also decreased wind competence that results in more clay in the primary airborne particles.

In the preceeding discussion, we emphasized that loess deposition is not limited to glacial periods. Having said this, it is also apparent from detailed stratigraphic records that span the last interglacial-glacial cycle (Fig. 7), as well as those that span several interglacial-glacial cycles (Fig. 15), that the amount of loess deposition in most areas is greater during glacial periods than during interglacial or interstadial periods. Even in those areas where loess was not derived exclusively from glacial deposits, such as South America, China, and the Great Plains of North America, the amount of loess deposited during the last glacial period was much greater than during the Holocene (Fig. 16). Because loess deposition in these areas was not dependent exclusively on glaciogenic silts, the higher rates of sedimentation during the last glacial period must have been a function of other factors. Mahowald et al. (1999) reviewed some of the possible causes of high rates of loess flux during the last glacial period. These include stronger or more persistent winds, greater aridity, decreased intensity of the hydrological cycle, decreased vegetation cover, and increased sediment availability. It is possible that all these factors combined to produce the sedimentary extreme of rapid and dramatic loess deposition during the last glacial period. In this regard, we agree with Wright (2001a) that regardless of the process of origin, much loess can be considered to be "glacial" in the sense that the optimum climatic and geomorphic conditions for loess formation in many regions occurred during glacial periods.

Variability in Loess Sedimentation within a Single Glacial Period

In the past, loess was perceived to be a relatively uniform sediment that, at least within one depositional package, showed little variability over the period of sedimentation. Studies over

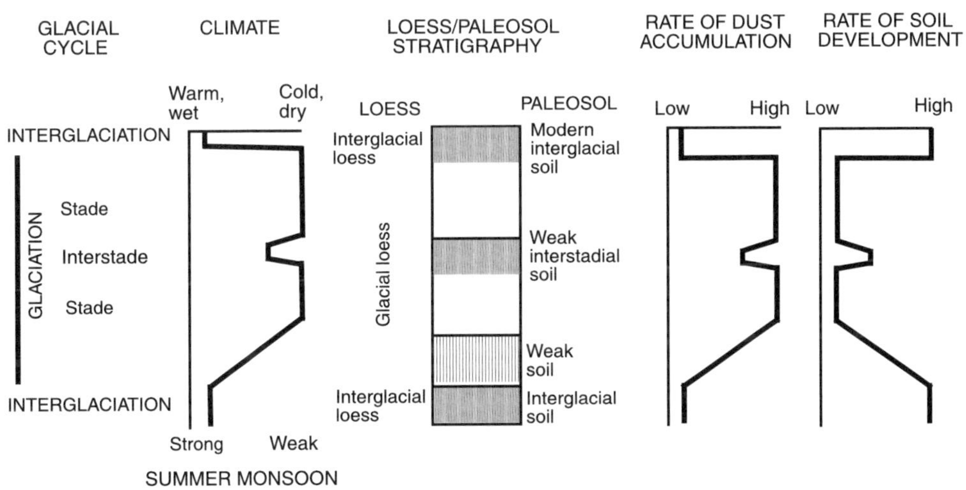

Figure 14. Model of loess deposition and soil formation cycles in China as a function of glacial-interglacial cycles and relative strengths of winter and summer monsoons. Redrawn from Porter et al. (1992).

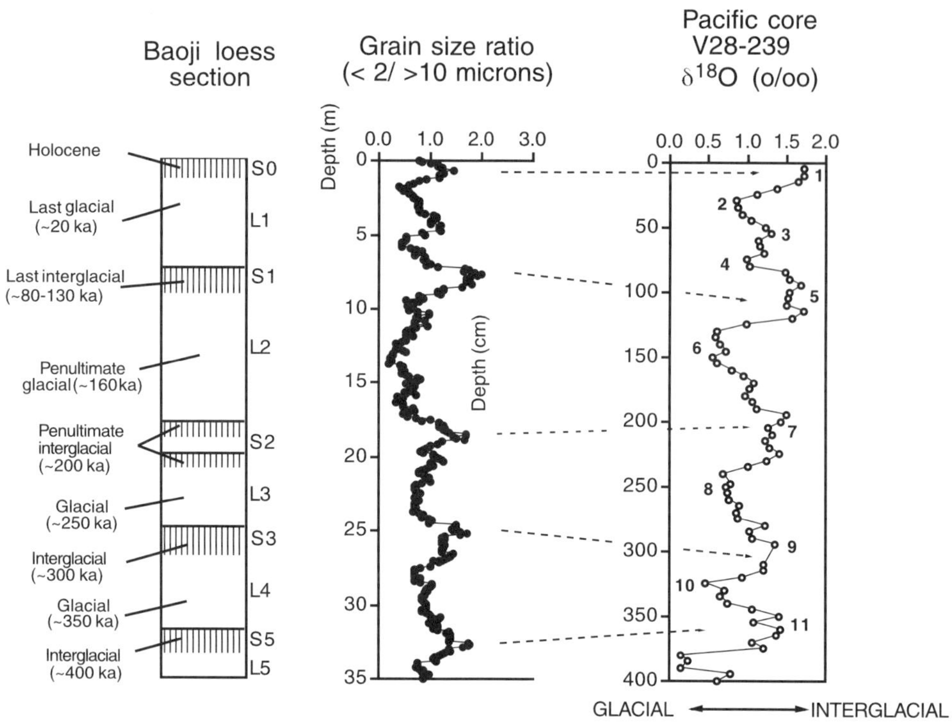

Figure 15. Illustration of relation of loess-paleosol sequences with glacial-interglacial cycles: stratigraphy, age estimates, and clay to medium and coarse silt ratios of upper part of loess section at Baoji, China, and proposed correlation with deep-sea oxygen isotope foraminiferal record of Pacific core V28-239 over the past five interglacial-glacial cycles. Loess units indicated by "L" prefix; paleosols indicated by "S" prefix. Oxygen isotope stages (bold numbers) indicate glacial periods (even numbers) or interglacial periods (odd numbers) Loess data from Ding et al. (1994); oxygen isotope data from Shackleton and Opdyke (1976). Correlations are based on age estimates in core V28-239, in turn derived from identification of the Brunhes-Matuyama boundary at 726 cm, an age for this boundary of ca. 780 ka (Spell and McDougall, 1992), and an assumed long-term average sedimentation rate of ~0.93 cm/ka.

the past 3 decades have shown that loess deposition rates can vary markedly within a single period of sedimentation, and subtle changes in loess properties can yield considerable information about changes in climate conditions and source areas within a glacial period. In several parts of North America (Iowa, Illinois, and Indiana) detailed stratigraphic studies also show that last-glacial (Peoria) loess sedimentation rates were not constant and are thought to reflect changes in source sediment supply and/or climatic conditions that influence source sediment availability (Ruhe et al., 1971; Hayward and Lowell, 1993; Wang et al., 2000).

Recent detailed studies of loess in the Rhine Valley of Germany indicate that loess in Europe, like that in North America, was not constantly deposited during the last glacial period (Antoine et al., 2001; Rousseau et al., 2002). Stratigraphic studies show that periods of loess sedimentation were separated by brief intervals of tundra soil formation (Fig. 17). The gleyed tundra soils are not well developed, indicating that periods of pedogenesis were brief and that loess sedimentation probably occurred contemporaneously, albeit at a greatly reduced rate. Furthermore, particle size analyses show that the periods of low sedimentation rate, which are marked by tundra soils, were characterized by much finer-grained loess than the periods of more rapid sedimentation. Rousseau et al. (2002) interpret these data to mean that wind competence was lower during periods of lower sedimentation rate. An alternative interpretation is that loess sources changed during the periods of differing sedimentation rate, as at Loveland, Iowa (Muhs and Bettis, 2000).

Magnetic Susceptibility in Loess-Paleosol Sequences

One of the primary means of verifying the presence of paleosols, correlating them from section to section and quantifying the degree of soil development in Chinese loess sequences, has been measurement of bulk magnetic susceptibility. A full discussion of this property and other magnetic mineralogical properties is beyond the scope of this paper, and Singer et al. (1996) and Maher (1998) provide useful reviews. Nevertheless, some discussion of this method is critical, because it has become the most commonly measured property in loess-paleosol sequences in China and in many other regions. Kukla et al. (1988) showed that bulk magnetic susceptibility in Chinese loess is relatively low, whereas the intercalated paleosols have relatively high values (Fig. 7). Heller and Liu (1984) interpreted the higher magnetic susceptibility in paleosols to be the result of concentration of magnetic minerals by sediment compaction and carbonate leaching in the soils. Kukla et al. (1988) interpreted these trends to be the result of quartz-dominated-dilution of a small component of detrital magnetic minerals in loess. Later workers (Zheng et al., 1991; Verosub et al., 1993; Maher and Thompson, 1995) proposed that much of the magnetic mineral enhancement in Chinese loess-derived paleosols is due to production of ferrimagnetic minerals, such as maghemite and fine-grained magnetite, during pedogenesis. This finding has led to the use of magnetic susceptibility not only to identify paleosols and correlate them between sections, but also to quantify paleoclimate at the time of soil formation (e.g., Maher and Thompson,1995; Maher, 1998).

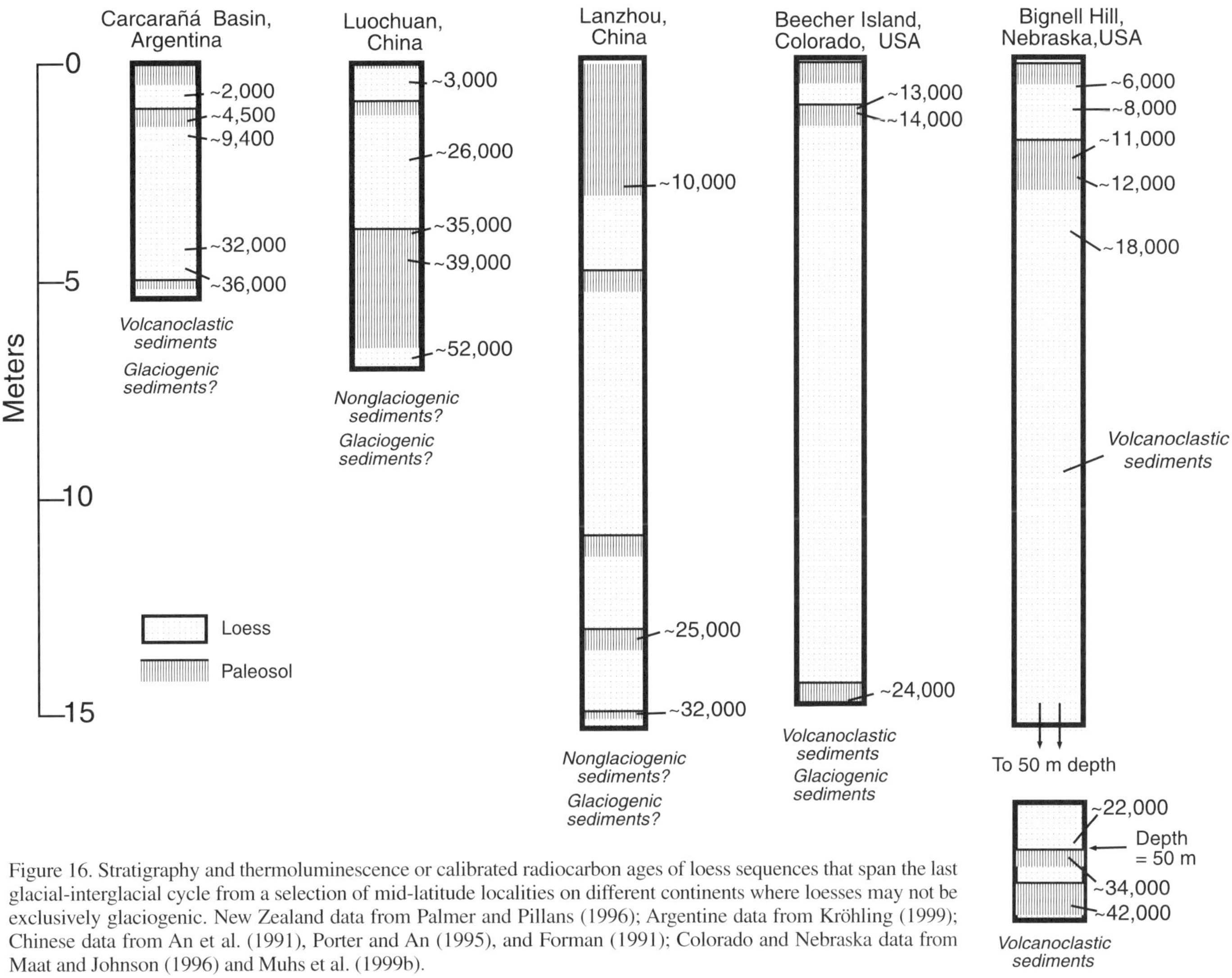

Figure 16. Stratigraphy and thermoluminescence or calibrated radiocarbon ages of loess sequences that span the last glacial-interglacial cycle from a selection of mid-latitude localities on different continents where loesses may not be exclusively glaciogenic. New Zealand data from Palmer and Pillans (1996); Argentine data from Kröhling (1999); Chinese data from An et al. (1991), Porter and An (1995), and Forman (1991); Colorado and Nebraska data from Maat and Johnson (1996) and Muhs et al. (1999b).

The widespread use of magnetic susceptibility in loess studies is likely due, at least in part, to the ease, rapidity, and inexpensive nature of the analysis. Magnetic susceptibility can be measured in the field rapidly with relatively high precision and accuracy. Nevertheless, interpreting magnetic susceptibility data is not simple, and not all loess sequences show the same trends as in China. For example, in Alaska and Siberia, magnetic susceptibility is not highest in soils and lowest in loess, but highest in loess and lowest in soils (Begét, 1990; Chlachula et al., 1997). Begét (2001) has summarized the current hypotheses for magnetic susceptibility variations in Alaskan loess. The trend of high susceptibility in loess and low susceptibility in paleosols has been interpreted to be a function of wind competence, with magnetic susceptibility as a proxy for amount of detrital magnetite, which is in turn a proxy for abundance of heavy minerals. Detailed particle size analyses of loess-paleosol sequences in Alaska support this interpretation, as loess has higher amounts of sand and coarse silt than intercalated paleosols, indicating stronger winds during periods of relatively high sedimentation rate (Fig. 18).

IMPLICATIONS FOR INTERPRETING LOESSITE IN THE ROCK RECORD

Loess deposits are not limited to the Quaternary. Ding et al. (1999) have shown that the Chinese loess record extends well back into the Tertiary period. As we alluded to at the beginning of this review, there has been an increasing recognition that loess, in the form of loessites, may be a more important part of the sedimentary rock record than previously thought (Johnson, 1989; Soreghan, 1992; Chan, 1999). A number of the concepts and observations discussed for the Quaternary loess-paleosol record are important for recognizing and interpreting loessites in the

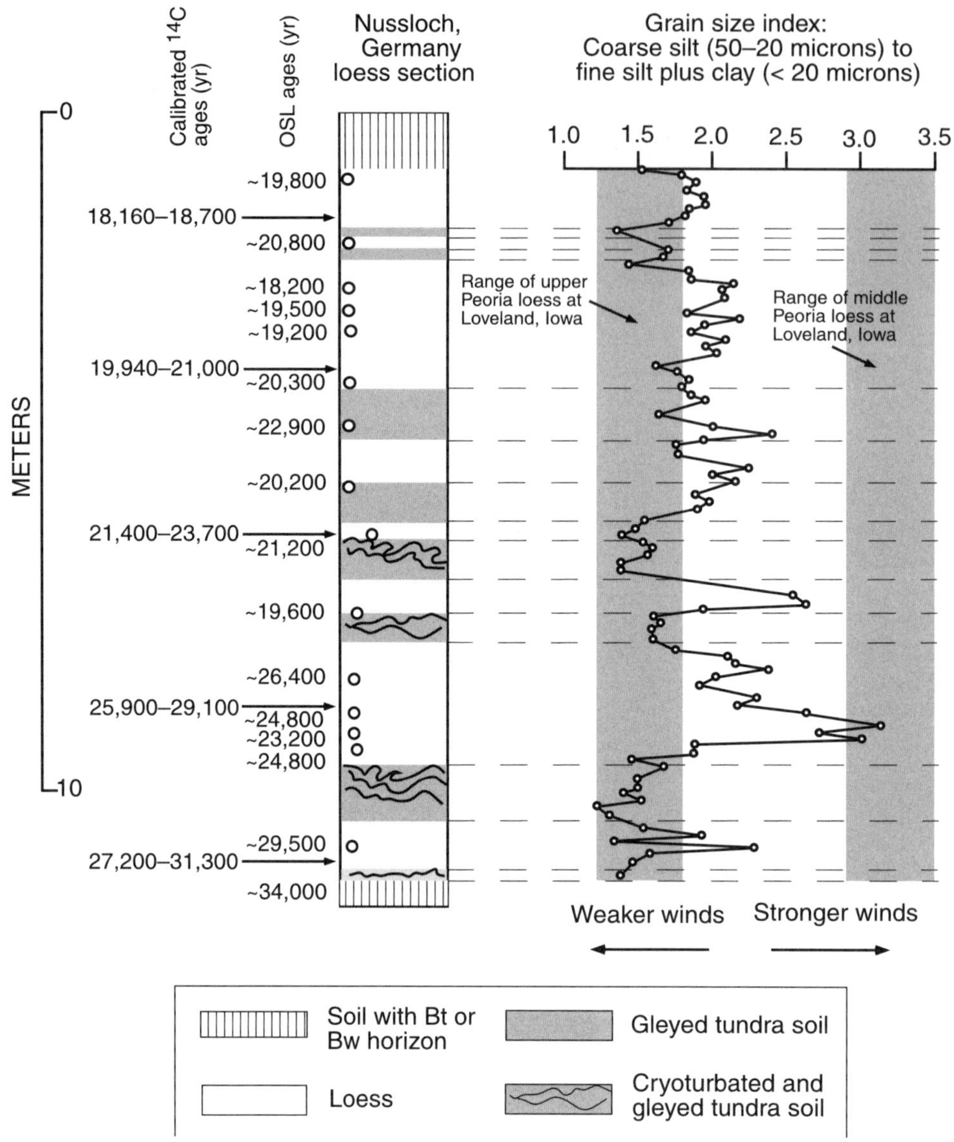

Figure 17. Stratigraphy, radiocarbon, and optically stimulated luminescence (OSL) ages and variations in loess particle size at Nussloch, Germany, loess section as an illustration of loess sedimentation variability within a single glacial period. Shown for comparison are similar particle size data for Loveland, Iowa, loess section. Nussloch data from Antoine et al. (2001) and Rousseau et al. (2002); Loveland data from Muhs and Bettis (2000).

rock record. The silts comprising the bulk of most widely distributed and thick loess deposits appear to be derived from glacial grinding, silt-rich protoliths, volcanic ash, or some combination of these sources. There is little evidence for *abundant* primary production of silt from sand in desert regions. Regional thickness, grain-size, geochemical, and isotopic trends of loess sediments permit identification of source areas, and by inference, the directions of transporting winds. Secondary alterations of loess, including soil development, can potentially provide key information for interpreting sedimentation history, as well as past climatic and vegetation conditions. However, as shown with the example from Greenbay Hollow, separating the effects of climate and time on pedogenesis is not easy, and care must be taken in paleoclimatic interpretations of loess-derived paleosols.

SUMMARY

Loess is a terrestrial eolian deposit that records climatically driven sedimentary extremes and may cover as much as 10% of the Earth's surface. It is dominated by silt-sized particles with a

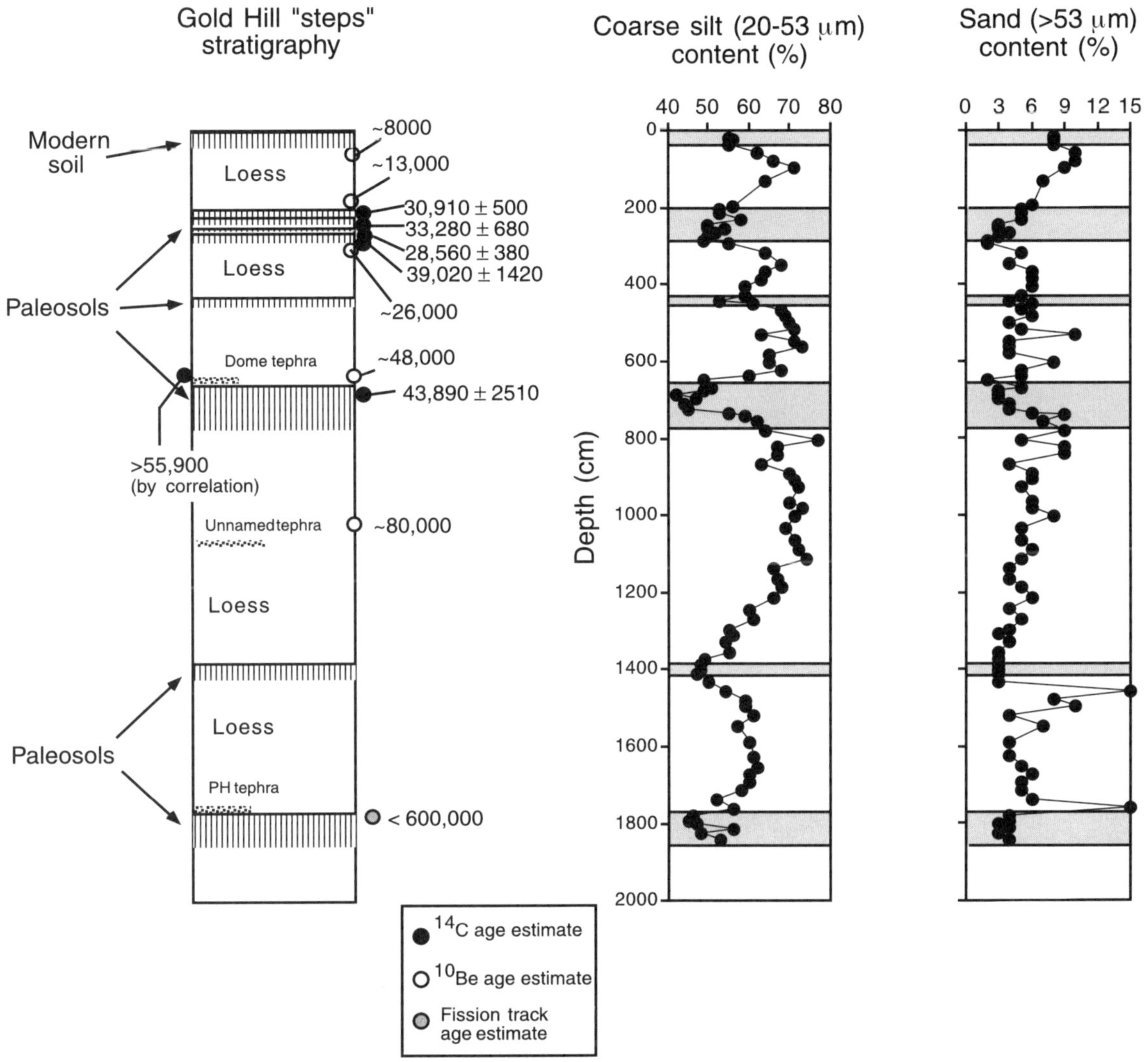

Figure 18. Stratigraphy, age estimates, and coarse silt and sand contents of loess and intercalated paleosols at the Gold Hill "steps" section near Fairbanks, Alaska. Stratigraphy and ages are from Muhs et al. (2003); coarse silt and sand contents are previously unpublished data obtained by the authors through conventional sieve and pipette methods.

majority of grains comprised of quartz, feldspars, and clay minerals. In many regions, loess also has varying amounts of carbonate minerals (calcite and dolomite). The geochemistry of loess varies from region to region, depending on source area, but it generally has a composition that resembles the bulk composition of upper crustal rocks. Trends in loess away from source areas include decreasing thickness, decreasing amounts of sand and coarse silt, and increasing amounts of fine silt and clay. Loess particle size also varies at a given locality over time and may be a function of varying wind strengths, changing source sediments or a combination of these two factors.

Traditionally, loess has been viewed primarily as a product of glacial grinding, with subsequent entrainment by wind from the surfaces of outwash deposits. Numerous studies have shown that this is an oversimplified concept and that other processes contribute to silt particle formation and loess accumulation. Recognition of these processes has led to the concept of "desert," nonglaciogenic loess, which is widespread in some regions, including China and the semiarid Great Plains of North America, and has limited occurrences elsewhere, such as Africa, Australia, and arid North America. Despite challenges to the importance of the role of glacial grinding, a review of the evidence suggests that glacial processes may still be much more important in the *new* formation of silt-sized particles. However, the importance of inheritance of silt-sized particles in loess from siltstones, mudstones, shales, and volcanic ash, whether in glaciated regions or elsewhere, probably has not been appreciated.

Loess is often geographically associated with eolian sand. Transects in loess bodies typically show decreasing amounts of sand downwind. At some localities, sand and loess are interbedded, which indicates that they are facies of the same deposit. However, in other regions, geochemical and isotopic analyses show that while sand may contribute to the sediment population of a loess body, the majority of loess particles are derived from a different and more distant source.

Buried soils, or paleosols, are important components of loess stratigraphy. They can be recognized by their distinctive morphological features and by systematic changes in particle size, chemistry and mineralogy. Buried soils formed during past periods when loess sedimentation rates either dropped to zero or at least slowed significantly enough that soil formation could keep ahead of loess deposition. Thus, loess and soils represent opposite members of a continuum of sedimentary extremes: high rates of sedimentation yield relatively unaltered loess in a stratigraphic record, whereas low rates of sedimentation leave a record of buried soils. This swing between sedimentary extremes can be recognized in the long-term glacial-interglacial record of the Quaternary. In some regions such as China and Alaska, loess deposition continues today. However, in most regions, including China, loess sedimentation rates were much higher during glacial periods than during interglacial periods. The sedimentary extreme of high loess sedimentation rates during the last glacial period on many continents was probably due to a cold, dry climate with strong winds, a decreased intensity of the hydrologic cycle, decreased vegetation cover, and increased sediment supplies, whether from glacial or nonglacial sources.

ACKNOWLEDGMENTS

We thank Margie Chan (University of Utah) and Allen Archer (Kansas State University) for inviting us to contribute this review and for helpful editing. It is a pleasure to extend our appreciation to the landowners at Elba, Lincoln, and Plattsmouth, Nebraska; Loveland, Iowa; and Morrison, Rapid City, and Greenbay Hollow, Illinois, for access to their property for our studies. Thanks also go to Ed Hajic (Illinois State Museum) for providing us with the Greenbay Hollow core and Josh Been (U.S. Geological Survey) for assistance with fieldwork and processing many samples discussed in this paper. Ralph Shroba (U.S. Geological Survey), Phillip Heckel (University of Iowa), George Kukla (Columbia University), and Jim Begét (University of Alaska) read an earlier version of the paper and made many helpful comments for its improvement. This research was supported by the Earth Surface Dynamics Program of the U.S. Geological Survey (to Muhs) and in part by National Science Foundation Grant EAR-00-87572 (to Bettis).

REFERENCES CITED

Ahlbrandt, T.S., and Fryberger, S.G., 1980, Eolian deposits in the Nebraska Sand Hills: U.S. Geological Survey Professional Paper 1120-A, 24 p.

Aleinikoff, J.N., Muhs, D.R., and Fanning, C.M., 1998, Isotopic evidence for the sources of late Wisconsin (Peoria) loess, Colorado and Nebraska: Implications for paleoclimate, *in* Busacca, A.J., ed., Dust aerosols, loess soils and global change: Washington State University College of Agriculture and Home Economics Miscellaneous Publication MISC0190, p. 124–127.

Aleinikoff, J.N., Muhs, D.R., Sauer, R., and Fanning, C.M., 1999, Late Quaternary loess in northeastern Colorado, Part II—Pb isotopic evidence for the variability of loess sources: Geological Society of America Bulletin, v. 111, p. 1876–1883.

An, Z., Kukla, G.J., Porter, S.C., and Xiao, J., 1991, Magnetic susceptibility evidence of monsoon variation on the loess plateau of central China during the last 130,000 years: Quaternary Research, v. 36, p. 29–36.

Antoine, P., Rousseau, D.-D., Zöller, L., Lang, A., Munaut, A.-V., Hatté, C., and Fontugne, M., 2001, High-resolution record of the last Interglacial-glacial cycle in the Nussloch loess-paleosol sequences, upper Rhine area, Germany: Quaternary International, v. 76/77, p. 211–229.

Baosheng, L., Zhang, D.D., Heling, J., Zheng, W., Mancun, Y., Wu, S., Yizhi, Z., and Donghuai, S., 2000, Paleo-monsoon activities of Mu Us Desert, China since 150 ka B.P.—A study of the stratigraphic sequences of the Milanggouwan section, Salawusu River area: Palaeogeography, Palaeoclimatology, Palaeoecology, v. 162, p. 1–16.

Begét, J.E., 1990, Middle Wisconsin climate fluctuations recorded in central Alaskan loess: Géographie Physique et Quaternaire, v. 44, p. 3–13.

Begét, J.E., 1996, Tephrochronology and paleoclimatology of the last interglacial-glacial cycle recorded in Alaskan loess deposits: Quaternary International, v. 34–36, p. 121–126.

Begét, J.E., 2001, Continuous Late Quaternary proxy climate records from loess in Beringia: Quaternary Science Reviews, v. 20, p. 499–507.

Bettis, E.A., III, Muhs, D.R., Roberts, H.M., and Wintle, A.G., 2003, Last glacial loess in the conterminous U.S.A.: Quaternary Science Reviews, (in press).

Bruins, H.J., and Yaalon, D.H., 1979, Stratigraphy of the Netivot section in the desert loess of the Negev (Israel): Acta Geologica Academiae Scientarum Hungaricae, v. 22, p. 161–169.

Bryan, K., 1945, Glacial versus desert origin of loess: American Journal of Science, v. 243, p. 245–248.

Butler, B.E., 1974, A contribution towards the better specification of parna and some other aeolian clays in Australia: Zeitschrift für Geomorphologie Supplement, v. 20, p. 106–116.

Chan, M.A., 1999, Triassic loessite of north-central Utah: Stratigraphy, petrophysical character, and paleoclimate implications: Journal of Sedimentary Research, v. 69, p. 477–485.

Chlachula, J., Rutter, N.W., and Evans, M.E., 1997, A late Quaternary loess-paleosol record at Kurtak, southern Siberia: Canadian Journal of Earth Science, v. 34, p. 679–686.

Clague, J.J., 1989, Character and distribution of Quaternary deposits (Canadian Cordillera), *in* Fulton, R. J., ed., Quaternary geology of Canada and Greenland: Geological Survey of Canada, Geology of Canada 1, p. 34–38.

Conant, L.C., and Goudarzi, G.H., 1964, Geologic map of Kingdom of Libya: U.S. Geological Survey Miscellaneous Geological Investigations Map I-350A, scale 1: 2,000,000, 1 sheet.

Condie, K.C., 1993, Chemical composition and evolution of the upper continental crust: Contrasting results from surface samples and shales: Chemical Geology, v. 104, p. 1–37.

Curry, B.B., and Pavich, M.J., 1996, Absence of glaciation in Illinois during marine isotope stages 3 through 5: Quaternary Research, v. 46, p. 19–26.

Derbyshire, E., 1983, On the morphology, sediments, and origin of the Loess Plateau of central China, *in* Gardner, R., and Scoging, H., eds., Mega-geomorphology: Oxford, Clarendon Press, p. 172–194.

Derbyshire, E., Meng, X., and Kemp, R.A., 1998, Provenance, transport and characteristics of modern aeolian dust in western Gansu Province, China, and interpretation of the Quaternary loess record: Journal of Arid Environments, v. 39, p. 497–516.

Dijkmans, J.W.A., and Tornqvist, T.E., 1991, Modern periglacial eolian deposits and landforms in the Søndre Strømfjord area, West Greenland and their palaeoenvironmental implications: Meddelelser om Grønland, Geoscience, v. 25, p. 3–39.

Ding, Z.L., Xiong, S.F., Sun, J.M., Yang, S.L., Gu, Z.Y., and Liu, T.S., 1999, Pedostratigraphy and paleomagnetism of a ~7.0 Ma eolian loess-red clay sequence at Lingtai, Loess Plateau, north-central China and the implications for paleomonsoon evolution: Palaeogeography, Palaeoclimatology, Palaeoecology, v. 152, p. 49–66.

Dutta, P.K., Zhou, Z., and dos Santos, P.R., 1993, A theoretical study of mineralogical maturation of eolian sand: Geological Society of America Special Paper 284, p. 203–209.

Flint, R.F., 1971, Glacial and Quaternary geology: New York, John Wiley and Sons, 892 p.

Forman, S.L., 1991, Late Pleistocene chronology of loess deposition near Luochuan, China: Quaternary Research, v. 36, p. 19–28.

Gallet, S., Jahn, B., and Torii, M., 1996, Geochemical characterization of the Luochuan loess-paleosol sequence, China, and paleoclimatic implications: Chemical Geology, v. 133, p. 67–88.

Gallet, S., Jahn, B., Van Vliet-Lanoe, B., Dia, A., and Rossello, E.A., 1998, Loess geochemistry and its implications for particle origin and composition of the upper continental crust: Earth and Planetary Science Letters, v. 156, p. 157–172.

Goudie, A.S., Cooke, R.U., and Doornkamp, J.C., 1979, The formation of silt from quartz dune sand by salt-weathering processes in deserts: Journal of Arid Environments, v. 2, p. 105–112.

Graham, I.J., Ditchburn, R.G., and Whitehead, N.E., 2001, Be isotope analysis of a 0–500 ka loess-paleosol sequence from Rangitatau east, New Zealand: Quaternary International, v. 76/77, p. 29–42.

Grimley, D.A., Follmer, L.R., and McKay, E.D., 1998, Magnetic susceptibility and mineral zonations controlled by provenance in loess along the Illinois and central Mississippi River Valleys: Quaternary Research, v. 49, p. 24–36.

Hajic, E.R., 1990, Late Pleistocene and Holocene landscape evolution, depositional subsystems, and the stratigraphy in the lower Illinois Valley and adjacent central Mississippi River Valley [Ph.D. dissertation]: Urbana, University of Illinois, 307 p.

Hallet, B., Hunter, L., and Bogen, J., 1996, Rates of erosion and sediment evacuation by glaciers: A review of field data and their implications: Global and Planetary Change, v. 12, p. 213–235.

Harrison, S.P., Kohfeld, K.E., Roelandt, C., and Claquin, T., 2001, The role of dust in climate changes today, at the last glacial maximum and in the future: Earth-Science Reviews, v. 54, p. 43–80.

Hayward, R.K., and Lowell, T.V., 1993, Variations in loess accumulation rates in the mid-continent, United States, as reflected by magnetic susceptibility: Geology, v. 21, p. 821–824.

Heller, F., and Liu, T.S., 1984, Magnetism of Chinese loess deposits: Geophysical Journal of the Royal Astronomical Society, v. 77, p. 125–141.

Hesse, P.P., and McTainsh, G.H., 1999, Last glacial maximum to Early Holocene wind strength in the mid-latitudes of the southern hemisphere from aeolian dust in the Tasman Sea: Quaternary Research, v. 52, p. 343–349.

Hovan, S.A., Rea, D.K., Pisias, N.G., and Shackleton, N.J., 1989, A direct link between the China loess and marine $\delta^{18}O$ records: Aeolian flux to the north Pacific: Nature, v. 340, p. 296–298.

Jahn, B., Gallet, S., and Han, J., 2001, Geochemistry of the Xining, Xifeng and Jixian sections, Loess Plateau of China: Eolian dust provenance and paleosol evolution during the last 140 ka: Chemical Geology, v. 178, p. 71–94.

Johnson, S.Y., 1989, Significance of loessite in the Maroon Formation (Middle Pennsylvanian to Lower Permian), Eagle Basin, northwest Colorado: Journal of Sedimentary Petrology, v. 59, p. 782–791.

Kemmis, T.J., Hallberg, G.R., and Lutenegger, A.J., 1981, Depositional environments of glacial sediments and landforms on the Des Moines Lobe, Iowa: Iowa Geological Survey Guidebook Series 6, 132 p.

Kohfeld, K.E., and Harrison, S.P., 2000, How well can we simulate past climates? Evaluating the models using global paleoenvironmental datasets: Quaternary Science Reviews, v. 19, p. 321–346.

Kohfeld, K.E., and Harrison, S.P., 2001, DIRTMAP: the geological record of dust: Earth-Science Reviews, v. 54, p. 81–114.

Kröhling, D.M., 1999, Upper Quaternary geology of the lower Carcarañá Basin, North Pampa, Argentina: Quaternary International, v. 57/58, p. 135–148.

Kukla, G., Heller, F., Ming, L.X., Chun, X.T., Sheng, L.T., and Sheng, A.Z., 1988, Pleistocene climates in China dated by magnetic susceptibility: Geology, v. 16, p. 811–814.

Lea, P.D., 1990, Pleistocene periglacial eolian deposits in southwestern Alaska; sedimentary facies and depositional processes: Journal of Sedimentary Petrology, v. 60, p. 582–591.

Leigh, D.S., 1994, Roxana silt of the Upper Mississippi Valley: Lithology, source, and paleoenvironment: Geological Society of America Bulletin, v. 106, p. 430–442.

Liu, C.-Q., Masuda, A., Okada, A., Yabuki, S., and Fan, Z.-L., 1994, Isotope geochemistry of Quaternary deposits from the arid lands in northern China: Earth and Planetary Science Letters, v. 127, p. 25–38.

Liu, T., 1985, Loess in China (2nd ed.): Beijing, China Ocean Press, and Berlin, Springer-Verlag, 224 p.

Maat, P.B., and Johnson, W.C., 1996, Thermoluminescence and new ^{14}C age estimates for Late Quaternary loesses in southwestern Nebraska: Geomorphology, v. 17, p. 115–128.

Maher, B.A., 1998, Magnetic properties of modern soils and Quaternary loessic paleosols: Paleoclimatic implications: Palaeogeography, Palaeoclimatology, Palaeoecology, v. 137, p. 25–34.

Maher, B.A., and Thompson, R., 1995, Paleorainfall reconstructions from pedogenic magnetic susceptibility variations in the Chinese loess and paleosols: Quaternary Research, v. 44, p. 383–391.

Mahowald, N., Kohfeld, K., Hansson, M., Balkanski, Y., Harrison, S.P., Prentice, I.C., Schulz, M., and Rodhe, H., 1999, Dust sources and deposition during the last glacial maximum and current climate: A comparison of model results with paleodata from ice cores and marine sediments: Journal of Geophysical Research, v. 104, p. 15,895–15,916.

Markewich, H.W., Wysocki, D.A., Pavich, M.J., Rutledge, E.M., Millard, Jr., H.T., Jr., Rich, F.J., Maat, P.B., Rubin, M. and McGeehin, J.P., 1998, Paleopedology plus TL, ^{10}Be, and ^{14}C dating as tools in stratigraphic and paleoclimatic investigations, Mississippi River Valley, U.S.A.: Quaternary International, v. 51/52, p. 143–167.

McFadden, L.D., Wells, S.G., and Dohrenwend, J.C., 1986, Influences of Quaternary climatic changes on processes of soil development on desert loess deposits of the Cima volcanic field, California: Catena, v. 13, p. 361–389.

McKay, E.D., 1979, Wisconsinan loess stratigraphy of Illinois, *in* Follmer, L.R., McKay, E.D., Lineback, J.A., and Gross, D.L., eds., Wisconsinan, Sangamonian, and Illinoian stratigraphy in central Illinois: Illinois State Geological Survey Guidebook 13, p. 95–108.

McTainsh, G., 1987, Desert loess in northern Nigeria: Zeitschrift für Geomorphologie, v. 31, p. 145–165.

McTainsh, G.H., 1989, Quaternary aeolian dust processes and sediments in the Australian region: Quaternary Science Reviews, v. 8, p. 235–253.

Muhs, D.R., 1983, Airborne dust fall on the California Channel Islands, U.S.A.: Journal of Arid Environments, v. 6, p. 223–238.

Muhs, D.R., and Bettis III, E.A., 2000, Geochemical variations in Peoria Loess of western Iowa indicate paleowinds of midcontinental North America during last glaciation: Quaternary Research, v. 53, p. 49–61.

Muhs, D.R., and Zárate, M., 2001, Eolian records of the Americas and their paleoclimatic significance, *in* Markgraf, V., ed., Interhemispheric Climate Linkages: San Diego, Academic Press, p. 183–216.

Muhs, D.R., Swinehart, J.B., Loope, D.B., Aleinikoff, J.N., and Been, J., 1999a, 200,000 years of climate change recorded in eolian sediments of the High Plains of eastern Colorado and western Nebraska: Geological Society of America Field Guide, v. 1, p. 71–91.

Muhs, D.R., Aleinikoff, J.N., Stafford Jr., T.W., Kihl, R., Been, J., Mahan, S.A., and Cowherd, S., 1999b, Late Quaternary loess in northeastern Colorado: Part I—Age and paleoclimatic significance: Geological Society of America Bulletin, v. 111, p. 1861–1875.

Muhs, D.R., Bettis III, E.A., Been, J., and McGeehin, J., 2001, Impact of climate and parent material on chemical weathering in loess-derived soils of the Mississippi River Valley: Soil Science Society of America Journal, v. 65, p. 1761–1777.

Muhs, D.R., Ager, T.A., Bettis III, E.A., McGeehin, J., Been, J.M., Begét, J.E., Pavich, M.J., Stafford Jr., T.W., and Pinney, D., 2003, Stratigraphy and paleoclimatic significance of Late Quaternary loess-paleosol sequences of the last interglacial-glacial cycle in central Alaska: Quaternary Science Reviews, (in press).

Palmer, A.S., and Pillans, B.J., 1996, Record of climatic fluctuations from ca. 500 ka loess deposits and paleosols near Wanganui, New Zealand: Quaternary International, v. 34-36, p. 155–162.

Pésci, M., 1990, Loess is not just the accumulation of dust: Quaternary International, v. 7/8, p. 1–21.

Pettijohn, F.J., Potter, P.E., and Siever, R., 1972, Sand and sandstone: New York, Springer-Verlag, 618 p.

Péwé, T.L., and Journaux, A., 1983, Origin and character of loesslike silt in unglaciated south-central Yakutia, Siberia, U.S.S.R.: U.S. Geological Survey Professional Paper 1262, 46 p.

Porter, S.C., 2001, Chinese loess record of monsoon climate during the last glacial-interglacial cycle: Earth-Science Reviews, v. 54, p. 115–128.

Porter, S.C., and An, Z., 1995, Correlation between climate events in the North Atlantic and China during the last glaciation: Nature, v. 375, p. 305–308.

Porter, S.C., An, Z., and Zheng, H., 1992, Cyclic Quaternary alluviation and terracing in a nonglaciated drainage basin on the north flank of the Qinling Shan, central China: Quaternary Research, v. 38, p. 157–169.

Prospero, J.M., Bonatti, E., Schubert, C., and Carlson, T. N., 1970, Dust in the Caribbean atmosphere traced to an African dust storm: Earth and Planetary Science Letters, v. 9, p. 287–293.

Pye, K., 1987, Aeolian dust and dust deposits: San Diego, California, Academic Press, 334 p.

Pye, K., 1995, The nature, origin and accumulation of loess: Quaternary Science Reviews, v. 14, p. 653–657.

Reheis, M.C., Goodmacher, J.C., Harden, J.W., McFadden, L.D., Rockwell, T.K., Shroba, R.R., Sowers, J.M., and Taylor, E.M., 1995, Quaternary soils and dust deposition in southern Nevada and California: Geological Society of America Bulletin, v. 107, p. 1003–1022.

Roberts, H.M., Wintle, A.G., Maher, B.A., and Hu, M., 2001, Holocene sediment-accumulation rates in the western Loess Plateau, China, and a 2500-year record of agricultural activity, revealed by OSL dating: The Holocene, v. 11, p. 477–483.

Rousseau, D.D., Antoine, P., Hatté, C., Lang, A., Zöller, L., Fontugne, M., Ben Othman, D., Luck, J.M., Moine, O., Labonne, M., Bentaleb, I., and Jolly, D., 2002, Abrupt millennial climatic changes from Nussloch (Germany) Upper Weichselian eolian records during the Last Glaciation: Quaternary Science Reviews, v. 21, p. 1577–1582.

Rozycki, S.Z., 1991, Loess and loess-like deposits. Warsaw, Ossolineum Press, Polish Academy of Sciences, 187 p.

Ruddiman, W.F, 1997, Tropical Atlantic terrigenous fluxes since 25,000 yrs B.P.: Marine Geology, v. 136, p. 189–207.

Ruhe, R.V., 1983, Depositional environment of Late Wisconsin loess in the mid-continental United States, *in* Wright Jr., H.E., and Porter, S.C., eds., Late-Quaternary environments of the United States, The Late Pleistocene: Minneapolis, University of Minnesota Press, v. 1, p. 130–137.

Ruhe, R.V., Miller, G.A., and Vreeken, W.J., 1971, Paleosols, loess sedimentation and soil stratigraphy, in Yaalon, D.H., ed., Paleopedology—Origin, nature and dating of paleosols: Jerusalem, Israel Universities Press, p. 41–59.

Shackleton, N.J., and Opdyke, N.D., 1976, Oxygen-isotope and paleomagnetic stratigraphy of Pacific core V28–239 Late Pliocene to latest Pleistocene: Geological Society of America Memoir 145, p. 449–464.

Singer, M.J., Verosub, K.L., Fine, P., and TenPas, J., 1996, A conceptual model for the enhancement of magnetic susceptibility in soils: Quaternary International, v. 34–36, p. 243–248.

Smalley, I.J., 1995, Making the material: The formation of silt-sized primary mineral particles for loess deposits: Quaternary Science Reviews, v. 14, p. 645–651.

Smalley, I.J., and Krinsley, D.H., 1978, Loess deposits associated with deserts: Catena, v. 5, p. 53–66.

Smalley, I.J., and Smalley, V., 1983, Loess material and loess deposits: Formation, distribution and consequences, in Brookfield, M.E., and Ahlbrandt, T.S., eds., Eolian sediments and processes: Amsterdam, Elsevier, p. 51–68.

Smalley, I.J., Jefferson, I.F., Dijkstra, T.A., and Derbyshire, E., 2001, Some major events in the development of the scientific study of loess: Earth-Science Reviews, v. 54, p. 5–18.

Smith, G.D., 1942, Illinois loess: Variations in its properties and distribution, a pedologic interpretation: University of Illinois Agricultural Experiment Station Bulletin, 490, p. 139–184.

Soreghan, G.S., 1992, Preservation and paleoclimatic significance of eolian dust in the ancestral Rocky Mountain Province: Geology, v. 20, p. 1111–1114.

Spell, T.L., and McDougall, I., 1992, Revisions to the age of the Brunhes-Matuyama boundary and the Pleistocene geomagnetic polarity timescale: Geophysical Research Letters, v. 19, p. 1181–1184.

Sun, J., Yin, G., Ding, Z., Liu, T., and Chen, J., 1998, Thermoluminescence chronology of sand profiles in the Mu Us Desert, China: Palaeogeography, Palaeoclimatology, Palaeoecology, v. 144, p. 225–233.

Swineford, A., and Frye, J.C., 1951, Petrography of the Peoria Loess in Kansas: Journal of Geology, v. 59, p. 306–322.

Taylor, S.R., and McLennan, S.M., 1985, The continental crust: Its composition and evolution: Oxford, Blackwell Scientific Publications, 312 p.

Tegan, I., Lacis, A.A., and Fung, I., 1996, The influence on climate forcing of mineral aerosols from disturbed soils: Nature, v. 380, p. 419–422.

Thorp, J., 1945, Discussion of "Glacial versus desert origin of loess": American Journal of Science, v. 243, p. 246–247.

Tsoar, H., and Pye, K., 1987, Dust transport and the question of desert loess formation: Sedimentology, v. 34, p. 139–153.

Verosub, K.L., Fine, P., Singer, M.J., and TenPas, J., 1993, Pedogenesis and paleoclimate: Interpretation of the magnetic susceptibility record of Chinese loess-paleosol sequences: Geology, v. 21, p. 1011–1014.

Wang, H., Follmer, L.R., and Liu, J.C., 2000, Isotope evidence of paleo-El Niño-Southern Oscillation cycles in loess-paleosol record in the central United States: Geology, v. 28, p. 771–774.

Whalley, W.B., Marshall, J.R., and Smith, B.J., 1982, Origin of desert loess from some experimental observations: Nature, v. 300, p. 433–435.

Willman, H.B., and Frye, J.C., 1970, Pleistocene stratigraphy of Illinois: Illinois State Geological Survey Bulletin 94, 204 p.

Wright, J.S., 1995, Glacial comminution of quartz sand grains and the production of loessic silt: A simulation study: Quaternary Science Reviews, v. 14, p. 669–680.

Wright, J.S., 2001a, "Desert" loess versus "glacial" loess: quartz silt formation, source areas and sediment pathways in the formation of loess deposits: Geomorphology, v. 36, p. 231–256.

Wright, J., 2001b, Making loess-sized quartz silt: data from laboratory simulations and implications for sediment transport pathways and the formation of "desert" loess deposits associated with the Sahara: Quaternary International, v. 76/77, p. 7–19.

Wright, J., and Smith, B., 1993, Fluvial comminution and the production of loess-sized quartz silt: A simulation study: Geografiska Annaler, v. 75A, p. 25–34.

Wright, J., Smith, B.J., and Whalley, W.B., 1998, Mechanisms of loess-sized quartz silt production and their relative effectiveness: Laboratory simulations: Geomorphology, v. 23, p. 15–34.

Xiao, J., Porter, S.C., An, Z., Kumai, H., and Yoshikawa, S., 1995, Grain size of quartz as an indicator of winter monsoon strength on the Loess Plateau of central China during the last 130,000 yr: Quaternary Research, v. 43, p. 22–29.

Xiao, J.L., An, Z.S., Liu, T.S., Inouchi, Y., Kumai, H., Yoshikawa, S., and Kondo, Y., 1999, East Asian monsoon variation during the last 130,000 years: Evidence from the Loess Plateau of central China and Lake Biwa of Japan: Quaternary Science Reviews, v. 18, p. 147–157.

Zárate, M., and Blasi, A., 1993, Late Pleistocene-Holocene eolian deposits of the southern Buenos Aires Province, Argentina: A preliminary model: Quaternary International, v. 17, p. 15–20.

Zheng, H., Oldfield, F., Yu, L., Shaw, J., and An, Z., 1991, The magnetic properties of particle-sized samples from the Luo Chuan loess section: Evidence for pedogenesis: Physics of the Earth and Planetary Interior, v. 68, p. 250–258.

Manuscript Accepted by the Society January 22, 2003

Printed in the USA

Geological Society of America
Special Paper 370
2003

Lessons from large lake systems— Thresholds, nonlinearity, and strange attractors

Kevin M. Bohacs*
ExxonMobil Upstream Research Company, 3120 Buffalo Speedway, Houston, Texas 77098, USA

Alan R. Carroll
Department of Geology and Geophysics, University of Wisconsin, 1215 W. Dayton Avenue, Madison, Wisconsin 53706 USA

Jack E. Neal
ExxonMobil Exploration Company, 233 Benmar Street, Houston, Texas 77060-2598, USA

ABSTRACT

Lake systems are the largest integrated depositional complexes in the continental realm: modern lakes have areas up to 374,000 km^2, and ancient lake strata extend up to 300,000 km^2 in the Cretaceous systems of the south Atlantic and eastern China and the Permian system of western China. The largest lakes do not appear to form a significantly different population in many of their attributes. Their area, maximum depth, and volume closely follow power-law distributions with fractional exponents (–1.20, –1.67, –2.37 respectively), with minimal breaks between the largest lakes and the majority of lakes.

Controls on lake size and stratigraphic extent are not straightforward and intuitively obvious. For example, there is little relation of modern lake area, depth, and volume, with origin, climatic conditions, mixis, or water chemistry. Indeed, two-thirds of the largest-area lakes occur in relatively dry climates (precipitation-evaporation ratio [P/E] <1.6). In ancient lake strata, deposits of largest areal extent and thickness tended to form mostly under relatively shallow-water, evaporitic conditions in both convergent and divergent tectonic settings.

Geometric and dynamical thresholds appear to govern lake systems as complex, sensitive, nonlinear dynamical systems. Phanerozic examples worldwide indicate that the existence, character, and stacking patterns of lake strata are a function of the interaction of rates of supply of sediment + water and potential accommodation change. Lake-system behavior reflects interactions of four main state variables: sediment supply, water supply, sill height, and basin-floor depth. The stratal record ultimately records five main modes of behavior indicating that nonmarine basin dynamical systems are governed by two fundamental bifurcations and five strange attractors in the sediment + water supply – potential accommodation phase plane: fluvial, overfilled lake, balanced filled lake, underfilled lake, and aeolian/playa. Thus, extremely large lakes are highly dependent on intricate convolutions of climatic and tectonic influences and occur in a variety of settings and climates.

Keywords: aeolian, Caspian Sea, climate, climate change, continental depositional systems, contingency, fluvial, fractal, lacustrine, lake, lake size, mixis, nonlinear dynamics, nonmarine depositional systems, paleoclimate, paleolatitude, playa, rift basin, rivers, sag basin, tectonics, topography, water chemistry.

*Kevin.M.Bohacs@exxonmobil.com

Bohacs, K.M., Carroll, A.R., and Neal, J.E., 2003, Lessons from large lake systems—Thresholds, nonlinearity, and strange attractors, *in* Chan, M.A., and Archer, A.W., eds., Extreme depositional environments: Mega end members in geologic time: Boulder, Colorado, Geological Society of America Special Paper 370, p. 75–90.

INTRODUCTION

Large lakes have intrigued and fascinated humans since time immemorial; indeed, a significant portion of early hominid evolution transpired around the large lakes of east Africa. As with any familiar object, there are numerous clashes between facts and hypotheses, between what "everybody knows" and comprehensive observations. Ptolemy, in about 154 A.D., reported quite accurately on the location and sizes of the large east African lakes, Victoria, Tanganyika, and Malawi. Later, natural philosophers of the Enlightenment era thought these huge inland seas to be fabulous, for they reasoned that no large bodies of water could exist beneath the scorching equatorial sun (Guadalupi and Shugaar, 2001). It took the intensive efforts of mid-nineteenth century European explorers to "discover" and map these lakes, complete with native towns and fishing and trading fleets (Burton, 1860; Speke, 1863; Baker, 1866). Analogous clashes of thought and observations about lakes persist even into our day; although very large lakes are commonly termed "inland seas," they do not behave like small oceans, at least in the stratigraphic or geochemical sense (e.g., Bohacs, 1999, Bohacs et al., 2000c; Buoniconti, 2001). Additionally, one might tend to think of large lakes as somehow distinctive and special in their origin and behavior. This thought does not, however, stand up under close scrutiny of either modern or ancient lakes (Bohacs et al., 2000b).

Numerous observations of modern lakes and Phanerozoic lacustrine strata strongly indicate that lake size, shape, chemistry, and ecology are not simple functions of climatic humidity or tectonic subsidence. All lakes owe their existence and character to the nonlinear interaction of rates of potential accommodation increase and supply of sediment and water; potential accommodation is the space available for sediment accumulation below a lake's sill or spillpoint (Carroll and Bohacs, 1999). A lake's size, chemistry, and biota can vary rapidly, and lake strata are punctuated by many widespread breaks (Gierlowski-Kordesch and Kelts, 1994; Sladen, 1994; Bohacs et al., 2000a). Rates of change can be extreme: lake level changes of 300 m in 10–20 k.y. are common throughout the Pleistocene (e.g., Street-Perrott and Roberts, 1983; Street-Perrott and Harrison, 1984, 1985; Contreras and Scholz, 2001). For example, Lake Victoria changed from totally desiccated to the largest area lake in Africa, populated with over 300 fish species in less than 25 k.y. (Johnson et al., 1996).

The responses of lake systems to climatic, tectonic, and other forcing functions are complex, and their stratigraphic records can be difficult to interpret. Lacustrine strata record various modes of lake response to changes in forcing conditions as a function of climate, tectonics, sediment supply, and inherited topography. As with so many other geological phenomena, the origin and existence of large lakes are contingently conditioned, and these "inland seas" can form in a wide variety of settings: convergent and divergent tectonics and both dry and wet climates.

Our paper presents the size and character of lake systems in the modern and ancient, investigates the ultimate controls of large lakes, and discusses insights that large lakes provide for all lacustrine and continental depositional systems.

Our main data sources cover 253 modern lakes from an extensive compilation by Herdendorf (1984) of modern lakes greater than 500 km^2 in extent, and 211 ancient lakes from Cambrian to Pleistocene (from Gierlowski-Kordesch and Kelts, 1994, 2000; Carroll and Bohacs, 1999; and Bohacs et al., 2000a, along with other ExxonMobil proprietary studies). All of the data on modern systems are available in GSA Data Repository item 9916[1]; most of the ancient examples are covered to some degree in the publications cited.

Based on these observations, our hypotheses are that large lakes do not form a particularly distinctive population and that lake size is not a particularly strong function of climate, latitude, altitude, origin, mixis, or water chemistry. We do see that lake size is an intricate function of four main factors: sill height, basin-floor depth, water supply, and sediment supply. We suggest that a variety of combinations of these factors can yield large lakes—that bigness is an accidental, and not an essential, attribute of a lake system.

MODERN LACUSTRINE SYSTEMS

When discussing large lakes, the first issue to be addressed is the measure of size, or which dimension of the lake is to be considered: volume, depth, area, or some combination of these measures (e.g., Hutchinson, 1957; Cole, 1979). The answer depends on the scope and focus of one's investigation: in general, volume is of interest for investigating system hydrology and ecosystem issues; depth correlates mainly with mixis, distribution of biota, and water-sediment interactions; and area relates to trophic state, energy influx, and the lateral extent of lacustrine strata.

There is only a weak relation among these three measures of lake size. Table 1 lists the largest 10 in each size category and shows an imperfect correspondence among rankings. For instance, Lake Baikal is second in volume, first in depth, but seventh in area. Analysis of the lake data reveals a key insight: the deepest lakes are not necessarily the lakes with largest areal extent (Fig. 1A). Also, it appears that maximum lake depth is the strongest factor in determining lake volume, and area is only a weak factor: $r^2 = 0.64$ for depth versus volume, but $r^2 = 0.03$ for area versus volume (even allowing for the auto-correlation inherent in these relations; Figs. 1B and C). Areal extent is the main focus of this paper because it forms the strongest link between depositional environment and the extent of its stratal record, but first we examine the other two measures of lake size.

The volumes of modern lakes follow a very strong trend with rank along a power-law distribution with a fractional exponent or dimension of 2.37 ($r^2 = 0.97$, n = 253; Fig. 2A). Depth follows a reasonably consistent trend with rank ($r^2 = 0.83$, n = 253; Fig. 2B) There appear to be several statistical populations in

[1]GSA Data Repository item 9916, Modern Lakes Data, is available on request from Documents Secretary, GSA, P.O. Box 9140, Boulder, CO 80301-9140, USA, editing@geosociety.org, or at www.geosociety.org/pubs/ft1999.htm.

TABLE 1. SIZE RANKING OF LARGEST MODERN LAKES ACCORDING TO VOLUME, DEPTH, AND AREAL EXTENT

	Volume (km^3)	Depth (m)	Area (km^2)
Largest	Caspian	Baikal	Caspian
	Baikal	Tanganyika	Superior
	Tanganyika	Caspian	Victoria
	Superior	Nyasa	Aral*
	Nyasa	Issykkul	Huron
	Michigan	Great Slave	Michigan
	Huron	Toba	Tanganyika
	Victoria	Tahoe	Baikal
	Great Bear	Kivu	Great Bear
Smallest	Great Slave	Great Bear	Great Slave

Note: Data from Herdendorf (1984).
* As of 1984.

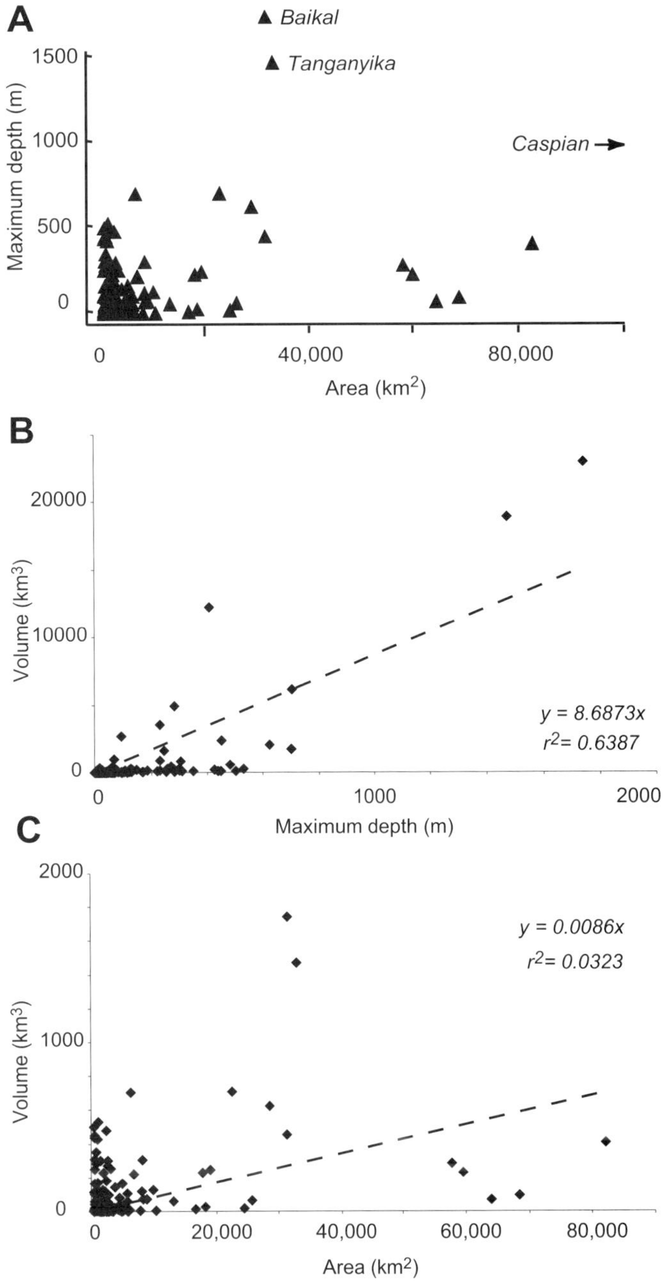

Figure 1. Relations of surface area, maximum depth, and volume for 253 modern lakes >500 km^2 in area. Note lack of any correlation and presence of significant outliers in depth (Baikal and Tanganyika) and in area (Caspian Sea plots far off the chart).

depth with two distinct breaks; however, all lakes deeper than 200 m follow a single trend. Finally, looking at area, the dimension on which we concentrate in this paper, we observe a very strong trend with area rank along another power-law distribution with an exponent of 1.20 (Fig. 3).

These distributions reveal several interesting insights about whether very large lakes are somehow distinctive: All of the distributions are relatively continuous with no major gap between the largest lakes and the rest of the population, indicating that large lakes do not form an inherently distinct population. Furthermore, the strong correlation along a power-law distribution, especially for volume and area, suggests the existence of underlying controls that operate across the broad scale of lake sizes (cf. Goodings and Middleton, 1991; Turcotte, 1997).

The shapes of the distributions, however, do show breaks in slope at about the upper decile (the top 20); the breaks in slope are distinct for depth but rather subtle for volume and area (Figs. 2 and 3). These breaks suggest a potentially different population that spans the top 20 largest lakes. The following observations test whether there is a distinctively different population of large lakes and seek to reveal fundamental controls on lake size.

We examined a broad range of factors commonly suspected of controlling lake behavior to search for distinctive characteristics of large lakes and to reveal potential fundamental controls. We started with climate, specifically the precipitation-evaporation ratio (P/E).

Figure 4, with the 253 modern lakes in area rank order versus P/E, shows no obvious trend of larger lakes in wetter climates. Figure 5 shows P/E along with five other climate parameters, comparing the 20 largest lakes with all others; it also shows no significant difference between the largest lakes and the rest. In general, these relations indicate that lake size is not a strong function of climate parameters such as precipitation-evaporation ratio, annual precipitation, evaporation potential, actual evaporation, transpiration, or annual runoff. However, two-thirds of the entire population of modern lakes and the bulk of the 20 largest area lakes do occur in relatively dry areas with P/E between 0.5 and 1.6.

Latitude, a commonly used proxy for climate in studies of the ancient (e.g., Barron, 1990; Katz, 1990; Smith, 1990; Sladen, 1994), also shows no strong relation with lake size, either for

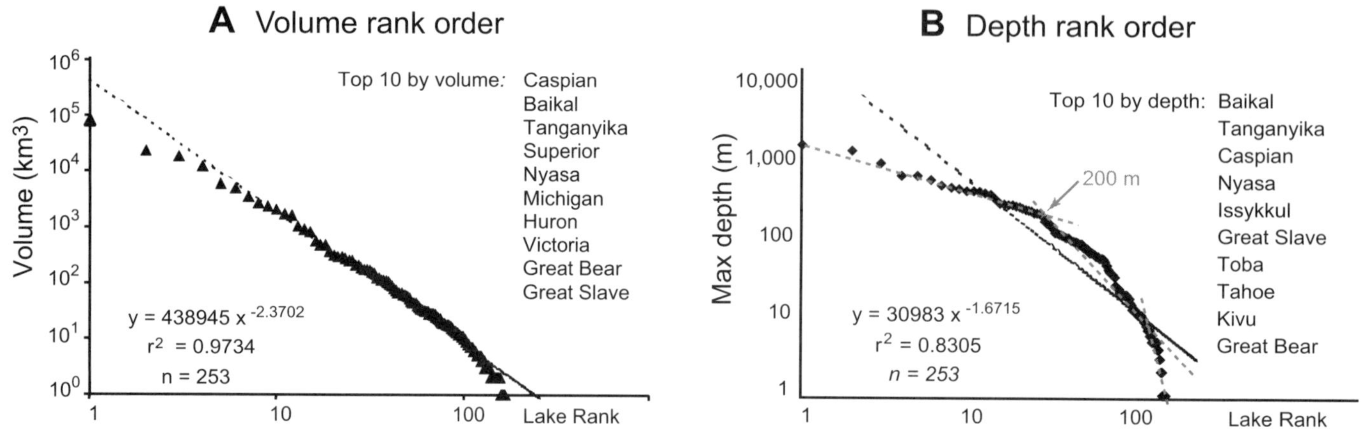

Figure 2. A: Volume-size distribution of 253 modern lakes (>500 km^2 in area) and listing of 10 largest volume lakes. Relation closely follows a power-law distribution with an exponent of 2.37 over five orders of magnitude of volume. This apparent fractal character suggests some unifying influence on lake volume. B: Depth-size distribution of 253 modern lakes (>500 km^2 in area) and listing of 10 deepest lakes. Relation overall follows a power-law distribution with an exponent of 1.67 over three orders of magnitude of depth. The overall distribution has two distinct breaks at 8 m and 200 m that segment it into three populations, suggesting a multifractal character.

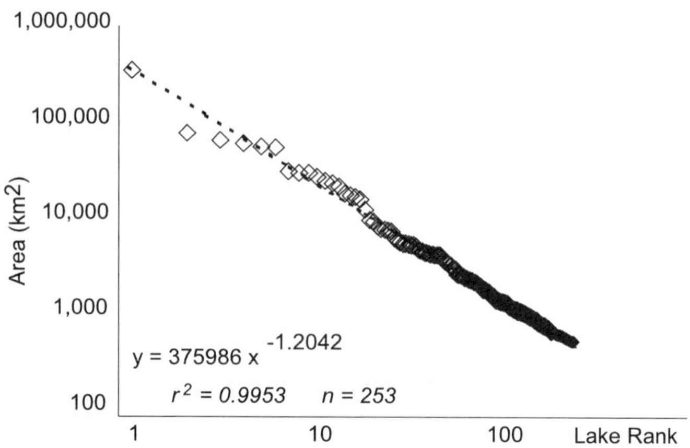

Figure 3. Surface-area-size distribution of 253 modern lakes (>500 km^2 in area). Relation very closely follows a power-law distribution with an exponent of 1.20 over more than three orders of magnitude, suggesting lake systems are scale invariant within this range. This fractional distribution has same exponent as that of erosional topography (Huang and Turcotte, 1989; Turcotte, 1997).

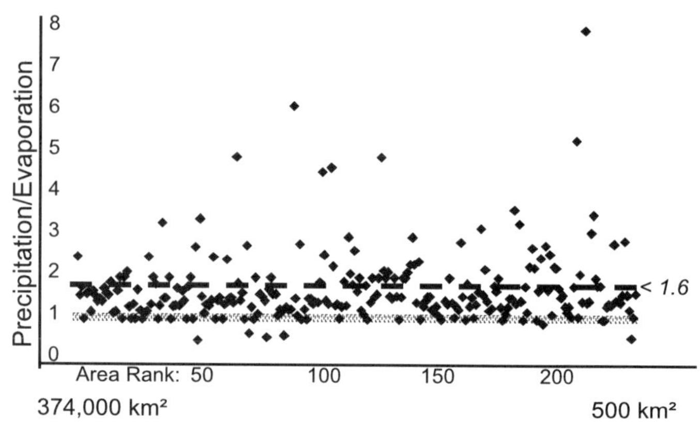

Figure 4. Relation of precipitation/evaporation ratio with lake-area size rank for 253 modern lakes >500 km^2 in area. There is no obvious trend of larger lakes in wetter climates, but most lakes are distinctly clustered between precipitation/evaporation ratio of 0.5 and 1.6.

modern or ancient systems. Modern lake distribution more or less mirrors the present-day latitudinal distribution of land area (Fig. 6). The distribution of ancient lake strata shows a similar lack of strong latitudinal control (Fig. 7).

Other physical parameters of lake settings in the modern data also show no strong relation of lake size to any single parameter, including altitude (linear regression $r^2 = 0.01$) and drainage basin size (linear regression $r^2 = 0.16$).

In a converse sense, neither water chemistry (Fig. 8A) nor mixis (Fig. 8B) shows a strong relation with lake size, except for the influence of the very largest modern lake, the Caspian Sea, which is brackish and meromictic. The Caspian Sea is five times larger than the next largest lake and hence significantly skews any statistical analysis of modern systems. Large ancient examples, however, also tend to record brackish, meromictic conditions, as discussed later in this paper.

Finally, the origin of a lake appears to have little control on size for most modern lakes (Fig. 9). Of the 20 largest modern lakes, 11 are glacial in origin and eight are tectonic in origin, a legacy of the Pleistocene. Tectonic lakes, however, can be very significant: the Caspian Sea is as large as the next six largest modern lakes combined. Also, most geologically significant lake strata are tectonic in origin because of lake persistence in areas of

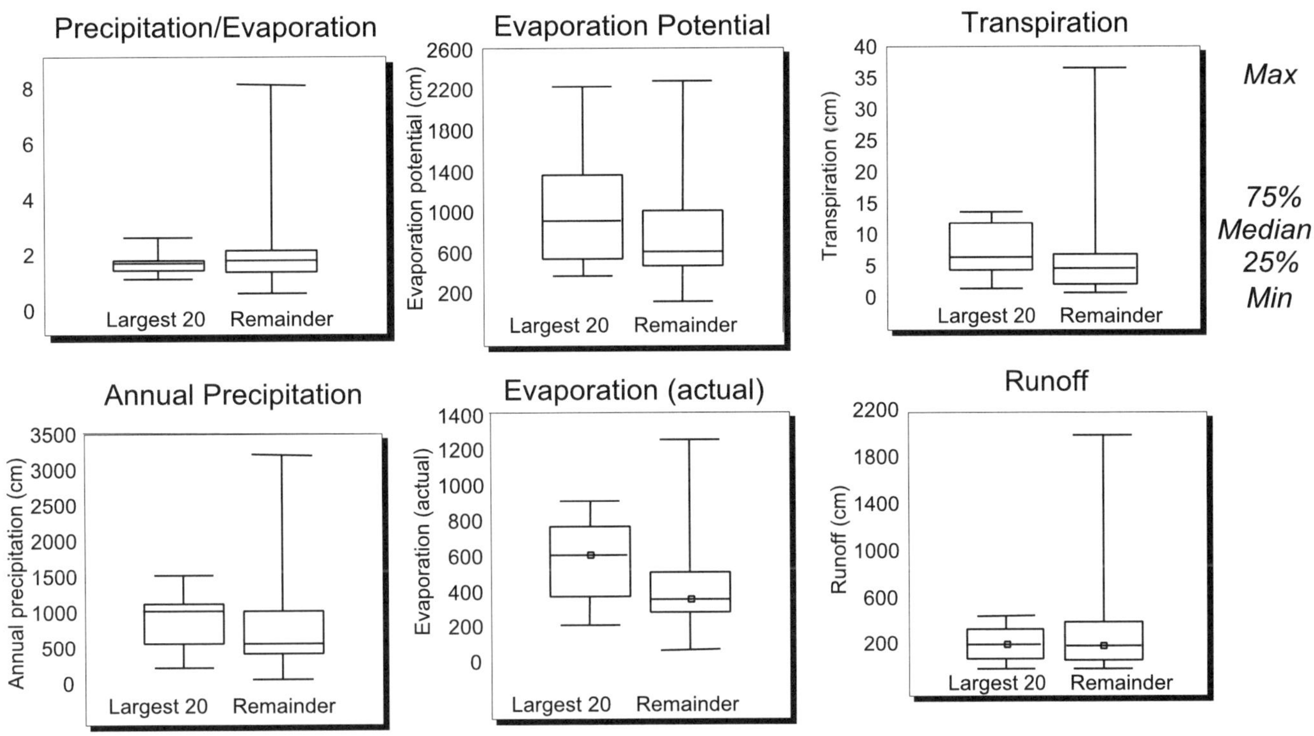

Figure 5. Distribution of various climate parameters according to lake-area size rank for 253 modern lakes >500 km^2 in area. Note lack of significant differences between 20 largest lakes and remaining 233 lakes. Box-and-whisker plots range from minimum to maximum, with central two quartiles denoted by box and median by central line.

long-lived accommodation, which confers higher preservation potential. Ancient lake deposits of tectonic origin include the Irati Formation (Permian, Brazil), Qinshankou 1 and 2 (Cretaceous, China), Termit Graben (Cretaceous, Niger), Green River Formation (Eocene, Utah and Colorado, United States), Junggar and Jianghan Basins (Permian, China), Waltman Shale (Paleocene, Wyoming, United States), Karoo Basin (Permian, South Africa), Doba and Doseo Basins (Cretaceous, Chad), Rubielos de Mora (Miocene, Spain), Cantwell Shale (Devonian, United Kingdom), Panonian Basin (Miocene, Hungary), Shahejie Formation (Eo-Oligocene, China), Lagoa Feia Formation (Cretaceous, Brazil), Rundle Formation (Eocene, Australia), and the Brown Shale (Oligocene, Indonesia) (Fig. 10).

Two components of the tectonic setting influence lake character: lake-floor subsidence and sill uplift. In the modern data it appears that uplift of the sill or spill point is as common a control as lake-floor subsidence (Fig. 11). All lakes owe their origin to some impediment to the free through-flow of water through the depositional system (e.g., Hutchinson, 1957); in tectonically active settings, structural uplift of a sill or spillpoint appears to be very effective in forming lakes. In an analogous manner, ancient examples indicate that convergent settings that impede drainage can form lakes as large, or larger than, divergent or rift settings—that extensional subsidence of a lake floor is not the exclusive mechanism that produces potential accommodation (Bradley, 1925; Carroll et al., 1995; Gierlowski-Kordesch and Kelts, 1994, 2000; Bohacs et al., 2000a).

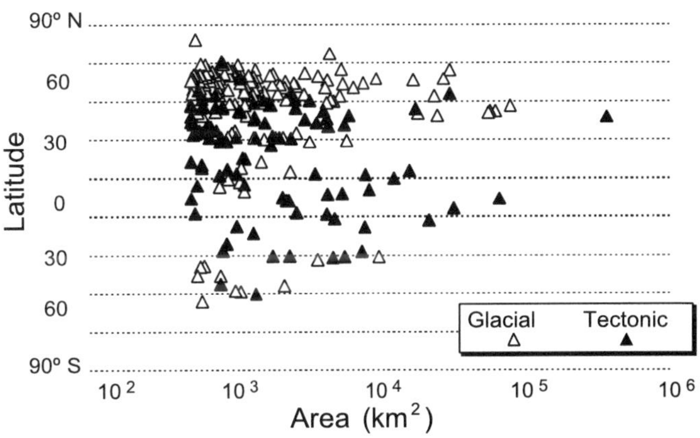

Figure 6. Cross plot of latitude with lake-area size for 253 modern lakes (>500 km^2 in area). There is no strong relation of lake size with latitude even for lakes of glacial origin—the distribution mostly mirrors present-day distribution of land area.

In summary, the modern data shows no strong relation of lake size with climate, latitude, mixis, water chemistry, drainage-basin area, or altitude. Significantly, in all of the data, there are no

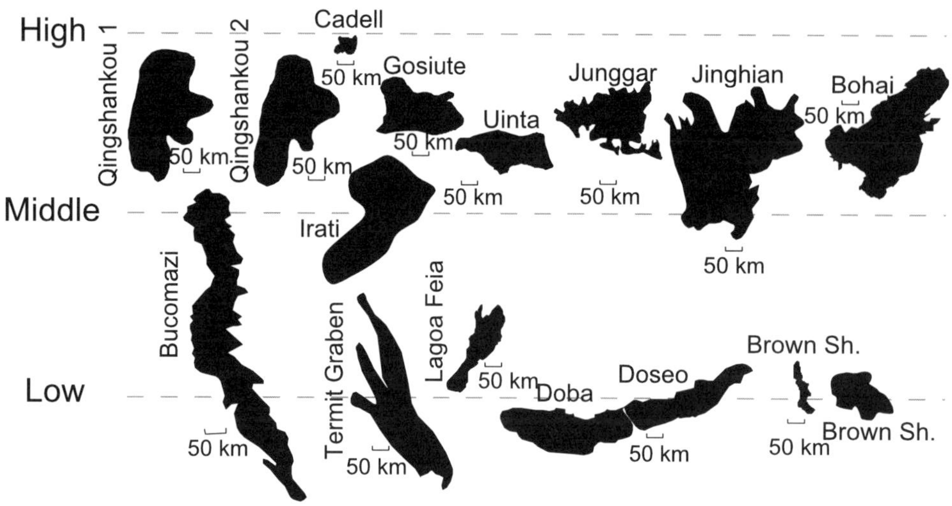

Figure 7. Distribution of selected ancient lake strata versus paleolatitude. As with present-day lakes, there is no strong relation of lake size or existence with latitude. Ancient lake strata are as prevalent in middle to high paleolatitudes as in low paleolatitudes. These ancient examples range in age from Devonian to Miocene.

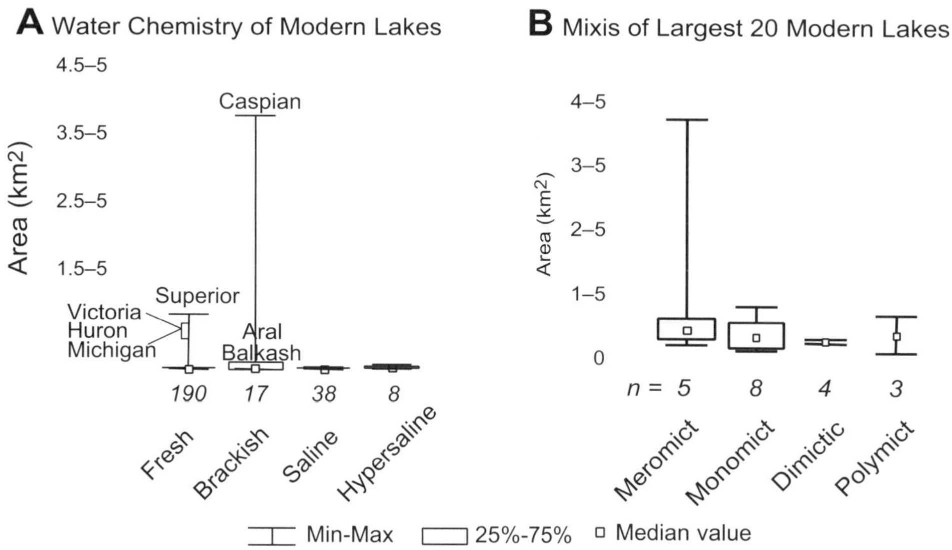

Figure 8. A: Distribution of lake-area size according to lake-water chemistry for 253 modern lakes >500 km^2 in area. Fresh-water lakes are not significantly larger than other lake types. The largest modern lake, the Caspian Sea, has brackish waters. B: Distribution of lake-area size according to lake mixis state for 20 largest area modern lakes. Note lack of significant differences among 20 largest area modern lakes, and that the largest modern lake, the Caspian Sea, is meromictic. Box-and-whisker plots range from minimum to maximum, with central two quartiles denoted by box and median by central line.

systematic or significant differences between the 20 largest lakes and the rest of the 233 modern lakes larger than 500 km^2 in area. The largest modern lake, the Caspian Sea, is tectonic in origin, controlled mainly by sill uplift, and occurs in a convergent setting under brackish, meromictic waters. It has much in common with the large, ancient lacustrine systems discussed below.

ANCIENT LACUSTRINE SYSTEMS

Pleistocene glacial lakes and pre-Pleistocene ancient examples lie close to the same power-law distribution as modern lakes over four orders of magnitude of area, despite progressively lower spatial resolution of the data (Fig. 12). This again points to fundamental controls that operate on all lakes of all ages. When searching for these fundamental controls, we commonly find relations such as those demonstrated by the Green River and associated formations in the Eocene of Wyoming (Fig. 13). There, the largest areal extent occurs in a brackish, shallow lake; the thickest cumulative stratal record results from the most shallow and evaporitic lake; and the smallest areal extent represents a freshwater lake with the thickest individual depositional sequences. All of these changes occurred under relatively stable climate conditions (Horsfield et al., 1995; Carroll and Bohacs, 1999; Wilf, 2000). Similar trends are seen in many other basins (Bohacs et al., 2000a), for example, the Permian Hongyanchi, Jingjingzagao, and Lucagao Formations of western China discussed by Carroll (1998).

Looking at the place of large lakes in basin-fill evolution, several investigators observe that the deepest lake strata form commonly at mid-rift phase and the largest-area lake strata occur

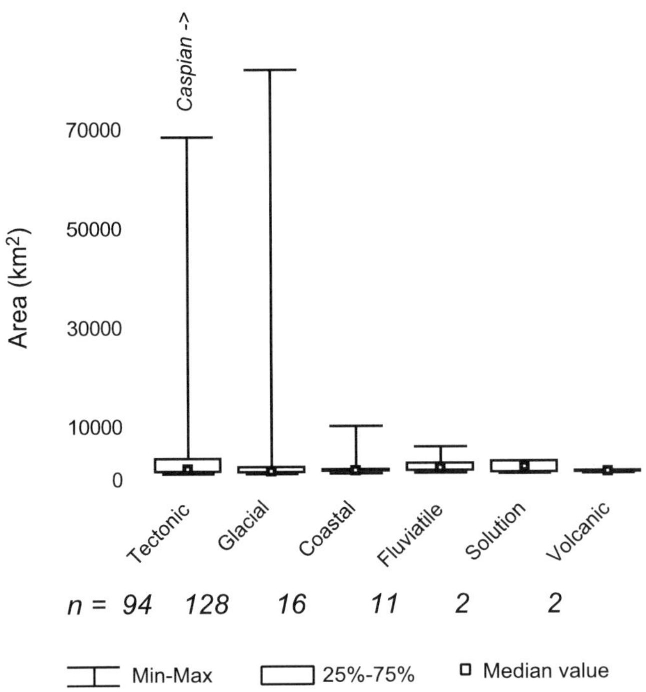

Figure 9. Distribution of lake-area size according to lake origin for 253 modern lakes >500 km^2 in area. There are minimal differences among most modern lakes, although Caspian Sea of tectonic origin is main outlier. Box-and-whisker plots range from minimum to maximum, with central two quartiles denoted by box and median by central line.

at the transition from rift to sag phase or early sag phase (Ebinger et al., 1987; Lambiase, 1990; Bohacs et al., 2000a). Figure 14 illustrates a representative example from the Cretaceous strata of the Songliao Basin, China. Seismic sections with successively younger datums show that the thickest sequences in the Quantou 3 and 4 Formations formed in active half-grabens and the most areally extensive Qinshankou 1 Formation in the overlying sag phase (Schwans et al., 1997).

These ancient examples, along with many others (e.g., Rosendahl, 1987; Schlische and Olsen, 1990; Strecker et al., 1999), show that the largest-area lakes tend to form at moderate subsidence rates under brackish, meromictic waters developed under intermittently open hydrologies. Overall, the size and evolution of the container is a very strong control on lake character; the basin accommodation affects the hydrologic balance.

SEARCH FOR FUNDAMENTAL CONTROLS

The data considered so far show no obvious relations among modern lake size and any of the usually considered controls. Indeed, 15 of the largest 20 modern lakes have freshwater, open hydrologies, but the very largest (by a factor of 4.5) has a brackish, closed hydrology. Ancient lake strata lie along the same areal-size distribution as modern lakes with no significant break. The larger ones do, however, tend to indicate deposition under intermittently open hydrologic conditions. To move beyond these simple descriptions to a deeper understanding, it would be useful to develop insights into the root causes or controls on lake size—the interrelation of lake genesis, character, size, and the nature of lake-system dynamics. One possible path towards this goal is indicated by a closer examination of the size distributions and detailed behavior of lake hydrology.

The distributions of volume, depth, and area along power-law trends with fractional exponents suggest a self-similar or fractal character of lake size. They further suggest the existence of unifying controls underlying these attributes (e.g., Goodings and Middleton, 1991; Turcotte, 1997). Fractal geometry is an effective way of describing natural objects, where traditional Euclidean geometry is found wanting. ("Clouds are not spheres, mountains are not cones, coastlines are not circles, and bark is not smooth, nor does lightning travel in a straight line," Mandelbrot, 1982, p. 1.) Extensive work has shown that landscapes can be characterized well by fractal geometries and their evolution analyzed profitably by using concepts of nonlinear dynamics (e.g., Turcotte, 1997). Lakes are an important component of landscapes, and both are constructed by tectonic, erosive, and depositional processes. Many of the tools used in these analyses of landscape formation may also be useful for explaining lake size distributions and understanding their formation.

Volume closely follows a power-law distribution with a fractional exponent or dimension of 2.37 (r^2 = 0.97, n = 253; Fig. 2A). Lake area also has a strong power-law trend with an exponent of 1.20, which is equivalent to the fractal dimension of the lateral distribution of erosional topography (Huang and Turcotte, 1989; Turcotte, 1997). This trend covers more than four orders of magnitude when ancient lake examples are included (Fig. 3). This is a satisfying result, as the area of a lake effectively defines a closed topographic contour.

Depth also follows a power-law distribution with an overall fractional dimension of 1.67 (Fig. 2B), similar to that of elevation hypsometry (e.g., Turcotte, 1997). There are, however, two distinct breaks in the distribution, at 200 m and 8 m depth. This agrees with the conclusions of workers who analyzed topographic fields and observed intertwined fractal subsets with different distributions or scaling exponents (e.g., Lavallée et al., 1993). These breaks suggest a multifractal character of lake bathymetry (Benzi et al., 1984; Frisch and Parisi, 1985; Feder, 1988; Stanley and Meakin, 1988) and the existence of several fundamental controls or processes that influence maximum depth.

Thus, all of these strong trends indicate that lake systems are scale invariant and bear the spatial signature of self-similar fractal objects (Bak et al., 1987, 1988; also see Rodríguez-Iturbe and Rinaldo, 1997). Nonlinearity is a necessary condition for scale invariance and fractal distributions (Turcotte, 1997). All geologists are familiar with the concept of scale invariance, that without an object for scale, it is commonly impossible to determine whether a photograph of a geological feature covers 10 cm or 10 km. The concept of self-similar fractal

Figure 10. Areal extents of selected ancient lake strata of tectonic origin. Tectonic lakes are geologically important because of persistent accommodation that allows accumulation of significant volumes of lacustrine strata. These ancient examples range in age from Devonian to Miocene.

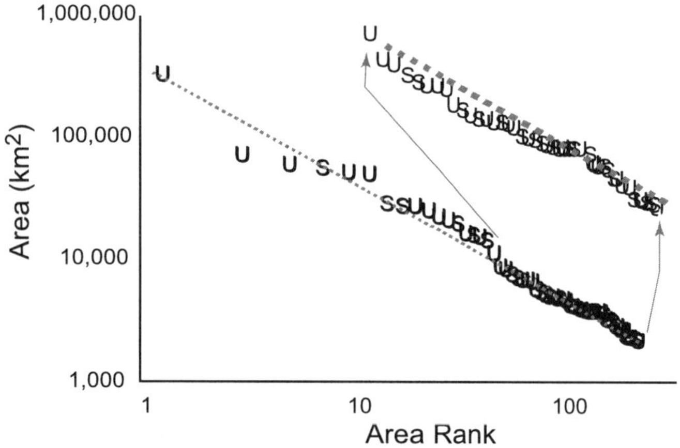

Figure 11. Surface-area-size distribution of 253 modern lakes >500 km^2 in area according to major control on origin: sill uplift (U) or lake-floor subsidence (S). Note that uplift or convergence is as common as lake-floor subsidence or divergence in influencing lake formation.

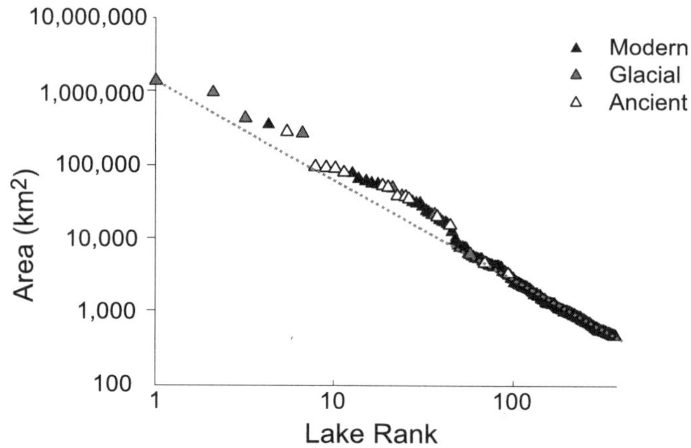

Figure 12. Surface-area size distribution of 253 modern lakes >500 km^2 in area, with nine Pleistocene glacial lakes and 14 pre-Pleistocene ancient examples. Data still closely follow power-law distribution for modern lakes alone, with an exponent of 1.20 over more than four orders of magnitude in area, indicating lake systems are scale invariant and self-similar throughout this range. This suggests that fundamental controls operate on lakes of all ages.

geometry is closely related: such fractal objects look similar at any scale of magnification. Unlike the surface of a three-dimensional Euclidean sphere, which looks less curved with increasing magnification until it resembles a two-dimensional plane, a self-similar fractal object shows continuously more detail with increasing magnification. The area-size distribution indicates that lakes have self-similar fractal geometries; without a scale, it is difficult to determine the altitude from which an aerial photograph of a lake was taken (Figs. 3 and 15). Other consequences of nonlinearity not quite so familiar to many geologists also provide valuable insights into the genesis of large lakes. The size distributions and attributes suggest the existence of underlying controls that operate across a broad scale of lake sizes. The nonlinear approach also helps us reconcile how relatively subtle differences among lake attributes can result in such a large range in lake sizes.

Integrating information from all the modern and ancient examples allows us to postulate these underlying controls on lake systems. At the most fundamental level, it appears that two factors are essential for the existence of a lake: water is necessary but not sufficient, and there also must be a hole in the ground to contain the water and sediment (Hutchinson, 1957; Cole, 1979). The hole in the ground or lake basin has two key controls: lake-floor subsidence and sill uplift, and it is the integrated effect of the height of the sill relative to the lake floor that controls the existence and nature of the lake. The space below the sill/thresh-

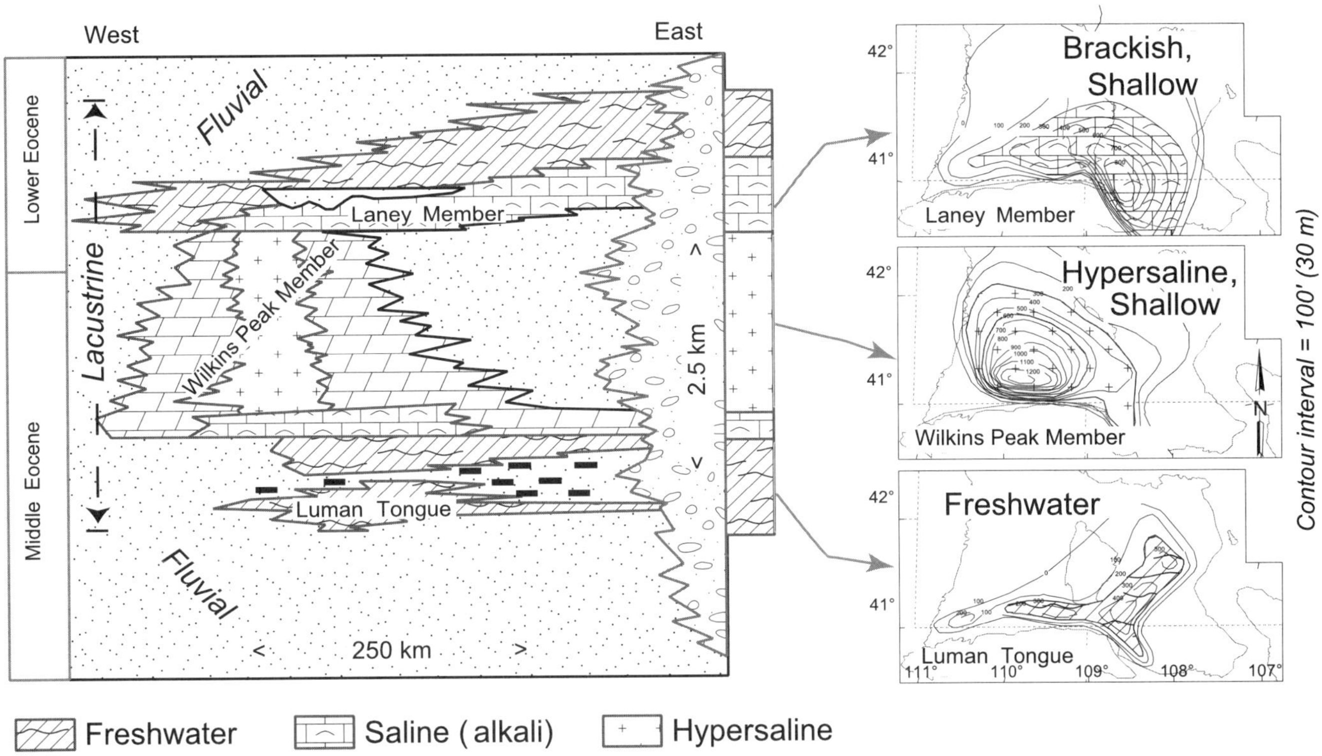

Figure 13. Representative example of relations among lacustrine stratal areal extent, thickness, and lake character taken from Eocene Green River and associated formations of Greater Green River Basin, southwestern Wyoming. Largest areal extent occurs in a brackish, shallow lake; thickest cumulative stratal record results from the most shallow and evaporitic lake; smallest areal extent represents a freshwater lake with the thickest individual depositional sequences. All of these changes occur under relatively stable climate. Cross section modified from Roehler, 1993; maps modified from Sullivan, 1980.

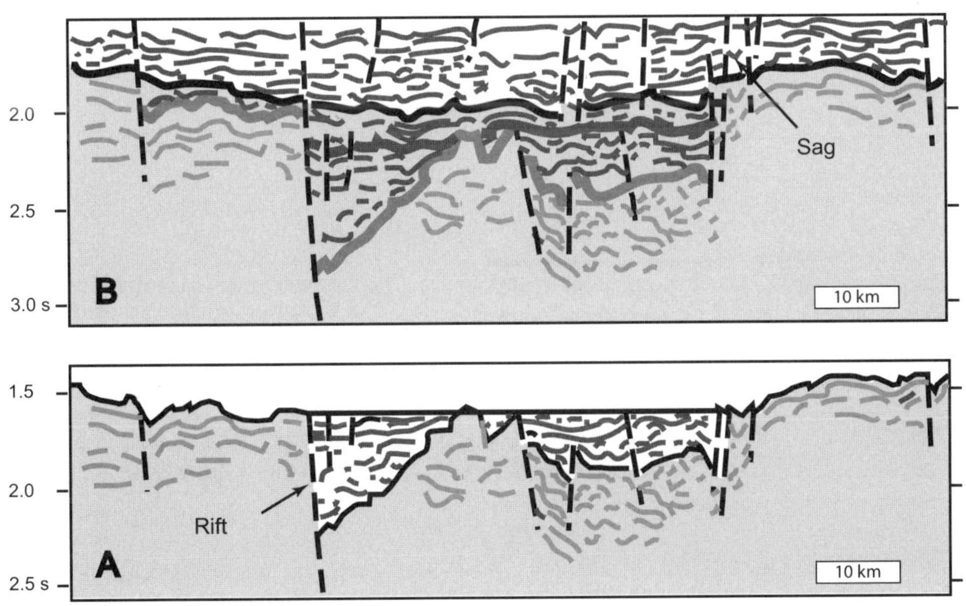

Figure 14. Illustration of relation of thickness and areal extent of lacustrine strata to basin evolution in Cretaceous strata of Songliao Basin, China. Line drawings of seismic sections with successive datums show thickest sequences in Quantou 3 and 4 Formations formed in (A) active half-grabens and (B) the most areally extensive Qinshankou 1 Formation in overlying sag phase. Figure courtesy of Peter Schwans.

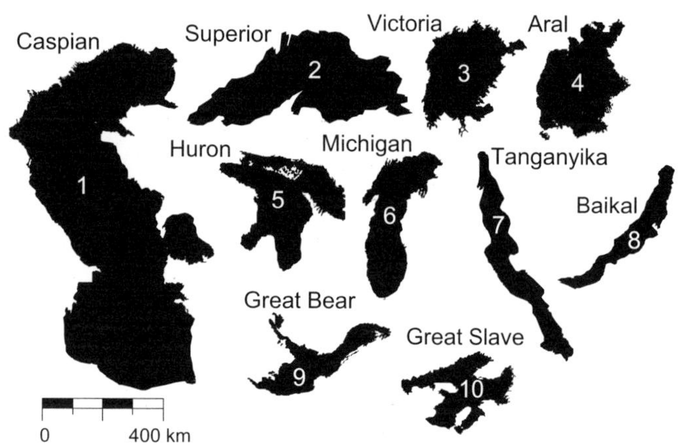

Figure 15. Map patterns of 10 largest area modern lakes at same map scale. Note nonlinear distribution of sizes. Although most of these lakes have shapes that are distinctly elongate and not equant, the distribution of their areas indicate that they are scale invariant and self-similar despite varied origins and tectonic settings.

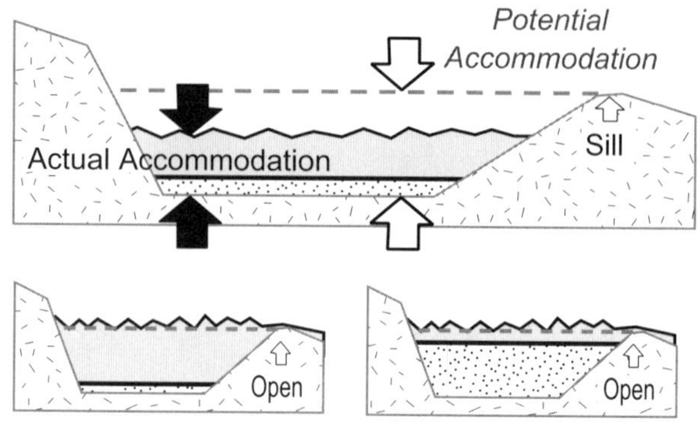

Figure 16. Schematic diagram of key controls on lake formation and character: basin-floor subsidence relative to sill-height uplift. Lakes contrast with oceanic systems because there is a distinct upper limit to available accommodation, defined as potential accommodation (volume below spillpoint or sill of lake basin). This potential accommodation can be filled with various combinations of sediment and water to yield an open (overflowing) hydrology. Thus, rate of potential accommodation change relative to supply of sediment and water controls the lake's very existence as well as its character and areal distribution through evolving hydrology.

old (or potential accommodation) can be filled with any combination of sediment and water (Fig. 16).

Lake-basin volume or potential accommodation relative to the supply of sediment and water controls the very existence of a lake, along with its character and areal distribution through evolving hydrology (see discussion in Carroll and Bohacs, 1999). If the lake level is at the sill elevation on average, the lake will have a dominantly open hydrology and will frequently overflow, with minimal changes in lake level, area, depth, or volume (Fig. 17A). If lake level is near sill elevation on average, but intermittently drops below the sill, the hydrology will vary between open and closed, and the lake level curve will be a clipped waveform (Fig. 17B). At the extreme, if lake level is always below sill elevation, the lake will have a closed hydrology and will never overflow, but this is the only way to get a fully unconstrained cycle of lake level and widely varying lake area (Fig. 17C). These considerations give us a very direct indication of the nonlinear nature of lake systems—the height of the sill relative to lake level controls how changes in water input due to climate, for instance, are felt and recorded by the lake. The system response is strongly influenced by a physical threshold: the sill or spillpoint.

We interpret that these three modes of lake-level response—dominantly open, intermittently open, and dominantly closed—are recorded in lacustrine strata that can be grouped into only three main facies associations, based on all parameters: stratigraphy, lithology, paleontology, and organic and inorganic geo-

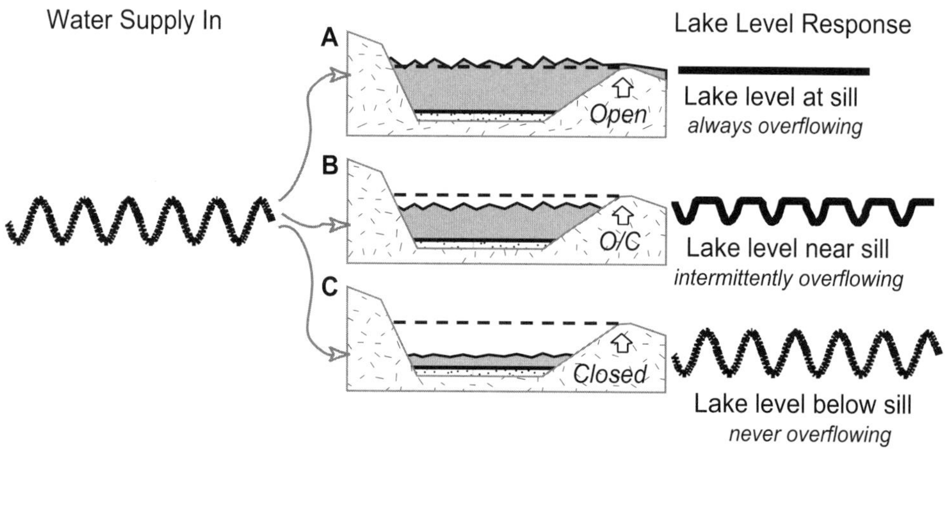

Figure 17. Height of sill relative to lake level controls how changes in water input (due to climate, for instance) are felt and recorded by a lake. For same amount of variation in sediment + water supply, three distinct responses of lake level are possible, depending on preexisting condition of lake system. System response is influenced very strongly by a physical threshold—the sill or spillpoint. This sensitive dependence on initial conditions and non-unique response to similar inputs give direct indication of nonlinear nature of lake systems.

chemistry (Carroll and Bohacs, 1999; Bohacs et al., 2000a). The lacustrine facies associations and their interpreted lake-basin types are summarized in Table 2. The mostly open hydrology lake with fresh waters (due to continual flushing) corresponds to the overfilled lake basin type marked by the fluvial-lacustrine facies association (Carroll and Bohacs, 2001). Its stratal record is dominated by progradational stacking of mostly clastic lithologies. The intermittently open hydrology lake, whose waters typically fluctuate between alkaline/saline and fresh, corresponds to the balanced-fill lake basin type marked by the fluctuating profundal facies association (Carroll and Bohacs, 2001). Its stratal record is a mixture of progradational and aggradational stacking of both clastic and biogenic/chemical lithologies; this is enhanced by concentration of solutes during its closed hydrology phases. These facies are commonly the most laterally extensive in a particular lake system. The dominantly closed hydrology lake, with saline to hyper-saline waters, corresponds to the underfilled lake basin type marked by the evaporative facies association (Carroll and Bohacs, 2001). Its stratal record is dominated by aggradational stacking of widely varying mixtures of rock types—clastic, biogenic, and chemical—with a significant component of evaporative lithologies.

We observe very few intermediate cases, which indicates that this system is almost intransitive and may represent self-organized criticality (Bak et al., 1987, 1988). For example, the strata of different lake-basin types are typically sufficiently distinct that they are the basis for dividing many formations that record lake deposition into subunits; each member, tongue, or bed commonly represents deposition in a different lake-basin type (Table 3). Hence, it is appropriate to use the terminology of nonlinear system dynamics, and it may be helpful to consider each lake-basin association as a strange attractor or a pattern of behavior that complex systems more or less replicate over the long term (Smale, 1967; Ruelle and Takens, 1971; Ruelle, 1980; Glieck, 1987).

We represent these lake-basin types on a phase diagram to indicate where each lacustrine and nonmarine environment is

TABLE 2. REPRESENTATIVE ATTRIBUTES OF THREE MAJOR LACUSTRINE FACIES ASSOCIATIONS

Lacustrine facies association; lake basin type	Stratigraphy	Stratal stacking patterns	Sedimentary structures	Lithologies	Organic matter
Fluvial-lacustrine; overfilled lake basin	Maximum progradation: 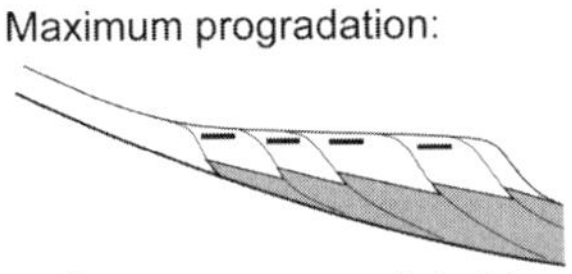 ▪ Parasequences related to lateral progradation ▪ Maximum fluvial input	Dominantly progradation Indistinctly expressed parasequences	Physical transport: ripples, dunes, flat bed Root casts Burrows (in- and epi-faunal)	Mudstone, marl Sandstone Coquina Coal, coaly shale	Freshwater biota Land-plant, charophytic and aquatic algal OM Low to moderate TOC Terrigenous and algal biomarkers
Fluctuating profundal; balanced-fill lake basin	Mixed progradation and desiccation: 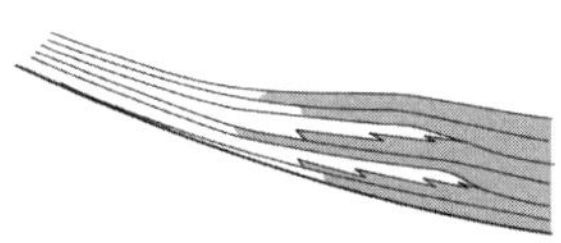 ▪ Distinct shoaling cycles common ▪ Fluvial input variable	Mixed progradation and aggradation Distinctly expressed parasequences	Physical and biogenic: flat bed, current, wave, and wind ripples; stromatolites, pisolites, oncolites Mudcracks Burrows (epifaunal)	Marl, mudstone Siltstone, sandstone Carbonate grainstone, wackestone, micrite Kerogenite	Salinity tolerant biota Aquatic algal OM Minimal land plant Moderate to high TOC Algal biomarkers
Evaporative; underfilled lake basin	Maximum desiccation: 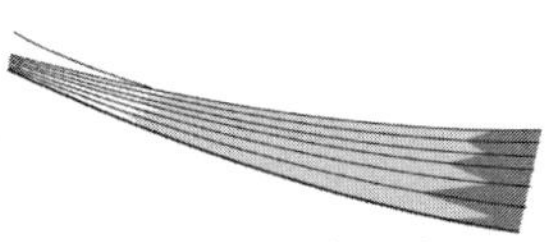 ▪ Closely spaced packages of wet-dry lithologies ▪ Minimum fluvial input	Dominantly aggradation Distinctly to indistinctly expressed parasequences	Physical, biogenic, and chemical: climbing current ripples, flat bed, stromatolites, displacive fabrics, cumulate textures	Mudstone, kerogenite Evaporite Siltstone, sandstone Grainstone, boundstone, flat-pebble conglomerate	Low-diversity, halophytic biota Algal-bacterial OM Low to high TOC Hypersaline biomarkers

Note: OM—organic matter; TOC—total organic carbon. Table adapted from Bohacs et al., 2000b.

TABLE 3. EXAMPLES OF DISTINCT LAKE BASIN TYPES WITHIN SINGLE BASIN

Basin: Age:	Chinle, USA Triassic	Hartford, USA Jurassic	Recôncavo, Brazil Cretaceous	Green River, USA Eocene	
Fluvial		upper Portland Formation	São Sebastião Formation	Washakie Formation	youngest
Overfilled	*(climate dries, system changes to aeolian dominated)*	middle Portland Formation	Ilhas Group	Upper LaClede Bed	
Balanced-fill		lower Portland Formation	Maracangalha Formation	Lower LaClede Bed	< Complete accommodation cycle >
Underfilled	Rock Point Member	*(climate too wet for development)*		Wilkins Peak Member	
Balanced-fill	Owl Rock Member	East Berlin Formation	Candeias Formation	Rife Bed	
Overfilled	Monitor Butte Member	Shuttle Meadow Formation	lower Salvador Formation	Luman Tongue–Scheggs Bed	
Fluvial	Shinarump Member	New Haven Formation	Agua Grande	Wasatch Formation	oldest
Reference:	Dubiel (1994)	Gierlowski-Kordesch and Rust (1994)	Figueiredo et al. (1994)	Roehler (1993)	

most likely in terms of rates of potential accommodation and sediment and water supply (Fig. 18). This shows that continental depositional systems possess two fundamental bifurcations, or breakpoints, in system behavior between the types of systems or strange attractors—perennially open hydrologies equating to fluvial systems and perennially closed hydrologies that favor only aeolian systems and playas. Lakes, with intermittently open and closed hydrologies, are most likely to form between these two bifurcation points. The concept of bifurcation can be seen as separating the continental sedimentary realm into three main types of systems that each have fundamentally different behaviors and sets of controls. Fluvial systems are open, dynamical systems with throughput of energy, sediment, and water. They provide only temporary storage of transport load and medium; they respond mainly to discharge rates, erodibility, and gradients, mostly the sediment + water variable (e.g., Leopold et al., 1964; Bull, 1991). Lakes are selectively open dynamical systems, intermittently bypassing energy and water, but permanently storing most sediment. They respond to rates of both sediment + water supply and potential accommodation subequally. Internally drained systems are permanently closed with no throughput of energy, water, or sediment. They receive minimal overland input of sediment; in the absence of overland water flow, other transport processes dominate, such as aeolian and slope failure. Along with direct precipitation and evaporation, they respond mainly to local land slope and relief, which are mostly potential accommodation changes (e.g., Blair and McPherson, 1994). This approach, therefore, can provide a unified framework for understanding the dynamical basis of continental deposits and the interrelations among and evolution of fluvial, lacustrine, and aeolian/playa strata.

These considerations of strange attractors and bifurcations—along with the empirical observations of the common occurrence of fractal relations and indications of scale invariance and self-similarity over four orders of magnitude in the modern lake data—-suggest it is appropriate and potentially very useful to treat modern and ancient lakes as nonlinear dynamical systems.

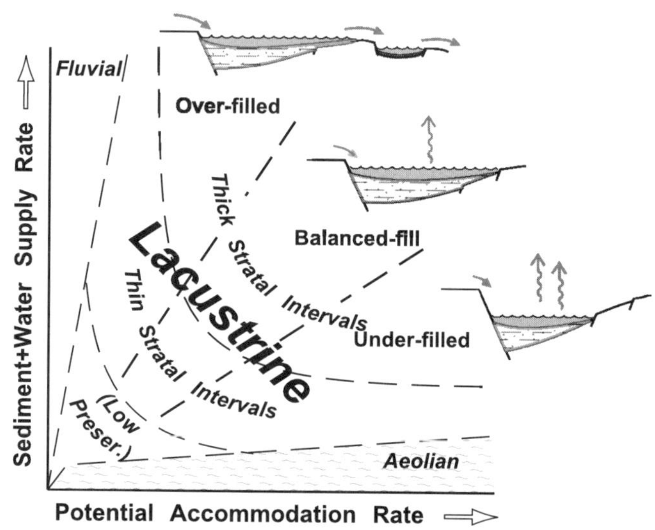

Figure 18. Lake-basin-type phase diagram, illustrating stability fields of major continental depositional systems in potential-accommodation–sediment + water supply space (after Carroll and Bohacs, 1999). The lack of significant intermediate cases between lake-basin types indicates that lake systems are almost intransitive in behavior and represent self-organized criticalities. It also suggests that lake-basin types might be considered as strange attractors.

The nonlinear approach is supported by well-documented cases and analyses of catastrophic shifts between alternative stable states in modern lake ecosystems (Scheffer et al., 1997; Carpenter and Pace, 1999). The most dramatic and well-studied case is the sudden loss of water transparency and lake vegetation that occurs in shallow lakes as a consequence of human-induced eutrophication (Scheffer et al., 1993; Jeppesen et al., 1999). Here, the pristine state of clear water and abundant submerged vegetation is altered abruptly by algal blooms fertilized by increased nutrient input. However, this change occurs only above a critical threshold in nutrient concentration when algal production exceeds the consuming capacity of endemic organisms, and the lake waters shift rapidly from clear to turbid, causing the submerged vegetation to largely disappear. The rapid disappearance of the Aral Sea and large fluctuations of the Caspian Sea, Lake Chad, and maritime Antarctic lakes are other clear examples of this nonlinear behavior (Mohler et al., 1995; Quayle et al., 2002).

The lake phase diagram highlights the position of the balanced-fill lake basin type, whose laterally extensive strata typically record brackish, meromictic conditions at intermediate rates of potential accommodation relative to sediment + water supply. This agrees with our observations of the most likely conditions for large lakes in the ancient. Balanced-fill conditions appear to represent an optimum hydrologic history for large lakes: closed long enough to pond abundant waters, but not too long to allow too much water to evaporate.

In summary, the size distributions suggest a nonlinear character of lake systems. Further examination reveals other signs of nonlinear behavior in lake systems: different responses to similar input depending on antecedent lake conditions, sensitive dependence on initial conditions of lake origin, and scale invariance. All of this indicates that the formation of large lakes is not deterministically controlled, but depends on convergences of causes: lake size is contingent on proper combinations of controls and not on unique factors of climate or tectonics alone.

INSIGHTS FOR INTERPRETATION, PREDICTION, AND FURTHER WORK

This is all very interesting, of course, but what does it teach us that we didn't know before? The insights and work of the nonlinear dynamics community (e.g., Middleton, 1991; Turcotte, 1997; Rodíguez-Iturbe and Rinaldo, 1997; Scheffer et al., 2001) provide potentially useful tools and approaches for interpreting lacustrine behavior and an appropriate path towards quantitative modeling. Their approach and results from analogous dynamical systems should help us appreciate what is and is not possible to extract from records of lake-systems behavior—what aspects of the system might be fruitful to pursue and which we can never know from the stratigraphic record. The world is fundamentally nonlinear, and our investigations, models, and conclusions must take that into account.

One possible application of this approach might be in estimating the range of areas covered by ancient lake strata in a basin, analogous to estimating sizes of oil fields or ore bodies (e.g., Turcotte, 1997). Different periods of lake expansion and contraction are commonly mapped as distinct members of a formation (e.g., Green River Formation, Fig. 13, Table 3). The distribution of estimated lake sizes of individual members can be compared to the expected distribution shown in Figure 3 as a check for reliability. The same distribution could be used to interpolate the sizes of incompletely sampled lake members, for instance, in the subsurface. This approach also allows determination of the parent-population characteristics from sampling truncated by seismic resolution or outcrop limitations (following the approach of Molz and Boman [1993] and Crovelli and Barton [1995]) and provides another, potentially higher resolution approach to estimating volumes of buried organic carbon, for example.

The nonlinear approach also indicates that, in the search for proximate causes of large lakes, it is not as straightforward as supposing "tectonics provides the hole in the ground and climate supplies the fill." Their complex interactions force us to obtain external data independent of stratal geometries and lithofacies to attribute a particular response or stratal character to climate or tectonics, for example, isotopic data about water input, direct dating of fault movements, or paleobotanical analysis of upland vegetation (Talbot and Kelts, 1989; Gawthorpe et al., 1994; Wilf, 2000; Cross et al., 2001; Wolfe et al., 2001; Rhodes et al., 2002).

For example, we may wish to investigate controls of the two fundamental state variables for lake behavior: potential accommodation and sediment + water supply. It is tempting to equate these with the more traditional controls of tectonics and climate, but a closer examination through the nonlinear approach reveals how intricately interwoven and non-independent these "controls" are. Potential accommodation is certainly a function of tectonics, but it has three distinct components: First, basin-floor subsidence is a function of tectonics and sediment load—but sediment load is strongly influenced by climate and geomorphology (e.g., Garner, 1957; Wilson, 1973; Fuller et al., 1998; Leeder et al., 1998). Sill movement is also influenced by tectonics, but also by erosion and stream piracy; thus, once again, it is affected by climate and geomorphology. Finally, basin shape is mostly controlled by tectonic evolution (e.g., Rosendahl et al., 1986; Withjack et al., 1995; Gawthorpe and Leeder, 2000), but also by inherited accommodation, which is a function of sediment supply over time. Similarly, sediment + water supply is not a simple function of climate, for climate itself is nonlinear (e.g., Lorenz, 1963), and there is a strongly nonlinear relation of water supply to sediment supply due to the nature of sediment transport, strong memory in the watershed system, and significant hysteresis and sensitivity to the direction of change (e.g., Garner, 1957; Wilson, 1973; Fuller et al., 1998; Leeder et al., 1998). For instance, a very dry climate may have abundant mechanically weathered sediment, but insufficient precipitation to provide persistent transport. Thus, it would yield little clastic sediment. Sufficient precipitation to provide persistent transport will also support abundant vegetation (in post-Devonian time) that acts as baffles and traps, and so it will also yield little clastic sediment. Most sediment yield to a lake tends to occur during changes between persistently wet and dry conditions, and the direction of change makes a difference: increasing precipitation from a very dry climate tends to yield abundant, mechanically weathered clastic sediments until pervasive plant growth occurs, whereas decreasing precipitation from a very wet climate will not yield abundant coarse clastic sediments until the system crosses the threshold for a significant decrease in vegetation and a change from the dominantly chemical weathering of the wet phase (e.g., Einsele and Hinderer, 1998).

CONCLUSIONS

Ultimately, lake size and character are functions of both current and inherited conditions. Lake-system responses and their stratigraphic record can be several steps removed from obvious causes. Nonlinear theory indicates that deconvolving climate and tectonic signals using stratal geometries and lithologies is not just hard in a practical sense, but theoretically impossible (e.g., Lorenz, 1963; Bergé et al., 1984; Middleton, 1991; Shaw, 1991). One must delve more deeply into aspects of the lake system that are directly sensitive to the climate or tectonic parameters sought, e.g., isotopes, organic matter evolution, and upland vegetation.

To understand how a lake system will react to and record a particular change in climate, one must have some insight into the preexisting state of the lake (e.g., Fig. 18). Apparent quasi-periodic variations in stratal character may be related to climate changes, but care must be taken to ensure that the lake system did not cross a critical threshold or change lake-basin type within the interval under consideration for deconvolution.

From this analysis, we see that lake systems are difficult to predict in detail, as they are nonlinear systems that are exponentially sensitive to boundary conditions. We see that lake size and character is a complex function of four main state variables: (1) basin-floor depth, (2) sill height, (3) water supply, and (4) sediment supply. Most importantly, modern and ancient examples demonstrate that a variety of combinations of factors can yield large lakes and that bigness is an accidental and not an essential attribute of lake systems (to use the precise and well-established terminology of Aristotle). In such systems, slightly different input can result in widely different results, and different inputs can result in generally similar results: the issues of divergence and convergence of causes and effects with which geomorphologists regularly wrestle (e.g., Schiedeggar, 1991). Hence, one cannot automatically assume that a significant change in lake size or character is due to large changes in forcing functions. It indicates that one must be extremely cautious when assigning extreme size to extreme causes or when interpreting quasi-periodic stratal changes in terms of periodic forcing functions (e.g., Lorenz, 1963; Olsen et al., 1978; Renaut and Tiercelin, 1994; Olsen and Kent, 1996; Cole, 1998; Prokopenko et al., 2002).

ACKNOWLEDGMENTS

We thank our many fellow fans of lakes for generously sharing their insights and data: Ken Stanley, George Grabowski Jr., Nilo de Azambuja, Filho, Terry Blair, Paul Bucheim, Lluis Cabrera, Andy Cohen, Beth Gierlowski-Kordesch, C.E. Herdendorf, Tom Johnson, Kerry Kelts, Feng-Chi Lin, Paul Olsen, David Reynolds, Gao Ruiqi, Chris A. Scholz, Peter Schwans, Randy Steinen, and J.-J. Tiercelin. Jie Huang helped with our foray into the nonlinear world of fractals. We especially thank Steve Colman and Phil Jewell for their thoughtful and thorough reviews that improved our manuscript immensely, and Margie Chan and Alan Archer for inviting us to the original symposium and for their editorial assistance and patience. We appreciate the continued support of the management of ExxonMobil Upstream Research Company and their permission to publish this work.

REFERENCES CITED

Bak, P.C., Tang, C., and Wiesenfeld, K., 1987, Self-organized criticality: An explanation of 1/*f* noise: Physics Review Letters, v. 59, p. 381–385.

Bak, P.C., Tang, C., and Wiesenfeld, K., 1988, Self-organized criticality: Physics Reviews A, v. 38, no. 1, p. 364–374.

Baker, S., 1866, The Albert Nyanza, great basin of the Nile: New York, Dover edition 1964, 235 p.

Barron, E.J., 1990, Climate and lacustrine petroleum source prediction, *in* Katz, B.J., ed., Lacustrine basin exploration; case studies and modern analogs: American Association of Petroluem Geologists Memoir 50, p. 1–8.

Benzi, R., Paladin, G., Parisi, G., and Vulpiani, A., 1984, On the multifractal nature of fully developed turbulence and chaotic systems: Journal of Physics A, v. 17, p. 3521–3531.

Bergé, P., Pomeau, Y., and Vidal, Ch., 1984, Order within chaos: New York, Wiley, and Paris, Hermann, 183 p.

Blair, T.C., and MacPherson, J.G., 1994, Historical adjustments by Walker River to lake-level fall over a tectonically tilted half-graben floor, Walker Lake Basin, Nevada: Sedimentary Geology, v. 92, p. 7–16.

Bohacs, K.M., 1999, Sequence stratigraphy of lake basins; unraveling the influence of climate and tectonics: American Association of Petroleum Geologists Bulletin, v. 83, p. 1878.

Bohacs, K.M., Carroll, A.R., and Neal, J.E., 2000a, Lessons from large lake systems—Thresholds, nonlinearity, and strange attractors: Geological Society of America Abstracts with Programs, v. 32, no. 7, p. A-312.

Bohacs, K.M., Carroll, A.R., Neal, J.E., and Mankiewicz, P.J., 2000b, Lake-basin type, source potential, and hydrocarbon character: An integrated sequence-stratigraphic–geochemical framework, *in* Gierlowski-Kordesch, E., and Kelts, K., eds., Lake basins through space and time: American Association of Petroleum Geologists Studies in Geology v. 46, p. 3–37.

Bohacs, K.M., Neal, J.E., Carroll, A.R., Reynolds. D.J., 2000c, Lakes are not small oceans! Sequence stratigraphy in lacustrine basins [abs.]: American Association of Petroleum Geologists Annual Meeting Expanded Abstracts, v. 9, p. 14.

Bohacs, K.M., Grabowski, G.J., Jr, Carroll, A.R., Miskell-Gerhardt, K.J. 2001, Non-marine sequence stratigraphy field workshop: Guidebook to Wyoming & Colorado outcrops: Rocky Mountain Section, SEPM Field Trip Guidebook, Denver Colorado, 102 p.

Bradley, W.H., 1925, A contribution to the origin of oil shale: Bulletin of the American Association of Petroleum Geologists, v. 9, p. 247–262.

Bull, W.B., 1991, Geomorphic responses to climatic change: London, Oxford University Press, 312 p.

Buoniconti, M.R., 2001, The role of lake level dynamics in lacustrine stratal architecture and sequence development: Evidence for the Songwe delta, Lake Malawi, east Africa [abs.]: American Association of Petroleum Geologists Abstracts with Program, v. 10, p. A-25.

Burton, R., 1860, The lake region of central Africa: London, Tylston and Edwards, 298 p.

Carpenter, S.R., and Pace, M.L., 1999, Dystrophy and eutrophy in lake ecosystems: Implications of fluctuating inputs: Oikos, v. 78, p. 3–14.

Carroll, A.R., 1998, Upper Permian lacustrine organic facies evolution, southern Junggar Basin, NW China: Organic Geochemistry, v. 28, p. 649–667.

Carroll, A.R., and Bohacs, K.M., 1999, Stratigraphic classification of ancient lakes: Balancing tectonic and climatic controls: Geology, v. 27, p. 99–102.

Carroll, A.R., and Bohacs, K.M., 2001, Lake-type controls on petroleum source rock potential in nonmarine basins: American Association of Petroluem Geologists Bulletin, v. 85, p. 1033–1053.

Carroll A.R., Graham, S.A., Hendrix, M.S., Ying, D., and Zhou, D., 1995, Late Paleozoic tectonic amalgamation of northwestern China: Sedimentary record of the northern Tarim, northwestern Turpan, and southern Junggar Basins: Geological Society of America Bulletin, v. 107, p. 571–594.

Cole, G.A., 1979, Textbook of limnology (2nd ed.): St. Louis, C.V. Mosby, 426 p.

Cole, R.D., 1998, Possible Milankovitch cycles in the lower Parachute Creek Member of Green River Formation (Eocene), north-central Piceance Creek basin, Colorado; an analysis, *in* Pitman, J.K., and Carroll, A.R., eds., Modern & ancient lake systems; new problems and perspectives: Utah Geological Association Publication, v. 26, p. 233–259.

Contreras, J., and Scholz, C.A., 2001, Evolution of stratigraphic sequences in multisegmented continental rift basins; comparison of computer models with the basins of the East African rift system: American Association of Petroleum Geologists Bulletin, v. 85, p. 1565–1581.

Cross, S.L., Baker, P.B., Seltzer, G.O., Fritz, S.C., and Dunbar, R.B., 2001, Late Quaternary climate and hydrology of tropical South America inferred from an isotopic and chemical model of Lake Titicaca, Bolivia and Peru: Quaternary Research, v. 56, p. 1–9.

Crovelli, R.A., and Barton, C.C., 1995, Fractals and the Pareto distribution applied to petroleum accumulation-size distributions, *in* Barton, C.C., and LaPointe, P.R., eds., Fractals in petroleum geology and Earth processes: New York, Plenum Press, p. 59–72.

Dubiel, R.F., 1994, Lacustrine deposits of the Upper Triassic Chinle Formation, Colorado Plateau, USA, *in* Gierlowski-Kordesch, E., and K. Kelts, eds., Global geological record of lake basins, volume 1: Cambridge, Cambridge University Press, p. 151–154.

Ebinger, C.J., Rosendahl, B.R., and Reynolds, D.J., 1987, Tectonic model of the Malawi Rift, Africa, *in* Ben Avraham Zvi, ed., Sedimentary basins within the Dead Sea Rift and other rift zones: Tectonophysics, v. 141, p. 215–235.

Einsele, G., and Hinderer, M., 1998, Quantifying denudation and sediment accumulation systems (open and closed lakes): Basic concepts and first results: Palaeogeography, Palaeoclimatology, Palaeoecology, v. 140, p. 7–21.

Feder, J., 1988, Fractals: New York, Plenum, 492 p.

Figueiredo, A.M.G., Braga, J.A.E., Zabalaga, H.M.C., Oliveira, J.J., Aguiar, G.A., Silva, O.B., Mato, L.F., Daniel, L.M.F., Magnavita, L.P., and Bruhn, C.H.L., 1994, Recôncavo Basin: A prolific intracontinental rift basin, *in* Landon, S.M., ed., Interior rift basins: American Association of Petroleum Geologists Memoir 59, p. 157–203.

Frisch, U., and Parisi, G., 1985, Fully developed turbulence and intermittency, *in* Ghil, M., Benzi, R., and Parisi, G., eds., Turbulence and predictability in geophysical fluid dynamics: New York, North Holland, p. 84–92.

Fuller, I.C., Macklin, M.G., Lewin, J., Passmore, D.G., and Wintle, A.G., 1998, River response to high-frequency climate oscillations in southern Europe over the past 200 k.y.: Geology, v. 26, p. 275–278.

Garner, H.F., 1957, Stratigraphic-sedimentary significance of contemporary climate and relief in four regions of the Andes Mountains: Geological Society of America Bulletin, v. 70, p. 1327–1368.

Gawthorpe, R.L., Fraser, A.J., and Collier, R.E.L.I., 1994, Sequence stratigraphy in active extensional basins: Implications for the interpretation of ancient basin fills: Marine and Petroleum Geology, v. 11, p. 642–658.

Gawthorpe, R.L., and Leeder, M.R., 2000, Tectono-sedimentary evolution of active extensional basins, *in* Guptta, S., and Cowie, P.A., eds., Processes and controls in the stratigraphic development of extensional basins: Basin Research, v. 12, p. 195–218.

Gierlowski-Kordesch, E., and Kelts, K., 1994, Global geological record of lake basins, volume 1: Cambridge, Cambridge University Press, 427 p.

Gierlowski-Kordesch, E., and Kelts, K., editors, 2000, Lake basins through space and time: American Association of Petroluem Geologists Studies in Geology 46, p. 3–37.

Gierlowski-Kordesch, E., and Rust, B.R., 1994, The Jurassic East Berlin Formation, Hartford Basin, Newark Supergroup (Connecticut and Massachusetts); a saline lake-playa-alluvial plain system, *in* Renaut, R.W., and Last, W.M., eds., Sedimentology and geochemistry of modern and ancient saline lakes: Society for Sedimentary Geology (SEPM) Special Publication, v. 50, p. 249–265.

Glieck, J., 1987, Chaos: Making a new science: New York, Viking Penguin, 352 p.

Goodings, D., and Middleton, G., 1991, Fractals and fractal dimension, *in* Middleton, G., Nonlinear dynamics, chaos and fractals: Geological Association of Canada Short Course Notes, v. 9, p. 13–22.

Guadalupi, G., and Shugaar, A., 2001, Latitude zero: Tales of the equator: New York, Carroll and Graf, 258 p.

Herdendorf, C.E., 1984, Inventory of the morphometric and limnologic characteristics of the large lakes of the world: The Ohio State University Sea Grant Program, Technical Bulletin OHSU-TB-17, 78 p.

Horsfield, B., Curry, D.J., Bohacs, K.M., Carroll, A.R., Littke, R., Mann, U., Radke, M., Schaefer, R.G., Isaksen, G.H., Schenk, H.J., Witte, E.G., and Rullkötter, J., 1995, Organic geochemistry of freshwater and alkaline lacustrine environments, Green River Formation, Wyoming: Organic Geochemistry, v. 22, p. 415–450.

Huang, J., and Turcotte, D.L., 1989, Fractal mapping of digitized images: Application to the topography of Arizona and comparisons with synthetic images: Journal of Geophysical Research, v. 94, p. 7491–7495.

Hutchinson, G.E., 1957, A treatise on limnology, vol. 1, Geography, physics and chemistry: New York, John Wiley & Sons, 1015 p.

Jeppesen, E., Sondergaard, M., and Kristensen, P., 1999, Lake and catchment management in Denmark: Hydrobiologia, v. 396, p. 419–432.

Johnson, T.C., Scholz, C.A., Talbot, M.R., Kelts, K., Ricketts, R.D., Ngobi, G., Beuning, K., Ssemmanda, I., and McGill, J.W., 1996, Late Pleistocene desiccation of Lake Victoria and rapid evolution of cichlid fishes: Science, v. 273, p. 1091–1093.

Katz, B.J., editor, 1990, Lacustrine basin exploration; case studies and modern analogs: American Association of Petroluem Geologists Memoir 50, 340 p.

Lambiase, J.J., 1990, A model for tectonic control of lacustrine stratigraphic sequences in continental rift basins, *in* Katz, B.J., Lacustrine basin exploration; case studies and modern analogues: American Association of Petroleum Geologists Memoir 50, p. 265–276.

Lavallée, D., Lovejoy, S., Schertzer, D., and Ladoy, P., 1993, Nonlinear variability of landscape topography: Multifractal analysis and simulation, *in* Lam, N.S., and De Cola, L., eds., Fractals in geography: Englewood Cliffs, N.J., Prentice Hall, p. 158–192.

Leeder, M.R., Harris, T., and Kirkby, M.J., 1998, Sediment supply and climate change: Implications for basin stratigraphy: Basin Research, v. 10, p. 7–18.

Leopold, L.B., Wolman, M.G., and Miller, J.P., 1964, Fluvial processes in geomorphology: San Francisco, Freeman, 365 p.

Lorenz, E.N., 1963, Deterministic nonperiodic flow: Journal of Atmospheric Sciences, v. 20, p. 130–141.

Mandelbrot, B.B., 1982, The fractal geometry of nature: San Francisco, Freeman, 468 p.

Middleton, G.V., 1991, Nonlinear dynamics, chaos, and fractals with applications to geological systems: Geological Association of Canada Short Course Notes, v. 9, 235 p.

Mohler, R.R., Wilkinson, M.J., and Giardino, J.R., 1995, The extreme reduction of Lake Chad surface area: Input to paleoclimatic reconstructions [abs.]: Geological Society of America Abstracts with Programs, v. 27, no. 6, p. A-265.

Molz, F.J, and Boman, G.K., 1993, A fractal-based stochastic interpolation scheme in subsurface hydrology: Water Resources Research, v. 29, p. 3769–3774.

Olsen, P.E., and Kent, D.V., 1996, Milankovitch climate forcing in the tropics of Pangaea during the Late Triassic: Palaeogeography, Palaeoclimatology, Palaeoecology, v. 122, p. 1–26.

Olsen, P.E., Remington, C.L., Cornet, B., and Thomson, S., 1978, Cyclic change in Late Triassic lacustrine communities: Science, v. 201, p. 729–733.

Prokopenko, A.A., Karabanov, E.B., and Williams, D.F., 2002, Age of long sediment cores from Lake Baikal: Nature, v. 415, p. 976.

Quayle, W.C., Peck, L.S., Peat, H., Ellis-Evans, J.C., and Harrigan, P.R., 2002, Extreme responses to climate change in Antarctic lakes: Science, v. 295, p. 645.

Renaut, R.W., and Tiercelin, J.J., 1994, Lake Bogoria, Kenya Rift valley; a sedimentological overview, *in* Renaut, R.W., and Last, W.M., eds., Sedimentology and geochemistry of modern and ancient saline lakes: SEPM (Society for Sedimentary Geology) Special Publication 50, p. 101–123.

Rhodes, M.K., Carroll, A.R., Pietras, J.T., Beard, B.L, and Johnson, C.M., 2002, Strontium isotope record of paleohydrology and continental weathering, Eocene Green River Formation, Wyoming: Geology, v. 30, p. 167–170.

Rodríguez-Iturbe, I., and Rinaldo, A., 1997, Fractal river basins: Cambridge, Cambridge University Press, 547 p.

Roehler, H.W., 1993, Correlation, composition, areal distribution, and thickness of Eocene stratigraphic units, greater Green River basin, Wyoming, Utah, and Colorado: U.S. Geological Survey Professional Paper 1506E, 49 p.

Rosendahl, B.R., 1987, Architecture of continental rifts with special reference to East Africa: Annual Review of Earth and Planetary Sciences, v. 15, p. 445–503.

Rosendahl, B.R., Reynolds, D.J., Lorber, P.M., Burgess, C.F., McGill, J., Scott, D., Lambiase, J.J., and Derksen, S.J., 1986, Structural expressions of rifting; lessons from Lake Tanganyika, Africa, *in* Frostick, L.E., Renaut, R.W., and Tiercelin, J.J., eds, Sedimentation in the African rifts: London, Geological Society Special Publication, v. 25, p. 29–43.

Ruelle, D., 1980, Strange attractors: Mathematics Intelligencer, v. 2, p. 1267.

Ruelle, D., and Takens, F., 1971, On the nature of turbulence: Communications in Mathematical Physics, v. 23, p. 343.

Scheffer, M., Carpenter, S., Foley, J.A., Folke, C., and Walker, B., 2001, Catastrophic shifts in ecosystems: Nature, v. 413, p. 591–596.

Scheffer, M., Hosper, S.H., Meijer, M.L., and Moss, B., 1993, Alternative equilibria in shallow lakes: Trends in Ecological Evolution, v. 8, p. 275–279.

Scheffer, M., Rinaldi, S., Gragnani, A., Mur, L.R., and Van Nes, E.H., 1997, On the dominance of filamentous cyanobacteria in shallow, turbid lakes: Ecology, v. 78, p. 272–282.

Schiedeggar, A.E., 1991, Theoretical geomorphology, 3rd edition: Berlin, Springer Verlag, 434 p.

Schlische, R.W., and Olsen, P.E., 1990, Quantitative filling model for continental extensional basins with applications to Early Mesozoic rifts of eastern North America: Journal of Geology, v. 98, p. 135–155.

Schwans, P., Bohacs, K.M., Hohensee, V.K., Lin, F.C., Gao, R.Q., Qiu, J.Y., Sun, S.H., and Li, F.J., 1997, Nonmarine sequence stratigraphy; impact on source, reservoir, and seal distribution in a rift setting; Cretaceous Songliao Basin, China: American Association of Petroleum Geologists Bulletin, v. 81, p. 1411.

Shaw, H.R., 1991, Magmatic phenomenology as nonlinear dynamics: Anthology of some relevant experiments and portraits, *in* Middleton, G., Nonlinear dynamics, chaos and fractals: Geological Association of Canada Short Course Notes, v. 9, p. 97–150.

Sladen, C.P., 1994, Key elements during the search for hydrocarbons in lake systems, *in* Gierlowski-Kordesch, E., and Kelts, K., Global geological record of lake basins, vol. 1: Cambridge, Cambridge University Press, p. 3–17.

Smale, S., 1967, Differentiable dynamical systems: Bulletin of the American Mathematical Society, v. 13, p. 747.

Smith, M.A., 1990, Lacustrine oil shale in the geological record, *in* Katz, B.J., ed., Lacustrine basin exploration; case studies and modern analogs: American Association of Petroleum Geologists Memoir 50, p. 43–60.

Speke, J.H., 1863, Journal of the Discovery of the Source of the Nile: London, Tylston and Edwards, 255 p.

Stanley, H.E., and Meakin, P., 1988, Multifractal phenomena in physics and chemistry: Nature, v. 335, p. 405–409.

Strecker, U., Steidtmann, J.R., and Smithson, S.B., 1999, A conceptual tectonostratigraphic model for seismic facies migrations in a fluvio-lacustrine extensional basin: American Association of Petroleum Geologists Bulletin, v. 83, p. 43–61.

Street-Perrott, F.A., and Harrison, S.P., 1984, Temporal variations in lake levels since 30,000 yr B.P.—An index of the global hydrological cycle: American Geophysical Union Geophysical Monograph 29, p. 118–129.

Street-Perrott, F.A., and Harrison, S.P., 1985, Lake levels and climate reconstructions, *in* Hecht, A.D., ed., Paleoclimate analysis and modeling: New York, Wiley, p. 291–304.

Street-Perrott, F.A., and Roberts, N., 1983, Fluctuations in closed basin lakes as indicators of past atmospheric circulation patterns, *in* Street-Perrott, F.A., Beran, M., and Ratcliffe, R.A.S., eds., Variations in the global water budget: Dordrecht, Reidel, p. 331–345.

Sullivan, R., 1980, A stratigraphic evaluation of the Eocene rocks of southwestern Wyoming: Geological Survey of Wyoming Report of Investigations 20, 50 p.

Talbot, M.R., and Kelts, K., editors, 1989, The Phanerozoic record of lacustrine basins and their environmental signals: Palaeogeography, Palaeoclimatology, Palaeoecology, v. 70, p. 121–137.

Turcotte, D.L., 1997, Fractals and chaos in geology and geophysics, second edition: Cambridge, Cambridge University Press, 370 p.

Wilf, P., 2000, Late Paleocene–early Eocene climate changes in southwestern Wyoming: Paleobotanical analysis: Geological Society of America Bulletin, v. 112, p. 292–307.

Wilson, L., 1973, Variations in mean annual sediment yield as a function of mean annual precipitation: American Journal of Science, v. 273, p. 335–349.

Withjack, M.O., Olsen, P.E., and Schlische, R.W., 1995, Tectonic evolution of the Fundy rift basin, Canada; Evidence of extension and shortening during passive margin development: Tectonics, v. 14, p. 390–405.

Wolfe, B.B., Aravena, R., Abbott, M.B., Seltzer, G.O., and Gibson, J.J., 2001, Reconstruction of paleohydrology and paleohumidity from oxygen isotope records in the Bolivian Andes: Palaeogeography, Palaeoclimatology, Palaeoecology, v. 176, p. 177–192.

Manuscript Accepted by the Society January 22, 2003

Printed in the USA

Geological Society of America
Special Paper 370
2003

Organic carbon burial by large Permian lakes, northwest China

Alan R. Carroll
Marwan A. Wartes
Department of Geology and Geophysics, University of Wisconsin, 1215 W. Dayton Street, Madison, Wisconsin 53711, USA

ABSTRACT

Permian strata of the Junggar-Turpan-Hami Basins represent one of the thickest and most laterally extensive lacustrine deposits in the world, yet they are very poorly known outside of China. Deposition spanned approximately 30 m.y., from the Sakmarian through Changhsingian epochs. Continuous intervals of organic-rich lacustrine mudstone may exceed 1000 m, and the total thickness of lacustrine and associated nonmarine strata locally exceeds 4000 m. Early Permian basin subsidence coincided with regional normal faulting and associated volcanism, interpreted to result from extension or transtension of newly amalgamated accretionary crust. In contrast, relatively uniform regional subsidence occurred during the Late Permian, most likely due to flexure caused by renewed regional compression. The maximum expansion of Permian lakes during the Wordian to Capitanian postdates any evidence for significant normal faulting or volcanism. Organic-rich mudstone facies cover an area at least 900 × 300 km, indicating that at their maximum, the Permian lakes were comparable in size to the Caspian Sea. An organic-rich profundal mudstone section in the south Junggar Basin has been ranked as the thickest and richest petroleum source rock interval in the world, with total organic carbon content averaging 4% and commonly exceeding 20%. Total Late Permian carbon burial is estimated at 10^{19} gC. Maximum organic carbon burial rates are estimated at 4×10^{12} gC/yr, equivalent to approximately 4–8% of estimated global carbon burial rates during this time. Permian lakes in northwestern China were broadly synchronous with other large lakes (or inland seas) in South America and Africa that also formed due to the amalgamation of Pangea. Collectively, these basins represent a large and heretofore unrecognized organic carbon sink that may have influenced atmospheric CO_2 concentrations.

Keywords: Junggar, Turpan, Hami, lacustrine, climate, stratigraphy.

INTRODUCTION

Organic carbon burial in modern lakes may rival the magnitude of burial in many nearshore marine settings (Dean and Gorham, 1998; Einsele et al., 2001), suggesting that these deposits should be considered an important carbon sink. Dean and Gorham (1998) concluded that Holocene carbon burial in small lakes accounts for the majority of this carbon burial, based on estimates derived from post-glacial lakes in Minnesota. A smaller, but still significant rate of carbon burial was estimated to occur in large lakes (>5000 km^2). Carbon burial by lakes and inland seas was estimated to represent about one-third of the total global burial rate.

Collectively, nonmarine depositional environments, including reservoirs, account for approximately 75% of present day organic carbon burial. Einsele et al. (2001) pointed to rapid accumulation rates and high preservation factors in lakes as causal factors and noted that burial of organic carbon increases with increased drainage basin area and change to wetter and warmer climate.

Ancient lacustrine stata deposited in tectonically subsided basins have long been recognized to include thick intervals of highly organic-rich mudstone, which is often economically

Carroll, A.R., and Wartes, M.A., 2003, Organic carbon burial by large Permian lakes, northwest China, *in* Chan, M.A., and Archer, A.W., eds., Extreme depositional environments: Mega end members in geologic time: Boulder, Colorado, Geological Society of America Special Paper 370, p. 91–104.

exploited as oil shale. Three of the world's five thickest and richest petroleum source rock intervals were deposited in lake basins (Demaison and Huizinga, 1991). Many of these deposits also cover very large areas. For example, the Neocomian-Barremian organic-rich lacustrine mudstone strata associated with the opening of the south Atlantic extend at least 3000 km along the African and South American continental margins across numerous basins (cf., Ojeda, 1982; Mello and Maxwell, 1990; Katz and Mello, 2000). In contrast to the deposits of relatively ephemeral Holocene post-glacial lakes, carbon buried by large tectonic lake basins is typically removed from the atmospheric reservoir for millions to hundreds of millions of years. However, carbon buried in tectonic lake basins has generally been ignored in global mass balance calculations. These deposits represent potentially important carbon sinks that may have helped to ameliorate past episodes of global warming.

The Permian Junggar-Turpan-Hami Basins in northwest China collectively represent one of the largest known Phanerozoic lake basins (Wartes et al., 2000, 2002) and include the world's thickest and richest petroleum source rock interval (Demaison and Huizinga, 1991). Organic-rich lacustrine mudstone is distributed over an area roughly equivalent to that of the modern Caspian Sea. These strata also span an important tectonic transition, from late Paleozoic continental amalgamation to recurrent intraplate collisional deformation (Allen and Windley, 1993; Allen et al., 1991, 1995; Carroll et al., 1990, 1992, 1995; Hendrix et al. 1992; Graham et al., 1993). This paper summarizes what is known about the origin of the Junggar-Turpan-Hami deposits and provides the first estimates of the magnitude and rate of carbon burial they represent.

Tectonic Setting of Northwest China

The Junggar, Turpan, and Hami Basins (Figs. 1 and 2) overlie oceanic materials that have been interpreted as part of the Paleozoic Altaid accretionary complex (Şengör et al., 1993). Exposed basement lithologies consist chiefly of highly deformed Ordovician through Carboniferous volcanogenic turbidites, with local occurrences of mafic and ultramafic igneous rocks and chert (e.g., Coleman, 1989; Feng et al., 1989; Carroll et al., 1990, 1995). Precambrian rocks are limited to the central Tian Shan and areas to the south. Isotopic ($^{87}Sr/^{86}Sr$ and εNd) evidence from late Paleozoic granitic plutons that pierce basement lithologies confirms their oceanic affinity and reflect an increased influence of Precambrian continental crust to the south (Hopson et al., 1989, 1998; Wen, 1991). The nature of the substrate hidden beneath the Junggar Basin sedimentary strata is unknown, but it is inferred to be similar to the basement lithologies that crop out on all sides of the basin. Total crustal thicknesses in this region have been estimated at approximately 40 km (Lee, 1985), which locally includes 15 km or more of relatively undeformed Carboniferous through Cenozoic sedimentary rocks (Hendrix et al., 1992; Graham et al., 1993; Carroll et al., 1995). Uplift of the Bogdashan Range (Fig. 2) first occurrred in Late Triassic to Early Jurassic

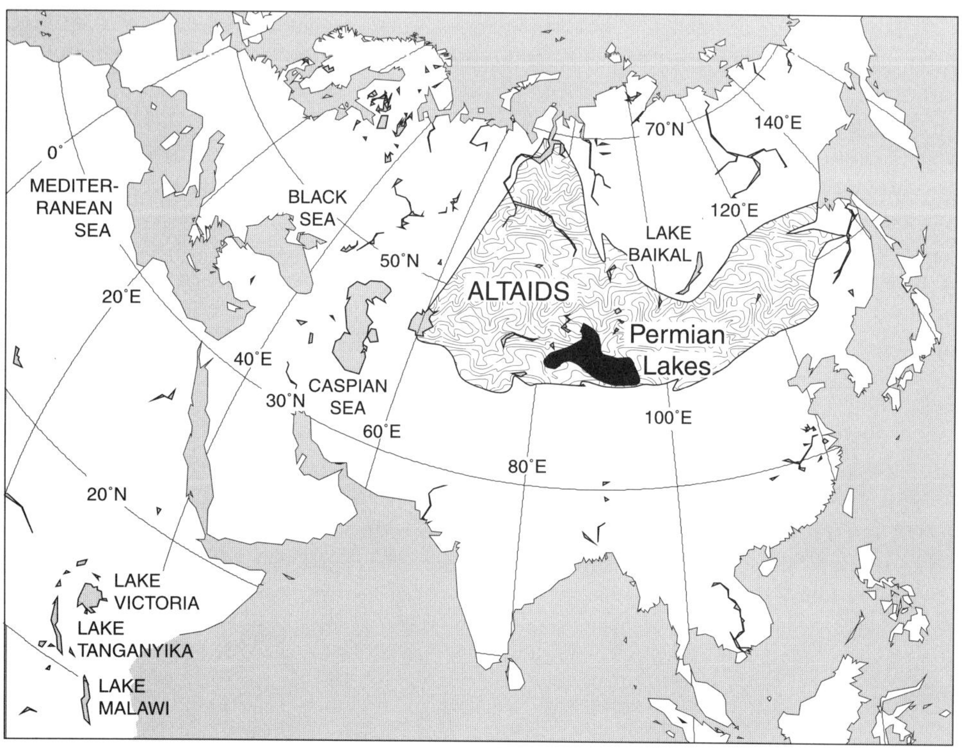

Figure 1. Location of the Junggar-Turpan-Hami Permian lake deposits and the Altaid orogenic complex (area of Altaids modified from Şengör et al., 1993).

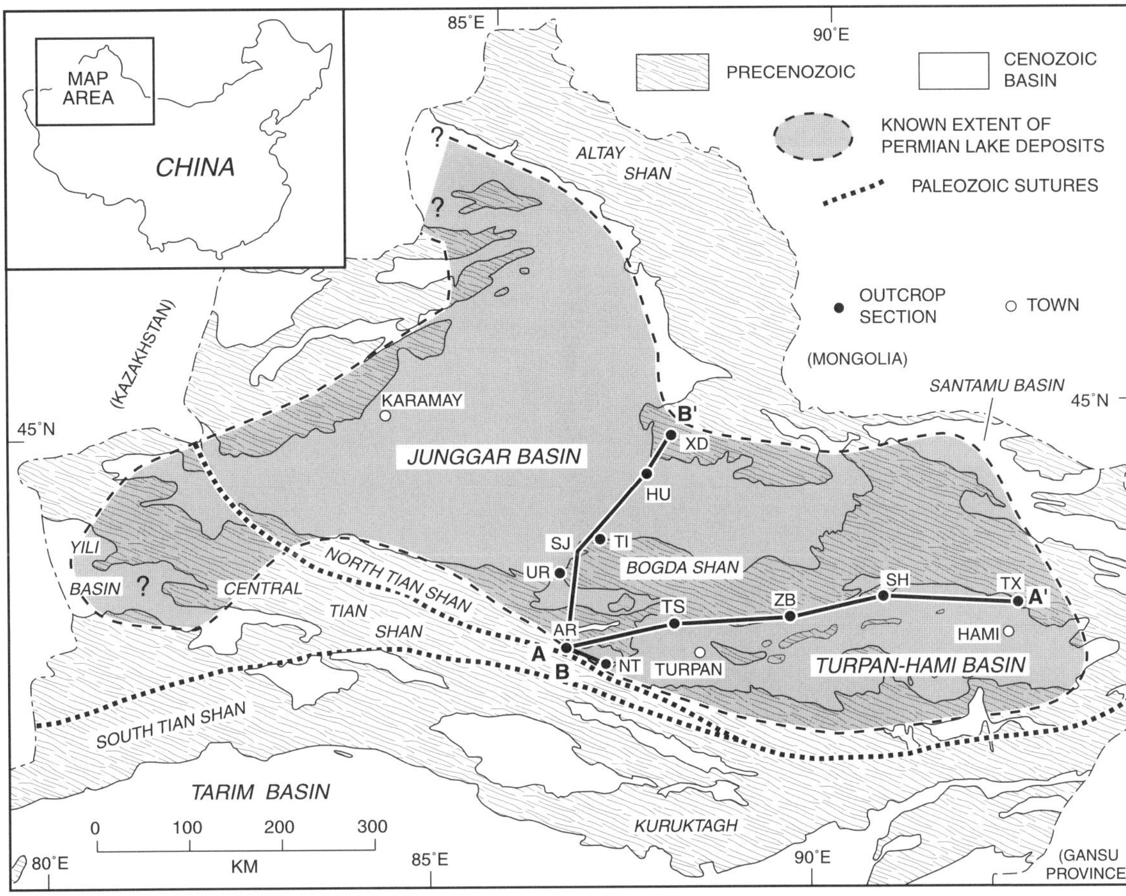

Figure 2. Location of Junggar and Turpan-Hami Basins, with maximum known extent of Upper Permian lake deposits (modified from Wartes et al., 2002). Field localities: AR—Aiweiergou, HU—Huoshaoshan, NT—North Tian Shan, SH—Shisanjifang, SJ—South Junngar Composite Section, TI—Tianchi, TS—Taoshuyuan, TX—Tian Shan Xiang, UR—Urumqi, XD—Xidagou, ZB—Zaobishan. A–A′ is line of section in Figure 3; B–B′ is line of section in Figure 4.

(Hendrix et al., 1992; Greene et al., 2001). Prior to that time, the Junggar, Turpan, and Hami Basins were united, although a partial drainage divide may have existed between Junggar and Turpan-Hami Basins (Wartes et al., 2002).

The unified Junggar-Turpan-Hami Basin occupies the site of a relict Early to Middle Carboniferous sea that filled with marine deep water through shelf facies following the extinction of arc magmatism in the Late Carboniferous (Carroll et al., 1990, 1995). The tectonic setting of this area prior to the Permian is ambiguous and may be interpreted as either a remnant ocean basin or backarc basin (Carroll et al., 1990; Windley et al., 1990; Allen et al., 1991) that became tectonically isolated from the sea. A relatively continuous shoaling-upward succession in excess of 2-km-thick records the Late Carboniferous through Early Permian retreat of marine waters from the southern Junggar Basin (Carroll et al., 1995). Overlying Permian nonmarine strata total over 4 km in thickness. A similar marine to nonmarine transition is recorded in southern Bogdashan exposures of the Turpan-Hami Basin, although these strata are generally thinner and change more abruptly.

Stratigraphy and Depositional Environments

Absolute chronostatigraphic control for nonmarine Permian facies in the Junggar-Turpan-Hami Basins is poor due to faunal and floral endemism. However, relative age assignments at approximately the stage level are possible through comparison of faunal, floral, and palynolomorph assemblages reported from different units (Wartes et al., 2000; Wartes et al., 2002). Radioisotopic dating of volcanic units is limited to the Early Permian, but the results are consistent with the established biostratigraphic framework (Wartes et al., 2000). These data permit regional correlations between Permian nonmarine stratigraphic units exposed in the southern Bogdashan, and between surface and subsurface lithologies within and adjacent to the Junggar Basin.

Lacustrine facies assemblages range from fluvial-lacustrine to evaporative, based on physical features and biomarker geochemistry, and have been previously described by Carroll et al. (1992), Carroll (1998), Wartes et al. (2000), Bohacs et al. (2000), and Carroll and Bohacs (2001). Outcrop exposures of Lower Permian lacustrine facies are restricted to the southern Bogdashan, and are represented by portions of the Zaobishan and Yierxitu Formations (Fig. 3). These fluvial-lacustrine to fluctuating profundal facies (cf., Carroll and Bohacs, 1999) are interbedded with mafic to intermediate volcanic rocks, and deposition was localized by Early Permian normal faulting. They typically lie unconformably on underlying Carboniferous marine facies. In contrast, Lower Permian facies of the Junggar Basin are dominantly marine and generally conformable with underlying strata (Fig. 4). The thickness of Lower Permian strata in the subsurface was determined using a combination of well logs and reflection seismic data, and facies similar to those in outcrop were verified by core examination (Greene et al., 2001).

Upper Permian facies in both basins are exclusively nonmarine and dominantly siliciclastic and record a wide range of lacustrine depositional environments. Average carbonate contents in the southern Junggar Basin range between 20% and 30%, with dolomite common in the most organic-rich facies (unpublished X-ray diffraction and X-ray fluorescence spectrometry data). Upper Permian strata overlie a major regional unconformity in southern Bogdashan exposures, but they appear to be conformable with Lower Permian nonmarine strata in the northwestern Bogdashan (Figs. 3 and 4). The Tarlong Formation in the southern Bogdashan and northern Tian Shan consists dominantly of fluctuating profundal facies, with mudstone, sandstone, and minor limestone cyclically interbedded on the scale of meters to

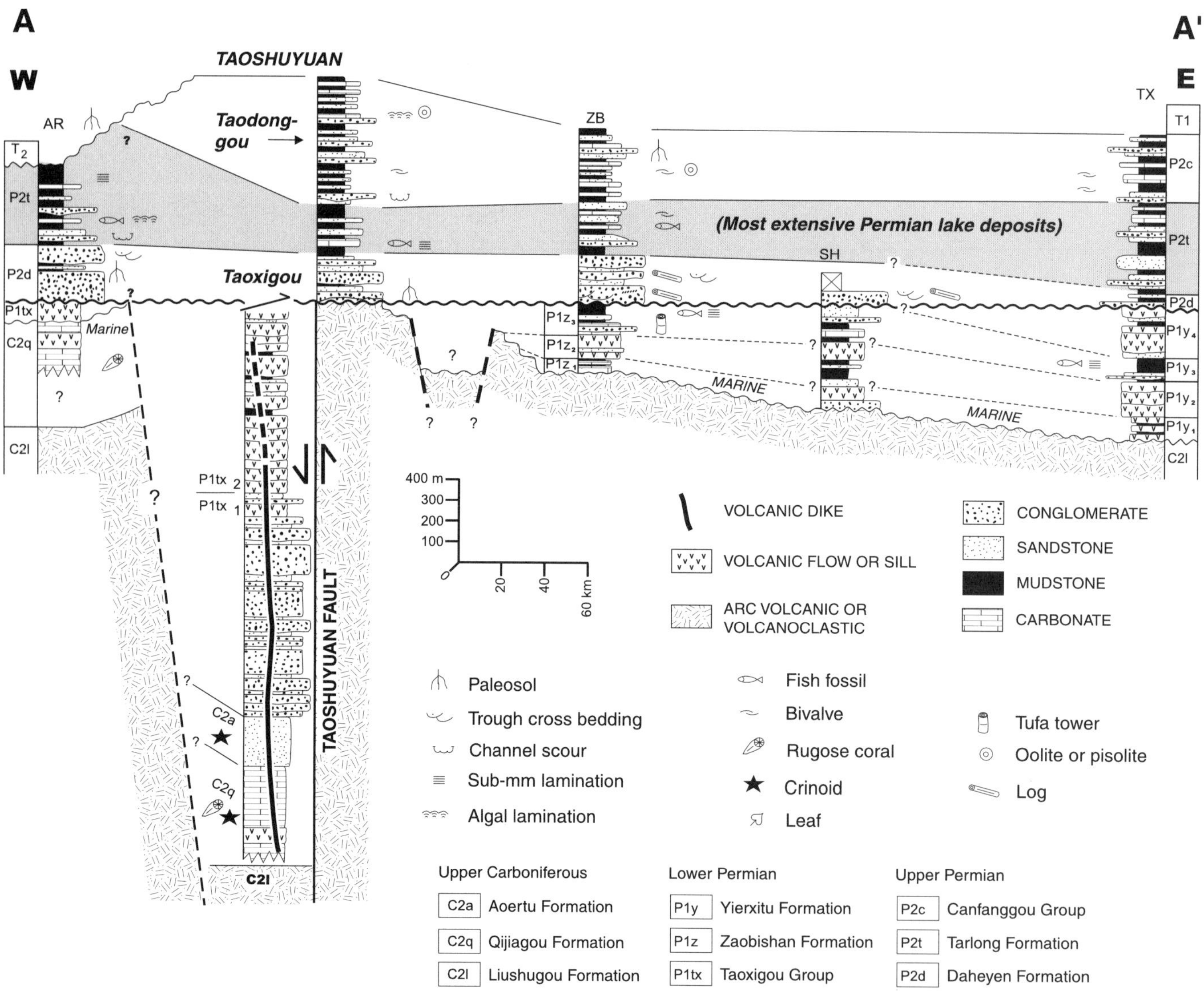

Figure 3. East-west stratigraphic cross section of Upper Carboniferous through Lower Triassic strata outcrops on southern flank of Bogdashan (modified from Wartes et al., 2002; see Fig. 2 for location). AR—Aiweiergou, SH—Shisanjifang, TX—Tian Shan Xiang, ZB—Zaobishan, T1—Lower Triassic, T2—Middle Triassic.

tens of meters (Figs. 5A and 5B). Laminated mudstone and well-preserved fish fossils attest to deposition under deep, low oxygen conditions (cf., Olsen 1986), mostly likely in a stratified lake, whereas mudcracks indicate periodic dessication. The Tarlong correlates with the Lucaogou and Hongyanchi Formations of the Junggar Basin (Fig. 4). These units record a gradual progression from evaporative (Fig. 6A) to fluvial-lacustrine facies (Fig. 6C) in exposures on the northwestern flank of the Bogdashan, based on physical evidence for desiccation and synaeresis (mudcracks) and biological marker compounds in mudrock extracts (Carroll, 1998). Fluctuating profundal facies of the Lucaogou Formation reach a thickness of approximately 1300 m (Figs. 6B and 7). The lower Lucaogou Formation is commonly dolomitic and contains possible synaeresis cracks (Fig. 5D), which suggests fluctuating salinity. The upper Lucaogou Formation includes at least 500 continuous meters of laminated mudstone with no evidence of subaerial exposure. Laminae thicknesses vary cyclically from ~0.1 to 2.0 mm over a vertical scale of several meters. Variations in laminae thickness and organic matter content were interpreted by Carroll (1998) to result from changes in the rate of clastic sediment supply in a deep, sediment-starved lake basin. Finally, the Cangfanggou Group records a gradation to fluvial-lacustrine facies in both basins (Figs. 3 and 4). Fluvial sandstone and lacustrine mudstone are interbedded on the scale of meters with no apparent cyclicity, and plant fossils, freshwater molluscs, and evidence for paleosols are common.

Permian Basin Evolution

After retreat of marine waters, Early Permian extension led to the creation of localized nonmarine depocenters that were controlled, in part, by normal faults. Normal faults have also been

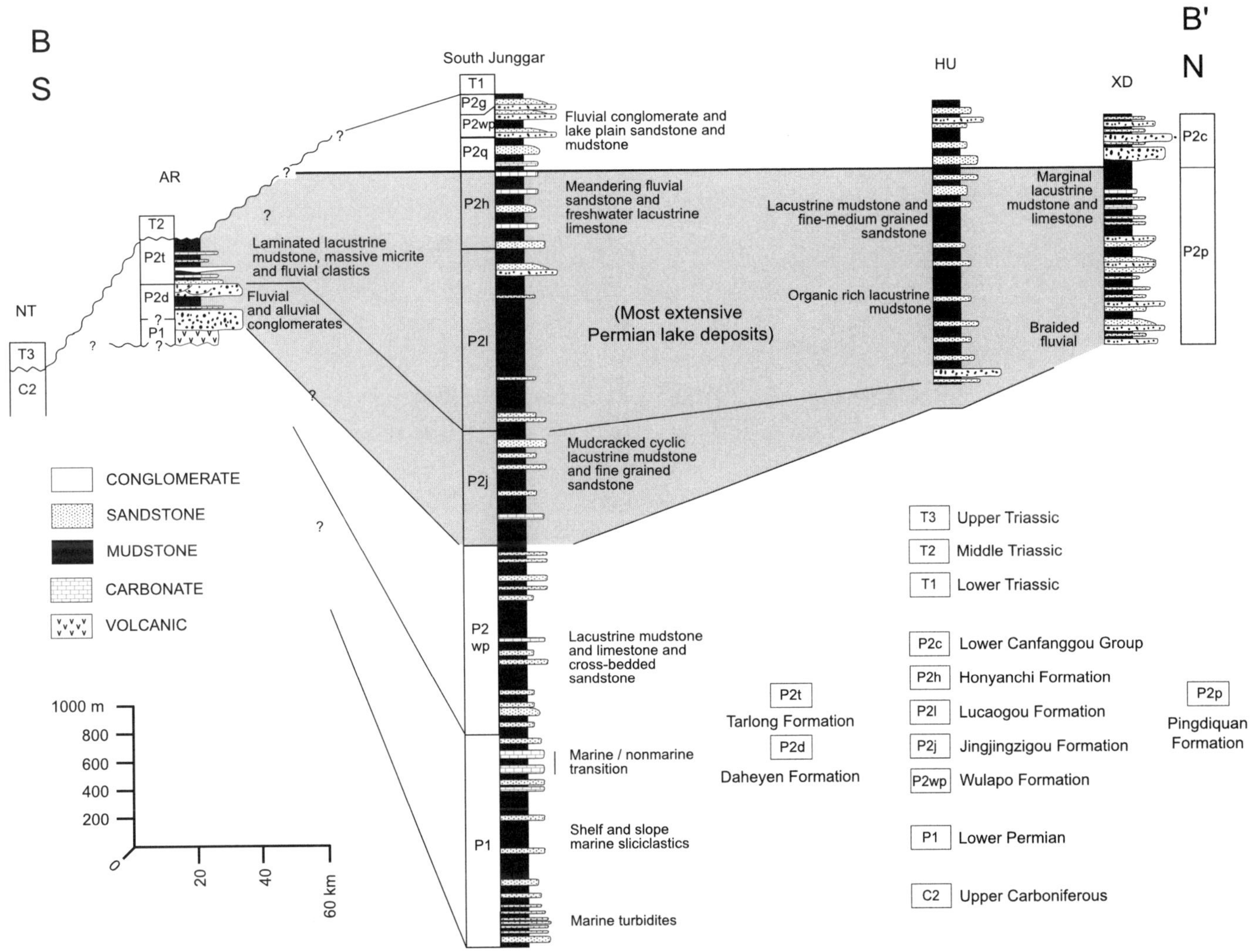

Figure 4. North-south stratigraphic cross section of Upper Carboniferous through Triassic outcrops (NT, AR, and South Junggar) and wells (HU, XD) (modified from Wartes et al., 2002; see Fig. 2 for location).

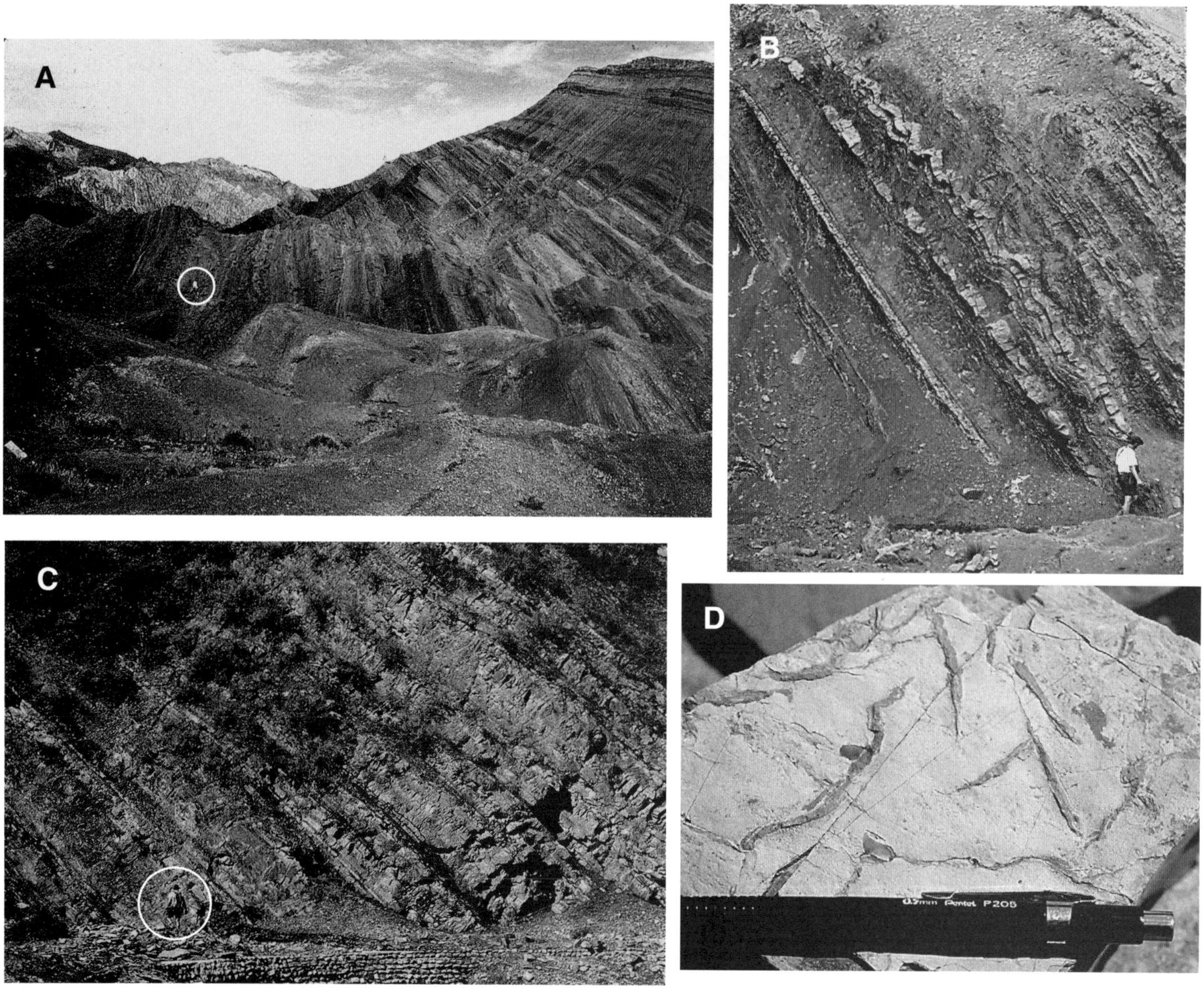

Figure 5. Outcrop photographs of Upper Permian lacustrine facies. A: Tarlong Formation mudstone with interbedded sandstone at Aierwaigou (see "AR" in Fig. 2 for location). Person circled for scale. B: Detail of Tarlong Formation at Aierwaigou (note slumped sandstone bed). C: Lucaogou Formation at Tianchi aqueduct excavation (see "SJ" in Fig. 2 for location). Person circled for scale. D: Detail of possible synaeresis cracks in Lucaogou Formation at Tianchi aqueduct excavation.

postulated to exist beneath the Junggar Basin (Bally et al., 1986; Liu, 1986; Peng and Zhang, 1989), but the precise timing of these structures is unconstrained. Allen et al. (1995) and Şengör and Natal'in (1996) hypothesized that these structures and the origin of the Alakol, Junggar, and Turpan Basins were all related to Late Permian-Early Triassic sinistral shear. However, field relationships near Taoshuyuan in the southern Bogdashan show that a major north-south trending normal fault there is actually Early Permian (Wartes et al., 2002; Figs. 2 and 3). This fault and associated volcanic rocks are roughly coeveal with Early Permian basaltic magmatism in the northwest Tarim Basin and to the east in the Beishan, suggesting a widespread episode of extension following the late Paleozoic amalgamation of this region. Carroll et al. (1995) and Wartes et al. (2002) interpreted this extension to have been dominantly east-west oriented, occurring during ongoing north-south compression (present orientation). However, regional shear remains a viable alternative hypothesis.

Early Permian paleogeography was characterized by a complex series of fault-related sub-basins containing mostly nonmarine fill in the Turpan-Hami area and a relict marine basin in the Junggar area (Fig. 8). The nature of the drainage divide between these areas is unclear, but it may be topography inherited from

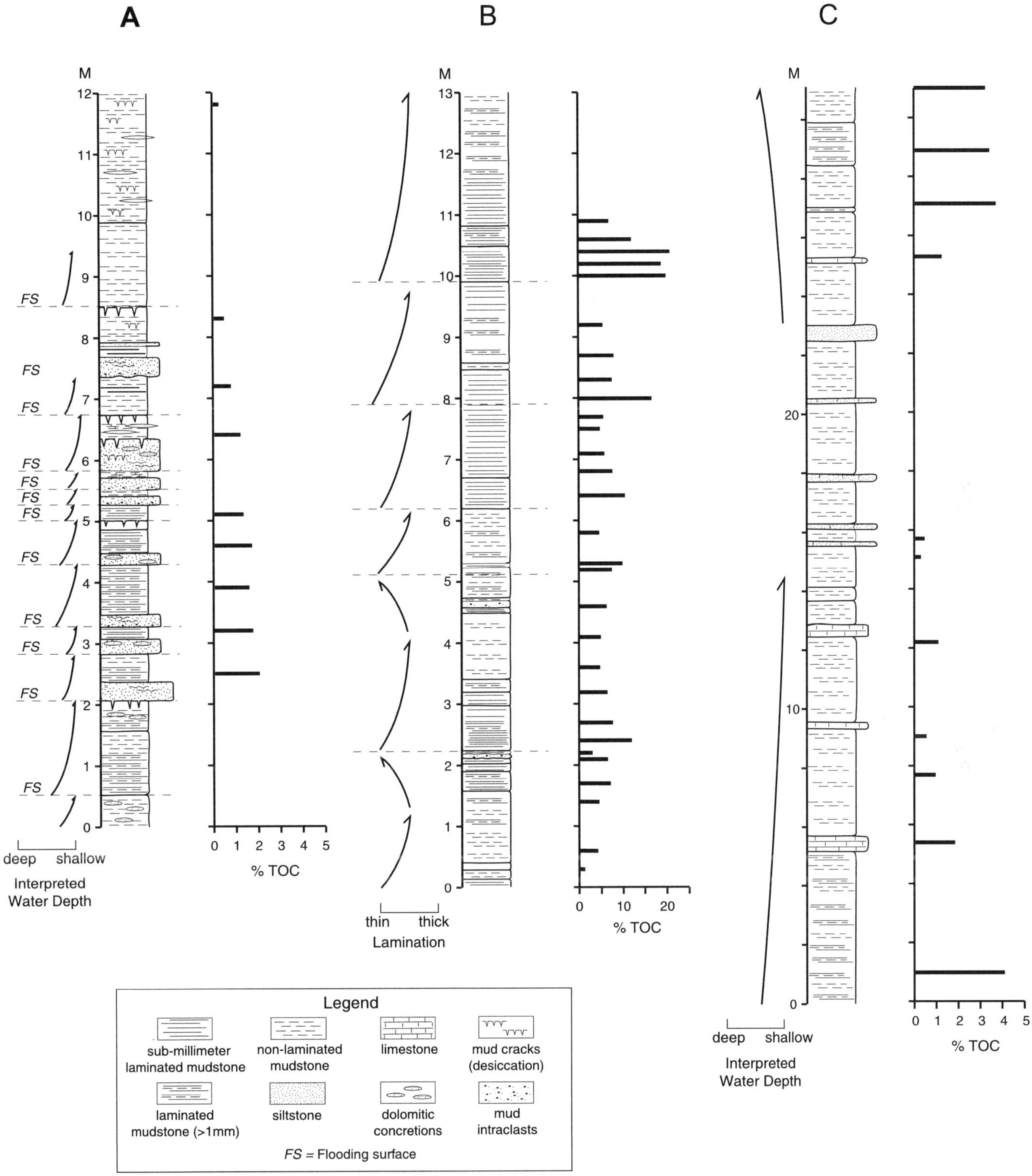

Figure 6. Representative detailed measured sections of Upper Permian lacustrine and associated alluvial facies exposed in western Bogdashan (modified from Carroll, 1998). Arrows indicate deepening and shallowing trends. A: Jingjingzigou Formation at Tianchi aqueduct excavation. B: Lucaogou Formation at Tianchi aqueduct excavation. C: Hongyanichi Formation at Urumqi. TOC—total organic content.

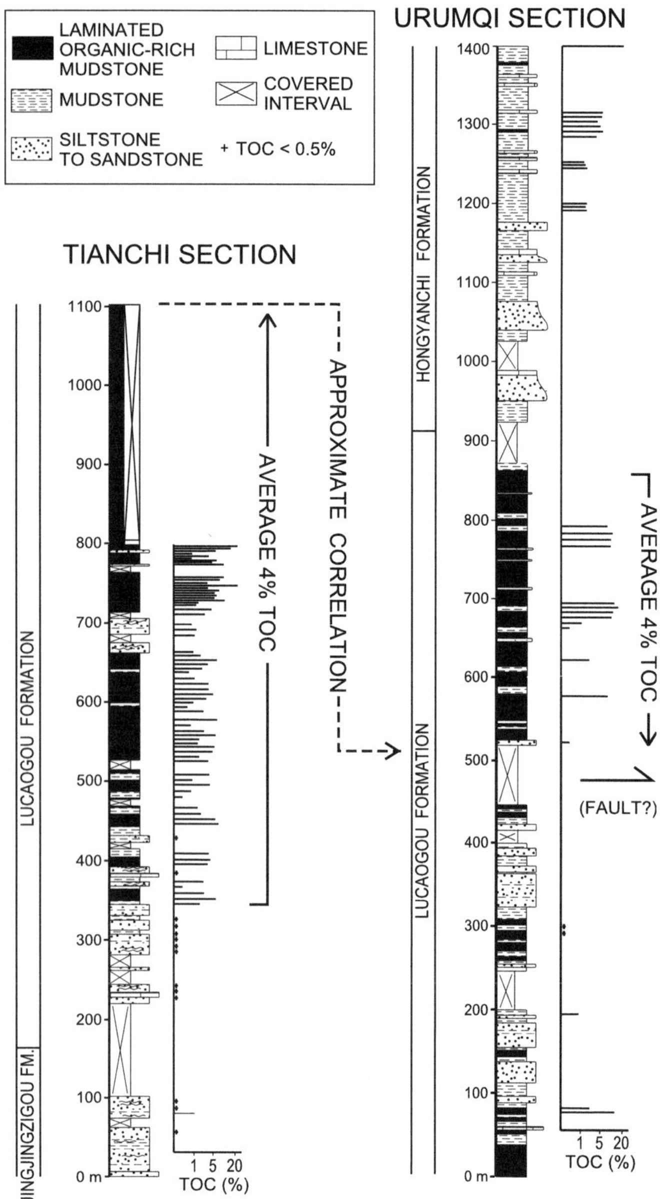

Figure 7. Master measured sections showing total organic carbon (%TOC) at Tianchi aqueduct excavation and at Urumqi (see Fig. 2 for location). Together these sections constitute south Junggar composite section, based on approximate correlation shown.

the late stages of a Carboniferous magmatic arc co-located with the Bogdashan (c.f., Coleman, 1989; Carroll et al., 1990; Windley et al., 1990; Allen et al., 1991). In contrast, late Permian subsidence was more uniformly distributed, resulting in widespread deposition of relatively continuous lacustrine facies and alluvial facies (Fig. 9). Geohistory analysis of outcrop sections from the northwestern Bogdashan indicates that subsidence rates increased markedly during the Late Permian, most likely due to lithospheric flexure in response to uplift of an ancestral Tian Shan Range (Watson et al., 1987; Graham et al., 1990; Carroll et al., 1990, 1995; Allen and Windley, 1993). Greene et al. (2001) and Wartes et al. (2002) proposed that the Turpan-Hami Basin occupied a wedge-top position, which would account for the thinner Upper Permian deposits preserved there and the development of a basal Upper Permian unconformity. In contrast, the southern Junggar Basin is interpreted as a flexural foredeep, which received over 4 km of Upper Permian nonmarine basin fill. The maximum extent of lakes occurred during deposition of the Lucaogou, Tarlong, and equivalent formations.

Organic Matter Burial

High enrichments of organic carbon in the Lucaogou Formation were documented by Graham et al. (1990), who reported values up to 34% total organic carbon and Type I kerogen. Vitrinite reflectance values for the same samples indicated that these rocks have reached the early stages of petroleum generation (Graham et al., 1993; Carroll et al., 1992; Carroll, 1998), so the original total organic carbon values may have been slightly higher prior to onset of oil generation and migration. Oils on the northwestern, northern, and northeast flanks of the Junggar Basin have biomarker distributions indicating derivation from Upper Permian lacustrine facies (Clayton et al., 1997), and at least two fields in the Turpan-Hami Basin also produce similar oils (Greene, 2000). Outcropping mudstone facies in the southern Bogdashan are relatively weathered and appear to have been exposed to higher levels of thermal maturation, but preliminary data suggest that they also had high original total organic carbon contents (Greene, 2000).

To determine the average total organic carbon of the Lucaogou Formation mudstone facies, which is also referred to as "oil shale," Carroll et al. (1992) conducted sampling at fixed, regular intervals from an aqueduct excavation in the northwestern Bogdashan (Fig. 7). They reported that total organic carbon values of 20% or more are common, and that total organic carbon values over a continuous 800-m interval average 4%. Our subsequent investigations have shown that similar but less well-exposed organic rich mudstone facies continue stratigraphically above the aqueduct excavation for at least another 500 m. The total thickness of the Lucaogou Formation averaging 4% total organic carbon at this locality is taken to be approximately 1300 m.

Subsurface mapping of the Junggar Basin indicates several depocenters where the Upper Permian mudstone reaches thicknesses of up to 2000 m (Fig. 10). The precise age of these deposits is uncertain, but their indicated thickness is broadly consistent with that of the measured outcrop exposures in the Bogdashan. Furthermore, biomarker distributions in oils produced from the northwestern and eastern flanks of the Junggar Basin correlate closely with bitumen extracts from the outcropping Lucaogou Formation (Carroll et al., 1992; Clayton et al., 1997; Carroll, 1998; Fig. 10). Oils with similar biomarker characteristics are also produced from two fields in the Turpan-Hami Basin (Greene, 2000), suggesting that correlative facies also occur

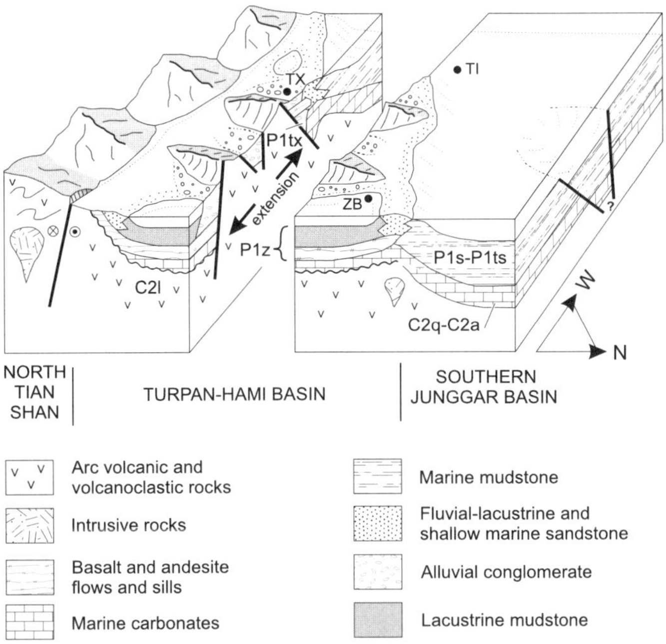

Figure 8. Schematic block diagram showing Early Permian evolution of Junggar and Turpan-Hami Basins (modified from Wartes et al., 2002). TI—Tianchi, TX—Tian Shan Xiang, ZB—Zaobishan. C2a—Aoertu Formation, C2l—Liushugou Formation, C2q—Qijiagou Formation, P1s—Lower Permian Shirenzigou Formation, P1ts—Lower Permian Taoshuyuan Formation, P1tx—Taoxigou Group, P1z—Zaobishan Formation. See Figures 3 and 4 for stratigraphy.

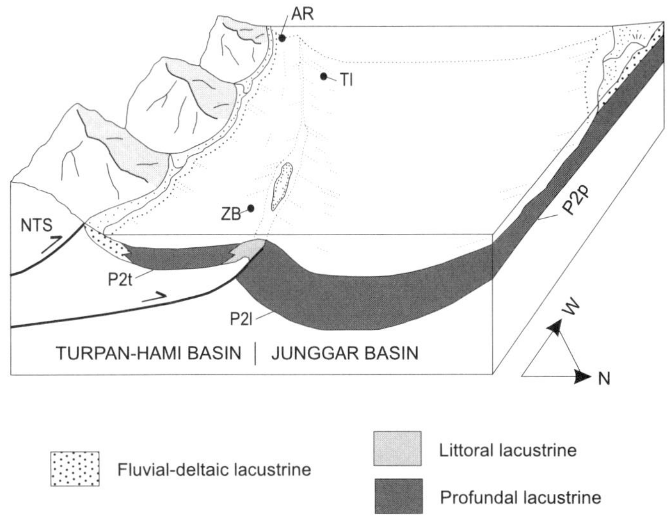

Figure 9. Schematic block diagram showing Late Permian (Wordian) evolution of Junggar and Turpan-Hami Basins (modified from Wartes et al., 2002). AR—Aiweiergou, TI—Tianchi. ZB—Zaobishan. See Figures 3 and 4 for stratigraphy. NTS—North Tian Shan, P2t—Tarlong Formation, P2l—Lucaogou Formation, P2p—Pingdiquan Formation.

south of the Bogdashan. Permian-sourced oils are generally absent in areas with thick accumulations of Mesozoic strata (the southern Junggar and much of the Turpan-Hami Basins), apparently due to overmaturation of Permian source facies during deep burial (Carroll et al., 1992).

The absolute limit for the distribution of the Lucaogou Formation and its equivalents is uncertain, due to very incomplete mapping of areas surrounding the Junggar-Turpan-Hami Basins. However, similar facies have been reported from an area that greatly exceeds the boundary of these basins. For example, intervals of organic-rich, Upper Permian mudstone ~100 m thick have been reported in the Yili Basin to the west of Junggar and in the Santamu Basin to the east (Wang, 1992; Cheng Keming, personal commun., 1998). The Yili Basin rocks are offset by Cenozoic dextral strike-slip faulting between central and northern Tian Shan (Tapponier and Molnar, 1979), suggesting that they may have originally been deposited adjacent to the Turpan-Hami Basin. The possibility that Permian lake deposits continued across the border into Mongolia is tantalizing but entirely untested. Finally, organic-rich Permian mudstone facies have been reported in the Irtysh-Zayshan area of Kazahkstan, adjacent to the northwest corner of the Junggar Basin (J. Degenstein, personal commun., 1993). In total, these known occurrences are distributed over an area roughly 900 × 300 km.

We estimate an average thickness for Lucaogou Formation and equivalent strata of 400 m for the area outlined in Figure 2. This assumption most likely leads to an underestimate of total mud rock volume due to the preferential exposure of these strata in areas with relatively less Permian subsidence and greater post-Permian uplift. It is likely that the thickest intervals of mud rock remain buried beneath thick Mesozoic and Cenozoic cover strata. The ultimate limit of mud rock deposition is unknown due to incomplete preservation and the rudimentary state of geological mapping, but it is inferred to exceed the outlined area. Based on the above assumptions, we estimate total carbon burial within this interval of approximately 10^{19} grams (Table 1).

DISCUSSION

Based on the thickness and average richness of organic-rich rocks in the south Junggar Basin composite section (Fig. 7), Demaison and Huizinga (1991) ranked the Junggar Basin as having the highest cumulative hydrocarbon potential of any petroleum-producing basin in the world (Table 2). This ranking is based on source potential index, a parameter that is based on average Rock Eval S_1 (thermally distilled hydrocarbons) plus S_2 (hydrocarbons derived from pyrolytic breakdown of kerogen; see Espitalié et al., 1977, and Peters, 1986, for further explanation of Rock Eval techniques). Source potential index values thus understate the total magnitude of carbon burial, since they do not

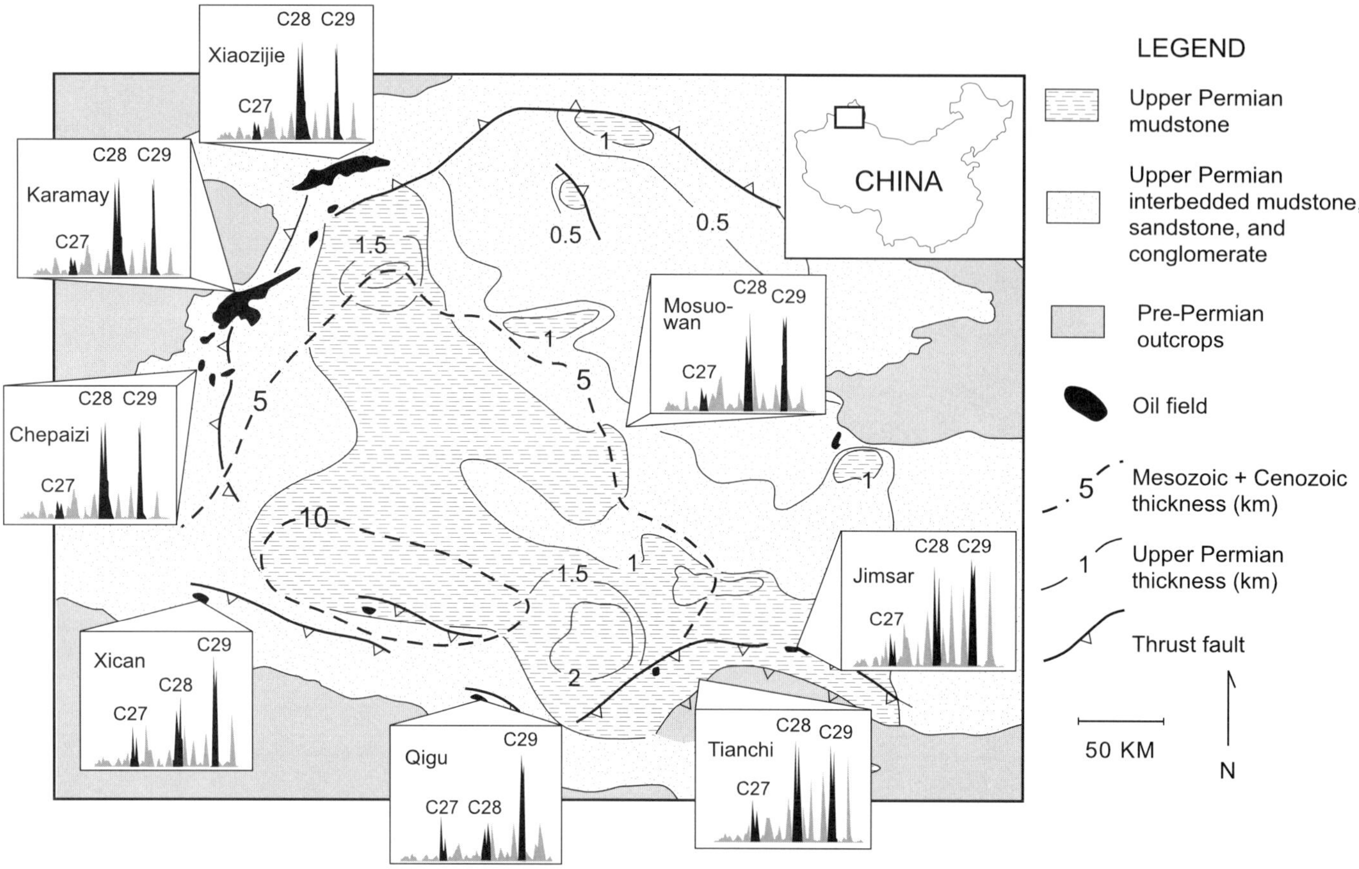

Figure 10. Distribution of Upper Permian lacustrine petroleum source facies in the Junggar Basin (modified from Carroll et al., 1992) and biomarker "fingerprints" of Permian-source oils. Inset diagrams show *m/z* 218 mass fragmentograms for alkane fractions of Junggar oils and for one rock extract from outcropping organic-rich mudstone ("Tianchi"). Each peak on fragmentograms corresponds to a different steroidal compound. Black peaks indicate specific groups of sterane isomers and highlight relative distribution of compounds with 27, 28, or 29 carbon atoms in each sample. Note that Permian-sourced oils from rock extract and from oils produced in western, central, and eastern Junggar Basin all have similar sterane distributions, characterized by low C_{27} and subequal C_{28} and C_{29} homologues. In contrast, Jurassic-sourced oils from southern Junggar Basin have low C_{27} and C_{28} but high C_{29}, typical of coaly source facies. Potential Upper Permian source facies in southern Junggar Basin are buried by up to 11 km of post-Permian strata and are thus overmature with respect to oil generation.

TABLE 1. JUNGGAR-TURPAN-HAMI ORGANIC CARBON CONTENT

Area of lake deposits	3×10^5 km^2
Average interval thickness	4×10^2 m
Duration of deposition	3×10^6 yr
Average total organic carbon	4%
Total carbon burial	1.2×10^{19} gC

account for carbon contained in refractory compounds. Nonetheless, Demaison and Huizinga's (1991) summary provides a useful means of comparing the relative importance of various intervals of organic-rich rock. It is interesting to note that three of the top five such intervals reported represent lacustrine basins that preserve Type I kerogen. A different ranking might result if the areal extent could be integrated with the SPI values to obtain an estimate of total hydrocarbon potential in each of these basins. In particular, Lower Cretaceous lacustrine strata associated with various south Atlantic margin rift basins, for example, the Bucomazi and Marnes Noires Formations in offshore west Africa and equivalent units in various Brazillian basins, could potentially exceed the total carbon burial of the Permian Junggar-Turpan-Hami Basin due to the large area of the Lower Cretaceous basins.

The lacustrine Green River Formation (Eocene of Colorado, Utah, and Wyoming) has long been recognized as the world's largest oil shale deposit, representing an estimated resource of 1×10^{12} bbls of oil in Colorado alone (Pitman et al., 1989) and 1.5×10^{12} bbls total (J. R. Dyni, personal commun., 2002). These estimates are determined primarily from Fischer assay measurements and thus cannot be directly compared with the total organic carbon measurements from the Junggar-Turpan-Hami Basins.

TABLE 2. SOURCE POTENTIAL INDEX (SPI) VALUES FOR SELECTED PETROLEUM SOURCE ROCK INTERVALS

Basin (Country)*	Source-rock sequence	Kerogen type	Average SPI (t HC/m^2)†
Junggar (China)	Upper Permian	I	65
Lower Congo (Cabinda)	Lower Cretaceous	I	46
Santa Barbara Channel (USA)	Miocene	II	39
San Joaquin (USA)	Miocene	II	38
Central Sumatra (Indonesia)	Miocene	I	34§
Eastern Venezuela (Venezuela)	Middle to Upper Cretaceous	II	27**
Offshore Santa Maria (USA)	Miocene	II	21
Middle Magdalena (Colombia)	Middle to Upper Cretaceous	II	16
North Sea (UK)	Upper Jurassic	II	15
Central Arabia (Saudi Arabia)	Upper Jurassic	II	14
Green River Formation (Eocene)‡			
Washakie (USA)	Laney Member	I	20
Washakie (U.S.)	Luman Tongue	I–III	7
Green River (U.S.)	Wilkins Peak Member	I	3

* Petroleum-producing basins (Demaison and Huizinga, 1991).
† SPI = $h(S_1+S_2)\rho/1000$, h = thickness, (S1+S2) = averaged Rock Eval S_1+S_2 in kg/t, ρ = rock density.
§ Estimated.
** Fold and thrust belt.
‡ Carroll and Bohacs (2001).

However, 1.5×10^{12} barrels equates to approximately 1.6×10^{17} g of oil, which is two orders of magnitude less than the estimated total carbon buried in the Junggar-Turpan-Hami Basins. This comparison fails to account for refractory organic carbon compounds in the Green River Formation, but the higher level of thermal maturity in the Junggar-Turpan-Hami deposits at least partially offsets the deficit. Source potential index values offer an alternative means of comparison. The Laney Member of the Green River Formation in Wyoming, which is the stratigraphic equivalent of the Parachute Creek Member in Colorado, has an SPI value that is about one third that of the Upper Permian in the Junggar-Turpan-Hami Basin (Table 2). Total source potential index for the Green River Formation may locally equal or exceed that of the south Junggar Basin, but the Green River Formation Lakes at their maximum extent covered only about one-third of the area of the Junggar-Turpan-Hami Lakes.

Although it is clear that Permian lacustrine mudstone in western China basins contains a very large mass of organic carbon, calculation of accurate carbon burial rates is difficult due to poor age control. One typical approach to this problem is to assume that mudstone laminations are annual in nature and therefore represent varves. If this is the case and sedimentation rates were reasonably constant, then the Lucaogou Formation and its equivalents were deposited over a period of approximately 3 m.y., based on counts of its finest-scale laminations. This duration is broadly consistent with other constraints on the age of the Lucaogou interval and is reasonable when compared with mudstone accumulation rates measured in other nonmarine basins (Table 3). However, we observed considerable variation in lamination thickness in outcrop, ranging between ~0.1 and 2.0 mm. Thicker laminae appear to be systematically associated with lower % total organic carbon, suggesting that considerable variation occurred in the flux of inorganic sediment (Carroll, 1998). If so, then the actual duration of the Lucaogou interval may be less than 3 m.y.

TABLE 3. COMPARATIVE NET SEDIMENT ACCUMULATION AND ORGANIC CARBON BURIAL RATES

Net sediment accumulation rates	μmyr^{-1}
Junggar-Turpan-Hami Basin	133
Green River Formation*	70–150
Pannonian basin†	96–205
Organic carbon burial rates	gCm^{-2}yr^{-1}
Junggar-Turpan-Hami Basin	13.3
Modern lakes (range)§	5–72
Modern lakes (average)§	13.5
Black Sea**	2–5
Total organic carbon burial	10^{12}gCyr^{-1}
Junggar-Turpan-Hami Basin	4
Total Permian‡	50–100
Modern lakes§	54

*Smith et al. (2001).
†Horváth and Pogácsás (1988).
§Dean and Gorham (1998).
**Arthur and Dean (1998).
‡Berner and Canfield (1989).

Based on the above assumptions, the calculated organic carbon burial rate per unit area for the Junggar-Turpan-Hami

Lake is comparable to the average rate for modern lakes and much lower than the maximum seen in modern eutrophic lakes (Dean and Gorham, 1998; Table 3). If valid, this calculated rate suggests that the preservation of large amounts of organic matter corresponded to a period of relatively modest primary aquatic productivity. Carroll (1998) argued on the basis of textural and geochemical evidence that high %total organic carbon values in the Lucaogou Formation resulted, in part, from low rates of inorganic sedimentation in a deep, chemically stratified lake with oxygen-depleted bottom water. This conclusion is also consistent with Dean and Gorham (1998) and others' observation that large modern lakes and inland seas typically bury organic carbon at much lower rates than do small eutrophic lakes, due to relatively lower nutrient concentrations in the surface waters of the larger bodies.

Despite the relatively modest rate of organic carbon burial per unit area beneath the Junggar-Turpan-Hami Lake, its extreme size resulted in a very large magnitude of total organic carbon burial per year. This mass has not been included in previous physical assessments of organic carbon burial. For example, Berner and Canfield (1989) stated that "The organic carbon content of non-coal-containing continental clastics is essentially zero." They estimated that global organic carbon burial rates decreased from approximately 100×10^{18} to 50×10^{18} gC/yr during the Permian. Depending on the actual timing of the Junggar-Turpan-Hami Lake, its organic carbon burial therefore represented 4–8% of global organic carbon burial. This amount is clearly not zero, suggesting that large lakes may be more important geological reservoirs of organic carbon than previously thought.

The occurrence of other large, inland water bodies in the southern hemisphere at about the same time helps to reinforce the hypothesis that such features were globally significant sinks for organic carbon. An inland water body that was apparently several times the size of the Junggar-Turpan-Hami Lake covered parts of Brazil and southern Africa during the latest Early Permian or early Late Permian, and its deposits were preserved in the Paraná and Great Karoo Basins (Oelofsen, 1987; França et al., 1995; Ziegler et al., 1996; Fig. 11). The precise size of this water body is uncertain because these deposits are incompletely exposed, but it may have been large enough to ameliorate the otherwise harsh climatic conditions that would have prevailed in the interior of Gondwana (Yemane, 1993; Kutzbach and Ziegler, 1993). The Irati and Whitehill Formations (South America and southern Africa) contain organic-rich mudstone facies similar to the Lucaogou Formation, commonly with greater than 10% total organic content (e.g., Correa da Silva and Cornford, 1985; Oelofsen, 1987). These somewhat controversial deposits have been variously interpreted to record either a fresh to brackish-water lake (e.g., Correa da Silva and Cornford, 1985; Faure and Cole, 1999) or else an inland sea (e.g., Teichert, 1974; Mello et al., 1993; Visser, 1994; Goldberg, 2001). The precise age relationship of these units to the Lucaogou Formation is unknown. In contrast to the Lucaogou Formation, the oil shale facies of the Irati and Whitehill Formations are relatively thin (meters to tens of meters).

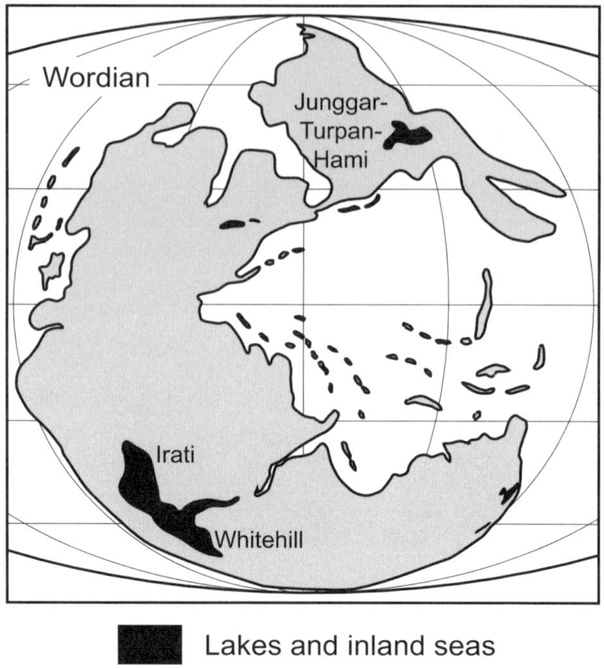

Figure 11. Plate tectonic reconstruction of Pangea during Late Permian (Wordian), indicating approximate position of Junggar-Turpan-Hami Lakes and of the lake or inland sea that deposited the Irati-Whitehill Formations in Brazil and southern Africa (modified from Ziegler et al., 1996).

The occurrence of large lakes as discussed above is directly related to processes of continental convergence and orogenesis by the entrapment of thinned continental or oceanic crust within collisional zones and by the development of internal drainages behind rising mountain ranges. Late Cenozoic examples of this process may be found in the southern Caspian and Pannonian Basins; both are former marine realms that were trapped within the developing Alpine orogenic zone (Zonenshain and Le Pichon, 1986; Mattick et al., 1988; Geary et al., 1989). The Black Sea and parts of the Mediterranean will likely suffer the same fate in the future. Significant organic matter burial is common in such basins, but the mechanisms of organic matter preservation appear to be complex. In some cases, the burial rates of organic carbon may reach a peak during transition between marine and lacustrine conditions. For example, Arthur and Dean (1998) noted that organic carbon-rich sapropel in the Black Sea was deposited during the initial incursion of marine waters through the Bosporus during the Early Holocene. These authors suggested that marine spillover into the previous freshwater lake resulted in vertical mixing of nutrients and that subsequent salinity stratification helped to maintain bottom-water anoxia. However, in the case of the Junggar-Turpan-Hami deposits, no evidence for marine incursions has been reported, either because such incursions did not occur or else because these vast deposits have been inadequately investigated. Regardless of the causes of organic matter preservation, lakes and inland seas within convergent zones are capable of burying large total quantities of carbon due to frequent, high rates of tectonic subsidence. In contrast to the much smaller eutrophic lakes that are often geomorphic in origin, large tectonic lakes can permanently isolate organic carbon from the atmosphere over time periods lasting up to hundreds of millions of years.

ACKNOWLEDGMENTS

We thank K. Cheng, J. Degenstein, T. Hu, S. A. Graham, T. J. Green, and Y. Liang for helpful discussions of parts of this study. Financial support has been provided by the Donors of the Petroleum Research Fund of the American Chemical Society, Conoco, Texaco, the Graduate School of the University of Wisconsin, and the Stanford-China Industrial Affiliates. We are grateful to Charles Oviatt and Walter Dean for constructive reviews.

REFERENCES CITED

Allen, M.B., and Windley, B.F., 1993, Evolution of the Turfan Basin, Chinese central Asia: Tectonics, v. 12, p. 889–896.

Allen, M.B., Windley, B.F., Zhang, C., Zhao, Z.Y., and Wang, G.R., 1991, Basin evolution within and adjacent to the Tien Shan range, NW China: Journal of the Geological Society, London, v. 148, p. 369–378.

Allen, M.B., Şengör, A.M.C., and Natal'in, B.A., 1995, Junggar, Turfan, and Alakol basins as Late Permian to Early Triassic extensional structures in a sinistral shear zone in the Altaid orogenic collage, Central Asia: Journal of the Geological Society, London, v. 152, p. 327–338.

Arthur, M.A., and Dean, W.E., 1998, Organic-matter production and preservation and evolution of anoxia in the Holocene Black Sea: Paleoceanography, v. 13, p. 395–411.

Bally, A.W., Chou, I.M., Clayton, R., Eugster, H.P., Kidwell, S., Meckel, L.D., Ryder, R.T., Watts, A.B., and Wilson, A.A., 1986, Notes on sedimentary basins in China—Report of the American Sedimentary Basins Delegation to the People's Republic of China: United States Geological Survey Open-File Report 86-327, 108 p.

Bohacs, K.M., Carroll, A.R., Neal, J.E., and Mankiewicz, P.J., 2000, Lake-basin type, source potential, and hydrocarbon character: An integrated sequence-stratigraphic geochemical framework, *in* Gierlowski-Kordesch, E.H., and Kelts, K., eds., Lake basins through space and time: American Association of Petroleum Geologists Studies in Geology 46, p. 3–33.

Berner, R.A., and Canfield, D.E., 1989, A new model for atmospheric oxygen over Phanerozoic time: American Journal of Science, v. 289, p. 333–361.

Carroll, A.R., and Bohacs, K.M., 2001, Lake type control on hydrocarbon source potential in nonmarine basins: American Association of Petroleum Geologists Bulletin, v. 85, p. 1033–1053.

Carroll, A.R., 1998, Upper Permian lacustrine organic facies evolution, southern Junggar Basin, NW China: Organic Geochemistry, v. 28, p. 649–667.

Carroll, A.R., and Bohacs, K.M., 1999, Stratigraphic classification of ancient lakes: Balancing tectonic and climatic controls: Geology, v. 27, p. 99–102.

Carroll, A.R., Liang, Y., Graham, S.A., Xiao, X., Hendrix, M.S., Chu, J., and McKnight, C.L., 1990, Junggar Basin, northwest China: Trapped late Paleozoic ocean: Tectonophysics, v. 181, p. 1–14.

Carroll, A.R., Brassell, S.C., and Graham, S.A., 1992, Upper Permian lacustrine oil shale of the southern Junggar Basin, northwest China: American Association of Petroleum Geologists Bulletin, v. 76, p. 1874–1902.

Carroll, A.R., Graham, S.A., and Hendrix, M.S., 1995, Late Paleozoic tectonic amalgamation of northwestern China: Sedimentary record of the northern Tarim, northwestern Turpan, and southern Junggar Basins: Geological Society of America Bulletin, v. 107, p. 571–594.

Clayton J.L., Yang, J., King, J.D., Lillis, P.G., and Warden, A., 1997, Geochemistry of oils from the Junggar Basin, northwest China. American Association of Petroleum Geologists Bulletin, v. 81, p. 1926–1944.

Coleman, R.G., 1989, Continental growth of northwest China: Tectonics, v. 8, p. 621–635.

Correa da Silva, Z.C., and Cornford, C., 1985, The kerogen type, depositional environment and maturity, of the Irati Shale, Upper Permian of Paraná Basin, southern Brazil: Organic Geochemistry, v. 8, p. 399–411.

Dean, W.E., and Gorham, E., 1998, Magnitude and significance of carbon burial in lakes, reservoirs, and peatlands: Geology, v. 26, p. 535–538.

Demaison, G., and B.J. Huizinga, 1991, Genetic classification of petroleum systems: American Association of Petroleum Geologists Bulletin, v. 75, p. 1626–1643.

Einsele, G., Yan, J., and Hinderer, M., 2001, Atmospheric carbon burial in modern lake basins and its significance for the global carbon budget: Global and Planetary Change, v. 30, p. 167–195.

Espitalié, J., Laporte, J.L., Madec, M., Marquis, F., Lepat, P., Paulet, J., and Boutefeu, A., 1977, Méthode rapide de caractérisation des roches mères, de leur potentiel petrolier et leur degré d'évolution: Review Institut Français du Pétrole, v. 32, p. 23–42.

Faure, K., and Cole, D., 1999, Geochemical evidence for lacustrine microbial blooms in the vast Permian Main Karoo, Paraná, Falkland Islands and Huab Basins of southwestern Gondwana: Palaeogeography, Palaeoclimatology, Palaeoecology, v. 152, p. 189–213.

Feng, Y., Coleman, R.G., Tilton, G., and Xiao, X., 1989, Tectonic evolution of the west Junggar region, Xinjiang, China: Tectonics, v. 8, p. 729–752.

França, A.B., Milani, E.J., Schneider, R.L., Lopez-Paulsen, M., Suárez-Soruco, R., Santa Ana, H., Wiens, F., Ferreiro, O., Rossello, E.A., Bianucci, H.H., Flores, R.F.A., Vistalli, M.S., Fernandez-Seveso, F., Fuenzalida, R.P., Munoz, N., 1995, Phanerozoic correlation in southern South America, *in* Tankard, A.J., Suarez-Soruco, R., and Welsink, H.J. eds., Petroleum basins of South America: Tulsa, Oklahoma, American Association of Petroleum Geologists Memoir 62, p. 129–161.

Geary, D.H., Rich, J., Valley, J.W., and Baker, K., 1989, Stable isotopic evidence of salinity change: Influence on the evolution of melanopsid gastropods in the late Miocene Pannonian basin: Geology, v. 17, p. 981–985.

Goldberg, K., 2001, The paleoclimatic evolution of the Permian in the Paraná Basin in southern Brazil [Ph.D. dissertation]: Chicago, University of Chicago, 267 p.

Graham, S.A., Brassell, S., Carroll, A.R., Xiao, X., Demaison, G., McKnight, C.L., Liang, Y., Chu, J., and Hendrix, M.S., 1990, Characteristics of selected petroleum source rocks, Xinjiang Uygur Autonomous Region, China: American Association of Petroleum Geologists Bulletin, v. 74, p. 493–512.

Graham, S.A., Hendrix, M.S., Wang, L.B., and Carroll, A.R., 1993, Collisional successor basins of western China: Impact of tectonics inheritance of sand composition: Geological Society of America Bulletin, v. 105, p. 323–344.

Greene, T.J., 2000, Tectonics, basin analysis and organic geochemical attributes of Permian through Mesozoic deposits and their derivative oils of the Turpan-Hami Basin, northwestern China: [Ph.D. Dissertation], Stanford, Stanford University 392 p.

Greene, T.J., Carroll, A.R., Hendrix, M.S., and Graham, S.A., 2001, Sedimentary record of intra-Mesozoic deformation and Middle to Late Triassic inception of the Turpan-Hami Basin, northwest China, *in* Hendrix, M.S., and Davis, G., eds., Paleozoic and Mesozoic tectonic evolution of central and eastern Asia—from continental assembly to intracontinental deformation: Geological Society of America Memoir 194, p. 317–340.

Hendrix, M.S., Graham, S.A., Carroll, A.R., Sobel, E.R., McKnight, C.L., Schulein, B.J., and Wang, Z., 1992, Sedimentary record and climatic implications of recurrent deformation in the Tian Shan: Evidence from Mesozoic strata of the north Tarim, south Junggar, and Turpan Basins, northwest China: Geological Society of America Bulletin, v. 104, p. 53–79.

Hopson, C., Wen, J., Tilton, G., Tang, Y., Zhu, B., and Zhao, M., 1989, Paleozoic plutonism in east Junggar, Bogdashan, and eastern Tianshan, northwest China: Eos, v. 70, p. 1403–1404.

Hopson, C.A., Wen, J., and Tilton, G.R., 1998, Isotopic variation of Nd, Sr, and Pb in Paleozoic granitoid plutons along an east Junggar-Bogdashan-Tianshan transect, NW China [abs.]: Geological Society of America Abstracts with Programs, v. 30, p. 20.

Horváth, F., and Pogácsás, Gy., Contribution of seismic reflection data to chronostratigraphy of the Pannonian basin, *in* Royden, L.H., and Horváth, F., eds., The Pannonian Basin; a study in basin evolution :American Association of Petroleum Geologists Memoir 45, p. 97–105.

Katz, B.J., and Mello, M.R., 2000, Petroleum systems of South Atlantic marginal basins—An overview, *in* Katz, B.J., and Mello, M.R., eds., Petroleum systems of south Atlantic margins: American Association of Petroleum Geologists Memoir 73, p. 1–13.

Kutzbach, J.E., and Ziegler, A.M., 1993, Simulation of Late Permian climate and biomes with an atmosphere-ocean model: comparison with observations: Philosophical Transactions of the Royal Society of London, B, v. 341, p. 327–340.

Lee, K.Y., 1985, Geology of the petroleum and coal deposits in the Junggar Basin, Xinjiang Uygur Zizhiqu, northwest China: U.S. Geological Survey Open-File Report 85-230, 53 p.

Liu, H., 1986, Geodynamic scenario and structural styles of Mesozoic and Cenozoic basins in China: American Association of Petroleum Geologists Bulletin, v. 70, p. 377–395.

Mattick, R.E., Phillips, R.L., and Rumpler, J., 1988, Seismic stratigraphy and depositional framework of sedimentary rocks in the Pannonian Basin in southeastern Hungary, *in* Royden, L.H., and Horvath, F., eds., The Pannonian Basin; a study in basin evolution: American Association of Petroleum Geologists Memoir 45, p. 117–145.

Mello, M.R., and J.R. Maxwell, 1990, Organic geochemical and biological marker characterization of source rocks and oils derived from lacustrine environments in the Brazilian continental margin, *in* Katz, B.J., ed., Lacustrine basin exploraion: American Association of Petroleum Geologists Memoir 50, p. 77–97.

Mello, M.R., Koutsoukos, E.A.M., Santos Neto, E.V., and Silva Tellas, J.A.C., 1993, Geochemical and micropaleontological characterization of lacustrine and marine hypersaline environments from Brazilian sedimentary basins: American Association of Petroleum Geologists Studies in Geology 37, p. 17–34.

Oelofsen, B.W., 1987, The biostratigraphy and fossils of the Whitehill and Irati Shale Formations of the Karoo and Paraná Basins, *in* McKenzie, G.D., ed., Gondwana six: Stratigraphy, sedimentology, and paleontology: American Geophysical Union Monograph 41, p. 131–138.

Ojeda, H.A.O., 1982, Structural framework, stratigraphy, and evolution of Brazilian marginal basins: American Association of Petroleum Geologists Bulletin, v. 66, p. 732–749.

Olsen, P.E., 1986, A 40-million year lake record of Early Mesozoic orbital climatic forcing: Science, v. 234, p. 842–848.

Peng, X., and Zhang, G., 1989, Tectonic features of the Junggar Basin and their relationshiop with with oil and gas distribution, *in* Zhu, X., ed., Chinese sedimentary basins: Amsterdam, Elsevier, p. 17–31.

Peters, K.E., 1986, Guidelines for evaluating petroleum source rock using programmed pyrolysis: American Association of Petroleum Geologists Bulletin, v. 70, p. 318–329

Pitman, J.K., Pierce, F.W., and Grundy, W.D., 1989, Thickness, oil-yield, and kriged resource estimates for the Eocene Green River Formation, Piceance Creek Basin, Colorado: U.S. Geological Survey Oil and Gas Investigations Chart OC-0132, 6 sheets, 4 p.

Şengör, A.M.C., and Natal'in, B.A., 1996, Paleotectonics of Asia: Fragments of a synthesis, *in* Yin, A., and Harrison, M.T., eds., The tectonic evolution of Asia: New York, Cambridge, p. 486–640.

Şengör, A.M.C., Natal'in, B.A., and Burtman, V.S., 1993, Evolution of the Altaid tectonic collage and Paleozoic crustal growth in Eurasia: Nature, v. 364, p. 299–307.

Smith, M.E., Pietras, J.T., Carroll, A.R., and Singer, B , 2001, Laser fusion $^{40}Ar/^{39}Ar$ geochronology and cyclostratigraphy of the Green River Formation: Testing micro- and meso-scale climate forcing [abs.]: Eos (Transaction, American Geophysical Union), v. 82, p. F1141.

Tapponnier, P., and Molnar, P., 1979, Active faulting and Cenozoic tectonics of the Tien Shan, Mongolia, and Baykal regions: Journal of Geophysical Research, v. 84, p. 3425–3459.

Teichert, C., 1974, Marine sedimentary environments and their faunas in Gondwana area: American Association of Petroleum Geologists Memoir 23, p. 43–145.

Visser, J.N. J., 1994, A Permian argillaceous syn- to post-glacial foreland sequence in the Karoo Basin, South Africa, *in* Deynoux, M., Miller, J.M.G., Domack, E.W., Eyles, N., Fairchild, I.J., and Young, G.M., eds., Earth's glacial record: Cambridge, Cambridge University Press, p. 193–203.

Wang, X., 1992, Petroleum geological properties of the Permo-Carboniferous in the Ili Basin: Oil and Gas Geology, v. 13, p. 219–228 (in Chinese with English abstract).

Wartes, M.A., Carroll, A.R., and Greene, T.J., 2002, Permian stratigraphic evolution of the Turpan-Hami Basin and adjacent regions, northwest China: Constraints on post-amalgamation tectonic evolution: Geological Society of America Bulletin, v. 114, p. 131–152.

Wartes, M.A., Carroll, A.R., Greene, T.J., Cheng, K., and Hu, T., 2000, Permian lacustrine deposits of northwest China, *in* Gierlowski-Kordesch, E.H., and Kelts, K., eds., Lake basins through space and time: American Association for Petroleum Geologists Studies in Geology 46, p. 123–132.

Watson, M.P., Hayward, A.B., Parkinson, D.N., and Zhang, Z.M., 1987, Plate tectonic history, basin development and petroleum source rock deposition in onshore China: Marine and Petroleum Geology, v. 4, p. 205–225.

Wen, J., 1991, Geochronological and isotopic tracer studies of some granitoid rocks from Xinjiang, China: Constraints on Paleozoic crustal evolution and granitoid petrogenesis [Ph. D. thesis]: Santa Barbara, University of California, 163 p.

Windley, B.F., Allen, M.B., Zhang, C., Zhao, Z.Y., and Wang, G.R., 1990, Paleozoic accretion and Cenozoic redeformation of the Chinese Tien Shan range, central Asia: Geology, v. 18, p. 128–131.

Yemane, K., 1993, Contributions of Late Permian paleogeography in maintaining a temperate climate in Gondwana: Nature, v. 361, p. 51–54.

Ziegler, A.M., Hulver, M.L., and Rowley, D.B., 1996, Permian world topography and climate, *in* Martini, I.P. ed., Late glacial and postglacial environmental changes-Quaternary, Carboniferous-Permian and Proterozoic: New York, Oxford University Press, p. 111–142.

Zonenshain, L.P., and Le Pichon, X., 1986, Deep basins of the Black Sea and Caspian Sea as remnants of Mesozoic back-arc basins: Tectonophysics, v. 123, p. 181–211.

Manuscript Accepted by the Society January 22, 2003

Printed in the USA

Geological Society of America
Special Paper 370
2003

Features and origin of the giant Cucomungo Canyon alluvial fan, Eureka Valley, California

Terence C. Blair
Blair & Associates LLC, 1949 Hardscrabble Place, Boulder, Colorado 80305, USA

ABSTRACT

With an area of 119 km² and a radial length of 17.1 km, the Cucomungo Canyon alluvial fan of Eureka Valley, California, is notably larger than all previously documented examples. This fan singularly covers 29% of the Eureka Valley, as compared with 54% of the basin covered by about 110 other fans, and 17% by a mid-basin channel, playa, and eolian erg. Analyses of 1–6-m-thick surface exposures at 62 stations spanning the fan reveal that it is built predominantly (88.3%) of debris-flow deposits. Most cuts (average 79.3%) consist of amalgamated beds of clast-poor debris flows (mudflows) typically 10–40 cm thick (Facies A). Cobbly and pebbly clast-rich debris flow beds 10–260 cm thick (Facies B) are most prevalent in the proximal fan (20.8%) and less so distally (3.5%) for an average of 9.0%. The fan radial slope decreases from 4 to 5° in the proximal zone to 2 to 3° distally coincident with a decrease in Facies B content. The remainder of the fan exposures consists of clast-supported and imbricated pebble cobble gravel deposited by rare water flows in large channels (Facies C, 6.4% of cuts), waterlaid laminated granular sand on the beds of abundant shallow channels (Facies D, 5.0%), and laminated eolian sand present as sparse coppice dunes (Facies E, 0.3%).

Fan sediment is derived from an 86.8 km² catchment in the Sylvania Mountains that is traversed by a transpressive sector of the Furnace Creek strike-slip fault. Granitic bedrock in this fault zone is widely crushed from tectonic shearing that promotes weathering, producing extraordinarily high volumes of sediment that accumulate as a thick colluvial mantle. This colluvium is rich in coarse silt, sand, and fine pebbles but is abnormally deficient in coarser gravel. When saturated by thunderstorm rainfall, these slopes fail and typically transform into mudflows of extraordinary volume (200,000–600,000 m³). The large volume and deficiency of coarse clasts create high-runout mudflows capable of depositing 100-m-wide tracts that span the 13–17-km-long radius of the Cucomungo fan, through time building this rare giant.

Keywords: alluvial fan, mudflow processes, faulted catchment, Eureka Valley, California.

INTRODUCTION

Alluvial fans constitute typically gravelly coalesced to uncoalesced semiconical accumulations at a mountain front built by processes that transfer sediment from an upland catchment to the adjoining valley (Surrell, 1841; Drew, 1873; Bull, 1972). Alluvial fans are differentiated from other environments such as rivers and river deltas by their unique form that radiates from a range front point commonly at slopes of 2–15°, and also by their distinctive principal processes and facies such as catastrophic rock avalanches, debris flows, and sheetfloods (Blair, 1987, 1999a, 1999b, 1999c, 1999d; Blair and McPherson, 1994a, 1994b, 1998). Sediment transferred from the catchment by these processes accumulates at the mountain front in response to flow

*tcblair@aol.com

Blair, T.C., 2003, Features and origin of the giant Cucomungo Canyon alluvial fan, Eureka Valley, California, *in* Chan, M.A., and Archer, A.W., eds., Extreme depositional environments: Mega end members in geologic time: Boulder, Colorado, Geological Society of America Special Paper 370, p. 105–126.

expansion, producing deposits with slopes that manifest the angle at which particle transport is no longer possible (Blair and McPherson, 1994b).

Past research on the size of alluvial fans, especially the radial length and area, is of four types. Fan area has been related by many to be a function of catchment area (e.g., Bull, 1962; Whipple and Traylor, 1996; Allen and Densmore, 2000). Although it is intuitive that a large fan cannot be derived from a small catchment, equating the planview area of three-dimensional features lacking a constant vertical value has long been identified as mathematically invalid (Lustig, 1965), as shown by the wide scatter in a cross-plot of previous fan area-catchment area data (Blair and McPherson, 1994a). A second approach to fan size involves relating their length to the locations of active faulting. Fans lining half grabens commonly are observed to be smallest on the faulted margin due to high vertical accommodation caused by tectonic subsidence (Hunt and Mabey, 1966). In the type case of central Death Valley, fans along the passive margin have radial lengths of 5–10 km and those along the tectonically active margin 0.5–3 km. A third approach to the analysis of fan size was provided by Anstey (1965, 1966), who did a systematic study of the radial length of over 4000 fans in the western United States and Pakistan using large scale maps. Anstey determined from this dataset that most fans have radial lengths of less than 8 km, and rarely exceed 10 km. A fourth and most recent perspective on fan size involved the reclassification of rivers hundreds of kilometers long and with areas of >10,000 km^2 as alluvial fans or megafans, starting with the Kosi and Gandak rivers of India and now including rivers from around the globe (e.g., Schumm, 1977; Horton and DeCelles, 2001; Shukla et al., 2001). As discussed by Blair and McPherson (1994a, 1994b), this reclassification is unscientific because these features lack a fan form or slope, and have channel and floodplain processes and resultant facies that are unlike alluvial fans but typical of rivers, as they were originally identified. Excluding these misclassified rivers, all fans documented to date are consistent in size with those of Anstey's survey. The largest of these fans for which sedimentologic studies have also been made are the Anvil, Warm Springs, Hell's Gate, and Tuttle fans of southeast California, which have radial lengths of 8.1–11.8 km and areas of 25.4–49.5 km^2 (Blair, 1999a, 1999b, 1999c, 2000, 2001).

The Cucomungo Canyon alluvial fan of Eureka Valley, California, with a length of 17.1 km and an area of 119 km^2, is notably larger than other documented fans and is the largest fan known to the author (Figs. 1 and 2A). Despite occurring along the tectonically active side of the Eureka Valley half graben, the Cucomungo dwarfs the other ~110 fans in the valley, having a radial length nearly 3 times longer and an area 10 times greater than the next largest fans (Table 1). The purpose of this paper is to document the form and sedimentology of the Cucomungo fan, and to elucidate the origin of its abnormally large size. The study methods involved: (1) characterizing the fan and its catchment using aerial photographs and topographic maps (1:24,000 scale; 12.2 m contour interval), and by field reconnaissance; (2) describing surface

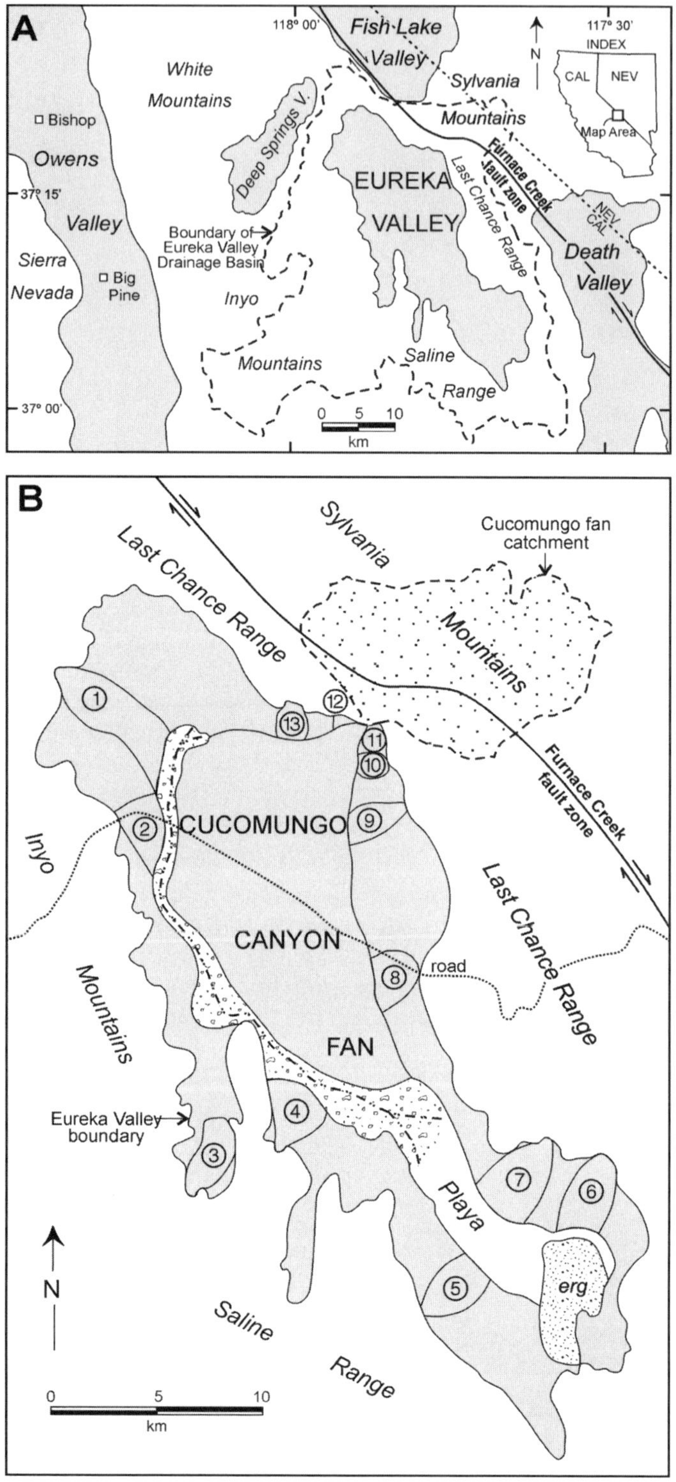

Figure 1. A: Location map of Eureka Valley, California, and its drainage basin. B: Map of sedimentary environments of Eureka Valley, including basin-rimming piedmont alluvial fans (shaded) separated by an ephemeral channel tract that leads to a playa and eolian erg. The Cucomungo and other individual fans referred to in the text are delineated and numbered (see Table 1).

Figure 2. A: Aerial photograph taken 21 June 1981 of the Cucomungo fan; a road (*r*) crosses the medial fan. The photograph covers an area of 14 × 31 km and is orthogonally oriented with north to the left. Note the darker desert-varnished fan sectors, a lighter mudflow tract (*m*), and the mid-basin channel (arrows) at the fan toe. Catchment bedrock is of Pliocene basin fill (*b*), Jurassic granite (*g*), and Paleozoic strata (*s*). Smaller adjoining fans also are visible. B: Across-valley view of the sparsely vegetated Cucomungo fan; *A* is immediately left of the apex. C: Westward view in catchment down the yucca-covered Cucomungo Canyon to the fan feeder channel (arrow). Dark Paleozoic strata (left) and light granite (right) are separated in the canyon by the Furnace Creek fault. D: View of catchment bedrock showing a serrated and rilled morphology caused by tectonic shearing and weathering of the granite. Forested catchment crests also are visible (arrows). E: Photomicrograph 4.5 × 6.4 mm in area of catchment granite at station #1 showing crystals >1 cm across of plagioclase (*p*), potassium feldspar (*k*), and quartz (*q*) that are tectonically crushed into fragments <0.2 mm across.

TABLE 1. FEATURES OF EUREKA VALLEY FANS

Fan	Fan length (km)	Fan area (sq km)	Fan avg. radial slope (°)	Catchment area (sq km)	Main deposits
Cuc	17.1	119.2	1.8	86.8	Mudflow
1	6.8	12.3	2.3	6.6	Sheetflood
2	3.4	6.6	2.7	66.3	Sheetflood
3	4.5	5.6	2.6	279.0	?
4	3.1	7.3	3.6	1.9	?
5	3.4	5.1	2.8	13.3	?
6	3.7	5.9	7.4	5.6	Debris flow
7	4.3	9.3	3.6	33.9	?
8	2.4	4.1	4.0	17.6	Sheetflood?
9	2.6	2.9	6.4	1.4	Debris flow
10	1.4	0.8	9.2	0.7	Debris flow
11	1.0	0.4	8.3	0.4	Debris flow
12	0.3	0.1	22.5	0.2	Rockfall
13	1.8	2.1	6.7	1.7	Debris flow

Note: See Figure 1B for fan locations. Cuc—Cucomungo fan.

forms, texture, fabric, structures, bedding, and other features at 62 stratigraphic stations exposed at channel walls and in pits spanning the fan and lower catchment; (3) measuring surface and bedding slopes at the stations using a clinometer; (4) analyzing the matrix (<1.6 cm) fraction of 14 representative samples at 1/4 ϕ intervals using standard methods; (5) documenting gravel composition by ten pebble counts; (6) analyzing petrographically two bedrock samples from the catchment; (7) reconnoitering exposures of nine other fans in Eureka Valley to determine their facies; and (8) integrating available data to formulate a depositional model and to identify the causes of the large size of the Cucomungo fan. Fabric terms follow Walker (1975) and textural terms follow Folk (1974) as modified for the gravel fraction by Blair and McPherson (1999).

SETTING

Eureka Valley is a NNW-trending extensional basin 45 km long and 16 km wide located between Owens, Fish Lake, Deep Springs, and Death Valleys in east-central California (Fig. 1A). It is a little-studied, uninhabited basin recently added to Death Valley National Park and the Piper Mountain Wilderness. Like other valleys of the Basin and Range province, Eureka is lined by alluvial fans commonly 1–5 km long that collectively cover 83% of valley (Fig. 1B). The flanking piedmonts are separated on the basin floor by an ephemeral channel and terminal playa system 60 km^2 in area, and to the southeast by a 13 km^2 eolian erg. Eureka Valley is bounded by the Inyo Mountains on the west, Saline Range on the south, and the Sylvania Mountains and Last Chance Range on the north and east. These ranges rise 1100–1600 m above the 875 m elevation of Eureka playa. Saline Range consists of Neogene basalt and the other ranges of Paleozoic carbonate, quartzite, and shale with Neogene basalt and Mesozoic granitic intrusions that are most common along the northern valley margins (Strand, 1967). Eureka basin is forming by normal fault offset along the east margin as manifested by fault scarps 1–5 m high, whereas the west side constitutes the tectonically passive flank of this half graben.

Eureka Valley is a hydrologically closed basin draining an area of 1506 km^2, for which Eureka playa is the ultimate sink (Fig. 1). Most (72%) of the drainage area consists of uplands, the majority in the Inyo Mountains on the western margin, whereas the remaining 28% is in the valley proper. The highest peaks of the drainage basin are in the Inyo Mountains, where a maximum relief of 2516 m exists between the 3391-m-high Waucoba Mountain peak and Eureka playa. Little precipitation falls within Eureka Valley due to its position in the rain shadow of the Sierra Nevada and Inyo Mountains, creating desert conditions with a sparse (<10%) vegetative cover of desert grass, creosote, and other scrub (Fig. 2B). Precipitation is higher in the uplands where Joshua trees (yucca) appear at an elevation of about 1400 m and pine trees at 2000 m (Fig. 2C, D). Except for wind processes, most sedimentation in Eureka Valley results from overland flows triggered by thunderstorm precipitation in the adjoining mountains.

OVERVIEW OF THE CUCOMUNGO CANYON FAN AND CATCHMENT

The Cucomungo fan of northeastern Eureka Valley is derived from a catchment about 10 km long, 12 km wide, and 86.8 km^2 in area located in the Sylvania and Last Chance Ranges (Fig. 1B). The catchment backwall rises about 1000 m above the 1348 m elevation of the fan apex, with a maximum relief of 1146 m attained at Sylvania Peak. The catchment is underlain mainly by Jurassic orbicular granitic bedrock (78.3% of area) along with Paleozoic sedimentary rocks (21.3%) and a small (0.4%) zone of Plio-Pleistocene basin fill (Figs. 2A and 3; Strand 1967; Reheis et al., 1991; Reheis, 1992). The catchment is traversed by the northwest-trending, dextral strike-slip Furnace Creek fault, a major fault that accommodates extension in this part of the Basin and Range province (Brogan et al., 1991; Oldow, 1992). Within the catchment this fault zone is an active transpressive transfer segment ~10 km wide passing from the east side of Death Valley to the west side of Fish Lake Valley, and separating the granitic Sylvania Mountains from Paleozoic strata of the Last Chance Range (Figs. 1 and 3; Brogan et al., 1991). Granite within this fault zone is crushed along closely spaced shears that produce mangled exposures rather than cliffs and exfoliation surfaces more typical of disintegrating granitic bedrock in this region (Fig. 2C, D; Strand, 1967). Petrographic analysis of samples from this zone shows that the constituent coarse (2–20 mm) crystals of quartz, potassium feldspar, plagioclase, and accessory minerals are widely fragmented into angular grains of silt to fine sand size (<0.25 mm; Fig. 2E).

Hydrologically, the Cucomungo catchment is a seventh-order ephemeral basin consisting of four compartments drained southwestward by fifth-order channels (Fig. 3). The fifth-order channels flow into two orthogonally oriented sixth-order channels

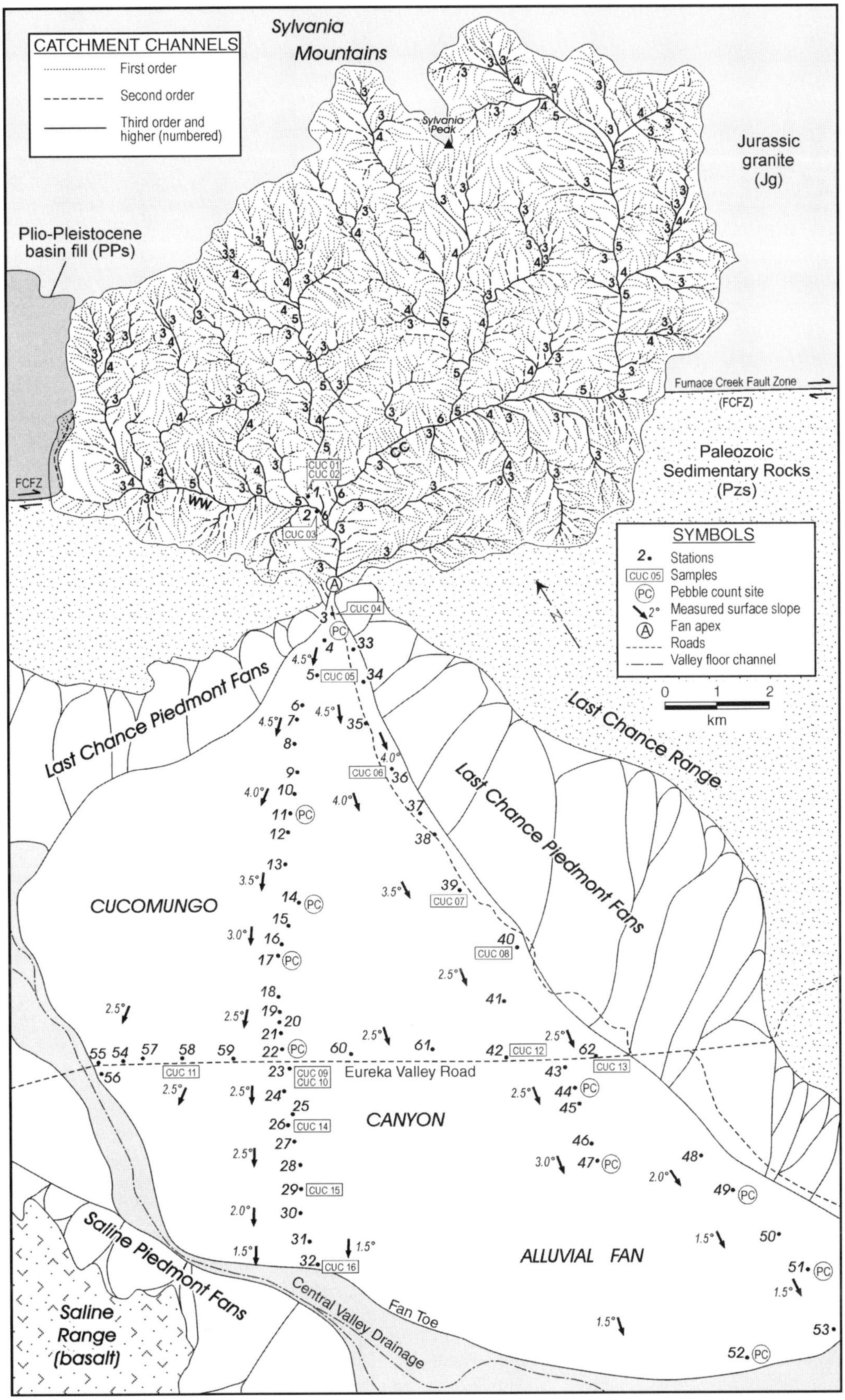

Figure 3. Map of the Cucomungo fan and catchment showing drainage net, fan stations, sample sites, radial slopes, and pebble-count sites. Catchment channels are delineated and numbered using 1:24,000 scale U.S. Geological Survey topographic maps with 12.2 m contours, and the stream ordering method of Horton (1945) and Strahler (1964). Bedrock is after Strand (1967) and Reheis (1992). The trace of the Furnace Creek fault is delineated by the Willow Wash (WW) and Cucomungo Canyon (CC) channels; it also is the contact between Jg and Pzs bedrock.

called Cucomungo Canyon and Willow Wash, which are developed upon the Furnace Creek fault (Figs. 2C and 3). Cucomungo Canyon and Willow Wash combine 1.3 km upslope from the fan apex to form the seventh-order channel leading to the fan. The average slope of the catchment drainage net is 6.6° for the Cucomungo Canyon sector and 5.4° for the Willow Wash sector. Lower-order channels also drain Paleozoic bedrock along the southwest side of the catchment. The catchment slopes, especially in the Furnace Creek fault zone, are widely mantled by sandy and muddy colluvium derived from the sheared granitic bedrock.

The Cucomungo alluvial fan spans southwestward from the apex to cover 29% of Eureka Valley, in contrast to 54% of the valley covered by the ~110 other Eureka fans (Figs. 1B and 3). The Cucomungo has an area of 119.2 km^2, a maximum width of 10.6 km, and a maximum radius of 17.1 km. The radius is restricted to a length of 13.5 km along the southwest margin by fans of the opposing piedmont, whereas the unrestricted south sector stretches to Eureka playa. The radial slope of the Cucomungo fan surface is comparatively low, averaging 1.8° along the bisectrix, in contrast to the more typical 3–9° radial slopes of other fans in this valley and elsewhere (Table 1; Blair and McPherson, 1994a, 1994b). Rather than having a constant value, the surface slope varies from 4–5° in the proximal zone, 2–3° in the medial zone, and 1.5° distally. Cross profiles display a broadly bowed-upward form with 40 m of relief proximally and 10 m distally. A weak to moderately developed desert pavement of sandy gravel covers the inactive parts of the fan, and hosts a desert varnish (Figs. 2A and 4A).

Constituent forms of the Cucomungo fan mostly are of low relief. The largest feature is the incised channel of the proximal fan, representing a continuation of the catchment feeder channel that extends for about 4.5 km down the fan. This channel is 10–30 m across and 3–6 m deep in the lower catchment and upper fan (Fig. 4B), and lessens to a width of 5 m and depth of 2 m near its terminus. It and fan gullies 1–3 m deep and across typically have floors that are flat in cross section and covered by imbricated pebbles and cobbles. More shallow (<1.5 m) and narrow (1–5 m) channels and rills with sandy to fine pebbly beds are common fanwide, separating pavement-covered tracts that locally are studded with boulders (Fig. 4A). Another feature common on the surface is radially aligned mudflow (clast-poor debris-flow) tracts 20–100 cm thick and 50–300 m wide that begin at the fan apex or at the end of the incised channel and span to the fan toe. Recent mudflow tracts are light in color, in contrast to older tracts that are darker due to oxidation. A light-colored tract extending to the fan toe is visible on the 1981 aerial photograph (Fig. 2A), and a mudflow tract with a similar extent was deposited on 2 September 1997, while this study was ongoing, in response to intense thunderstorms triggered by the passage of a weather front fed by subtropical moisture. The 1997 mudflow deposits were derived from colluvium mantling granitic bedrock in the two middle fifth-order compartments of the catchment (Fig. 3). The common presence of tree limbs and logs within this and older mudflow tracts indicates that they commonly originate by colluvial slips from hollows in the uppermost level of the compartments where the limited tree cover is developed, as confirmed for the 1997 mudflow by aerial photographs and field reconnaissance.

The 1997 mudflow tract, like the geomorphically intact older tracts on the fan surface, consists of broad lobes 10–100 cm thick, 10–30 m across, and 50 to 150 m long that are laterally amalgamated and radially telescoped to form a continuous expanse extending from the end of the incised channel to the fan toe (Fig. 4C). The lobes expand outward and downslope from a central channel where the mudflow was focused. The sharp and steep margins of the lobes typically are 10–40 cm high and have a sparse concentration of logs, cobbles, and boulders that were transported in the flow (Fig. 4D). The mudflow enveloped older elements on the fan surface such as plants and protruding clasts (Fig. 4E) and filled negative features such as channels. It has a smooth surface that contains abundant desiccation cracks. Deposits from the 1997 event also partially filled the incised channel and are present as lobes extending outward from this channel where the confining walls were topped (Fig. 4F). The within-channel mudflows and the central part of the lobate tract were partially eroded by waning stage water flow that winnowed mud and sand, and concentrated gravel (Fig. 4B, C, F). Using an average width of 80 m and thickness of 35 cm, the volume of the 1997 mudflow sediment on the fan is roughly estimated at 400,000 m^3.

SEDIMENTOLOGY OF THE CUCOMUNGO FAN

Sedimentologic features of Cucomungo fan exposures were described at 62 widely distributed stations grouped for discussion into a medial cross-fan transect and two radial transects, one radial transect comprising the fan bisectrix and the other located near the eastern margin (Fig. 3). Grain-size sample and pebble-count sites are divided amongst these stations. Exposure heights at the stations vary from 80 to 600 cm, generally lessening distally (Fig. 5A). The exposures consist of five associated sedimentary facies labeled A through E (Table 2).

Facies A and B: Clast-Poor and Clast-Rich Debris-Flow Deposits

Descriptions

Facies A consists of light pinkish to greenish gray, poorly to very poorly sorted, pebbly granular muddy sand and pebbly sandy mud in sharply bounded, matrix-rich and matrix-supported beds (Fig. 6A–F). Angular to subangular, coarse to very coarse pebbles, cobbles, and boulders rarely are present as scattered clasts, as are coniferous tree logs 50 cm in diameter in the proximal fan and 20 cm near the toe. Rare large clasts usually are present at the top of the beds (Fig. 6E). In contrast, granules and fine to medium pebbles are widespread in Facies A. They usually are evenly distributed within a given bed but in some cases are most abundant in the basal zones and less so upwards to display coarse-fraction normal grading (Fig. 6C, D). Besides grading, no other internal

Figure 4. A: Across-fan view near station #38 showing sparse vegetation, a shallow ephemeral channel (*c*), and scattered boulders. B: Upslope view at station #2 of the 6-m-deep and 15-m-wide feeder channel leading to the fan apex. Note the imbricated cobble pebble gravel (Facies C) on the channel bed. C: Downfan view near station #18 of a 60-cm-thick and 25-m-wide part of the mudflow tract deposited 2 September 1997. Recessional water flow through a central channel (*c*) concentrated coarse gravel. D: View of 1997 mudflow deposits 35 cm thick and >15 m wide near station #14; arrows point to steep margin. These deposits extend outward from a central channel (*c*), enveloping plants and aggrading a planar bed upon the smooth surface of older mudflows. E: View near station #11 of the margin of 1997 mudflow lobes with a concentration of tree logs and limbs, such as near the 20-cm-long fieldbook (arrow). F: Downfan view of 10-m-wide and 3.5-m-deep incised channel near station #6 showing 1997 mudflow deposits within the channel (*m*) and extending laterally outward from the crest of the channel walls (arrows at mudflow margins). Note the concentration of sparse cobbles near the margins of the overbank deposits (lower two arrows).

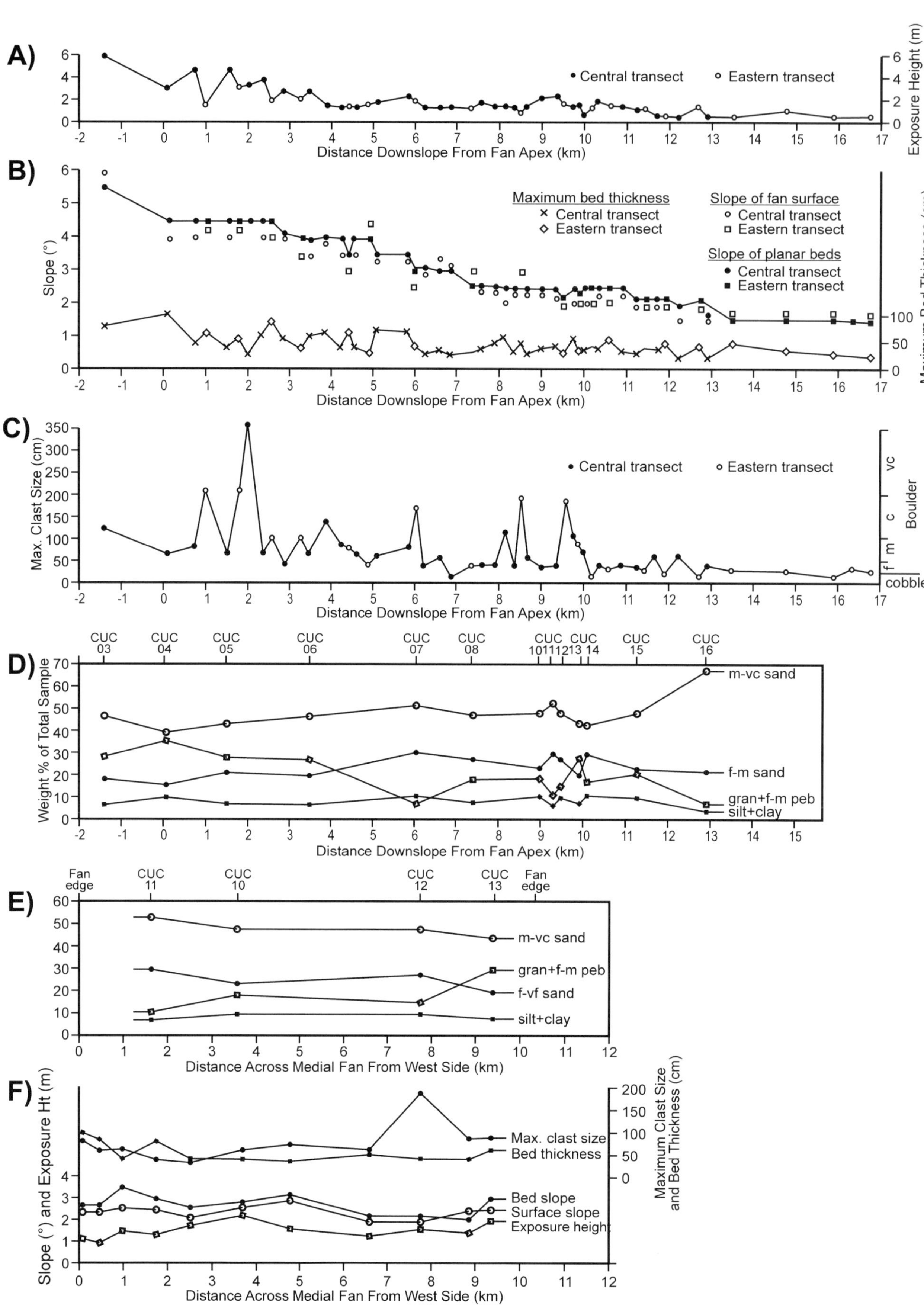

Figure 5. A–C: Plots versus radial distance from the fan apex of exposure height (A); measured surface slope, planar bedding slope, and maximum bed thickness (B); and maximum (*b* axis) clast size (C). D–E: Weight percent of the granule to medium pebble, medium to very coarse sand, fine to very fine sand, and mud fractions of analyzed Facies A and B samples versus radial distance (D) and versus medial cross-fan position (E, Table 3, Fig. 3). F: Medial cross-fan plots of maximum clast size (*b* axis), maximum bed thickness, measured bed and surface slopes, and exposure height.

TABLE 2. FACIES AND PROCESSES ON THE CUCOMUNGO CANYON FAN

Facies	Features	Processes	Facies abundance (%) at stations		
			Fanwide	Proximal	Distal
A	Unsorted matrix-supported pebbly mud and muddy sand in planar beds	Clast-poor debris flows across the fan surface	79.3	66.8	85.0
B	Unsorted matrix-supported cobbly sandy muddy pebble gravel in lenticular/planar beds	Clast-rich debris flows in or near fan channels	9.0	20.8	3.5
C	Imbricated clast-supported cobbly pebble gravel in lenticular beds	Coarse waterlaid deposits on channel beds	6.4	7.6	5.8
D	Laminated pebbly sandy granule gravel and pebbly sand	Waterlaid deposits on beds of gullies and rills	5.0	4.8	5.3
E	Laminated sand in discontinuous low-angle beds near plants	Eolian coppice dune accumulations	0.3	0.0	0.4

Note: Average fanwide facies abundance is based on data from each of the 60 Cucomungo fan stations (Fig. 3). Proximal average facies abundance is based on data from 19 stations located < 6.1 km from the fan apex. Distal average facies abundance is based on data from 41 stations located > 6.1 km downslope from the apex and extending to the fan toe.

features are apparent in this facies except for variations in the volume of fine gravel, which allow beds to be differentiated.

Individual beds of Facies A range in thickness from 1 to 70 cm but most typically are 10 to 40 cm, and beds are stacked in a pancake-like manner all across the fan (Fig. 6A, B). Bed thicknesses generally are similar fanwide, although the maximum thickness of exposed Facies A beds at the stations decreases distally from 70 to 20 cm (Fig. 5B). Successive beds either are amalgamated or are separated by partings or thin beds of Facies C or D (Figs. 6 and 7). The beds have sharp bases that usually are planar or slightly undulatory in form (<10 cm relief), although in strike sections bed relief increases in some cases to 50 cm, depicting a more trough-like form (Fig. 6F). The tops of the beds mostly are sharp and planar except locally where they are cut by surface channels 5–50 cm deep. In radial cuts the beds commonly are continuous for >100 m and have boundaries that slope 4–5° in the proximal stations and 2–3° distally, like the fan surface (Fig. 5B). Beds extend for 5–50 m in strike cuts, wherein the thickest beds rapidly thin with a lenticular geometry (Fig. 6F). The beds terminate both radially and laterally by wedging out into other beds (Fig. 6A, B). Roots and root casts are sparse in this facies, whereas older deposits have pedogenic carbonate coatings on clasts, cracks, or root walls. The 1997 mudflow deposits have textures, bedding types, and other features identical to and thus are a part of Facies A.

Facies B consists of light gray, bouldery cobbly sandy muddy pebble gravel to muddy cobble boulder gravel in extremely poorly sorted and matrix-supported beds typically 20–70 cm thick, and locally to 260 cm thick (Fig. 8A, B). Internally, coarse clasts in most of the beds appear to be randomly distributed, but less commonly they are concentrated in the lower part of the beds in a normally graded fashion (Fig. 8C, D). Beds either are amalgamated or are separated by partings or thin interbeds of other facies (Fig. 7). Individual beds are discernible due to graded fabric, variations in coarse gravel mode or volume, or to the presence of interbeds of other facies that tend to weather more recessively in exposures. Facies B beds extend radially for 100s of meters but pinch-out laterally over a distance of <5–20 m. The bed bases typically are planar in radial cuts but are more undulatory in strike cuts, whereas the tops usually are sharp and in some cases are studded by boulders or tree logs. Radial cuts show that these beds are not horizontally oriented but slope 4–6° in the proximal fan and 3–4° in the medial fan cuts, like the fan surface slope. This facies texturally is like Facies A but differs from it by thicker bedding, greater bed lenticularity, and by greater content of coarse gravel, including of coarse to very coarse pebbles, cobbles, and boulders with *b* axes as large as 365 cm.

Facies A and B are further characterized by grain-size data. Data from sieve and laser-particle analyses of the matrix fraction (<1.6 cm or <–4 ϕ) of 12 Facies A samples collected at sites spanning the fan and lower catchment, and of one Facies B sample (CUC-04) from the proximal fan, show that these samples are similar texturally, with modes ranging from medium sand to granules and sorting from poor to very poor (1.58–2.78 ϕ, Fig. 9, Table 3). Facies A samples mostly consist of sand (64%–88%; average 72.8%) that is evenly distributed amongst all five grades. Additionally, these samples contain an average of 10.3% granules, 8.6% fine and medium pebbles, 7.3% silt, and 1.0% clay. Their gravel-sand-mud ratios are similar near the average of 19-73-8, as are the normalized sand-silt-clay ratios at an average of 90-9-1 (Fig. 9F, G). The Facies B sample is like those of Facies A except for greater gravel content, consisting of 20.5% fine to medium pebbles, 15.0% granules, 53.4% sand, 9.9% silt, and 1.2% clay. Plots of the grades for all Facies A and B samples show a fanwide consistency irrespective of transport distance (Fig. 5D, E). The data also show that the two Facies A samples from the 1997 deposits (CUC-03 and -15) have textures that are the same as the older deposits (Fig. 9A; Table 3).

Maximum clast-size data and pebble compositional data for Facies A and B generally also are consistent fanwide. The largest

Figure 6. Photographs of Facies A mudflow deposits (fieldbook or pen scales are 20 cm long). A: Radial view at station #15 of stacked mudflow beds 1–30 cm thick, including the upper 15-cm-thick bed deposited in 1997. Bedding planes are sharp and delineated by partings to thin beds of Facies D (recessional areas). Some beds terminate laterally by wedging (arrow). B: Obliquely oriented view at station #29 of stacked mudflow beds 1–30 cm thick divided by Facies D partings. Beds have wedge-planar (arrow) and planar forms. C: Close-up radial view of mudflow strata at station #22 (pen for scale, arrow). Note variations in the content of granules and fine pebbles. D: Close-up view of stacked mudflow beds at station #42. Note the apparent coarse-fraction normal grading in the middle bed (between arrows). E: Stacked mudflow beds in a radial cut at station #18 containing an isolated boulder (near pen, *p*). Arrows define some of the sharp bedding planes. F: Cross-fan cut through mudflow deposits at station #61. Mudflows are delineated by variations in gravel content and by thin beds of Facies D (recessional areas). Note the trough-like cross-sectional geometry of some of the beds (arrows).

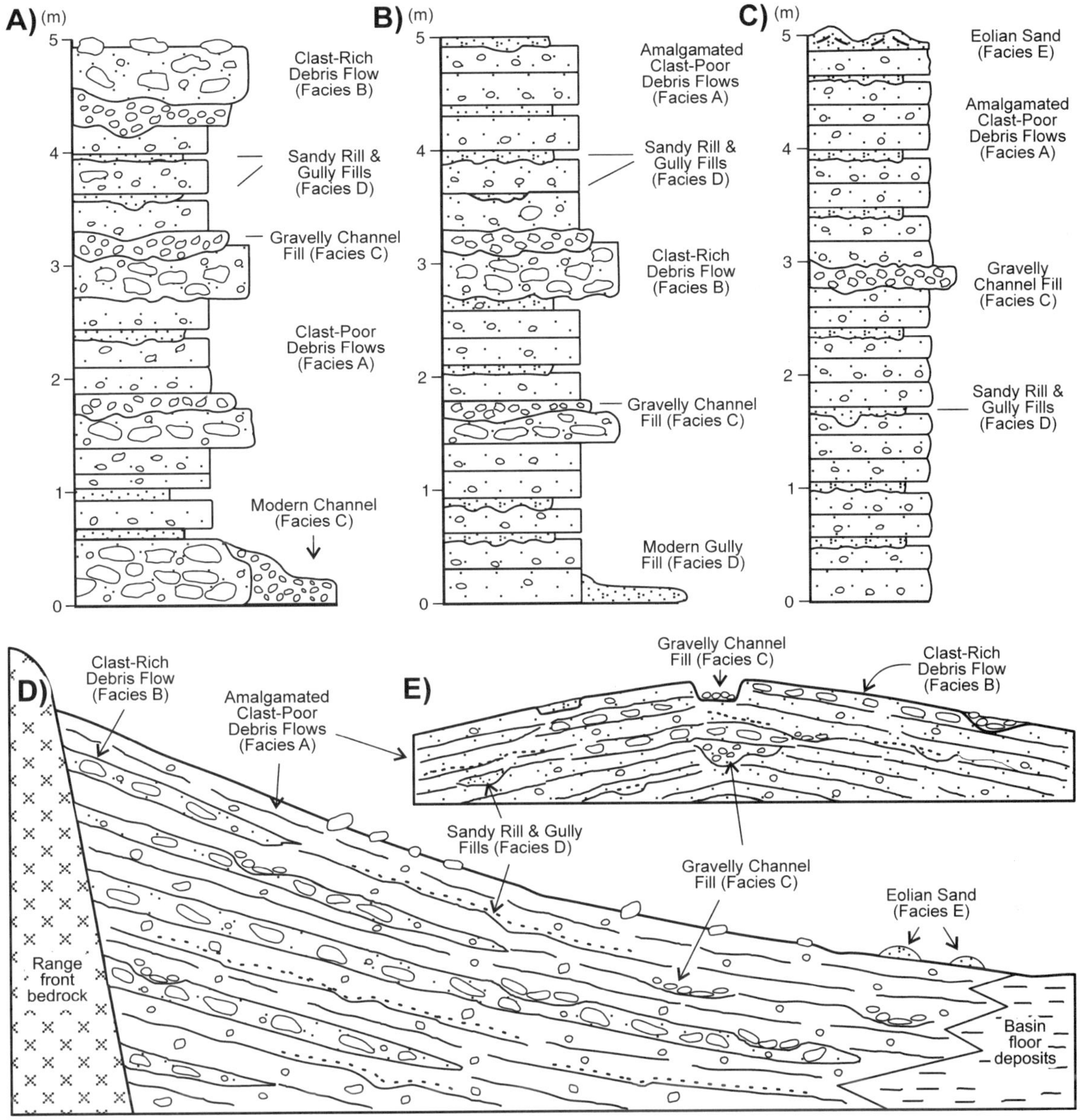

Figure 7. A–C: Stratigraphic summary of Cucomungo facies for the proximal (A), medial (B), and distal (C) fan. D–E: Stratigraphic model for radial (D) and transverse (E) transects of the Cucomungo fan.

clast (*b* axis) ranges from 10 to 100 cm (cobbles to medium boulders) at most stations, and reaches 150–365 cm (coarse to very coarse boulders) at seven stations. Plots of these values versus distance from the fan apex or across the medial fan show that they are independent of distance except that clasts coarser than fine boulders (>50 cm) are not present at sites >10 km from the apex (Fig. 5C, F). Seven compositional counts show that coarse to very coarse pebbles of Facies A and B are similar, averaging 86% granite, 7% quartzite, 4% carbonate rock, and 3% of volcanic rock (Table 4). Cobbles and boulders in either facies consist almost exclusively of granite.

Abundance data at the stations show that Facies A and B dominate the Cucomungo fan from apex to toe, together accounting for 80%–100% of the exposures, and averaging 88.3% (Fig. 10). Facies A is the dominant of the two, comprising 65%–95% of most stations for an average of 79.3% of the exposures fanwide. It is most prevalent distally, with the 41 stations located >6.1 km from the apex composed on average of 85.0% Facies A, and those at more proximal sites 66.8% (Fig. 7; Table 2). Facies B makes up an average of 9.0% of the station exposures, occurring most abundantly (average of 20.8%) in the proximal sites. It is less common in the medial and distal fan where it

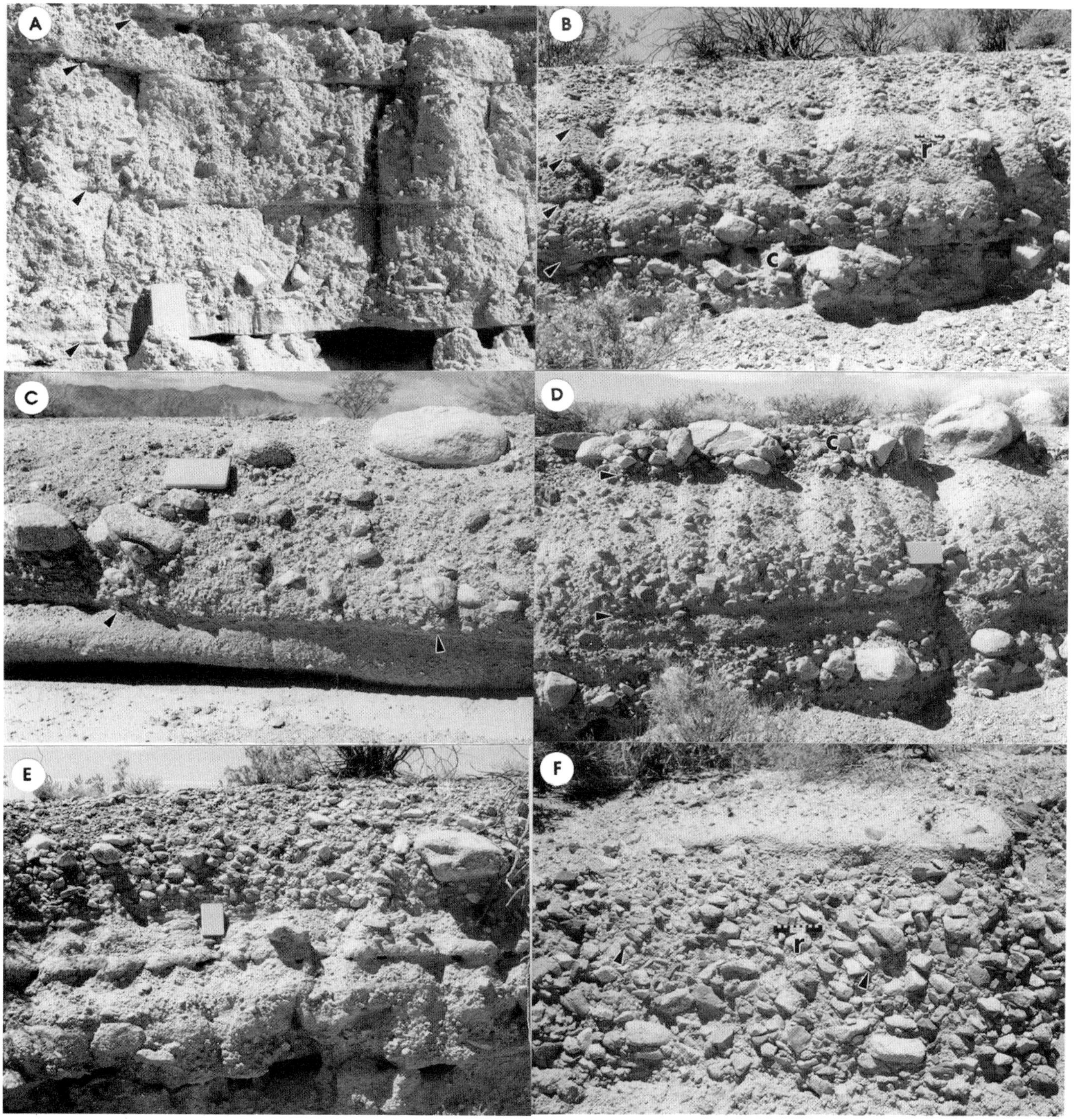

Figure 8. Photographs of Facies B and C. Arrows demarcate bedding planes and either a 20-cm-long fieldbook or a 15 cm ruler (*r*) provide scale. A: Obliquely oriented cut at station #5 of sharply bounded (arrows), stacked clast-rich debris-flow beds (Facies B) 20–50 cm thick. B: Radially oriented exposure at station #38 of stacked Facies B beds 20–60 cm thick with planar boundaries oriented at a 4° slope, parallel to the fan surface. The two lower beds are divided by a lenticle of clast-supported cobble boulder gravel (*c*) of Facies C. C: Radial cut at station #19 of 60-cm-thick Facies B bed (arrows at undulatory sharp base). Note the random distribution of matrix-supported clasts. D: Radial cut at station #13 of stacked clast-rich debris-flow beds capped by a Facies C unit (*c*). Note the apparent coarse-fraction normal grading in the 80-cm-thick bed delineated by arrows. E: Radial cut at station #39 of stacked clast-rich debris flows (below fieldbook) sharply overlain by clast-supported and well imbricated, coarse pebble to cobble gravel unit of Facies C. Subtle variations in clast modes define internal bedding in the Facies C unit. F: Radial cut of cobbly coarse to very coarse pebble gravel unit one m thick of Facies C at station #3. Note the well developed clast imbrication (arrows).

TABLE 3. SUMMARY OF GRAIN-SIZE ANALYSIS OF CUCOMUNGO FAN SAMPLES

Feature	CUC-03	CUC-04	CUC-05	CUC-06	CUC-07	CUC-08	CUC-09	CUC-10	CUC-11	CUC-12	CUC-13	CUC-14	CUC-15	CUC-16
Facies:	A	B	A	A	A	A	D	A	A	A	A	A	A	A
Station:	2	3	5	36	39	40	23	23	58	42	62	26	29	32
Distance:	–140	50	1540	3460	6050	7460	9000	9000	9290	9310	9910	10,030	11,280	12,960
Mode (ϕ):	0.38	–0.62	–0.62	0.38	–0.63	0.38	3.13	–0.63	1.38	0.38	–1.88	2.83	–0.63	0.37
Sort (ϕ):	2.36	2.78	2.35	2.38	2.03	2.17	1.12	2.31	1.86	2.20	2.32	2.35	2.28	1.58
G-S-M:	28-65-7	36-53-11	28-65-7	27-66-7	9-81-10	19-73-8	0-76-24	19-71-10	10-83-7	15-75-10	28-64-8	17-72-11	20-70-10	6-88-6
NmS-Z-C:	91-8-1	83-15-2	90-9-1	91-8-1	89-10-1	91-8-17	6-22-28	88-11-1	92-7-1	88-10-2	90-9-1	87-11-2	87-11-2	93-6-1
Spl Wt (g):	673.8	536.1	619.6	594.6	802.5	675.2	379.4	831.8	551.7	611.8	659.5	767.2	824.7	748.5
Grade							Weight percent of total sample							
m pebble	6.2	9.5	5.7	7.6	0.0	1.3	0.0	0.9	0.7	1.1	2.9	0.8	1.0	1.6
f pebble	9.9	11.0	8.9	8.2	0.9	6.4	0.0	7.1	3.3	4.0	9.8	6.5	6.6	1.8
Granule	11.7	15.0	13.0	11.7	8.1	11.6	0.0	11.3	6.0	9.9	15.8	9.8	12.9	2.8
vc sand	14.3	18.1	13.6	13.9	15.3	14.9	0.6	15.7	13.6	13.9	14.7	11.1	16.3	15.2
c sand	17.3	10.5	16.1	16.5	17.6	16.8	0.7	16.3	17.6	17.3	15.3	14.7	16.1	29.3
m sand	15.3	8.5	14.5	15.1	18.0	15.9	4.0	16.0	21.4	16.9	14.5	16.6	15.1	22.7
f sand	11.3	8.5	12.4	12.8	16.8	14.9	27.2	13.6	20.0	15.3	11.7	15.6	12.4	13.2
vf sand	7.4	7.8	8.8	7.7	13.2	10.6	43.5	9.3	10.4	11.8	7.8	14.2	9.6	7.3
c silt	3.0	5.1	3.6	3.4	4.3	3.1	10.9	4.0	2.8	4.1	2.8	4.5	4.5	2.8
m silt	1.5	2.4	1.5	1.4	2.9	1.9	6.3	2.3	1.5	2.4	1.8	2.6	2.4	1.4
f silt	0.8	1.4	0.7	0.7	1.2	1.0	3.1	1.5	1.0	1.3	1.2	1.5	1.4	0.7
vf silt	0.5	1.0	0.5	0.4	0.6	0.6	1.5	0.8	0.7	0.8	0.8	0.9	0.7	0.5
Clay	0.8	1.2	0.7	0.6	1.1	1.0	2.2	1.2	1.0	1.2	0.9	1.2	1.0	0.7

Note: Data are summaries of values collected at 1/4 ϕ intervals. Facies: A—mudflow, B—clast-rich debris flow, D—sandy washed debris-flow top. See Figure 3 for stations. Distance: radial distance downslope from fan apex to sample site. Sort: Sorting is Inclusive Graphic Standard Deviation (S_I) of Folk (1974). G-S-M—gravel-sand-mud weight percent; Nm S-Z-C—normalized sand-silt-clay weight percent; Spl Wt—sample weight. Grades: vc—very coarse, c—coarse, m—medium, f—fine, vf—very fine.

comprises 0%–35% of the stations and averages 3.5%. The lower volume of Facies A in the most proximal fan stations relative to the distal ones is due to the greater volume of Facies B (Fig. 10).

Depositional Processes

Both Facies A and B have features typical of debris-flow deposits, including extremely poorly sorted textures with grain sizes ranging from clay to boulders, lack of internal stratification, poorly organized and matrix-supported clast distribution, and recent deposits with lobate forms and sharp margins lined by the coarser clasts and plant debris moved in the flow (Beaty, 1963; Johnson, 1970, 1984; Blair and McPherson, 1994a, 1994b, 1998; Blair, 1999d). Historical debris flows on fans in the southwestern U.S.A., like the 2 September 1997 event on the Cucomungo fan, are mainly triggered by heavy rainfall on the steep colluvial slopes of the catchment usually from summer thunderstorms. These slopes destabilize from saturation and increased pore pressure, slip under the force of gravity, and then liquefy into a debris flow that moves to the alluvial fan developed at the catchment terminus (Blackwelder, 1928; Beaty, 1963; Johnson, 1970, 1984). The matrix support and distribution of clasts, as well as the sharp and steep margins of the modern flows, indicate that the Cucomungo Facies A and B debris flows were cohesive mixtures of wet sediment. Abundant conifer tree limbs, logs, and cones in many of beds reveals that these flows were in part derived from failed slopes of the upper catchment where the limited forests are developed. Clasts probably were supported during flow by a combination of matrix strength and buoyancy (Middleton and Hampton, 1976). Deposition of the Facies A and B debris flows on the fan likely was promoted by a lessening of slope and thinning as the flow expanded, which increase internal friction (e.g., Johnson, 1970, 1984; Bull, 1972). The stratigraphy of the Facies A and B beds reflects their aggradation as expansive lobes on the fan, with the planar beds accumulating on the generally smooth surface of older debris flows, and the beds with irregular bases filling pre-flow channels or gullies. The presence of the greatest bed lenticularity in transverse cuts reflects the pre-flow channel morphology. A consistency between radial fan slope and the slope of planar bedding planes in radial cuts illustrates the laminar aggradation of the debris flows on the smoother parts of the fan, with the smooth tops of the recent flows collectively dictating the slope of the fan surface (Fig. 4). Lateral wedging of beds results from the overlapping of debris-flow lobes.

The coarse-fraction normal grading observed in some of the debris-flow beds of both Facies A and B suggests that those flows lacked sufficient matrix strength to maintain support of the large clasts. This condition may have resulted from a loss of competency as the flows thinned or from an inadequate volume of mud needed for reducing permeability to maintain pore pressure (Hampton, 1975; Middleton and Hampton, 1976). The relatively low amount of clay and silt in many of the analyzed matrix samples (1%–2% and 5%–6%, respectively; Table 3) suggests that this second factor may have affected some debris flows, leading to normal grading. Alternately, the grading may have formed from overriding of the frontal snout of the debris flow, where clasts become concentrated, by the more clast-deficient body and tail of the flow. Clasts are known to become concentrated in the

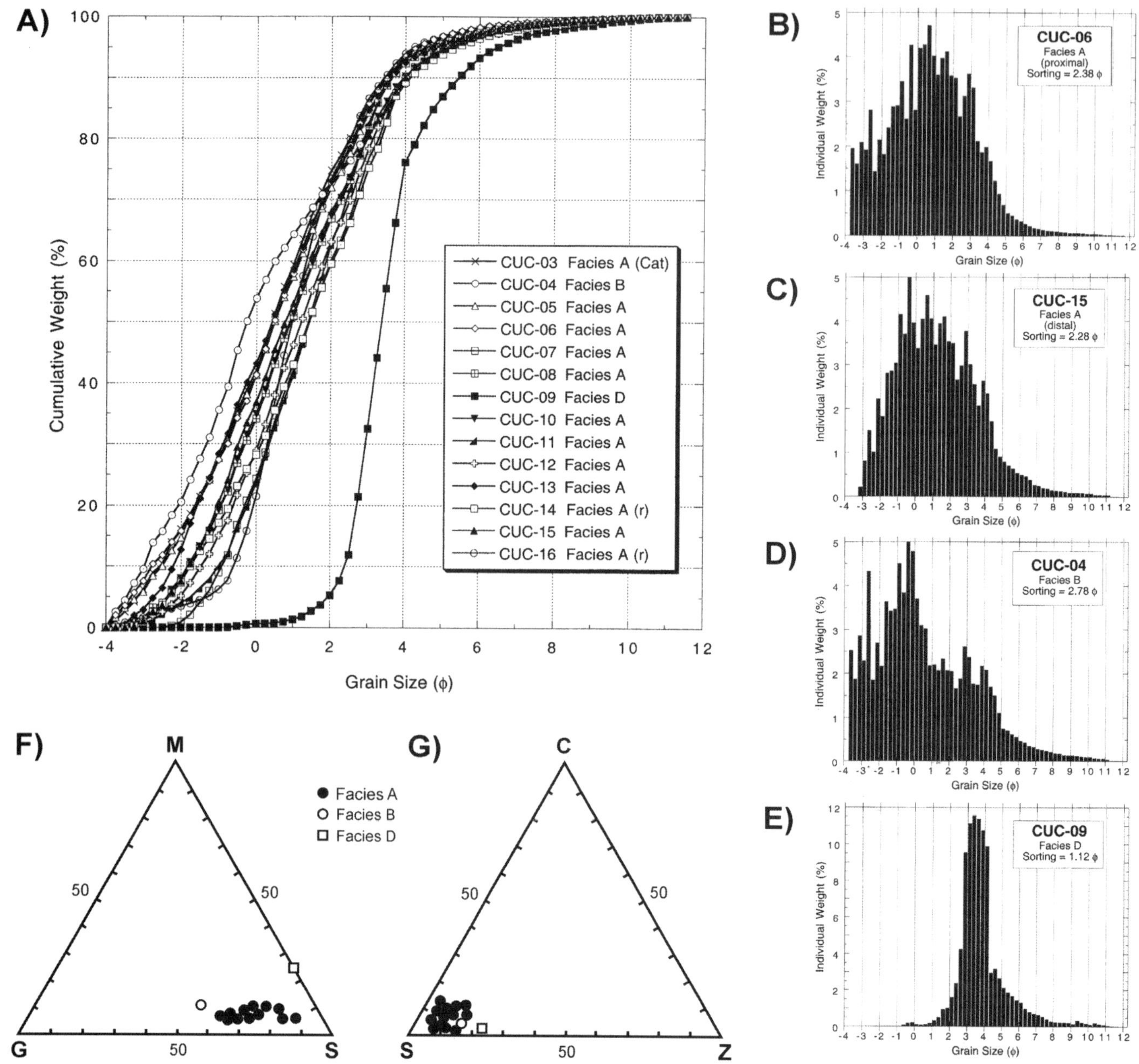

Figure 9. A: Grain-size cumulative curves of the finer than –4 ϕ fraction (clay through medium pebbles) of the analyzed Cucomungo samples (Fig. 3; Table 3); "*r*" specifies samples from the 1997 mudflow deposits. B–E: Representative histograms of samples from Facies A, B, and D, as labeled. F, G: Ternary gravel-sand-mud (F) and normalized sand-silt-clay (G) plots of the analyzed samples.

upper zone of a moving debris flow by buoyancy forces, and then are moved to the flow front because of greater velocity of the upper tread as a result of friction near the base (Johnson, 1970, 1984). High concentrations of clasts in the snout can retard the flow as a result of clast interlocking and loss of pore pressure (Johnson, 1970). In the case of the graded Cucomungo beds, clasts probably were of an insufficient concentration to halt movement of the body of the debris flow through interlocking, which would have passed over the clast-rich front as it moved downslope to form the apparent grading.

The key difference between the two debris-flow facies on the Cucomungo fan, and the basis for their differentiation, is that Facies B contains abundant coarse to very coarse pebbles and cobbles with some boulders whereas Facies A is deficient in this coarse fraction. Although the graded beds are arguably gradational in texture between clast-rich and clast-poor debris flows, nearly

TABLE 4. PEBBLE COMPOSITION OF FAN DEPOSITS

Station	Facies	Distance from apex (m)	# of counts	Composition (%)			
				Granite	Quartzite	Carbonate	Volcanic
3	B	50	358	91.9	1.7	5.0	.1.4
11	B	4250	353	87.4	7.6	2.2	2.8
14	B	5810	423	87.0	7.6	2.8	2.6
17	B	6840	383	85.1	7.5	3.7	3.7
22	A	8710	378	90.4	2.1	2.7	4.8
44	A	10,200	366	81.2	8.7	5.2	4.9
47	A	11,760	409	81.9	10.0	5.4	2.7
49	C	13,580	393	58.8	11.7	23.7	5.8
51	C	15,840	439	31.2	30.5	28.9	9.4
52	C	16,560	323	83.0	5.6	3.1	8.3

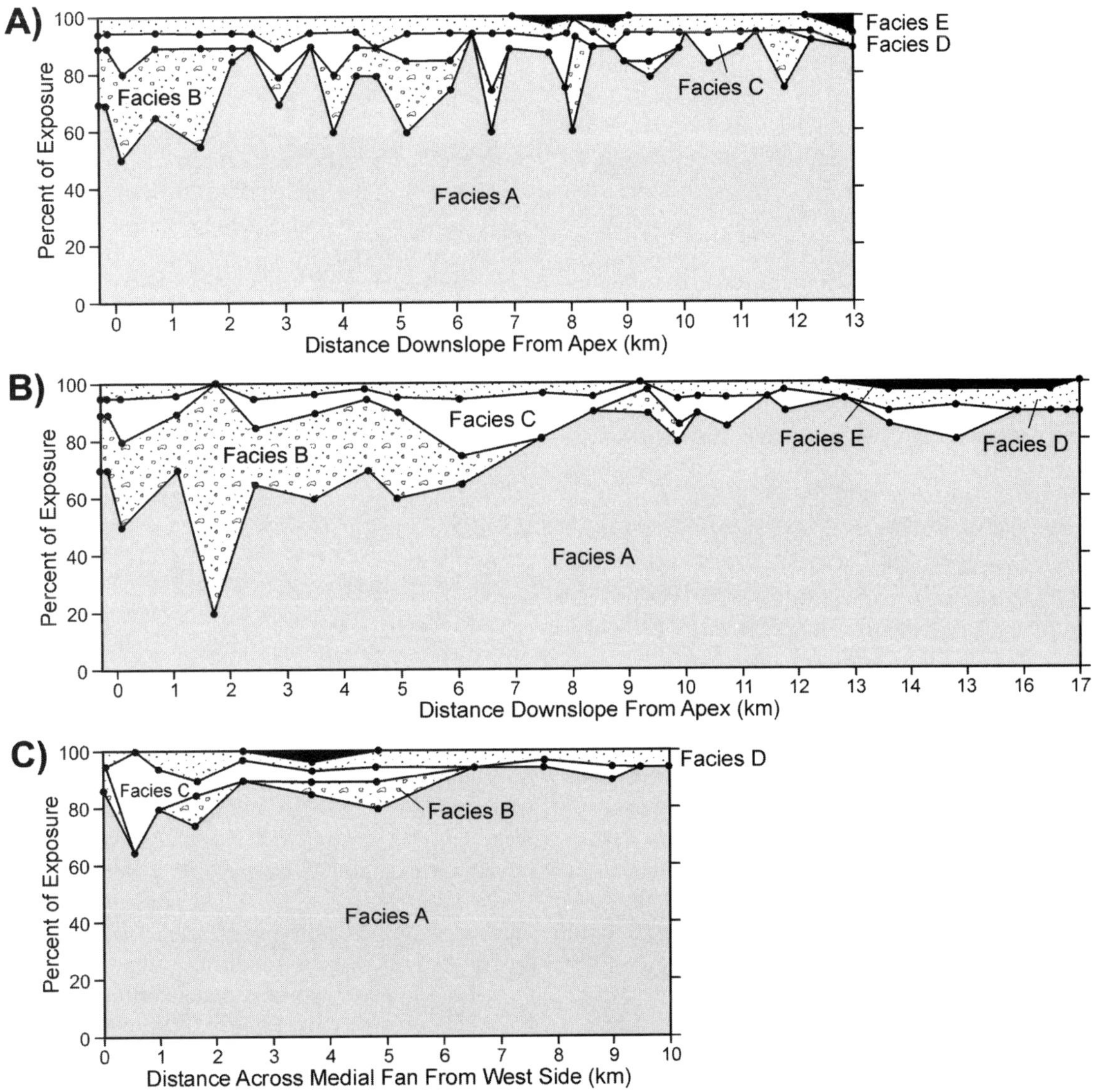

Figure 10. Facies abundance at fan stations plotted for the central (A) and eastern (B) radial transects, and for the medial cross-fan transect (C). Facies abundance values are the percentage of the total exposures at the individual stations in transects as long as 50 m.

all of the exposures can easily be assigned to either Facies A or B. The dominance of mudflow (clast-poor debris-flow) beds in the exposures relative to clast-rich debris-flow beds (79.3% versus 9.0%) suggests either that most debris flows from the catchment lacked the competency to carry coarse clasts or that clasts of this size typically were of limited availability in the failed colluvial masses spawning the flows. The presence of cobbles and boulders in the upper part of many mudflow beds (Fig. 6E) shows that their lack of coarse clasts is not due to low competency. An abundance of mudflow remnants throughout the catchment indicates that coarse clast availability in the colluvial slopes is the key factor promoting the development of mudflows rather than clast-rich debris flows during failure of the catchment colluvium.

Pebble count data from Facies A and B show that pebble composition is similar across the fan, consisting mostly (86%) of clasts derived from granitic bedrock with a small amount (14%) of quartzite, carbonate, and volcanic rock derived from the 21% of the catchment underlain by Paleozoic strata (Fig. 3; Table 4). The absence of debris-flow beds with clasts mostly derived from the Paleozoic strata indicates either that the southwest part of the catchment does not shed debris flows or that they do not reach the fan. Examination of this part of the catchment reveals that small fans and cones composed of clast-rich debris flows have been built along and are derived from the eastern flank of the Last Chance Range (Fig. 2C), and that these flows do not extend far down Cucomungo Canyon or Willow Wash, where remnants of debris flows with granitic provenance are widespread. The limited runout of the Paleozoic debris flows probably is a function of their clast richness, which promotes gravel interlocking. Their mobility may also be limited by the particle-size suite of the parent colluvium, which contains abundant clay and calcareous clay. Unlike the Paleozoic-derived debris flows, remnants of flows derived from the granitic terrain are widespread in the catchment, with flows readily passing through the drainage network and then many more kilometers down the fan. The small amount of quartzite, carbonate, and volcanic clasts present in Facies A and B implies that such clasts were incorporated as the granitic debris flows moved along Last Chance colluvium in the catchment en route to the fan.

The fanwide distribution of Facies A in the Cucomungo exposures versus the general restriction of Facies B to the proximal third of the fan (Fig. 7) shows that runout distances for the mudflows are 2 to 3 times greater than the typical runout distance of the clast-rich debris, despite their identical granitic provenance. Runout of Facies B debris flows likely was limited relative to the mudflows due to the greater abundance of clasts that hinder flow by interlocking as they become concentrated at the flow fronts and pushed laterally to the margins (Johnson, 1970; Blair and McPherson, 1998). Another factor affecting runout of the mudflows is the volume of sediment mobilized during a slip event. The recent mudflows on the Cucomungo fan are able to span 13–17 km from apex because of the high volume (~400,000 m^3) of sediment. The large sediment volume promotes runout by continuously supplying the flow on the fan, inhibiting flow thinning that commonly is responsible for deposition.

Facies C and D: Channel and Winnowed Debris-Flow Deposits

Descriptions

Facies C consists of light gray, poorly to very poorly sorted, cobble pebble gravel or pebble cobble gravel in clast-supported beds 10–100 cm thick and in sets to 20–250 cm thick (Figs. 8E, F, and 11A). Fine to medium boulders are widely scattered within these beds. Clasts typically are tightly packed and interclast pores are filled with sand. Clusters of elongate clasts display an imbricated fabric having their *a* axes aligned parallel to strike and the *a-b* planes dipping upfan. Internal stratification is visible due to changes in modal clast size (Fig. 11A). Both the bases and tops of the beds are sharp, delineating either a planar or undulatory geometry. Bedding planes in radially-oriented cuts slope 3–6° (Fig. 8E), whereas they are more irregular in strike cuts (Fig. 11B). In radial exposures the beds extend for tens of meters but in transverse cuts span just 1 to 5 m before pinching-out against debris-flow deposits. Most Facies C sets are vertically isolated by interstratified debris-flow beds, especially of Facies B, with which it is associated. Facies C comprises 0%–20% of the stations and has a fanwide average of 6.4%. It is more abundant (7.6%) in the proximal stations than in the distal ones (5.8%, Fig. 10; Table 2). It also is present on the fan surface as the bed of the largest channels (Fig. 4A).

Pebble counts of Facies C at three stations identify two compositional suites. Pebbles at station #52 consist of 83% granite, 6% quartzite, 3% carbonate, and 8% volcanic rock, similar to the counts from Facies A and B (Table 4). The two counts from stations #49 and #51 are similar to each other with an average of 45% granite, 21% quartzite, 26% carbonate, and 8% volcanic rock. Unlike the other pebble-count sites, these stations are located along the eastern margin of the Cucomungo fan near the toes of the Last Chance piedmont fans, which are rich in carbonate, quartzite, and volcanic rock (Fig. 3).

Facies D consists of tan to light gray, slightly pebbly granular sand or silty sand present as partings to laminated sets 1–30 cm thick that usually are present between Facies A mudflow beds (Figs. 7 and 11D). This facies also is common on the fan surface within shallow channels 1–3 m wide that are incised within mudflow deposits (Fig. 11C). Facies D laminations are horizontally oriented in strike cuts and slope 2–4° in radial cuts. Grain-size analysis of the <1.6 cm fraction of one sample (CUC-09) shows that it consists mostly (70.7%) of fine to very fine sand, with the remainder including 5.3% medium to very coarse sand, 21.8% silt, and 2.2 clay (Table 3). Although poorly sorted, this sample is much cleaner than those from Facies A and B (Fig. 8A, E). The tops of the Facies D sets usually are sharp and planar, whereas the bases are sharp and typically undulatory with 5–20 cm of relief. Facies D commonly weathers recessively in stratigraphic cuts to delineate the mudflow interbeds (Fig. 6). This facies comprises 0%–10% of the exposures at the stations, with a fanwide average of 5.0% (Fig. 10; Table 2). It is slightly more common distally (5.3%) than in the proximal stations (4.8%).

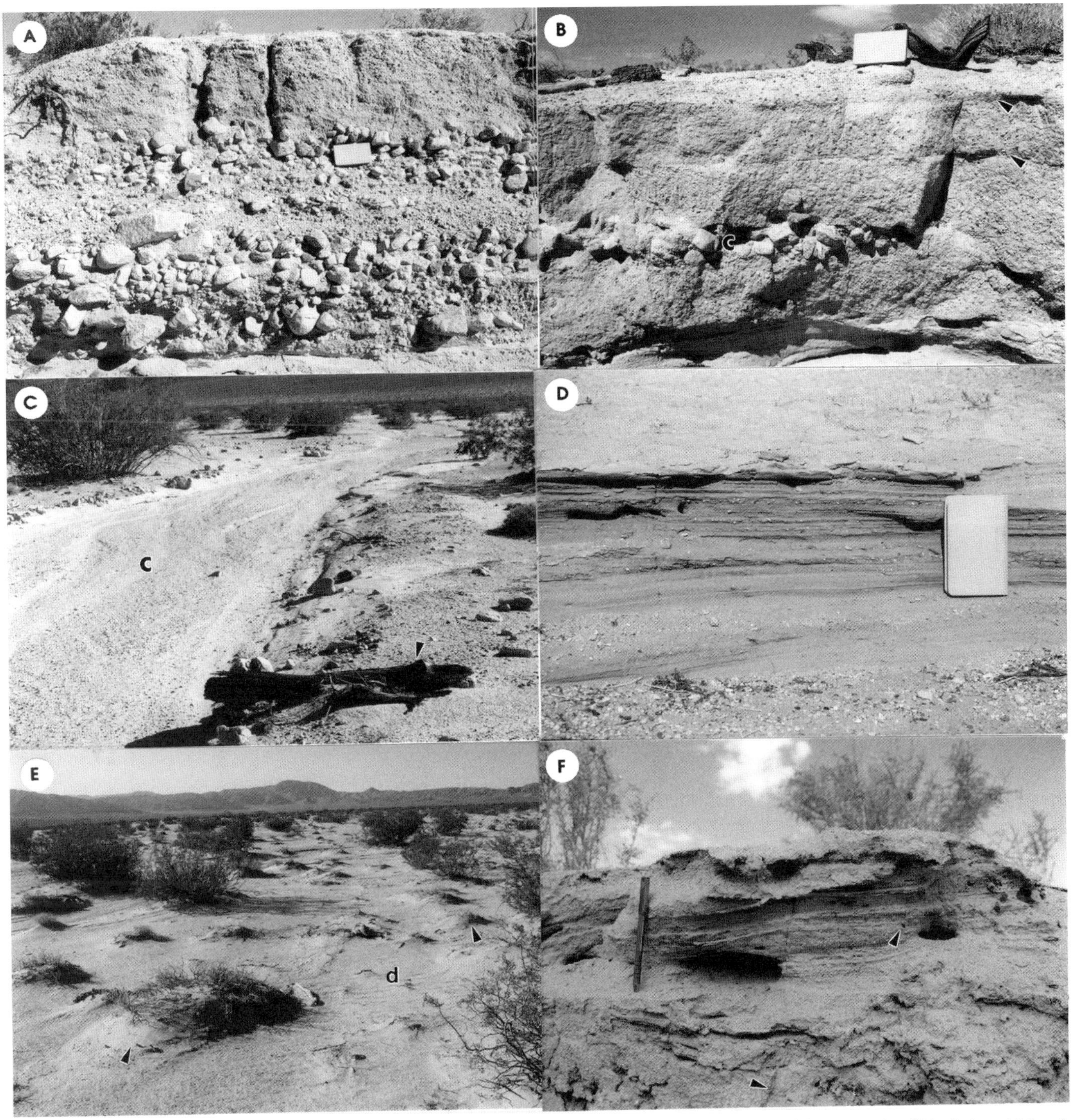

Figure 11. Photographs of Cucomungo deposits; a 20-cm-long fieldbook or pen are provided for scale. A: Radial cut at station #24 of >1-m-thick unit of clast-supported and imbricate bouldery cobble and pebble gravel (Facies C) capped (above fieldbook) by a 50-cm-thick mudflow bed. Variations in gravel modes in the Facies C unit define weak bedding. B: Obliquely oriented cut at station #18 consisting of a lenticle 20–30 cm thick of cobbly Facies C bed (*c*) within stacked Facies A mudflow units (arrows at bedding planes). C: Downslope view of distal fan (station #31) showing a 2 m wide and 30 cm deep ephemeral channel (*c*) with a smooth bed of fine pebbles, granules, and sand of Facies D. This channel is set within mudflow beds bearing tree logs (arrow). D: Radial exposure of Facies D at station #56 consisting of laminated fine pebbly granular sand. E: Across-fan view near station #32 of mudflows capped by rippled eolian drift (*d*) and low-relief eolian coppice dunes (e.g., arrows). F: Cross-fan cut through a coppice dune at station #49 showing laminated granular sand dipping at a low angle and containing finer-grained partings. Roots (lower arrow) and insect burrows (upper arrow) are common.

Depositional Processes

The textures, imbrication, and sedimentary structures indicate that Facies C was deposited by high-energy water flows down the fan slope (e.g., Walker, 1975; Blair, 1987, 1999c). The restriction of this facies to the bed of the larger channels on the fan surface, and their channel-like geometry in exposures, imply that older Facies C units also were deposited within main fan channels. The textures, sedimentary structures, and geometry of Facies D deposits also are indicative of unidirectional water flows. The match between these attributes and those of the granular sandy deposits flooring the small channels on the fan indicates that older Facies D units also were deposited in such channels. The lateral pinch out of both Facies C and D units against debris-flow deposits is consistent with incision of the surface channels into debris-flow deposits. These relationships indicate that Facies C and D develop through erosive winnowing during incision of Facies B clast-rich debris-flow deposits and Facies A mudflow deposits, respectively. In either case, coarser sediment is concentrated on the beds of the channels and the finer sediment is transported off the fan to Eureka playa. The dominance of planar stratification in Facies D and more crude planar bedding in Facies C denotes that sediment is not simply lagged during incision but is transported by the water flows, and that deposition subsequently occurs under upper-flow-regime plane-bed conditions, consistent with flows on fan slopes (Blair and McPherson, 1994b; Blair, 1999c). The dominance of planar stratification further indicates that sedimentation occurs by aggradation upon the relatively smooth channel floors. The presence of gravelly Facies C deposits in the distal fan as abundantly as in the proximal fan, where most of the clast-rich debris flows are restricted, attests to the ability of the channelized water flows to transport coarse clasts downfan. Scattered outsized boulders present within Facies C deposits (Fig. 8D, E) may represent a particle fraction that is lagged rather than transported by the water flows. The overall low volume of Facies C and D in the exposures indicates that surficial water-reworking of debris flows is a minor process on the Cucomungo fan. The common filling of both channel types by debris flows, as indicated by stratigraphic data, suggests that the fan channels are short lived because they become conduits for subsequent debris flows.

The contrasting modes of the two channel types denote that they form by flows of different strength. The greater depth (2–6 m) of the modern fan channels floored by Facies C gravel versus the shallow depth (<1 m) of channels with Facies D sand indicate that water flows producing Facies C were deeper and of greater competence than those producing Facies D. Although the channels may be activated by overland flow from rainfall on the fan or in the catchment, documented case studies in the area suggest that fan channels are most active during the recessional stage of a debris-flow event, when debris flows are no longer instigated but catchment drainage ensues (Beaty, 1963; Johnson, 1970; Blair and McPherson, 1998), as was the case for the 1997 event on the Cucomungo fan. In contrast, the wide distribution on the fan of shallow channels with Facies D deposits, many without direct connection to the catchment, suggests that they may be active during less catastrophic discharge caused by rainfall on the fan or in the nearby highlands. The greater richness of quartzite and carbonate pebbles within Facies D deposits in the vicinity of the Last Chance Range piedmont (Fig. 3; Table 4) attests to runoff from these adjoining fans as a source of water flowing through some of the Facies D channels.

Facies E: Eolian Deposits

Description

Facies E consists of moderately to poorly sorted medium to very fine sand, or of slightly granular medium to very coarse sand, that are present in laterally discontinuous, laminated sets 1–25 cm thick (Fig. 11F). Laminations vary from planar to undulatory, and generally slope at a low (0–10°) angle in multiple directions. Plant roots and insect burrows are common. This facies is present on the fan surface as low relief mounds 1–2 m across centered on and sloping away from plants (Fig. 11E). It is absent in the fan cuts except for some of the distal-fan stations where it comprises the surficial bed.

Depositional Processes

The grain size, laminations, and morphology of Facies E are typical of eolian drift and coppice dunes on sand-bearing fans that accumulate in sheltered zones, such as in rills or around plants (Blair, 2000). This facies is produced by wind reworking of the sand present in the Cucomungo channels or exposed mudflows. The flat laminations are typical of transport as wind ripples (Fryberger and Schenk, 1988), bedforms of which are visible on the fan surface (Fig. 11F). The poor sorting of these deposits relative to dune facies indicates that they are not highly reworked by the wind. Also, the low volume of this facies in the studied exposures, despite high sand availability, reveals that wind erosion is a minor secondary surface process on the Cucomungo fan. The absence of buried eolian beds suggests that these deposits may be readily reworked by passing flows. Other secondary processes noted on the Cucomungo fan are plant rooting, burrowing by insects and rodents, clast weathering, desert varnish development, wind and rainsplash erosion to form desert pavements, and soil development.

ORIGIN OF THE GIANT SIZE OF THE CUCOMUNGO FAN

In addition to sedimentologic and morphologic evidence, clues to the origin of the extraordinary size of the Cucomungo alluvial fan are provided by comparison with the other fans in Eureka Valley. Geomorphic data were collected from 13 additional fans chosen to represent the spectrum of the ~110 fans in the valley, and sedimentological data were obtained by examining the exposures of nine of these fans (Fig. 1B). The Cucomungo dwarfs these other fans, with a radial length 3 to 50 times as long and an area 10 to 1200 times as large (Table 1). The Cucomungo also has a lower average composite slope (1.8°) than the others (2.3–22.5°). Although the area of the Cucomungo catchment is large at 86.8 km^2, other fans have catchments with similar or greater area, including 66.3 km^2 for Fan #2 and 279.0 km^2 for Fan #3. These data show that the size of the Cucomungo fan cannot simply be attributed to catchment size. The large size of the Cucomungo fan likewise cannot be attributed to its position on the passive side of a half graben because it exists on the tectonically active side of Eureka Valley.

Exposures of other Eureka Valley fans reveal a variety of facies. Many of the fans of the Last Chance Range piedmont flanking eastern Eureka Valley are comparatively steep (6.4–9.2°; Table 1) and have surface features typical of clast-rich debris-flow levees on the proximal fan that are continuous downslope with debris-flow lobes (Fig. 12A). Their exposures consist of matrix-supported, muddy bouldery pebbly cobble gravel typical of clast-rich debris flows (Fig. 12B; Beaty, 1963; Blair and McPherson, 1998). They denote that clast-rich debris flows are triggered when heavy rainfall saturates the catchments due to the presence of colluvium rich in particles spanning in size from mud through boulders formed through weathering of the Paleozoic quartzite, mudrock, and carbonate rock of the Last Chance Range. The three fans in the dataset with lower slopes (2.3–4.0°; Fans #1, 2, and 8) have exposures consisting of interstratified couplets of pebbly granular sand and sandy pebble gravel with backset-bedded units typical of sheetflood-dominated fans (Fig. 12C, D; Blair, 1987, 1999c, 2000). Fans #1 and #2 are mainly derived from granitic bedrock along northwest Eureka Valley and #8 from Quaternary conglomeratic sandstone on the east side, both of which typically yield sandy and gravelly but mud-deficient colluvium known to incite sheetfloods when destabilized during rainfall (Blair, 1987, 1999a, 1999c). The steepest (22.5°) fan in the dataset, #12, has a composite and constituent morphology typical of a rockfall-dominated fan, as corroborated by exposures of unstratified chutes of angular gravel (Fig. 12E, F). This facies forms by the downward rolling and sliding of loose gravel liberated from fractured bedrock (Blair and McPherson, 1994b).

These reconnaissance studies reveal that the Cucomungo is the only fan in Eureka Valley containing mudflow deposits. Thus, its dominance by mudflows may explain the exceptional size of the Cucomungo fan because, with all other factors constant, mudflows have greater runout than clast-rich debris flows due to low clast interlocking at the margins. Yet, other mudflow fans in the region, such as in Death Valley, are not of giant size. Also, the presence of steep and sharp margins along the recent mudflow tracts on the Cucomungo fan indicates that they were cohesive flows, not more watery flows that would have greater mobility. The most critical evidence to explain the large runout of Cucomungo mudflows is the size of the recent mudflow tracts, which shows that the catchment typically supplies an enormous volume (~200,000–600,000 m^3) of sediment to the fan during a single flood event. This high sediment supply allows the mudflows to maintain thickness and inertia, promoting runout of 17 km or more down the fan slope. Mudflows of smaller volume would undergo deposition closer to the fan apex due to increased friction resulting from flow thinning, regardless of mobility. Thus, it likely is the extraordinary volume of a typical Cucomungo mudflow aided by low clast content that has caused the large fan size. A third factor permitting the fan to reach giant size is that space is available in Eureka Valley for progradation. The relevance of this factor is illustrated by the shorter (13.5 km) radial length of the western edge of the fan resulting from topographic restriction by the opposing fans of the Inyo piedmont. The maximum length attained by mudflow processes is indicated by the 17.1 km radius along the south margin, where the fan is bordered by a mid-basin channel and playa complex that does not impede fan progradation.

The abundance of mudflows on the Cucomungo fan and their extraordinary volume are attributes directly caused by conditions in the catchment and, more specifically, in the part of the catchment underlain by granitic rock, where a colluvial mantle generally deficient in coarse clasts is well developed. The recent Cucomungo mudflows attest that this colluvium can readily be mobilized during a heavy rainfall event. Although fans derived from granitic bedrock are common in the region, none besides the Cucomungo is known to consist mainly of mudflows, most likely because coarser clasts are commonly supplied by catchments. A comparison of the colluvium and deposits of fans derived from granitic bedrock in other parts of the region shows the abnormality of the particle suite yielded from the Cucomungo catchment. Colluvium in the granitic catchments of the Tuttle and Lone Pine fans in Owens Valley California, the Deadman Canyon fan in Walker Lake Basin, Nevada, the Smith Mountains and Anvil Canyon fans in Death Valley, and the Roaring River fan of Colorado all possess abundant coarse gravel, including pebbles, cobbles, boulders, and blocks (Blair, 1987, 1999c, 2001, 2002). The size suites of the <1.6 cm fraction of these other fans are similar to each other but differ from that of the Cucomungo samples. In the case of the Deadman, Lone Pine, and Tuttle fans, the first of which is dominated by clast-rich debris-flow deposits and the second two by noncohesive sediment-gravity-flow deposits, the matrix fraction is much richer in gravel (especially fine to medium pebbles) and in fine to very fine silt and clay relative to the Cucomungo samples (Fig. 13A). The matrix of samples from sheetflood-dominated fans in Death Valley and Colorado also are more gravelly than the Cucomungo samples, although their mud fraction is more similar (Fig. 13B).

Coarse to very coarse pebbles, cobbles, boulders, and blocks form in granitic uplands by jointing, fracturing, exfoliating, sheeting, and other forms of physical disintegration (e.g, Ritter, 1978; Blair and McPherson, 1994a). In contrast, silt and clay typically form through chemical weathering. Granitic bedrock in the Cucomungo catchment differs from the catchments of the other studied granitic fans by the lack of morphologic features reflecting large-scale physical disintegration, such as exfoliation domes (Fig. 2C, D). The one condition of the Cucomungo catchment

Figure 12. A: Photographs of other fans in Eureka Valley (Fig. 1B); fieldbook scale is 20 cm long. A: Overview of Fan #10. Light zone is a recent clast-rich debris-flow deposit with levees on the upper fan (arrows) changing distally to lobes. B: Cross-fan exposure near the distal end of Fan #10 consisting of clast-rich, matrix-supported debris flows; arrows mark bedding planes. C: View of Hanging Rock fan (Fan #8) probably built of sheetflood deposits. D: Radial exposure 7 m thick on Fan #1 showing alternating couplets of pebbly sand and sandy cobbly pebble gravel typical of sheetflood sequences. E: View of the steep (20–25°) Fan #12 composed of rockfall deposits. F: Radial exposure of Fan #12 consisting of matrix-free, clast-supported cobble pebble gravel in planar and lenticular beds typical of deposition in rockfall chutes.

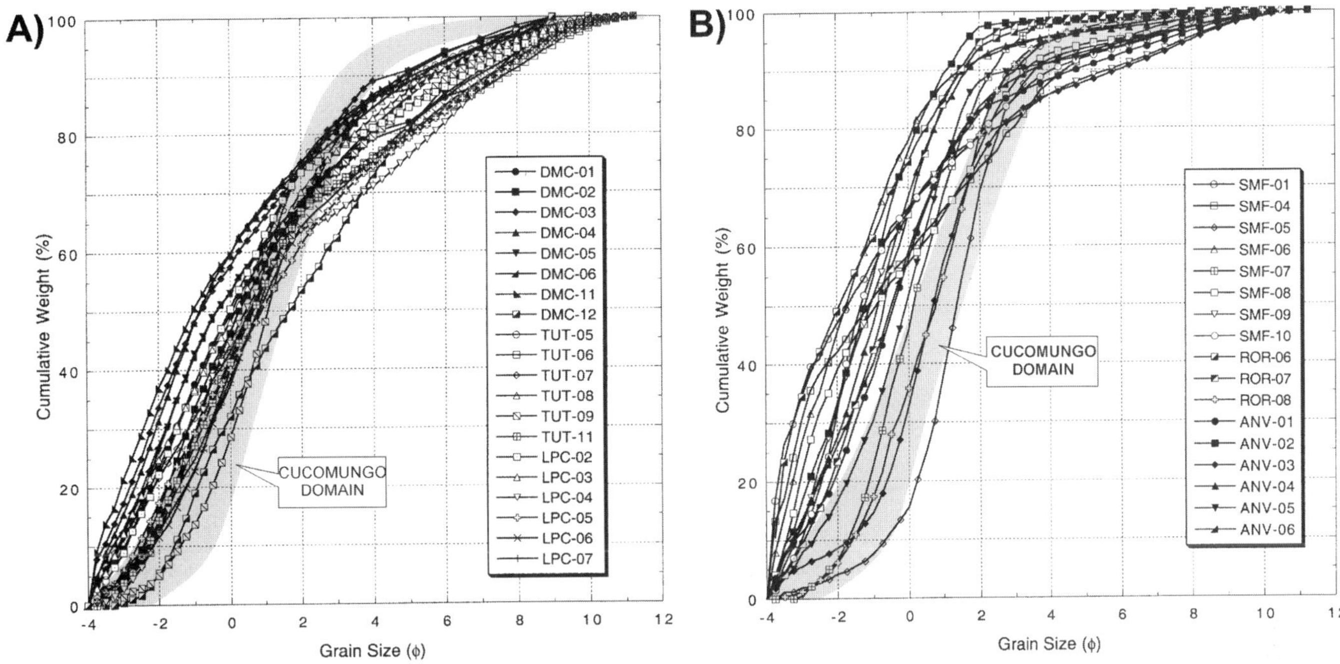

Figure 13. Grain-size cumulative curves of the <–4 ϕ fraction of samples from other fans in Nevada and California derived from catchments underlain by granitic bedrock. The domain of Cucomungo mudflow samples is shaded. A: Curves of samples from fans built by debris flows (DMC samples) or noncohesive sediment-gravity flows (TUT and LPC). B) Curves of samples from sheetflood fans. Abbreviations and sources: DMC—Deadman Canyon fan, Walker Lake, Nevada; TUT and LPC—Tuttle and Lone Pine Canyon fans, Owens Valley, California (Blair, 2001, 2002); SMF—Smith Mountain fans, southern Death Valley; ROR—Roaring River fan, Colorado (Blair, 1987); ANV—Anvil Canyon fan, Death Valley (Blair, 1999c).

that can account for all of its abnormalities is that the granitic bedrock there has been widely crushed from transpressive shear within the dextral strike-slip Furnace Creek fault zone (Figs. 1 and 3). Intense tectonic shear along closely spaced fractures within this major fault zone has resulted in a significantly higher sediment yield and greater abundance of silt, sand, and fine pebbles, at the expense of coarser clasts, than is typically yielded from catchments underlain by granite. This tectonic control on colluvial yield and grain size is the ultimate reason why the catchment exports high-volume mudflows responsible for building the giant Cucomungo fan.

CONCLUSIONS

The Cucomungo fan is notably larger than other fans documented around the globe. It is built mainly of mudflows of abnormally high volume (200,000–600,000 m^3). Less abundant facies are clast-rich debris flows, gravelly channel beds, sandy rill deposits, and coppice dunes. The high volume of clast-deficient colluvium yielded from the Cucomungo catchment is caused by extreme tectonic shearing of granitic bedrock along a transpressive segment of the dextral strike-slip Furnace Creek fault. Granite is widely crushed in this shear zone, causing it to yield large volumes of colluvium rich in silt, sand, and fine to medium pebbles, but not coarser clasts. The mobilization of this colluvium during heavy precipitation creates mudflows of extraordinary volume that are responsible for building the giant Cucomungo fan.

ACKNOWLEDGMENTS

I thank R. Anderson of the U.S. National Park Service for approving this research and the sampling permit. Reviewers H. Mills and J. O'Connor and editors A. Archer and M. Chan are thanked for critical comments and for handling this article. This project was funded by Blair & Associates.

REFERENCES CITED

Allen, P.A., and Densmore, A.L., 2000, Sediment flux from an uplifting fault block: Basin Research, v. 12, p. 367–380.

Anstey, R.L., 1965, Physical characteristics of alluvial fans: United States Army Natick Laboratory, Technical Report ES-20, 109 p.

Anstey, R.L., 1966, A comparison of alluvial fans in west Pakistan and the United States: Pakistan Geographical Review, v. 21, p. 14–20.

Beaty, C.B., 1963, Origin of alluvial fans, White Mountains, California and Nevada: Association of American Geographers Annals, v. 53, p. 516–535.

Blackwelder, E., 1928, Mudflow as a geologic agent in semi-arid mountains: Geological Society of America Bulletin, v. 39, p. 465–484.

Blair, T.C., 1987, Sedimentary processes, vertical stratification sequences, and geomorphology of the Roaring River alluvial fan, Rocky Mountain National Park, Colorado: Journal of Sedimentary Petrology, v. 57, p. 1–18.

Blair, T.C., 1999a, Cause of dominance by sheetflood versus debris-flow processes on two adjoining alluvial fans, Death Valley, California: Sedimentology, v. 46, p. 1015–1028.

Blair, T.C., 1999b, Form, facies, and depositional history of the North Long John rock avalanche, Owens Valley, California: Canadian Journal of Earth Sciences, v. 36, p. 855–870.

Blair, T.C., 1999c, Sedimentary processes and facies of the waterlaid Anvil Spring Canyon alluvial fan, Death Valley, California: Sedimentology, v. 46, p. 913–940.

Blair, T.C., 1999d, Sedimentology of the debris-flow-dominated Warm Spring Canyon alluvial fan, Death Valley, California: Sedimentology, v. 46, p. 941–965.

Blair, T.C., 2000, Sedimentology and progressive tectonic unconformities of the sheetflood-dominated Hell's Gate alluvial fan, Death Valley, California: Sedimentary Geology, v. 132, p. 233–262.

Blair, T.C., 2001, Outburst flood sedimentation on the proglacial Tuttle Canyon alluvial fan, Owens Valley, California, U.S.A.: Journal of Sedimentary Research, v. 71, p. 658–680.

Blair, T.C., 2002, Alluvial-fan sedimentation from a glacial-outburst flood, Lone Pine, California, and contrasts with meteorological flood deposits, *in* Martini, I.P., Baker, V.R., and Garzón, G., Flood and megaflood processes and deposits: International Association of Sedimentologists Special Publication 32, p. 113–140.

Blair, T.C., and McPherson, J.G., 1994a, Alluvial fan processes and forms, *in* Abrahams, A.D., and Parsons, A., eds, Geomorphology of desert environments: London, Chapman Hall, p. 354–402.

Blair, T.C., and McPherson, J.G., 1994b, Alluvial fans and their natural distinction from rivers based on morphology, hydraulic processes, sedimentary processes, and facies: Journal of Sedimentary Research, v. A64, p. 451–490.

Blair, T.C., and McPherson, J.G., 1998, Recent debris-flow processes and resultant form and facies of the Dolomite alluvial fan, Owens Valley, California: Journal of Sedimentary Research, v. 68, p. 800–818.

Blair, T.C., and McPherson, J.G., 1999, Grain-size and textural classification of coarse sedimentary particles: Journal of Sedimentary Research, v. 69, p. 6–19.

Brogan, G.D., Kellogg, K.S., Slemmons, D.B., and Terhune, C.L., 1991, Late Quaternary faulting along the Death Valley-Furnace Creek fault system, California and Nevada: U.S. Geological Survey Bulletin 1991, 23 p.

Bull, W.B., 1962, Relations of alluvial fan size and slope to drainage basin size and lithology in western Fresno County, California: U.S. Geological Survey Professional Paper 450-B, p. 51–53.

Bull, W.B., 1972, Recognition of alluvial fan deposits in the stratigraphic record, *in* Rigby, J.K., and Hamblin, W.K., eds., Recognition of ancient sedimentary environments: SEPM Special Publication 16, p. 63–83.

Drew, F., 1873, Alluvial and lacustrine deposits and glacial records of the Upper Indus Basin: Geological Society of London Quarterly Journal, v. 29, p. 441–471.

Folk, R.L., 1974, Petrology of sedimentary rocks: Austin, Texas, Hemphill's Publishing Company, 182 p.

Fryberger, S.G., and Schenk, C.J., 1988, Pin-stripe lamination: A distinctive feature of modern and ancient eolian sediments: Sedimentary Geology, v. 55, p. 1–15.

Hampton, M.A., 1975, Competence of fine-grained debris flows: Journal of Sedimentary Petrology, v. 45, p. 834–844.

Horton, B.K., and DeCelles, P.G., 2001, Modern and ancient fluvial megafans in the foreland basin system of the central Andes, southern Bolivia: Implications for drainage network evolution on fold-thrust belts: Basin Research, v. 13, p. 43–63.

Horton, R.E., 1945, Erosional development of streams and their drainage basins; hydrophysical approach to quantitative morphology: Geological Society of America Bulletin, v. 56, p. 275–370.

Hunt, C.B., and Mabey, D.R., 1966, General geology of Death Valley, California—Stratigraphy and structure: U.S. Geological Survey Professional Paper 494-A, 165 p.

Johnson, A.M., 1970, Formation of debris flow deposits, *in* Johnson, A.M., ed., Physical processes in geology: San Francisco, Freeman, Cooper, Co., p. 433–448.

Johnson, A.M., 1984, Debris flow, *in* Brunsden, D., and Prior, D.B., eds., Slope instability: New York, Wiley, p. 257–361.

Lustig, L.K., 1965, Clastic sedimentation in Deep Springs Valley, California: U.S. Geological Survey Professional Paper 352-F, 192 p.

Middleton, G.V., and Hampton, M.A., 1976, Subaqueous sediment transport and deposition by sediment gravity flows, *in* Stanley, D.J., and Swift, D.J.P., eds., Marine sediment transport and environmental management: New York, John Wiley & Sons, p. 197–218.

Oldow, J.S., 1992, Late Cenozoic displacement partitioning in the northwestern Great Basin: Reno, Nevada, Geological Society of Nevada, Walker Lake Symposium Proceedings, p. 17–52.

Reheis, M.C., 1992, Geologic map of late Cenozoic deposits and faults in parts of the Soldier Pass and Magruder Mountain 15' quadrangles, Inyo and Mono counties, California, and Esmeralda County, Nevada: U.S. Geological Survey Miscellaneous Investigations Series Map I-2268, scale 1:24,000.

Reheis, M.C., Sarna-Wojcicki, A.M., Burbank, D.M., and Meyer, C.E., 1991, The late Cenozoic section at Willow Wash, west-central California: A tephrochronologic Rosetta Stone: U.S. Geological Survey Open-File Report 91-290, p. 46–66.

Ritter, D.F., 1978, Process geomorphology: Dubuque, Iowa, W.C. Brown, 603 p.

Schumm, S.A., 1977, The fluvial system: New York, Wiley, 338 p.

Shukla, U.K., Singh, I.B., Sharma, M., and Sharma, S., 2001, A model of alluvial megafan sedimentation: Ganga megafan: Sedimentary Geology, v. 144, p. 243–262.

Strahler, A.N., 1964, Quantitative geomorphology of drainage basins and channel networks, *in* Chen, V.T., ed., Handbook of applied hydrology: New York, McGraw-Hill, p. 40–74.

Strand, R.G., 1967, Geologic map of California, Mariposa Sheet: Sacramento, California Division of Mines and Geology, scale 1:250,000.

Surrell, A., 1841, Etude sur les torrents des Hautes-Alps, 1st Edition: Paris, Imprimerie Cusset.

Walker, R.G., 1975, Conglomerate: Sedimentary structures and facies models, *in* Depositional environments as interpreted from primary sedimentary structures and stratification sequences: SEPM Short Course Notes 2, p. 133–161.

Whipple, K.X., and Traylor, C.R., 1996, Tectonic controls on fan size: The importance of spatially-variable subsidence rates: Basin Research, v. 8, p. 351–366.

Manuscript Accepted by the Society January 22, 2003

Printed in the USA

Geological Society of America
Special Paper 370
2003

Desmoinesian coal beds of the Eastern Interior and surrounding basins: The largest tropical peat mires in Earth history

Stephen F. Greb
William M. Andrews
Cortland F. Eble
Kentucky Geological Survey, University of Kentucky, Lexington, Kentucky 40506, USA

William DiMichele
Smithsonian Institution, National Museum of Natural History, Washington, D.C., USA

C. Blaine Cecil
U.S. Geological Survey, Reston, Virginia, USA

James C. Hower
Center for Applied Energy Research, University of Kentucky, Lexington, Kentucky, USA

ABSTRACT

The Colchester, Springfield, and Herrin Coals of the Eastern Interior Basin are some of the most extensive coal beds in North America, if not the world. The Colchester covers an area of more than 100,000 km^2, the Springfield covers 73,500–81,000 km^2, and the Herrin spans 73,900 km^2. Each has correlatives in the Western Interior Basin, such that their entire regional extent varies from 116,000 km^2 to 200,000 km^2. Correlatives in the Appalachian Basin may indicate an even more widespread area of Desmoinesian peatland development, although possibly slightly younger in age. The Colchester Coal is thin, but the Springfield and Herrin Coals reach thicknesses in excess of 3 m.

High ash yields, dominance of vitrinite macerals, and abundant lycopsids suggest that these Desmoinesian coals were deposited in topogenous (groundwater fed) to soligenous (mixed-water source) mires. The only modern mire complexes that are as widespread are northern-latitude raised-bog mires, but Desmoinesian Midcontinent paleomires were topogenous and accumulated within 10° of the paleo-equator.

The extent and thickness of Desmoinesian paleomires resulted from the coincidence of prime peat-forming factors, including a seasonally wet paleoclimate; cyclothemic transgressions and base-level rise above extensive, low-relief cratonic areas floored by vast, impermeable paleosols; broad floodplains along large rivers with a groundwater table high enough to hydrologically link peatlands and keep them wet; low, relatively uniform rates of tectonic subsidence; and accumulation in a basin surrounded by low relief, which led to minimal sediment input.

Keywords: Carboniferous, Carbondale Formation, Illinois Basin, Midcontinent, topogenous, peatlands.

Greb, S.F., Andrews, W.M., Eble, C.F., DiMichele, W., Cecil, C.B., Hower, J.C., 2003, Desmoinesian coal beds of the Eastern Interior and surrounding basins: The largest tropical peat mires in Earth history, *in* Chan, M.A., and Archer, A.W., eds., Extreme depositional environments: Mega end members in geologic time: Boulder, Colorado, Geological Society of America Special Paper 370, p. 127–150.

INTRODUCTION

Wanless (1975a) noted that the Colchester Coal of the Eastern Interior (Illinois) Basin, and its correlatives in the Western Interior Basin, combined to form the most widespread coal bed in North America and possibly the world. Currently there is no database or central source area for comparing global coal areas on a bed basis, but for the purpose of determining the most widespread coals of all time, basin size can be used as an initial limiting extent. Some of the largest coal basins in the world are the Bowen (Queensland) Basin of Australia; Karoo Basin of South Africa; and the Powder River, Williston, Appalachian, Western Interior, and Eastern Interior Basins of the United States (Fig. 1).

Many Bowen Basin coal seams split into multiple beds or benches, rather than occurring as single, widespread beds (Hower et al., 1995; Diessel, 1998). At least one Permian coal bed contains a tuff, which is extensive along the outcrop margin of the coal, suggesting coeval, basin-wide peat accumulation (Michaelsen et al., 2000). The Leichardt and Vermont "superseams" are extensive for 200 km along strike and may have areal distributions of 20,000 km^2 (Mallett et al., 1995; Michaelsen et al., 2000), which is vast, but less extensive than the most extensive Desmoinesian coals of the Eastern Interior Basin.

Permian coals are also extensive in the Karoo Basin of South Africa, but are restricted to three coal fields on the northern stable platform of the basin. The number 4 and 5 seams of the Ecca Group can be correlated for more than 250 km along the outcrop margin of the coal fields, but they are bisected by numerous paleochannels and split into multiple subseams across part of the Natal Coal Field (Cadle et al., 1993).

In North America, the Wyodak-Anderson coal, a Paleocene coal of the Powder River Basin (Fig. 1), is currently the largest coal producer in the United States, producing 320 million short tons in 2001 (U.S. Department of Energy, 2001). It has a total area of 24,000 km^2 (Ellis et al., 1999), which is less than the extent of the most extensive coals in the Eastern Interior Basin. Additionally, the Wyodak-Anderson is actually a zone of as many as 11 separate beds (Hardie and Van Gosen, 1986) in a coal zone rather than a single bed, such that it may be difficult to discern the distribution of any single coeval paleomire.

The Pittsburgh coal is the second largest producer in the United States, producing 81 million short tons in 2001 (U.S. Department of Energy, 2001). This Upper Pennsylvanian bed of the Northern Appalachian Basin (Fig. 1) covers an area of more than 21,450 km^2 (Northern and Central Appalachian Basin Coal Regions Assessment Team, 2001). Another extensive coal from the Northern Appalachian Basin is the upper Middle Pennsylvanian Upper Freeport coal. This coal produced 10.3 million short tons in 2001, and ranked nineteenth nationally (Northern and Central Appalachian Basin Coal Regions Assessment Team, 2001). The Upper Freeport coal covers an area of at least 27,000 km^2 (Northern and Central Appalachian Basin Coal Regions Assessment Team, 2001). Some of the top-producing coals of the Central Appalachian Basin (Fig. 1), such as the Fire Clay (Hazard No. 4) coal (18.5 million short tons, eleventh nationally, Northern and Central Appalachian Basin Coal Regions Assessment Team, 2001), cover areas of less than 17,000 km^2 (Northern and Central Appalachian Basin Coal Regions Assessment Team, 2001), and most of these Appalachian Basin coals split or develop into zones toward the foreland basin axis.

Eastern Interior Basin Coals

The largest producing coals in the Eastern Interior (Illinois) Basin are the Desmoinesian coals of the Carbondale Formation (Fig. 2). These coal beds have member status in Illinois and Indiana, but bed status in Kentucky. For the purposes of this report, Eastern Interior Basin coal beds will be treated as members for

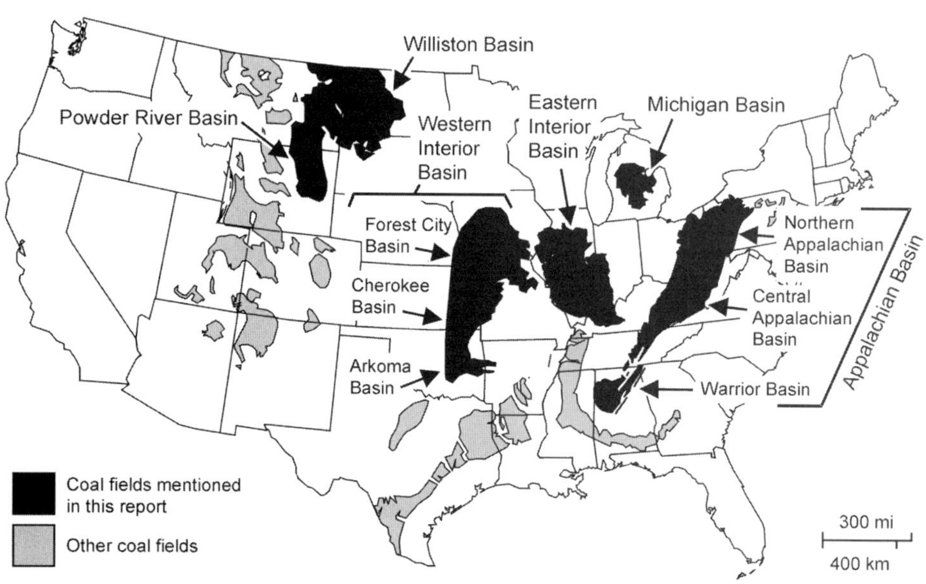

Figure 1. Map of coal fields in conterminous United States (after Tully, 1996).

Series	Stage	Illinois (coals are members)		Indiana (coals are members)		Kentucky (coals are beds)		Tri-state nomenclature (2001)	
Middle Pennsylvanian (part)	Desmoinesian (part)		Danville (No. 7)		Danville (VII)	McLeansboro Group (part)	Wheatcroft	*	Danville
		Carbondale Formation	———		———		Western Ky. No. 13		Baker
			Jamestown *Brereton Ls.*	Carbondale Formation	Jamestown *Providence Ls.*		Western Ky. No. 12 *L. Providence Ls.*	**	Paradise/ Jamestown
			Herrin (No. 6)		**Herrin (VI)**	Carbondale Formation	**Western Ky. No. 11**	Carbondale Formation	**Herrin**
			Briar Hill (No. 5A)		Bucktown (Vb)		Western Ky. No. 10		Briar Hill
			Springfield-Harrisburg (No. 5)		**Springfield (V)**		**Mulford-W. Ky. No. 9**		**Springfield**
			Summum (No. 4)		Houchin Creek (IVa)		Western Ky. No. 8b		Houchin Creek
			Survant		Survant (IV)		Western Ky. No. 8		Survant
			Colchester (No. 2)		**Colchester (IIIa)**		**"S" coal**		**Colchester**
		Tradewater Formation	Dekoven		———		Western Ky. No. 7		Dekoven
			Seelyville		Seelyville (No. III)		Western Ky. No. 6		Davis

Figure 2. Stratigraphic column for Eastern Interior Basin showing nomenclature for the basin's three states. Tri-state Committee on Correlation of the Pennsylvanian System in the Illinois Basin nomenclature (2001) is used in this report. Ls—limestone.

consistency. The two largest producers are the Springfield and Herrin Coals. The Springfield is the third largest producer in the nation at 42 million short tons in 2001, and the Herrin ranked tenth nationally, with 19.1 million short tons (U.S. Department of Energy, 2001). Figure 3 is a cross section of the Eastern Interior Basin showing the stratigraphic position and extent of the Carbondale Formation coals. More coals are preserved beneath the Colchester Coal in the deeper, southern part of the basin above depocenters called the Fairfield Basin (Fb in Fig. 3) and Moorman Syncline (Ms in Fig. 3). Only three coals, the Herrin, Springfield, and Colchester, have basin-wide extent. Extensive and uniform distribution of strata and facies are typical of Desmoinesian and younger Pennsylvanian strata in the basin (Wanless and Weller, 1932; Kosanke et al., 1960; Wanless et al., 1963; Wanless, 1975b; Wanless and Wright, 1978; Greb et al., 1992). In general, coal bed extent and uniformity decreases beneath the Colchester and above the Herrin Coal. Increasingly extensive coals in the basin seem to parallel a trend toward decreasing tectonic accommodation in the basin from Morrowan into Desmoinesian time (Greb et al., 2002).

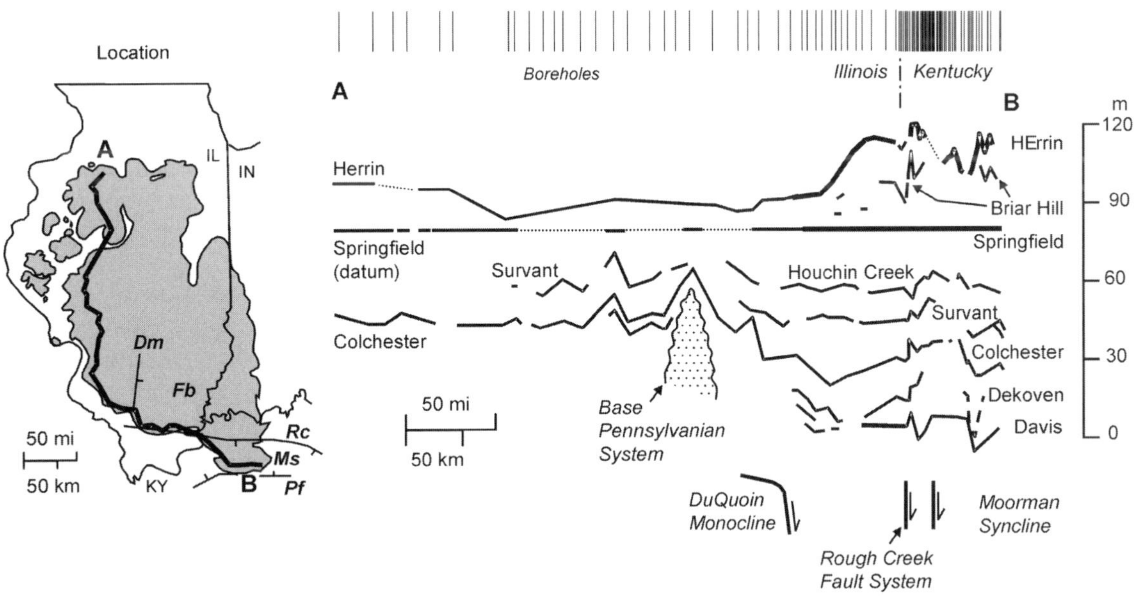

Figure 3. Cross section of Carbondale Formation (Davis to Herrin Coals) from the northern (A) to southern (B) margin of the Eastern Interior Basin (Borehole data from Smith, 1958, 1961; Smith and Berggren, 1963; and the Kentucky Geological Survey database). Dm—DuQuoin monocline, Fb—Fairfield Basin, Ms—Moorman Syncline, Pf—Pennyrile fault system, Rc—Rough Creek fault system, IL—Illinois, IN—Indiana, IA—Iowa, KY—Kentucky.

Cyclothems

Cyclothems are vertically repetitive successions of strata, including coals, clastics, and carbonates, which were named and first investigated in the Desmoinesian of the Eastern Interior Basin (Udden, 1912; Wanless and Weller, 1932). Similar groupings were also noted in the Western Interior Basin of the U.S. Midcontinent and eastward into the Appalachian Basin (Wanless and Weller, 1932; Wanless and Shepard, 1936; Wanless, 1939). Basin comparisons indicate a greater percentage of carbonate deposition in the Western Interior Basin and a greater percentage of clastic deposition in Appalachian Basin cyclothems (Wanless and Shepard, 1936; Wanless, 1975a, 1975b, 1975c, 1975d; Heckel, 1986, 1995, Heckel et al., 1998).

Figure 4 is a cross section of part of the Desmoinesian Carbondale Formation in the southern part of the Eastern Interior Basin. The only persistent marine carbonate in this part of the section, and in this part of the basin, is the Brereton (Providence) Limestone above the Herrin Coal. Most cycles consist of coarsening-upward sequences above the coals. Cyclothems, as defined by Wanless and Weller (1932), extend upward from the base of each scour-based sandstone to the next scour-based sandstone. These sand bodies, however, are not as continuous as the coals or underclays, which represent paleosols, in this part of the section; informally, it is easier to visualize the cyclicity as bounded by successive coal beds. In sequence analyses, (1) sequence boundaries are generally defined at paleosols within each cycle to mark lowstand surfaces of fourth-order sequences, or (2) in a genetic analysis, the base of marine-fossil bearing dark gray to black shales above coals is used to mark marine-flooding surfaces for delineation of transgressive-regressive (TR) cycles (e.g., Weibel, 1996).

Desmoinesian and younger cyclothemic deposition has been attributed to tectonic controls (e.g., Weller, 1930), delta switching (e.g., Ferm, 1970), glacio-eustacy (e.g., Wanless and Shepard, 1936), and combinations of glacio-eustacy and tectonics (e.g., Klein and Willard, 1989). Eustatic controls are most commonly inferred (Kosanke et al., 1960; Heckel, 1977, 1986, 1994, 1995; Ross and Ross, 1985). Using a mean duration for the Late Pennsylvanian, Heckel (1986) estimated that Midcontinent cyclothems and bundles of cyclothemic sequences fell within Milankovitch orbital parameters of 44 to 393 ka. Durations of 400 ka have been inferred for Appalachian Basin cyclothem-scale units (Chesnut, 1992), which have been analyzed as fourth-order sequences (Aitken and Flint, 1994).

Coalification

Coal is formed from peat, which accumulates in mires, where large amounts of plant material can accumulate and be buried without significant transport, degradation, or dilution by sedimentation. When peat is buried it undergoes physical and chemical changes during the process of coalification. One of the results of coalification is compaction of the peat. Compaction ratios of peat to bituminous coal generally range from 20:1 to 7:1 (Stach et al., 1982), although the degree of compaction may be minimal for some coals, or at least may have happened at or near the surface, prior to deep

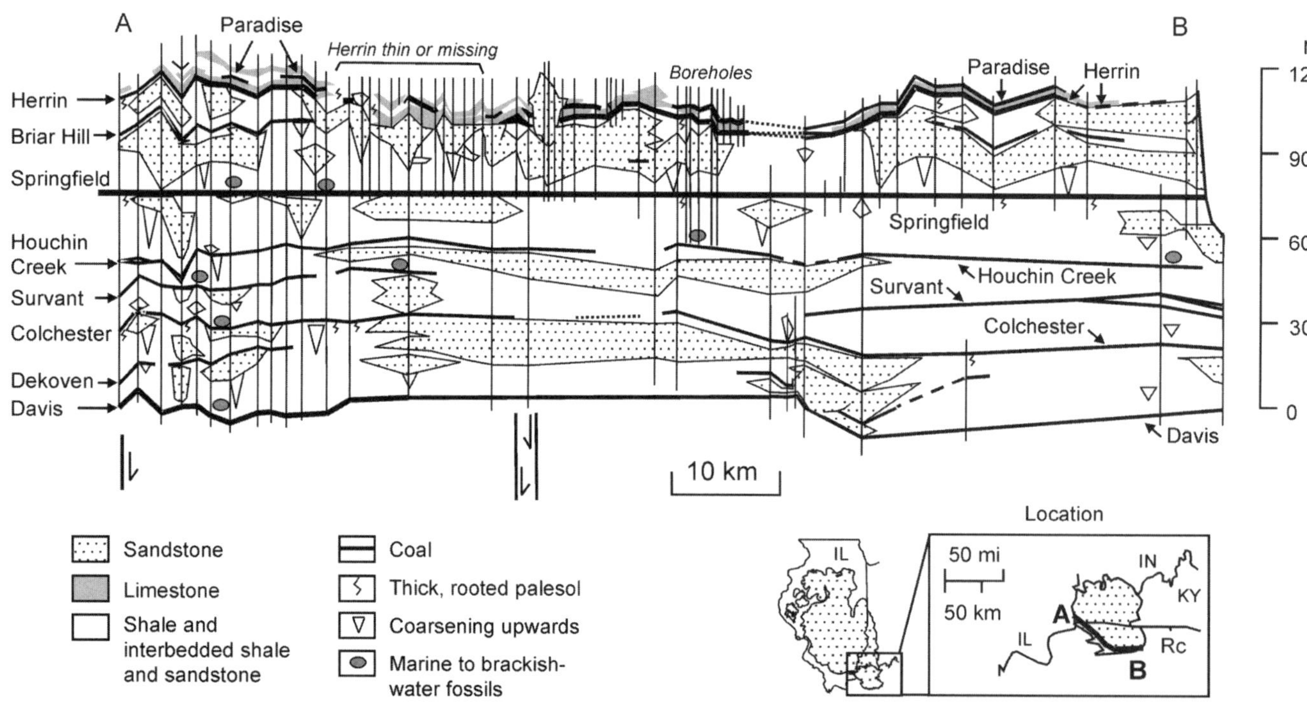

Figure 4. Detail of cross section shown in Figure 3 in southern part of the Eastern Interior Basin, showing characteristic Desmoinesian depositional cycles (cyclothems) of Carbondale Formation. Rc—Rough Creek Fault System. IL—Illinois, IN—Indiana, IA—Iowa, KY—Kentucky.

burial (Nadon, 1998). At a 10:1 compaction ratio, which is commonly used for Carboniferous coals, a 10-m-thick peat would be required to produce 1 m of bituminous coal. Likewise, not all peats become coals. Coals must have less than 50% mineral matter, and most economic deposits have less than 20% mineral matter. Different types of peats have characteristic amounts of mineral matter, water cover, and plant successions, which can be preserved and interpreted in coal beds. Understanding modern peat environments is important for interpreting analogues for coal beds.

Peat mires can be classified in many ways, but they are commonly classified based on the manner in which water enters the mire or by the nutrient content of the mire. Topogenous mires (also sometimes referred to as planar) get their water mostly from surface and ground water, soligenous mires get their water from mixed sources, and ombrogenous (also sometimes referred to as domed or raised) mires get their water from rainwater. Eutrophic mires are high-nutrient mires, mesotrophic mires have mixed nutrients, and oligotrophic mires are low-nutrient mires (Gore, 1983; Cecil et al., 1985; Moore, 1989).

Topogenous mires tend to fill depressions, while most ombrogenous mires dome upward above the groundwater table and receive all of their water from rain water. Mires may pass through a succession from topogenous (swamps, forest peats) to ombrogenous (bogs) stages (Gore, 1983; Clymo, 1987; Moore, 1989).

The broadest modern peats occur in northern-latitude mires of Siberia and Canada (Fig. 5; Walter, 1977; Gore, 1983; Clymo, 1987; Ziegler et al., 1987). The thickest peats occur in ombrogenous mires of equatorial Indonesia (Fig. 5; Anderson, 1983; Esterle et al., 1992; Cecil et al., 1993). Mire type, latitude, and climate are important considerations when interpreting analogues for coal beds.

PURPOSE

The purpose of this report is to summarize salient attributes of three extensive Desmoinesian coal beds of the Eastern Interior Basin, to compare those attributes with modern extensive peatlands, and then to demonstrate that the coals and their correlatives represent the largest tropical peatlands in Earth history. Aspects of Desmoinesian paleoclimate, eustacy, sedimentation, topography, and tectonics are examined to better understand the controls on these ancient giant paleomires.

DESMOINESIAN COALS OF THE EASTERN INTERIOR BASIN

Herrin Area and Thickness

Recent mapping of resources in the Herrin and Springfield Coals allows for accurate determination of the area and thickness of these coals in the basin (Hatch and Affolter, 2002). The Herrin Coal covers an area of 73,900 km^2 (Fig. 6). It is extensive across most of the basin and is the principal mined seam in Illinois. It reaches a maximum thickness of more than 4.3 m adjacent to the Walshville paleochannel (W in Fig. 6) in southern Illinois. The coal is thick in a belt 25–30 km wide on either side of the paleochannel for a distance of at least 350 km. The coal is more than 1.7 m thick across most of the present southwestern limit of the bed. The coal is at least 1.1 m thick across 32,950 km^2.

The Herrin Coal contains a 4–5-cm-thick claystone parting called the "blue band" in the lower third of the coal. Wanless (1939) noted the consistency of the parting throughout much of

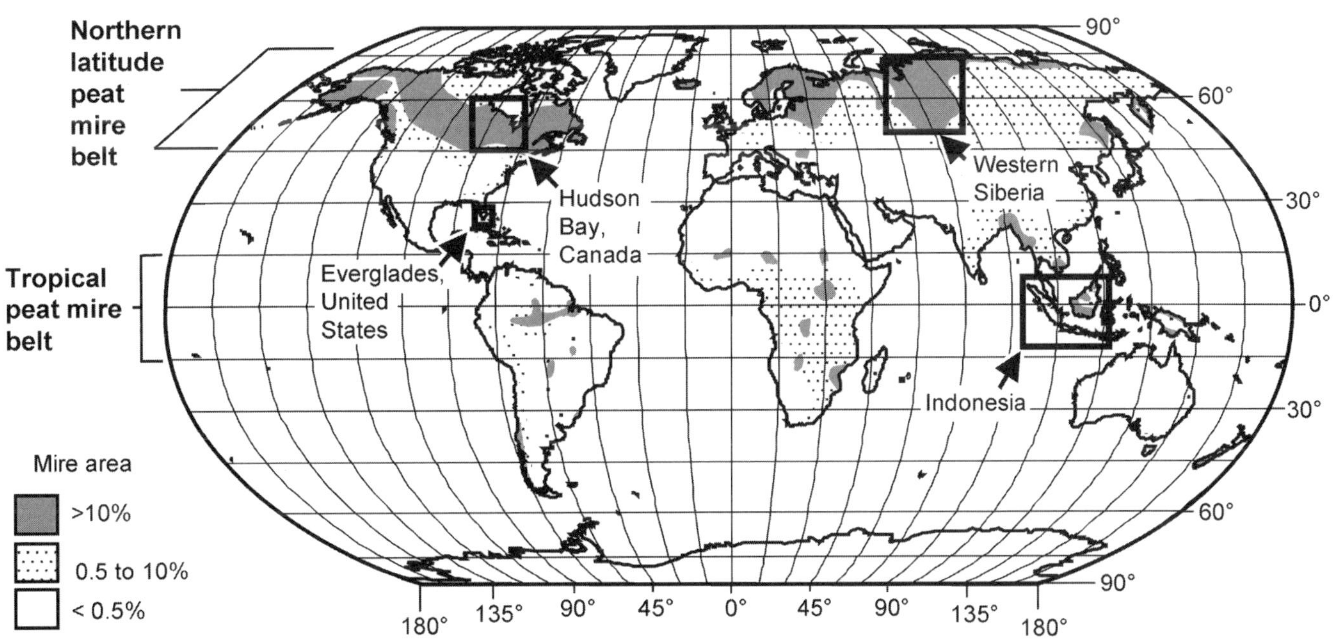

Figure 5. Peatlands of the world (modified from Gore, 1983). Specific areas outlined in diagram are shown in more detail in Figure 17.

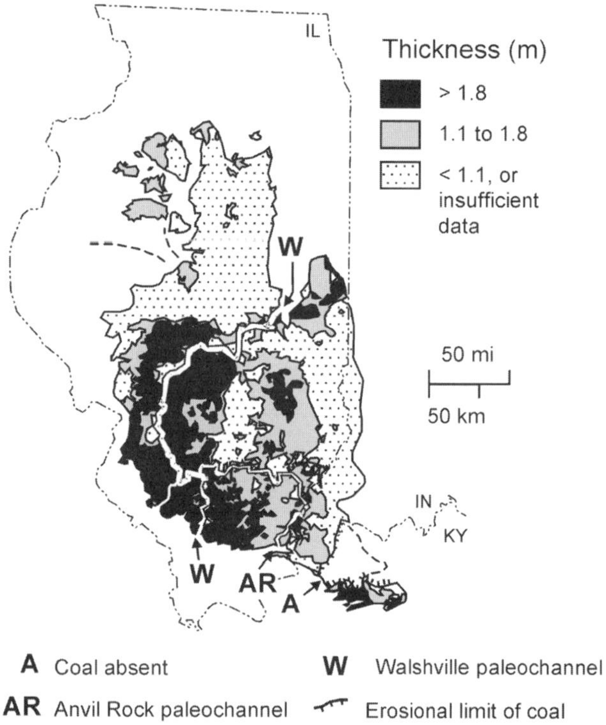

Figure 6. Isopach map of Herrin Coal (modified from Hatch and Affholter, 2002). A–Area where coal is absent in the southern part of the basin, AR—Anvil Rock paleochannel, W—Walshville paleochannel, IL—Illinois, IN—Indiana, IA—Iowa, KY—Kentucky. Dashed lines show possible trends of thicker coal beyond preserved basin margin.

the basin (Fig. 7). The parting locally thickens toward the Walshville paleochannel (W in Figs. 6 and 7) (Johnson, 1972). Several studies have also noted the consistency of smaller partings in the coal bench above the parting (e.g., Nelson, 1987).

The Herrin Coal thins to the north and northeast but thickens again on the northwest margin of the basin (Fig. 6). The coal splits and is truncated along the Walshville paleochannel (Fig. 8A; Hopkins, 1968; Gluskoter and Simon, 1968; Krausse et al., 1979; Hopkins et al., 1979; Nelson, 1983, 1987). Another elongate paleochannel, mapped as the Anvil Rock Sandstone (AR in Figs. 6 and 7), truncates the coal along an elongate belt in southern Illinois (Potter and Simon, 1961; Krausse et al. 1979; Nelson, 1983).

The Herrin Coal is overlain across much of its extent by thin black shale, the Anna Shale, and the Brereton (Providence) Limestone. The shale contains the bivalve *Dunbarella* and the inarticulate brachiopod *Orbiculoidea*. The limestone contains a more diverse marine assemblage, including brachiopods, bryozoans, crinoids, corals, and fusulinids (Utgaard, 1979). In the area of western Kentucky where the coal is absent (A in Figs. 6 and 7), the limestone thickens (Fig. 4). The coal has a sharp "ragged edge" or margin where it is missing (Fig. 8B). This ragged edge is accompanied by brecciation of the overlying limestone and, in some cases, a conglomeratic mudstone (Hower et al., 1987; deWet et al., 1997). The limestone is overlain by the Paradise Coal across much of western Kentucky, where the two coals are commonly mined together (Greb et al., 1992). South of the area of absent coal, the Herrin again thickens southward to the present outcrop margin of the basin (Figs. 6 and 7).

Springfield Area and Thickness

The Springfield Coal covers an area of 73,500 to 81,000 km^2 (Fig. 9). Uncertainty in the estimate is caused by thinning and possible nondeposition west of the Du Quoin Monocline (Dm in Fig. 9; maximum area of 7,500 km^2). The Springfield thickens eastward from the monocline and along the margins of the Galatia paleochannel (G in Fig. 9). The coal is uniformly more than 1.1 m across much of the southern basin, where it is the principal mined seam in western Kentucky. The Springfield averages more than a meter in thickness across an area of nearly 20,000 km^2. It reaches

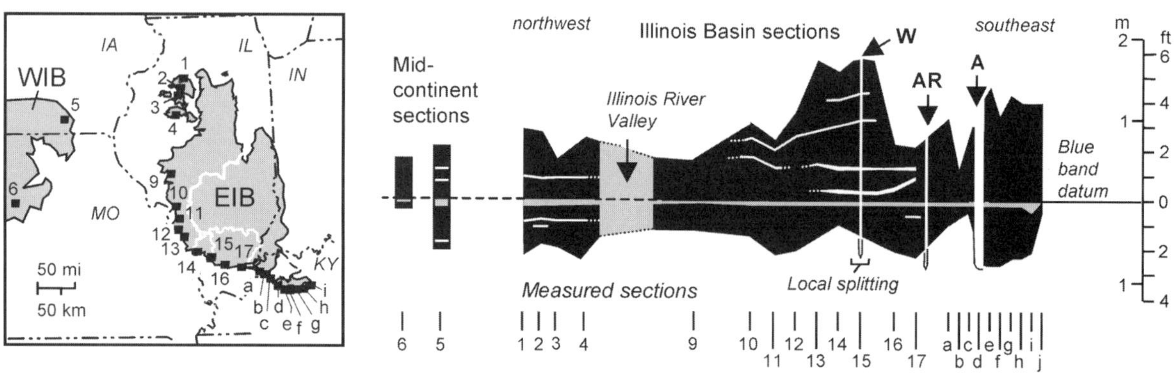

Figure 7. Cross section of Herrin Coal showing consistency of "blue band" parting across the Eastern Interior Basin (EIB) and westward into the Western Interior Basin (WIB). Numbered data from Wanless (1939), Figure 3, but adjusted to show spatial distribution and using blue band parting as datum. Data a through i are from Kentucky Geological Survey data. A—Area where coal is absent in southern part of the basin, AR—Anvil Rock paleochannel, W—Walshville paleochannel, IL—Illinois, IN—Indiana, IA—Iowa, KY—Kentucky, MO—Missouri.

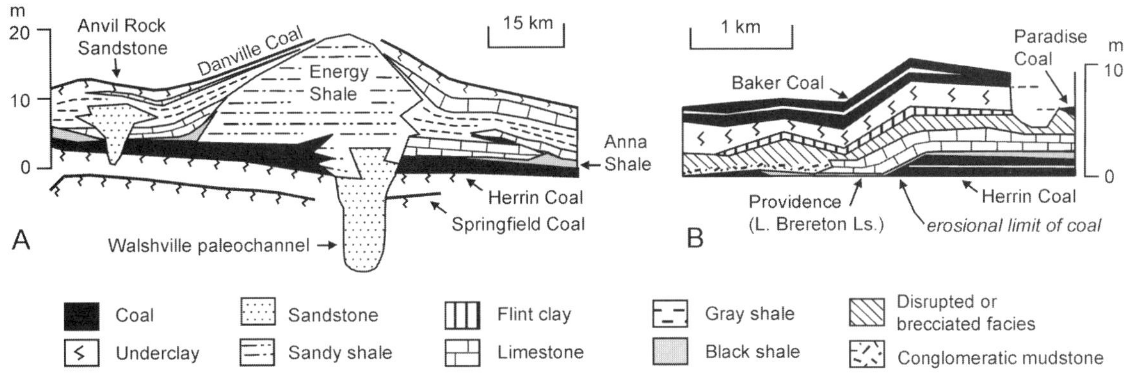

Figure 8. Interruptions in Herrin Coal continuity. A: Generalized diagram showing splitting of Herrin Coal and truncation beneath Walshville paleochannel in southern Illinois (after Nelson, 1987, Fig. 4, p. 7). B: Ragged edge of Herrin Coal in the barren area of western Kentucky (A in Fig. 6) showing truncation by carbonates and disrupted lithofacies (after deWet et al., 1997).

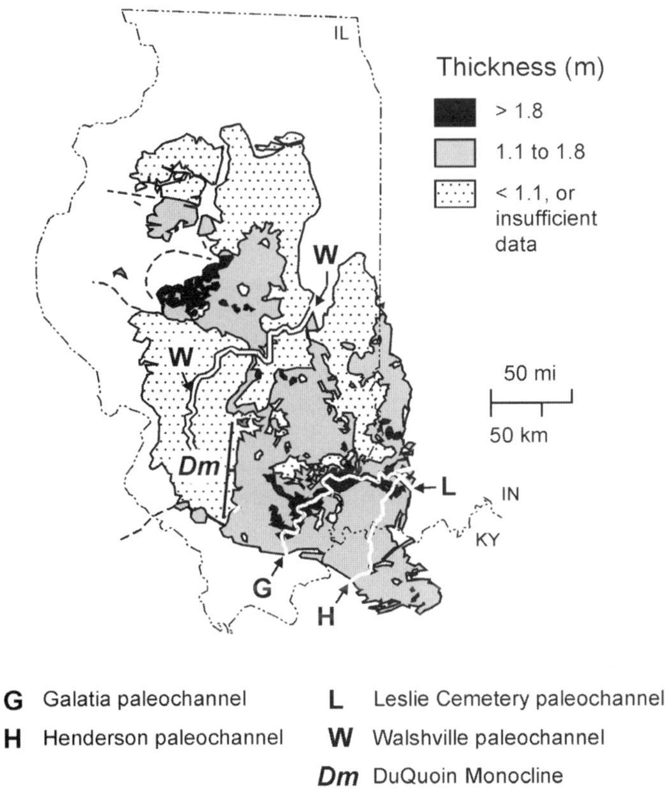

Figure 9. Isopach map of Springfield Coal (modified from Hatch and Affholter, 2002). IL—Illinois, IN—Indiana, KY—Kentucky. Dashed lines show possible trends of thicker coal beyond preserved basin margin.

a maximum thickness of more than 3 m along the Galatia paleochannel. The Springfield is thick in a belt 160 km-long along the paleochannel and is locally split and truncated along the margin of the channel (Hopkins et al., 1979; Nelson, 1983). On the eastern margin of the basin, a channel that is secondary to the Galatia paleochannel has been called the Leslie Cemetery paleochannel (L in Fig. 9). The Springfield is split along the Leslie Cemetery channel as well (Eggert, 1984, 1987). The coal is also truncated by overlying paleochannels, including the previously mentioned Walshville (W in Fig. 9) paleochannel in Illinois (Krausse et al., 1979; Nelson, 1983, 1987) and the Henderson (H in Fig. 9) paleochannel in western Kentucky (Beard and Williamson, 1979).

Colchester Area and Thickness

The Colchester Coal is generally thin (< 1 m), so it has not been the subject of regional resource analyses as were the previous two coals. It was historically mined in northern Illinois during the early 1900s because of its shallow depth. The Colchester appears to cover a slightly larger area than the Springfield Coal, perhaps more than 107,000 km^2 (Fig. 10). The coal reaches a maximum thickness of 1 m along the northern outcrop margin in Illinois (Hopkins et al., 1979). The thickness of the bed in the subsurface is uncertain because it is mostly known from geophysical well logs, although it appears to be thin across much of its extent. The Colchester is distinctive on subsurface geophysical logs because it is underlain by a thick underclay/paleosol and overlain by a carbonaceous dark gray to black shale, the Mecca Quarry Shale. The fact that the coal crops out along much of the basin's northern, western, and southern margins suggests that it is probably continuous into the deeper part of the basin, similar to the Herrin and Springfield Coals.

Coal Composition

The Herrin, Springfield, and Colchester Coals are all highly volatile bituminous coals. Each of the coals exhibits mean ash yields of 9%–12% and mean sulfur contents of 3%–5% (dry basis, Indiana Geological Survey, Illinois State Geological Survey, and Kentucky Geological Survey data). Each of the coals is dominated by vitrinite macerals (generally more than 80%), but some compositional differences are noted within and among coals. These differences are discussed in the following paragraphs.

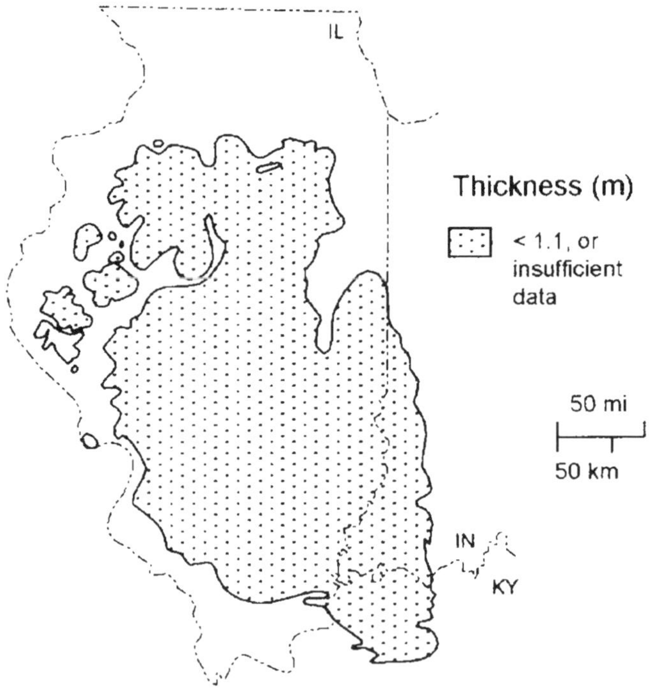

Figure 10. Isopach map of Colchester Coal (modified from Treworgy and Bargh, 1984). IL—Illinois, IN—Indiana, KY—Kentucky.

Herrin Coal

Of the three coals studied, the Herrin is probably the most uniform in composition. Austin (1979) noted that the lower bench of the Herrin, below the "blue band," was duller than the upper bench at sites in Muhlenberg County, Kentucky, but even these are still high in vitrinite content. At sites in Ohio and Hopkins County, Kentucky, Hower et al. (1987) found brecciated and oxidized coal, which was inferred to have oxidized in situ shortly after deposition of the peat. Further examples of the so-called "ragged edge" of the coal were found in western Hopkins County (de Wet et al., 1997; Schultz et al., 2002) and central Webster County, Kentucky (Hower and Williams, 2000). In western Hopkins County, lateral and vertical transitions from "normal," high-vitrinite to brecciated, inertinite-rich Herrin Coal occur along the southeastern limit of the coal (de Wet et al., 1997).

Another exception to the uniformity of Herrin composition is locally lower sulfur content (<2.5%) in the vicinity of the Walshville paleochannel (Fig. 8A; Hopkins, 1968; Gluskoter and Simon, 1968; Nelson, 1983, 1987). Low sulfur values occur beneath splay-form gray shale wedges (Energy Shale) that thicken toward the paleochannel. The Brereton Limestone, which overlies the Herrin Coal distal to the paleochannel, rises to more than 10 m above the coal where the Energy Shale is thick, resulting in low-sulfur coal beneath (Hopkins, 1968; Gluskoter and Simon, 1968; Krausse et al., 1979; Treworgy and Jacobson, 1979; Nelson, 1987). The Herrin Coal is also truncated by the Henderson paleochannel (H in Fig. 6) in western Kentucky (Beard and Williamson, 1979), but the coal retains high sulfur content adjacent to the Henderson paleochannel.

Springfield Coal

The Springfield Coal exhibits a dulling-upward trend, corresponding to an upward decrease in vitrinite content (Ault et al., 1979; Hower and Wild, 1982; Hower et al., 1990a), although even dull lithotypes have more than 65% vitrinite. Similar to the Herrin Coal, the Springfield shows a decrease in sulfur content beneath gray shale wedges (Fig. 11). These low-sulfur values occur beneath the Dykersburg Shale along the Galatia paleochannel (G in Fig. 9) and the Folsomville Shale adjacent to the Leslie Cemetery paleochannel (L in Fig. 9). The gray shale wedges formed a barrier to the downward percolation of sulfates from the marine-fossil bearing black shale and Alum Cave Limestone, which occur directly above the coal laterally (Hopkins, 1968; Treworgy and Jacobson, 1979; Eggert, 1984, 1987; Willard et al., 1995).

The Springfield also exhibits a generally higher rank and high chlorine (Cl) content in parts of southern Illinois and western Kentucky. The high Cl content is interpreted, at least in part, as a remnant of hydrothermal fluids that passed through the coal during diagenesis (Hower et al., 1990b, 1991). Similarly, enrichment of vanadium, zinc, nickel, and other trace metals near the top of the Springfield Coal in the same region may be a function of the same passage of Cl-rich, hydrothermal fluids through the coal (Zubovic, 1966; Hower et al, 1990a, 1990b, 1991; Hower and Gayer, 2002; Rowan et al., 2002). Some of the elements may have been remobilized from the overlying black shale.

Paleobotany

Palynological and coal-ball analyses indicate that all three coals contain similar arborescent lycopsids and tree ferns, although they differ in proportion (Phillips and Peppers, 1984; Phillips et al., 1985; DiMichele et al., 2002). The dominant lycopsid species is *Lepidophloios*, with lesser *Paralycopodites* (producers of *Lycospora* miospores), *Sigillaria* (producer of *Crassispora*), and *Diaphorodendron* and *Synchysidendron* (producers of *Granasporites* miospores). Tree ferns of several kinds were present, especially those producing the miospores *Thymospora*, *Laevigatosporites*, and *Punctatosporites*.

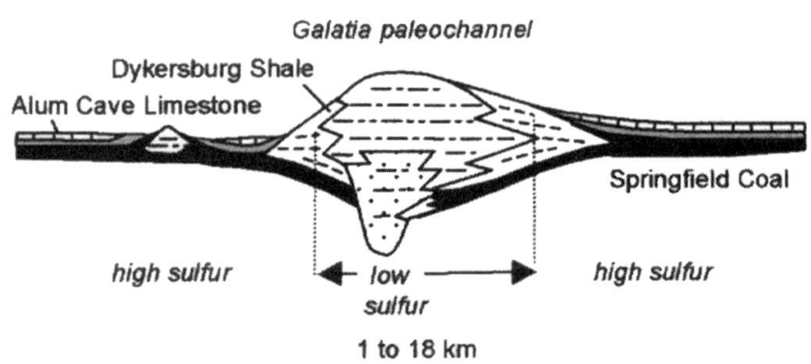

Figure 11. Generalized diagram of splits in Springfield Coal and areas of lower sulfur coal beneath gray shale wedges in Illinois and Indiana (modified from Eggert, 1987). See Figure 8 for explanation of symbols.

Most of the Desmoinesian coals of the Carbondale Formation, including the Herrin and Colchester Coals, are dominated by the lycopod tree spore *Lycospora* (Kosanke, 1950; Phillips et al., 1985; DiMichele and Phillips, 1988; Phillips and Cross, 1991). Analysis of Springfield Coal samples (Kosanke, 1950; Peppers, 1970; Mahaffy, 1988; Willard, 1993) shows that it exhibits greater palynologic diversity. Mahaffy (1988) studied bench samples of the Springfield Coal from three locations in Illinois and Indiana and established four miospore phases based on vertical abundance trends. Palynological studies of the Springfield Coal (Willard, 1993) from 10 locations in Illinois, Indiana, and Kentucky found most locations to be dominated by tree fern spores (47–69%) with subdominant lycopsid tree miospores (18–44%). Profiles collected in proximity to the Leslie Cemetery and Galatia paleochannels exhibited a greater diversity in miospore abundance among tree fern species and generally higher amounts of *Lycospora.* For example, profiles collected near the Leslie Cemetery paleochannel contained 54–62% lycopsid tree spores and 28–32% tree fern spores. Coal-ball studies indicate that the Springfield Coal near paleochannels exhibits larger ash values and is enriched in lycopsids rarely encountered in other late Westphalian coals of the basin, including *Lepidodendron hickii*, *L. mannabachense*, and *Sublepidophloios* (DiMichele and Nelson, 1989; Willard et al., 1995; Phillips and DiMichele, 1998). In contrast, *Thymospora* and *Anacanthotriletes spinosus*, additional types of lycopod spores, were found to be more abundant at localities distal to the paleochannels. In addition, *Granasporites medius, Crassispora kosankei*, and *Anacanthotriletes spinosus* were all found to be more abundant at northern locations (Willard, 1993).

Correlative Desmoinesian Coal Beds

Correlations and paleogeography of the Herrin, Springfield, and Colchester Coals across the Eastern and Western Interior Basins are discussed to demonstrate their possible extent. Figure 12 shows the correlation of the Colchester, Springfield, and Herrin Coals, as well as significant intervening units, with equivalents in surrounding basins. Coal correlations are based on palynology (Peppers, 1996; Eble, 2002). These correlations have been collaborated by correlations of conodonts in overlying marine black shales (Heckel, 1986, 1999). Many of the correlations have not changed significantly since interbasinal correlations were first attempted by Wanless and Weller (1932). Correlations between the Eastern and Western Interior Basins may indicate temporally equivalent units or slightly time-transgressive facies. Temporal equivalence versus lateral time-transgressive facies is more difficult to demonstrate between the Eastern Interior and Northern Appalachian Basins (discussed later).

Possible limits to the extent of individual facies include regional tectonic structures (Fig. 13). During the Desmoinesian,

Series	Stage	Eastern Interior Basin (Peppers, 1996)	Western Interior Basin (Wanless and Wright, 1978; Peppers, 1996; Ravn et al., 1984)			Northern Appalachian Basin	
			Oklahoma	Missouri and Kansas	Iowa	Peppers (1996)	Eble (2002)
Middle Pennsylvanian (part)	Desmoinesian (part)	Brereton/Providence Ls	Upper Oologah Ls	Myrick Station Ls	Myrick Station Ls		
		Anna Sh	Anna Sh.	Anna Sh	Anna Sh		
		Herrin c.	**Lexington c.**	**Lexington c.**	**Mystic c.**	Upper Kittanning c.	**Lower Freeport c.**
		Briar Hill c.	———	———	———	———	Upper Kittanning c.
		St. David/Alum Cave Ls	Higginsville Ls	Higginsville/Houx Ls	Houx Ls		
		Unnamed black shale	Little Osage Sh	Little Osage Sh	Little Osage Sh		
		Springfield c.	———	**Summit c.**	**Summit c.**	**Middle Kittanning c.**	**Middle Kittanning c.**
		Hannover Ls	Blackjack Creek Ls	Blackjack Creek Ls	Blackjack Creek Ls		
		Excello Sh	Excello Sh	Excello Sh	Excello Sh		
		Houchin Creek c.	Mulky c.	Mulky c.	Mulky c.	———	———
		———	Breezy Hill Ls ?	Breezy Hill Ls	Breezy Hill Ls		
		Survant c.	Bevier? c.	Bevier c.	Bevier c.	———	———
		Oak Grove Ls	Verdigris Ls	Verdigris/Ardmore Ls	Ardmore Ls		
		Mecca Quarry Sh	Oakley Sh	Oakley Sh	Oakley Sh		
		Colchester c.	**Croweburg c.**	**Croweburg c.**	**Whitebreast c.**	**Lower Kittanning c.**	**Lower Kittanning c.**

Figure 12. Correlation of coals (bold) and other significant rock units between Eastern Interior, Western Interior, and northern Appalachian Basins (based on data from Donaldson and Eble, 1991; Peppers, 1996; Eble, 2002). Ls—limestone, Sh—shale.

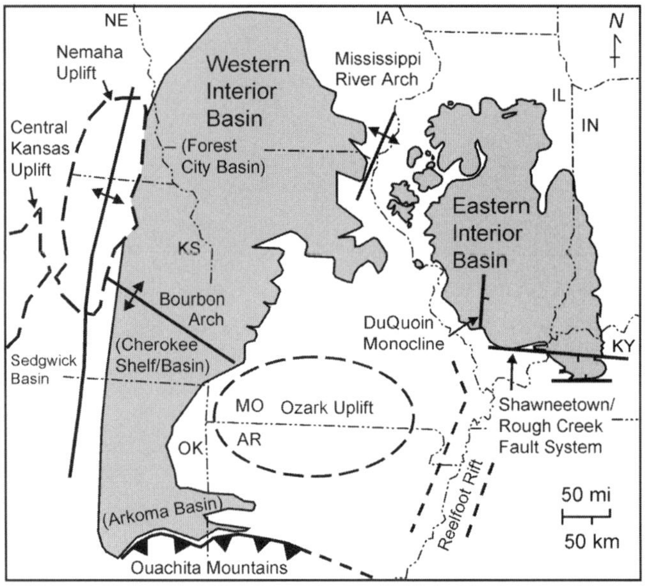

Figure 13. Structural features of U.S. Midcontinent. AR—Arkansas, IL—Illinois, IN—Indiana, IA—Iowa, KY—Kentucky, KS—Kansas, MO—Missouri, NE—Nebraska, OK—Oklahoma.

the Ouachita Mountains were uplifting along the southern margin of the craton. The Arkoma Basin was a foreland basin north of the Ouachitas (Rascoe and Adler, 1983; Houseknecht, 1986; Johnson et al., 1988; Thomas, 1989). The Ozark, Nemaha, and Central Kansas Uplifts (Fig. 13) are all inferred to have been positive, low-relief features (Wanless and Wright, 1978; Bunker et al., 1988; Johnson et al., 1988).

Herrin Correlation

The Herrin Coal is equivalent to the Lexington coal in Kansas, Missouri, and Oklahoma; and the Mystic coal in Iowa (Figs. 12 and 14A; Wanless and Weller, 1932; Wanless et al., 1969; Heckel, 1986, 1995; Peppers, 1996). These coals were the first to be correlated between basins on the basis of spores (Schopf, 1938). The Lexington contains a claystone "blue band" parting in the lower part, and it underlies the Anna Shale, the black and fissile marine shale that is named for a village in Kansas and traced eastward through the Eastern Interior Basin above the Herrin Coal. The Lexington coal is one of Missouri's most important coal resources, but it thins and becomes discontinuous westward (Robertson, 1971) into Oklahoma and Kansas (Friedman, 1977, Fig. 14A). The Lexington and Mystic coals cover a combined area of approximately 30,000 km^2. The combined area of these coals and the Herrin Coal is 73,900 km^2. Correlative coals do not occur west of the Nemaha Uplift, where marine and marginal marine facies dominate (Wanless and Wright, 1978, Fig. 14A).

The black, phosphatic, carbonaceous Anna Shale overlies the Herrin, Lexington, and Mystic coals from Illinois to Kansas (Wanless and Weller, 1932; Wanless et al., 1969), and southward into Oklahoma (Heckel, 1999). The shale is interpreted as a condensed section (Heckel, 1986, 1994, 1999). It is overlain by the Brereton (Providence) Limestone in the Eastern Interior Basin (Fig. 12) and the Myrick Station Limestone in the Midcontinent (Fig. 14B; Wanless and Wright, 1978; Thompson, 1995). Southward, the Myrick Station Limestone may pinch out into deeper-water, black shale facies, or it may be replaced by the overlying Labardie Limestone. The Myrick Station and Labardie Limestone are equivalent to the Oologah Limestone in Oklahoma (Heckel, 1999). The continuity of marine facies across the region indicates connection between basins during deposition of marine facies. If there was connection between basins during deposition of marine facies, it seems possible that there was also connection during initial base level rise when peat was accumulating. If peat accumulation was continuous between basins (dashed lines in Fig. 14A), the total area of Herrin-equivalent peatlands may have exceeded 116,000 km^2.

Marine connection appears likely to have occurred northward across the Mississippi River Arch, and possibly from the south and east of the Ozark Uplift. Paleocurrents from inferred fluvial paleochannels throughout the Carboniferous of the Eastern Interior Basin show southwesterly flow toward the present-day Mississippi River Embayment (Fig. 14A; Siever, 1957; Potter and Simon, 1961; Sedimentation Seminar, 1978; Greb, 1989). The embayment occurs above a Precambrian rift called the Reelfoot Rift (Soderberg and Keller, 1981; Hildenbrand et al., 1982). Subsidence above the rift during the Pennsylvanian caused lowstand valleys to trend along the valley before ultimately depositing their sediments in the Arkoma Basin (Sedimentation Seminar, 1978; Greb, 1989; Thomas, 1989; Donaldson and Eble, 1991). The Herrin Coal is thickest in a belt along the Walshville paleovalley, which drained southward into the embayment area. The Herrin is still thick at the present outcrop margin of the coal, and it is possible that this trend continued southward into the embayment (dashed lines in Fig. 6). During transgression, it is likely that these valleys would have been converted to estuaries and transgression would have moved up-dip into the Eastern Interior Basin. Also, tidal facies have been noted in gray-shale wedges that followed peat accumulation and preceded Anna Shale deposition in parts of the Galatia paleochannel (Archer and Kvale, 1993). It seems likely that rising base level in the floodplains of these valleys would have led to additional peat accumulations that were not preserved because they accumulated outside of the basin.

Eastward from the Eastern Interior Basin, Peppers (1996) correlated the Herrin Coal to the Upper Kittanning coal in the Northern Appalachian Basin (Fig. 12). Locally in Ohio, the Upper Kittanning may contain a parting in the lower part, similar to the parting in the Herrin Coal (J. Nelson, 2002, personal commun.). It has also been suggested that the Herrin Coal may correlate with the Lower Freeport coal (Fig. 12), just above the Upper Kittanning coal (Eble, 2002).

Springfield Correlation

The Summit coal of the Western Interior Basin in Missouri and eastern Kansas is the probable equivalent of the Springfield Coal (Figs. 12 and 15A; Heckel, 1994; Peppers, 1996). The

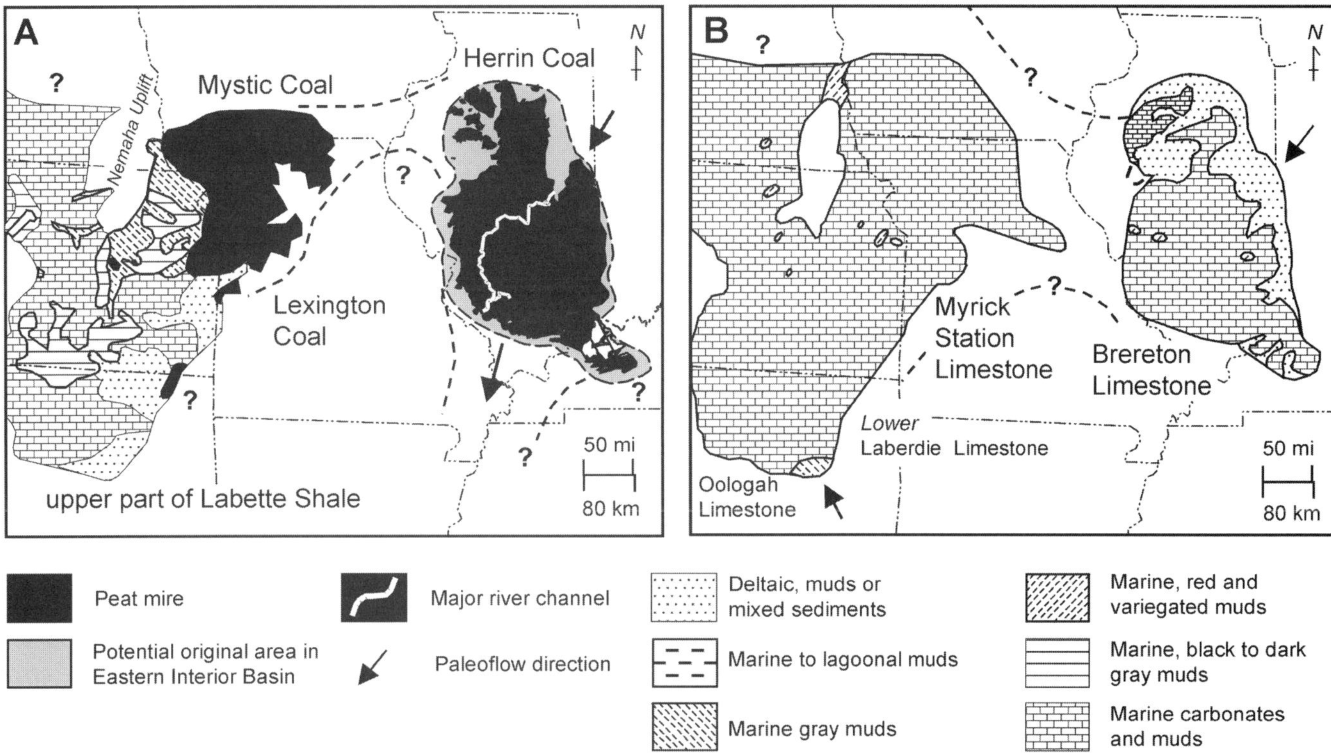

Figure 14. Generalized paleogeographic maps of key Desmoinesian beds across U.S. Midcontinent. A: Herrin Coal and its correlatives. Dashed lines show trend of possible original connections of coal/peat between preserved basins (modified from Wanless and Wright, 1978, Fig. 48, with outcrops from Robertson, 1971; Hatch and Affholter, 2002). B: Brereton Limestone and its correlatives. Dashed lines show possible limit of limestone between preserved basins (modified from Wanless and Wright, 1978, Fig. 51, with correlations from Heckel, 1994, 1999).

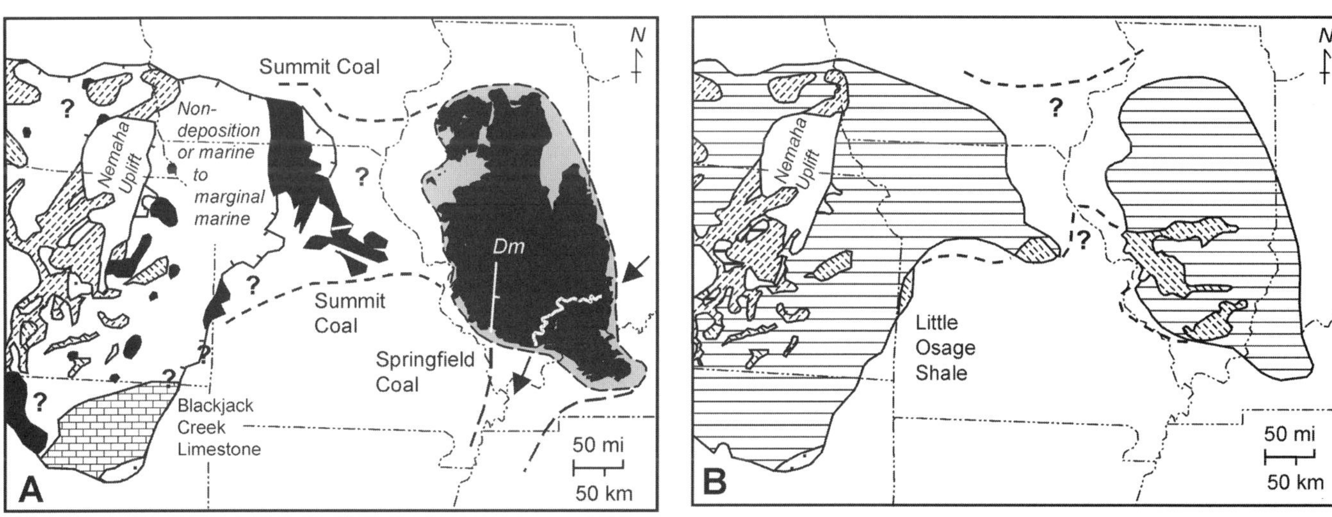

Figure 15. Generalized paleogeographic maps of key Desmoinesian beds across U.S. Midcontinent. A: Springfield Coal and its correlatives. Dashed lines show trend of possible original connections of coal/peat between preserved basins (modified from Wanless and Wright, 1978, Fig. 34, with outcrops from Robertson, 1971; Hatch and Affholter, 2002). B: Little Osage Shale and its correlatives. Dashed lines show possible path of connection between preserved basins (modified from Wanless and Wright, 1978, Fig. 35).

correlation is based on palynology (Peppers, 1996) and regional tracing of key beds that bracket the Summit coal, including the black, phosphatic Excello Shale and the Little Osage Shale above (Heckel, 1986, 1994). The Little Osage Shale in the western interior is correlated with the unnamed black shale above the Springfield (Fig. 15B; Wanless and Wright, 1978). The Little Osage Shale represents transgression of the Desmoinesian seas north and westward, again indicating connection between the Eastern and Western Interior Basins during transgression.

In Missouri, the Summit is best developed in the east-central part of the state and thins into a 1–2-cm-thick "smut zone" or rash above its underclay across much of the rest of the state (Robertson, 1971). The Summit is generally not a significant economic resource, and it is missing across large parts of the Western Interior Basin where nondeposition has been inferred (Wanless and Wright, 1978). Some of the area of inferred nondeposition includes area in which rashy layers above the underclay may indicate wetland, although non-peat-forming, environments. It is also possible that marine shales above the paleosol were partly coeval with lateral peat accumulations. In Oklahoma, Heckel (1999) inferred that the sequence from the Blackjack Creek Limestone to the overlying Little Osage Shale is entirely marine.

Regional estimates of areas for these correlative coals are more uncertain than for the Springfield Coal, but they appear to cover an area of 18,000 km^2. Hence, the combined peatland complex covered an area of 91,500–99,000 km^2. If the basins were connected, the Springfield-Summit paleomires may have covered an area of 121,500 km^2. Correlative coals do not occur west of the Nemaha Uplift, where marine and marginal marine facies dominate (Fig. 15A).

Eastward in the Appalachian Basin, Peppers (1996) correlated the Springfield Coal (Harrisburg, No. V) to the Middle Kittanning coal of Ohio and Pennsylvania and the Princess (No. 7) coal bed of eastern Kentucky based on spores (Fig. 12). This correlation is supported by preliminary correlation of ammonoid fauna between the Washingtonville shale above the Middle Kittanning coal and the Little Osage Shale above the Summit coal in the Midcontinent (D.M. Work, 2002, personal commun.).

Colchester Correlation

Wanless and Weller (1932) noted that the Croweburg coal of Kansas, Missouri, and Oklahoma occupies a similar stratigraphic position and shared a geophysical signature similar to that of the Colchester (No. 2) Coal in the Eastern Interior Basin (Fig. 12). Wanless (1975a) inferred that the Colchester-Croweburg coal may have formed a peat mire 960 km across as part of the Liverpool cyclothem (Fig. 16A; Wanless and Wright, 1978). The Croweburg extends across large areas of northern and western Missouri (Robertson, 1971), southeastern Kansas, and northeastern Oklahoma. Reserves in Oklahoma include coal containing less than 1% sulfur and ranging from 0.3 to 1.1 m thick (Friedman, 1977). The Croweburg occurs within the cyclothem containing the Verdigris Limestone, which can be correlated northward with lithologies above the Whitebreast coal (Heckel, 1986, 1995, 1999). Ravn (1986) correlated the upper Atokan (Westphalian C, Bolsovian)–Desmoinesian (Wesphalian D) boundary with the Whitebreast coal of Iowa, and thereby its equivalents, the Croweburg and Colchester. These coals all occur within the *Shopfites colchesternsis-Thymospora pseudothiessenii* (CP) spore assemblage zone (Peppers, 1996), in which one of the spores is named after the Colchester Coal. The Whitebreast and Croweburg coals may have covered an area of 105,000 km^2 (Fig. 16A). If continuous with the Colchester, as inferred by Wanless and Wright (1978), the total original peatland area exceeded 200,000 km^2 (Fig. 16A). Correlative coals do not occur west of the Nemaha Uplift, where marine and marginal marine facies dominate (Fig. 16A).

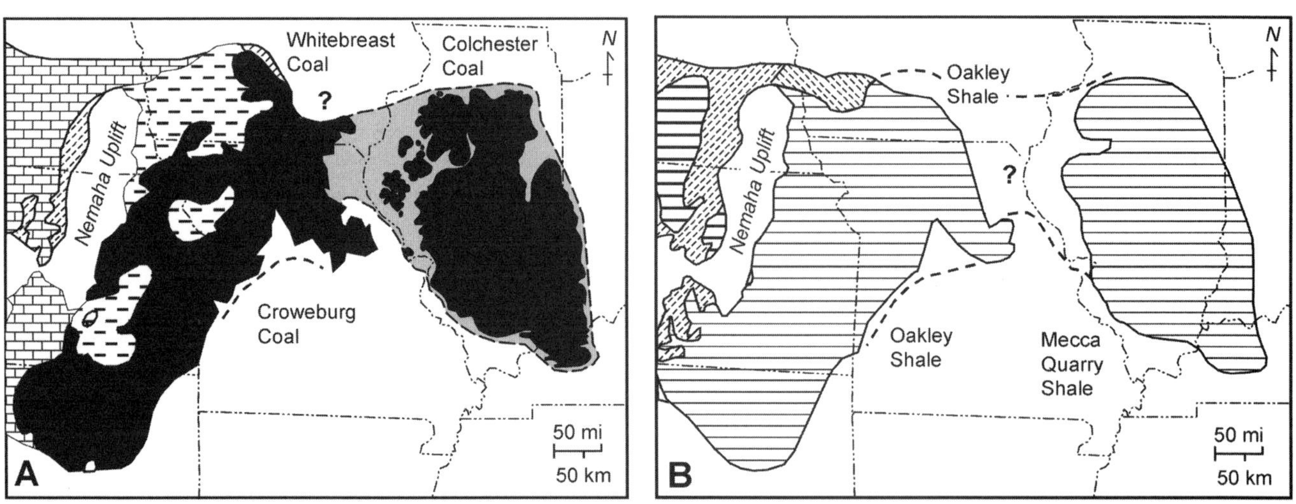

Figure 16. Generalized paleogeographic maps of key Desmoinesian beds across U.S. Midcontinent. A: Colchester Coal and its correlatives. Shaded area shows trend of likely original connections of coal/peat between preserved basins (modified from Wanless and Wright, 1978, Fig. 9, with outcrops from Robertson, 1971; Treworgy and Bargh, 1984). B: Mecca Quarry Shale and its correlatives. Dashed lines show possible path of connection between preserved basins (modified from Wanless and Wright, 1978, Fig. 11).

Each of the coals is overlain by a thick, brackish-water to marine unit: the Mecca Quarry and equivalent black shales in the Eastern Interior Basin, and the Oakley Shale (Raven et al, 1984) and Ardmore Limestone of the Verdigris cyclothem in the Western Interior Basin (Fig. 16B; Wright, 1975; Wanless and Wright, 1978; Heckel, 1986, 1999). Correlations of these marine to marginal-marine units were confirmed by conodont studies (see references in Heckel, 1986, 1999). In northern Illinois, the Colchester Coal is locally overlain by the Francis Creek Shale, which contains the famous Mazon Creek fossils, a mixed terrestrial, marginal marine, and marine fauna (Nitecki, 1979; Baird et al., 1985). Much of the Mazon Creek biota was buried in tidal environments, with evidence for increasing marine conditions to the west and southwest (Kuecher et al., 1990).

Eastward in the Appalachian Basin, the Colchester was first correlated with the Lower Kittanning coal by Wanless (1939). Subsequent spore analysis confirms that correlation in Ohio and Pennsylvania, as well as with the Princess No. 6 coal in eastern Kentucky and the No. 6 Block coal in West Virginia (Fig. 12; Peppers, 1996). Each of these coals is underlain by a thick paleosol across part of its extent. In fact, the paleosols may be more extensive than the coal beds. Spodosols, Ultisols, and gleyed Vertisols are interpreted for the claystone beneath the Lower Kittanning coal in the Northern Appalachian Basin (Cecil and Dulong, 2003). The paleosol beneath the Colchester Coal shares characteristics with gleyed Vertisols. The paleosol beneath the Croweburg coal of the Western Interior Basin is a moderately gleyed Vertisol. Mid-Desmoinesian Aridosols were also noted in several western basins, which may be equivalent to the paleosols beneath the Croweburg-Colchester coal but are not overlain by coal (Cecil et al., 2003).

INTERPRETATION

Types of Paleomires

For the purpose of interpreting modern analogues of comparable scale to the Desmoinesian coals studied, interpretations of the types of paleomires are required. Additionally, care is needed in interpreting Desmoinesian paleomires because of the strong diagenetic overprint, which resulted in locally high rank, chlorine, and trace metals in some areas (e.g., Hower et al., 1990b) and low-sulfur values beneath gray shale wedges in others (e.g., Gluskoter and Simon, 1968).

Comparison of available palynologic and coal-ball studies suggests that the Desmoinesian coals of the Eastern Interior Basin accumulated in three principal plant assemblages (DiMichele and Phillips, 1988, 1996). The first assemblage is very common and is enriched in species of the lycopsid tree *Lepidophloios*, particularly *L. hallii*. This assemblage exhibits low species richness, with few ground cover plants and vines or tree ferns. These features are consistent with an interpretation of a flooded peat surface. In addition, the vegetative and reproductive attributes of arborescent lycopods are distinctive and suggestive of semi-aquatic life-habitats, particularly the aerenchymatos (air-chambered) tissues in the roots and dispersal units that appear to be adapted to flotation and dispersal in water (Phillips et al., 1977; DiMichele and Phillips, 1985, 1988, 1994).

A second common assemblage is also rich in lycopsids, especially *Diaphorodendron scleroticum* and *Synchysidendron resinosum*, but contains abundant tree ferns of the genus *Psaronius*, many species of which produced a variety of miospore types; variable amounts of medullosan pteridosperms; and many kinds of small, ground-cover ferns and vines associated with limited amounts of clastic matter and fusain. *Diaphorodendron* has been interpreted as occupying saturated peat substrates that were occasionally covered with water (DiMichele and Phillips, 1994). Likewise, ferns require some period of substrate exposure to complete their life cycles (DiMichele and Phillips, 1994).

A third type of assemblage, rich in medullosans and the small lycopsid tree *Paralycopodites brevifolius*, is only locally abundant in habitats rich in clastics and, in some cases, fusain. Such environments were likely in ecotonal (not conducive to diverse flora) areas, intermediate between peat and clastic substrates. Other types of rare, but distinctive, assemblages also have been encountered. These include assemblages dominated almost entirely by the small shrubby or scrambling plant *Sphenophyllum*, or mixed assemblages of *Sphenophyllum* and the small lycopsid *Chaloneria*. Such assemblages have been interpreted as characteristic of disturbed or marsh-like vegetation (DiMichele et al., 1979). Each of these assemblages suggests topogenous to possibly soligenous mires.

Tree ferns in the late Middle Pennsylvanian were largely small, opportunistic weedy forms, contrasting with the large trees of the Late Pennsylvanian (Lesnikowska, 1989). These plants had little woody tissue and would not have been unexpected in low-nutrient habitats with exposed peat surfaces. Although the basic species pool of the Springfield Coal is similar to that of the Herrin, Baker, and Danville coals (DiMichele et al., 1996), tree-fern enrichment, in some areas associated with *Sigillaria* (a lycopsid associated with highly decayed, possibly exposed peats) (Willard, 1993), is a distinctive attribute of the Springfield paleomires.

In the area of the Galatia (G in Fig. 9A) and Leslie Cemetery (L in Fig. 9A) channels, lycopsid spores are more prominent than in other parts of the Springfield Coal (Willard, 1993). A number of these trees have been associated with either standing water or periodically inundated habitats, areas likely to have been associated with topogenous peat deposits. In areas of split coal, especially where clastic enrichment in the coal is significant, coal-ball (Phillips and DiMichele, 1998) and palynological (Willard et al., 1995) studies reveal dominance by medullosan pteridosperms and the small lycopsid tree *Paralycopodites brevifolius*. This same combination of dominant taxa has been identified in the Herrin Coal (DiMichele and Phillips, 1988) and Secor coal of Oklahoma (DiMichele et al., 1991) and in the Hamlin coal of eastern Kentucky (Phillips et al., 1985) in association with mineral matter and fusain. Such assemblages may have preferred organic mucks and other environments transitional between peats and oxygenated clastic environments. Vitrinite dominance in each of the Desmoinesian coals also suggests topogenous to soligenous mires as vitrinite preservation is enhanced in aqueous conditions (Teichmüller, 1989). Relatively high ash contents in each of the coals suggest that mineral matter periodically flooded the mires, which is also suggestive of topogenous mires.

Mahaffy (1988) noted the greater diversity of palynomorphs in the Springfield compared to the younger, *Lycospora*-dominant Herrin Coal. It was suggested that this may have been the result of the Springfield paleomire having more area with exposed substrates (if even temporary), which potentially could support a more diverse flora, versus those with mostly standing water cover that would be dominated by lycopsid trees. The Colchester paleomire shows similar high diversity and presumably had large areas of at least temporarily (possibly seasonally) exposed substrates (Eble et al., 2001).

Likewise, the Springfield shows some distinct spatial patterns in average plant compositional variation, based both on coal-ball and palynological data (Mahaffy, 1988; Willard, 1993; Willard et al., 1995). Palynological analyses suggest that areas near contemporaneous paleochannels differed from areas distal to paleochannels. Tree fern composition of the two areas differed, and in the channel areas, the proportion of lycopsid trees was elevated. These differences represent the influence of contemporaneous flooding and standing water near channels and are common in topogenous to soligenous mires. These are also the areas of thickest peat accumulations (Figs. 6 and 9).

DISCUSSION

Extent of Desmoinesian Midcontinent Mires

In some thick coal deposits, the coal represents a succession of different peat stages and in some cases entirely different mires, each succeeding the previous to form a thick peat, and thereby coal. Stacked mires produce vertical successions of changing petrography and palynology within coals. These successions are often bound by regionally extensive partings or high-ash layers (Shearer et al., 1994; Greb et al., 1999, 2002). No extensive partings or durains have been noted within the Colchester Coal, such that the coal at any of the locations sampled may represent the accumulation of a single, temporally extensive mire rather than stacked mires through time. Likewise, the Springfield Coal does not contain a persistent parting or durain layer, although it is more variable in composition, especially near contemporaneous paleochannels (DiMichele and Nelson, 1989; Willard, 1993; Willard et al., 1995; Phillips and DiMichele, 1998).

The Herrin Coal consists of two benches, but the benches are extremely uniform in petrographic, palynologic, and paleobotanical composition so that at least the upper bench could represent a single paleomire developed above the "blue band" detrital incursion. The extent and uniformity of the parting (Fig. 7) indicates that it was deposited in water above a relatively flat surface. This flat surface was formed when the underlying lower bench peat filled in the pre-Herrin peat paleotopography. Relative uniformity in the composition of the upper bench of the Herrin suggests an extensive, interconnected topogenous mire rather than individual mires separated by local paleotopography and wide drainages. Likewise, the persistence of several thin partings in the upper bench across much of the western margin of Illinois indicates that the upper bench may have been a persistent coeval peat on either side of the Walshville paleochannel. The paleochannel divides the coal area roughly in half, such that two peat mires, each possibly in excess of 35,000 km^2, could have constituted the upper Herrin paleomire.

There was undoubtedly some lateral translation of the mires as the Desmoinesian seas transgressed, but the persistence of partings in the Herrin Coal suggests that the bed was not wholly time transgressive. If the Herrin paleomire was a narrow coastal deposit that shifted laterally with transgression, partings in the coal would occur at different positions in the seam concurrent with that transgression. The extent of the blue band and overlying partings indicates that whatever paleotopographic variation existed in the pre-peat surface was infilled during the blue band clastic incursion. Post blue-band peat accumulation was broadly blanketform. Additionally, a similar parting in the coeval Mystic coal of the Western Interior Basin suggests exceptionally widespread, coeval peat accumulation in at least the northern part of the Western Interior Basin.

If the Herrin and Mystic coals represent an extensive, coeval mire complex, it is possible that the other widespread Desmoinesian coals of the Eastern Interior Basin also may have comprised one. Moreover, all three coals were extensive beyond the limits of the present basin. Wanless and Wright (1978) inferred continuity across the Mississippi River Arch between the Colchester and equivalent coals in the Western Interior Basin. The Springfield Coal was not drawn as continuous, but outliers in western Illinois are still more than 1-m thick, and so it must have extended beyond the present limit of the basin (Figs. 9 and 15A). Likewise, the Herrin is very thick along almost the entire southwestern margin of the basin (Figs. 6, 7, and 14A) and obviously was greater in extent than is presently preserved. Even where the coal is missing is western Kentucky (B in Fig. 14A), brecciation and conglomerates in overlying strata (Fig. 8B) indicate that the Herrin Coal may have been removed by post-peat erosion, rather than absence due to nondeposition (Hower et al., 1987; deWet et al., 1997).

The extent of these coals shows that the paleomire complexes that formed them may have covered areas of more than 70,000 km^2 in the Eastern Interior Basin alone. The Colchester and equivalent paleomires may have covered an area of more than 200,000 km^2 across the Midcontinent. The Springfield and Herrin Coals can be demonstrated to have covered slightly smaller areas, but they may have covered total areas much greater than are presently preserved. The Desmoinesian paleomires may have continued southward for an unknown extent above the Reelfoot Rift (Figs. 14A and 15A). The rift area was probably a broad lowland connecting the Illinois Basin southward to the Ouachita foreland.

Comparison to Modern Vast Peatlands

In the modern world, the most widespread peatlands occur in northern, cold-temperate to subpolar latitudes (Fig. 5), and are dominated by *Sphagnum* moss, but they also contain sedges, heath, and pines (e.g., Gore, 1983). Peatlands composed of topogenous to ombrogenous peats cover more than 1.29 million km^2 in northern Canada (Zoltai and Pollett, 1983) and 1.39 million km^2 in West Siberia (Yefremov and Yefremova, 2001).

In terms of individual peatlands, Ziegler et al., (1987) reported a 300,000 km^2 area of peat on the southwest margin of Hudson Bay, Canada (Fig. 17A), as the largest continuous area

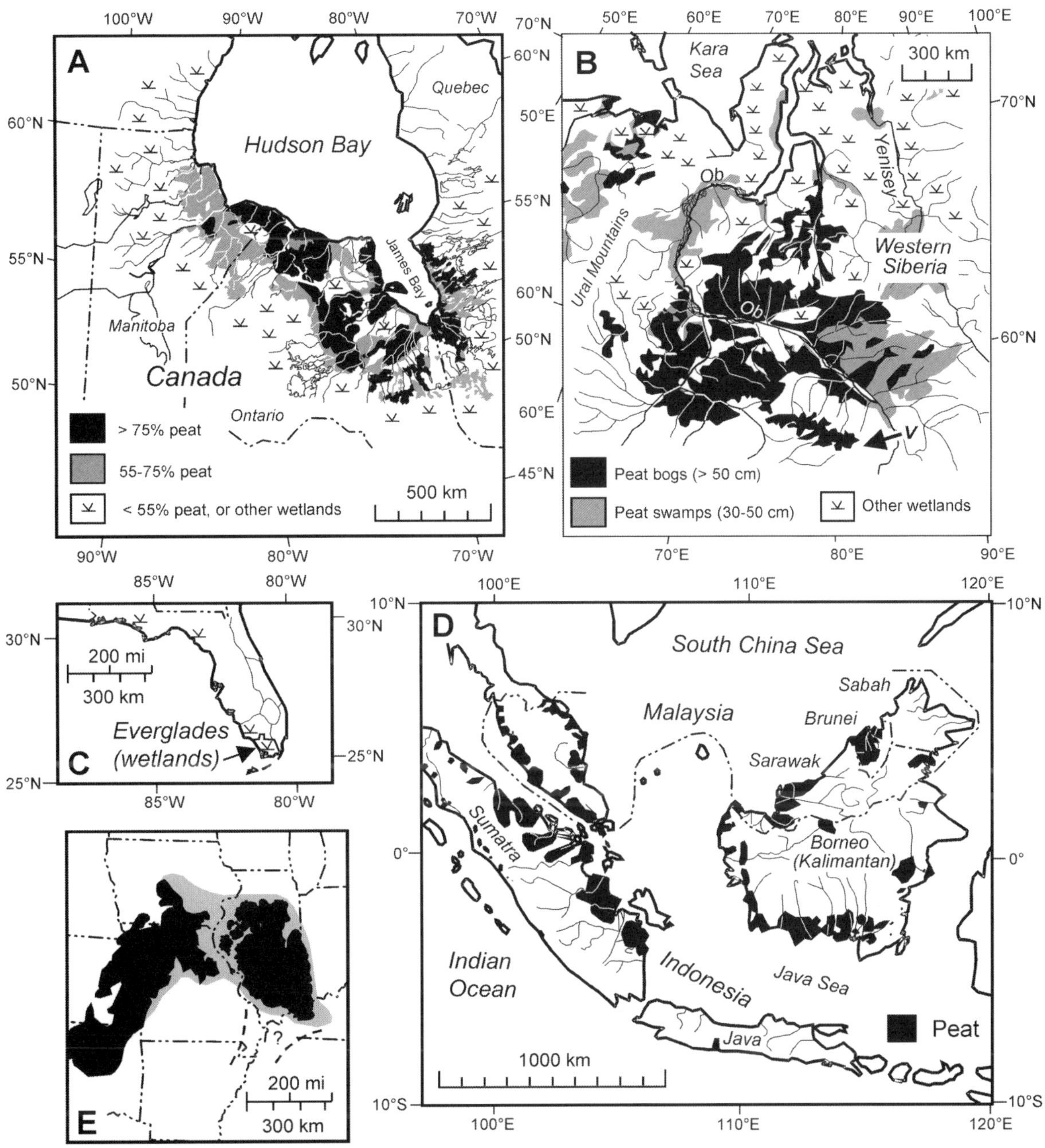

Figure 17. Comparison of scales of modern peatlands (see Fig. 5 for locations) with Desmoinesian peatlands. A: Hudson Bay Lowlands, Canada. Not all of black area is one continuous peat body (modified from Tarnocai et al., 2000, 2002). B: West Siberian peatlands (after Stolbovoi and McCallum, 2002). V—Vasyugan bog complex. C: Everglades area. Very little of this area is actually peatland; most is non–peat-accumulating wetland. D: Indonesian peatlands (after Anderson, 1983, Fig. 6.1). E: Desmoinesian peatlands of U.S. Midcontinent. All to same approximate scale.

of peat accumulation in the world. The Hudson Bay Lowland consists of a complex of coastal marshes, swamps, vast fens, and raised bogs (Martini and Glooschenko, 1985; Martini, 1989; Tarnocai et al., 2000, 2002). Walter (1977) inferred that a 1.4 million km^2 area of western Siberia, including raised peat bogs, hollows, and thousands of lakes, was joined into a single hydrological system during seasonal flooding of the Ob River (Fig. 17B). One of the bogs along this drainage is the Vasyugan bog complex (V in Fig. 17B). It covers 51,000 km^2 and may be the largest individual, undrained peat bog in the world (Botch and Massing, 1983; Bleuten et al., 2000; Lapshina et al., 2001). The total potential area of preserved Desmoinesian Midcontinent peatlands (Fig. 17E) is similar in scale to the main areas of thickest peat in the Hudson Bay Lowland and West Siberian peatlands (Fig. 17, A and B). These modern peatlands illustrate that ancient peatlands need not have been confined to individual basins. If the Desmoinesian peatlands were connected between basins, their total area may have rivaled these modern giant peatlands. In fact, individual mires during Colchester and Herrin peat accumulation may have been similar in size or even exceeded the Vasyugan bog complex.

Extensive, warm-temperate climate mires also occur in the modern world and include two of the most-studied modern mires, the Okefenokee and Everglades of North America (Hofsetter, 1983). The Okefenokee covers an area of 1600 km^2, and the Everglades 10,000 km^2 (Fig. 17C), both significantly less than the area covered by the Desmoinesian coals of the Eastern Interior Basin. Additionally, the Everglades are dominated by non-peat accumulating wetlands, and the peats that do exist are thin, discontinuous, and high in ash content.

Comparison to Modern Thick Peatlands

Although the most widespread modern peats occur in northern latitudes, the thickest modern peats occur at low latitudes. Plant production and net primary productivity is greater in tropical climates than in temperate (Clymo, 1987; Ziegler et al., 1987). Tropical forest peats may accumulate at rates of 3 to 4.8 mm/yr (Anderson, 1983), whereas raised bogs of northern latitudes accumulate at 1–2 mm/yr, and temperate climate topogenous peats accumulate at only 0.5–1 mm/yr (Stach et al., 1982). Most northern latitude peatlands have average thicknesses of 1 to 5 m, with maximum local thicknesses of 7 to 11 m in peats of west Siberia (Kazakov, 1954; Botch and Masing, 1983). Some of these thick northern latitude peats are valley fills, but the thickest are raised bogs (soligenous to ombrogenous peats), like the Vasyugan bog complex (Bleuten et al., 2000; Lapshina et al., 2001). Ombrogenous peatlands of Indonesia (Fig. 17D) may exceed 10 m in thickness for areas of hundreds of square kilometers, and may reach thicknesses of more than 15 m toward the center of peat domes (Anderson, 1983; Esterle et al., 1992; Cecil et al., 1993). These low-latitude peat domes are the thickest peats in the modern world. Individual peat domes are mostly coastal peats and are not as extensive as their northern latitude counterparts. Indonesian peatlands are limited in extent by the width of the coast and the distances between streams, estuaries, and marine straits (Fig. 17D).

The Desmoinesian coals of the Eastern Interior Basin accumulated at tropical latitudes (Witzke, 1990; Heckel, 1994, 1995) and appear to have consisted of widespread topogenous to soligenous mires or mire complexes, relatively undivided by extrabasinal secondary and tertiary drainages. If a 10:1 peat-to-coal compaction ratio is assumed, the only modern peats with comparable thicknesses to these Desmoinesian paleomires are the ombrogenous mires of Indonesia. Like the Desmoinesian paleomires, Indonesian peats occur within 10° of the equator. As stated previously, however, there is no evidence that the Desmoinesian peats were domed, as are modern Indonesian peats. Even if ombrogenous conditions were indicated, the maximum thickness of modern Indonesian peats is less than that indicated for the decompacted thickness of the Springfield peat, which could have been more than 30 m thick, and Herrin peat, which could have been more than 40 m thick (at a 10:1 peat-to-coal compaction ratio). The Desmoinesian coals studied herein not only represent more continuous, widespread, topogenous tropical paleomires than modern tropical mires, but they possibly were thicker.

Extent of Desmoinesian Mires in North America

If examined in terms of total peatland area, the Eastern Interior Basin coals can be looked at not solely by their present extent, but relative to the total potential peat-covered area of adjacent basins, in essence, as a vast Desmoinesian peatland (Fig. 18). The interconnection of peatlands and roof facies between the Eastern and Western Interior Basins has been previously discussed. Likewise, each of the coals has a correlative in the Northern Appalachian Basin, and each is underlain by a thick paleosol. In addition, highstands and lowstands appear to have been near-contemporaneous between basins, as indicated by available biostratigraphy and correlation of cyclothems (discussed in Heckel, 1994).

Although cyclothems appear to be near-contemporaneous, it is difficult to confirm temporal equivalence of coals between basins versus temporal shifts, within the limits of biostratigraphic control, accompanying transgression of the Desmoinesian seas. Correlative coals in the Northern Appalachian Basin cover areas of 20,000 to 30,000 km^2. If Desmoinesian peatlands in the Northern Appalachian Basin accumulated at the same time as those in the Eastern Interior Basin, then the combined peatlands area could have been more than 100,000 km^2 in the basins alone, and if connected by wetlands and peatlands, much more.

Although possibly coeval accumulations, there is evidence to suggest a broad, ramp-like Desmoinesian paleoslope between the Eastern Interior and Northern Appalachian Basins because the number of marine units increases westward from the Appalachians to Iowa, and then southward from Iowa into Oklahoma (Heckel, 1995; Heckel et al., 1998). Also, Ting (1989) provided evidence for a change in coal lithotypes within the upper parts of the Lower Kittanning coal related to the inferred salinity of the roof fauna. This suggests lateral translation of at least the late stages of

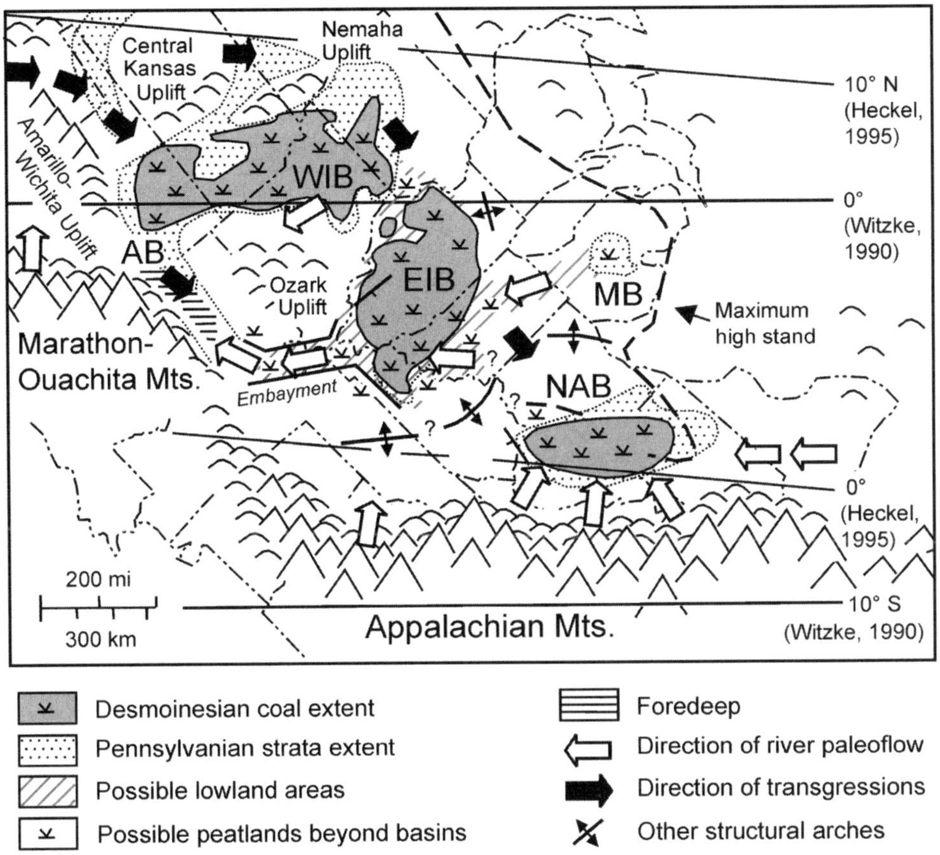

Figure 18. Paleogeography of Desmoinesian peatlands in eastern and middle United States (modified from information in Heckel, 1980; Thomas, 1989; Johnson et al., 1988; Donaldson and Eble, 1991). Peats may not have been completely coeval. Westphalian (Morrowan-Early Desmoinesian) paleoequator from Witzke (1990). Desmoinesian–Virgilian paleoequator from Heckel (1995). AB—Arkoma Basin, EIB—Eastern Interior Basin, MB—Michigan Basin, NAB—Northern Appalachian Basin, WIB—Western Interior Basin.

the northern Appalachian paleomires directly related to rising sea level. Ultimate drowning of the mires reflects the easternmost extent (highstands) of the Desmoinesian Midcontinent seas (Wanless et al., 1969; Heckel, 1995; Heckel et al., 1998). If transgression were needed for the base-level rise that initiated Appalachian peat accumulation, then Appalachian Basin peats would have been slightly younger than "correlative" Eastern and Western Interior Basin peats.

In the modern world, extensive peatlands can form between basins along low-lying coastal plains (Fig. 5). In west Siberia, peats are extensive 1000 km inland from the coast. Since (1) giant-peat-forming conditions seem to have existed in each of the Desmoinesian basins, and (2) extensive interbasinal paleosols indicate widespread exposure and weathering between basins, it seems probable that the areas between the present outcrop limits of the Desmoinesian coals would also have been low-relief, heavily-weathered areas. If low relief areas were extensive between basins, then Desmoinesian peatlands could have been extensive beyond the present limit of the basins, especially in low-lying areas such as might have existed in the embayment above the Reelfoot rift (Fig. 18). It seems likely that coastal mires could also have developed along much of the coast as the Desmoinesian seas transgressed to their maximum highstands (dashed line in Fig. 18), even if outside the present outcrop limit of the basins.

Controls on Desmoinesian Peatlands

Eustatic Controls

The extent of Desmoinesian coals in Midcontinent and eastern North American basins is generally attributed to glacioeustacy, as previously noted (see discussion in Heckel, 1994). Coals occur in well-developed cyclothems and are overlain by dark shales that are attributed in part to condensed sections of marine transgressions and then overlying regressive deposits. The most laterally and vertically consistent coal-bearing cyclothems in the Desmoinesian of the Eastern Interior Basin occur in the stratigraphic interval between the Colchester and Herrin Coals, including the Springfield Coal (Fig. 4). The similarity of cyclothemic deposition across the central and eastern United States strongly supports eustatic controls on Desmoinesian coals. Heckel (1995) and Heckel et al. (1998) inferred that eustatic rise could have created the accommodation space for the development of the thick late Middle and Upper Pennsylvanian coals.

Not all of the Desmoinesian coals are widespread. The Davis, Dekoven, Houchin Creek, Survant, and Briar Hill Coals (Figs. 3 and 4) are not as extensive as the Colchester, Springfield, and Herrin. A consistent difference between the three coals studied and the other coals mentioned is that the three coals studied are underlain by widespread paleosols and overlain by more marine-influenced (at least stenohaline) roof strata than the other coals are. Each is also overlain by gray shale wedges that locally contain tidal stratification (Kuecher et al., 1990; Archer and Kvale, 1993). Marine strata and tidal facies accumulated during fourth-order transgressive systems tracts. The Mecca Quarry, Little Osage, and Anna Shales (Fig. 12) are all marine carbonaceous shales across much of their extent, containing orbiculoids, *Dunbarella*, and conodonts. These shales represent widespread condensed sections (Heckel, 1986, 1999). The Herrin Coal is overlain by the Anna Shale and the Brereton Limestone (Figs. 12 and 14B); the Colchester Coal is overlain by the Mecca Quarry Shale and Oak Grove Limestone. That these marine zones are more extensive than the marine facies above less extensive Desmoinesian coals illustrates the importance of the extent of transgression to Desmoinesian peat extent, and possibly their importance to thickness.

The correlation between the extent of post-peat marine influences and peat extent may reflect greater relative height of base-level rise and a more widespread base-level rise. It may also represent a longer duration of base-level rise or an optimal rate of base-level rise in which peats could accumulate and keep pace with the transgression.

Climatic Controls

A critical factor in modern peat accumulation is climate. In modern, extensive peat-forming mires, humid climates are important for rapid growth of vegetation and peat formation (Kylcyński, 1949; Pearsall, 1950; Clymo, 1987). The extensive west Siberian peatlands (Semenova and Lapshina, 2001; Lapshina et al., 2001) and Hudson Bay Lowland (Zoltai and Pollett, 1983; Martini and Glooschenko, 1985; Martini, 1989) show north-south vegetational gradients related to climate and coastal proximity, as well as temporal changes in vegetation related to Holocene climate changes. The thickest modern peats occur within the Intertropical Convergence Zone within 10° of the equator. It has been postulated that when the Intertropical Convergence Zone is narrow, as it is today, rain forests occur at tropical latitudes (Ziegler et al., 1987). During the Desmoinesian, the coals of middle and eastern North America accumulated within the Intertropical Convergence Zone, within 10° of the equator (Fig. 18).

Desmoinesian paleoclimates in the Eastern Interior Basin are inferred to have had a long wet season and short dry season (Cecil and Dulong, 2003). Comparison of mid-Desmoinesian paleosols suggests a westward change from humid to arid paleoclimates west of the Western Interior Basin. Spodosols, Ultisols, and gleyed Vertisols beneath the Lower Kittanning coal in the Appalachian Basin are soil types that indicate humid to perhumid climates (Retallack, 1990). Gleyed Vertisols beneath the Colchester Coal in the Eastern Interior Basin also suggest humid conditions. Moderate gleying of the paleosol beneath the Croweburg coal of the Western Interior Basin suggests a moist, subhumid paleoclimate. Mid-Desmoinesian Aridosols in western basins demonstrate more arid climates westward in the Desmoinesian (Cecil et al., 2003). A lack of correlative coals in the western United States north of 10° from the paleoequator may also suggest a narrow Intertropical Convergence Zone during the Desmoinesian.

Since Desmoinesian coal beds are found only east of the Nemaha Uplift (Fig. 18) it can be concluded that at least seasonally humid paleoclimates were critical to the establishment of vast mid-Desmoinesian peatlands (Cecil et al., 2003). Some seasonality was also probably needed to preclude development of ombrogenous mires. Indonesian equatorial peats receive rain all year, allowing domed peat to accumulate. Precipitation continuity is critical to establishing tropical ombrogenous peats (Ziegler et al., 1987).

Additional evidence for the importance of climate controls on the regional development of Desmoinesian peatlands is that cyclothemic sedimentation patterns continued into the Late Pennsylvanian in middle and eastern North America (e.g., Heckel, 1986; Greb et al., 1992), but coal beds became less widespread in the Western and Eastern Interior Basins. There is also a sharp change in palynology from Middle to Upper Pennsylvanian coals. Upper Pennsylvanian coals contain more fern-dominated mires, which is suggestive of increased seasonality and drying (Cecil et al., 1985; 2003; Phillips and DiMichele, 1985; Cecil, 1990; DiMichele and Phillips, 1996; Eble et al., 2001). Seasonality and drying climate would have precluded widespread inter-basinal peatland development.

Topographic and Hydrologic Controls

Another important reason for the broad extent of the Colchester and later coals is related to infilling of the last remnants of the sub-Pennsylvanian unconformity along the margins of the Midcontinent and Eastern Interior Basins (Wanless and Wright, 1978). Older coals were restricted to a smaller area, and the unconformity was still an exposure surface along the basin margins with topographic relief. The Colchester Coal and younger deposits formed above that surface, across a broader possible area for deposition (Wright, 1975; Wanless and Wright, 1978).

Wanless et al. (1969) inferred that Desmoinesian peats accumulated on widespread, low-relief delta platforms of the Eastern Interior (Illinois) Basin. These platforms and correlative environments were mapped across the Eastern and Western Interior Basins (Wright, 1975; Wanless and Wright, 1978). The widespread Desmoinesian coals of middle and eastern North America are underlain by well-developed underclays, which are typically more extensive than the coals themselves. These underclays represent widespread paleosols, which formed due to weathering of sediments from underlying paleoenvironments rather than just weathering of the delta platforms. As extensive paleosols, they most likely represent fourth-order lowstands (e.g., Weibel, 1996).

The extensive northern-latitude peats of our modern world have developed on widespread low-relief, glacially-scoured, coastal plains with little topographic relief. The Hudson Bay Lowland peats (Fig. 17A) are developed on a vast, essentially flat, marine platform that has isostatically rebounded since the last glacial retreat (Martini and Glooschenko, 1985; Ziegler et al., 1987; Martini, 1989). In the Hudson Bay Lowland, slopes average 0.5m/km (Martini et al., 1980). Likewise, the vast, low-relief cratonic setting of the West Siberian peatlands, has gradients between 0.5 to 1.5 degrees (Semenova and Lapshina, 2001). This type of setting is undoubtedly crucial for the development of widespread peatlands.

In west Siberia, cold, humid, maritime-influenced climates with low evapotranspiration because of cold average annual temperatures led to paludification of the extensive lowlands. Paludification involves the development of soil gleying or podzolization (Pearsall, 1950), which results in limited infiltration capacity and substrate permeability, impeded drainage, seasonal to permanent waterlogging of the impermeable substrates, and peat accumulation (Heinselman, 1963; Walter, 1977; Tallis, 1983). Frenzel (1983) inferred that modern flat-lying basins may be prone to mire formation through paludification and may become self-perpetuating ecosystems. Examination of key areas in the West Siberian peatlands indicates that peats began in topographic depressions (lake depressions and poorly drained basins) and then spread laterally, ultimately fusing into expansive peat massifs and bog complexes (Lapshina et al., 2001). Surplus water running from the growing mires, either superficially or as groundwater, resulted in downslope paludification and lateral extension of the peat. This resulted in impeded water flow and ponding of the water table, which allowed upslope expansion of the peat (Heinselman, 1963; Walter, 1977; Tallis, 1983). In fact, Romanova (1967) and Frenzel (1983) inferred that geomorphic and geologic conditions may have been more important to paludification and extensive mire development in west Siberia than climate.

The widespread paleosols that developed prior to the Colchester, Springfield, and Herrin peats probably formed a vast, low-relief surface that would have been susceptible to paludification and peat accumulation. These paleo-peats were significantly more expansive than modern tropical peats, such as the mires in Indonesia. One of the reasons that tropical mires in Indonesia are not greater in expanse is that they occupy relatively narrow coastlines on a series of volcanic islands (Fig. 17D). In fact, coastal position is not necessary for the development of peatlands. Widespread northern-climate peatlands are more extensive and thicker away from the coast. In the Hudson Bay Lowland, the thickest and most extensive peats occur 50–300 km inland (Fig. 17A; Martini and Glooshenko, 1985; Martini, 1989). The largest West Siberian peatlands are essentially vast floodplain deposits developed in flat, cratonic settings, and they are extensive for more than 800 km inland from the coast.

Much of the West Siberian peatlands is drained by a single large anastomosing to sinuous river, the Ob (Fig.17B), along which the peatlands are most continuous. The Springfield and Herrin Coals are likewise bisected by single, large contemporaneous paleochannels and are best developed on either side of these paleochannels in a cratonic floodplain setting (Figs. 6, 8, and 9), inland from the encroaching Desmoinesian seaway (Fig. 18). Seasonal flooding from the large paleochannels that bisected each of the Desmoinesian paleomires supplied nutrients and water to adjacent peatlands. Seasonal floods may have been impeded by peat infilling of some secondary and tertiary drainage pathways, as happens along the modern Ob River. Because of resulting poor drainage, individual mire complexes are linked into single hydrological systems across broad areas. This is significant because the modern analogue indicates that Desmoinesian peats could develop and become widespread independent of a eustatic rise. The combination of water table rises along the floodplain may have reinforced regional base-level rise and widespread peat saturation, leading to extremely thick and widespread topogenous peats. The major water sources of the ancient peatlands also greatly affected their botanical makeup. The Herrin exhibits greater palynologic diversity than the Springfield and Colchester Coals, and it is associated with the Walshville paleochannel, which is the largest syndepositional channel system of the three coals. Higher water tables across larger areas could have inhibited the proliferation of ferns and thereby decreased average plant diversity in the Springfield and Colchester peatlands (Eble et al., 2001).

Too much flooding from fluvial sources can lead to oxidation and preclude peat accumulation in tropical settings (Ziegler et al., 1987). In both the Hudson Bay Lowland and West Siberian peatlands, winter freezing restricts sedimentation during parts of the year. The Hudson Bay Lowland is drained only by small rivers, which carry small sediment loads (Martini and Glooshenko, 1985). Too much clastic influx would preclude peat accumulation and result in deposition of carbonaceous shales rather than coals. Since widespread peats developed, it can be inferred that the Desmoinesian rivers did not excessively flood the adjacent mires with coarse clastics but did keep water tables high enough for thick peat accumulation. Coarse clastics were confined to narrow zones along the margins of syndepositional channels in the Herrin and Springfield peatlands. Because the Herrin and Springfield mires were only bisected by single large contemporaneous river systems, much of the peatlands may have been protected from coarse clastic influx during flooding. Multiple rock partings in the coals in channel areas (see Fig. 7) attest to periodic flooding. Widespread, high ash yields (9%–12%) in the coals also partly resulted from flooding and may attest to the regional extent of flooding influences.

The relative lack of additional drainages in each of the three Desmoinesian coal beds studied in the Eastern Interior Basin indicates not just poor drainage and low relief, but also low relief in subadjacent areas. Low relief outside of the depositional basin would have contributed to poor drainage within the basin and to the lack of dissected topography. Additionally, extra-basinal low relief might have led to a much lower total detrital sediment volume, which would aid in widespread peat accumulation. In contrast, the splitting and zoning of correlative coals in the Appalachian Basin indicates more influx of detrital clastics and more relief adjacent to the basin in the Appalachian orogen.

Tectonic Controls

A comparison of the Western Kentucky Coal Field (southern Eastern Interior Basin) and Eastern Kentucky Coal Field (central Appalachian Basin) shows that although tectonic accommodation was greater in the Appalachian Basin throughout most of the Pennsylvanian, there was a general decrease in tectonic accommodation from the Morrowan to Desmoinesian in both basins (Greb et al., 2002). Minimal tectonic subsidence coincided with mid-Desmoinesian cyclothem development. Areas of thick Herrin Coal away from the Walshville paleochannel correspond to the Fairfield Basin depocenter in Illinois (Fig. 6), suggesting at least small tectonic influences. Earlier in the Pennsylvanian, increased tectonic subsidence toward basin depocenters in both the Eastern Interior and Appalachian Basins led to splitting, changes in coal thickness, and diversions of paleodrainages due to structural influences (Greb et al., 2002), which probably precluded widespread giant paleomire development.

Intrabasinal tectonics were also primary controls on limiting the extent of pre-Colchester peatlands in the Eastern Interior Basin, especially along the Du Quoin monocline (Fig. 3). Updip thinning of the Springfield Coal and concomitant thinning of the interval between the Springfield and Herrin Coals (Figs. 3 and 8A) suggest that the monocline created a low accommodation area on the western shelf of the basin, which negatively influenced peat accumulation. The Walshville paleochannel (W in Fig. 6) was apparently deflected southward toward the Reelfoot rift by the monocline. Stacking of channel sandstones, informally termed the "highlands fluvial complex," has been attributed to subsidence along the structure (Palmer et al., 1979). Hence, tectonics indirectly affected the accumulation of thick Herrin peat mires by influencing the position of the dominant paleodrainage and, thereby, the pathways for sedimentation. The absence of the Herrin Coal along part of the Rough Creek Graben (B in Fig. 6), and correspondence of part of the mire to local faults (Hower et al., 1987; deWet et al., 1997), may indicate that tectonics influenced the trend of post-Herrin transgression as well.

SUMMARY

Wanless (1975a) inferred that the Colchester Coal of the Eastern Interior (Illinois) Basin and its correlatives in the Western Interior Basin formed the most widespread coals in Earth's history. The Springfield and Herrin Coals and their correlatives in the Western Interior Basin also represent vast Desmoinesian peatlands. These coal beds all have correlatives, although possibly slightly younger, in the Northern Appalachian Basin as well. Herein, the coals were examined as part of vast paleopeatlands. Petrography, palynology, and ash contents of the coals indicate that each was deposited in topogenous to soligenous mires or mire complexes, which may have covered areas in excess of 200,000 km^2.

Desmoinesian giant-topogenous mire complexes were deposited in an equatorial setting and were thicker and much more extensive than modern tropical topogenous mires. They may represent the most extensive tropical topogenous mires in Earth history. The only modern peatlands similar in extent to the Desmoinesian peatlands are the northern-latitude peatlands of the Hudson Bay Lowland and west Siberia. Although these modern peatlands accumulated under a much colder climatic regime, some of the controls on their extent are applicable to understanding the development of giant Desmoinesian paleomires. Paludification above vast paleosols, infilling and damming of early peat paleotopography, and widespread seasonal flooding that links separated mire complexes into single hydrological systems, all may be applicable to the development of the tropical Desmoinesian paleomires.

The superposition of favorable Desmoinesian humid paleoclimates, broad cratonic and impermeable paleosol substrates, strong eustatic influences, basin-wide decreased tectonic subsidence, low relief and low-sediment yield in areas updip of the peat mires all combined to form some of the most widespread peats in Earth history.

ACKNOWLEDGMENTS

We are thankful for the collective works of Harlold R. Wanless, who first correlated these coal beds across the middle and eastern United States. Thanks also to Garland Dever, Phil Heckel, and Erik Kvale for their helpful reviews. Data for Eastern Interior Basin coal isopach maps were funded by the U.S. Geological Survey National Coal Resource Assessment.

REFERENCES CITED

Aitken, J.F., and Flint, S.S., 1994, High-frequency sequences and the nature of incised-valley fills in fluvial systems of the Breathitt Group (Pennsylvanian), Appalachian foreland basin, eastern Kentucky, *in* Dalrymple, R., Boyd, R., and Zaitlen, B., eds., Incised valley systems–Origin and sedimentary sequences: SEPM (Society for Sedimentary Geology) Special Publication 51, p. 353–368.

Anderson, J.A.R., 1983, The tropical pet swamps of western Malesia, *in* Gore, A.J.P., ed., Mires: Swamps, bog, fen, and moor: Ecosystems of the World: New York, Elsevier, v. 4B, p. 181–199.

Archer, A.W., and Kvale, E.P., 1993, Origin of gray-shale lithofacies ("clastic wedges") in U.S. Midcontinental coal measures (Pennsylvanian), *in* Cobb, J.C., and Cecil, C.B., eds., Modern and ancient coal-forming environments: Geological Society of America Special Paper 286, p. 181–192.

Ault, C.H., Carr, D.D., Chen, P.Y., Eggert, D.L., Hassenmueller, W.A., and Hutchinson, H.C., 1979, Geology of the Springfield Coal Member (V) in Indiana–A review, *in* Palmer, J.E., and Dutcher, R.R., eds., Depositional and structural history of the Pennsylvanian system in the Illinois Basin. Part 2: Invited papers: Field trip 9, 9th International Congress of Carboniferous Stratigraphy and Geology, Urbana, Illinois: Illinois State Geological Survey, p. 43–49.

Austin, S.A., 1979, Depositional environment of the Kentucky No. 12 coal bed (Middle Pennsylvanian) of western Kentucky, with special reference to the origin of coal lithotypes [Ph.D. thesis]: University Park, The Pennsylvania State University, 390 p.

Baird, G.C., Shabica, C.W., Anderson, J.L., and Richardson Jr., E.S., 1985, Biota of a Pennsylvanian muddy coast; habitats within the Mazonian delta complex, northeast Illinois: Journal of Paleontology, v. 59, p. 253–281.

Beard, J.G., and Williamson, A.D., 1979, A Pennsylvanian channel in Henderson and Webster Counties, Kentucky: Kentucky Geological Survey, ser. 11, Information Circular 1, 12 p.

Bleuten, W., Vasilev, S.V., and Lapshina, E.D., 2000, The scientific relevance of the greatest raised bog of the world: Vasuganskoe bog (west Siberia): Quebec, Canada, 11th International Mire Conservation Group Peat Congress, Program and Abstracts, p. 251.

Botch, M.S., and Masing, V.V., 1983, Mire ecosystems in the USSR, *in* Gore, A.J.P., ed., Mires: Swamps, bog, fen, and moor: Ecosystems of the World: New York, Elsevier, v. 4B, p. 95–152.

Bunker, B.J., Witzke, B.J., Watney, W.L., and Ludvigson, G.A., 1988, Phanerozoic history of the central Midcontinent, U.S., *in* Sloss, L.L., ed., Sedimentary cover—North American craton: Boulder, Colorado, Geological Society of America, The Geology of North America, v. D-2, p. 243–260.

Cadle, A.B., Cairncross, B., Christie, A.D.M., and Roberts, D.L., 1993, The Karoo Basin of South Africa: Type basin for the coal-bearing deposits of southern Africa: International Journal of Coal Geology, v. 23, p. 117–157.

Cecil, C.B., 1990, Paleoclimate controls on stratigraphic repetition of chemical and siliciclastic rocks: Geology, v. 18, p. 533–536.

Cecil, C.B., Stanton, R.W., Neuzil, S.G., Dulong, F.T., Ruppert, C.F., and Pierce, B.S., 1985, Paleoclimate controls on Late Paleozoic sedimentation and peat formation in the central Appalachian Basin (U.S.A.): International Journal of Coal Geology, v. 5, p. 195–230.

Cecil, C.B., Dulong, F.T., Cobb, J.C., and Supardi, S., 1993, Allogenic and autogenic controls on sedimentation in the central Sumatra basin as an analogue for Pennsylvanian coal-bearing strata in the Appalachian basin, *in* Cobb, J.C., and Cecil, C.B., eds., Modern and ancient coal-forming environments: Geological Society of America Special Paper 286, p. 3–22.

Cecil, C.B., and Dulong, F.T., 2003, Precipitation models for sediment supply in warm climates, *in* Cecil, C.B. and Edgar, N.T., eds., Climate Controls on Stratigraphy: Society of Sedimentary Petrology, SEPM (Society for Sedimentary Geology) Special Publication 77.

Cecil, C.B., Dulong, F.T., West, R.R., Stamm, R., Wardlaw, B., and Edgar, N.T., 2003, Climate controls on the stratigraphy of a middle Pennsylvanian cyclothem in North America, *in* Cecil, C.B. and Edgar, N.T., eds., Climate Controls on Stratigraphy: Society of Sedimentary Petrology, SEPM (Society for Sedimentary Geology) Special Publication 77.

Chesnut, D.R., Jr., 1992, Stratigraphic and structural framework of the Carboniferous rocks of the central Appalachian Basin in Kentucky: Kentucky Geological Survey, Bulletin 3, Series 11, 42 p.

Clymo, R.S., 1987, Rainwater-fed peats as a precursor of coal, *in* Scott, A.C., ed., Coal and coal-bearing strata—Recent advances: Geological Society of America Special Publication 32, p. 7–23.

de Wet, C.B., Moshier, S.O., Hower, J.C., de Wet, A.P., Brennan, S., Helfrich, C.T., and Raymond, A.L., 1997, Disrupted coal and carbonate facies within two Pennsylvanian cyclothems, Southern Illinois Basin, USA: Geological Society of America Bulletin, v. 109, p. 1231–1248.

Diessel, C.F.K., 1998, Sequence stratigraphy applied to coal seams: Two case histories, in Shanley, K.W., and McCabe, P.J., eds., Relative role of eustacy, climate, and tectonism in continental rocks: SEPM (Society for Sedimentary Geology) Special Publication 59, p. 151–173.

DiMichele, W.A., and Nelson, W.J., 1989, Small-scale spatial heterogeneity in Pennsylvanian-age vegetation from the roof-shale of the Springfield Coal: PALAIOS, v. 4, p. 276–280.

DiMichele, W.A., and Phillips, T.L., 1985, Arborescent lycopod reproduction and paleoecology in a coal-swamp environment of late Middle Pennsylvanian age (Herrin Coal, Illinois, U.S.A): Reviews Palaeobotany and Palynology, v. 44, p. 1–26.

DiMichele, W.A., and Phillips, T.L., 1988, Paleoecology of the Middle Pennsylvanian-age Herrin Coal swamp near a contemporaneous river system, the Walshville paleochannel: Review of Palaeobotany and Palynology, v. 56, p. 157–176.

DiMichele, W.A., and Phillips, T.L., 1994, Paleobotanical and paleoecological constraints on models of peat formation in the Late Carboniferous of Euramerica: Palaeogeography, Palaeoclimatology, Palaeoecology, v. 106, p. 39–90.

DiMichele, W.A., and Phillips, T.L., 1996, Climate change, plant extinctions, and vegetational recovery during the Middle-Late Pennsylvanian transition: The case of tropical peat-forming environments in North America, *in* Hart, M.L., ed., Biotic recovery from mass Extinctions: London, Geological Society Special Publication 102, p. 201–221.

DiMichele, W.A., Mahaffy, J.F., and Phillips, T.L., 1979, Lycopods of Pennsylvanian age coals: *Polysporia*: Canadian Journal of Botany, v. 57, p. 1740–1753.

DiMichele, W.A., Phillips, T.L., and McBrinn, G.E., 1991, Quantitative analysis and paleoecology of the Secor coal and roof-shale floras (Middle Pennsylvanian, Oklahoma): PALAIOS, v. 6, p. 390–409.

DiMichele, W.A., Phillips, T.L. and Nelson, W.J., 2002, Place vs. time and vegetational persistence: a comparison of four tropical mires from the Illinois Basin during the height of the Pennsylvanian Ice Age: International Journal of Coal Geology, v. 50, p. 43–72.

DiMichele, W.A., Pfefferkorn, H.W., and Phillips, T.L., 1996, Persistence of Late Carboniferous tropical vegetation during glacially driven climatic and sea-level fluctuations: Paleoclimatology, Paleogeography, and Paleoecology, v. 125, p. 105–128.

Donaldson, A.C., and Eble, C.F., 1991, Pennsylvanian coals of central and eastern United States, *in* Gluskoter, H.J., Rice, D.D., and Taylor, R.B., eds., Economic geology, U.S.: Boulder, Colorado, Geological Society of America, Geology of North America, v. P-2, p. 523–546.

Eble, C.F., 2002, Palynology of late-Middle Pennsylvanian coal beds in the Appalachian Basin: International Journal of Coal Geology, v. 50, in press.

Eble, C.F., and Grady, W.C., 1990, Paleoecological interpretation of a Middle Pennsylvanian coal bed in the central Appalachian Basin, U.S.A.: International Journal of Coal Geology, v. 16, p. 255–286.

Eble, C.F., Greb, S.F., and Williams, D.A., 2001, The geology and palynology of Lower and Middle Pennsylvanian strata in the Western Kentucky coal field: International Journal of Coal Geology, v. 47, p. 189–205.

Eggert, D.L., 1984, The Leslie Cemetery and Francisco distributary fluvial channels in the Petersburg Formation (Pennsylvanian) of Gibson County, Indiana, U.S.A., *in* Rahmnai, R.A., and Flores, R.M., eds., Sedimentology of coal and coal-bearing sequences: Special Publications of the International Association of Sedimentologists, v. 7, p. 309–315.

Eggert, D.L., 1987, Earlier differential compaction in Gibson County, Indiana: International Journal of Coal Geology, v. 8, p. 305–334.

Ellis, M.S., Gunther, G.L., Ochs, A.M., Roberts, S.B., Wilde, E.M., Schuenemeyer, J.H., Power, H.C., Stricker, G.D., and Blacke, D., 1999, Coal resources, Powder River basin, *in* Fort Union Coal Assessment Team, 1999 resource assessment of selected tertiary coal beds and zones in the northern Rocky Mountains and Great Plains region: U.S. Geological Survey Professional Paper 1625 A, Chapter PN, version 1.2, disc 1.

Esterle, J.S., Gavett, K.L., and Ferm, J.C., 1992, Ancient and modern environments and associated controls on sulfur and ash in coal, *in* Platt, J., Price, J.P., Miller, M., and Suboleski, S., eds., 1.2–New perspectives on central Appalachian low-sulfur coal supplies: Fairfax, Virginia, Techbooks, Coal Decisions Forum Publication, p. 55–76.

Ferm, J.C., 1970, Allegheny deltaic deposits, *in* Morgan, J.P., ed., Deltaic sedimentation, modern and ancient: Society of Economic Paleontologists and Mineralogists Special Publication 15, p. 246–255.

Frenzel, B., 1983, Mires—Repositories of climatic information or self-perpetuating ecosystems?, *in* Gore, A.J.P., ed., Mires: Swamp, bog, fen and moor: Ecosystems of the world: New York, Elsevier, p. 35–66.

Friedman, S.A., 1977, Investigation of the coal reserves in the Ozarks section of Oklahoma and their potential uses: Final report to the Ozarks Regional Commission, July 10, 1974, Norman, Oklahoma, Oklahoma Geological Survey Special Publication 74-2, 117 p.

Gluskoter, H.J., and Simon, J.A., 1968, Sulfur in Illinois coals: Illinois State Geological Survey Circular 432, 28 p.

Gore, A.J.P., ed., 1983, Mires: Swamp, bog, fen and moor: Ecosystems of the world: New York, Elsevier, 440 p.

Greb, S.F., 1989, Structural controls on the formation of the sub-Absaroka unconformity in the U.S. Eastern Interior Basin: Geology, v. 17, p. 889–892.

Greb, S.F., Williams, D.A., and Williamson, A.D., 1992, Geology and stratigraphy of the western Kentucky coal field: Kentucky Geological Survey Bulletin 2, 77 p.

Greb, S.F., Eble, C.F., and Hower, J.C., 1999, Depositional history of the Fire Clay coal bed (late Duckmantian), eastern Kentucky, USA: International Journal of Coal Geology, v. 40, p. 255–280.

Greb, S.F., Eble, C.F., and Chesnut Jr., D.R., 2002, Comparison of the eastern and western Kentucky coal fields, U.S.A—Why are coal distribution patterns and sulfur contents so different in these coal fields?: International Journal of Coal Geology, v. 50, p. 89–118.

Greb, S.F., Eble, C.F., Hower, J.C., and Andrews, W.M., 2002, Multiple-bench architecture and interpretations of original mire phases–Examples from the

Middle Pennsylvanian of the central Appalachian Basin, U.S.A.: International Journal of Coal Geology, v. 49, p. 147–175.

Hardie, J.K., and Van Gosen, B.S., 1986, Fence diagram showing coal bed correlations within Upper Fort Union Formation in and adjacent to the eastern part of the Kaycee 30′ × 60′ quadrangle, Johnson and Campbell Counties, Wyoming: U.S. Geological Survey Coal Investigations Map C-107, 1 sheet.

Hatch, J.R., and Affolter, R.H., 2002, Resource assessment of the Springfield, Herrin, Danville, and Baker coals in the Illinois Basin: U.S. Geological Survey Professional Paper 1625-D, CD-ROM.

Heckel, P.H., 1977, Origin of black phosphatic shale facies in Pennsylvanian cyclothems of the Midcontinent, North America: American Association of Petroleum Geologists Bulletin, v. 61, p. 1045–1068.

Heckel, P.H., 1980, Paleogeography of eustatic model for deposition of Midcontinent Upper Pennsylvanian cyclothems, *in* Pouch, T.D., and Magathan, E.R., eds., Paleozoic paleogeography of the west-central United States: Society of Economic Paleontologists and Mineralogists, Rocky Mountain Section, Rocky Mountain Paleogeography Symposium, p. 197–215.

Heckel, P.H., 1986, Sea-level curve for Pennsylvanian eustatic marine transgressive-regressive depositional cycles along Midcontinent outcrop belt, North America: Geology, v. 14, p. 330–334.

Heckel, P.H., 1994, Evaluation of evidence for glacio-eustatic control over marine Pennsylvanian cyclothems in North America and consideration of possible tectonic effects, *in* Dennison, J.M., and Ettensohn, F.R., eds., Tectonic and eustatic controls on sedimentary cycles: SEPM (Society for Sedimentary Geology) Concepts in Sedimentology and Paleontology, v. 4, p. 65–87.

Heckel, P.H., 1995, Glacial-eustatic base-level-climatic model for late Middle to Late Pennsylvanian coal-bed formation in the Appalachian basin: Journal of Sedimentary Research, v. B65, p. 348–356.

Heckel, P.H., 1999, Field Trip No. 8: Middle and Upper Pennsylvanian Cyclothem Succession in Midcontinent Basin, U.S.A.: Kansas Geological Survey Open-file Report 99-27, 236 p.

Heckel, P.H., Gibling, M.R., and King, N.R., 1998, Stratigraphic model for glacial-eustatic Pennsylvanian cyclothems in highstand nearshore detrital regimes: Journal of Geology, v. 106, p. 373–383.

Heinselman, M.L., 1963, Forest sites, bog processes, and peatland types in the glacial Lake Agassiz region, Minnesota: Ecology Monographs, v. 33, p. 327–374.

Hildenbrand, T.G., Kane, M.F., and Hendricks, J.D., 1982, Magnetic basement in the Upper Mississippi Embayment region and preliminary report: U.S. Geological Survey Professional Paper 1236, 53 p.

Hofsetter, R.H., 1983, Wetlands in the U.S., *in* Gore, A.J.P., ed., Mires: Swamps, bog, fen, and moor: Ecosystems of the World: New York, Elsevier, v. 4B, p. 201–244.

Hopkins, M.E., 1968, Harrisburg (No. 5) coal reserves of southeastern Illinois: Urbana, Illinois State Geological Survey Circular 431, 25 p.

Hopkins, M.E., Nance, R.B., and Treworgy, C.G., 1979, Mining geology of Illinois coal deposits, *in* Palmer, J.E., and Dutcher, R.R., eds., Depositional and structural history of the Pennsylvanian system of the Illinois Basin: Field trip 9, Ninth International Congress of Carboniferous Stratigraphy and Geology, Part 2: Invited Papers, p. 142–151.

Houseknecht, D.W., 1986, Evolution from passive margin to foreland basin—The Atokan Formation of the Arkoma Basin, south-central U.S.A., *in* Allen, P.A., and Homewood, P., eds., Foreland basins: Special Publications of the Industrial Association of Sedimentologists, v. 8, p. 327–345.

Hower, J.C., and Gayer, R.A., 2002, Mechanisms of coal metamorphism: Case studies from Paleozoic coalfields: International Journal of Coal Geology, v. 50, p. 215–245.

Hower, J.C., and Wild, G.D., 1982, Petrographic variation in the Springfield (No. 9) coal in western Kentucky: International Journal of Coal Geology, v. 2, p. 17–30.

Hower, J.C., Riley, J.T., Thomas, G.A., and Griswold, T.B., 1991, Chlorine in Kentucky coals: Journal of Coal Quality, v. 10, p. 152–158.

Hower, J.C., and Williams, D.A., 2000, Further examination of the ragged edge of the Herrin Coal bed, Webster County, western Kentucky coal field: International Journal of Coal Geology, v. 46, p. 145–155.

Hower, J.C., Trinkle, E.J., Graese, A.M., and Neuder, G.L., 1987, Ragged edge of the Herrin (No. 11) coal, western Kentucky: International Journal of Coal Geology, v. 7, p. 1–20.

Hower, J.C., Wild, G.D., Pollock, J.D., Trinkle, E.J., Bland, A.E., and Fiene, F.L., 1990a, Petrography, geochemistry, and mineralogy of the Springfield (Western Kentucky No. 9) coal bed: Journal of Coal Quality, v. 9, p. 90–100.

Hower, J.C., Rimmer, S.M., Williams, D.A., and Beard, J.G., 1990b, Coal rank trends in the western Kentucky coal field and relationship to hydrocarbon occurrence, *in* Nuccio, V.F., and Barker, C.E., eds., Applications of thermal maturity studies to energy exploration: Tulsa, Oklahoma, Rocky Mountain Section, SEPM (Society for Sedimentary Geology), p. 127–138.

Hower, J.C., Scott, A.C., Hutton, A.C., Pareleh, B.K., and Dowley, B., 1995, Coal: Availability, mining, and preparation, *in* Encyclopedia of energy technology and the environment: John Wiley and Sons, p. 603–684.

Johnson, D.O., 1972, Stratigraphic analysis of the interval between the Herrin (No. 6) coal and the Palsa limestone in southwestern Illinois [Ph.D. thesis]: Urbana, University of Illinois, 105 p.

Johnson, K.S., Amsden, T.W., Denison, R.E., Dutton, S.P., Goldstein, A.G., Rascoe Jr., B., Sutherland, P.K., and Thompson, D.M., 1988, Southern Midcontinent region, *in* Sloss, L.L., ed., Sedimentary cover–North American craton, U.S.: Boulder, Colorado, Geological Society of America, Geology of North America, v. D-2, p. 307–359.

Kazakov, G., 1954, Soviet peat resources: A descriptive study: New York, Edwards Brothers, Research Program on the U.S.S.R., 201 p.

Klein, G.D., and Willard, D.A. 1989, Origin of the Pennsylvanian coal-bearing cyclothems of North America: Geology, v. 17, p.152–155.

Kosanke, R.M., 1950, Pennsylvanian spores of Illinois and their use in correlation: Illinois State Geological Survey Bulletin v. 74, 128 p.

Kosanke, R.M., Simon, J.A., Wanless, H.R., and Willman, H.B., 1960, Classification of the Pennsylvanian strata of Illinois: Urbana, Illinois State Geological Survey Report of Investigations 214, 84 p.

Krausse, H.-F., Damberger, H.H., Nelson, W.J., Hunt, S.R., Ledvina, C.T., Treworgy, C.C., and White, W.A., 1979, Roof strata of the Herrin (No. 6) Coal and associated rocks in Illinois—a summary report: Illinois State Geological Survey Illinois Minerals Notes 72, 54 p.

Kuecher, G.J., Woodland, B.G., and Broadhurst, F.M., 1990, Evidence of deposition from individual tides and tidal cycles from the Francis Creek Shale (host rock to the Mazon Creek Biota), Westphalian D (Pennsylvanian), northeastern Illinois: Sedimentary Geology, v. 68, p. 211–221.

Kylcyński, S., 1949, Peat bogs of Polesie: Memoirs of the Academy of Polish Science, B15, p. 1–356.

Lapshina, E.D., Mouldiyarov, E. Ya., and Vasilev, S.V., 2001, Analyses of key area studies, *in* Bleuten, W., and Lapshina, E.D., eds., Carbon storage and atmospheric exchange by West Siberian Peatlands: Utrecht, The Netherlands, Utrecht University, Department of Physical Geography, p. 23–37.

Lesnikowska. A.D., 1989, Anatomically preserved Marattiales from coal swamps of the Desmoinesian and Missourian of the mid-continent United States: Systematics, ecology, and evolution: [PhD dissertation] University of Illinois, Urbana, 227 p.

Mahaffy, J.F., 1988, Vegetational history of the Springfield Coal (Middle Pennsylvanian of Illinois) and distribution patterns of a tree-fern miospore, *Thymospora pseudothiessenii*, based on miospore profiles: International Journal of Coal Geology, v. 10, p. 239–260.

Mallett, C.W., Pattison, P., McLennan, C., Balfe, P., and Sullivan, D., 1995, Bowen Basin, *in* Ward, C.W., Harrington, H.J., Mallett, C.W., and Beeston, J.W., ed., Geology of Australian coal basins: Geological Society of Australia, Coal Geology Group, Special Publication 1, p. 299–339.

Martini, I.P, 1989, The Hudson Bay lowland: Major geologic features and assets: Geologie en Mijnbouw, v. 68, p. 25–34.

Martini, I.P., and Glooschenko, W.A., 1985, Cold climate peat formation in Canada and its relevance to Lower Permian coal measures of Australia: Earth-Science Reviews, v. 22, p. 107–140.

Martini, I.P, Morrison, R.I.G., Glooshenko, W.A., and Protz, R., 1980, Coastal studies in James Bay, Ontario: Geoscience Canada, v. 7, p. 11–21.

Michaelsen, P., and Henderson, R.A., 2000, Facies relationships and cyclicity of high-latitude, Late Permian coal measures, Bowen Basin, Australia: International Journal of Coal Geology, v. 44, p. 19–48.

Michaelsen, P., Henderson, R.A., Crosdale, P.J., and Mikkelsen, S.O., 2000, Facies architecture and depositional dynamics of the Upper Permian Rangal Coal Measures, Bowen Basin, Australia: Journal of Sedimentary Research, v. 70, p. 879–895.

Moore, P.D., 1989, The ecology of peat-forming processes—A review: International Journal of Coal Geology, v. 12, p. 89–103.

Nadon, G.C., 1998, Magnitude and timing of peat-to-coal compaction: Geology, v. 26, p. 727–730.

Nelson, W.J., 1983, Geologic disturbances in Illinois coal seams, Illinois State Geological Survey Circular 530, 47 p.

Nelson, W.J., 1987, The Hornsby District of low-sulfur Herrin Coal in central Illinois (Christian, Macoupin, Montgomery, and Sangamon Counties): Illinois State Geological Survey Circular 540, 40 p.

Nitecki, M.H., ed., 1979, Mazon Creek fossils: New York, Academic Press, 581 p.

Northern and Central Appalachian Basin Coal Regions Assessment Team, 2001, 2000 resource assessment of selected coal beds and zones in the northern and central Appalachian Basin coal regions: U.S. Geological Survey Professional Paper 1625-C, CD-ROM, Discs 1 and 2, Version 1.0.

Palmer, J.E., Jacobson, R.J., and Task, C.B., 1979, Depositional environments of strata of Late Desmoinesian age overlying the Herrin (No. 6) Coal Member in southwestern Illinois, *in* Palmer, J.E., and Dutcher, R.R., eds., Depositional and structural history of the Pennsylvanian system of the Illinois Basin, Ninth International Congress of Carboniferous Stratigraphy and Geology, Field Trip 9, p. 86–92.

Pearsall, W.H., 1950, Mountains and moorlands: London, Collins Publishing, New Naturalist Series, 32 p.

Peppers, R.A., 1970, Correlation and palynology of coals in the Carbondale and Spoon Formations (Pennsylvanian) of the northeastern part of the Illinois Basin: Illinois State Geological Survey Bulletin 93, 173 p.

Peppers, R.A., 1996, Palynological correlation of major Pennsylvanian (Middle and Upper Carboniferous) chronostratigraphic boundaries in the Illinois and other coal basins: Boulder, Colorado, Geological Society of America Memoir 188, 111 p.

Phillips, T.L. and Cross, A.T., 1991, Paleobotany and paleoecology of coal, *in* Gluskoter, H.J., Rice, D.D., and Taylor, R.B., eds., Economic geology, U.S.: Boulder, Colorado, Geological Society of America, Geology of North America, v. P-2, p. 483–502.

Phillips, T.L. and DiMichele, W.A., 1998, A transect through a clastic-swamp to peat-swamp ecotone in the Springfield Coal, Middle Pennsylvanian age of Indiana, U.S.A.: PALAIOS, v. 13, p. 110–125.

Phillips, T.L., and Peppers, R.A., 1984, Changing patterns of Pennsylvanian coal swamp vegetation and implications of climate control on coal occurrence: International Journal of Coal Geology, v. 3, p. 205–255.

Phillips, T.L., Kunz, A.B., and Mickish, D.J., 1977, Paleobotany of permineralized peat (coal balls) from the Herrin (No. 6) Coal Member of the Illinois Basin, *in* Given, P.H., and Cohen, A.D., eds., Interdisciplinary studies of peat and coal origins: Boulder, Colorado, Geological Society of America Microform Publication 7, p. 18–49.

Phillips, T.L., Peppers, R.A., and DiMichele, W.A., 1985, Stratigraphic and interregional changes in Pennsylvanian-age coal-swamp vegetation: Environmental inferences: International Journal of Coal Geology, v. 5, p. 43–109.

Potter, P.E., and Simon, J.A., 1961, Anvil Rock sandstone and channel cutouts of Herrin (No. 6) coal in west-central Illinois: Illinois State Geological Survey Circular 314, 12 p.

Rascoe, B., Jr., and Adler, F.J., 1983, Permo-Carboniferous hydrocarbon accumulations, Midcontinent, U.S.A.: American Association of Petroleum Geologists Bulletin 67, p. 979–1001.

Ravn, R.C., 1986, Palynostratigraphy of the Lower and Middle Pennsylvanian coals of Iowa: Iowa Geological Survey Technical Paper 7, 245 p.

Ravn, R.C., Swade, J.W., Howes, M.R., Gregory, J.L., Anderson, R.R., and Van Dorpe, P.E., 1984, Stratigraphy of the Cherokee Group and revision of Pennsylvanian stratigraphic nomenclature in Iowa: Iowa Geological Survey, Technical Information Series 12, 76 p.

Retallack, G.J., 1990, Soils of the past: Boston, Unwin Hyman Publishing, 520 p.

Robertson, C.E., 1971, Evaluation of Missouri's coal resources: Missouri Geological Survey, Report of Investigation 48, 92 p.

Romanova, E.A., 1967, Nekotorye morfologicheskie charakteristiki oligotrofnykh bolotnykh landshaftov zapadnosibirskoi nizmennosti kak osnova ikh tipologii I rayonirovannii, *in* Nítsenko, A.A., ed., Priroda bolot i metody ikh issledovaniya: Lenningrad, Nauka, p. 63–67.

Ross, C.A., and Ross, J.R.P., 1985, Late Paleozoic depositional sequences are synchronous and worldwide: Geology, v. 13, p. 27–30.

Rowan, E.L., Goldhaber, M.B., and Hatch, J.R., 2002, The role of regional fluid flow in the Illinois Basin's thermal history: Constraints from fluid inclusions and the maturity of Pennsylvanian coals: American Association of Petroleum Geologists Bulletin, v. 86, p. 257–277.

Schopf, J.M., 1938, Spores from the Herrin (No. 6) coal bed in Illinois: Illinois State Geological Survey Report of Investigations 50, 73 p.

Sedimentation Seminar, 1978, Sedimentology of the Kyrock sandstone (Pennsylvanian) in the Brownsville paleovalley, Edmonson and Hart Counties, Kentucky: Kentucky Geological Survey, ser. 10, Report of Investigations 21, 24 p.

Semenova, N.M., and Lapshina, E.D., 2001, Description of the West Siberian Plain, *in* Bleuten, W., and Lapshina, E.D., eds., Carbon Storage and Atmospheric Exchange by West Siberian Peatlands: FGUU Scientific Reports, The Netherlands, Utrecht University, Department of Physical Geography, Utrecht, p. 10-22.

Shearer, J.C., Staub, J.R., and Moore, T.A., 1994, The conundrum of coal bed thickness—A theory for stacked mire sequences: Journal of Geology, v. 102, p. 611–617.

Shultz, M., Klein, J., McKenzie, F.M., Rimmer, S.M., Hower, J.C., and Popp, J.T., 2002, On the road to Paradise: Depositional setting of the Herrin and Paradise coals, western Kentucky coalfield: Nineteenth Annual International Pittsburgh Coal Conference, University of Pittsburgh, Sept. 23-27, 2002, CD-ROM.

Siever, R., 1957, Pennsylvanian sandstones of the Eastern Interior Basin: Journal of Sedimentary Petrology, v. 27, p. 227–250.

Smith, W.H., 1958, Strippable coal resources of Illinois—Part 2, Jackson, Monroe, Perry, Randolph, and St. Clair Counties: Illinois State Geological Survey Circular 260, 35 p.

Smith, W.H., 1961, Strippable coal resources of Illinois—Part 3, Madison, Macoupin, Jersey, Greene, Scott, Morgan, and Cass Counties: Illinois State Geological Survey Circular 310, p. 3–40.

Smith, W.H., and Berggren, D.J., 1963, Strippable coal resources of Illinois—Part 5A, Fulton, Henry, Knox, Peoria, Stark, Tazewell, and parts of Bureau, Marshall, Mercer, and Warren Counties: Illinois State Geological Survey Circular 348, 59 p.

Soderberg, R.K., and Keller, G.R., 1981, Geophysical evidence for deep basins in western Kentucky: American Association of Petroleum Geologists Bulletin 65, p. 226–234.

Stach, E., Mackowsky, M.-Th., Teichmüller, M., Taylor, G.H., Chandra, D., and Teichmüller, R., 1982, Stach's textbook of coal petrology: Berlin, Gebrüder Borntraeger, 535 p.

Stolbovoi, V., and McCallum, I., 2002, Land resources of Russia: Laxenburg, Austria, International Institute for Applied Systems Analysis and the Russian Academy of Sciences, CD-ROM.

Tallis, J.H., 1983, Changes in wetland communities, *in* Gore, A.J.P., ed., Mires: swamps, bog, fen, and moor: Ecosystems of the World: New York, Elsevier, v. 4A, p. 311–318.

Tarnocai, C., Kettles, I.M., and Lacelle, B., 2000, Peatlands of Canada: Geological Survey of Canada, Open File 3834 (1:6,500,000 map).

Tarnocai, C., Kettles, I.M., and Lacelle, B., 2002, Peatlands of Canada Database: Geological Survey of Canada, Open File 4002 (Digital data).

Treworgy, C.G., and Jacobson, R.J., 1979, Paleoenvironments and distribution of low-sulfur coal in Illinois, *in* Cross, A.T., ed., Economic geology: Coal, oil, and gas: Carbondale, Southern Illinois Press, Compte Rendu of Ninth International Congress on Carboniferous Stratigraphy and Geology, v. 4, p. 349–359.

Teichmüller, M., 1989, The genesis of coal from the viewpoint of coal petrography: International Journal of Coal Geology, v. 12, p. 1–87.

Thomas, W.A., 1989, The Appalachian-Ouachita orogen beneath gulf coastal plain between the outcrops in the Appalachian and Ouachita Mountains, *in* Hatcher Jr., R.D., Thomas, W.A., and Viele, G.W., eds., Appalachian-Ouachita orogen in the United States: Boulder, Colorado, Geological Society of America, Geology of North America, v. F-2, p. 537–553.

Thompson, T.L., 1995, The stratigraphic succession in Missouri: Missouri Department of Natural Resources, Division of Geology and Land Survey, v. 40 (revised), 189 p.

Ting, F.T.C., 1989, Facies in the Lower Kittanning coal bed, Appalachian Basin (U.S.A.): International Journal of Coal Geology, v. 12, p. 425–442.

Treworgy, J.D., and Bargh, M.H., 1984, Coal resources of Illinois: Illinois State Geological Survey, 1:500,000, 5 map sheets.

Tri-state Committee on Correlation of the Pennsylvanian System in the Illinois Basin, 2001, Toward a more uniform stratigraphic nomenclature for rock units (formations and groups) of the Pennsylvanian system in the Illinois Basin: Illinois Basin Consortium Study 5: Illinois State Geological Survey, Indiana Geological Survey, and Kentucky Geological Survey, 26 p.

Tully, J., 1996, Coal fields of the conterminous United States: U.S. Geological Survey Open-File Report OF96–92, 1 map.

Udden, J.A., 1912, Geology and mineral resources of the Peoria quadrangle, Illinois: U.S. Geological Survey Bulletin 506, 103 p.

United States Department of Energy, 2001, Energy Information Administration Annual Coal Report, Web data tables, <www.eia.doe.gov/cneaf/coal/page/acr/acr_html_tabs.html>.

Utgaard, J., 1979, Paleoecology and depositional history of rock strata associated with the Herrin (No. 6) Coal Member, Delta Mine, southern Illinois, *in* Palmer, J.E., and Dutcher, R.R., eds., Depositional and structural history of the Pennsylvanian System of the Illinois Basin, Ninth International Congress of Carboniferous Stratigraphy and Geology, Field Trip 9: Urbana, Illinois State Geological Survey, p. 86–92.

Walter, H., 1977, The oligotrophic peatlands of western Siberia: The largest peino-heiobiome in the world: Vegetatio, v. 34, p. 167–178.

Wanless, H.R., 1939, Pennsylvanian correlation in the Eastern Interior and Appalachian coal fields: Boulder, Colorado, Geological Society of America Special Paper 17, 130 p.

Wanless, H.R., 1975a, Distribution of coals in the United States, *in* McKee, E.D., and Crosby, E.J. (coords.), Paleotectonic investigations of the Pennsylvanian system in the United States; Part II, Interpretive summary and special features of the Pennsylvanian system: U.S. Geological Survey Professional Paper 853, p. 33–47.

Wanless, H.R., 1975b, Illinois Basin region, *in* McKee, E.D., and Crosby, E.J. (coords.), Paleotectonic investigations of the Pennsylvanian system in the United States; Part I Introduction and regional analyses of the Pennsylvanian system: U.S. Geological Survey Professional Paper 853, p. 71–95.

Wanless, H.R., 1975c, Missouri and Iowa, *in* McKee, E.D., and Crosby, E.J. (coords.), Paleotectonic investigations of the Pennsylvanian system in the United States; Part I Introduction and regional analyses of the Pennsylvanian system: U.S. Geological Survey Professional Paper 853, p. 97–114.

Wanless, H.R., 1975d, Appalachian region, *in* McKee, E.D., and Crosby, E.J. (coords.), Paleotectonic investigations of the Pennsylvanian system in the United States; Part I Introduction and regional analyses of the Pennsylvanian system: U.S. Geological Survey Professional Paper 853, p. 17–62.

Wanless, H.R., and Shepard, F.P., 1936, Sea level and climatic changes related to Late Paleozoic cycles: Geological Society of America Bulletin, v. 47, p. 1177–1206.

Wanless, H.R., and Weller, J.M., 1932, Correlation and extent of Pennsylvanian cyclothems: Geological Society of America Bulletin, v 43, p. 1003–1016.

Wanless, H.R., and Wright, C.R., 1978, Paleoenvironmental maps of Pennsylvanian rocks, Illinois basin and northern Midcontinent region: Geological Society of America Map and Chart Series, MC-23, 1:3,937,000, 165 plates, 32 p.

Wanless, H.R., Tubb, J.B., Jr., Gednetz, D.E., and Weiner, J.L., 1963, Mapping sedimentary environments of Pennsylvanian cycles: Geological Society of America Bulletin, v. 74, p. 437–486.

Wanless, H.R., Baroffio, J.R., and Trescott, P.C., 1969, Conditions of deposition of Pennsylvanian coal beds: Geological Society of America Special Paper 114, p. 105–142.

Weibel, C.P., 1996, Applications of sequence stratigraphy to Pennsylvanian strata in the Illinois basin, *in* Witzke, B.J., Ludvigson, G.A., and Day, J., eds., Paleozoic sequence stratigraphy, Views from the North American Craton: Geological Society of America Special Paper 306, p. 331–339.

Weller, J.M., 1930, Cyclical sedimentation of the Pennsylvanian period and its significance: Journal of Geology, v. 38, p. 97–135.

Willard, D.A., 1993, Vegetational patterns in the Springfield Coal (middle Pennsylvanian): Comparison of miospore and coal-ball records, *in* Cobb, J.C., and Cecil, C.B., eds., Modern and ancient coal-forming environments: Geological Society of America Special Paper 286, p. 139–152.

Willard, D.A., DiMichele, W.A., Hower, J.C., Eggert, D.L., Rexroad, C.B., and Scott, A.C., 1995, Paleoecology of the Springfield Coal member (middle Pennsylvanian, Illinois Basin) near the Leslie Cemetery Paleochannel, southwestern Indiana: International Journal of Coal Geology, v. 27, p. 59–98.

Witzke, B.J., 1990, Paleoclimatic constraints for Paleozoic palaeolatitudes of Laurentia and Euramerica, *in* McKerrow, W.S., and Scotese, C.R., eds., Palaeozoic palaeogeography, and biogeography: Geological Society [London] Memoir 12, p. 57–73.

Wright, C.R., 1975, Environments within a typical Pennsylvanian cyclothem, *in* McKee, E.D., and Crosby, E.J., eds., Paleotectonic investigations of the Pennsylvanian system in the United States, Part II, interpretive summary and special features of the Pennsylvanian system: United States Geological Survey Professional Paper 853, p. 73–84.

Yefremov, S.P., and Yefremova, T.T., 2001, Present stocks of peat and organic carbon in bog ecosystems of west Siberia, *in* Bleuten, W., and Lapshina, E.D., eds., Carbon storage and atmospheric exchange by West Siberian peatlands: Utrecht University, The Netherlands, Department of Physical Geography, p. 23–37.

Ziegler, A.M., Scotese, C.R., McKerrow, W.S., Johnson, M.E., and Bambach, R.K., 1987, Paleozoic paleogeography: Annual Review of Earth and Planetary Science, v. 7, p. 473–502.

Zoltai, S.C., and Pollett, F.C., 1983, Wetlands in Canada: Their classification, distribution, and use, *in* Gore, A.J.P., ed., Mires: swamps, bog, fen, and moor: Ecosystems of the World: New York, Elsevier, v. 4B, p. 245–268.

Zubovic, P., 1966, Physico-chemical properties of certain minor elements as controlling factors in their distribution in coal, *in* P.H. Given, ed., Coal science: American Chemical Society Advances in Chemistry Series 55, p. 211–231.

Manuscript Accepted by the Society January 22, 2003

Printed in the USA

Geological Society of America
Special Paper 370
2003

Highest tides of the world

Allen W. Archer
Mary S. Hubbard
Department of Geology, Kansas State University, Manhattan, Kansas 66506, USA

ABSTRACT

Based upon analyses of global navigational tidal tables, 13 areas in the world have spring tidal ranges that commonly exceed 8 m. In these areas, specific coastal geometries result in tidal amplification. The amplified tides can occur hundreds of kilometers inland from open coasts. Here, a simple approach for discussion of tidal amplification based upon linear transects is used to demonstrate how macrotidal conditions can occur within restricted and inland settings. This approach, although very simple as compared to dynamical tidal theory, is useful for comparing modern to less-well constrained ancient depositional systems. Many misconceptions persist regarding the distribution of tidal ranges and controls on sedimentation, particularly within inland tidal systems. Understanding the larger-scale patterns of tidal ranges and very high tides can assist in the development of more realistic models of sedimentation within ancient inland seas and cratonic seaways.

In order of highest to lowest ranges, the areas include: (1) Bay of Fundy in eastern Canada; (2) Bristol Channel in southwest England; (3) Ungava Bay area in northeastern Canada; (4) Gulf of St. Malo in northwestern France; (5) Straits of Magellan of southernmost South America; (6) Cook Inlet in Alaska, USA; (7) Sea of Okhotsk near Siberia; (8) eastern English Channel in northeastern France; (9) northwest Australia area; (10) west-central England; (11) Gulf of Cambay in eastern India; (12) Gulf of Mezan in northwestern Russia; and (13) the Yellow Sea of China and Korea.

Tectonic controls on regional topography and basinal geometry affect the occurrence of very high tidal ranges. Nine (~70%) of these settings occur on present-day passive continental margins. Four of these passive-margin settings are related to funnel- or trumpet-shaped estuaries, whereas the other five relate to straits or seaways that connect the major oceans to more inland seas. Only four of the highest-tide settings occur on active continental margins, and these are primarily related to foreland-basin subsidence within convergent zones.

Keywords: tides, tidal amplification, estuary, macrotidal systems, epicontinental seas, tidal limit.

INTRODUCTION

A variety of terms have been applied to inland and shelf seas. These include seaway, epeiric, continental, epicontinental, and cratonic seas. Historically, some depositional models of ancient inland seas suggested tideless conditions. For example, a highly influential book on stratigraphy and depositional facies by Shaw (1964, p. 7) stated:

> I am not enough of a mathematical oceanographer to be able to demonstrate the impossibility of having normal diurnal tides over the entire extent of an epeiric sea, but it appears that this conclusion is sufficiently self-evident so that rigorous quantitative proof is not required.

Subsequently, quantitative analyses and observational data indicated that Shaw's notion of tide-free inland seas was not actually "sufficiently self-evident." Redfield (1958) and Klein and

Archer, A.W., and Hubbard, M.S., 2003, Highest tides of the world, *in* Chan, M.A., and Archer, A.W., eds., Extreme depositional environments: Mega end members in geologic time: Boulder, Colorado, Geological Society of America Special Paper 370, p. 151–173.

Ryer (1978) documented that areas with wide continental shelves exhibited higher coastal tides than areas with narrow shelves. In addition, documentation of tidal facies within ancient Carboniferous-age inland seas of the U.S. midcontinent contradicts earlier theoretical models (see review in Archer, 1998). These analyses provide a strong theoretical and practicable basis for understanding inland tidal processes that occur in modern analogs of ancient bays, gulfs, and continental seaways.

This discussion of the world's highest tides will focus upon tidal amplification that can occur within inland seas, gulfs, bays, estuaries, and similar settings throughout the world (Fig. 1). Herein, areas are compared that have maximum tidal ranges (during perigean spring tides) that exceed 8 m in height. All data, unless otherwise discussed, has been compiled from tidal tables published by the U.S. National Oceanic and Atmospheric Administration (NOAA, 1998). These tables are designed for navigational purposes and include only a selected suite of tidal stations. Despite these shortcomings, the NOAA data are consistent and globally distributed. Other published information on highest tides contains unsubstantiated or even anecdotal estimates of tidal ranges. Such problems do not appear to occur within the more conservative NOAA tidal-range database.

A primary intent of this paper is to provide comparative information for ancient settings. In most cases, details of paleocoastal geometries and basinal paleodepths are largely unknown. Thus, to make the following discussion germane to ancient settings, coverage of the modern oceanic tidal systems is highly simplified. Complicating effects of amphidromic systems and bathymetry are not discussed here. This approach, although obviously simplistic, can provide information for applications such as paleo-drainage basin analyses (see Feldman et al., 1995). Thus, analyses of tidal heights are compared along simple and direct linear transects. These transects are measured from a "zero line" that is arbitrarily designed near the mouth of the bay, gulf, or seaway. This zero line is set to be essentially perpendicular to the longitudinal axis of the tidal-system amplification. Such an approach allows computation of the variations in tidal range (i.e., effects of tidal amplification) along a transect that is perpendicular to the zero line.

Throughout the following discussion, tidal ranges are referred to as: (1) "microtidal" that have ranges from 0 to 2 m, (2) "mesotidal" for ranges of 2–4 m, and (3) "macrotidal" for ranges of greater than 4 m. Discussion of tidal ranges within inland systems is complex, because low-water conditions cannot be measured in an estuary with fluvial throughput. In such cases, the low-tide waterline may be many kilometers seaward. Thus, the descriptions of tidal range, as used herein, are based on the actual tidal-related changes in water level that can be observed at specific sites. This usage does not conform to NOAA-type measurements based on a consistent datum.

In general, open-oceanic tidal ranges are microtidal, especially for islands that are in the centers of the great oceans of the world. For example, the maximal tidal range at the remote Easter Island is only 58 cm. For the Island of Hawaii, the maximum range is 76 cm. On the Pacific atoll of Kwadjalen, somewhat higher ranges of 107 cm occur. As a generalization, remote and isolated islands commonly have tidal ranges of 1 m or less. Continental coastlines can also have microtidal ranges, but tidal ranges can be orders of magnitude higher. Regionally, tidal effects can be significantly amplified, and ranges that exceed 8 m occur within 13 regions of the world. Characteristics of these regions will be discussed in subsequent sections.

The spring tidal ranges of areas with the world's highest tides can be ordered (Fig. 2). It should be pointed out, however, that this ordering is based upon the somewhat limited data available in the NOAA tidal tables. Use of all available data would result in a different ordering, but some of this non-NOAA data is not particularly well verified. In addition, some published data on extreme tidal ranges refers to unique and unusual tides that were related to meteorological effects, such as combined conditions of spring tides and strong onshore winds. These unusual and unique conditions are not characteristic of the normal highest ranges.

For these 13 locations, there are general similarities and differences in plate-tectonic settings and geographical locations. Nine (~70%) of the settings are on passive continental margins. As compared to active margins, passive margins have generally very low slopes and thus much greater potential for coastal flooding and the development of inland seas. Four of the passive-margin settings are related to large, flooded river valleys, and the highest tides occur in estuarine settings. Funnel- and trumpet-shaped geometries tend to characterize these systems, which include the Bay of Fundy, Bristol Channel, west-central England, and the Gulf of Cambay. The other five of the passive-margin settings occur in seaways that connect open oceans with inland seas. Such areas commonly have complex geometries, but many have a funnel-shaped component where the highest tides occur. These include the Ungava Bay area, Gulf of St. Malo, Straits of Magellan, eastern English Channel, and Gulf of Mezan.

The five remaining settings occur near active continental margins, particularly subduction zones. Three of these settings—Cook Inlet, the Sea of Okhotsk, and the Yellow Sea—occur within inlet or sea related to backarc foreland-basin subsidence. Along the northwest coast of Australia, the amplification is related both to continental shelf width and the development of funnel-shaped geometries of the coast of Australia and islands of Indonesia that include a convergent-plate boundary.

PASSIVE-MARGIN ESTUARIES

Bay of Fundy (1)

The highest tidal ranges in the world occur within the Bay of Fundy, where a tidal range of as much as 16.1 m was reported by Dawson (1902). The bay is located in eastern Canada, between the maritime provinces of Nova Scotia and New Brunswick (Fig. 3A). This area has a relatively humid climate, and rainfall can occur throughout the warmer parts of the year. The winters can be cold, and the rivers freeze. The depositional systems have been greatly

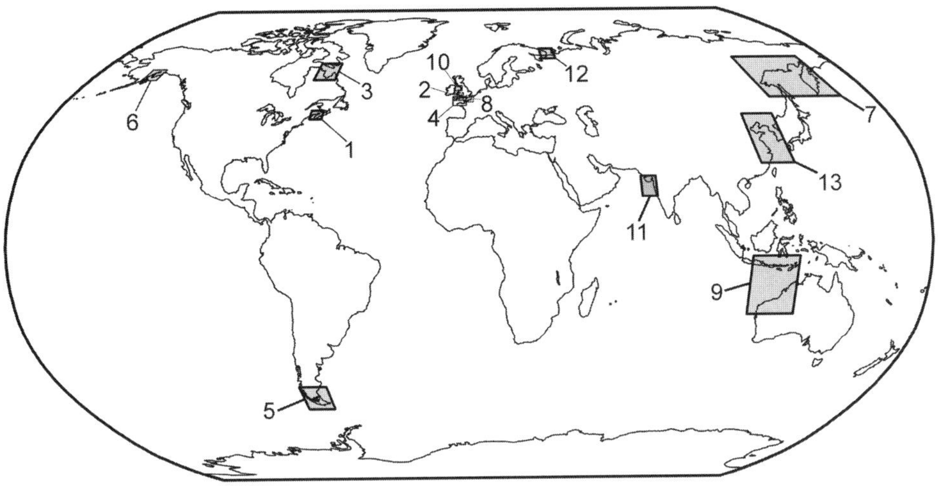

Figure 1. Location of areas with tidal ranges that normally exceed 8 m. In order of highest to lowest ranges these areas are: (1) Bay of Fundy in eastern Canada, (2) Bristol Channel in southwest England, (3) Ungava Bay area in northeastern Canada, (4) Gulf of St. Malo in northwestern France, (5) Straits of Magellan of southernmost South America, (6) Cook Inlet in Alaska, USA, (7) Sea of Okhotsk in Siberia, (8) eastern English Channel in northeastern France, (9) northwest Australia area, (10) Liverpool Bay, west-central England, (11) Gulf of Cambay in eastern India, (12) Gulf of Mezan in northwestern Russia, and (13) Yellow Sea of China and Korea.

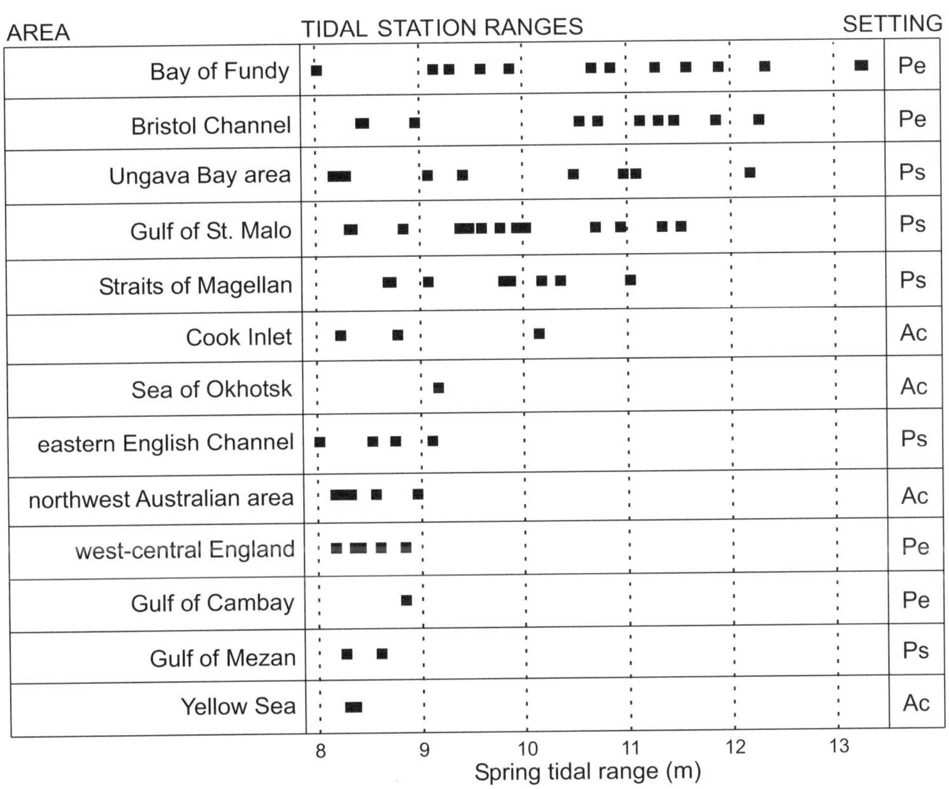

Figure 2. Spring tidal ranges that exceed 8 m as reported by NOAA (1988). Codes for settings include passive-margin estuary (Pe), passive-margin seaway (Ps), and active-margin settings (Ac).

affected by anthropogenic changes. Much of the surrounding land is affected by agriculture, particularly dairy farming. The shorelines of the inner part of the system are extensively diked, and areas of salt marsh behind the dikes have been reclaimed as pasture.

Tidal sedimentation within the Bay of Fundy has been extensively studied (Klein, 1970; Dalrymple et al., 1991). Within the zone of increasing tidal range (bay mouth to ~325 km inland), the system is primarily erosional and the coastline is characterized by sea cliffs and gravelly beaches. Mudflats do occur in the upper intertidal, but they are seasonal, very highly bioturbated, and apparently have little to no long-term preservation potential. The effects of winter storms and wave erosion dominate over summer accumulation rates. Deposition only occurs within the zone characterized by decreasing tidal range.

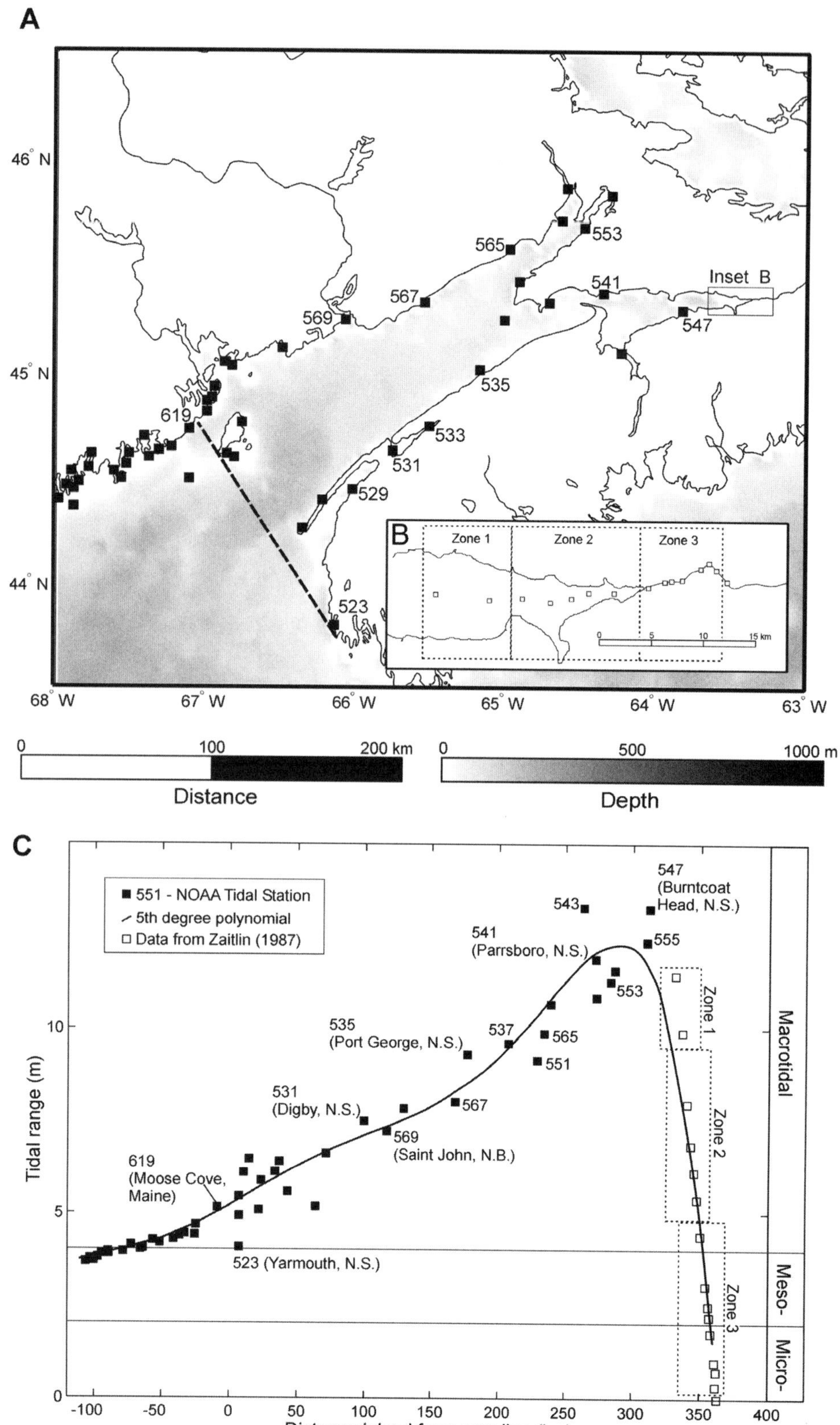

Figure 3. A: Location of NOAA tidal stations in Bay of Fundy, Canada, area showing "zero line" (heavy dashed line) that is used to measure inland distances. B: Inset shows area and Zones 1, 2, and 3, which were described in detail by Zaitlin (1987) and Dalrymple et al. (1991). C: Tidal amplification within Bay of Fundy system. Note that highest tidal ranges occur approximately 300 km inland from the zero line (thick dashed line). Further inland, the tidal flux undergoes a dramatic decrease; however, tidal effects occur to ~370 km inland from the zero line.

The morphology of the Bay of Fundy is the result of an episode of Triassic-Early Jurassic rifting that coincided with the earliest phase of the opening of the Atlantic Ocean Basin (Keppie, 1982). By the Middle Jurassic, rifting was focused south and east of present-day Nova Scotia. The Bay of Fundy is a half-graben structure with the bounding fault, known as the Fundy Fault, located on the northern side of the bay (Keen et al., 1991).

For an analysis of tidal amplification, a zero line was designated as running between Yarmouth, Nova Scotia (523), and Moose Cove, Maine, USA (619). Because of the funnel-shaped geometry of the bay, there is a progressive inland amplification of the tides (Fig. 3B). The highest tidal ranges occur within the two inland arms of the bay, which are the southern Minas Basin and northern Chignecto Bay. NOAA tidal data only extend eastward to Burntcoat Head (547), where the tidal range is 13.26 m. Tidal height data for the area to the east, and delineation of zones, were obtained from Zaitlin (1987). Inland, there is a rapid decrease in actual tidal flux, with the inland macro- and mesotidal zones having a longitudinal extent of only a few tens of kilometers.

Bristol Channel (2)

In southwestern England, the system defined by the Bristol Channel to Severn Estuary and River exhibits very high tidal ranges (Fig. 4A). This area has a humid, maritime climate with relatively warm winters. The degree of anthropogenic influences is very high, and the system has been extensively diked along all of the inland reaches of the estuary and river. A number of changes within the channels were related to improvements to navigation. Today, however, all river traffic is within canals that parallel the Severn.

Tidal bores occur within the Severn and can achieve heights of 2 m at Gloucester (station 4, Fig. 4A). Bores move at about 16 km/hr. Currently, the bore forms only during the highest tides, and not all spring tides produce bores. At Gloucester, the inland progressive of the bores, which formerly reached nearly to Worcester (station 5), is now impended by small weirs (dams).

In the western parts of this area, high tidal ranges occur even on the open coast, where spring tidal ranges exceed 8 m (Fig. 4B). There is progressive inland amplification with maximal ranges, based upon the NOAA data, of 13.26 m at Avonmouth (997). Tidal ranges have been reported to "reach 50 ft" (15.2 m) (Rowbothham, 1983). Further upstream, tidal range undergoes a dramatic decrease, and a range of 8.44 m occurs near Wellhouse Rock (1001) within the Severn River. Tides continue within the Severn for more than 50 km inland (Rowbotham, 1983). Tidal effects can extend to Worcester, where a 30 cm rise in the Severn River is related to the tides impeding the free outflow of water.

Rift-related extension, associated with the Late Paleozoic-Mesozoic early opening of the Atlantic Ocean Basin, was responsible for North-South extension across southern England (Chadwick, 1986). The western end of the resultant graben system is the present-day Bristol Channel Basin (Nemcock and Gayer, 1996). The Bristol Channel Basin was a half-graben structure that received dominantly shallow-water calcareous sediment during the Late Paleozoic and Mesozoic. Although there was some Early Tertiary basin inversion, the region has remained a topographic low (Nemcock et al., 1995).

West-central England (10)

Along the northwestern coast of England and southwestern coast of Scotland, a number of macrotidal estuaries occur near, and to the north of, Liverpool. Important estuarine systems are associated with the Mersey, Ribble, and Dee Rivers. High tidal ranges also occur within the Solway Firth, which is at the border of England and Scotland. These systems all occur within a large-scale indentation of the coastline, which is the eastward extent of the Irish Sea (Fig. 5A). This area has a generally cool and humid climate. Significant anthropogenic changes have modified the estuarine systems, particularly those near the larger urban areas, such as Liverpool (1057). The rivers have all been diked in this area.

The sea off the coast of west-central England is part of the East Irish Sea Basin. This basin was formed by multiple phases of extension, subsidence, and deposition related to the break-up of Laurasia in the Late Paleozoic (Rowley and White, 1998). These episodes of extension created several sub-basins within the east Irish Sea. The sub-basins contribute to the irregular coastline in this area.

A zero line for the mouth of this complex system can be defined as making landfall near Holyhead (1043) to the south and Drummore, Scotland (1089), to the north (Fig. 5B). This zero line also passes near the most westward part of the Isle of Man, which includes a tidal station at Peel (1083). These three stations have tidal ranges that are approximately 5 m.

Stations to the east have progressively higher tidal ranges. In the Liverpool area, this includes Eastham (1059) and Liverpool (1057) on the Mersey Estuary. These two stations have ranges of 8.84 and 8.38 m, respectively. A tidal range of 8.38 m occurs at Silloth in the Solway Firth (1077). Ranges of nearly 8 m occur in the estuaries of River Dee (1053) and River Ribble (1063). Tens of kilometers inland, within the fluvio-estuarine system, the tidal range is greatly reduced. This is apparent for the data from Preston on the River Ribble (1061), Chester on the River Dee (1055), and the estimates made by the senior author for other rivers in this area. Tidal effects do not propagate very far inland in the fluvial systems. This is due, in part, to the relatively high longitudinal gradients and topographic relief of the area. The polynomial fit for the tidal-range versus inland distance data is rather remarkable because it includes at least four major estuaries that are separated, in a north-to-south direction, by distances of hundreds of kilometers.

Gulf of Cambay (11)

The Gulf of Cambay (Khambhat) is a trumpet-shaped gulf of the Arabian Sea located in eastern India (Fig. 6A). The city of

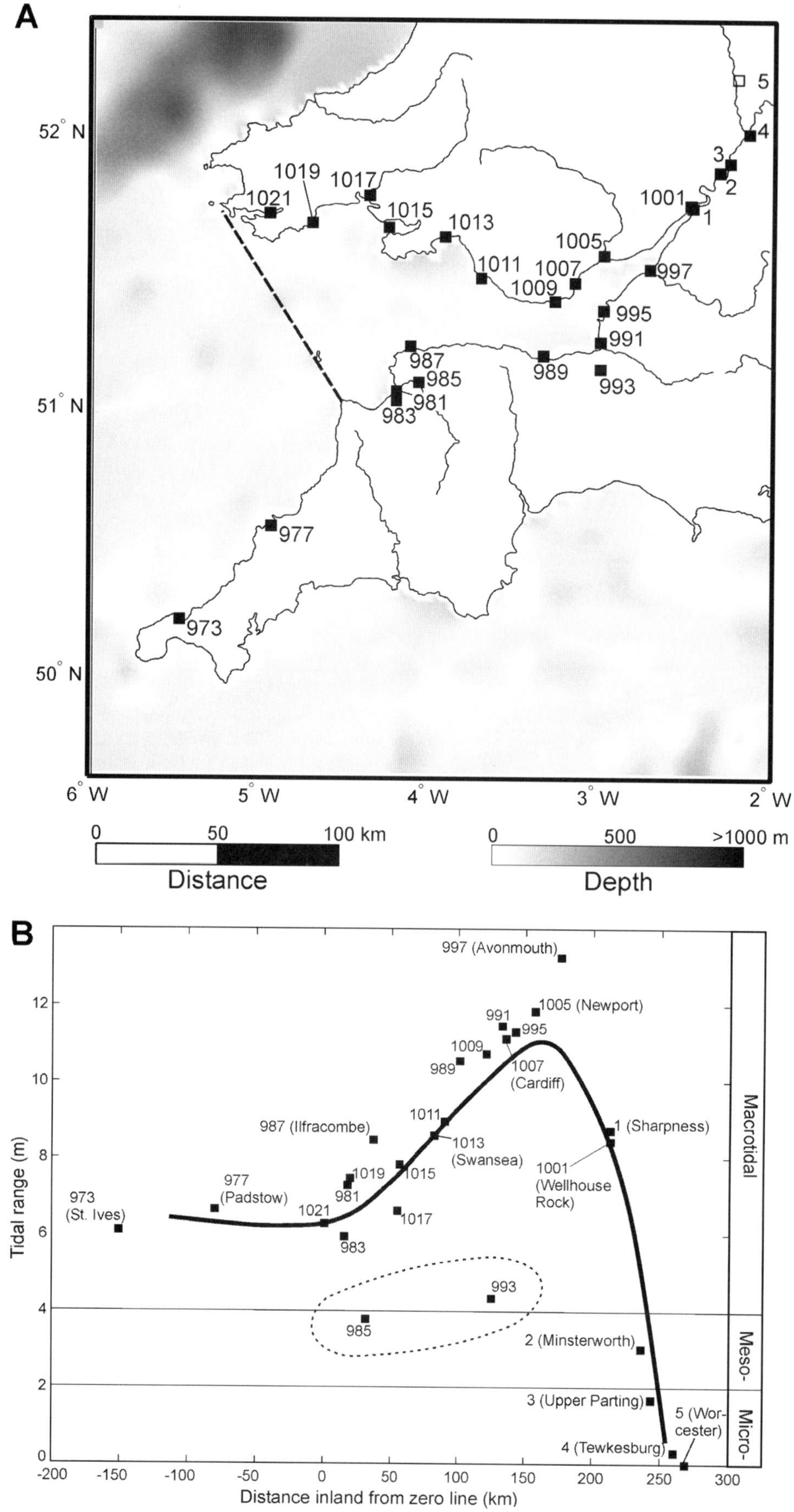

Figure 4. A: Location of NOAA tidal stations (973 through 1021) within the Bristol Channel and ranges described by Rowbotham (1983; stations 1–5) in the Severn Estuary and River, in southwestern England. Heavy dashed line delineates arbitrary "zero line." B: Tidal amplification with Bristol Channel system illustrates that highest ranges occur near Station 997 (Avonmouth) with progressively inland stations exhibiting decreased tidal range. According to Rowbotham (1983), tidal effects formerly extended to Worcester (station 5) prior to anthropogenic channel modifications. Stations 985 and 993, which are enclosed in a dashed oval, are noncoastal stations located upstream within small, inland rivers. Thus, their tidal range is significantly reduced.

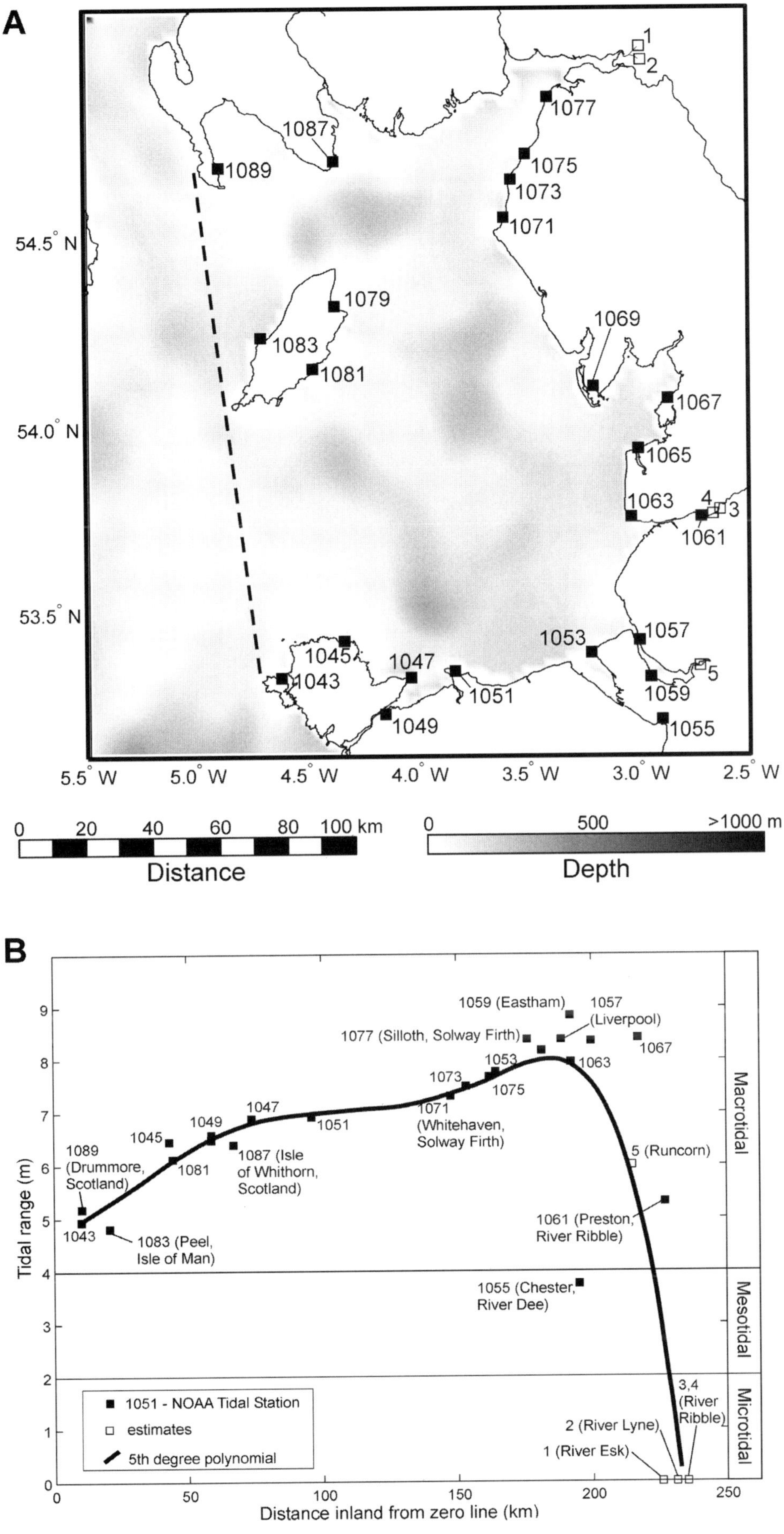

Figure 5. A: Location of NOAA tidal stations (1043 through 1089) combined with personal observations (1 through 5) in west central England. The "zero line" (heavy dashed line) encompasses a complex system, which includes several estuary and river systems. B: Despite its complexity, all stations fit closely to fifth degree polynomial regression. Highest tides occur nearly 200 km inland from zero line and more inland ranges dramatically decrease.

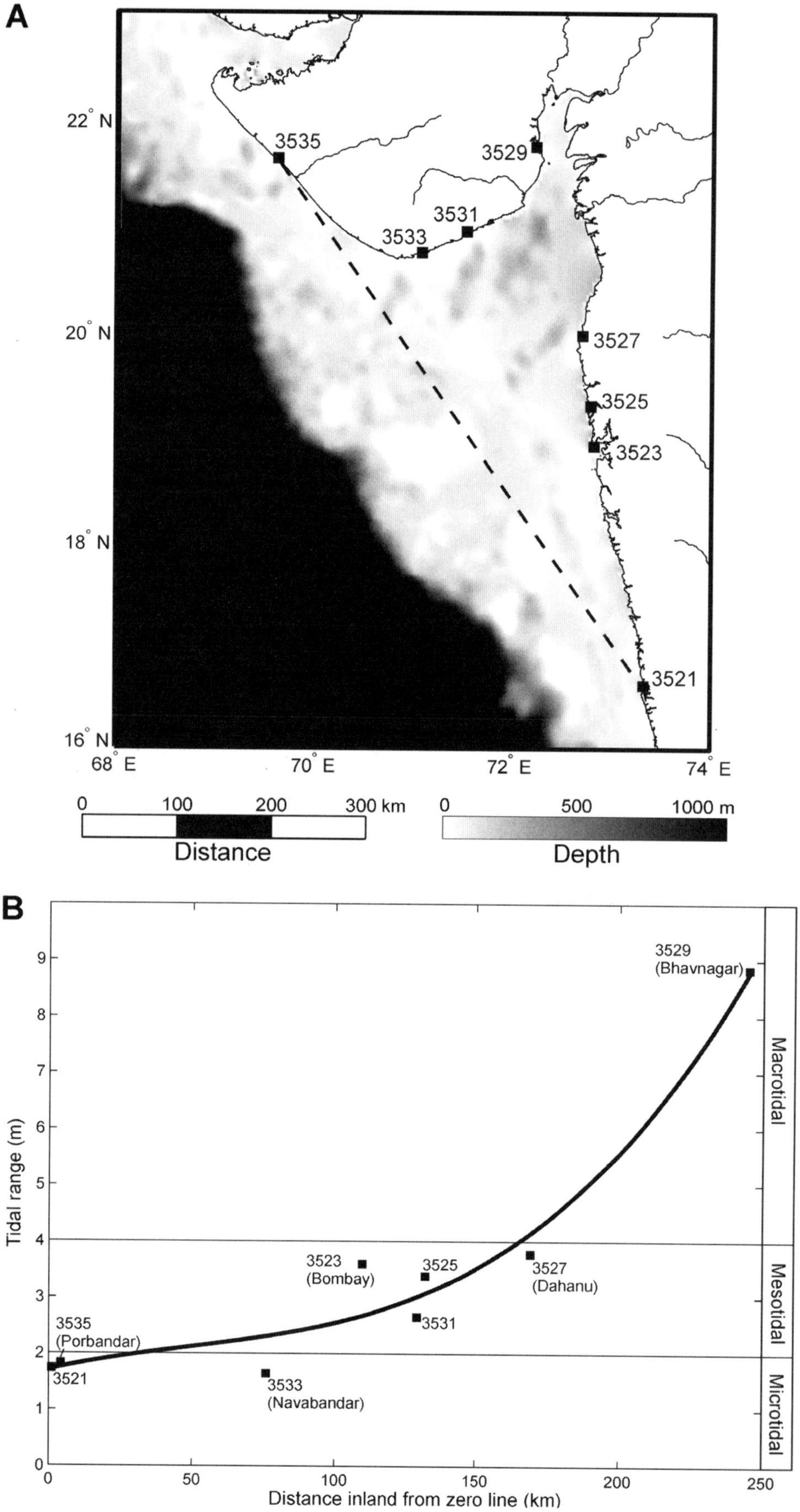

Figure 6. A: Location of NOAA tidal stations within Gulf of Cambay in western India showing "zero line" (heavy dashed line). B: Despite small number of stations, dramatic amplification of tides occurs, with maximal tides occurring nearly 250 km inland from zero line. Estuaries north and northeast of Bhavagar (3529) include major historic ports. Over past several centuries, rapid siltation has forced their closure.

Bombay (3523) is located on the eastern side of the gulf. Rivers that enter the gulf include the Sabarmati, Mahi, Narmada (Narbaba), and Tapti. The climate is tropical savanna, and warm temperatures persist throughout the year. The orientation of the gulf to the southwest monsoons and its trumpet-shaped geometry yield very high tidal ranges.

Shoals and sandbanks present a number of problems to navigation, and a number of harbors in the northern reaches of the gulf have severe silting problems. At the northern end of the gulf, the former port at Cambay (Khambhat) on the Mahi River has been largely abandoned owing to siltation (Bradnock and Bradnock, 1997), and NOAA does not provide tidal data for this area. The highest ranges occur at Bhavnagar (3529). About 50 m south of Bhavnagar, the world's largest scrapyard for ships is at Alang Beach. Because of the high tidal range, ships can be beached well onshore during spring tides and then salvaged over the intervening 2-week period (Bradnock and Bradnock, 1997).

The Gulf of Cambay is on the passive margin of southern India and is located approximately equidistant from the convergent margin of the Himalayas to the northeast and the divergent Carlsburg Ridge to the southwest. The Cambay graben is a north-south-trending rift that formed when India separated from Africa and Madagascar in the Jurassic. The southern portion of the Cambay graben forms the Gulf of Cambay (Gombos et al., 1995).

The gulf exhibits a prominent funnel-shaped geometry that flares to the south toward the Arabian Sea. Because of the gulf, the continental shelf is very wide. In this area, the coast-to-shelf edge distances are approximately 350 km (Fig. 6A). The NOAA data are limited and include only eight tidal stations. The zero line was defined as connecting stations at the mouth of the Rajapur River (3521) and Porbandar (3535), which are on the open coast (Fig. 6B). The outermost three of the stations (3521, 3533, and 3535) are microtidal, with ranges from 1.6 to 1.8 m. Microtidal conditions are characteristic for much of the eastern coast of India. The four stations within the mid portion of the gulf are mesotidal; this includes Bombay (3523), with a range of 3.60 m, and Dahanu (3527), with a range of 3.78 m. The innermost station is Bhavnagar (3529), with a tidal range of 8.84 m. When compared to areas with higher data densities, such as the Bay of Fundy, Bhavnagar could well be seaward of the zone of maximum tidal amplification. Thus, significantly higher ranges probably exist in the northern extents of the gulf, where several large fluvio-estuarine systems are evident in the Mahi, Narmada, and Tapi Rivers. Defant (1961) reports that large tidal bores occur on the Narmada River.

PASSIVE-MARGIN SEAWAYS

Ungava Bay Area (3)

Ungava Bay is located in the northern part of the province of Quebec in Canada. It lies along the southern part of Hudson Strait, which leads westward to Hudson Bay (Fig. 7A). This entire area is mostly tundra and is only sparsely populated, and the number of tidal stations is very low. Wide tidal flats and marshes are common within bays along the shoreline. Discontinuous bodies of permafrost, which reach thicknesses of 5 m, occur within the silty sediments within the upper part of the intertidal zone (Sequin and Champagne, 1979; Michel et al., 1992).

Ungava Bay is located at the southwest end of a fracture zone that formed during the Cretaceous opening of the Labrador Sea. At Ungava Bay this fracture zone truncates a northwest-southeast-trending fault system in Hudson Strait. Localized extensional deformation associated with these fault systems is the likely cause of the topographic depression that makes Ungava Bay (Adams and Basham, 1989; Grant and Sanford, 1988). Post-glacial isostatic rebound has resulted in an emergent coast (Gray et al., 1980). Seismic and core-based biostratigraphic analyses have been used to delineate the Late Cenozoic sedimentation within the Bay and surrounding areas (McLean et al., 1992).

A well-developed amplification is developed within the Ungava Bay system when distances and ranges are compared based on a zero line that connects stations 151 and 167 (Fig. 7B). Stations on the more open Atlantic Ocean coastline, such as Sorry Harbor on Resolution Island (123), have ranges of only 5.36 m. The highest tidal ranges are within the estuaries at the southwestern limit of the bay. The highest range occurs at Leaf Lake (161), where there is a spring tidal range of 12.19 m. A range of 11.09 m occurs at the mouth of the Koksoak River (163). At Leaf Bay (159), the range is 10.97 m, and at Hopes Advance Bay (157), the range is 10.49 m. Conversely, at the mouth of the bay, the tidal range is much reduced, although it remains macrotidal. At the eastern margin of Ungava Bay, the range is 4.69 m at Button Islands (167). At the western margin, the range is 8.17 m at Diana Bay (155).

There are sites with tidal ranges that exceed 9 m in this region on Baffin Island, and these, although not actually within Ungava Bay, are included as part of this region but were not used for the amplification computations. Two such stations lie on the southeast coast of Baffin Island and the southern coast of Baffin Island. Northwest of Ungava Bay and on the northern side of Hudson Strait, tides attain ranges of 9.42 m at Ashe Inlet on Big Island (127). At the head of Frobisher Bay (119), in the southeastern part of Baffin Island, the tidal range is 9.08 m.

To the west is the extensive, but shallow, Hudson Bay, which is characterized by low tidal ranges. This relates to frictional damping and distances that exceed the potential for developing tidal amplification.

Gulf of St. Malo (4)

Very high tidal ranges occur throughout the English Channel (La Mance) that separates the southern coast of England from the northern coast of France. This seaway and the North Sea to the east are large areas of flooded continental shelf and are useful analogs for understanding ancient inland tidal systems.

High tidal ranges are well developed within the Gulf of St. Malo (Fig. 8A), which is east of a line connecting St. Malo (725)

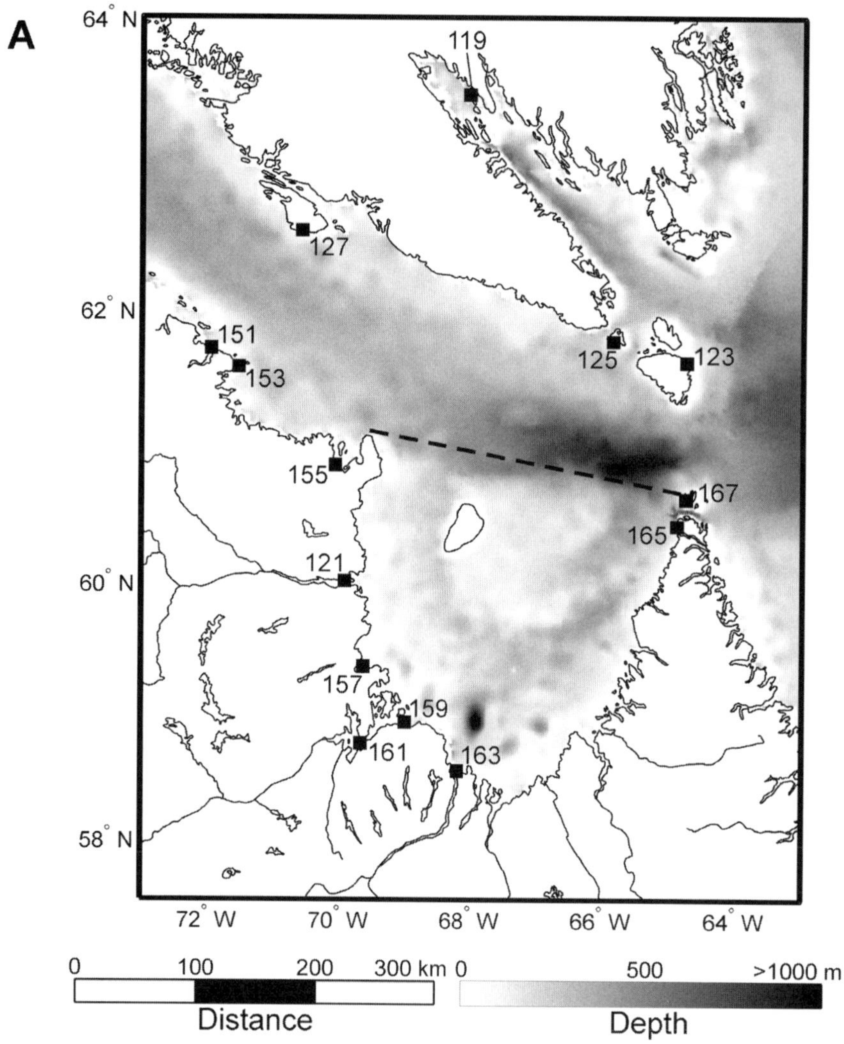

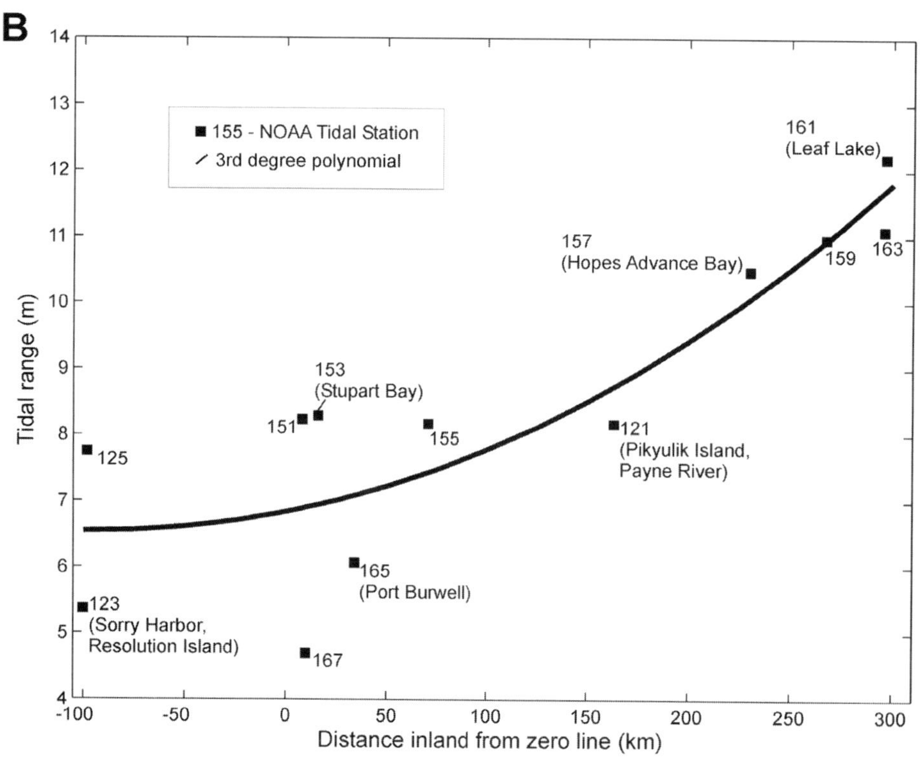

Figure 7. A: Location of NOAA station in Ungava Bay area of northeastern Canada showing "zero line" (heavy dashed line). In addition to high tidal ranges in Ungava Bay, very high tides also occur at stations 119 and 127; however, there is insufficient data for computation of amplification. B: Within Ungava Bay, there is significant tidal amplification inland of zero line. Significant tidal effect probably extends for considerable distances into fluvial-estuarine systems; however, data on these inland tidal ranges were not available for analysis.

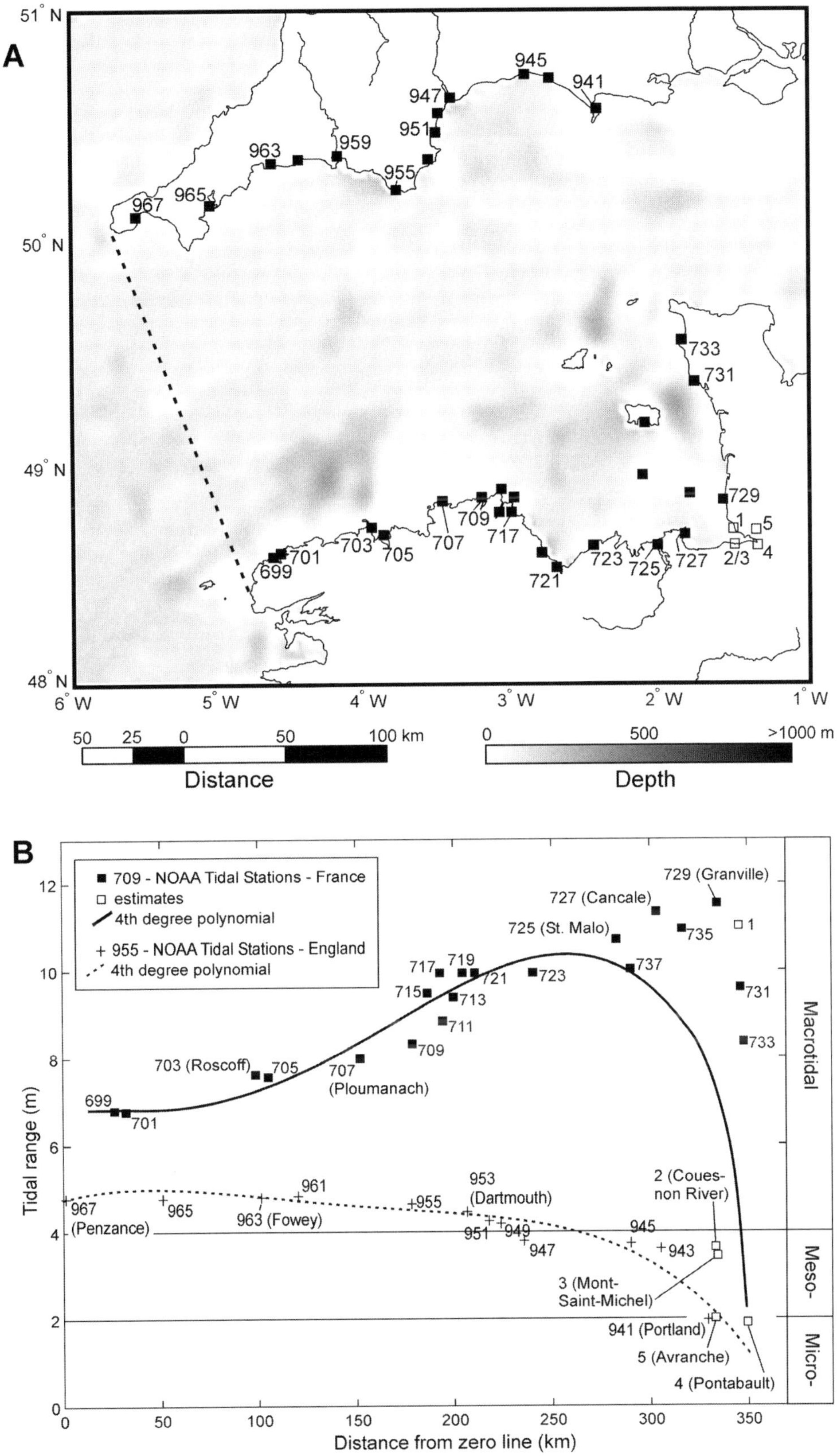

Figure 8. A: Gulf of St. Malo and surrounding area in western parts of English Channel that separates England and France, showing locations of NOAA tidal data. "Zero line" is defined as the channel's western end. B: Highest tides occur at Granville (Station 729). Estimated ranges in Mont-Saint-Michel area (stations 1 through 5) were made in 1999 by the senior author. Because of Coriolis effects, tidal ranges are considerably higher on French coast than on English coast.

and Granville (729). To the east, extensive tidal flats surround the famous Mont-Saint-Michel Abbey (3).

This area has a maritime climate with relatively cool summers and warm winters. Large-scale winter storms can cause significant amounts of coastal erosion throughout the English Channel. Anthropogenic changes have greatly affected this region, particularly around Mont-Saint-Michel. Large areas of salt marsh have been reclaimed by the construction of dikes. The fluvio-estuarine systems within the inland portions of the system are diked and commonly protected by barriers that prevent saltwater intrusion into the freshwater systems. The salt marshes that persist are used extensively for grazing sheep. Aquiculture of oysters and mussels is carried on throughout the intertidal zones of the area.

Larsonneur (1989) and Tessier et al. (1989) have studied sedimentary facies within the Bay of Mont-Saint-Michel extensively. The finer-grained sediments and sedimentary structures share many similarities with equivalent facies of the Bay of Fundy. Siltation has been very rapid, and this has reduced the tidal ranges and rates of sedimentation within the system. The area remains a valuable modern analog for tidal rhythmite production (see Tessier, 1993); a number of great similarities exist between rhythmites of Mont-Saint-Michel and the Douglas Group (Carboniferous) from the U.S. state of Kansas (Tessier et al., 1995).

The Gulf of St. Malo is adjacent to the English Channel, which began a north-south rifting phase during the Early Mesozoic (Zeigler, 1982).

Within the western part of the English Channel, a progressive tidal wave is formed that moves from west to east. Throughout this entire area, coastal tidal ranges are macrotidal. Owing to the Coriolis deflection, which is toward the right in the northern hemisphere, higher tidal ranges occur along the coast of France as compared to similar longitudinal positions on the English coast. This can be seen in a comparison of the higher ranges at the Morlaix River (705) and Paimpol (717) in France as compared to the lower ranges at Fowey (963) and Salcombe (955) in England (Fig. 8B). The ranges along the French coast best express the effects of tidal amplification. Conversely, tidal ranges along the coast of England show a general decrease from west to east (Fig. 8B).

Although it is not a simple funnel-shaped system, there are funnel-shaped components within the Gulf of St. Malo. The world's best-known tidal-power generating station is on the Rance Estuary at St. Malo and was built between 1961 and 1967 (Frau, 1993).

Within this complex system, the tidal ranges are highest at Granville (729) and Cancale (727), which have spring tidal ranges of 11.52 and 11.34 m, respectively. Tides within this area are reported to range as high as 15 m (Tessier, 1990). At St. Malo, spring tidal ranges have been reported to be 13.3 m (Macmillan, 1966). Tidal ranges decrease dramatically to the east, and the tidal flux at Mont-Saint-Michel (3) is less than 4 m. At Avranche (5) and Pontabault (4), tidal flux falls into the microtidal range (2 m and less), and the inland tidal limit is within a few kilometers to the east.

Straits of Magellan (5)

This passage (Estrecho de Magallanes) connects the Atlantic and Pacific Oceans near the southern tip of South America. This region includes tidal stations along the Atlantic coast of Argentina and inland and Pacific coast stations of Chile (Fig. 9A). The Strait of Magellan is 560 km long and ranges from ~3 to 32 km in width. The area has a generally cold and commonly foggy climate, and the strait has numerous channels separated by islands. Formerly an important sailing-ship route for passages from the Atlantic to Pacific Oceans, its importance was greatly reduced by the opening of the Panama Canal in 1914.

This region is particularly interesting because of the differences between active and passive continental margins. The western coast of South America is an active convergent margin where the Pacific plate is being subducted under the continental plate. The continental shelf is very narrow, and the tidal ranges are low. Conversely, the eastern coast of South America is passive and also has a very wide continental shelf, which includes the Falkland Islands. This results in very high coastal tides, which are in part propagating northward along the shelf, and this creates even higher inland tidal ranges.

The Strait exists because of a series of northwest-southeast-trending rift valleys in the foreland basin immediately east of the southernmost Andes. The kinematics of the northeast-southwest extension that formed these rift valleys is related to the left-lateral strike-slip system that links subduction west of the Andes to the Scotian Arc to the east. Based on the ages of offset units, timing of rift formation is Neogene and is consistent with plate motion changes that occurred at that time as the Scotian system developed (Diraison et al., 1997).

For an analysis of amplification, a zero line was set that extends from near Santa Cruz, Argentina (5259), southward to Bahia Thetis, Argentina (5275). As thus defined, this includes not only the Straits of Magellan but also other rivers and coastal stations in Argentina (Fig. 9B). Near the zero line, the tides are mesotidal with a tidal range of 3.23 m at Bahia Thetis (5275). The tidal ranges increase westward, and maximum amplification within the Straits of Magellan occur at Banco Direccion, Chile (13), which has a tidal range of 10.36 m. The tidal range is also very high within the Rio Gallegos, Argentina (5265), which has a spring tidal range of 11.03 m.

To the west, the tidal ranges decrease dramatically within the Straits of Magellan. At Punta Arenas, Chile (25), the range has been reduced to microtidal (1.49 m). Microtidal conditions persist to the west and characterize the Pacific coast of Chile in this area.

Eastern English Channel (8)

In the western part of the English Channel, the Gulf of St. Malo, as discussed above, includes tidal ranges that approach 12 m. The eastern part of the English Channel exhibits a prominent west-to-east narrowing and serves as another, but lesser, area of tidal amplification (Fig. 10A). Much of the coast in this part of

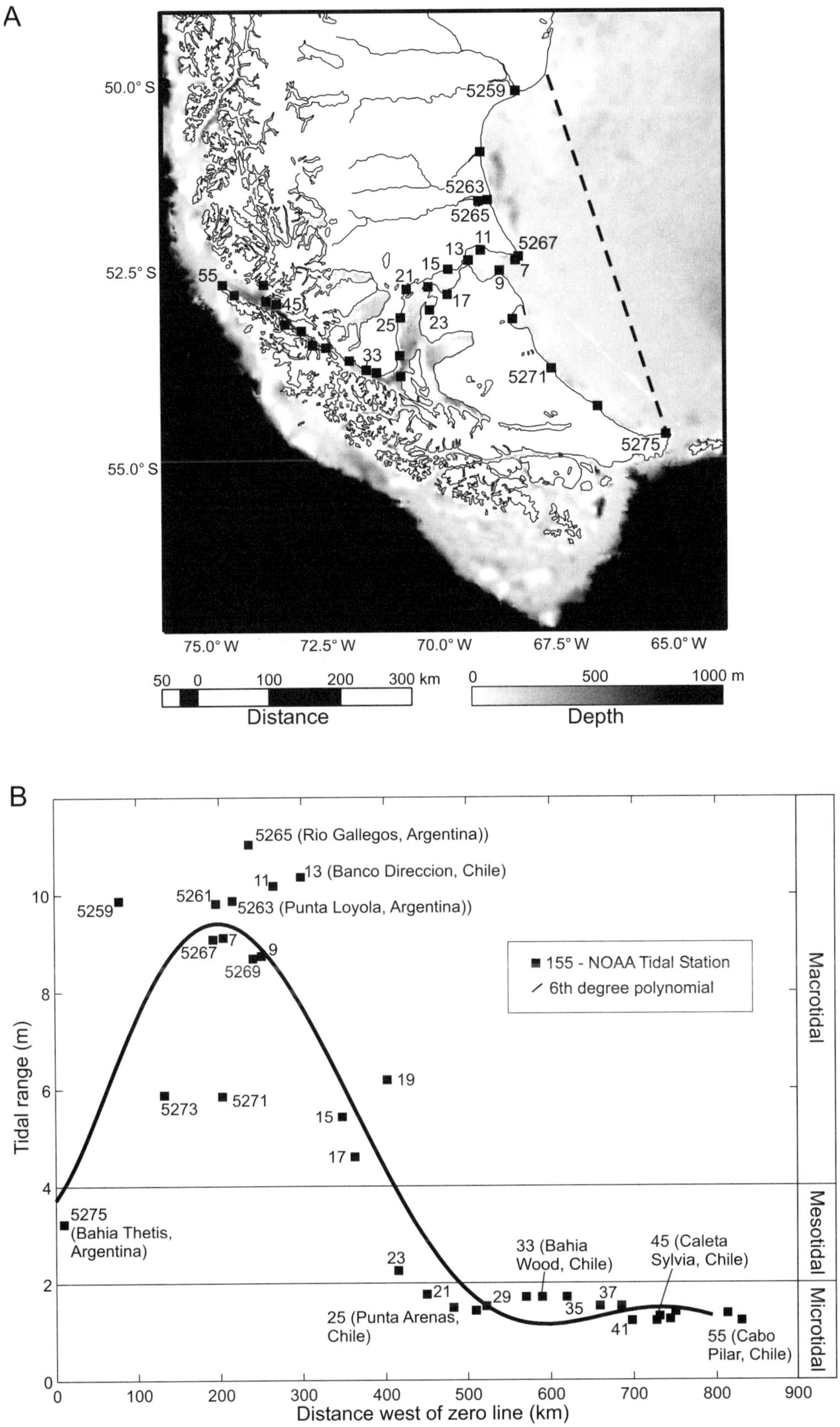

Figure 9. A: Location of NOAA tidal data in Straits of Magellan area of southern South America and location of "zero line" used for computing tidal amplification. B: Wide shelf along Atlantic coast results in considerably higher tides than those that occur on Pacific coast's narrow shelf. Highest tidal ranges occur within embayments and estuaries along Atlantic coast.

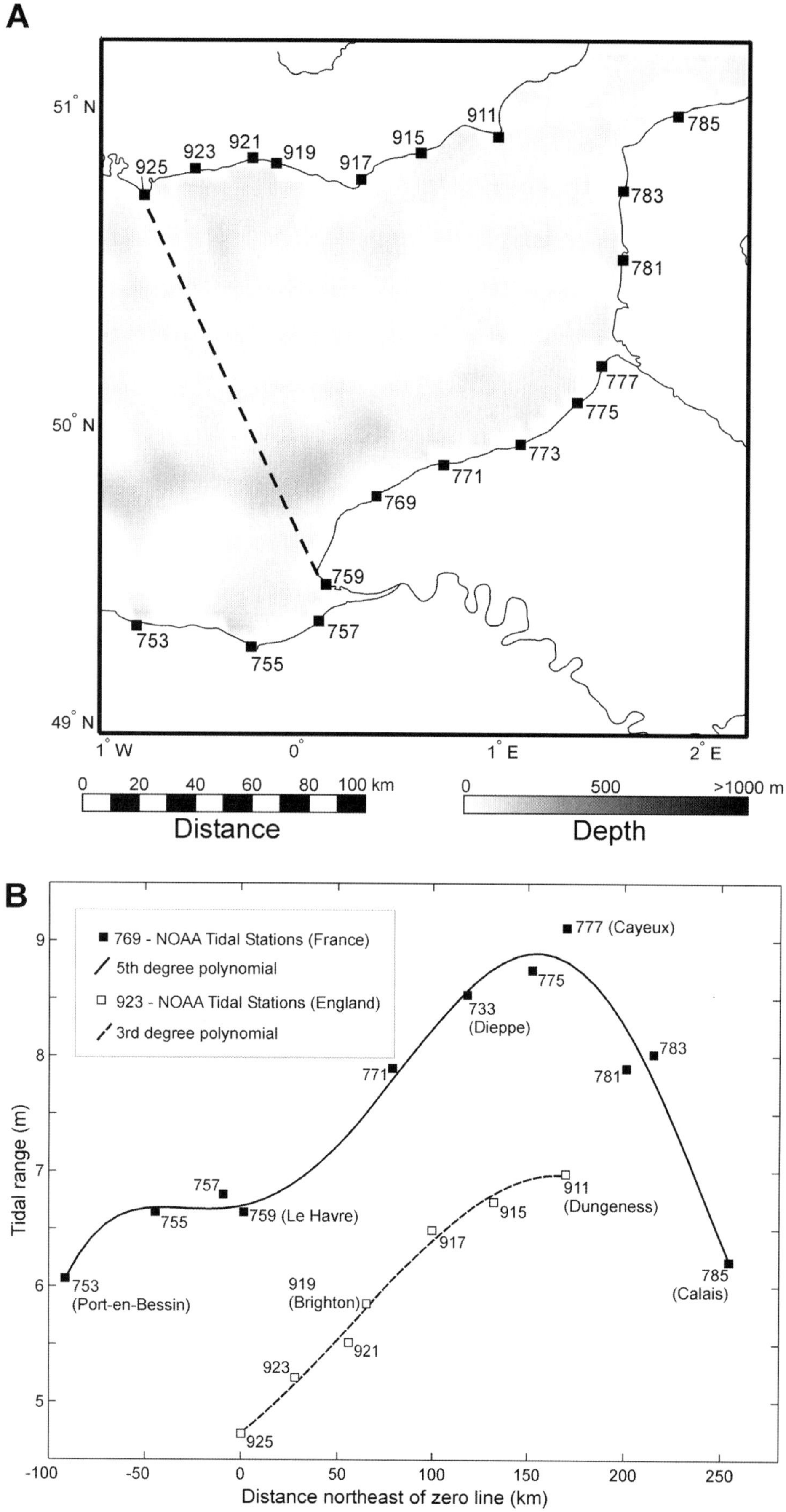

Figure 10. A: Location of NOAA tidal data within eastern English Channel. "Zero line" is defined as running from La Havre (station 759) on French coast to station 925 on English coast. B: Tidal amplification occurs west-to-east, with highest tides occurring near Cayeux (station 777). Because of Coriolis effects, tidal ranges on English Coast are considerably lower than on French coast.

the channel consists of sea cliffs of Cretaceous-age chalk. Much of the subsidence of the English Channel formed during rifting that coincided with the early opening of the Atlantic Ocean during the Triassic period (Zeigler, 1982).

Tidal ranges along both the English and French coasts exhibit a prominent increase in an eastern direction. The entire English Channel develops a progressive tidal wave that moves to the east. Owing to Coriolis deflection, significantly higher tides occur along the coast of France. The zero line is defined as running from Selsey Bill (925) in England in a southeasterly direction to Le Havre (759) in France (Fig. 10B). All the stations in this area exhibit macrotidal ranges. The highest of these high tides occurs at Cayeux (777) in France, where spring tidal ranges attain 9.11 m. Further to the east, along the coast of France, the tidal ranges undergo a marked reduction. The range at Calais (785), for example, is reduced to only 6.21 m.

Gulf of Mezan (12)

The nearly landlocked inland White Sea (Beloye More) is an extension of the Arctic Ocean into northeastern Russia. This area has a tidal power station that was completed in 1969. The highest tidal ranges in the White Sea occur in the Gulf of Mezan, which is located between the Kanin Peninsula to the east and Kola Peninsula to the west (Fig. 11A). It has a generally funnel-shaped geometry; however, this simple geometry is interrupted along the western side by a seaway that connects the more inland parts of the White Sea. This is an important area of shipping and has cold winters due to its subarctic climate. The use of icebreakers, however, allows year-long navigation. The land surrounding the gulf is extensively forested and low in population density.

Waters within the Gulf of Mezan are shallow, and sandy ridges occur near the mouth of the Mezan River. Within the Mezan estuary, there is a regular increase in mean water levels, which varies with neap-spring tidal periods. This tidal pumping results in residual water transport within the fluvio-estuarine system (Lupachev, 1989).

The Baltic Shield underlies the gulf, and the entire area is on a passive-margin continental shelf that is located at great distances from any active margins. The floor of the gulf is relatively irregular because of extensive faulting. The north-south orientation of the Gulf of Mezan parallels the Precambrian terrane boundaries of the Baltic Shield region (Gorbatschev and Bogdanova, 1993). The east-west entrance to the White Sea marks the southern limit of shield exposures and the northern limit of Paleozoic rocks (Lidmar-Bergstrom, 1993). The morphology of this region can likely be attributed to a combination of this bedrock geology and the erosion by Quaternary glaciers.

A zero line can be defined at the mouth of the Gulf of Mezan, which connects Savika Bay (1641) on the west to Cape Kanin (1801) on the east (Fig. 11A). These stations have ranges of 4.51 and 2.53 m, respectively. The outer stations along the eastern side (1801, 1799, and 1797) are all mesotidal, whereas the outer stations on the western side (1641, 1643, 1645, and 1647) have macrotidal ranges. This is the result of Coriolis deflection of the incoming tidal wave. There is a gradual increase in tidal range to the south to Morzhovetz Island (1777), which is within the seaway that connects to the more inland parts of the White Sea. In this area, tidal ranges of 5 m can occur. Near Cape Abramov (1781), the system develops a funnel-shaped geometry, and this results in a significant inland amplification of tides. The highest tidal range of 8.6 m occurs at the Semzha River mouth (1787). A similar range of 8.26 m occurs at Nerninski Point (1783). Upstream within the Semzha River, the tidal flux begins to decay, and the range is reduced to mesotidal (3.41 m) at Kamenka (1791).

ACTIVE MARGINS

Cook Inlet (6)

Cook Inlet is in the southern coastal area of the state of Alaska in the United States (Fig. 12A). The inlet connects to the Gulf of Alaska in the northern Pacific Ocean. This is an area of coniferous forest with a humid, subarctic climate. There is extensive formation of winter ice, particularly in the more inland parts of the inlet. This is an area of generally low population density and anthropogenic alterations are minimal, except in the region of Anchorage (2055).

At Anchorage, the system bifurcates into the northern Knik Arm and the southern Turnagain Arm. Tidal bores are well developed within Turnagain Arm. Aspects of sedimentation and tidal regimes are discussed in detail by Sharma (1969) and Bartsch-Winkler (1988).

Cook Inlet formed within a backarc extensional zone related to subduction of the Pacific plate under the North American plate. To the west, this boundary has resulted in the Aleutian Islands, which form an island arc that extends westward toward Asia. This is a seismically active area and Anchorage (2055), which is near the landward limit of the inlet, is particularly earthquake prone. In 1964, a quake in this area registered a Richter magnitude of 8.4 and caused about $500 million in property damage (Eckel, 1970). The seismicity of the area has had profound effects on long-term deposition and tidal-channel stability (Bartsch-Winkler, 1988; Atwater et al., 2001).

The entire inlet is relatively shallow and has a macrotidal regime. A zero line can be defined as connecting Port Chatham (2021) to the south with Oil Bay (2067) to the north (Fig. 12A). The tidal range near the zero line is low mesotidal (ranging from 4.2 to 5 m). There is a gradual tidal amplification in an inland direction; at Nikiski (2047), the tidal range is 6.25 m. Strong amplification occurs where Cook Inlet bifurcates into the northern Knik Arm and southern Turnigan Arm. This bifurcation occurs near Anchorage (2055), which has a tidal range of 8.78 m. Tides attain ranges of 10.15 m within Turnagain Arm (2053). Within Turnagain Arm, tidal ranges have been reported to be "about 11.5 m" (Bartsch-Winkler, 1988), which would be about 13% higher than the information presented by NOAA.

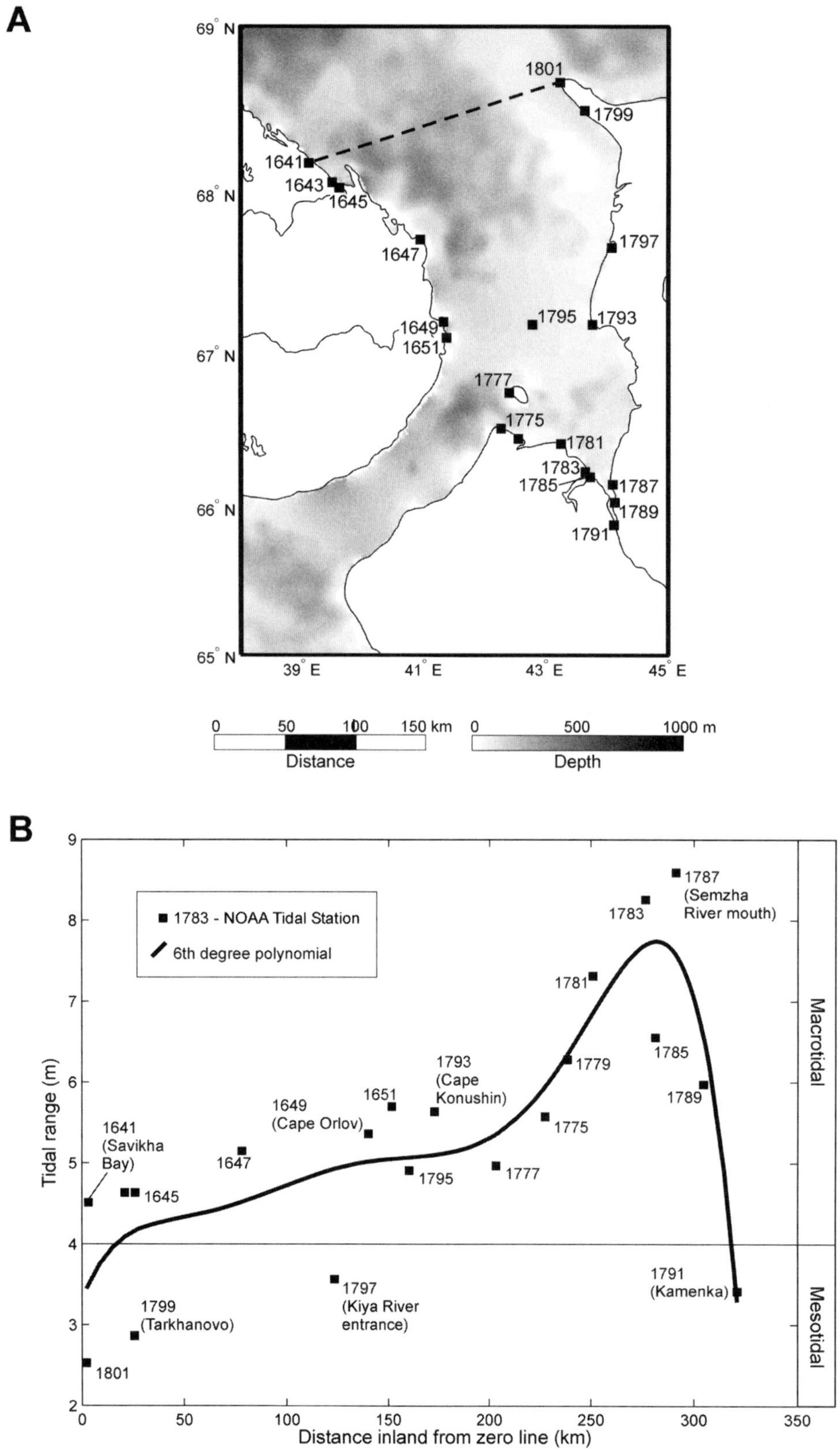

Figure 11. A: Gulf of Mezan area in northeastern Russia showing location of NOAA tidal data. "Zero line" (heavy dashed line), used to analyze effects of inland tidal amplification, is defined at gulf's mouth. B: Based upon distances from zero line, highest tidal ranges occur nearly 300 km inland. Further inland, within estuary, tidal ranges decrease rapidly, however mesotidal ranges are still reported at approximately 320 km from zero line.

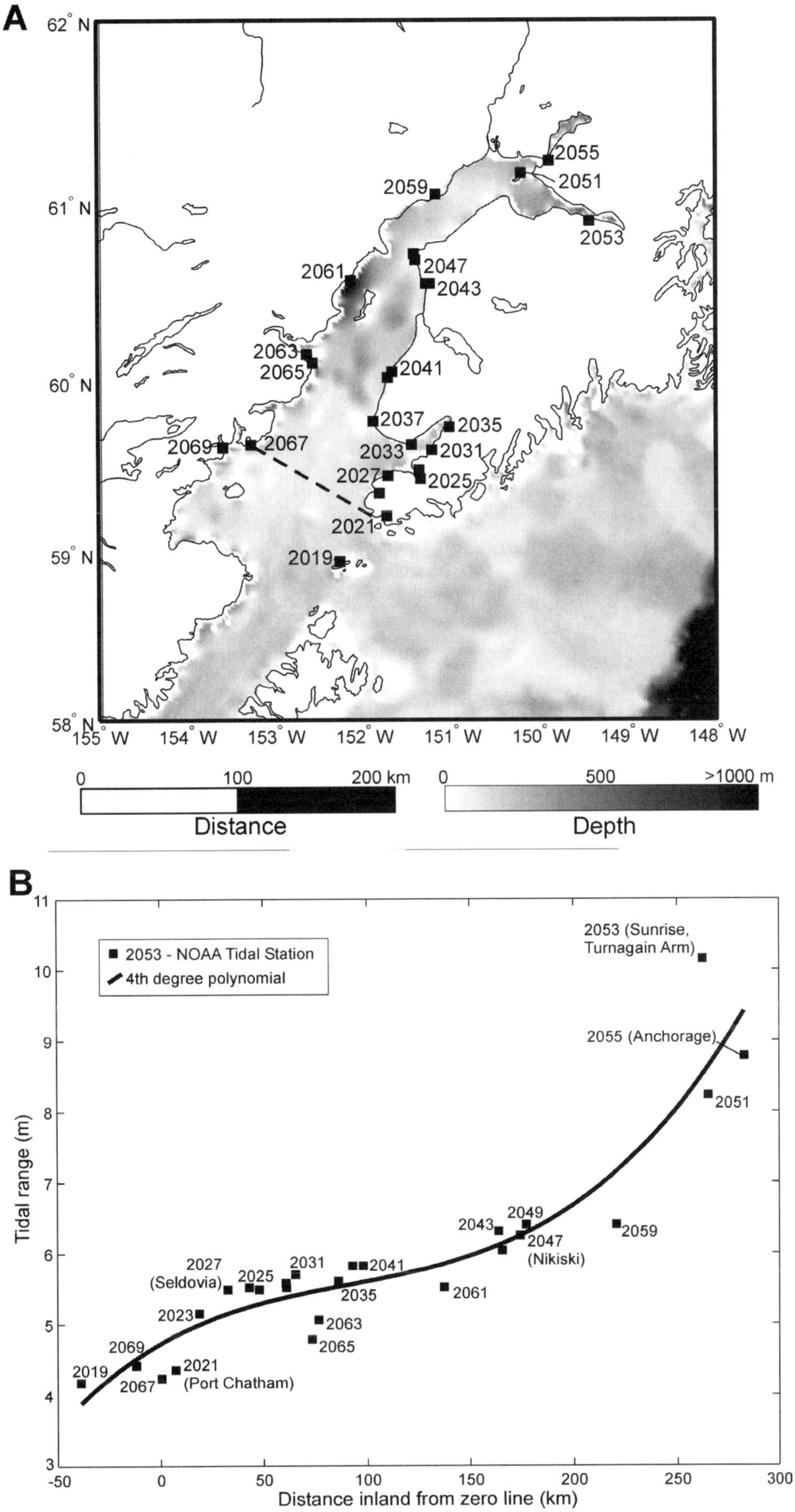

Figure 12. A: Location of NOAA tidal data within Cook Inlet area of Alaska, USA. "Zero line" is defined near seaward mouth of inlet. B: Tidal ranges are maximized near Anchorage (station 2055) and within Turnigain Arm (station 2053). Inland tidal limits are not known, but must extend over 300 km inland from zero line.

Sea of Okhotsk (7)

West of the Kamchatka Peninsula of Siberia is the Sea of Okhotsk (Fig. 13A). This is an area of subarctic climate with coniferous forests. The population density is low and anthropogenic effects are minimal. Parts of this sea relate to a backarc extensional zone developed west of the convergent boundary of the Pacific plate with the Eurasian plate. The Kuril trench occurs in the southern part of this area. Although the coastal outline is somewhat irregular, it exhibits a generalized funnel-shaped geometry, which opens to the south. The Sea of Okhotsk consists of three parts: northern, central, and southern. The northern and central parts consist of subsided continental crust and basin-fill sediments. The southern part is floored by oceanic crust that formed in a backarc setting (Maeda, 1990).

This sea is of sufficient width to produce amphidromic (rotary) tidal systems. For purposes of this simple analysis, however, the tidal ranges will be compared to their inland distance from a "zero line" extending from near the southern tip of the Kamchatka Peninsula (near 127) to the city of Okhotsk (149). Tidal ranges are high microtidal on the tip of the Kamchatka Peninsula. At Okhotsk, the range is mesotidal (2.71 m). There is a gradual increase in tidal ranges to the northeast. Where the Sea constricts at Udacha Bay (143), macrotidal ranges become pervasive. Tidal amplification occurs in the northeastern part of the northern coast of the Sea of Okhotsk, where ranges are commonly about 6 m (137, 139, and 141). Within the restricted embayment in the vicinity of Cape Astronomicheski (135), the spring tidal ranges are 9.17 m, which is nearly 1500-km inland from the "zero line." Higher tides may occur farther inland.

Northwest Australian area (9)

This area includes the northwest coast of Australia and adjacent parts of Indonesia (Fig. 14A). The distance between the islands of Indonesia and the Australian coast decreases in a west-to-east direction. This combination of Indonesian and Australian tidal stations results in two converging coasts. Together with the parts of South America, this area is one of the few examples of very high tidal ranges within the southern hemisphere. The high tides in Australia occur upon a passive coastal margin, and a unique aspect of this area is that the amplification appears to relate, at least in part, to increasing shelf width and also potentially to propagation along the shelf. As such, there is not a significant inland component of this system.

A major convergent boundary occurs within this combined system. Volcanism associated with a subduction zone is prominent throughout the islands of Indonesia. The Sunda Trench, to the south of the island, has depths that exceed 6000 m (see Fig. 14A). The narrowing of the seaway in this region is due to the encroachment of the Australian continent on the island arc complex of Indonesia. This site is the locus of a possible future continental collision.

The zero line for this system runs from Semerang, Java (1865), south to near Learmouth, Australia (3347) (Fig. 14A). At the western end of the system, as thus defined, tidal ranges are microtidal to low mesotidal. Along the Australian coast to the east, the system becomes meso- and then macrotidal. At Port Hedland (3359), the tidal range is 5.79 m. Between Port Hedland and Broome (3361) is one of the longest continuous stretches of beaches in the world, the Eighty Mile Beach. To the east of this beach, the coastline becomes more complex and embayed, and the tides achieve ranges that exceed 7 m. At Broome (3361) the range is 8.29 m. At the eastern end, the tidal range is nearly 9 m at Hall Point on Kid Islet (3373). Farther east, the tidal ranges decrease. At Baudin Island (3377), the range has been reduced to 5.73 m and at Geranium Harbor (3381) the range is low mesotidal (2.29 m). This decrease in range is of interest because the width of the continental shelf offshore from these stations is somewhat wider when compared to shelf widths offshore from the stations with the highest ranges. Thus, shelf width is not the only control on the highest ranges in this region. This is confirmed by examination of the ranges for the Indonesian stations (Fig. 14B). Although shelf widths are minimal, because of the subduction zone, there is a slight west-to-east increase in tidal ranges. Near the zero line, the Indonesia stations generally have low microtidal ranges. About 1000 km east of this zero line, some stations exhibit mid-mesotidal ranges. This is particularly true of stations on the southern coasts of the Indonesian islands.

Yellow Sea (13)

The Yellow Sea connects to the south to the East China Sea, which is landward of the islands of Japan (Fig. 15A). The Korean peninsula lies along the eastern margin, and the northern and western areas are in China. A variety of tidally influenced systems occur in the Yellow Sea and surrounding areas. To the south, the mouth of Chiang Jiang (Yangtze) River (near 1409) has a tide-dominated estuary. The highest reported tidal bores in the world occur within the Hangzhou Estuary (1421), where bore heights approach 3 m (Bartsch-Winkler and Lynch, 1988).

Tides are complex within the Yellow Sea and associated inland seas. The highest tidal ranges do not occur at the most inland positions but instead occur along the west coast of Korea, where very wide muddy tidal flats are produced by tides that exceed 8 m. Frey et al. (1989) has described biogenic and sedimentary structures within the wide tidal flats. The source of the vast amounts of mud that comprise the flats is problematic, and the mineralogy has been discussed by Shin et al. (1993). Wells and Adams (1988) discuss the controls on tidal channels within these dynamic flats.

The Yellow Sea has a geological setting that is similar to that of the Sea of Okhotsk. Both of these areas are on the active margin of the east coast of Asia. The convergent boundary here is related to subduction of the Philippine plate under the Eurasian plate. The area is seismically active, and deep-focus earthquakes are common along the seaward end of the system. The Yellow

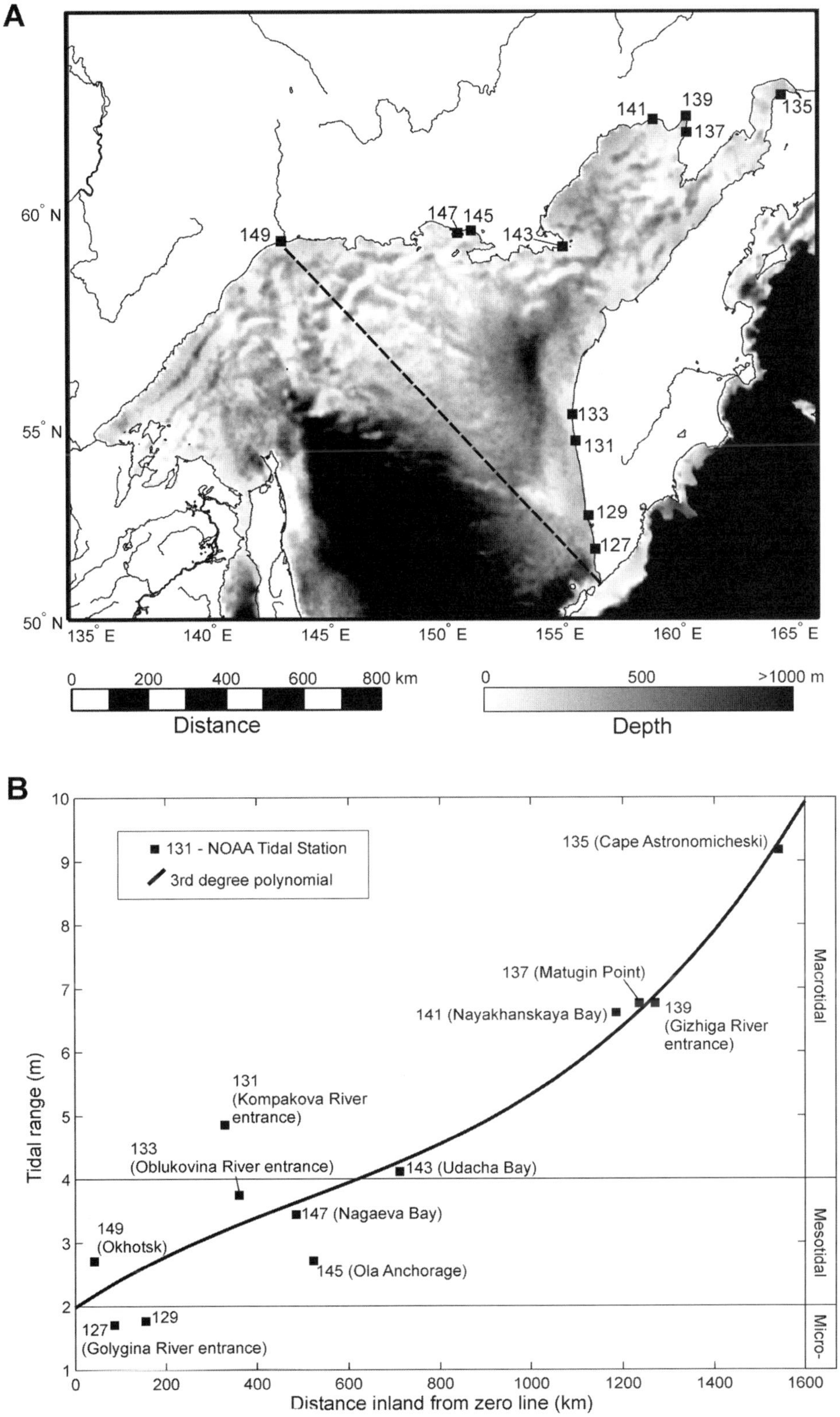

Figure 13. A: Tidal data, as listed by NOAA, for Sea of Okhotsk in Siberia. "Zero line" is defined at sea's southern end. B: Highly amplified tides occur within sea's most inland portions, with ranges exceeding 9 m at Cape Astronomicheski (station 135). Data regarding tidal limit within fluvio-estuarine systems to the north was not available for analysis.

A

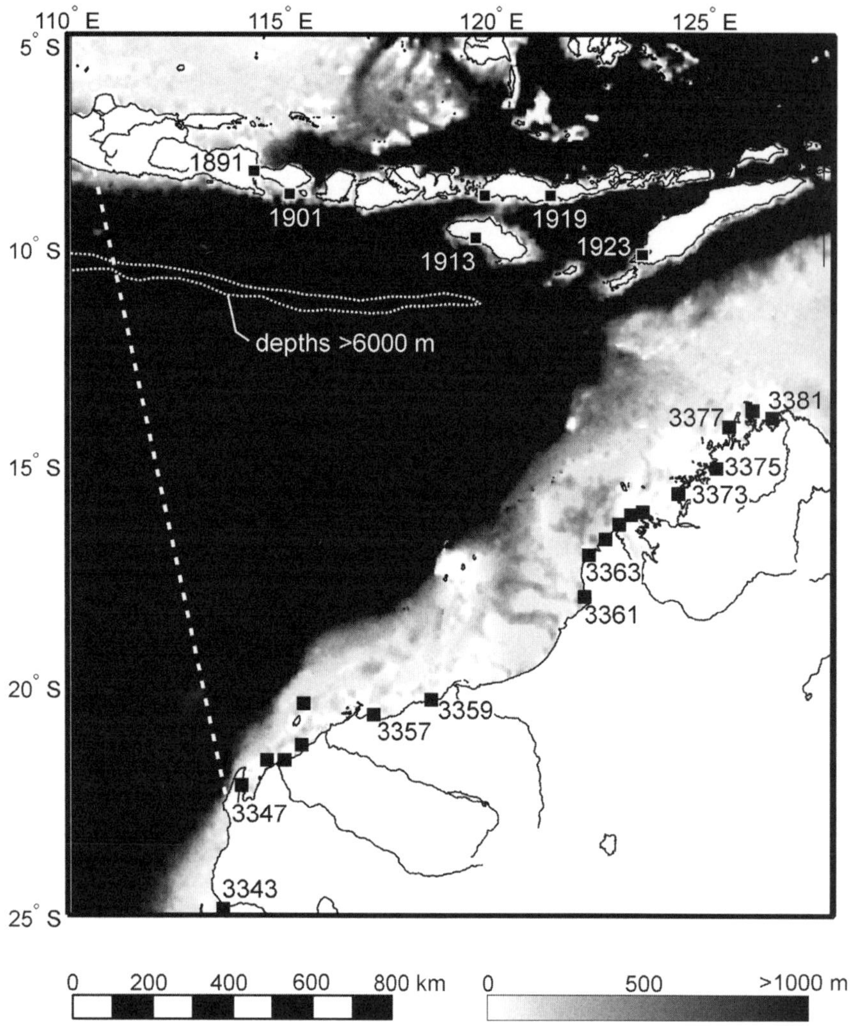

Figure 14. A: NOAA tidal stations for northwest Australia coastal area and adjacent areas of Indonesia. Dashed "zero line" is used for analysis of lateral amplification of tidal ranges. Deep-sea trench in this area exceeds 6000 m in depth. B: Because of a considerably wider shelf, tidal ranges along coast of Australia are considerably higher than ranges on islands of Indonesia. Tidal ranges exceed 9 m near Hall Point (station 3373) and subsequently decrease eastward.

B

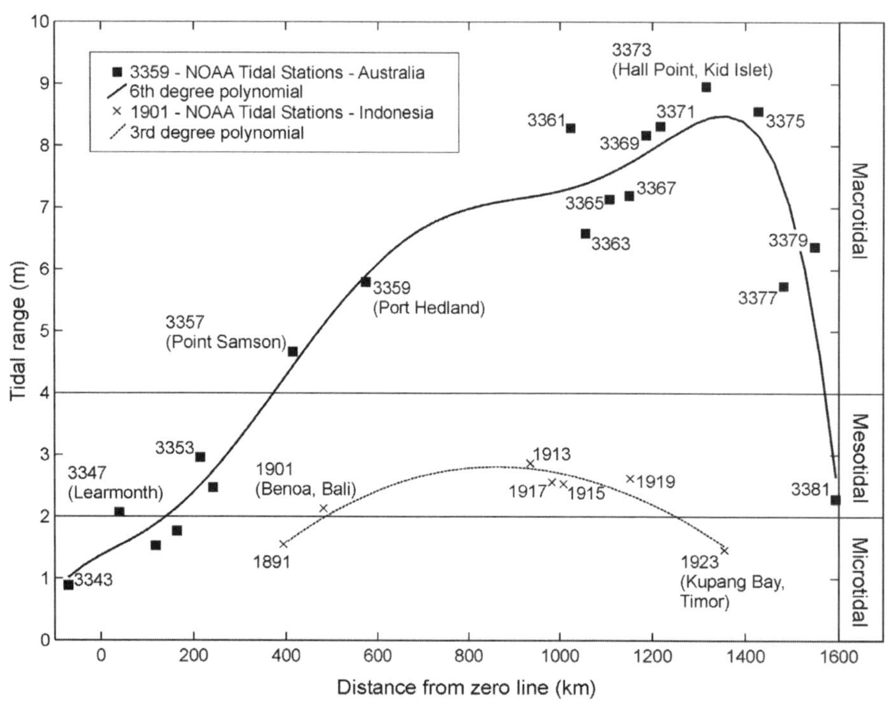

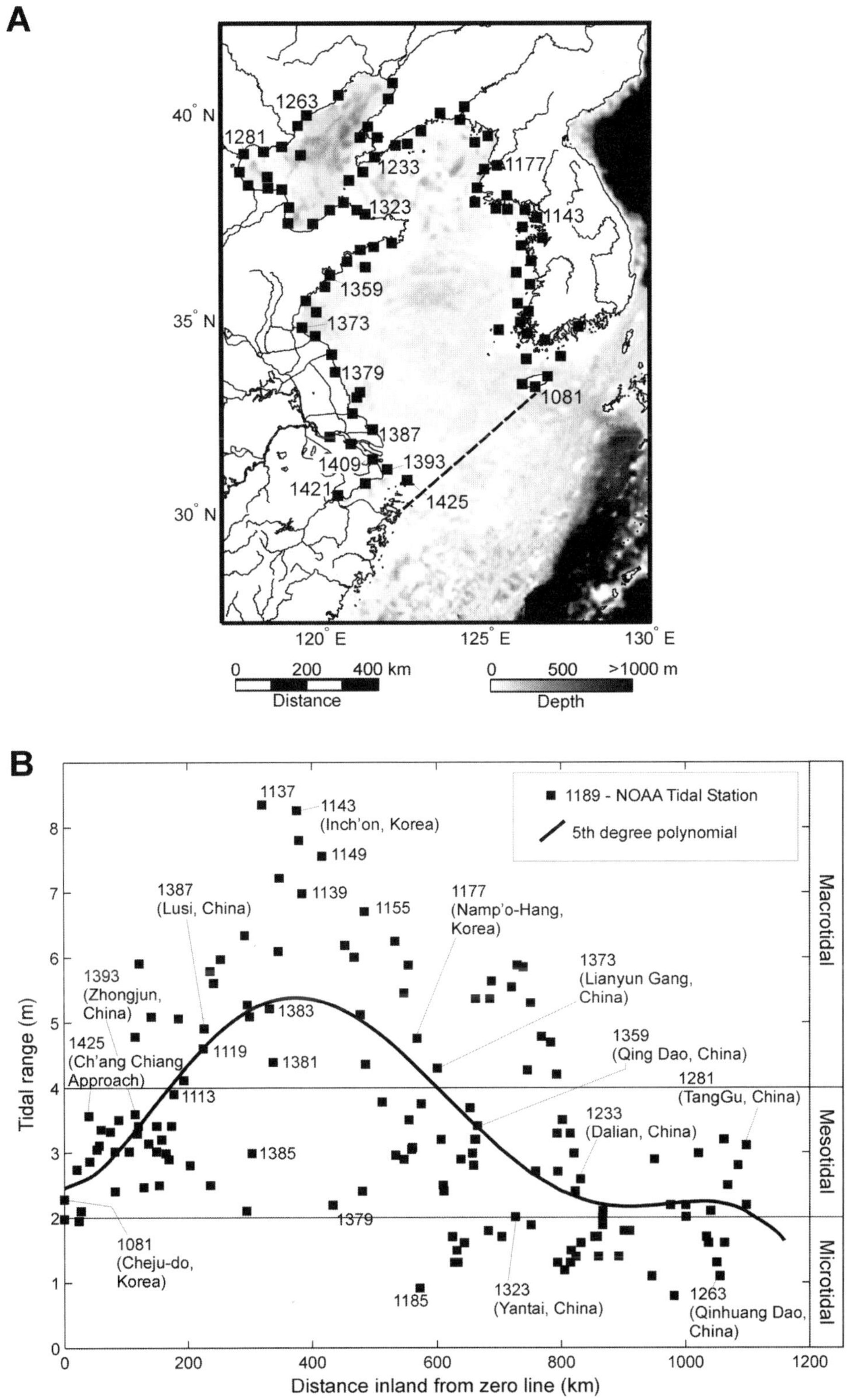

Figure 15. A: Tidal data from NOAA within Yellow Sea area and location of "zero line" used for analysis of tidal amplification. B: This complex system is composed of several different seas. Highest reported tides occur along coast of Korea near Inch'on (stations 1137 and 1143).

Sea is a backarc extensional setting that began in the Tertiary with the onset of subduction of the Pacific plate in this region (Wu and Murray, 1998).

Because of complex geometries and a great number of tidal stations, it is not easy to clearly define a simple amplification curve. Nonetheless, a zero line can be defined running from Sogwi-p'o on the Korean island of Cheju-do (1081) southwest to the Chinese coast near Ningpo (Fig. 15A). As a generalization, the tides near the zero line are low mesotidal with ranges from about 2 to 3.7 m.

Unlike most of the other systems discussed above, the highest tides do not occur at the most inland positions. Instead, they occur along the west coast of Korea. At Asan (1137), the tidal range is 8.34 m, and at Inch'on (1143), the range reaches 8.26 m. These stations are nearly 400-km inland from the zero line. Macrotidal ranges persist to about 800 km inland from the zero line. Mesotidal and microtidal ranges occur farther inland.

DISCUSSION OF INLAND TIDAL AMPLIFICATION

Funnel-shaped or flaring systems can develop strong tidal amplification. In such settings, more open coastal areas may have only microtidal conditions whereas more inland settings can develop a macrotidal regime. Not only do wider shelves results in higher coastal tidal ranges (Redfield, 1958; Klein and Ryer; 1978), but also specific coastal geometries can significantly affect tidal ranges. Inland seaways, gulfs, and bays may have much higher tidal ranges than nearby open-coastal settings. Funnel-shaped settings, particularly bays, in which the width exhibits a continuously decreasing width toward the inland direction, exhibit the highest tidal ranges in the world. This includes the world's highest tidal range, which occurs within the Bay of Fundy. Maximal heights of the highest tides within a particular bay are not easy to quantify, because storms and onshore winds can further heighten individual, astronomically driven tides. Thus, the most seaward settings can exhibit less tidal influences than contemporaneous, more inland systems. The amplified, macrotidal systems can be more than 1000 km in an inland direction from the open ocean.

CONCLUSIONS

Herein, a compilation of the world's highest tidal ranges indicates that 13 regions of the world include spring-tidal ranges that can regularly exceed 8 m in height. Most of these settings (~70%) occur on passive continental margins, many of which consist of "failed-rift" structures. Upon such low-slope coastlines, marine incursions and flooding of low-lying continental margins can result in propagation of tidal influences many hundreds of kilometers inland and into bays, gulfs, and seaways. Even within areas of microtidal open coasts, the more inland areas can experience macrotidal conditions. These simple observations have profound implications for the study of ancient inland tidal systems, because the "most marine" parts of the systems may be less influenced by tide than the "most inland" portion of the same depositional basin.

REFERENCES CITED

Adams, J., and Basham, P.W., 1989, The seismicity and seismotectonics of Canada east of the Cordillera: Geoscience Canada, v. 16, no. 1, p. 3–16.

Archer, A.W., 1998, Hierarchy of controls on cyclic rhythmite deposition: Carboniferous basins of eastern and mid-continental, U.S.A.: Tidalites: Processes and products: SEPM (Society for Sedimentary Geology) Special Publication 61, p. 59–68.

Atwater, B.F., Yamaguchi, D.K., Bondevik, S., Barnhardt, W.A., Amidon, L.J., Benson, B.E., Skjerdal, G., Shulene, J.A., and Nanayama, F., 2001, Rapid resetting of an estuarine recorder of the 1964 Alaska earthquake: Geological Society of America Bulletin, v. 113, p. 1193–1204.

Bartsch-Winkler, S., 1988, Cycle of earthquake-induced aggradation and related tidal channel shifting, upper Turnagain Arm, Alaska, USA: Sedimentology, v. 35, p. 621–628.

Bartsch-Winkler, S., and Lynch, D.K., 1988, Catalog of worldwide tidal bore occurrences and characteristics: U.S. Geological Survey Circular 1022, 17 p.

Bradnock, R., and Bradnock, R., 1997, India handbook: Bath, England, Footprint Handbooks, 1440 p.

Chadwick, R.A., 1986, Extensional tectonics in the Wessex Basin, southern England: Journal Geological Society London, v. 143, p. 465–488.

Dalrymple, R.M., Makino, Y., and Zaitlin, B.A., 1991, Temporal and spatial patterns of rhythmite deposition on mud flats in the macrotidal Cobequid Bay-Salmon River Estuary, Bay of Fundy, Canada, *in* Smith, D.G., Reinson, G.E., Zaitlin, B.A., and Rahmani, R.A., eds., Clastic tidal sedimentology: Canadian Society of Petrololeum Geologists Memoir 16, p. 137–160.

Dawson, R., 1902, Tides of the Bay of Fundy: Nature, v. 66, p. 85.

Defant, A., 1961, Physical oceanography, Volume 2: New York, Pergamon Press, 598 p.

Diraison, M., Cobbold, P.R., Gapais, D., and Rossello, E.A., 1997, Magellan Strait: Part of a Neogene rift system: Geology, v. 25, p. 703–706.

Eckel, E.B., 1970, The Alaska earthquake March 27, 1964: Lessons and conclusions: U.S. Geological Survey Professional Paper 546, p. 57.

Feldman, H.R., Gibling, M.R., Archer, A.W., Wightman, W.G., and Lanier, W.P., 1995, Stratigraphic architecture of the Tonganoxie paleovalley fill (Lower Virgilian) in northeastern Kansas: American Association of Petroleum Geologists Bulletin, v. 79, p. 1019–1043.

Frau, J.P., 1993, Tidal energy: Promising projects—La Rance, a successful industrial-scale experiment: IEEE Transaction on Energy Conversion, v. 8, p. 552–558.

Frey, R.W., Howard, J.D., Han, S.J., and Park, B.K., 1989, Sediments and sedimentary sequences on a modern macrotidal flat, Inchon, Korea: Journal Sedimentary Petrology, v. 59, p. 28–44.

Gombos, A.M., Powell, W.G., and Norton, I.O., 1995, The tectonic evolution of western India and its impact on hydrocarbon occurrences: An overview, *in* Davies, T.A., Coffin, M.F., and Wise, S.W., eds., Selected topics related to the Indian Ocean basins and margins: Sedimentary Geology, v. 96, p. 119–129.

Gorbatschev, R., and Bogdanova, S., 1993, Frontiers in the Baltic Shield: Precambrian Research, v. 64, p. 3–21.

Grant, A.C., and Sanford, B.V., 1988, Bedrock geological mapping and basin studies in the Hudson Bay region: Geological Survey of Canada, Current Research, Part B, Paper 88-1B, p. 287–296.

Gray, J., DeBoutray, B., Hillaire, M.C., and Lauriol, B., 1980, Post glacial emergence of the west coast of Ungava Bay, Quebec: Arctic and Alpine Research, v. 12, p. 19–30.

Keen, C.E., Kay, W.A., Keppie, D., Marillier, F., PePiper, G., and Waldron, J.W.F., 1991, Deep seismic reflection data from the Bay of Fundy and Gulf of Maine: Tectonic implications for the northern Appalachians: Canadian Journal Earth Science, v. 28, p. 1096–1111.

Keppie, J.D., 1982, The Minas Geofracture, *in* St. Julien, P., and Bérland, J., eds., Major structural zones and faults of the northern Appalachians: Geological Association of Canada Special Paper 24, p. 263–280.

Klein, G.deV., 1970, Depositional and dispersal dynamics of intertidal sand bars: Journal of Sedimentary Petrology, v. 40, p. 1095–1127.

Klein, G.deV., and Ryer, T.A., 1978, Tidal circulation patterns in Precambrian, Paleozoic, and the Cretaceous epeiric and mioclinal shelf seas: Geological Society America Bulletin, v. 89, p. 1050–1058.

Larsonneur, C., 1989, La baie du Mont-Saint-Michel: Un modele de sedimentation en zone temperee: Bulletin de l'Institute Géologique du Bassin d' Aquitaine, v. 46, p. 5–74.

Lidmar-Bergstrom, K., 1993, Denudation surfaces and tectonics in the southernmost part of the Baltic Shield: Precambrian Research, v. 64, p. 337–345.

Lupachev, Y.V., 1989. The effect of resultant tidal pumping in estuaries: Meteorologiva I Hidrologiya, v. 9, p. 78–82.

Macmillan, D.H., 1966, Tides: London, Companion Resources Books, 240 p.

Maeda, J., 1990, Opening of the Kuril Basin deduced from the magmatic history of central Hokkaido, North Japan: Tectonophysics, v. 174, p. 235–255.

McLean, B., Vilks, G., and Bhan, D., 1992, Depositional environments and history of Late Quaternary sediments in Hudson Strait and Ungava Bay; further evidence from seismic and biostratigraphic data: Geographie Physique et Quaternaire, v. 46, p. 311–329.

Michel, A., Allard, M., and Sequin, M.K., 1992, The thermal regime of intertidal permafrost, George River Estuary, Ungava Bay, Quebec: Canadian Journal Earth Science, v. 29, p. 249–259.

National Oceanographic and Atmospheric Administration (NOAA), 1998, Tide and current tables CD-ROM, 1999 Predictions: Riverdale, NOAA.

Nemcock, M., and Gayer, R., 1996, Modelling palaeostress magnitude and age in extensional basins: A case study from the Mesozoic Bristol Channel Basin, U.K: Journal of Structural Geology, v. 18, p. 1301–1314.

Nemcock, M., Gayer, R.A., and Miliorizos, M., 1995, Structural analysis of the inverted Bristol Channel Basin: Implications for the geometry and timing of the fracture permeability, *in* Buchanan, J.G., and Buchanan, P.G., eds., Basin inversion: Geological Society [London] Special Publication 88, p. 355–392.

Redfield, A.C., 1958, The influence of continental shelf on the tides of the Atlantic coast of the United States: Journal of Marine Research, v. 17, p. 432–448.

Rowbotham, F.W., 1983, The Severn Bore: London, David & Charles, 104 p.

Rowley, E., and White, N., 1998, Inverse modelling of extension and denudation in the East Irish Sea and surrounding areas: Earth and Planetary Science Letters, v. 161, p. 57–71.

Sequin, M.K., and Champagne, P., 1979, Ungava, pays du pergelisol continu (Ungava, land of continuous permafrost): Ressources Quebec, v. 3, p. 18–24.

Sharma, G.D., 1969, Sediments and tidal regimes in Cook Inlet: Eos (Transactions, American Geophysical Union), v. 50, p. 636.

Shaw, A.B., 1964, Time in stratigraphy: New York, McGraw Hill, 365 p.

Shin, D.H., Yoon, H.I., Han, S.J., and Oh, J.K., 1993, Distribution and provenance of clay minerals in tidal-flat sediments of the west coast of Korea: Ocean Research, v. 15, p. 123–136.

Tessier, B., 1993, Upper intertidal rhythmites in the Mont-Saint-Michel Bay (NW France): Perspectives for paleoreconstructions: Marine Geology, v. 110, p. 355–367.

Tessier, B., Archer, A.W., Lanier, W.P., and Feldman, H.R., 1995, Comparison of ancient tidal rhythmites (Carboniferous of Kansas and Indiana, USA) with modern analogues (the Bay of Mont-Saint-Michel, France): Oxford, England, International Association of Sedimentologists Special Publication 24, p. 259–271.

Tessier, B., Monfort, Y., Gigot, P., and Larsonneur, C., 1989, Enregistrement des cycles tidaux en acretion verticale, ples en baie du Mont-Saint-Michel et dans la molasse marine miocene du bassin de Digne: Bulletin Société Géologie de France, v. 5, p. 1029–1041.

Wells, J.T., and Adams, C.E., 1988, Speed, scale, and variability of tidal channel processes along the west coast of South Korea [abs.]: Eos (Transactions, American Geophysical Union), v. 69, p. 1086.

Wu, C., and Murray, G., 1998, Tectonics and stratigraphy of the greater Bohai Bay Basin, China [abs.]: American Assocation of Petroleum Geologists Bulletin, v. 82, p. 1983.

Zaitlin, B.A., 1987, Sedimentology of the Cobequid Bay-Salmon River estuary, Bay of Fundy, Canada [Ph.D. thesis]: Kingston, Ontario, Queen's University, 391 p.

Ziegler, P.A., 1982, Geological Atlas of western and central Europe: The Hague, Netherlands, Shell Internationale Petroleum Maatschappij, 130 p.

Manuscript Accepted by the Society January 22, 2003

Printed in the USA

Geological Society of America
Special Paper 370
2003

Giant submarine canyons: Is size any clue to their importance in the rock record?

William R. Normark
Paul R. Carlson
U.S. Geological Survey, 345 Middlefield Road, Menlo Park, California 94025, USA

ABSTRACT

Submarine canyons are the most important conduits for funneling sediment from continents to oceans. Submarine canyons, however, are zones of sediment bypassing, and little sediment accumulates in the canyon until it ceases to be an active conduit. To understand the potential importance in the rock record of any given submarine canyon, it is necessary to understand sediment-transport processes in, as well as knowledge of, deep-sea turbidite and related deposits that moved through the canyons. There is no straightforward correlation between the final volume of the sedimentary deposits and size of the associated submarine canyons. Comparison of selected modern submarine canyons together with their deposits emphasizes the wide range of scale differences between canyons and their impact on the rock record.

Three of the largest submarine canyons in the world are incised into the Beringian (North American) margin of the Bering Sea. Zhemchug Canyon has the largest cross-section at the shelf break and greatest volume of incision of slope and shelf. The Bering Canyon, which is farther south in the Bering Sea, is first in length and total area. In contrast, the largest submarine fans—e.g., Bengal, Indus, and Amazon—have substantially smaller, delta-front submarine canyons that feed them; their submarine drainage areas are one-third to less than one-tenth the area of Bering Canyon. Some very large deep-sea channels and turbidite deposits are not even associated with a significant submarine canyon; examples include Horizon Channel in the northeast Pacific and Laurentian Fan Valley in the North Atlantic. Available data suggest that the size of turbidity currents (as determined by volume of sediment transported to the basins) is also not a reliable indicator of submarine canyon size.

Keywords: submarine canyons, turbidite systems, submarine fans, turbidity currents, submarine mass wasting.

INTRODUCTION

Submarine canyons continued to be a major area of interest for Francis P. Shepard throughout his career. His *Submarine Geology* textbooks, which were revised over the years, devoted substantial space to a review of the world's canyons that were known at the time; these reviews were a standard resource for several decades of marine geologists (e.g., Shepard, 1973). Shepard and his coworkers observed that canyon size was not directly related to the size of the rivers that fed them. In fact, many major canyons were not fed directly from rivers but fed from littoral drift transport of beach sediment, for example, the La Jolla Canyon (Shepard, 1973; Shepard and Dill, 1966). As our ability to map the deep-sea floor has improved since Shepard's pioneering work, it has become clear that the size of submarine canyons also shows no simple relation to the size of the turbidite systems—submarine fans, abyssal plains, slope basin deposits, etc.—fed through the canyons.

Normark, W.R., and Carlson, P.R., 2003, Giant submarine canyons: Is size any clue to their importance in the rock record? *in* Chan, M.A., and Archer, A.W., eds., Extreme depositional environments: Mega end members in geologic time: Boulder, Colorado, Geological Society of America Special Paper 370, p. 175–190.

In this review, we accept the characteristics of submarine canyon presented by Shepard (1963) to distinguish canyons from the channels that form their seaward extension. Canyons are cut into shelf and slope bedrock and sediment, typically have a V-shaped profile, and have a steeper gradient than the channels formed in the depositional areas associated with the canyons. Canyons lack levee morphology commonly found along channels extending from mouths of the canyons. Shepard (1973) did distinguish delta-front troughs, which are "closely related to submarine canyons." For the sake of brevity, we use the term *canyon* for both types of features in this review.

Before declaring the "world's largest submarine canyon," we need to consider what criteria should be used to define largest. Should the criteria be the maximum submarine "drainage" area, the largest cross section, the maximum volume, or the greatest length? Just based on these four characteristics, Table 1 shows that there are several candidates for "largest canyon."

Because submarine canyons are bypass zones in the transport of sediment from continents to the oceans, one must look at the deposits in the adjacent basins to understand the importance of a submarine canyon. As a consequence, this review will compare a limited number of candidates for largest canyon (together with a few of the smaller, but perhaps more thoroughly studied canyons) and will also consider the size of the associated deposits. It is also instructive to look at those canyon systems that have delivered the greatest volume of terrigenous sediment to the oceans. The examples of smaller canyons selected for inclusion in this review reinforce the importance of physical-scale considerations and emphasize that the rock record is much better represented by the small canyons and their deposits rather than the giant canyons.

LARGEST SUBMARINE CANYON AND LARGEST SUBMARINE FAN

The areas shown in Figure 1 encompass the submarine canyon and the area of its associated deposit(s). The drainage areas of the canyons are given in Table 1, and the cross-sections of these canyons or their major fan valleys are compared in Figure 2. It can by seen that large canyons feeding turbidite systems

TABLE 1. AREA OF SUBMARINE CANYONS

No.	Canyon	Area (km^2)	Comments	Reference
1	Zhemchug	11,350	Largest cross section (Fig. 2) and largest volume of incision on continental margin but only third in total area (see Table 2).	Carlson and Karl (1988)
2	Bering	30,800	Longest canyon (495 km); only large Beringian margin canyon that has associated submarine fan	Carlson and Karl (1988)
3	Navarin	14,600	Second largest in area (see Table 2) and in volume of incision on continental margin	Carlson and Karl (1988)
4	Monterey	2380	Cross section at shelf edge is greater than Grand Canyon width and relief (Fig. 2)	Normark et al. (1985)
5	La Jolla	33	Size is typical of canyons that provide sediment to continental margin basins that are analogs for many hydrocarbon resource areas	Buffington (1964); Shepard and Dill (1966)
6	Horizon Channel	n.a.	The channel, which is more than 1200 km long, is not connected to a submarine canyon but to slope gullies, possibly associated with glacial trough	Stevenson and Embley (1987)
7	Swatch of No Ground	9000	Current (modern) canyon for Bengal Fan, the world's largest fan with channels extending more than 2000 km (Fig. 6)	Shepard (1973); Smith and Sandwell (1997)
8	Swatch	1700	Current (modern) canyon for the Indus Fan	Shepard and Dill (1966)
9	Amazon	2250	Current (modern) canyon associated with the river having largest discharge	Damuth and Kumar (1975); Milliman (1979); Damuth and Flood (1985)
10	Zaire (Congo)	4470	Canyon extends 30 km into the Zaire River Estuary and feeds a fan nearly 1000 km in length	Heezen et al. (1964); Droz et al. (1996)
11	Laurentian Fan Valley	n.a.	Largest levee/channel floor relief (Fig. 2); the fan valley heads in a gullied upper slope below the mouth of a glacial trough cut 300 m below the shelf	Piper and Normark (1982); Piper et al. (1985)

Note: Numbers refer to map in Figure 1. Area given represents total submarine drainage area of canyon and tributaries. If canyon area is not provided directly by the reference(s) listed, it is derived from maps presented in these references.

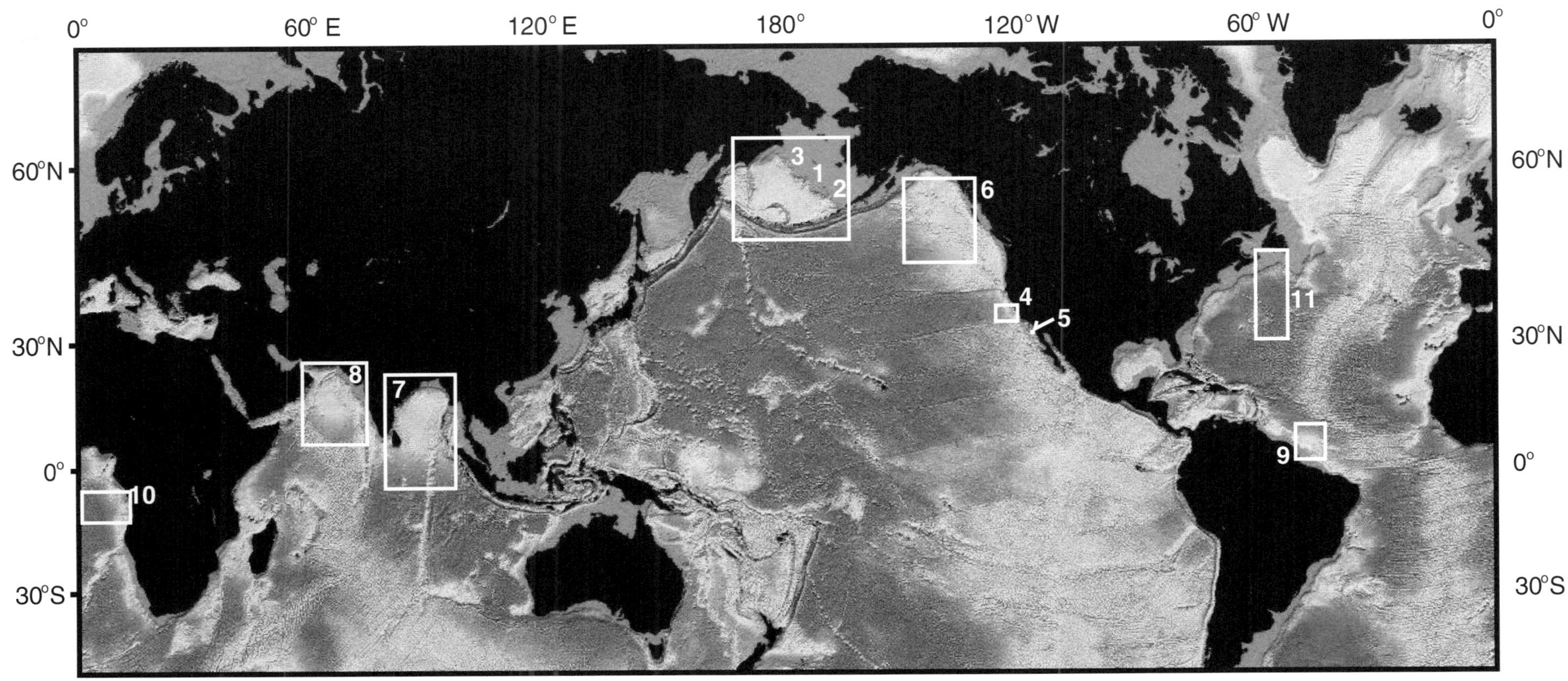

Figure 1. Shaded world relief map with locations of submarine canyons and associated turbidite deposits mentioned in text (shaded relief from Miller et al., 2001). Canyon-fan systems are: (1) Zhemchug, (2) Bering, (3) Navarin, (4) Monterey, (5) La Jolla, (6) Horizon Channel (leading to Tufts Abyssal Plain in Gulf of Alaska), (7) Swatch of No Ground (Bengal), (8) The Swatch (Indus), (9) Amazon, (10) Zaire (Congo), and (11) Laurentian Fan Valley and Sohm Abyssal Plain. At this "world view" scale, only three large submarine canyons can be distinguished on Bering Sea margin. Light shading in boxes denoting canyon-fan systems generally marks areas covered by turbiditic fan deposits. Several larger submarine fan channels are shown for scale comparison. See text for discussion and references.

on the deep ocean floor are typically about 1 km deep in profiles across the canyon near the shelf break. The Zhemchug Canyon, which is twice as deep as the other large canyons depicted, is equaled in width only by Navarin Canyon, which has less than half of the vertical relief of Zhemchug. The La Jolla Canyon, which empties into a small borderland basin on the continental margin, is less than 300 m deep; its size is typical for canyons that feed small basins within continental margin settings. To help visualize the immense size of large submarine canyons, Figure 2 includes a section across the Grand Canyon of the Colorado River in the southwestern United States. The Grand Canyon is deeper than all of the submarine canyons depicted except for the Zhemchug; in cross-sectional area, however, the Grand Canyon is nearly an order of magnitude smaller than the Zhemchug Canyon and falls in the middle of the range of canyons shown in Figure 2.

The Swatch of No Ground, which is the equivalent of a submarine canyon on the delta of the Ganges-Brahmaputra Rivers, is only about a tenth of the cross-sectional area of the Zhemchug Canyon (Fig. 2). The Swatch of No Ground, however, feeds the Bengal Fan, which is the largest submarine fan in the ocean (Emmel and Curray, 1985). The size of a submarine canyon is, thus, not a reliable indicator of the volume of sediment that has moved through the conduit.

Using the characteristics given in Table 1, we will review the best candidates for largest submarine canyon, which are all from the Bering Sea margin (Figs. 1 and 2). We then consider the relation of canyon size relative to their associated turbidite deposits.

Zhemchug Canyon

The three largest submarine canyons, based on drainage area (Table 1) and cross-sectional area (Fig. 2), are all from the North American margin of the Bering Sea; Zhemchug Canyon is the largest of these canyons (Carlson and Karl, 1988). Zhemchug Canyon has a volume of at least 5800 km^3, nearly double the volume of the Swatch of No Ground, which is 2950 km^3 (see Table 1 in Carlson and Karl, 1988) and feeds the largest submarine fan, the Bengal.

Zhemchug Canyon, named for the Soviet expeditionary vessel *Zhemchug*, has two main branches, and each is larger than typical continental-margin canyons such as the Monterey (Fig. 2). A strong case might be made for each branch of the

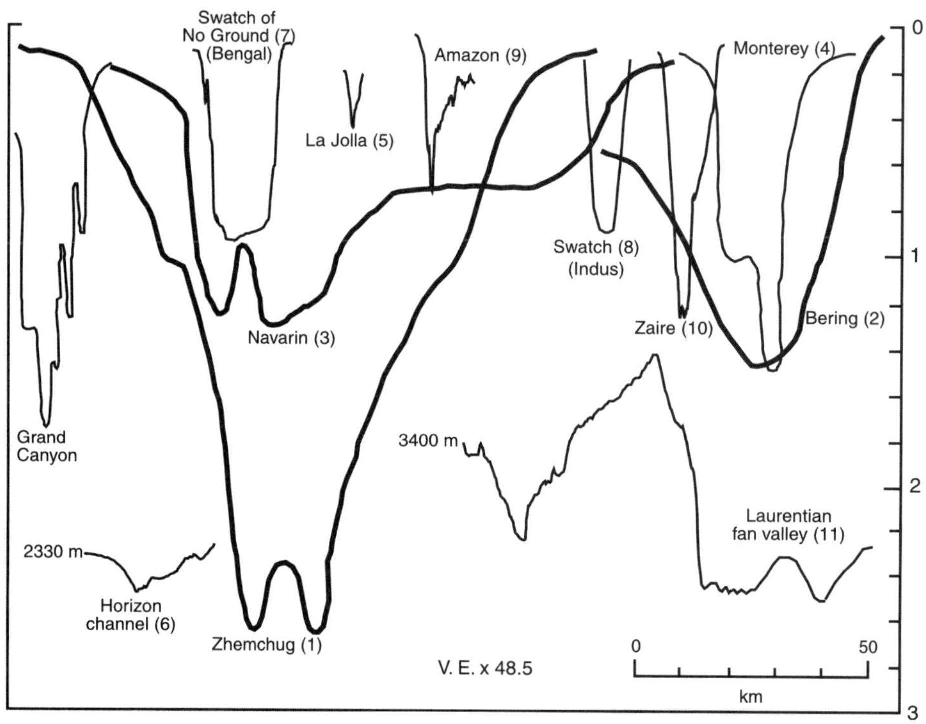

Figure 2. Comparison of canyon cross-sections near shelf edge for selected fans shown in Figure 1 and Table 1. Canyon sections are plotted at true depth; Horizon and Laurentian turbidite channels are plotted at same scale and water depth is given at left side of profile. Sections for three largest canyons are shown with bold lines. Sections for Bering, Monterey, Navarin, and Zhemchug submarine canyons and Grand Canyon are from Carlson and Karl (1988). Sections for other submarine canyons and fan channels are: Amazon Canyon (Damuth and Kumar, 1975); Horizon channel (Stevenson and Embley, 1987); La Jolla Canyon (Shepard and Buffington, 1968); Laurentian fan valley (Piper and Normark, 1982); Swatch (Shepard, 1973); Swatch of No Ground (Curray and Moore, 1974; Shepard, 1973); Zaire (formerly Congo, Heezen et al., 1964).

Zhemchug Canyon being separate canyons because both thalwegs traverse the entire slope before merging on the upper rise (Fig. 3; Carlson and Karl, 1988). These two branches occupy a 160-km-long, 30-km-wide, steep-walled trench that is incised into the shelf and oriented northwest-southeast, roughly parallel to the shelf-slope break. The canyon breaches an outer shelf structural high named Pribilof Ridge by Marlow et al. (1976). The canyon, at the regional shelf break, has cut a gorge 100 km wide and 2600 m deep (Figs. 3 and 4). The axial profiles of both branches steepen in a step-like fashion. Transverse profiles of the canyon are steep walled and V-shaped landward of the shelf break. Seaward of the shelf break, the walls are still steep and have great relief (2550 m), but the floor becomes flat and is as much as 10 km wide.

Of all the processes that have been instrumental in shaping the large submarine canyons of the Beringian margin, mass movement has been by far the most important agent, followed by density flows (Carlson et al., 1991). The imprint of tectonism controlled or influenced the locations of the canyons, and glacio-eustatic sea-level changes regulated the timing during which the canyon-cutting processes were most effective. Zhemchug Canyon has breached a structural basin that underlies the Bering shelf. The canyon is eroding into the basin fill, and the shape of the basins and bounding faults (Scholl et al., 1970) control the configuration of the developing canyon heads. The epicenter of a recent earthquake (Abers et al., 1993) is adjacent to the northwest branch of the canyon, which underscores the structural aspect of its formation.

Evidence of mass wasting of sediment in the Beringian margin canyons has been recognized from seismic-reflection profiles (Scholl et al., 1970, Carlson and Karl, 1984–1985; 1988). The GLORIA images collected in 1986 reveal that products of mass wasting are much more common than previously interpreted and that mass wasting is the dominant erosional process on the Beringian continental slope (Carlson et al., 1991). No discrete fan in the classical sense (e.g., Amazon Fan,

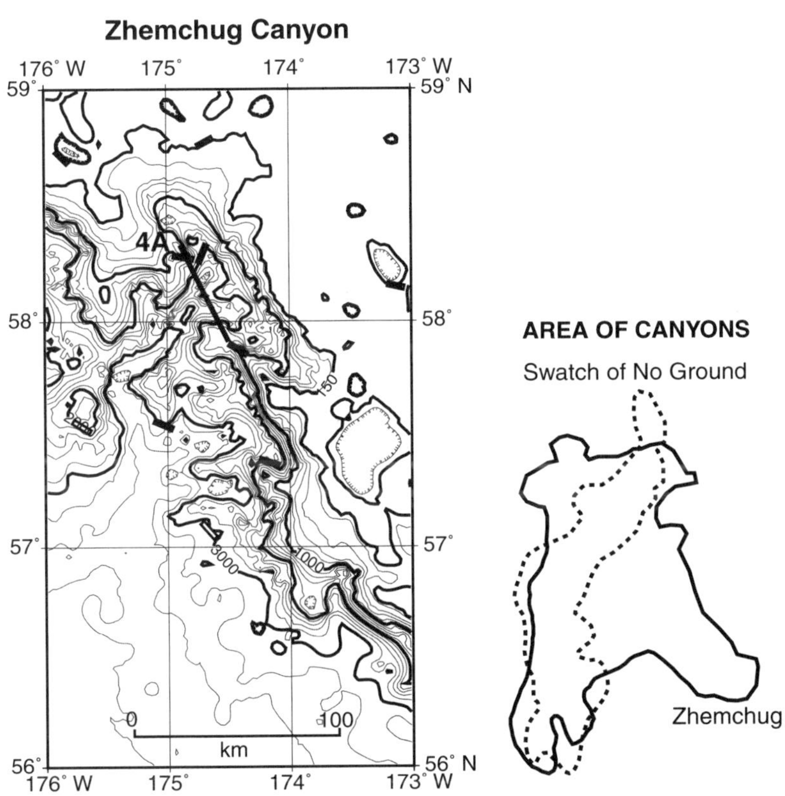

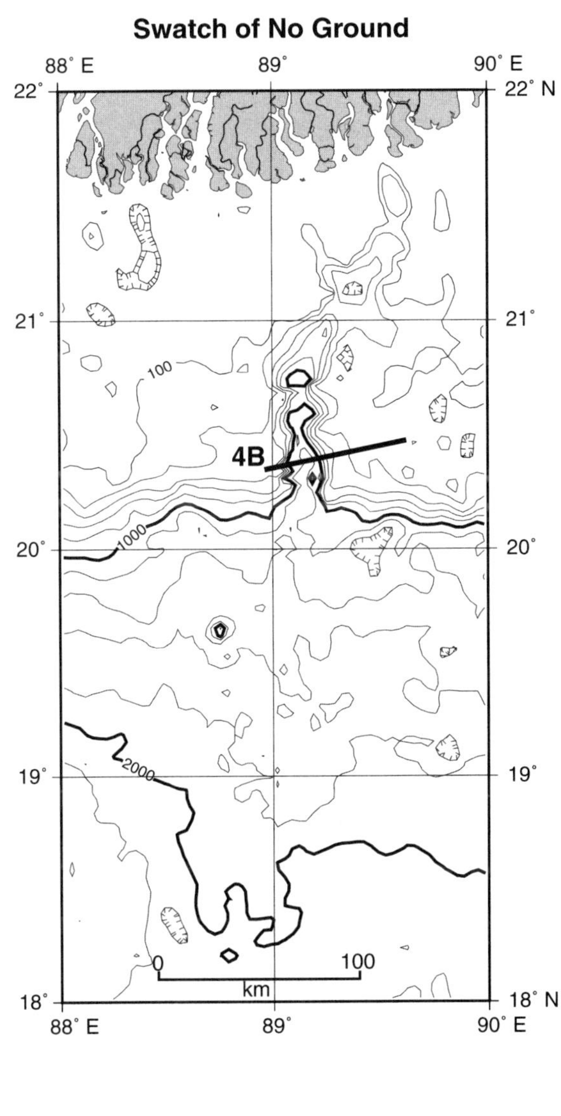

Figure 3. SEASAT/ETOPO2 bathymetry of largest submarine canyon (Zhemchug) and submarine canyon ("Swatch of No Ground") with largest fan (Bengal) derived from Smith and Sandwell (1997). Contour interval is 200 m. For Zhemchug Canyon area, 150 m contour is also shown, as is 100 m contour for Swatch of No Ground. Areas of two canyons are compared in center of figure. Locations of seismic-reflection profiles of Figure 4 are shown.

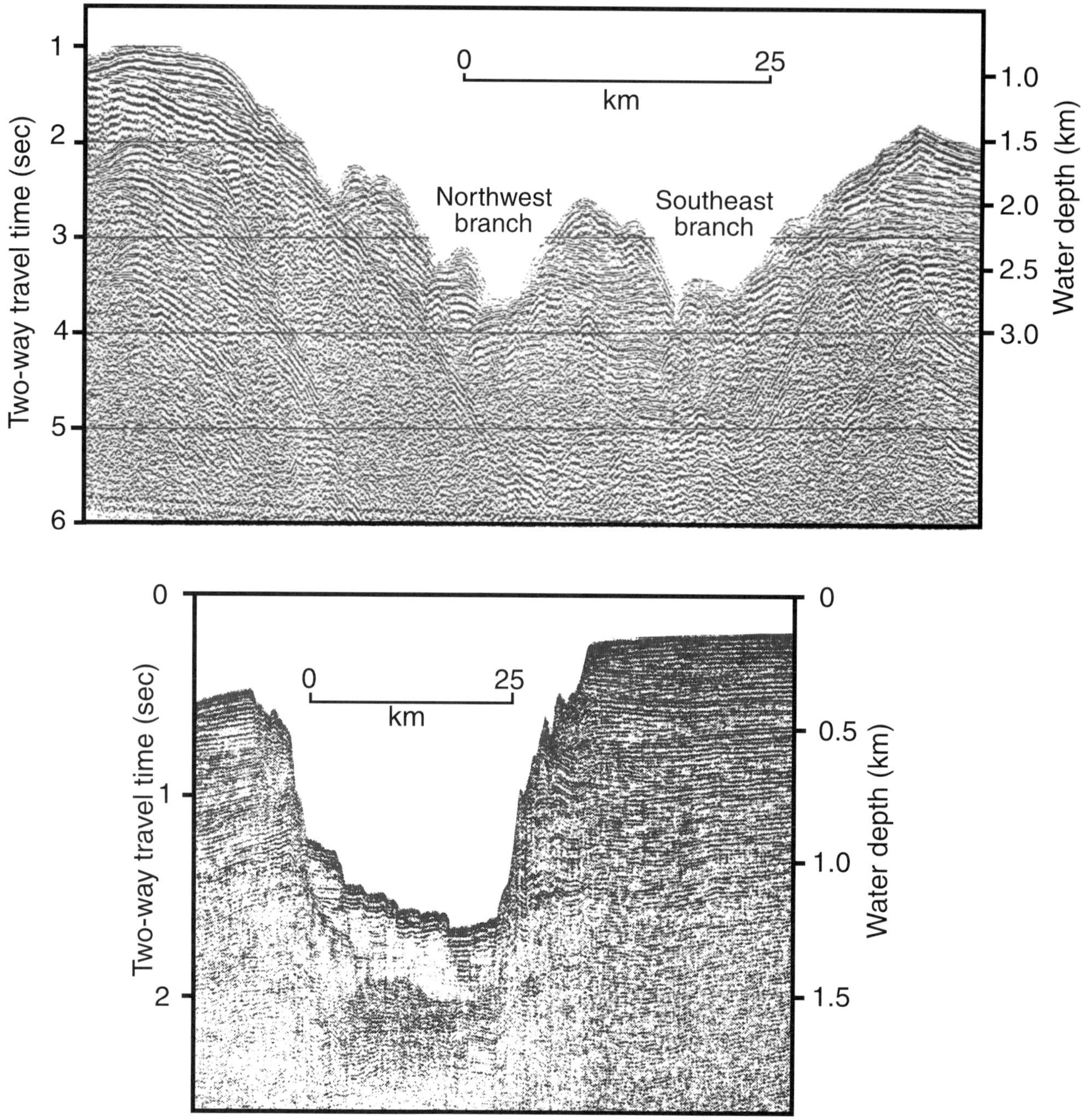

Figure 4. Seismic-reflection profiles across Zhemchug and Swatch of No Ground canyons. (A) Airgun seismic-reflection profile crossing rugged, slump-dominated walls of Zhemchug Canyon. This profile crosses two branches of this massive canyon slightly east of shelf edge. Profile collected on cruise F-3-86-BS (Bering Sea EEZ-Scan Scientific Staff, 1991). (B) Profile across Swatch of No Ground showing thick fill and/or slump deposits in canyon floor (adapted from Shepard; Fig. 11–21 *in* Shepard, 1973). See Figure 3 for location.

Monterey Fan, etc.) occurs at the mouth of Zhemchug Canyon. It appears that the canyon was not an important source of sediment during the latest Quaternary because a subtle channel extends only a relatively short distance onto the deep Aleutian Basin of the Bering Sea.

"Swatch of No Ground" (Bengal Fan)

The Bengal Fan is the largest submarine fan in the modern ocean. Its length is at least 2800 km and its width locally exceeds 1400 km (Emmel and Curray, 1985). The apex of the

Bengal Fan is offshore of the "Swatch of No Ground" Canyon, which feeds the most recently active channel on the fan (Figs. 2 and 3). Unlike the Zhemchug Canyon, which has no associated major fan or fan channel, the Bengal Fan Channel fed by the incised Swatch of No Ground is an elevated levee-channel system. This elevated channel begins at 1400 m water depth on the continental slope and extends more than 2300 km down fan (Curray and Moore, 1974; Emmel and Curray, 1985). This contrast in canyon morphology is exemplified in the bathymetric contours of Figure 3, where the Zhemchug has a broad, flat floor at 3600 m water depth and extends into the basin as a broad, shallow depression in the seafloor.

The Swatch of No Ground is a delta-front canyon cut in flat-lying sediment of the shelf (Fig. 4B) and as such is probably a short-lived feature compared to submarine canyons cut in older sedimentary and basement rocks. As is typical for most modern canyons, the Swatch of No Ground is inactive during the current high stand of sea level (Emmel and Curray, 1985). It is probable that the form and location of the Swatch of No Ground was different during previous periods of active sediment transport during glacial lowstands, reflecting changes in the distributary system on the delta. Older equivalents of the Swatch of No Ground were probably similar in size or smaller because it is one of the larger delta front canyons in the modern ocean, c.f., Amazon and Indus (Fig. 2).

Curray and Moore (1974) define the modern Bengal Fan to include the upper 4 km of sediment under the fan apex. The volume of the Eocene to Holocene section under the Bengal Fan is 12.5×10^6 km^3 (Curray, 1994). The volume of the Swatch of No Ground is 2.95×10^3 km^3 (Carlson and Karl, 1988). Therefore, the amount of sediment that is Eocene and younger on the Bengal Fan is equivalent to a volume that is 4200 times the volume of the Swatch of No Ground itself. In contrast, the estimated volume of sediment in the Aleutian Basin of the Bering Sea is 1.9×10^6 km^3 (derived from Fig. 4 *in* Cooper et al., 1987). If all of the sediment in the Aleutian Basin had been transported through the Zhemchug Canyon, it would equal only 325 times the volume of the canyon. Given that the enclosed Aleutian Basin (Fig. 1) can receive sediment from nearly all its margins, it is unlikely that the Zhemchug Canyon contributed more than a few percent of the equivalent volume that has passed through the "Swatch of No Ground" and its predecessors. Thus, it is clear that the smaller Swatch of No Ground has been a much more active funnel for bringing sediment to the ocean. The similarity of delta-fed canyons in Figure 2 suggests we can assume that ancestral canyons on the delta were roughly the same size.

OTHER LARGE SUBMARINE CANYONS AND TURBIDITE SYSTEMS

In this section, we look at examples of several other large submarine canyons and several additional large turbidite systems to emphasize the problems inherent in using canyon size to predict parameters for deposits that have formed on the seafloor as a result of processes within the canyons. We will start with other large canyons on the Beringian margin and then look at large turbidite systems, some of which are not associated with submarine canyons.

Bering Canyon and Fan

Bering Canyon is the longest of the Bering Sea canyons; it extends about 400 km across the Bering shelf and slope. It is confined at its eastern edge by the Aleutian Islands (Fig. 1). The width of the canyon at the shelf break is about 65 km, only about two-thirds that of the Zhemchug and Navarin Canyons (Fig. 2), but because of its great length, the Bering Canyon has the largest area (Table 1). At a depth of 3200 m, the Bering Canyon thalweg reaches the Aleutian Basin, where a low-relief submarine channel-lobe system has developed.

A fan channel extending basinward from Bering Canyon had been suggested on early bathymetric maps but was not clearly defined until the GLORIA survey (Bering Sea EEZ-SCAN, 1991). The fan channel extends several hundred kilometers into the Aleutian Basin as a low-relief (10–20 m), very broad (20 km), flat-floored turbidite channel (Figs. 5 and 6B). The Bering Channel has a low levee on its north side. The form of the channel is distinctly different from the elevated leveed channel systems found on large, delta-fed fans (Fig. 6A). The Bering Channel terminates in an area of very low-relief branching channels or channel remnants in a channel-lobe transition area (Figs. 5 and 6C; Karl et al., 1996). Bering Fan lacks the distinctive upper, middle, and lower fan morphologic expression that is generally present on other fans (Karl et al., 1996); instead, it forms a relatively thin veneer of sediment in the Aleutian Basin. The turbidites fed by Bering Canyon underlie debris flow facies to the south and are indistinguishable from the Aleutian Basin fill in front of the Zhemchug Canyon. The latter observation also applies to the Aleutian Basin fill between Zhemchug Canyon and Navarin Canyon farther north.

Navarin Canyon

Navarin Canyon is the third largest submarine canyon that cuts the Beringian margin. It is the second largest in area, behind only the Bering Canyon, and its width at the shelf break is nearly the same as that of the Zhemchug Canyon (Fig. 2; Carlson and Karl, 1988). Navarin is also similar to the Zhemchug in that it has two main branches and it does not lead to any distinct submarine fan morphology. The lack of distinctive submarine fan morphologies for both the Zhemchug and Navarin Canyons, together with the very subdued relief of the Bering Fan, suggests that these canyons have not been particularly effective as major conduits for sediment transport from the continent.

The large canyons of the Beringian margin are not directly related to large rivers. During low stands of sea level, however, when the Alaskan shoreline was near the present 150-m isobath, the Yukon and Kuskokwim Rivers must have meandered across much of the emergent Bering shelf and, perhaps, influenced the

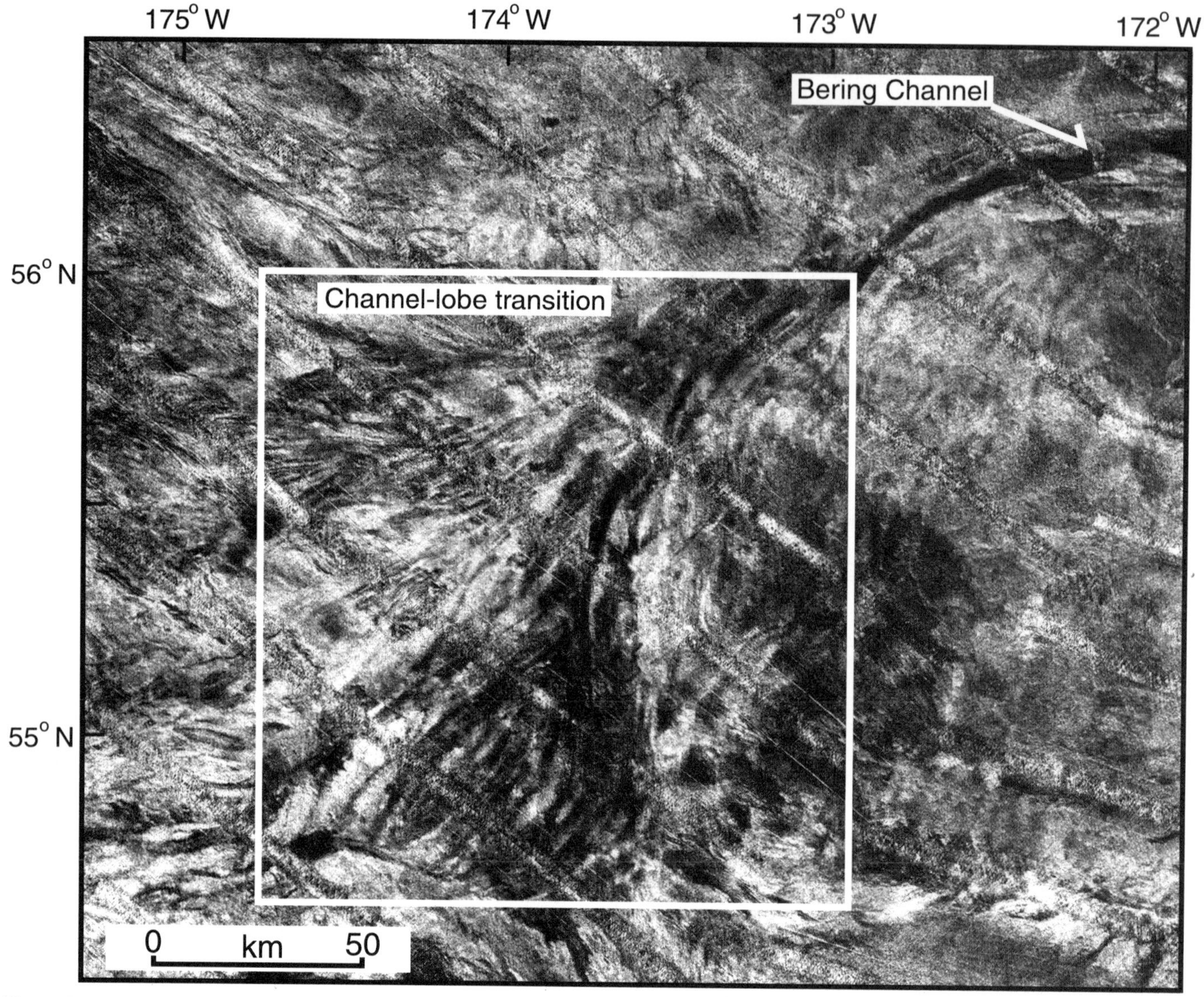

Figure 5. Long-range side-scan sonar image of area of channel-lobe transition seaward of Bering Channel from GLORIA survey of Aleutian Basin (adapted from Fig. 17-6 *in* Karl et al., 1996).

development of the present-day heads of the Bering, Navarin, and Zhemchug Canyons. Today, the upper part of the Bering shelf valley can be traced to a position offshore of the Kuskowim River. Buried channels located in the outer shelf along the northern Bering margin suggest that streams that were, perhaps, ancestral to the present Yukon River must have meandered across the shelf, thus affecting the present day dual heads of Navarin and Zhemchug Canyons (Carlson and Karl, 1984).

Amazon Canyon and Fan

The Amazon Fan, which is one of the largest modern submarine fans (Bouma et al., 1985a), is one of the better documented submarine turbidite systems because of extensive mapping of the fan surface (Damuth and Flood, 1985; Pirmez and Flood, 1995) and the extensive scientific drilling during Ocean Drilling Program (ODP) Leg 155 (Flood et al., 1995, 1997; Normark et al., 1997). The canyon feeding the Amazon Fan is much less well documented. It is thought that the modern Amazon Canyon probably formed as a result of mass failures and then was modified by subsequent erosion by turbidity currents. The cross-sectional shape of the Amazon Canyon is similar to the other delta-fed canyon-fan systems, although it is smaller than either the "Swatch of No Ground" or "Swatch" Canyons (Fig. 2). The smaller size of the associated fan, however, has allowed for a more complete mapping of the surface morphology, and the Amazon Fan provides the best-

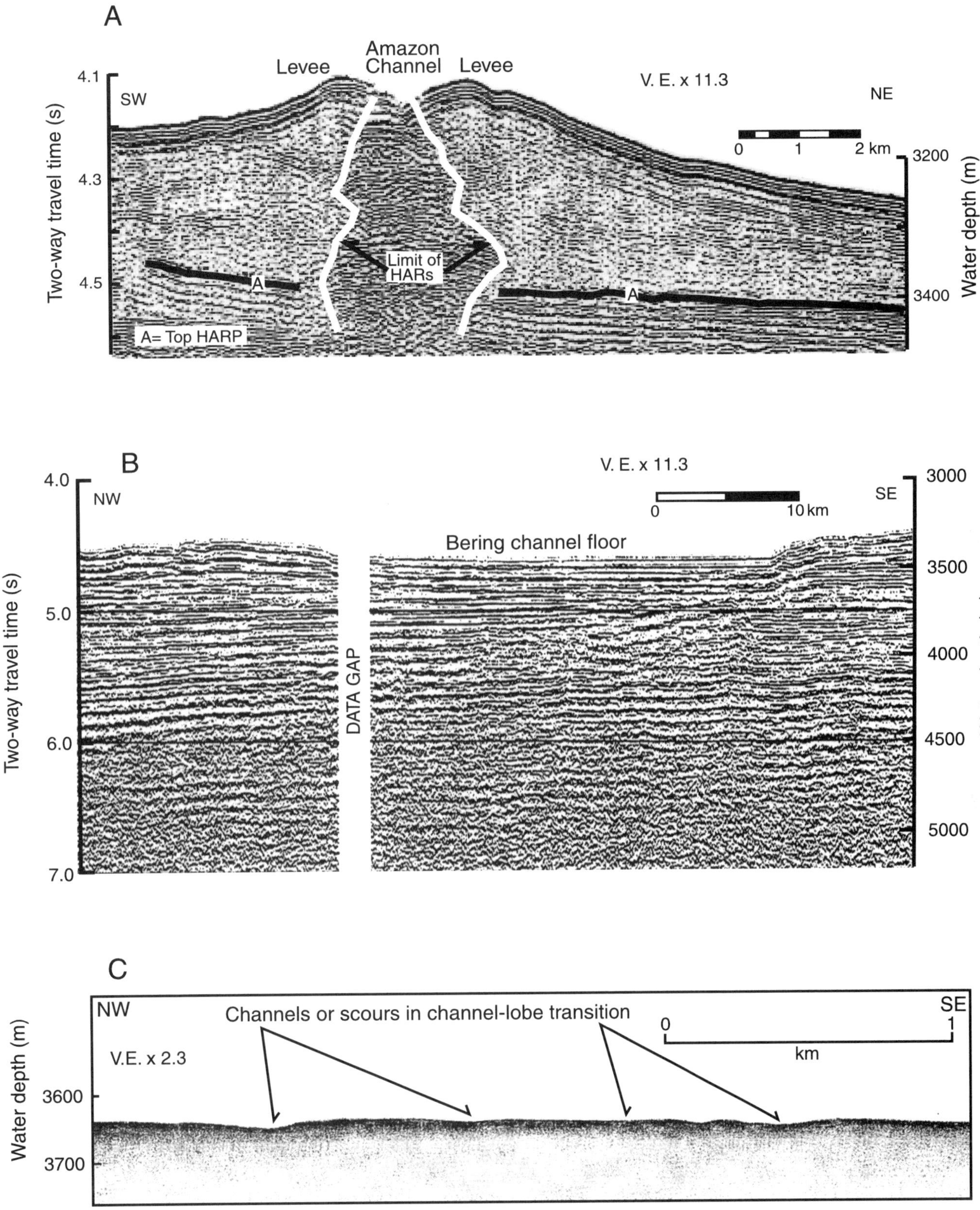

Figure 6. Comparison of submarine fan channels. A: Leveed channel typical of large, delta-fed submarine fans (modified from Flood et al., 1995). Leveed-channel complex overlies High Amplitude Reflection Packet (HARP), denoted A. Wide levees are deposited on low-relief, gently sloping HARP contact, where channel floor was initially incised before aggradation of entire levee-channel complex. B: Seismic-reflection profile across Bering Fan Channel just upstream (east) from channel shown in northeast corner of Figure 5. C: Small Bering Fan channels resolvable in 3.5-kHz profiles across low-relief, channel-lobe transition (adapted from Fig. 17-14 *in* Karl et al., 1996).

documented examples of large, sinuous, leveed valleys that are common on delta-fed submarine fans (Fig. 6A).

The upper part of Amazon Fan comprises stacked channel-levee systems, which have a sinuous planform. The channel representation in Figure 7A gives the highly sinuous nature of the youngest leveed channel on the fan (Pirmez and Flood, 1995). The aggrading levees attain thicknesses of hundreds of meters in proximal parts of the fan and, as a result, both the channel floor and levee crest are elevated well above the surface of the adjacent fan (Fig. 6A). The highest rates of sedimentation on the Amazon Fan are on levee crests, locally as great as 25 m/k.y., and channel-floor deposits aggrade at rates in excess of 15 m/k.y. (Flood et al., 1997). The fan grew rapidly during Pleistocene low stands of sea level, accumulating some 500 m of sediment in the past 0.5 Ma. Channel width varies only slightly through time, as shown by the ≤2 km width of the high-amplitude reflectors under the channel floor (Fig. 6A). Thus, the sinuosity and position of the channel, which forms the conduit of an elevated 10–20-km-wide levee-channel system, appears to remain fairly constant during levee aggradation, and there is little evidence for significant lateral migration of meanders (Flood et al., 1997).

Major channels on the upper fan may persist for tens of thousands of years, with avulsion taking place more frequently on the middle fan and most commonly near the distal end of channels on the lower part of the middle fan (Pirmez and Flood,

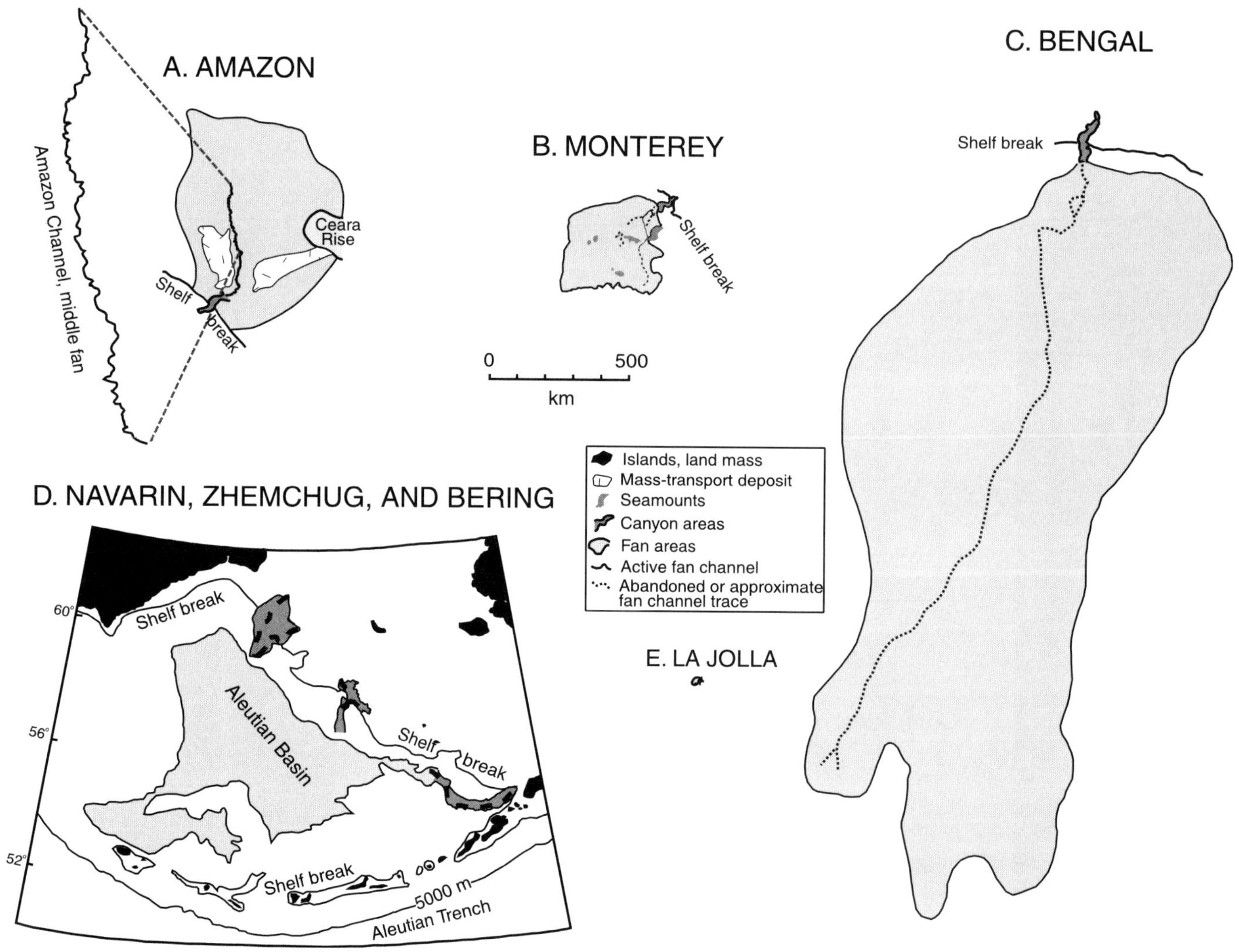

Figure 7. Comparison of submarine canyon area and area of related submarine fan. Examples are constructed using the following references: (A) Amazon Canyon Fan (Milliman, 1979; Damuth and Flood, 1985); sinuous Amazon Channel on middle fan is shown in blowup to left (Pirmez and Flood, 1995); (B) Monterey Canyon Fan (Normark et al., 1985; Fildani et al., 1999); (C) "Swatch of No Ground" and Bengal Fan (Shepard, 1973; Emmel and Curray, 1985; Smith and Sandwell, 1997); (D) Zhemchug, Navarin, and Bering Canyons and Aleutian Basin (Carlson and Karl, 1988; Karl et al., 1996); (E) La Jolla Canyon and Fan (adapted from Moore, 1972).

1995). The sampling from ODP Leg 155 suggests that only one channel system is active at any given time. Avulsion apparently results from autocyclic controls and generally occurs as a result of either sediment failure on the levee or erosion by large turbidity currents. Avulsion events do not appear to be controlled by sea-level change.

Apparently, a single submarine canyon feeds the shifting channel system during each lowstand of sea level (Flood et al., 1995, 1997). Changes in the position of this feeding canyon during each lowstand produce a levee complex made up of a series of stacked channel-levee deposits. Successive levee complexes are separated by a thin layer of pelagic sediment that accumulates during sea-level highstand conditions (Flood et al., 1997; Normark et al., 1997).

Zaire Canyon and Fan

The Zaire Canyon, with its main fan-channel extension, was one of the first large modern canyon/fan systems to be described from the river source to the deep ocean basin (Heezen et al., 1964). The Zaire Canyon (Fig. 2) and Fan were originally called the Congo (Heezen et al., 1964), but recent workers use Zaire (Droz et al., 1996). The Zaire Canyon is somewhat unusual for a modern system in that the canyon head not only extends across the shelf (which in itself is uncommon during the present high stand of sea level); it also continues 30 km into the river estuary (Heezen et al., 1964). As a result, the canyon easily receives sediment from the river, and its current activity is attested to by frequent telecommunication cable breaks.

The canyon feeds a major fan valley system that is quite similar to those on the Amazon Fan. The channels are highly sinuous and have broad levees (Droz et al., 1996; Savoye et al., 2000). The channels can be traced about 900 km across the fan to the Angola Abyssal Plain. Droz et al. (1996) further observe that the Zaire Fan is unique in that it is a large turbidite system that remains active today under high stand conditions and because it is a large muddy turbidite system.

"Swatch" (Indus Fan)

The "Swatch" is the modern submarine canyon for the Indus Fan, which is the second largest turbidite system in the modern ocean (Bouma et al., 1985a). It is smaller than the Swatch of No Ground but has a similar shape (although narrower; Fig. 2) and relation to its fan (Kolla and Coumes, 1985). Similar to the other large, delta-fed fans such as the Amazon, the Indus Fan is characterized by large, leveed valley systems (Droz and Bellaiche, 1991). Long-range side-scan sonar images show that the leveed channel systems are sinuous and similar to the Amazon channels (Kenyon et al., 1995).

As noted for the Swatch of No Ground, the Swatch is a delta-front canyon that is probably a short-lived feature compared to submarine canyons cut in older sedimentary and basement rocks. McHargue (1991) mapped a series of leveed channels on the inner Indus Fan that were fed by a shelf canyon that was wider than the Swatch and lies about 100 km to the northwest.

Monterey Canyon

The Monterey Canyon has been one of the most studied modern submarine canyons since Shepard and Emery (1941); numerous workers have compared its width and depth to that of the Grand Canyon, e.g., Shepard (1963) and this paper, Figure 2. The Monterey Canyon is the largest and deepest of several canyons that lead to a 400-km-long submarine turbidite system called the Monterey Fan (Normark et al., 1985; Greene and Hicks, 1990). As a result of these multiple canyons feeding the fan, the history of development of leveed channel systems on the upper fan is relatively complex (Fildani et al., 1999).

Along its course to the fan, Monterey Canyon cuts through granitic basement and through Miocene and younger sedimentary rocks (Shepard and Dill, 1966; Greene and Hicks, 1990). Early work concluded that the canyon is probably of Late Neogene age, and Greene and Hicks (1990) suggest that an ancestral Monterey Canyon originated by Early Miocene. Recent work on the age of sediment on Monterey Fan, however, indicates that the canyon, at least in its present form, may be no more than several million years old (Normark, 1999). Much of the older sediment underlying Monterey Fan apparently came from sources farther north than Monterey Bay (Fildani, 1993). In addition, the morphology of the largest leveed channel systems of the modern Monterey Fan show that they are probably not related to the current Monterey Canyon but to Ascension Canyon along the central California margin north of Monterey Bay. The main growth of the large leveed valley from the Ascension Canyon occurred in Late Pleistocene, and its present connection to the Monterey Canyon may have happened only within the last few hundred thousand years (Normark, 1999).

Horizon Channel

Horizon Channel extends nearly 1400 km southwest of Baranof Island, southeastern Alaska (Fig. 1). Figure 2 shows the cross section of this turbidite channel near the base of the continental slope. It is one of several very long channels in the Gulf of Alaska that feed the Tufts Abyssal Plain. Horizon Channel, with subparallel Mukluk Channel, are responsible for the major portion of the >200,000 km^3 of Late Miocene to Holocene sediment that has built the Baranof Fan complex (Stevenson and Embley, 1987). The ultimate sediment sources are the glaciated mountain ranges of southeastern Alaska and western Canada. At present, there is no submarine canyon that connects to the Horizon Channel. Rather, it is hypothesized (Stevenson and Embley, 1987) that as the Pacific Plate (Yakutat Block) moved northward, a series of small canyons and gulleys along the slope successively supplied sediment to the abyssal plain. In the process, three major fan channel systems were formed that presently make up most of the abyssal plain of the Gulf of Alaska, and none can be linked to prominent submarine canyons.

Laurentian Channel and Fan

The Laurentian Fan offshore eastern Canada is another example of a large turbidite system that is not fed by a prominent incised submarine canyon (Fig. 1; Piper et al., 1985). During low stands of sea level, sediment reached the fan through the Laurentian Channel, an 80-km-wide glacial trough that is incised 300 m deeper than the regional shelf depth; the shelf break at the end of the Laurentian Channel is about 400 m deep. The continental slope seaward of the Laurentian Channel is extensively gullied as a result of erosion of Late Quaternary sediment on the slope. The slope gullies transition downslope into several major fan valleys in water depths between 2000 m and 3000 m (see Fig. 2 in Piper and Normark, 1982). Two of these fan valleys extend for about 400 km to a low-relief sandy depositional lobe area, which extends another 400 km to the south (Piper et al., 1985). Turbidite sedimentation continues more than 500 km to the south beyond the fan margin onto the Sohm Abyssal Plain.

The Eastern Valley of Laurentian Fan, which is the largest of the fan channels on the upper fan, is also one of the largest turbidite channels yet documented. The Eastern Valley locally reaches a vertical relief of almost 1000 m from the channel floor to the crest of the eastern levee (Fig. 2). The valley floor adjacent to this area of high levee relief is nearly 20 km wide. Thus, in cross section, at its area of greatest relief, the Eastern Valley of Laurentian Fan is basically the same size as the Bering Canyon where it is measured at the shelf break (Fig. 2).

Although there is no major canyon associated with the fan, the Laurentian Channel was probably the main outlet for much of the ice in southern Quebec and the Atlantic Provinces of Canada during glacial periods. Similar to most modern submarine fans, the Laurentian Fan is basically inactive during the current high stand of sea level. The 1929 Grand Banks earthquake, however, generated a turbidity current event that disrupted telecommunication cables more than 500 km from the earthquake epicenter (see review in Heezen and Hollister, 1971). Sediment deposited from these turbidity currents has been recovered on the Sohm Abyssal Plain, indicating that the 1929 turbidity current flowed more than 800 km from the initiation zone near the earthquake epicenter. The flow was generated when the earthquake caused a series of mass failures on the upper slope in many of the gullies that feed into the channels on the Laurentian Fan (Normark and Piper, 1991). The volume of sediment carried in the 1929 Grand Banks turbidity current was about 160 km^3 (Piper and Aksu, 1987).

La Jolla Canyon and Fan

The La Jolla Canyon and Fan are included in this discussion to emphasize the physical-scale relationship between the biggest submarine canyons and turbidite fans and those canyons and fans that are more typical of those mapped in outcrop and borehole studies (Figs. 2 and 7). Large fans built on oceanic crust are rarely preserved in the rock record except as small, highly deformed slivers in subduction-zone complexes. The La Jolla Fan is one of several small turbidite systems off southern California that are contributing to the fill of San Diego Trough west of San Diego, California (Bachman and Graham, 1985). The La Jolla Canyon is the largest of these small systems (75 km^2), but, like the others, sediment is supplied predominantly from beach sources and not by rivers. Until recently, the La Jolla Canyon was the best-documented submarine canyon for morphology, sediment distribution, and sedimentary processes (Shepard, 1963, 1973; Buffington, 1964; Shepard and Dill, 1966; Shepard et al., 1969; Bachman and Graham, 1985).

The La Jolla Fan is about 30 km in length from the head of the fan to the termination of the fan valley in San Diego Trough. A length of 20–50 km is typical for modern fans in basins of the California Borderland; e.g., the Navy Fan in South San Clemente Basin is about 30 km in length from its apex to the ponded basin plain, and Hueneme Fan in Santa Monica Basin is about 40 km long (Normark and Piper, 1985; Normark et al., 1998). Thus, analogs for understanding ancient turbidite systems are more likely to be based on La Jolla Canyon and fan sized-features than on the examples presented to define the world's largest systems.

DISCUSSION

The identification of the largest modern submarine canyon is straightforward if the criteria for selection are agreed upon, and Table 1 illustrates the choices available. The primary goal of this volume is to look at the largest sedimentary deposits that have formed in a variety of environmental settings and that represent a range of depositional processes. There is no simple relationship, however, between submarine canyon size and the size of the deposits that might be associated with these large canyons. For this reason, we chose to include a few smaller submarine canyons in our review to emphasize the physical scale range of submarine canyons that is involved as a backdrop for understanding a few observations concerning their deposits and, therefore, their legacy in the rock record. We do not attempt a comprehensive review; we have only selected a few examples to examine the differences between large canyons, large deposits of terrigenous sediment on the seafloor, and large depositional events.

Largest Canyon

The Zhemchug Canyon, which is the largest submarine canyon, is one of the best examples of a submarine canyon formed by repeated mass failures (Carlson and Karl, 1988; Karl et al., 1996). The volume of the Zhemchug Canyon is 5800 km^3; this is comparable to the largest submarine landslides that have been documented, e.g., the major collapses of the volcanoes of the Hawaiian Ridge, some of which exceed 5000 km^3 in volume and 200 km in length (Moore et al., 1994). Comparably sized landslides have been documented from continental margins as well, e.g., the Storrega slide on the Norwegian margin, which exceeds 5500 km^3 (Kenyon, 1987).

The Beringian continental margin, which is underlain by an outer shelf structural high, is incised by three humungous submarine canyons. The present morphology of these canyons is primarily the product of mass sediment failures (Carlson, Karl, and Edwards, 1991). Where the Beringian margin now exists, the Pacific (Kula) plate was possibly subducted beneath the North American plate, according to Scholl et al. (1975), contributing to weakness and resulting in subsequent canyon erosion of the margin. This subduction, together with subsequent Cenozoic glaciation, may have influenced the development of the large canyons (Karl et al., 1996). Buried canyons and channels suggest intermittent channel development as the Beringian margin evolved during the Cenozoic. For example, the shelf landward of the margin contains evidence of buried channels that crossed the shelf from rivers such as the Yukon and Kuskokwim. The shelf then may have been destabilized by earthquakes and or large storm waves, for which the Bering Sea is famous, and moved downslope as large slump or slide blocks, mass flows, or turbidity currents, continuing to incise the canyons into the Beringian continental slope.

Large Canyons and the Rock Record

The relationship between size of modern canyons and the size of their associated deposits is illustrated in Table 2. The examples selected are ranked (from large to small) by area of the canyon within two groups: those cut into the continental margin and those that are on river deltas. As noted earlier, the largest canyons, all from the Beringian margin, are incising the edge of the continent. Table 2 includes an estimate of the area of the submarine fan and associated turbidite and mass-transport deposits that have been fed by the canyon. The ratio of the deposit area to the canyon area is also shown, from which it is clear that the bedrock-cutting canyons have much smaller deposit areas than do the generally smaller canyons formed on major deltas. The ratios for bedrock canyons are less than 100 while the ratios for the largest fans range from about 150 to nearly 650. If the Laurentian Fan, which does not have a canyon but multiple slope gullies, is excluded, then the difference in ratios between the two groups is more pronounced.

As a check on this general relationship, we compare two small, well-studied turbidite systems from offshore California formed in tectonically active inner basins of the California Borderland. The La Jolla and Hueneme Canyons are similar in size, but the ratio of the canyon to deposit area of the delta-fed Hueneme Fan and basin plain is twice the area of the La Jolla deposits (Table 2), despite the fact that the Hueneme Fan is in a completely enclosed basin.

Except for canyons that feed sediment to basins formed on continental crust, the submarine canyons themselves have a better chance of surviving in the rock record than do the deposits they feed (Normark et al., 1993). Much of the record of submarine fan deposition on the deep ocean floor is eventually lost when the oceanic crust on which submarine fans are formed is ultimately subducted. In general, only small remnants of these deep water systems are incorporated into the continent. The canyons, which are cut in the continental margin or formed on delta fronts, may

TABLE 2. COMPARISON OF THE AREAS OF CANYONS AND OF THEIR DEPOSITS

Canyon/Fan (if not same name)	Canyon area (km^2)	Depositional area (km^2)	Area ratio	References*
Incised margin canyons				
Bering	30,800	190,000	6	Carlson and Karl (1988); Karl et al. (1996)
Navarin	14,600	~250,000	17	Carlson and Karl (1988); Karl et al. (1996)
Zhemchug	11,350	~250,000	22	Carlson and Karl (1988); Karl et al. (1996)
Zaire (Congo)	4470	330,000	74	Heezen et al. (1964); Droz et al. (1996); Savoye et al. (2000)
Laurentian Fan Valley †	3800	375,000	99	Piper and Normark (1982); Piper et al. (1985)
Monterey	2380	95,600	40	Normark et al. (1985); Fildani et al. (1999); or Reid et al. (1999)?
La Jolla	33	880	27	Buffington (1964); Shepard and Dill (1966)
Delta-front canyons				
Swatch of No Ground/Bengal	9000	2,040,000	227	Shepard (1973); Smith and Sandwell (1997);
Amazon	2250	330,000	147	Damuth and Kumar (1975); Milliman (1979); Damuth and Flood (1985)
Swatch/Indus	1700	1,100,000	647	Shepard and Dill (1966); Kolla and Coumes (1985)
Hueneme	25	1400	56	This study based on Normark et al. (1998)

Note: Canyon Area refers to the submarine-drainage footprint of the canyon (from Table 1). Depositional Area includes mass-wasted deposits, turbidite fans, and any associated basin plain or abyssal plain (where these latter elements have been distinguished by previous work).

* Interpretation of the canyon and fan areas is based on data provided in the references listed for cases where that value was not provided in references directly.

† Interpretation of the Laurentian "canyon" area is based on the sum of the areas of the multiple slope gullies that lead to the main leveed valleys on the fan.

be preserved, but—especially in the case of the bedrock-eroded canyons—not necessarily with sediment fill that is representative of their primary activity as a conduit for terrigenous sediment to reach the seafloor. Submarine canyons that have been preserved in the rock record are generally small in area (500 km^2 to 2000 km^2) and typically are filled with mudstone underlain by minor amounts of coarser-grained facies (see comparisons in Williams et al., 1998). Of the modern examples presented herein, it is the smaller systems formed in basins on continental crust, e.g., the Hueneme and La Jolla systems, that are likely to remain as part of the rock record (Normark et al., 1993).

Submarine fans fed by large rivers—e.g., Bengal, Amazon, Indus, etc.—have broad, leveed-channel systems. The levee sequences appear relatively acoustically transparent on seismic-reflection profiles and are assumed to be generally muddy. Scientific drilling on the Mississippi and Amazon Fans has confirmed the muddy nature of these levees (Bouma et al., 1985b, 1985c; Flood et al., 1995, 1997). As a result, these large, delta-fed fan systems have become known as muddy turbidite systems. Drilling away from the levees on these muddy fans, however, has shown that sand is a major component of much of the rest of the fan and is not that different from supposedly sand-rich systems fed by canyons cut in continental margins (see review in Piper and Normark, 2001).

Large Depositional Events

As already noted in the discussion on Laurentian Fan Valley, large turbidity current events are not necessarily related to the largest canyons. One of the largest turbidite deposits documented in the modern ocean was generated by the catastrophic floods resulting from the rapid draining of glacial Lake Missoula. The floodwaters kept flowing as a hyperpycnally generated turbidity upon reaching the ocean at the mouth of the Columbia River (Zuffa et al., 2000). The turbidity current initially moved through Astoria Canyon, which is only 2000 km^2 in drainage area and 425 km^3 in volume (Carlson and Karl, 1988). Part of the deposit from the floods was trapped in the Escanaba Trough nearly 1000 km from the river mouth and was cored during ODP Leg 169. Using the data from Zuffa et al. (2000), it can be shown that the largest turbidite bed left by the flood-generated turbidity currents exceeded 80 km^3. In total, there are about 175 km^3 of Missoula flood sediment in Escanaba Trough. These authors also showed that it is unlikely that Escanaba Trough contains more than a few percent of the total sediment transported. Thus, it appears that the volume of sediment in the largest turbidity current generated by the Missoula floods is an order of magnitude larger than the volume of the canyon. This suggests that the flows generated by glacial-lake floods may have taken many days to transit through the canyon and/or overwhelmed the canyon and flowed down the adjacent continental slope as well.

Many of the submarine canyons discussed here are formed to a lesser or greater extent by mass failures. The extent of mass failures for the largest canyons are on the order of the largest slumps and debris avalanches found on continental margins and oceanic volcanoes.

CONCLUSIONS

The task of determining *the* largest submarine canyon in the world is difficult because one must decide which physical parameter—length, relief of incision, cross-sectional area, or volume—is the most important criterion. Ultimately, we have discovered that no single canyon leads the candidates in all four categories; however, the Bering Sea margin has been sculpted by three canyons that are collectively the leaders in all four physical parameters. Zhemchug Canyon has the greatest relief (2600 m, measured at the shelf break) and the largest volume, 5800 km^3 and is our choice as the largest modern canyon. Zhemchug and Navarin Canyons share the honors of being the widest at the shelf break (~100 km). Bering Canyon is the longest, stretching 400 km, with the greatest area of incision (30,000 km^2) from shelf to abyssal plain.

By contrast, the largest submarine fans—Bengal, Indus, and Amazon—are all fed by small canyons incised into their respective deltas that are generally an order of magnitude smaller than those cut in older sediment or basement. The rivers that feed these largest fans are all associated with significant mountain ranges, the Himalayas and Andes, which provide substantial sediment to the rivers. In general, the deposits related to delta-front canyons are much more extensive than those related to canyons that incise the bedrock of continental margins (Table 2).

Despite the extensive area of the seafloor covered by sediment that has been transported through the larger canyons and the delta-front troughs off the larger rivers, in the end, it is the deposits of small submarine canyons and fans formed on continental crust that have the greatest potential for being preserved in the rock record.

ACKNOWLEDGMENTS

We hope that in writing this review paper, we have correctly acknowledged and properly used the many references required. It came clear in reviewing the literature on submarine canyons that we, as well as all fans of submarine canyons, owe a debt to Francis P. Shepard. The paper has been improved by reviews from H. A. Karl, S. L. Eittreim, D. R. Lowe, and G. Shanmugam.

REFERENCES CITED

Abers, G.A, Ekstrom, G., Marlow, M.S, and Geist, E.L, 1993, Bering Sea earthquake of February 21, 1991; active faulting along the Bering shelf edge: Journal of Geophysical Research, B, Solid Earth and Planets, v. 98, p. 2155–2165.

Bachman, S.B., and Graham, S.A., 1985, La Jolla Fan, Pacific Ocean, *in* Bouma, A.H., Normark, W.R., and Barnes, N.E, eds., Submarine fans and related turbidite systems: New York, Springer-Verlag, p. 65–70.

Bering Sea EEZ-Scan Scientific Staff, 1991, Atlas of the U.S. Exclusive Economic Zone, Bering Sea: U.S. Geological Survey Miscellaneous Investigations Series I-2053, 152 p.

Bouma, A.H., Normark, W.R., and Barnes, N.E., eds., 1985a, Submarine fans and related turbidite systems: New York, Springer-Verlag, 351 p.

Bouma, A.H., Stelting, C.E., and Coleman, J.M., 1985b, Mississippi Fan, Gulf of Mexico, *in* Bouma, A.H., Normark, W.R., and Barnes, N.E., eds., Submarine fans and related turbidite systems: New York, Springer-Verlag, p. 143–150.

Bouma, A.H., Coleman, J.M., and Meyer, A.W., 1985c, Mississippi Fan: Leg 96 program and principal results, *in* Bouma, A.H., Normark, W.R., and Barnes, N.E., eds., Submarine fans and related turbidite systems: New York, Springer-Verlag, p. 247–252.

Buffington, E.C., 1964, Structural control and precision bathymetry of La Jolla submarine canyon, California: Marine Geology, v. 1, p. 44–58.

Carlson, P.R., and Karl, H.A., 1984, Discovery of two new large submarine canyons in the Bering Sea: Marine Geology, v. 56, p. 159–179.

Carlson, P.R., and Karl, H.A., 1984–1985, Mass movement of fine-grained sediment to the basin floor, Bering Sea, Alaska: Geo-Marine Letters, v. 4, p. 221–225.

Carlson, P.R., and Karl, H.A., 1988, Development of large submarine canyons in the Bering Sea indicated by morphologic, seismic, and sedimentologic characteristics: Geological Society of America Bulletin, v. 100, p. 1594–1615.

Carlson, P.R., Karl, H.A., and Edwards, B.D., 1991, Mass sediment failure and transport features revealed by acoustic techniques, Beringian Margin, Bering Sea, AK: Marine Geotechniques, v. 10, p. 33–51.

Cooper, A.K., Scholl, D.W., and Marlow, M.S., 1987, Structural framework, sedimentary sequences, and hydrocarbon potential of the Aleutian and Bowers Basins, Bering Sea, *in* Scholl, D.W., Grantz, A., and Vedder, J.G., eds., Geology and resource potential of the continental margin of western North America and adjacent ocean basins—Beaufort Sea to Baja California: Houston, Texas, Circum-Pacific Council for Energy and Mineral Resources, Earth Science Series, v. 6, p. 473–502.

Curray, J.R., 1994, Sediment volume and mass beneath the Bay of Bengal: Earth and Planetary Science Letters, v. 125, p. 371–383.

Curray, J.R., and Moore, D.G., 1974, Sedimentary and tectonic processes in the Bengal deep-sea fan and geosyncline, *in* Burk, C.A., and Drake, C.L., eds., The geology of continental margins: New York, Springer-Verlag, p. 617–627.

Damuth, J.E., and Flood, R.D., 1985, Amazon Fan, Atlantic Ocean, *in* Bouma, A.H., Normark, W.R., and Barnes, N.E., eds., Submarine fans and related turbidite systems: New York, Springer-Verlag, p. 97–106.

Damuth, J.E., and Kumar, N., 1975, Amazon cone: Morphology, sediments, age, and growth pattern: Geological Society of America Bulletin, v. 86, p. 863–878.

Droz, L., and Bellaiche, G., 1991, Seismic facies and geologic evolution of the central portion of the Indus Fan, *in* Weimer, P., and Link, M.H., eds., Seismic facies and sedimentary processes of submarine fans turbidite systems: New York, Springer-Verlag, p. 383–402.

Droz, L., Rigaut, F., Cochonat, P., and Tofani, R., 1996, Morphology and recent evolution of the Zaire turbidite system (Gulf of Guinea): Geological Society of America Bulletin, v. 108, p. 253–269.

Emmel, F.J., and Curray, J.R., 1985, Bengal Fan, Indian Ocean, *in* Bouma, A.H., Normark, W.R., and Barnes, N.E, eds., Submarine fans and related turbidite systems: New York, Springer-Verlag, p. 107–112.

Fildani, A., 1993, Evoluzione deposizionale e significato geodinamico delle torbiditi del Monterey Fan: California Centrale (U.S.A.) [Ph.D. Thesis]: Universita degli Studi di Roma "La Sapienza," 169 p.

Fildani, A., Normark, W.R., and Reid, J.A., 1999, Which came first: Monterey Canyon or fan? Elemental architecture of central California turbidite systems: American Association Petroleum Geologists Bulletin, v. 83, p. 687.

Flood, R.D., Piper, D.J.W., Klaus, A., et al., eds., 1995, Initial Reports of the Ocean Drilling Program, v. 155: College Station, Texas, Ocean Drilling Program, 1233 p.

Flood, R.D., Piper, D.J.W., Klaus, A., and Peterson, L.C., eds., 1997, Proceedings of the Ocean Drilling Program, Scientific Results Leg 155: College Station, Texas, Ocean Drilling Program, 695 p.

Greene, H.G., and Hicks, K.R., 1990, Ascension-Monterey canyon system: History and development, *in* Garrison, R.E., Greene, H.G., Hicks, K.R., Weber, G.E., and Wright, T.L., eds., Geology and tectonics of the central California coast region, San Francisco to Monterey Bay, Pacific Section: American Association Petroleum Geologists Volume and Guidebook, GB 67, Bakersfield, p. 229–249.

Heezen, B.C., and Hollister, C.D., 1971, The face of the deep: London, Oxford University Press, 659 p.

Heezen, B.C., Menzies, R.J., Schneider, E.D., Ewing, W.M., and Granelli, N.C.L., 1964, Congo submarine canyon: American Association Petroleum Geologists Bulletin, v. 48, p. 1126–1149.

Karl, H.A., Carlson, P.R., and Gardner, J.V., 1996, Aleutian Basin of the Bering Sea: Styles of sedimentation and canyon development, *in* Gardner, J.V., Field, M.E., and Twichell, D., eds., The geology of the United States' seafloor: The view from Gloria: New York, Cambridge Press, p. 305–332.

Kenyon, N.H., 1987, Mass-wasting features on the continental slope of northwest Europe: Marine Geology, v. 74, p. 57–77.

Kenyon, N.H., Amir, A., and Cramp, A., 1995, Geometry of the younger sediment bodies of the Indus Fan, *in* Pickering, K.T., Hiscott, R.N., Kenyon, N.H., Ricci Lucchi, F., and Smith, R.D.A., Atlas of Deep Water Environments: London, Chapman & Hall, p. 89–93.

Kolla, V., and Coumes, F., 1985, Indus Fan, Indian Ocean, *in* Bouma, A.H., Normark, W.R., and Barnes, N.E, eds., Submarine fans and related turbidite systems: New York, Springer-Verlag, p. 130–136.

Marlow, M.S., Scholl, D.W., Cooper, A.K., and Buffington, E.C., 1976, Structure and evolution of Bering Sea shelf south of St. Lawrence Island: American Association Petroleum Geologists Bulletin, v. 60, p. 161–183.

McHargue, T.R., 1991, Seismic facies, processes, and evolution of Miocene inner fan channels, Indus submarine fan, *in* Weimer, P., and Link, M.H., eds., Seismic facies and sedimentary processes of submarine fans turbidite systems: New York, Springer-Verlag, p. 403–413.

Miller, M., Smith, W.H.F., Kuhn, J., and Sandwell, D.T., 2001, An interactive global map of sea floor topography based on satellite altimetry & ship depth soundings: NOAA Laboratory for Satellite Altimetry, http://ibis.grdl.noaa.gov/cgi-bin/bathy/bathD.pl (accessed July 2001).

Milliman, J.D., 1979, Morphology and structure of Amazon upper continental margin: American Association Petroleum Geologists Bulletin, v. 63, p. 934–950.

Moore, D.G., 1972, Reflection profiling studies of the California continental borderland: structure and Quaternary turbidite basins: Boulder, Colorado, Geological Society of America Special Paper 107, 142 p.

Moore, J.G., Normark, W.R., and Holcomb, R.T, 1994, Giant Hawaiian landslides, Annual Reviews of Earth and Planetary Sciences, v. 22, p. 119–144.

Normark, W.R., 1999, Late Pleistocene channel-levee development on Monterey submarine fan, central California: Geo-Marine Letters, v. 18, p. 179–188.

Normark, W.R., and Piper, D.J.W., 1985, Navy Fan, Pacific Ocean, *in* Bouma, A.H., Normark, W.R., and Barnes, N.E, eds., Submarine fans and related turbidite systems: New York, Springer-Verlag, p. 87–94.

Normark, W.R., and Piper, D.J.W., 1991, Initiation processes and flow evolution of turbidity currents: Implications for the depositional record, *in* Osborne, R.H., ed., From shoreline to abyss: SEPM (Society for Sedimentary Geology) Special Publication 46, p. 207–230.

Normark, W.R., Gutmacher, C.E., Chase, T.E., and Wilde, P., 1985, Monterey Fan, Pacific Ocean, *in* Bouma, A.H., Normark, W.R., and Barnes, N.E, eds., Submarine fans and related turbidite systems: New York, Springer-Verlag, p. 79–86.

Normark, W.R., Posamentier, H., and Mutti, E., 1993, Turbidite systems: State of the art and future directions: Reviews of Geophysics, v. 31, p. 91–116.

Normark, W.R., Damuth, J.E., and the Leg 155 Sedimentology Group, 1997, Sedimentary facies and associated depositional elements of the Amazon Fan, *in* Proceedings of the Ocean Drilling Program Scientific Results, v. 155: College Station, Texas, Ocean Drilling Program, p. 611–651.

Normark, W.R., Piper, D.J.W., and Hiscott, R.N., 1998, Sea level controls on the textural characteristics and depositional architecture of the Hueneme and associated submarine fan systems, Santa Monica Basin, California: Sedimentology, v. 45, p. 53–70.

Piper, D.J.W., and Aksu, A.E., 1987, The source and origin of the 1929 Grand Banks turbidity current inferred from sediment budgets: Geo-Marine Letters, v. 7, p. 177–182.

Piper, D.J.W., and Normark, W.R., 1982, Acoustic interpretation of Quaternary sedimentation and erosion on the channeled Upper Laurentian Fan, Atlantic margin of Canada: Canadian Journal of Earth Sciences, v. 19, p. 1974–1984.

Piper, D.J.W. and Normark, W.R., 2001, Sandy fans—from Amazon to Hueneme and beyond: American Association Petroleum Geologists Bulletin, v. 85, p. 1407–1438.

Piper, D.J.W., Stow, D.A.V., and Normark, W.R., 1985, The Laurentian Fan, Atlantic Ocean, *in* Bouma, A.H., Normark, W.R., and Barnes, N.E, eds., Submarine fans and related turbidite systems: New York, Springer-Verlag, p. 137–142.

Pirmez, C., and Flood, R.D., 1995, Morphology and structure of Amazon Channel, *in* Initial Reports of the Ocean Drilling Program, v. 155: College Station, Texas, Ocean Drilling Program, p. 23–45.

Reid, J.A., Fildani, A., and Normark, W.R., 1999, Monterey East turbidite deposit: A proto channel/levee complex or a zone of turbulent flow stripping? [abs.]: Eos (Transactions, American Geophysical Union), v. 80, p. F560.

Savoye, B., and 37 others, 2000, Structure et evolution recente de l'eventail turbiditique du Zaire: Premiers resultats scientifiques des missions d'exploration ZaiAngo 1 & 2 (Marge Congo-Angola): Comtes Rendus de l'Academie des Sciences, Serie II, Science de la Terre et des Planetes, Paris, v. 331, p. 211–220.

Scholl. D.W., Buffington, E.C., and Marlow, M.S., 1975, Plate tectonics and the structural evolution of the Aleutian-Bering Sea region, *in* Forbes, R.B. ed., Contributions to the geology of the Bering Sea Basin and adjacent regions: Geological Society of America Special Paper 151, p. 1–32.

Scholl, D.W., Buffington, E.C., Hopkins, D.M., and Alpha, T.R., 1970, The structure and origin of the large submarine canyons of the Bering Sea: Marine Geology, v. 8, p. 187–210.

Shepard, F.P., 1963, 1973, Submarine geology: New York, Harper & Row, 517 p.

Shepard, F.P., and Buffington, E.C., 1968, La Jolla submarine fan-valley: Marine Geology, v. 6, p. 107–143.

Shepard, F.P., and Dill, R.F., 1966, Submarine canyons and other sea valleys: Chicago, Rand McNally, 381 p.

Shepard, F.P., and Emery, K.O., 1941, Submarine topography off the California coast: Canyons and tectonic interpretation: Geological Society of America Special Paper 31, 171 p.

Shepard, F.P., Dill, R.F., and von Rad, U., 1969, Physiography and sedimentary processes of La Jolla submarine fan and fan-valley, California: American Association Petroleum Geologists Bulletin, v. 53, p. 390–420.

Smith, W.H.F., and Sandwell, D.T., 1997, Global sea floor topography from satellite altimetry and ship depth soundings: Science, v. 277, p. 1956–1962.

Stevenson, A.J., and Embley, R., 1987, Deep-sea fan bodies, terrigenous turbidite sedimentation, and petroleum geology, Gulf of Alaska, *in* Scholl, D.W., Grantz, A., and Vedder, J.G., eds, Geology and resource potential of the continental margin of western North America and adjacent ocean basins—Beaufort Sea to Baja California: Houston, Texas, Circum-Pacific Council for Energy and Mineral Resources, Earth Science Series, v. 6, p. 503–522.

Williams, T.A., Graham, S.A., and Constenius, K.N., 1998, Recognition of a Santonian submarine canyon, Great Valley Group, Sacramento Basin, California: American Association of Petroleum Geologists Bulletin, v. 82, p. 1575–1595.

Zuffa, G.G., Normark, W.R., Serra, F., and Brunner, C.A., 2000, Turbidite megabeds in an oceanic rift valley recording jokulhlaups of Late Pleistocene glacial lakes of the western United States: Journal of Geology, v. 108, p. 253–274.

Manuscript Accepted by the Society January 22, 2003

Geological Society of America
Special Paper 370
2003

Remnant-ocean submarine fans: Largest sedimentary systems on Earth

Raymond V. Ingersoll*
Department of Earth and Space Sciences, University of California, Los Angeles, California 90095-1567, USA

William R. Dickinson*
Department of Geosciences, University of Arizona, Tucson, Arizona 85721, USA

Stephan A. Graham*
Department of Geological and Environmental Sciences, Stanford University, Stanford, California 94305, USA

ABSTRACT

The final stages of continental collision result in the development of remnant ocean basins, within which accumulate the most voluminous sedimentary deposits on Earth. The rapid uplift of continental crust during attempted subduction of buoyant crust results in huge sediment flux from the collision zone into nearby remnant ocean basins, where collision has yet to occur. Sediment is transported across and through foreland basins to deltas, which feed submarine fans. As diachronous suture zones migrate along strike, older remnant-ocean sediments are uplifted and recycled into younger remnant ocean basins.

The largest two active sedimentary systems on Earth are the Bengal and Indus fan systems, each of which easily surpasses the size of all other modern submarine fans, deltas, or alluvial/fluvial systems. These fans are fed by the Ganges/Brahmaputra and Indus River systems, respectively, which derive their immense sediment loads from the greatest uplifted continental crust on Earth, the Himalaya and Tibetan Plateau.

Ancient analogous remnant-ocean fan systems include the Triassic Songpan-Ganzi complex of northern Tibet and the Carboniferous-Permian Ouachita-Marathon flysch deposits of Arkansas, Oklahoma, and Texas. The diachronous collision of the south and north China blocks provided voluminous detritus to the Songpan-Ganzi remnant ocean basin, which was finally closed when the North Tibet block collided. The uplifted southern Appalachians provided most of the sediment fed westward into the Ouachita remnant ocean basin; sequential suturing from northeast to southwest recycled sediment westward into remnant ocean basins, culminating in Permian terminal suturing of Laurasia and Gondwana and development of the Permian basin of west Texas.

There is no known or suggested mechanism that can equal the volume of sediment deposited in remnant ocean basins during suturing of continents. Ancient remnant-ocean deposits are preserved as highly deformed accretionary masses that contribute significantly to the building of continental crust. Their recognition as such is a significant step in our understanding of the creation and recycling of continental crust.

Keywords: remnant ocean basins, submarine fans, Bengal fan, Songpan-Ganzi Complex, Ouachita Mountains.

*E-mails: ringer@ess.ucla.edu; wrdickin@geo.arizona.edu; graham@pangea.stanford.edu

Ingersoll, R.V., Dickinson, W.R., and Graham, S.A., 2003, Remnant-ocean submarine fans: Largest sedimentary systems on Earth, *in* Chan, M.A., and Archer, A.W., eds., Extreme depositional environments: Mega end members in geologic time: Boulder, Colorado, Geological Society of America Special Paper 370, p. 191–208.

INTRODUCTION

The most voluminous sedimentary systems and deposits on Earth result directly from the combined forces of tectonic uplift, erosion, transport, and deposition during continental collision (Graham et al., 1975; Ingersoll et al., 1995, 2000). As the broadest uplift on Earth today, the Himalayan/Tibetan system provides the greatest flux of sediment into the oceans, with most sediment being transported by the Ganges/Brahmaputra and Indus river systems into their respective deltaic and submarine-fan systems. Thus, the linked systems of orogenic uplift, erosion, transport, and deposition produce giant submarine fans (Bengal and Indus, respectively) that accumulate in remnant ocean basins adjacent to the Himalayan/Tibetan orogen. Such giant submarine fans and their related systems are the subject of this contribution.

A remnant ocean basin is a shrinking ocean basin, which is flanked by at least one convergent margin and whose floor is typically covered by turbidites derived predominantly from associated suture zones (Ingersoll et al., 1995). Graham et al. (1975) developed a general model for this class of basin (Fig. 1) and used the Cenozoic development of the Himalayan-Bengal system as an analog for the late Paleozoic development of the Appalachian-Ouachita system. Ingersoll et al. (1995) discussed remnant ocean basins in the broader context of evolution of continental crust during crustal collisions; much of the text and figures of the present contribution is modified from our 1995 chapter.

Intraplate (rifted) continental margins commonly consist of alternating coastal promontories and reentrants, formed from linked continental rifts, hot spots, transforms, and failed rifts (Dewey and Burke, 1974; Şengör, 1976; Ingersoll and Busby, 1995). As an intraplate continental margin approaches (generally obliquely) a subduction zone, coastal promontories collide first, resulting in diachronous orogenic uplift and erosion. Adjoining remnant ocean basins are the natural repositories for voluminous detritus eroded from the growing orogenic belts. As sequential suturing progresses, the flux of sediment eroded from the growing accretionary orogen increases at the same time that the repository (the remnant ocean basin) is shrinking due to continued plate convergence (Fig. 2). The common result is extremely rapid sedimentation, followed quickly by flexural loading to form proforeland basins (peripheral foreland basins of Dickinson, 1976) (Miall, 1995), terminal suturing, and ultimately, plate reorganization (e.g., Cloos, 1993). During the sequential suturing of supercontinents (e.g., Laurasia and Gondwanaland) regionally diachronous closure of remnant ocean basins usually occurs (e.g., the Silurian through Permian Caledonide-Mauritanide-Appalachian-Ouachita-Marathon system) (Dewey and Kidd, 1974; Graham et al., 1975).

Details of collision processes and orogenesis are highly variable (Ingersoll et al., 1995). Major continent-continent collisions (e.g., Himalayan-Bengal and Appalachian-Ouachita systems)

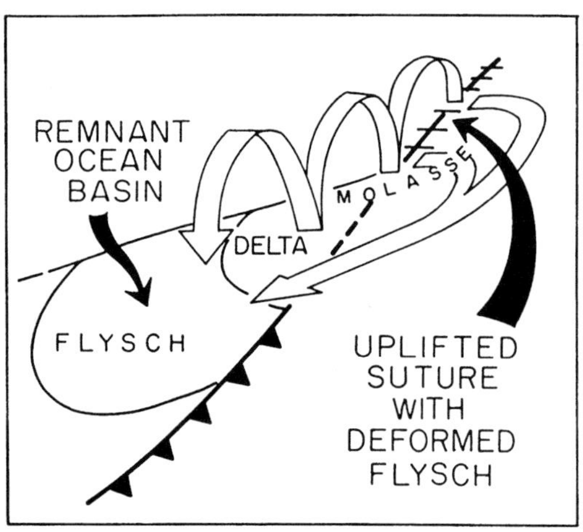

Figure 1. Conceptual diagram to illustrate progressive incorporation of synorogenic flysch within an orogenic suture belt by sequential closure of remnant ocean basin. From Graham et al. (1975).

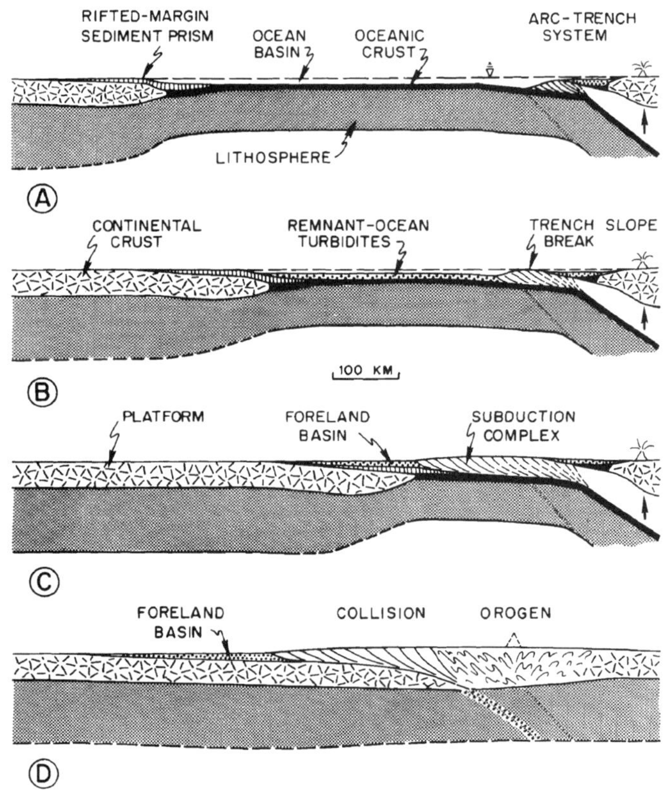

Figure 2. Idealized true-scale diagrams showing inferred evolution (A to D) of sedimentary basins associated with crustal collision to form cryptic intercontinental suture belts within collisional orogens. Diagrams represent sequence in time at one place along a developing collisional orogen, or coeval events at different places along suture belt marked by diachronous closure. Hence, erosion in one segment (D), where suture has formed, provides sediment that is dispersed longitudinally through a proforeland basin, past a migrating transition point (B to C), to feed subsea fans in a remnant ocean basin (B) along tectonic strike. From Dickinson (1976).

result in the largest orogenic and sediment-dispersal systems on Earth (Bouma et al., 1985; Dickinson, 1988; Ingersoll et al., 1995, 2000) (Figs. 3 and 4). In contrast, collision of two intraoceanic arcs (e.g., Molucca Sea; Silver and Moore, 1978; Moore and Silver, 1983) produces less detritus. Nonetheless, significant accretionary wedges are formed and plate boundaries change as a result of any of these collisions. Between these end-member examples are several variants of collisional orogenic events involving intraplate or transform margins and intraoceanic arcs (e.g., Taiwan [Teng, 1990] and Papua New Guinea [Crook, 1989]). These examples have in common the rapid accumulation of turbidites, commonly called "flysch," in remnant ocean basins immediately prior to accretion and suturing. All that usually remains of remnant ocean basins are the accretionary wedges formed by deformation of their sedimentary fill.

The early history of remnant ocean basins is as variable as the history of all oceanic crust. Subducting oceanic lithosphere underlying remnant ocean basins may be young or old. As collision

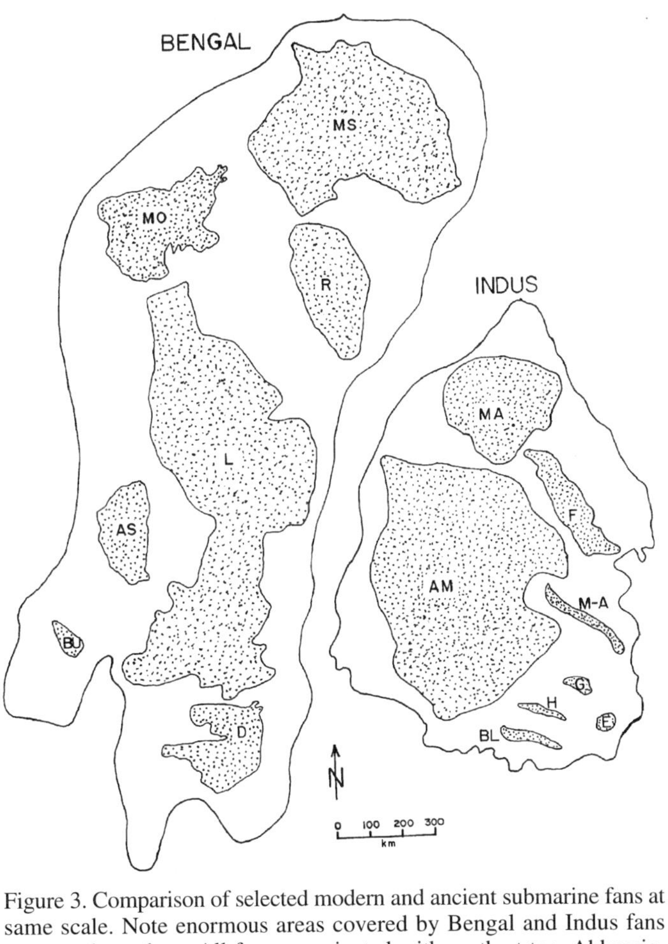

Figure 3. Comparison of selected modern and ancient submarine fans at same scale. Note enormous areas covered by Bengal and Indus fans compared to others. All fans are oriented with north at top. Abbreviations for other fans: AM—Amazon; AS—Astoria; BL—Blanca; BU—Butano; D—Delgado; E—Ebro; F—Ferrelo; G—Gottero; H—Hecho; L—Laurentian; MA—Magdalena; MO—Monterey; MS—Mississippi; M-A—Marnoso-Arenacea; R—Rhone. After Bouma et al. (1985).

begins along one or more parts of a suture zone, clastic sediments eroded from uplifting areas flood into transversely adjoining proforeland basins, transitional deltaic complexes, and laterally adjoining remnant ocean basins (Figs. 1, 4, and 5). Rates of subsidence and sedimentation increase rapidly as the tectonic and sedimentary loads of the growing accretionary wedge and sedimentary pile, respectively, affect the subducting continental margin (Fig. 2).

The transition from subduction of oceanic lithosphere (remnant ocean basin) to attempted subduction of continental crust (proforeland basin) is complex in three dimensions. Transitional continental crust typically underlies large deltaic complexes formed in this setting (e.g., Ganges-Brahmaputra; Alam, 1989; Lindsay et al., 1991; Reimann, 1993; Goodbred and Kuehl, 2000). Proforeland basins receive voluminous detritus shed from the growing orogenic belts (e.g., Indo-Gangetic; Burbank et al., 1986; Johnson et al., 1986; DeCelles and Cavazza, 1999). Big rivers (e.g., the Ganges and Indus) drain along suture zones, across fold-thrust belts either at strike-slip fault zones or at syntaxes, and longitudinally through foreland basins to build deltas. The deltas, which are transitional in both time and space, feed submarine fans (e.g., the Bengal and Indus fans) within remnant ocean basins. The four-dimensional nature of laterally suturing remnant ocean basins and proforeland basins is illustrated by the fact that cross sections (e.g., Fig. 2) can be viewed as sequential diagrams at any one location or as adjoining locations at one time; in either case, dominant sediment flux is perpendicular to the cross sections (e.g., Ganges-Brahmaputra-Bengal system).

DEPOSITIONAL SYSTEMS AND SEDIMENT DISPERSAL

Remnant ocean basins are dominated by turbidites and other submarine gravitites. These deposits represent the largest sedimentary accumulations on Earth (e.g., Bengal and Indus fans; Fig. 3); they are fed by the largest deltaic and fluvial systems, e.g., Ganges-Brahmaputra and Indus (Goodbred and Kuehl, 2000) and derived from the greatest orogenic systems, e.g., Himalayas and Tibetan Plateau (see below).

Sediment in remnant ocean basins usually is incorporated into collision suture belts, which constitute major segments of modern and ancient continental crust, e.g., Indoburman Range, Songpan-Ganzi terrane, and Ouachita Mountains (see below). Most "flysch" deposits of orogenic belts were deposited as turbidites in remnant ocean basins (Graham et al., 1975; Ingersoll et al., 1995).

The dominant source areas for sediment fed into remnant ocean basins are rapidly uplifted and recycled sedimentary and metasedimentary strata of associated foreland foldthrust belts (Figs. 1, 4, and 5) (Graham et al., 1976; Dickinson and Suczek, 1979; Ingersoll and Suczek, 1979). This recycled-orogenic provenance generates quartzolithic sand and sandstone, in strong contrast with quartzofeldspathic compositions derived from basement uplifts and feldspatholithic compositions derived from

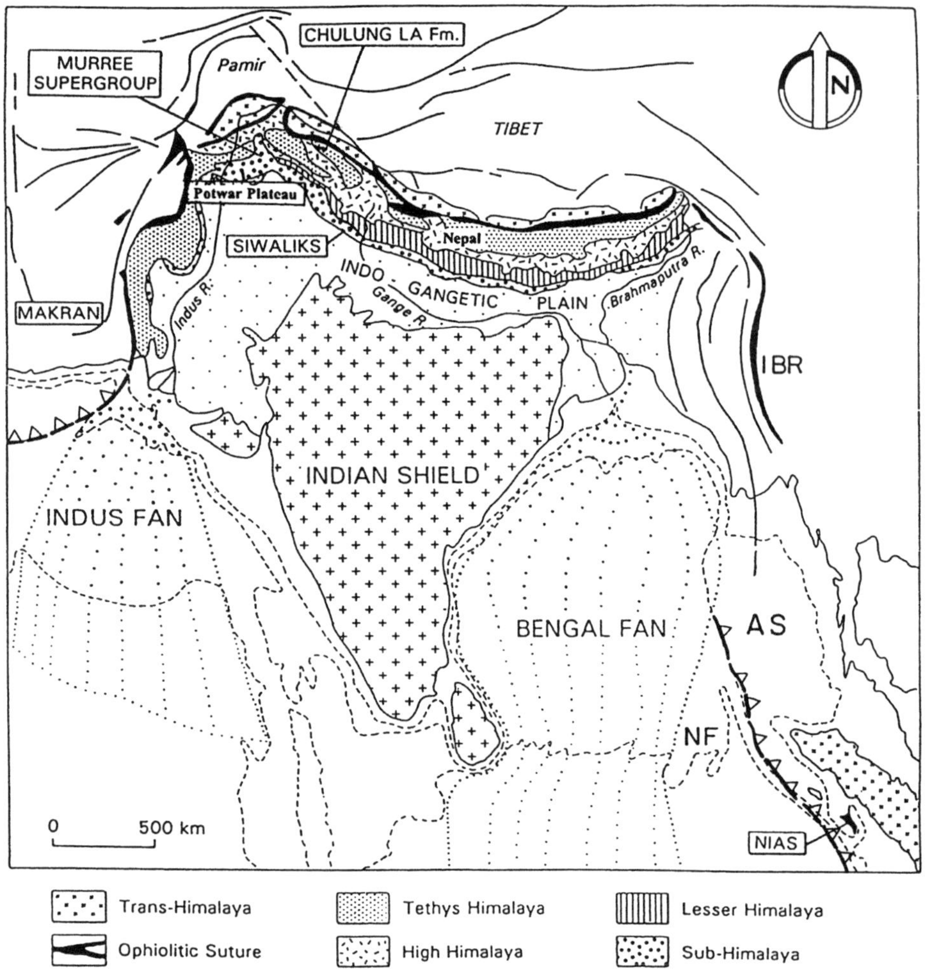

Figure 4. Indian-Asian collision zone. Since initial collision during Eocene, enormous quantities of detritus have been shed from Himalayas into ancient and modern proforeland basins (Murees and Siwaliks, deformed and uplifted in Sub-Himalayan belt; modern Indo-Gangetic Plain) and remnant ocean basins (Indus and Bengal fans). Older parts of the Indus and Bengal fans have been (and are being) deformed at subduction zones to form Makran and Sunda accretionary complexes, respectively. Abbreviations: IBR, Indoburman Ranges; AS, Andaman Sea; NF, Nicobar fan (subdivision of Bengal fan). After Critelli and Garzanti (1994).

magmatic arcs (Dickinson and Suczek, 1979; Dickinson, 1985; Ingersoll et al., 1995).

SUBSIDENCE HISTORY AND OROGENIC DEVELOPMENT

Lithosphere beneath remnant ocean basins has previously formed at spreading ridges in deep ocean basins, in contrast to thermally subsiding intraplate margins and lithospherically loaded foreland basins (e.g., Ingersoll and Busby, 1995). Initial depth of oceanic crust underlying remnant ocean basins is dependent on crustal age since formation at spreading ridges (e.g., Parsons and Sclater, 1977; Cloos, 1993); this crust will still be thermally subsiding (very slowly if old) when it becomes a remnant ocean basin (Ingersoll et al., 1995). Remnant ocean basins, therefore, are preexisting depressions in Earth's surface, relative to continental source areas. Rapid subsidence does not initiate due to tectonic processes such as stretching or flexure of the lithosphere; rather, subsidence is driven primarily by sedimentary loading as the oceanic lithosphere is confined between colliding buoyant crust and sediment flux to the basin increases. Reconstruction of subsidence history is inherently difficult because of uncertainties in paleobathymetric determinations of abyssal and bathyal strata and the likelihood that the strata will be deformed during suturing. Accurate chronostratigraphy is also problematic in most deformed "flysch" sequences. Palinspastic reconstruction of remnant ocean basins is a daunting challenge.

Suture zones are inherently complex (Dewey, 1977; Dewey et al., 1989; Şengör, 1990, 1991). Continental crust is compressed, thickened, and shifted laterally as subduction of oceanic lithosphere forces nonsubductable blocks together (also, see Cloos, 1993). Significant strike-slip motion is likely within both continental and oceanic crust associated with suture zones, especially where two continents collide, e.g., most of central to southern Asia. Where oblique oceanic convergence occurs, such as on most of the Bengal fan, which is converging obliquely with southeast Asia (Fig. 4), strike-slip motion is likely in intraarc and backarc segments of the overriding plate (e.g., Andaman Sea; Hamilton, 1979). During terminal suturing, deformation of remnant ocean turbidites commonly is extreme, to the point where detailed sedimentologic and paleoenvironmental studies may be impossible.

VARIANTS

Remnant ocean basins develop between nonsubductable crustal blocks of any dimension. At one extreme is the collision of major continents to form supercontinents, e.g., middle to late Paleozoic suturing of Laurasia and Gondwanaland. Major continents may also collide with subcontinents or intraoceanic arcs, e.g., modern Asia-India or Asia-Taiwan, respectively. At the other extreme are collisions involving two intraoceanic arcs (e.g., Molucca Sea collision zone; Silver and Moore, 1978; Hamilton, 1979; Moore et al., 1981; Moore and Silver, 1983). The history of remnant ocean basins associated with continent-continent collisions may span many geological periods, e.g., Silurian through Permian for the Laurasian-Gondwanan system, although individual basins have shorter life spans. Most collisions involving intraoceanic arcs are short-lived, lasting just a few million years, so that associated remnant ocean basins are destroyed soon after their initiation. Sediment flux to remnant ocean basins is directly dependent upon the rate and dimension of uplift in source areas in suture zones, so that overall sediment mass is huge for continent-continent collisions (e.g., Bengal and Indus fans) and relatively small for continent-arc collisions (e.g., Huon Gulf of the Solomon Sea; Crook, 1989; Silver et al., 1991). Accretionary wedges may grow to the point where their volume exceeds that of associated intraoceanic arcs (e.g., Taiwan orogen versus Luzon arc; Teng, 1990; Huang et al., 1992).

Collision of mid-sized continents (e.g., India) with large continents (e.g., Asia) is the ideal way to produce long-lived and voluminous sedimentary accumulations in remnant ocean basins fed by extremely indented and uplifted orogenic belts (e.g., Himalayas and Tibetan Plateau). The subcontinent of India is a relatively small part of the Indian oceanic plate, which is being pulled by subduction of dense lithosphere northward under Asia (e.g., Cloos, 1993). The modest size of India means that its buoyancy cannot overcome the negative buoyancy of the Indian oceanic lithosphere, and it continues to be driven northward as an indenter (e.g., Tapponnier et al., 1986; Dewey et al., 1989). In addition, the modest surface area of India has resulted in two large, remnant ocean basins, the Bay of Bengal and the Arabian Sea, immediately adjacent to the rapidly growing orogen.

Climate may modify rates of erosion and sediment production. Accelerated uplift of the Himalayas and the Tibetan Plateau during the Miocene may have resulted in development of, or intensification of, the Asian monsoon system (e.g., Quade et al., 1989; Ruddiman and Kutzbach, 1989; Harrison et al., 1993; Goodbred and Kuehl, 2000). Major collisional orogens may create their own climates as well as modify global climate, which, in turn, can affect tectonic development (e.g., Hoffman and Grotzinger, 1993). The greatest sediment accumulations probably form(ed) in tropical to subtropical settings, where sediment flux would be greatest.

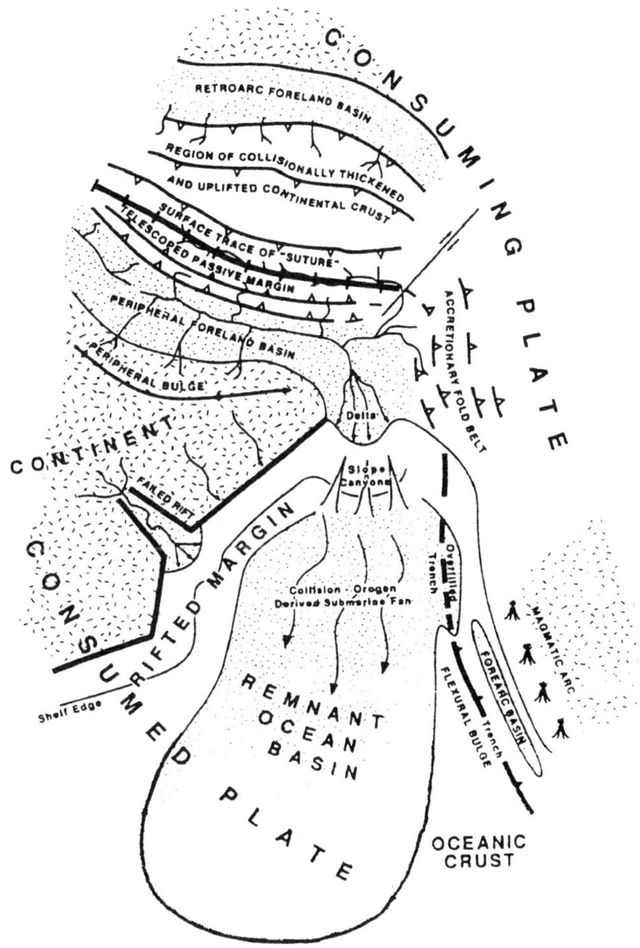

Figure 5. Morphotectonic map showing most components of typical remnant ocean basins closing sequentially between colliding continents. Illustrated geography is analogous to modern Bay of Bengal and surrounding areas and to inferred Carboniferous of Ouachita area. Details of components may be modified in many ways, as discussed in text. From Ingersoll et al. (1995).

The greatest accretionary bodies in the geologic record should have resulted from collision of a large continent with a moderate-sized continent, preferably in tropical to subtropical paleolatitudes. The Triassic Songpan-Ganzi terrane of central China is an excellent example (see Şengör and Okurogullari, 1991; Yin and Nie, 1993; Zhou and Graham, 1993, 1996; and below). Collision of two moderate-sized continents also may create large accretionary bodies, especially when long-term, complex multi-plate interactions are involved (e.g., Altaids; Şengör et al., 1993). In contrast, collision of major continents (e.g., Proterozoic Grenville and Paleozoic Appalachian-Ouachita systems) usually results in terminal suturing, which eliminates most remnant ocean basins early in the process. Preservation of accretionary complexes derived from remnant ocean basins along colliding irregular continental margins, such as the Ouachita-Marathon example, is the exception rather than the rule.

MODERN EXAMPLES

Bengal/Nicobar and Indus Submarine-Fan Systems

Accumulation of the immense Bengal/Nicobar and Indus submarine fans (Figs. 3 and 4) since the Eocene has been in direct response to the collision-induced uplift of the Himalayan suture belt (Curray and Moore, 1971, 1974). Recognition of this causal relationship between continental collision and synorogenic sedimentation in adjacent oceanic basins led to the first statement of the general model for remnant ocean basins by Graham et al. (1975). Remnant ocean basins both west and east of southern India are receiving voluminous sediment derived from the Himalayan suture belt at the same time that their flanks are being accreted along subduction zones (Makran and Indoburman Ranges, respectively). Both systems are fed by immense fluvial and deltaic systems, the Indus and Ganges-Brahmaputra, respectively. Some of the sediment derived from the Himalayas has accumulated in proforeland basins of the Indo-Gangetic Plain (e.g., Muree Supergroup and Siwalik Group), but most modern sediment bypasses these filled forelands and accumulates in the deltas and submarine fans along the continental margins.

The Bengal fan is approximately 16 km thick at the head of the Bay of Bengal; with a width of approximately 1000 km and a length of over 4000 km, the total volume of sediment is approximately 3×10^7 km^3 (Curray and Moore, 1971; Curray, 1991). The Indus fan is approximately 10 km thick at its head and consists of approximately 5×10^6 km^3 of sediment (Clift et al., 2001).

The timing of continental collision between India and southern Asia is constrained by the following observations (Graham et al., 1975; Ingersoll et al., 1995): (1) The northward movement of India is tracked by magnetic anomalies of the Indian Ocean (e.g., McKenzie and Sclater, 1971; Johnson et al., 1976; Norton and Sclater, 1979; Patriat and Achache, 1984; Dewey et al., 1989). (2) The intraplate margin of northern India experienced thermally influenced subsidence following rifting from Gondwanaland and prior to collision with Asia (Garzanti et al., 1987; Searle et al., 1987; Gaetani and Garzanti, 1991; Brookfield, 1993). (3) A proforeland basin formed as the northern edge of India was pulled below the growing Himalayan orogen (Critelli and Garzanti, 1994). (4) The south-facing magmatic arc of southern Asia became inactive as the subduction zone was stifled by the attempted subduction of buoyant Indian continental crust (Graham et al., 1975). (5) Massive quantities of sediment were deposited along the flanks of India as the Himalayas were uplifted and the foreland basins filled (Graham et al., 1975; Curray, 1991).

By the early Eocene (ca. 55 Ma), the continental rise of northern India began to enter the Transhimalayan subduction zone, causing initial flexural bulging of the margin (Garzanti et al., 1987). The remnant ocean basin of Neotethys was bounded on the north by a trench, accretionary wedge, and forearc basin (Garzanti et al., 1987), thus trapping most volcaniclastic detritus along the margin; to the south, Neotethys was bordered by the intraplate margin of northern India, which supplied limited quartzose detritus to the basin. As a result, no "flysch" analogous to the modern Bengal or Indus fans formed at this stage. Upon initial collision and uplift of the Himalayan margin in the Eocene, synorogenic sediment began to fill the evolving proforeland basin and associated wedge-top (piggy-back) basins (Garzanti et al., 1987; Critelli and Garzanti, 1994); sediment was also transported east and west into remnant ocean basins away from the initial point of contact in the Ladakh area (e.g., Patriat and Achache, 1984; Qayyum et al., 1996, 1997). These early flysch deposits were quickly deformed and incorporated into the lengthening suture zone. Some of the older foreland deposits, fluvio-deltaic molasse (e.g., Critelli and Garzanti, 1994), and (possibly) older flysch, are volcaniclastic, and thus represent early recycling of forearc and subduction-complex detritus. Other than these oldest syn-collisional deposits, however, both molasse and flysch in the Himalayan system are overwhelmingly dominated by sedimentary and metasedimentary detritus, which is characteristic of recycled-orogenic provenance (i.e., Dickinson and Suczek, 1979; Ingersoll and Suczek, 1979; Suczek and Ingersoll, 1985; Garzanti et al., 1987, 1996; Critelli and Garzanti, 1994). Once Neotethys was destroyed north of the Indian margin and proforeland basins began to fill, the only remaining repositories for detritus eroded from the growing Himalayan suture were the remnant ocean basins west and east of India. Thus, the pre-Eocene continental-rise deposits of the intraplate margins of west and east India are overlain by the Eocene to Holocene Indus and Bengal fans, respectively, and predecessors in Pakistan and the Indoburman Range (Curray and Moore, 1971, 1974; Kolla and Coumes, 1987; Curray, 1991; Uddin and Lundberg, 1998; Clift et al., 2001). Initial deposition was relatively slow, but by the early Miocene, vigorous uplift of the Himalayas and Tibet resulted in rapid building of the submarine fans (Alam, 1989; Cochran, 1990; Copeland and Harrison, 1990; Klootwijk et al., 1992; Harrison et al., 1993; Qayyum et al., 1996, 1997; Clift et al., 2001). Since the Miocene, most sediment has been transported by the Indus and Ganges/Brahmaputra river systems to their respective deltas (e.g., Uddin and Lundberg, 1998); much of the fluvial sediment bypasses the deltas and is transported, in places, over 3000 km as turbidity currents to form the Indus and Bengal/Nicobar fans (Curray and Moore, 1971, 1974; Bowles et al., 1978; Kuehl et al., 1989; Cochran, 1990; Lindsay et al., 1991; Clift et al., 2001).

The northwestern edge of the older part of the Indus fan is currently being deformed into the Makran accretionary wedge (Fig. 4) (Farhoudi and Karig, 1977; Critelli et al., 1990; Clift et al., 2001); almost symmetrically, the eastern edge of the Bengal/Nicobar fan is being deformed into the Sunda accretionary wedge (Curray and Moore, 1971, 1974; Graham et al., 1975; Moore, 1979). The Makran is transitional westward into the continental-collision zone of the Zagros and the proforeland basin of the Persian Gulf (Farhoudi and Karig, 1977), whereas the Sunda accretionary zone is transitional southeastward into the oceanic subduction zone of the Sunda arc (Hamilton, 1979). The Indoburman Ranges represent accreted proto-Bengal fan (Graham et al., 1975; Uddin and Lundberg, 1998). All evidence points to the ultimate derivation of most Indus and Bengal/Nicobar sediments from the Himalaya and bordering ranges (Ingersoll and Suczek, 1979; Suczek and Ingersoll, 1985). Most of the Makran accretionary wedge also has Himalayan sources, although younger, shallower parts contain volcaniclastic detritus derived from the magmatic arc to the north (Critelli et al., 1990). Himalayan detritus constitutes most of the Neogene to Holocene Sunda accretionary wedge, although older parts of the accretionary wedge include considerable volcaniclastic detritus derived from the Sunda arc (Ingersoll and Suczek, 1979; Moore, 1979; Moore et al., 1982). Both the Indus and Bengal/Nicobar fan systems are likely to continue receiving detritus from the Himalayas and Tibetan Plateau until their respective remnant ocean basins are closed by subduction. Final closure and destruction of these two remnant ocean basins is not likely to occur until additional continental fragments or arcs move northward relative to India in the distant future. Alternatively, the Indian subcontinent could rotate clockwise to close the Arabian Sea, thus emplacing the Makran/Indus accretionary wedge over the west coast of India, or counterclockwise to close the Bay of Bengal, thus emplacing the Sunda/Bengal/Nicobar accretionary wedge over the east coast of India. The final result in any of these scenarios is the destruction of oceanic crust between colliding continents and the creation of accretionary masses along suture zones (i.e., Graham et al., 1975; Şengör and Okurogullari, 1991). In rare cases, dormant ocean basins may persist, completely surrounded by collided continental blocks, but with no active plate boundaries, e.g., Black Sea and North Caspian Depression (see Ingersoll and Busby, 1995).

Other Modern Examples

Additional examples of modern remnant ocean basins were discussed by Ingersoll et al. (1995). Although several of these basins are as tectonically active as the Bay of Bengal and Arabian Sea examples discussed above, the total volume and flux of sediment, and the scale of depositional systems of these other modern remnant ocean basins are considerably smaller. Nonetheless, these more modest examples of modern remnant ocean basins provide considerable insights into tectonic and sedimentologic dynamics of these systems in general.

Ingersoll et al. (1995) discussed the following modern remnant ocean basins: (1) The Huon Gulf (HG) at the western extremity of the Solomon Sea (Papua New Guinea) provides a modern example of clastic sedimentation in a remnant ocean basin being closed by the sequential collision of the western end of the Bismarck volcanic arc with the northern fringe of the Australian continental block (Crook, 1989; Silver et al., 1991). (2) The late Cenozoic collision of Asia with the Luzon arc (Chai, 1972; Bowin et al., 1978; Teng, 1990) illustrates transitions from south to north from pre-collisional settings to remnant ocean basin to uplifted collisional orogen. (3) The eastern Mediterranean Sea is a complex remnant ocean basin receiving sediment primarily from young suture zones, e.g., erosion of the Apennines provides sediment to the Ionian Sea (Critelli and Le Pera, 1998) and intraplate margins, e.g., north Africa, especially the Nile Delta (Bartolini et al., 1975). (4) The extraordinarily complex area between southeastern Asia and Australia includes several present or future remnant ocean basins (e.g., Hamilton, 1979). (5) The only documented example of two intraoceanic arcs in the process of terminal suturing is the Talaud-Mindanao collision zone in the Molucca Sea, where the Sangihe and Halmahera arcs face each other and their accretionary wedges battle for supremacy (Silver and Moore, 1978; Moore et al., 1981; Moore and Silver, 1983; Silver et al., 1991).

ANCIENT EXAMPLES

Ouachita Remnant Ocean

Along the southern margin of the North American craton, the curvilinear Ouachita orogenic belt (Fig. 6) stretches more than 2000 km, mostly in the subsurface, westward from the southern limit of the Appalachian orogen into northern Mexico (Arbenz, 1989; Viele, 1989). The chief lithic assemblage of the Ouachita system, as exposed in the Ouachita Mountains of Arkansas and Oklahoma, and in the Marathon region of west Texas, is a thick, allochthonous succession of Paleozoic turbidites, chert, and mudrock that is thrust over coeval but contrasting platformal strata fringing the craton. The overthrust Ouachita-Marathon sedimentary assemblage is interpreted as the fill of a remnant ocean basin that was destroyed as the accreted assemblage was thrust cratonward over the continental margin during collisional orogenesis (Graham et al., 1975; Ingersoll et al., 1995). The alternate hypotheses that the Ouachita succession was deposited in a backarc basin (Morris, 1974) or within a failed rift trough (Lowe, 1985, 1989) do not account as well for overall geologic relationships (Viele and Thomas, 1989).

The Ouachita and Appalachian margins of the North American craton were delineated by Neoproterozoic-Cambrian rifting (Thomas, 1991). Several failed rifts oriented at high angles to the Ouachita continental margin extended into the adjacent craton (Fig. 6). Sedimentation that initiated along the nascent continental margin, within the failed rifts, and on the floor of the adjacent ocean basin by latest Cambrian time continued without significant tectonic interruption through much of Paleozoic time. In the interval from middle Pennsylvanian to earliest Permian time, however, diachronous thrusting, beginning first in the east and ending last in the west, carried the Ouachita-Marathon oceanic assemblage perhaps 75 km over shelf strata deposited along the continental margin. This terminal orogenesis reflected the attempted subduction of the continental margin beneath the flank of an arc-trench system that arrived from the south and faced the continental block (Graham et al., 1975; Wickham et al., 1976). The landmass upon which the arc was built is often termed Llanoria, and may be represented, in part, by the basement of Yucatan, which rifted away from Texas in Mesozoic time to form the Gulf of Mexico (e.g., Dickinson and Lawton, 2001).

Voluminous Carboniferous flysch was deposited by gravity flows down the axial trough of the Ouachita remnant ocean basin from sources lying dominantly to the east near the syntaxis

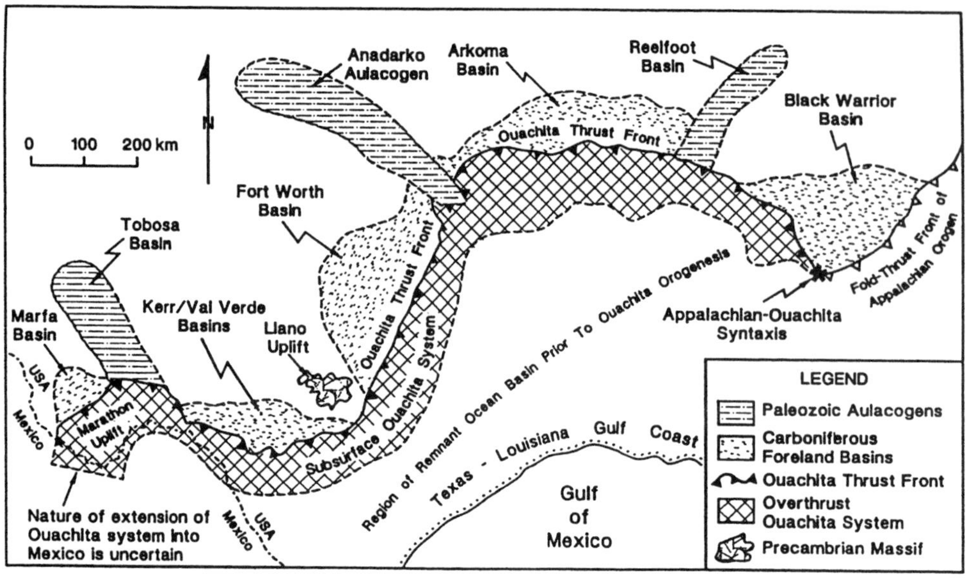

Figure 6. Regional tectonic relations of Ouachita orogenic belt. See text for discussion. Modified after Thomas (1989).

between the Appalachian and Ouachita orogenic belts (Graham et al., 1975, 1976). As thrusting proceeded, synorogenic basins that developed in front of advancing allochthonous masses evolved from a remnant ocean basin to a string of discrete proforeland basins floored by continental crust (Fig. 6). As basin architecture changed, turbidite (flysch) sedimentation in the remnant ocean basin was succeeded by deltaic and nonmarine (molasse) deposition in the foreland basins. Intermediate phases of basin evolution and the transition from flysch to molasse deposition involved transient hybrid basins floored by partly oceanic and partly continental crust (Houseknecht, 1986). The nature of basement beneath sediment deposited during the transitional stages of tectonic evolution is uncertain, because synorogenic sedimentary sequences were detached from underlying basement as the Ouachita allochthon overrode the continental margin.

Summary accounts (i.e., McBride, 1970, 1989; Morris, 1974, 1989; Ethington et al., 1989; Lowe, 1989) of the overthrust Ouachita-Marathon sedimentary assemblage permit the recognition of successive stratigraphic intervals that represent sequential depositional phases. Sub-Carboniferous strata are preorogenic deposits of an ocean basin lying south of the Ouachita intraplate continental margin, whereas Carboniferous strata are synorogenic deposits of the remnant ocean basin and related depocenters that developed just before and during arc-continent collision in the Ouachita region. The schematic geohistory diagram of Figure 7 depicts inferred patterns of tectonic and isostatic subsidence during evolution of the Ouachita system, as exposed in the Ouachita Mountains.

Carboniferous strata of the Ouachita remnant ocean basin and successor troughs reach an aggregate thickness of 12,000 to 15,000 m in the Ouachita Mountains; analogous strata in the Marathon region are less than 5000 m thick (McBride, 1989; Morris, 1989). Deformation was complete before the end of Pennsylvanian time in the Ouachita Mountains (Viele and Thomas, 1989), and by the end of Wolfcampian (earliest Permian) time in the Marathon region (Ross, 1986). Sediment accumulation rates were perhaps 125 m/m.y. during the initial phases of turbidite sedimentation, but increased to almost 1000 m/m.y. before deposition was terminated by thrust-dominated deformation.

Mississippian depositional systems record the earliest delivery of sediment derived from Llanoria. Widespread turbidite paleocurrent indicators of paleoflow to the northwest are dominant in the Stanley Group of the Ouachita Mountains and in the Tesnus Formation of the Marathon region. Lateral facies patterns in both units are more proximal in the southeast and more distal in the northwest (Niem, 1976; Flores, 1977). Both units also contain ashfall and subaqueous ashflow tuffs, apparently derived from arc eruptions farther south (Niem, 1977; Imoto and McBride, 1990).

Paleocurrent indicators in Pennsylvanian flysch (Jackfork and Atoka formations) of the Ouachita Mountains record dominantly longitudinal paleoflow toward the west along the axis of the trough that formed as the remnant ocean basin narrowed. Regional tectonic relationships (Graham et al., 1975) and the distribution of turbidite facies (Moiola and Shanmugam, 1984; Link and Roberts, 1986) imply that the main depositional system was an elongate submarine fan built westward from an initial apex near the southern end of the Appalachian orogenic belt. The Ouachita flysch fan was analogous to the modern Bengal fan built into the Bay of Bengal from an apex near the eastern end of the Himalayan orogen and was similar in size (Graham et al., 1975) (Fig. 8).

The immense volume of the Ouachita flysch and molasse requires a provenance able to supply sediment rapidly to the remnant ocean basin and its derivative troughs. The quartzose to quartzolithic character of the bulk of the Carboniferous sandstones (Graham et al., 1976; Ingersoll et al., 1995) implies recycling of

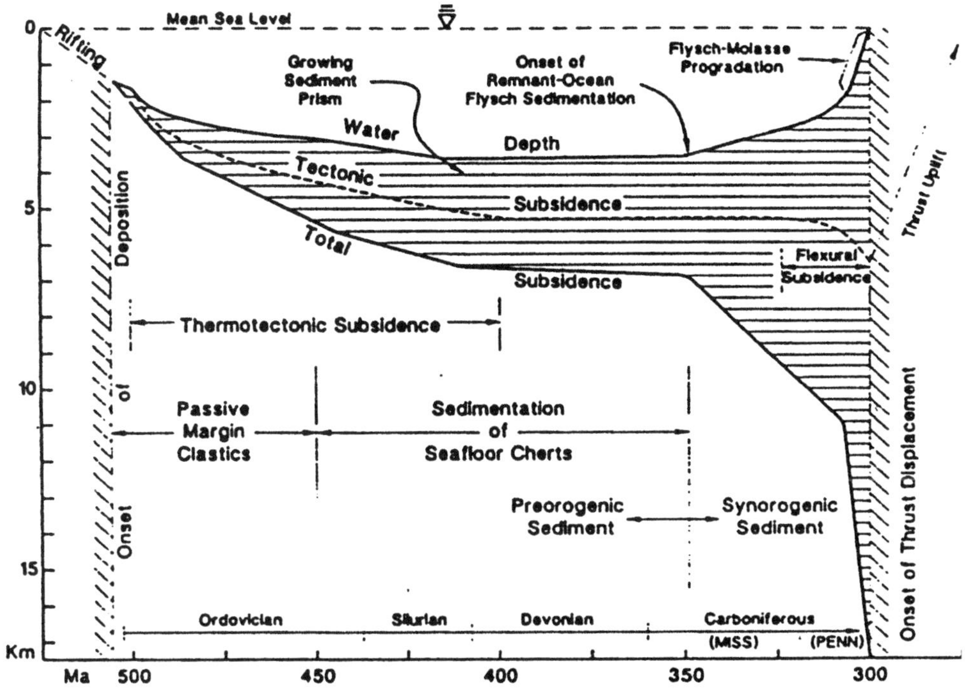

Figure 7. Hypothetical geohistory diagram for Ouachita succession of Ouachita Mountains. Thicknesses from Lowe (1989) and Morris (1989), and ages from Ethington et al. (1989) using Decade of North American Geology (DNAG) time scale (Palmer, 1983). Thermotectonic subsidence constrained to rates established for cooling of oceanic lithosphere (350 m times square root of elapsed time in million years after rifting) for a period of 100 m.y. after rifting, with no tectonic subsidence thereafter until onset of flexural subsidence under influence of structural loading by thrust sheets of Ouachita allochthon. Flexural subsidence constrained with flexural geometry inferred by Goebel (1991), assuming allochthon movement of 10 km/m.y. (faster rate of 100 km/m.y. would confine flexural subsidence to last 2.5 m.y. of depositional history and would smooth elbow of water-depth curve at transition from aggradational to progradational phases of flysch sedimentation, but would also sharpen corresponding elbow in curve for total subsidence of substratum). Backstripping constrained by net sediment densities inferred from equations for depth-porosity relations given by Dickinson et al. (1987). From Ingersoll et al. (1995).

sedimentary and metasedimentary detritus without unroofing of deep crustal basement, although significant feldspathic components in some units reflect additional contributions from igneous sources. Isotopic studies by Gleason et al. (1994, 1995) and Patchett et al. (1999) demonstrated the regional dominance of sediment dispersal from the Appalachian collision orogen into the Ouachita-Marathon remnant ocean basin. Given the regional tectonic setting of the Ouachita system, we inferred (Graham et al., 1976) that the primary source of the voluminous detritus was the vigorously uplifted older components of the Appalachian-Ouachita orogen, which developed diachronously through sequential closure of the oceanic region that lay between Laurentian and Gondwanan segments of Pangea. In this view, the longitudinal dispersal of turbidites down the axis of a closing remnant ocean basin systematically preceded orogenic deformation along each increment of the evolving orogenic belt. On a regional scale, sequential initiation of coarse-clastic sedimentation within proforeland depocenters along the cratonal flank of the Appalachian-Ouachita belt indicates diachronous suturing, from northeast to southwest, of Laurentian and Gondwanan continental blocks (Fig. 9).

Songpan-Ganzi Complex

Prominent on any geologic or tectonic map of Asia is a vast triangular region in central China underlain by deformed Triassic deep-marine strata (i.e., Zhang et al., 1984; Zhu, 1989). These rocks comprise the Songpan-Ganzi Complex, long recognized as an accretionary complex of deep-marine "flysch," trapped between cratonal blocks in the tectonic collage of Asia (Fig. 10) (Klimetz, 1983; Şengör, 1984; Ji and Coney, 1985; Watson et al., 1987; Chang, 2000). Following Yin and Nie (1993), Zhou and Graham (1993, 1996), and Nie et al. (1994), we interpret the Songpan-Ganzi Complex to be the largest (present surface area of 220,000 km^2; Huang and Chen, 1987) and best-preserved record of sedimentation and accretion of the fill of an ancient remnant ocean basin. If so, it may hold important answers about the evolution of remnant ocean basins in general, as well as the evolution of the Asian tectonic mosaic.

The Songpan-Ganzi Complex is crushed among continental nuclei within the pan-Eurasian Cimmeride orogenic system (Klimetz, 1983; Şengör, 1984, 1987). To the north of the Songpan-Ganzi Complex are amalgamated crustal elements, including Tarim, Qaidam, and Sino-Korea, or following Yin and Nie (1993), the "north China block" (Fig. 10). The contact between the Songpan-Ganzi Complex and the north China block is primarily tectonic, and where relatively well described, as in the Qinling Mountains, it involves nappes of Triassic flysch in thrust contact with north China block basement and platform cover rocks (Hsu et al., 1987). To the southeast lies the south China block, cored by the Yangtze craton. The south China block and its Paleozoic–Lower Triassic carbonate platform cover (Yin and Nie, 1993) are in tectonic contact with the Songpan-Ganzi Complex

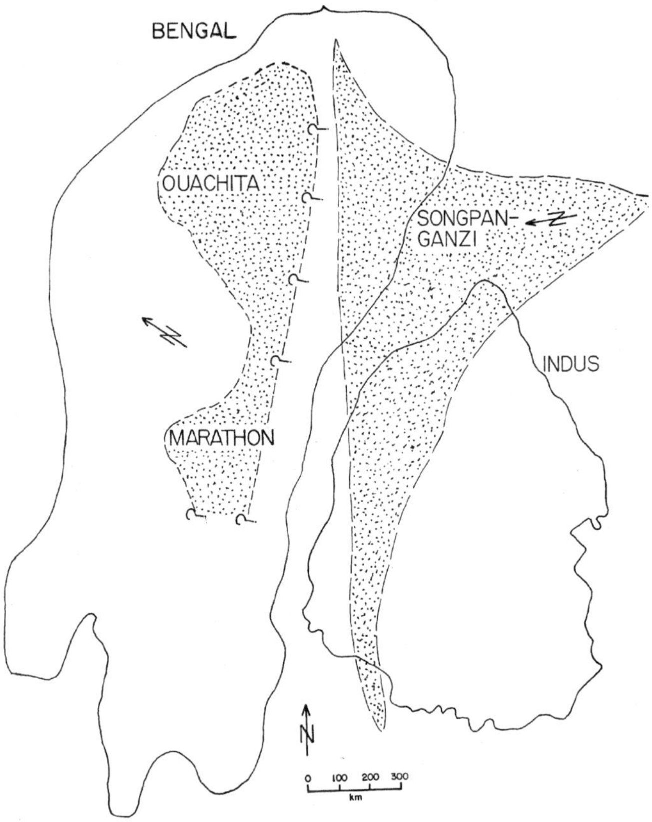

Figure 8. Comparison (at same scale) of two large modern remnant-ocean submarine-fan systems (Bengal and Indus from Fig. 3; after Bouma et al., 1985) with two large ancient remnant-ocean submarine-fan systems (Ouachita-Marathon, see Fig. 6, and Songpan-Ganzi, see Fig. 10). The four fan systems are of comparable size. All four systems are oriented with deltaic source areas toward top of figure. North arrows for two ancient systems show present orientations. For Ouachita-Marathon system, short dashes indicate inferred fan boundaries beneath younger cover; question marks denote unknown limits because of younger tectonic truncation. Long dashes for both ancient systems indicate present tectonic borders; no palinspastic reconstruction is attempted, so that areas shown are minima.

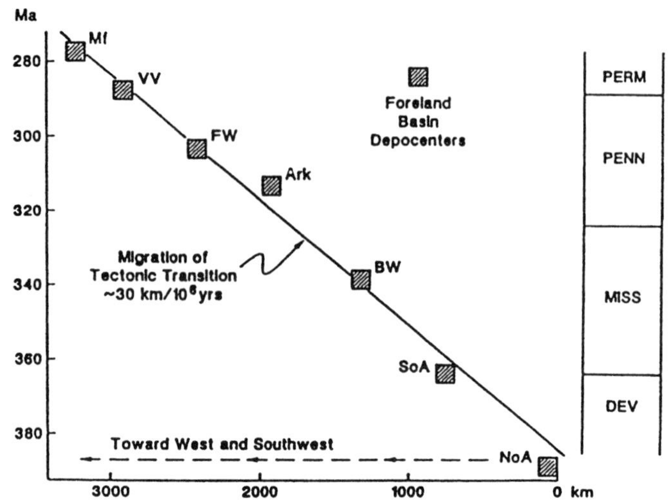

Figure 9. Approximate ages (plotted boxes) of oldest post-suture clastic strata in foreland basins along southeastern flank of North American craton (adjacent to Appalachian-Ouachita orogen): Mf, Marfa (Luff and Pearson, 1988); VV, Val Verde (Wuellner et al., 1986); FW, Fort Worth (T.J. Bornhorst, 1977, personal commun.); Ark, Arkoma (Houseknecht, 1986); BW, Black Warrior (Thomas, 1988); SoA and NoA, southern and northern Appalachian (Milici and deWitt, 1988). Indicated mean rate of migration of "flysch-molasse" transition is approximately 30 km/m.y. as Appalachian-Ouachita belt evolved. Modified after unpublished diagram by T.J.Bornhorst. From Ingersoll et al. (1995).

across the Longmen Shan fold-thrust belt (Fig. 10) (Şengör, 1984; Zhang et al., 1984; Klimetz, 1985), which evolved in Late Triassic time as a proforeland system during closure of the Songpan-Ganzi ocean basin (Huang and Chen, 1987; Lu et al., 1990).

The structure and timing of assembly of these continental blocks remain controversial. As the Songpan-Ganzi Complex is bounded on two sides by the north China block and south China block, and wedges out eastward between them (Fig. 10), it is the history of the collision of the north China block and south China block that is most significant. Conclusions regarding timing of this collision have revolved largely around the age and distribution of ophiolitic, fold-thrust, and magmatic belts of inferred subduction origin; the timing, nature, and distribution of metamorphic rocks; and paleomagnetic data. Disparate views have arisen about the closing of the ocean between the north China block and the south China block: middle Paleozoic (Mattauer et al., 1985) and late Paleozoic suturing (Zhang et al., 1984) have been advocated largely on geologic/geochronologic grounds, whereas Triassic closing has been inferred on the strength of paleomagnetic and geologic arguments (e.g., McElhinney et al., 1981; Klimetz, 1983; Şengör, 1984).

These opposing perspectives are partially reconciled in models (e.g., Watson et al., 1987; Eide, 1993; Yin and Nie, 1993; Zhou and Graham, 1993, 1996) that argue for diachronous ocean closure between the north China block and south China block from latest Paleozoic in easternmost China to Triassic in the west in the Qinling and Kunlun Mountains (Fig. 11). Diachronous closing of the north China block–south China block remnant ocean basin in late Paleozoic through early Mesozoic time is consistent with the following: (1) paleomagnetic data place the north China block and south China block apart in Permian time and together by Middle Jurassic time (McElhinny et al., 1981; Klimetz, 1983; Nie, 1991; Enkin et al., 1992); (2) coesite-and-diamond–bearing ultra-high-pressure metamorphic terranes within the nominal north China block–south China block suture zone in the Dabie Mountains are offset along the Tan-Lu strike-slip fault to the Shandong Peninsula (Wang and Liou, 1987, 1991; Yin and Nie, 1993; Nie et al., 1994); (3) strike-slip offset along the Tan-Lu fault is as great as 540 km, but dies to nothing within the south China block just south of the Dabie Mountains (Okay and Şengör, 1992; Yin and Nie, 1993); (4) metamorphic

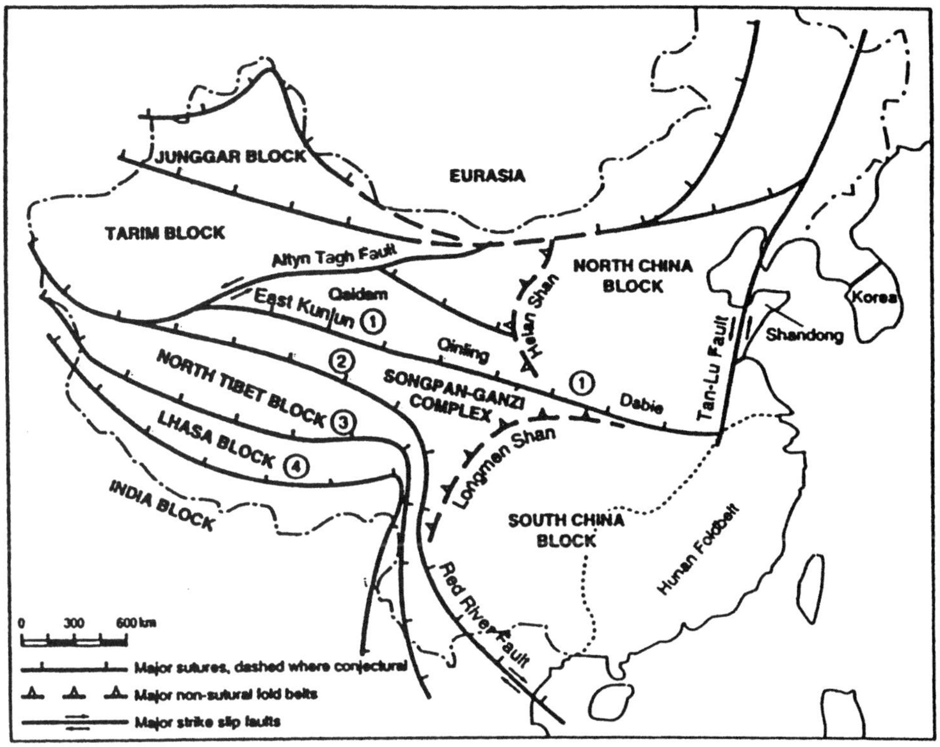

Figure 10. Selected tectonic features of China, emphasizing major accreted blocks, sutures between those blocks, and fault and fold systems discussed in text. Note triangular Songpan-Ganzi Complex, interpreted here as accreted fill of Triassic remnant ocean basin, surrounded by North China, South China, and North Tibet blocks. Sutures: 1—Kunlun-Qinling-Dabie, 2—Jinsha River, 3—Bangong Lake, and 4—Yarlung River. Map modified and simplified from Watson et al. (1987).

grade and probable age of metamorphism decrease from the Dabie Mountains westward toward the Qinling Mountains (Wang and Liou, 1991; Eide, 1993; Yin and Nie, 1993); (5) uplift cooling ages are only 5–35 m.y. younger than peak ultra-high-pressure metamorphism (Eide et al., 1992; Eide, 1993; Nie et al., 1994); (6) there was no early Mesozoic magmatic arc along the Dabie segment of the north China block (e.g., Zhang et al., 1985); (7) Songpan-Ganzi flysch is absent from the Dabie suture region (Zhang et al., 1984, 1985), although Hsu et al. (1987) noted that presumed lower Paleozoic metaflysch within the Qinling segment may be much younger; and (8) initiation of rapid clastic sedimentation in the south China block proforeland basins was younger westward (Yin and Nie, 1993).

Yin and Nie's (1993) and Zhou and Graham's (1993, 1996) hypothesis of diachronous ocean closure accounts for all of these phenomena and suggests that diachroneity of closure was related to irregularity of the northern intraplate margin of the south China block as it collided with the smoothly arcuate convergent margin of the north China block (Fig. 11) (i.e., Dewey and Burke, 1974; Graham et al., 1975; Şengör, 1976; Dewey, 1977; Ingersoll et al., 1995). In this scenario, a northeastern peninsula of the south China block impacted the north China block convergent margin in the east during the Permian in Shandong and the Early Triassic in the Dabie region farther west, driving a crustal wedge that produced ultra-high-pressure metamorphism, collisional strike-slip faulting, and thrusting analogous to the effects of the late Cenozoic collision of India and Asia (e.g., Molnar and Tapponier, 1975, 1977). The north China block–south China block collision elevated and erosionally exhumed high-grade metamorphic terranes; rapid erosion was enhanced by the tropical humid climate that prevailed by Late Triassic time (Wang, 1985; Huang and Chen, 1987). The diachronous collision also established the Songpan-Ganzi remnant ocean basin to the west as a receptacle for detritus eroded from the collision orogen (Fig. 11), which itself was deformed by continued ocean closure in the latest Triassic. The character, age, and origins of the rocks of the Songpan-Ganzi Complex are poorly known (Huang and Chen, 1987) due to geographic isolation, complex structure, and monotonous bedding style, but the term "flysch" has been widely applied to sedimentary elements of the complex (e.g., Wang, 1985). The Songpan-Ganzi flysch is extensively folded, faulted, and metamorphosed to varying degrees (Yang et al., 1986; Zhang et al., 1989). Reliable estimates of shortening across the complex are not available.

Within the Songpan-Ganzi Complex, flysch overlies Carboniferous pelagic limestone, which, in turn, overlies oceanic basement (Zhou and Graham, 1993, 1996). The stratigraphic column in the western Qinling Mountains generally consists of Lower Triassic pelagic carbonates and carbonate turbidites overlain by Middle Triassic siliciclastic turbidites (Zhou, 1987). The most striking deposits in the Middle Triassic section are olistostromes containing large blocks of carbonate-platform facies that were presumably derived from carbonate-dominated margins (Zhang et al., 1984; Zhou, 1987). Many blocks contain Devonian–Permian faunas that clearly predate deposition of the Triassic sediment-

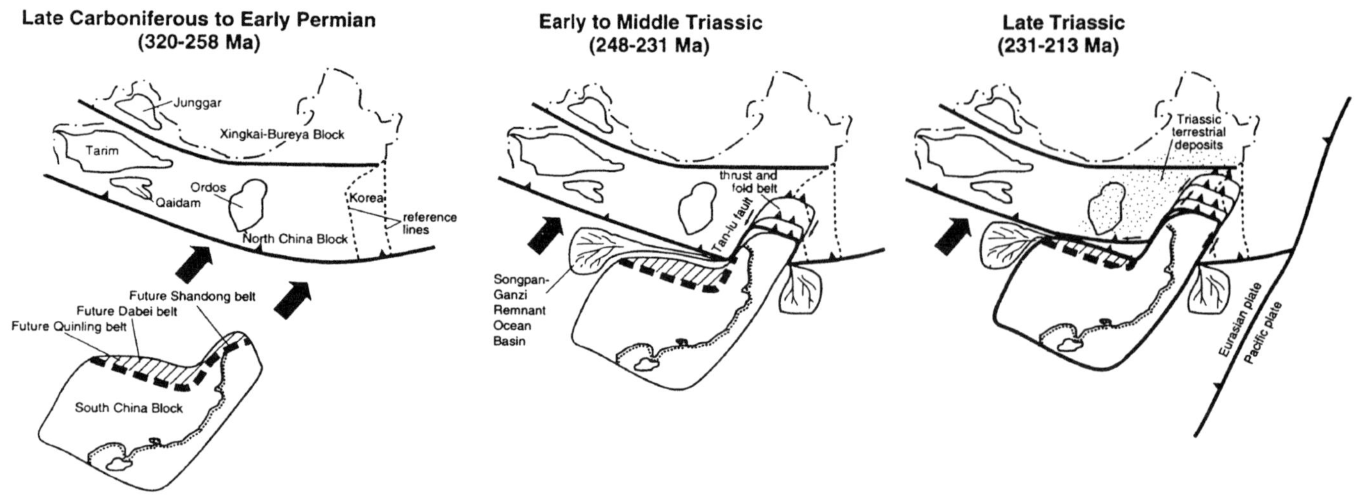

Figure 11. Development of Songpan-Ganzi remnant ocean basin and its flysch fill as a consequence of diachronous collision of North China and South China blocks. Simplified and modified slightly from Yin and Nie (1993).

gravity-flow deposits that contain them (Zhang et al., 1984; Zhou, 1987). Elsewhere in the Songpan-Ganzi Complex, the flysch sequence ranges into the Upper Triassic (Huang and Chen, 1987; Ren et al., 1987) and is overlain with angular unconformity by less deformed, nonmarine Jurassic strata (Huang and Chen, 1987). The total thickness of the section is poorly known, with intact sections greater than 5 km thick in the Qinling Mountains (Zhou, 1987), and total thickness variably estimated from 7 km (Yang et al., 1986) to 20 km (Huang and Chen, 1987).

Sources for Songpan-Ganzi detritus are poorly documented. Huang and Chen (1987), for instance, suggested Qinling, Kunlun, and Qilian sources, but stated that in view of the massive volume of the flysch, the source "remains an enigma." Şengör (Figure 16 *in* Şengör, 1984; Şengör and Okurogullari, 1991) suggested that the Songpan-Ganzi Complex accumulated as an accretionary wedge of the eastern Kunlun arc segment of the north China block during head-on subduction, bathymetrically isolated from sediment sources in the region of the north China block-south China block collisional orogen.

In contrast, reconstructions invoking east-to-west diachronous collision of the north China block and south China block predict that the voluminous flysch of the Songpan-Ganzi was the fill of a remnant ocean basin lying along tectonic strike to the west of the region of early collision (Fig. 11) (Watson et al., 1987; Yin and Nie, 1993; Zhou and Graham, 1993, 1996; Nie et al., 1994; Chang, 2000). Several lines of evidence favor, or are consistent with, this interpretation. First, the Triassic age of the flysch matches the timing of high-pressure and ultra-high–pressure metamorphism (244–201 Ma; Eide, 1993) and uplift cooling (230–195 Ma; Eide et al., 1992) of the Dabie suture zone. Second, the current 2.2×10^5 km^2 outcrop area of the Songpan-Ganzi complex (Huang and Chen, 1987), even without palinspastic reconstruction, compares favorably with the area of the largest modern remnant-ocean-basin, submarine-fan system, the Bengal fan, at 2.8–3.0×10^6 km^2 (Emmel and Curray, 1985) (Fig. 8). Furthermore, the 2.2×10^6 km^3 volume of the Songpan-Ganzi flysch (estimated by Huang and Chen, 1987, assuming 10 km average thickness) accommodates the 1.6×10^6km^3 of material estimated by Okay and Şengör (1992) to have been eroded from the Dabie ultra-high–pressure terrane during the Triassic–Jurassic. Some of the sediment eroded from the collision orogen may have been transported to the Pacific Basin (Okay and Şengör, 1992; Yin and Nie, 1993), and some is sequestered in Triassic–Jurassic nonmarine and shallow-marine siliciclastic sediments that supplanted carbonate deposition on the north China block–south China block platforms diachronously from east to west (Yin and Nie, 1993), but most detritus presumably passed westward through these forelands to the Songpan-Ganzi remnant ocean basin (Fig. 11).

Sparse detrital compositional data provide support for a source in the collision belt to the east for at least some of the Songpan-Ganzi turbidites. In general, Lower Triassic pelagic carbonate strata are overlain by carbonate turbidites, consistent with proximity to low-latitude carbonate–dominated continental margins. Olistostromes containing blocks of Carboniferous platform facies are interlayered with, and overlain by, siliciclastic turbidites. Our petrographic data confirm the arkosic composition of turbidite sandstones reported by Zhou (1987) from the Qinling area. Thus, the upward variation in sandstone compositions from carbonatoclastic to plutoniclastic may reflect unroofing of the Dabie suture region. Arkosic compositions are not typical of remnant ocean-basin sandstones (Graham et al., 1976; Dickinson and Suczek, 1979), but in this case, they are consistent with exposure of mid-crustal rocks during the Triassic in the focal region of collisional uplift (Fig. 11) (also, see Ingersoll and Suczek, 1979; Suczek and Ingersoll, 1985). It should also be noted that volcanic sandstones with possible Kunlun arc sources have been described, but not documented, from the western Songpan-Ganzi Complex (Zou et al., 1984). Paleocurrent and facies data are needed to document the westerly dispersal pathway that we infer from the Dabie suture to the Songpan-Ganzi remnant ocean basin. At present, paleocurrent data are available only from the western Qinling Mountains (Zhou, 1987), where they are directed toward the west and southwest, consistent with derivation of detritus from the north China block and the suture zone.

The timing of closure and mode of deformation of the Songpan-Ganzi remnant ocean basin are known generally, but not well documented in detail. Some deformation of the Songpan-Ganzi flysch was related to ongoing accretion at the southern margin of Eurasia (e.g., Şengör and Okurogullari, 1991). However, principal deformation of the flysch must have been associated with terminal suturing of the Songpan-Ganzi ocean, which is constrained to have occurred between deposition of the Upper Triassic deformed flysch and the Jurassic nonmarine coal-bearing molasse deposits that overlie the flysch with angular unconformity in many locations (Klimetz, 1983; Şengör and Hsu, 1984; Wang, 1985; Zhou and Graham, 1993, 1996). This deformation reflects not only the east-to-west diachronous closing of the remnant ocean basin, but later in the Triassic, the approach and collision of island arcs and the north Tibet terrane from the southwest (Fig. 10). Thus, incorporated by middle Mesozoic into the growing collage of terranes along the southern margin of Eurasia, the collapsed Songpan-Ganzi basin was redeformed and structurally reorganized by folding, thrusting, and strike-slip faulting during subsequent collisional events in the late Mesozoic, culminating in the collision of the Indian subcontinent and the Himalayan orogeny (e.g., Watson et al., 1987).

Other Ancient Examples

Examples of other ancient remnant ocean basins discussed by Ingersoll et al. (1995) include: (1) the Southern Uplands of Scotland, interpreted as an accretionary mass resulting from northward subduction of Iapetus and culminating in suturing in the Devonian (Mitchell and McKerrow, 1975; Mitchell, 1978; Leggett et al., 1982, 1983); (2) the Paleozoic Lachlan fold belt of southeastern Australia (Fergusson and Coney, 1992); (3) the Jurassic Mariposa Formation related to the Nevadan orogeny in California (Schweickert and Cowan, 1975; Ingersoll and Schweickert, 1986; Ingersoll, 2000); (4) the Siluro-Devonian Acadian orogeny in northern New England (Bradley, 1983); (5) the Cretaceous-Paleogene remnant ocean basin of western Iran (Cherven, 1986); (6) many of the flysch units of the Alps and Carpathians (e.g., Schwab, 1981; Homewood and Caron, 1982; Homewood, 1983; Caron et al., 1989); (7) the Paleogene Liguride Complex deposited between the European margin (represented by Corsica, Sardinia, and Calabria) and the Apulia margin (Adriatic foreland) (Boccaletti et al., 1990; Critelli, 1993); and (8) a small remnant ocean basin (which originated as a backarc basin) closed by transpression that has progressed from northwest to southeast in the northeastern Caribbean (Heubeck et al., 1991).

IMPLICATIONS FOR PALEOTECTONIC RECONSTRUCTIONS

Our actualistic model for the evolution of remnant ocean basins (Fig. 1) has significant implications for paleotectonic reconstruction and modeling of orogenic sedimentary basins. Most importantly, remnant ocean basins are inevitably destroyed during suturing, and their former presence must be inferred. Structurally complex accretionary wedges are all that remain following suturing; these accretionary wedges commonly contain deep-water turbidites (flysch) that have been thrust over intraplate continental margins or other accretionary wedges. The emplacement of these accretionary wedges usually results in tectonic flexure of adjoining continents and the formation of proforeland basins. The allochthonous nature of highly deformed flysch renders detailed palinspastic reconstruction difficult. Intricate structural relations must be determined in conjunction with sedimentologic and paleoecologic studies; for example, after approximately 200 years of study, the Alpine system is finally revealing much of its history (e.g., Homewood and Caron, 1982; Homewood, 1983; Caron et al., 1989).

Geosynclinal models for flysch and molasse (e.g., Aubouin, 1965) were essentially one-dimensional. The ophiolitic suite was overlain by the flysch facies, which was overlain by the molasse facies (Fig. 12). We now interpret the same vertical sequence, which is rarely completely preserved, as the result of seafloor spreading to form oceanic crust, followed by oceanic sedimentation, flysch sedimentation in a remnant ocean basin, and collision to form a proforeland basin with molasse sedimentation. This one-dimensional view of sedimentation related to sutures is useful as a first approximation, but it fails to account for the four-dimensional nature of these complex events. Especially significant are the time-transgressive nature of suturing (Figs. 2 and 9) and the lateral derivation of sediment derived from previously uplifted suture zones (Fig. 1). Two-dimensional models for foreland loading (e.g., Stockmal et al., 1986) provide important insights regarding flexure of continental margins as suturing occurs, but they fail to account for the dominant longitudinal flux of sediment. Useful mechanical models for the complex interactions of suturing, erosion, sedimentation, and accretion will require far more sophisticated treatment.

The most ambiguous aspect of our model (Fig. 1) is the nature of the deltaic transition. Modern deltaic systems, such as the Ganges-Brahmaputra, Indus, and Markham deltas, are relatively clear examples of tectonic and environmental transitions. In contrast, ancient examples of transitional deltas constitute volumetrically minor proportions of orogenic sequences (Fig. 12). The preservation potential for deltaic sequences at the tectonic transition is generally low because deformation occurs almost immediately after deposition as the suture migrates. In contrast, post-suturing foreland deposits (molasse) are commonly preserved because they primarily accumulate beyond the deformation front. Remnant-ocean submarine-fan deposits (flysch) are typically highly deformed during suturing, but their immense volume assures that a significant mass will be preserved in accretionary orogens.

Sutures that form in sediment-starved settings (intraoceanic settings, in general, and possibly continental settings during high sea level) tend to have minor deltaic deposits. In fact, the scarcity of sediment in such systems may delay the sedimentologic transition from flysch to molasse beyond the tectonic transition from remnant ocean basin to proforeland basin. In this case, early deposits within the proforeland will be turbidites (e.g., Miall, 1995); ambiguity as to whether the terms "flysch" and "molasse" refer to sedimentologic or tectonic features results (c.f., Miall, 1984).

Additional ambiguity results from the transition along tectonic strike of unfilled oceanic trenches to submarine fans that cover trenches (e.g., Java–Sumatra–Bay of Bengal transition). Subduction complexes of arc-trench systems are transitional to accretionary wedges of suture zones. A usable distinction in modern examples is that subduction complexes form by accretion of distantly derived oceanic and turbidite sediments, and local arc-derived sediment, whereas accretionary wedges of suture zones form by accretion of longitudinally derived, recycled-orogenic sediments. These distinctions are more difficult to make in highly deformed ancient examples, but provenance studies are the most promising method for making the distinction (e.g., Graham et al., 1976; Dickinson and Suczek, 1979; Ingersoll and Suczek, 1979; Dickinson, 1985; Critelli et al., 1990; Critelli, 1993).

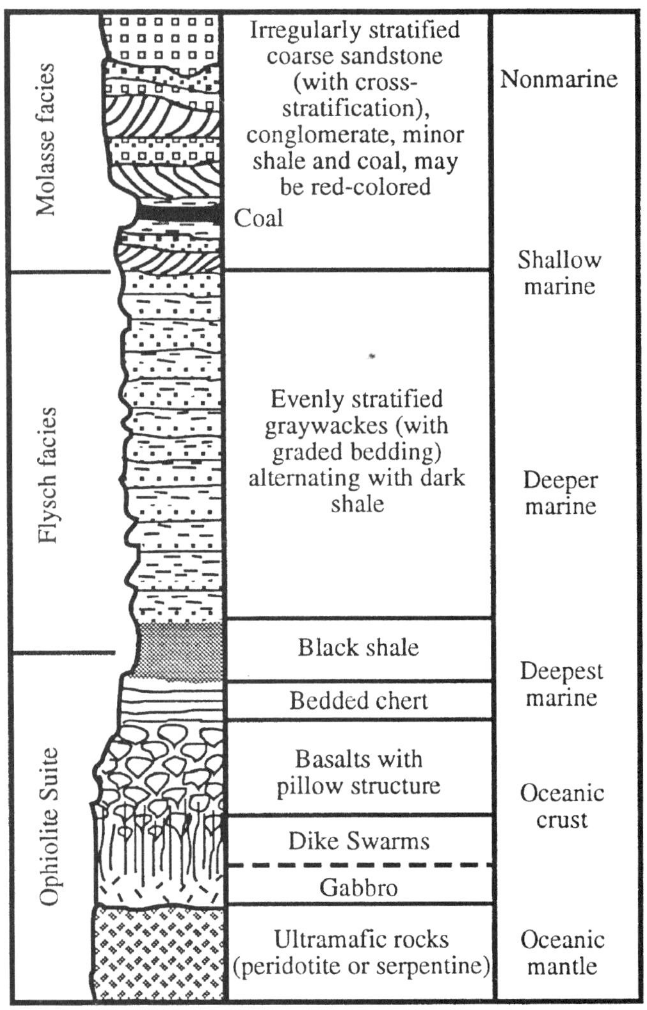

Figure 12. Idealized vertical sequence from oceanic igneous rocks and chert (ophiolite sequence) upward through hemipelagic to graywacke turbidite (flysch) strata to red clastic (molasse) deposits. Such successions are commonly seen in orogenic belts and represent progression from deep-marine to nonmarine conditions. We now interpret this succession as typically resulting from closing of remnant ocean basins, accretion of turbidites into suture zones, and creation of foreland basins adjacent to orogens. See text for discussion. Modified from Dott and Batten (1988).

CONCLUSIONS

Remnant ocean basins form during collisions between nonsubductable continental crustal elements of variable proportions. These elements include major and minor continents, small continental blocks, and intraoceanic magmatic arcs. At least one of the elements must include an arc-trench system, along which oceanic crust is subducted. Collisions usually involve irregular continental margins and oblique convergence, so that events are diachronous and complex; the result is uplift of orogens immediately adjacent to remnant ocean basins, which receive the bulk of their detritus.

Depositional systems of remnant ocean basins represent the largest sedimentary systems on Earth, both modern and ancient. Collisions involving one major continent and one moderate-sized continent, e.g., Asia and India, seem to produce the greatest volume of sediment, as a result of protracted collision and persistence of adjoining remnant oceans. Major sediment accumulations form rapidly at various stages during terminal suturing of supercontinents, but the remnant ocean basins are usually destroyed rapidly as suturing progresses. Collisions involving intraoceanic arcs produce less detritus.

Regardless of the dimensions of collision, most of the detritus eroded from collisional orogens accumulates in remnant ocean basins and is usually deformed in accretionary wedges during both oceanic subduction and terminal suturing. As a result, the sedimentary basins (floored by oceanic crust) are seldom preserved intact. Reconstruction of ancient remnant ocean basins is difficult, but it should be attempted because of their paleogeographic and paleotectonic significance. Geophysical models for the evolution of remnant ocean basins and their accretionary remains are nonexistent. Detailed structural, stratigraphic, sedimentologic, and petrologic studies are the primary methods through which remnant ocean basins must be reconstructed.

Much of Earth's continental crust has been created through the linked processes of uplift, erosion, deposition, accretion, and metamorphism during the evolution of collision orogens and remnant ocean basins. If we are to understand the development of sedimentary basins and the evolution and growth of continental crust in general, we must understand remnant ocean basins and the accretion of their sedimentary fill during crustal collision.

ACKNOWLEDGMENTS

We thank the following for help at various times concerning diverse aspects of this paper: T.J. Bornhorst, C.J. Busby, E.Z. Chang, S. Critelli, E. Garzanti, B.R. Haley, D.W. Houseknecht, E.F. McBride, W.R. Muehlberger, S. Nie, A.M.C. Şengör, C.G. Stone, W.A. Thomas, G.W. Viele, A. Yin, and D. Zhou. We thank A.W. Archer and M.A. Chan for organizing the Pardee Symposium and for inviting our participation. We also thank M.S. Hendrix and A.D. Miall for reviewing the manuscript.

REFERENCES CITED

Alam, M., 1989, Geology and depositional history of Cenozoic sediments of the Bengal basin of Bangladesh: Palaeogeography, Palaeoclimatology, Palaeoecology, v. 69, p. 125–139.

Arbenz, J.K., 1989, The Ouachita system, *in* Bally, A.W., and Palmer, A.R., eds., The geology of North America: An overview: Boulder, Colorado, Geological Society of America, Geology of North America, v. A, p. 371–396.

Aubouin, J., 1965, Geosynclines, *in* Developments in geotectonics 1: New York, Elsevier, 335 p.

Bartolini, C., Malesani, P.G., Manetti, P., and Wezel, F.C., 1975, Sedimentology and petrology of Quaternary sediments from the Hellenic Trench, Mediterranean Ridge and the Nile Cone from D.S.D.P., Leg 13, cores: Sedimentology, v. 22, p. 205–236.

Boccaletti, M., Ciaranfi, N., Cosentino, D., Deiana, G., Gelati, R., Lentini, F., Massari, F., Moratti, G., Pescatore, T., Ricci Lucchi, F., and Tortorici, L., 1990, Palinspastic restoration and paleogeographic reconstruction of the peri-Tyrrhenian area during the Neogene: Palaeogeography, Palaeoclimatology, Palaeoecology, v. 77, p. 41–50.

Bouma, A.H., Normark, W.R., and Barnes, N.E., eds., 1985, Submarine fans and related turbidite systems: New York, Springer–Verlag, 351 p.

Bowin, S., Lu, R.S, Lee, C.S., and Schouten, H., 1978, Plate convergence and accretion in the Taiwan Luzon region: American Association of Petroleum Geologists Bulletin, v. 62, p. 1645–1672.

Bowles, F.A., Ruddiman, W.F., and Jahn, W.H., 1978, Acoustic stratigraphy, structure, and depositional history of the Nicobar Fan, eastern Indian Ocean: Marine Geology, v. 26, p. 269–288.

Bradley, D.C., 1983, Tectonics of the Acadian orogeny in New England and adjacent Canada: Journal of Geology, v. 91, p. 381–400.

Brookfield, M.E., 1993, The Himalayan passive margin from Precambrian to Cretaceous times: Sedimentary Geology, v. 84, p. 1–35.

Burbank, D.W., Raynolds, R.G.H., and Johnson, G.D., 1986, Late Cenozoic tectonics and sedimentation in the north-western Himalayan foredeep, II, Eastern limb of the northwest syntaxis and regional synthesis: International Association of Sedimentologists Special Publication 8, p. 293–306.

Caron, C., Homewood, P., and Wildi, W., 1989, The original Swiss flysch: A reappraisal of the type deposits in the Swiss Prealps: Earth-Science Reviews, v. 26, p. 1–45.

Chai, B.H.T., 1972, Structure and evolution of Taiwan: American Journal of Science, v. 272, p. 389–442.

Chang, E.Z., 2000, Geology and tectonics of the Songpan-Ganzi fold belt, southwestern China: International Geology Review, v. 42, p. 813–831.

Cherven, V.B., 1986, Tethys-marginal sedimentary basins in western Iran: Geological Society of America Bulletin, v. 97, p. 516–522.

Clift, P.D., Shimizu, N., Layne, G.D., Blusztajn, J.S., Gaedicke, C., Schluter, H.-U., Clark, M.K., and Amjad, S., 2001, Development of the Indus fan and its significance for the erosional history of the Western Himalaya and Karakoram: Geological Society of America Bulletin, v. 113, p. 1039–1051.

Cloos, M., 1993, Lithospheric buoyancy and collisional orogenesis: Subduction of oceanic plateaus, continental margins, island arcs, spreading ridges, and seamounts: Geological Society of America Bulletin, v. 105, p. 715–737.

Cochran, J.R., 1990, Himalayan uplift, sea level, and the record of Bengal fan sedimentation at the ODP Leg 116 Sites: Proceedings of the Ocean Drilling Program, Scientific Results, v. 116, p. 397–414.

Copeland, P., and Harrison, T.M., 1990, Episodic rapid uplift in the Himalaya revealed by $^{40}Ar/^{39}Ar$ analysis of detrital K-feldspar and muscovite, Bengal fan: Geology, v. 18, p. 354–357.

Critelli, S., 1993, Sandstone detrital modes in the Paleogene Liguride Complex, accretionary wedge of the southern Apennines (Italy): Journal of Sedimentary Petrology, v. 63, p. 464–476.

Critelli, S., and Garzanti, E., 1994, Provenance of the Lower Tertiary Murree redbeds (Hazara-Kashmir syntaxis, northern Pakistan) and initial rising of the Himalayas: Sedimentary Geology, v. 89, p. 265–284.

Critelli, S., and Le Pera, E., 1998, Post-Oligocene sediment-dispersal systems and unroofing history of the Calabrian microplate, Italy: International Geology Review, v. 40, p. 609–637.

Critelli, S., DeRosa, R., and Platt, J.P., 1990, Sandstone detrital modes in the Makran accretionary wedge, southwest Pakistan: Implications for tectonic setting and long-distance turbidite transportation: Sedimentary Geology, v. 68, p. 241–260.

Crook, K.A.W., 1989, Suturing history of an allochthonous terrane at a modern plate boundary traced by flysch-to-molasse facies transitions: Sedimentary Geology, v. 61, p. 49–79.

Curray, J.R., 1991, Possible greenschist metamorphism at the base of a 22-km sedimentary section, Bay of Bengal: Geology, v. 19, p. 1097–1100.

Curray, J.R., and Moore, D.G., 1971, Growth of the Bengal deep-sea fan and denudation in the Himalayas: Geological Society of America Bulletin, v. 82, p. 563–572.

Curray, J.R., and Moore, D.G., 1974, Sedimentary and tectonic processes in the Bengal deep-sea fan and geosyncline, *in* Burk, C.A., and Drake, C.L., eds., The geology of continental margins: New York, Springer–Verlag, p. 617–627.

DeCelles, P.G., and Cavazza, W., 1999, A comparison of fluvial megafans in the Cordilleran (Upper Cretaceous) and modern Himalayan foreland basin systems: Geological Society of America Bulletin, v. 111, p. 1315–1334.

Dewey, J.F., 1977, Suture zone complexities: A review: Tectonophysics, v. 40, p. 53–67.

Dewey, J.F., and Burke, K., 1974, Hot spots and continental break-up: Implications for collisional orogeny: Geology, v. 2, p. 57–60.

Dewey, J.F., and Kidd, W.S.F., 1974, Continental collisions in the Appalachian-Caledonian orogenic belt: Variations related to complete and incomplete suturing: Geology, v. 2, p. 543–546.

Dewey, J.F., Cande, S., and Pitman III, W.C., 1989, Tectonic evolution of the India/Eurasia collision zone: Eclogae Geologicae Helvetiae, v. 82, p. 717–734.

Dickinson, W.R., 1976, Plate tectonic evolution of sedimentary basins: American Association of Petroleum Geologists Continuing Education Course Notes Series 1, 62 p.

Dickinson, W.R., 1985, Interpreting provenance relations from detrital modes of sandstones, *in* Zuffa, G.G., ed., Provenance of arenites: Dordrecht, D. Reidel Publishing, p. 333–361.

Dickinson, W.R., 1988, Provenance and sediment dispersal in relation to paleotectonics and paleogeography of sedimentary basins, *in* Kleinsehn, K.L., and Paola, C., eds., New perspectives in basin analysis: New York, Springer-Verlag, p. 3–25.

Dickinson, W.R., and Lawton, T.F., 2001, Carboniferous to Cretaceous assembly and fragmentation of Mexico: Geological Society of America Bulletin, v. 113, p. 1142–1160.

Dickinson, W.R., and Suczek, C.A., 1979, Plate tectonics and sandstone compositions: American Association of Petroleum Geologists Bulletin, v. 63, p. 2164–2182.

Dickinson, W.R., Armin, R.A., Beckvar, N., Goodlin, T.C., Janecke, S.U., Mark, R.A., Norris, R.D., Radel, G., and Wortman, A.A., 1987, Geohistory analysis of rates of sediment accumulation and subsidence for selected California basins, *in* Ingersoll, R.V., and Ernst, W.G., eds., Cenozoic basin development of coastal California, Rubey Volume VI: Englewood Cliffs, Prentice Hall, p. 1–23.

Dott Jr., R.H., and Batten, R.L., 1988, Evolution of the earth, 4th ed.: New York, McGraw-Hill, 643 p.

Eide, E.A., 1993, Petrology, geochemistry, and structure of high-pressure metamorphic rocks in Hubei Province, east central China, and their relationship to continental collision [Ph.D. Thesis]: Stanford, Stanford University, 235 p.

Eide, E.A., Liou, J.G., and McWilliams, M.D., 1992, Continent–continent collision and the tectonic history of the eastern Qinling–Dabie orogenic belt, east–central China: Eos (Transactions, American Geophysical Union), v. 73, p. 653.

Emmel, F.J., and Curray, J.R., 1985, Bengal fan, Indian Ocean, *in* Bouma, A.H., Normark, W.R., and Barnes, N.E., eds., Submarine fans and related turbidite systems: New York, Springer–Verlag, p. 107–112.

Enkin, R.J., Yang, Z., Chen, Y., and Courtillot, V., 1992, Paleomagnetic constraints on the geodynamic history of the major blocks of China from the Permian to the present: Journal of Geophysical Research, v. 97, p. 13953–13989.

Ethington, R.L., Finney, S.C., and Repetski, J.E., 1989, Biostratigraphy of the Paleozoic rocks of the Ouachita orogen, Arkansas, Oklahoma, west Texas, *in* Hatcher Jr., R.D., Thomas W.A., and Viele, G.W., eds., The Appalachian-Ouachita orogen in the United States: Boulder, Geological Society of America, Geology of North America, v. F-2, p. 563–574.

Farhoudi, G., and Karig, D.E., 1977, Makran of Iran and Pakistan as an active arc system: Geology, v. 5, p. 664–668.

Fergusson, C.L., and Coney, P.J., 1992, Implications of a Bengal fan-type deposit in the Paleozoic Lachlan fold belt of southeastern Australia: Geology, v. 20, p. 1047–1049.

Flores, R.M., 1977, Marginal marine deposits in the upper Tesnus Formation (Carboniferous), Marathon Basin, Texas: Journal of Sedimentary Petrology, v. 47, p. 582–592.

Gaetani, M., and Garzanti, E., 1991, Multicyclic history of the northern India continental margin (northwestern Himalaya): American Association of Petroleum Geologists Bulletin, v. 75, p. 1427–1446.

Garzanti, E., Baud, A., and Mascle, G., 1987, Sedimentary record of the northward flight of India and its collision with Eurasia (Ladakh Himalaya, India): Geodinamica Acta, v. 1, p. 297–312.

Garzanti, E., Critelli, S., and Ingersoll, R.V., 1996, Paleogeographic and paleotectonic evolution of the Himalayan Range as reflected by detrital modes of Tertiary sandstones and modern sands (Indus transect, India and Pakistan): Geological Society of America Bulletin, v. 108, p. 631–642.

Gleason, J.D., Patchett, P.J., Dickinson, W.R., and Ruiz, J., 1994, Nd isotopes link Ouachita turbidites to Appalachian sources: Geology, v. 22, p. 347–350.

Gleason, J.D., Patchett, P.J., Dickinson, W.R., and Ruiz, J., 1995, Nd isotopic constraints on sediment sources of the Ouachita-Marathon fold belt: Geological Society of America Bulletin, v. 107, p. 1192–1210.

Goebel, K.A., 1991, Paleogeographic setting of Late Devonian to Early Mississippian transition from passive to collisional margin, Antler foreland, eastern Nevada and western Utah, *in* Cooper, J.D., and Stevens, C.H., eds., Paleozoic paleogeography of the western United States II: Los Angeles, Pacific Section, Society of Economic Paleontologists and Mineralogists, Book 67, p. 401–418.

Goodbred Jr., S.L., and Kuehl, S.A., 2000, Enormous Ganges-Brahmaputra sediment discharge during strengthened Early Holocene monsoon: Geology, v. 28, p. 1083–1086.

Graham, S.A., Dickinson, W.R., and Ingersoll, R.V., 1975, Himalayan–Bengal model for flysch dispersal in Appalachian–Ouachita system: Geological Society of America Bulletin, v. 86, p. 273–286.

Graham, S.A., Ingersoll, R.V., and Dickinson, W.R., 1976, Common provenance for lithic grains in Carboniferous sandstones from Ouachita Mountains and Black Warrior Basin: Journal of Sedimentary Petrology, v. 46, p. 620–632.

Hamilton, W., 1979, Tectonics of the Indonesian region: U.S. Geological Survey Professional Paper 1078, 345 p.

Harrison, T.M., Copeland, P., Hall, S.A., Quade, J., Burner, S., Ojha, T.P., and Kidd, W.S.F., 1993, Isotopic preservation of Himalayan/Tibetan uplift, denudation, and climatic histories of two molasse deposits: Journal of Geology, v. 101, p. 157–175.

Heubeck, C., Mann, P., Dolan, J., and Monechi, S., 1991, Diachronous uplift and recycling of sedimentary basins during Cenozoic tectonic transpression, northeastern Caribbean plate margin: Sedimentary Geology, v. 70, p. 1–32.

Hoffman, P.F., and Grotzinger, J.P., 1993, Orographic precipitation, erosional unloading, and tectonic style: Geology, v. 21, p. 195–198.

Homewood, P., 1983, Palaeogeography of Alpine flysch: Palaeogeography, Palaeoclimatology, Palaeoecology, v. 44, p. 169–184.

Homewood, P., and Caron, C., 1982, Flysch of the western Alps, *in* Hsu, K.J., ed., Mountain building processes: New York, Academic Press, p. 157–168.

Houseknecht, D.W., 1986, Evolution from passive margin to foreland basin: The Atoka Formation of the Arkoma Basin, south-central USA.: International Association of Sedimentologists Special Publication 8, p. 327–345.

Hsu, K.J., Wang, Q., Li, J., Zhou, D., and Sun, S., 1987, Tectonic evolution of Qinling Mountains, China: Eclogae Geologicae Helvetiae, v. 80, p. 735–752.

Huang, C.-Y., Shyu, C.-T., Lin, S.B., Lee, T.-Q., and Sheu, D.D., 1992, Marine geology in the arc-continent collision zone off southeastern Taiwan: Implications for Late Neogene evolution of the Coastal Range: Marine Geology, v. 107, p. 183–212.

Huang, J., and Chen, B., 1987, The evolution of the Tethys in China and adjacent regions: Geologic Publishing House, Beijing, 109 p.

Imoto, N., and McBride, E.F., 1990, Volcanism recorded in the Tesnus Formation, Marathon uplift, Texas, *in* LaRoche, T.M., and Higgins, L., eds., Marathon thrust belt: Structure, stratigraphy, and hydrocarbon potential [Field Seminar]: Midland, West Texas Geological Society, p. 93–98.

Ingersoll, R.V., 2000, Models for origin and emplacement of Jurassic ophiolites of northern California, *in* Dilek, Y., Moores, E.M., Elthon, D., and Nicolas, A., eds., Ophiolites and oceanic crust: New insights from field studies and the Ocean Drilling Program: Geological Society of America Special Paper 349, p. 395–402.

Ingersoll, R.V., and Busby, C.J., 1995, Tectonics of sedimentary basins, *in* Busby, C.J., and Ingersoll, R.V., eds., Tectonics of sedimentary basins: Cambridge, Blackwell Science, p. 1–51.

Ingersoll, R.V., and Schweickert, R.A., 1986, A plate-tectonic model for Late Jurassic ophiolite genesis, Nevadan orogeny and forearc initiation, northern California: Tectonics, v. 5, p. 901–912.

Ingersoll, R.V., and Suczek, C.A., 1979, Petrology and provenance of Neogene sand from Nicobar and Bengal fans, DSDP sites 2ll and 218: Journal of Sedimentary Petrology, v. 49, p. 1217–1228.

Ingersoll, R.V., Graham, S.A., and Dickinson, W.R., 1995, Remnant ocean basins, *in* Busby, C.J., and Ingersoll, R.V., eds., Tectonics of sedimentary basins: Cambridge, Blackwell, p. 363–391.

Ingersoll, R.V., Graham, S.A., and Dickinson, W.R., 2000, Remnant-ocean turbidite fans: Largest sedimentary systems on Earth: Geological Society of America Abstracts with Programs, v. 32, no. 7, p. A-312.

Ji, X., and Coney, P.J., 1985, Accreted terranes in China, *in* Howell, D.G., ed., Tectonostratigraphic terranes of the Circum-Pacific region: Circum-Pacific Council for Energy and Mineral Resources, Earth Sciences Series 1, p. 349–361.

Johnson, B.D., Powell, C.McA., and Veevers, J.J., 1976, Spreading history of the eastern Indian Ocean and greater India's northward flight from Antarctica and Australia: Geological Society of America Bulletin, v. 87, p. 1560–1566.

Johnson, G.D., Raynolds, R.G.H., and Burbank, D.W., 1986, Late Cenozoic tectonics and sedimentation in the north-western Himalayan foredeep: I. Thrust ramping and associated deformation in the Potwar region: International Association of Sedimentologists Special Publication 8, p. 273–291.

Klimetz, M.P., 1983, Speculations on the Mesozoic plate tectonic evolution of eastern China: Tectonics, v. 2, p. 139–166.

Klimetz, M.P., 1985, An outline of the plate tectonics of China: Discussion: Geological Society of America Bulletin, v. 96, p. 407.

Klootwijk, C.T., Gee, J.S., Peirce, J.W., and Smith, G.M., 1992, Neogene evolution of the Himalayan-Tibetan region: Constraints from ODP Site 758, northern Ninetyeast Ridge; Bearing on climatic change: Palaeogeography, Palaeoclimatology, Palaeoecology, v. 95, p. 95–110.

Kolla, V., and Coumes, F., 1987, Morphology, internal structure, seismic stratigraphy, and sedimentation of Indus fan: American Association of Petroleum Geologists Bulletin, v. 71, p. 650–677.

Kuehl, S.A., Hariu, T.M., and Moore, W.S., 1989, Shelf sedimentation off the Ganges-Brahmaputra river system: Evidence for sediment bypassing to the Bengal fan: Geology, v. 17, p. 1132–1135.

Leggett, J.K., McKerrow, W.S., and Casey, D.M., 1982, The anatomy of a Lower Palaeozoic accretionary forearc: The Southern Uplands of Scotland: Geological Society of London Special Publication 10, p. 495–520.

Leggett, J.K., McKerrow, W.S., and Soper, N.J., 1983, A model for the crustal evolution of southern Scotland: Tectonics, v. 2, p. 187–210.

Lindsay, J.F., Holliday, D.W., and Hulbert, A.G., 1991, Sequence stratigraphy and the evolution of the Ganges-Brahmaputra delta complex: American Association of Petroleum Geologists Bulletin, v. 75, p. 1233–1254.

Link, M.H., and Roberts, M.J., 1986, Pennsylvanian paleogeography for the Ozarks, Arkoma, and Ouachita Basins in east-central Arkansas, *in* Stone, C.G., and Haley, B.R., eds., Sedimentary and igneous rocks of the Ouachita Mountains of Arkansas, Part 2, Guidebook 86-3: Little Rock, Arkansas Geological Commission, p. 37–60.

Lowe, D.R., 1985, Ouachita trough, part of a Cambrian failed rift system: Geology, v. 13, p. 790–793.

Lowe, D.R., 1989, Stratigraphy, sedimentology, and depositional setting of pre-orogenic rocks of the Ouachita Mountains, Arkansas and Oklahoma, *in* Hatcher Jr., R.D., Thomas, W.A., and Viele, G.W., eds., The Appalachian-Ouachita orogen in the United States: Boulder, Geological Society of America, Geology of North America, v. F-2, p. 575–590.

Lu, H., Dong, H., Deng, X., Li, P., and Yan, J., 1990, Preliminary study of the terranes in west Sichuan, *in* Wiley, T.J., Howell, D.G., and Wong, F.L., eds., Terrane analysis of China and the Pacific rim: Circum-Pacific Council for Energy and Mineral Resources, Earth Sciences Series 13, p. 261–267.

Luff, G.C., and Pearson, B.T., 1988, The Marfa Basin, *in* Sloss, L.L., ed., Sedimentary cover—North American craton: Boulder, Geological Society of America, Geology of North America, v. D-2, p. 295–299.

Mattauer, M., Matte, P., Malavieille, J., Tapponnier, P., Maluski, H., Xu, Z.Q., Lu, Y.L., and Tan, Y.Q., 1985, Tectonics of the Qinling belt: Build-up and evolution of eastern Asia: Nature, v. 317, p. 496–500.

McBride, E.F., 1970, Flysch sedimentation in the Marathon region, west Texas: Geological Association of Canada Special Paper 7, p. 67–83.

McBride, E.F., 1989, Stratigraphy and sedimentary history of pre-Permian Paleozoic rocks of the Marathon uplift, *in* Hatcher, R.D., Jr., Thomas, W.A., and Viele, G.W., eds., The Appalachian-Ouachita orogen in the United States: Boulder, Geological Society of America, The Geology of North America, v. F-2, p. 603–620.

McElhinny, M.W., Embleton, B.J.J., Ma, X.H., and Zhang, Z.K., 1981, Fragmentation of Asia in the Permian: Nature, v. 293, p. 212–216.

McKenzie, D., and Sclater, J.G., 1971, The evolution of the Indian Ocean since the Late Cretaceous: Geophysical Journal of the Royal Astronomical Society, v. 24, p. 437–528.

Miall, A.D., 1984, Flysch and molasse: The elusive models: Annales Societatis Geologorum Poloniae, v. 54, p. 281–291.

Miall, A.D., 1995, Collision-related foreland basins, *in* Busby, C.J., and Ingersoll, R.V., eds., Tectonics of sedimentary basins: Cambridge, Blackwell, p. 393–424.

Milici, R.C., and deWitt, W., Jr., 1988, The Appalachian Basin, *in* Sloss, L.L., ed., Sedimentary cover—North American craton: U.S.: Boulder, Geological Society of America, Geology of North America, v. D-2, p. 427–469.

Mitchell, A.H.G., 1978, The Grampian orogeny in Scotland: arc-continent collision and polarity reversal: Journal of Geology, v. 86, p. 643–646.

Mitchell, A.H.G., and McKerrow, W.S., 1975, Analogous evolution of the Burma orogen and the Scottish Caledonides: Geological Society of America Bulletin, v. 86, p. 305–315.

Moiola, R.J., and Shanmugam, G., 1984, Submarine fan sedimentation, Ouachita Mountains, Arkansas and Oklahoma: Gulf Coast Association of Geological Societies Transactions, v. 34, p. 175–182.

Molnar, P., and Tapponnier, P., 1975, Cenozoic tectonics of Asia: Effects of a continental collision: Science, v. 189, p. 419–426.

Molnar, P., and Tapponnier, P., 1977, Relation of the tectonics of eastern China to the India-Eurasia collision: Applications of slip-line field theory to large-scale continental tectonics: Geology, v. 5, p. 212–216.

Moore, G.F., 1979, Petrography of subduction zone sandstones from Nias Island, Indonesia: Journal of Sedimentary Petrology, v. 49, p. 71–84.

Moore, G.F., and Silver, E.A., 1983, Collision processes in the northern Molucca Sea: American Geophysical Union Geophysical Monograph 27, p. 360–372.

Moore, G.F., Kadarisman, D., Evans, C.A., and Hawkins, J.W., 1981, Geology of the Talaud Islands, Molucca Sea collision zone, northeast Indonesia: Journal of Structural Geology, v. 3, p. 467–475.

Moore, G.F., Curray, J.R., and Emmel, F.J., 1982, Sedimentation in the Sunda Trench and forearc region: Geological Society [London] Special Publication 10, p. 245–258.

Morris, R.C., 1974, Sedimentary and tectonic history of the Ouachita Mountains: Society of Economic Paleontologists and Mineralogists Special Publication 22, p. 120–142.

Morris, R.C., 1989, Stratigraphy and sedimentary history of post-Arkansas Novaculite Carboniferous rocks of the Ouachita Mountains, *in* Hatcher Jr., R.D., Thomas, W.A., and Viele, G.W., eds., The Appalachian-Ouachita orogen in the United States: Boulder, Geological Society of America, Geology of North America, v. F-2, p. 591–602.

Nie, S., 1991, Paleoclimatic and paleomagnetic constraints on the Paleozoic reconstruction of South China, North China, and Tarim: Tectonophysics, v. 196, p. 279–308.

Nie, S., Yin, A., Rowley, D.B., and Jin, Y., 1994, Exhumation of the Dabie Shan ultra-high-pressure rocks and accumulation of the Songpan-Ganzi flysch sequence, central China: Geology, v. 22, p. 999–1002.

Niem, A.R., 1976, Patterns of flysch deposition and deep-sea fans in the lower Stanley Group (Mississippian), Ouachita Mountains, Oklahoma and Arkansas: Journal of Sedimentary Petrology, v. 46, p. 633–646.

Niem, A.R., 1977, Mississippian pyroclastic flow and ash-fall deposits in the deep-marine Ouachita flysch basin, Oklahoma and Arkansas: Geological Society of America Bulletin, v. 88, p. 49–61.

Norton, I.O., and Sclater, J.G., 1979, A model for the evolution of the Indian Ocean and the breakup of Gondwanaland: Journal of Geophysical Research, v. 84, p. 6803–6830.

Okay, A.I., and Şengör, A.M.C., 1992, Evidence for intracontinental thrust–related exhumation of the ultra-high–pressure rocks in China: Geology, v. 20, p. 411–414.

Palmer, A.R., 1983, The Decade of North American Geology 1983 Time Scale: Geology, v. 11, p. 503–504.

Parsons, B., and Sclater, J.G., 1977, An analysis of the variation of ocean floor bathymetry and heat flow with age: Journal of Geophysical Research, v. 82, p. 803–827.

Patchett, P.J., Ross, G.M., and Gleason, J.D., 1999, Continental drainage in North America during the Phanerozoic from Nd isotopes: Science, v. 283, p. 671–673.

Patriat, P., and Achache, J., 1984, India-Eurasia collision chronology has implications for crustal shortening and driving mechanism of plates: Nature, v. 311, p. 615–621.

Qayyum, M., Niem, A.R., and Lawrence, R.D., 1996, Newly discovered Paleogene deltaic sequence in Katawatz basin, Pakistan, and its tectonic implications: Geology, v. 24, p. 835–838.

Qayyum, M., Lawrence, R.D., and Niem, A.R., 1997, Molasse-delta-flysch continuum of the Himalayan orogeny and closure of the Paleogene Katawaz remnant ocean, Pakistan: International Geology Review, v. 39, p. 861–875.

Quade, J., Cerling, T.E., and Bowman, J.R., 1989, Development of Asian monsoon revealed by marked ecological shift during the latest Miocene in northern Pakistan: Nature, v. 342, p. 163–166.

Reiman, K.-U., 1993, Geology of Bangladesh: Berlin, Gebruder Borntraeger, 160 p.

Ren, J., Jiang, C., Zhang, Z., and Qin, D., 1987, The geotectonic evolution of China: Berlin, Springer–Verlag, 203 p.

Ross, C.A., 1986, Paleozoic evolution of southern margin of Permian basin: Geological Society of America Bulletin, v. 97, p. 536–554.

Ruddiman, W.F., and Kutzbach, J.E., 1989, Forcing of Late Cenozoic northern hemisphere climate by plateau uplift in southern Asia and the American West: Journal of Geophysical Research, v. 94, p. 18409–18427.

Schwab, F.L., 1981, Evolution of the western continental margin, French-Italian Alps: Sandstone mineralogy as an index of plate tectonic setting: Journal of Geology, v. 89, p. 349–368.

Schweickert, R.A., and Cowan, D.S., 1975, Early Mesozoic tectonic evolution of the western Sierra Nevada, California: Geological Society of America Bulletin, v. 86, p. 1329–1336.

Searle, M.P., Windley, B.F., Coward, M.P., Cooper, D.J.W., Rex, A.J., Rex, D., Tingdong, L. Xuchang, X., Jan, M.Q., Thakur, V.C., and Kumar, S., 1987, The closing of Tethys and the tectonics of the Himalaya: Geological Society of America Bulletin, v. 98, p. 678–701.

Şengör, A.M.C., 1976, Collision of irregular continental margins: Implications for foreland deformation of Alpine-type orogens: Geology, v. 4, p. 779–782.

Şengör, A.M.C., 1984, The Cimmeride orogenic system and the tectonics of Eurasia: Geological Society of America Special Paper 195, 82 p.

Şengör, A.M.C., 1987, Tectonics of the Tethysides: Orogenic collage development in a collisional setting: Annual Reviews of Earth and Planetary Sciences, v. 15, p. 213–244.

Şengör, A.M.C., 1990, Plate tectonics and orogenic research after 25 years: A Tethyan perspective: Earth-Science Reviews, v. 27, p. 1–201.

Şengör, A.M.C., 1991, Plate tectonics and orogenic research after 25 years: Synopsis of a Tethyan perspective: Tectonophysics, v. 187, p. 315–344.

Şengör, A.M.C., and Hsu, K.J., 1984, The Cimmerides of eastern Asia: History of the eastern end of Palaeo-Tethys: Societe Geologique de France Memoires 147, p. 139–167.

Şengör, A.M.C., and Okurogullari, A.H., 1991, The role of accretionary wedges in the growth of continents: Asiatic examples from Argand to plate Tectonics: Eclogae Geologicae Helvetiae, v. 84, p. 535–597.

Şengör, A.M.C., Natal'in, B.A., and Burtman, V.S., 1993, Evolution of the Altaid tectonic collage and Palaeozoic crustal growth in Eurasia: Nature, v. 364, p. 299–307.

Silver, E.A., and Moore, J.C., 1978, The Molucca Sea collision zone, Indonesia: Journal of Geophysical Research, v. 83, p. 1681–1691.

Silver, E.A., Abbott, L.D., Kirchoff-Stein, K.S., Reed, D.L., and Bernstein-Taylor, B., 1991, Collision propagation in Papua New Guinea and the Solomon Sea: Tectonics, v. 10, p. 863–874.

Stockmal, G.S., Beaumont, C., and Boutilier, R., 1986, Geodynamic models of convergent margin tectonics: Transition from rifted margin to overthrust belt and consequences for foreland-basin development: American Association of Petroleum Geologists Bulletin, v. 70, p. 181–190.

Suczek, C.A., and Ingersoll, R.V., 1985, Petrology and provenance of Cenozoic sand from the Indus Cone and the Arabian Basin, DSDP sites 221, 222, and 224: Journal of Sedimentary Petrology, v. 55, p. 340–346.

Tapponnier, P., Peltzer, G., and Armijo, R., 1986, On the mechanics of the collision between India and Asia: Geological Society of London Special Publication 19, p. 115–157.

Teng, L.S., 1990, Geotectonic evolution of late Cenozoic arc-continent collision in Taiwan: Tectonophysics, v. 183, p. 57–76.

Thomas, W.A., 1988, The Black Warrior Basin, *in* Sloss, L.L., ed., Sedimentary cover—North American craton: Boulder, Geological Society of America, Geology of North America, v. D-2, p. 471–492.

Thomas, W.A., 1989, Tectonic map of the Ouachita orogen, *in* Hatcher Jr., R.D., Thomas, W.A., and Viele, G.W., eds., The Appalachian-Ouachita orogen in the United States: Boulder, Geological Society of America, Geology of North America, v. F-2, Plate 9.

Thomas, W.A., 1991, The Appalachian-Ouachita rifted margin of southeastern North America: Geological Society of America Bulletin, v. 103, p. 415–431.

Uddin, A., and Lundberg, N., 1998, Cenozoic history of the Himalayan-Bengal system: Sand composition in the Bengal Basin, Bangladesh: Geological Society of America Bulletin, v. 110, p. 497–511.

Viele, G.W., 1989, The Ouachita orogenic belt, *in* Hatcher Jr., R.D., Thomas, W.A., and Viele, G.W., eds., The Appalachian-Ouachita orogen in the United States: Boulder, Geological Society of America, Geology of North America, v. F-2, p. 555–561.

Viele, G.W., and Thomas, W.A., 1989, Tectonic synthesis of the Ouachita orogenic belt, *in* Hatcher Jr., R.D., Thomas, W.A., and Viele, G.W., eds., The Appalachian-Ouachita orogen in the United States: Boulder, Geological Society of America, Geology of North America, v. F-2, p. 695–728.

Wang, H.Z., chief compiler, 1985, Atlas of the paleogeography of China: Beijing, Beijing Institute of Geology and Wuhan Geological College, Cartographic Publishing House, 301 p.

Wang, X.M., and Liou, J.G., 1987, The large displacement of the Tanlu fault: Evidence from the distribution of coesite–bearing eclogite belt in eastern China: Eos (Transactions, American Geophysical Union), v. 70, p. 1312–1313.

Wang, X.M., and Liou, J.G., 1991, Regional ultrahigh-pressure coesite–bearing eclogite terrane in central China: Evidence from country rocks, gneiss, marble, and metapelite: Geology, v. 19, p. 933–936.

Watson, M.P., Hayward, A.B., Parkinson, D.N., and Zhang, Z.M., 1987, Plate tectonic history, basin development and petroleum source rock deposition onshore China: Marine and Petroleum Geology, v. 4, p. 205–225.

Wickham, J., Roeder, D.R., and Briggs, G., 1976, Plate tectonic models for the Ouachita foldbelt: Geology, v. 4, p. 173–176.

Wuellner, D.E., Lehtonen, L.R., and James, W.C., 1986, Sedimentary tectonic development of the Marathon and Val Verde Basins, west Texas, U.S.A.: A Permo-Carboniferous migrating foredeep: International Association of Sedimentologists Special Publication 8, p. 347–368.

Yang, Z., Cheng, Y., and Wang, H., 1986, The geology of China: Oxford Monographs on Geology and Geophysics 3, 303 p.

Yin, A., and Nie, S., 1993, An indentation model for the North and South China collision and the development of the Tan-Lu and Honam fault systems, eastern Asia: Tectonics, v. 12, p. 801–813.

Zhang, C., Ma, T., Song, Y., Wang, L., Jiang, H., and Wu, S., 1989, Tectonic systems and formation and evolution of the Qinghai–Xizang (Tibet) Plateau: Ministry of Geology and Mineral Resources, Geologic Memoir Series 5, no. 8, 161 p.

Zhang, Z.M., Liou, J.G., and Coleman, R.G., 1984, An outline of the plate tectonics of China: Geological Society of America Bulletin, v. 95, p. 295–312.

Zhang, Z.M., Liou, J.G., and Coleman, R.G., 1985, An outline of the plate tectonics of China: reply: Geological Society of America Bulletin, v. 96, p. 408.

Zhou, D., 1987, Triassic deep-sea sedimentation in west Qinling Mountians, and its plate tectonic setting [M.S. Thesis]: Beijing, Chinese Academy of Sciences, 128 p.

Zhou, D., and Graham, S.A., 1993, Songpan-Ganzi Triassic flysch complex as a remnant ocean basin along diachronous Qinling collisional orogen, central China (abs.): Geological Society of America Abstracts with Programs, v. 25, no. 6, p. A118.

Zhou, D., and Graham, S.A., 1996, Songpan-Ganzi complex of the west Qinling Shan as a Triassic remnant-ocean basin, in Yin, A., and Harrison, T.M., eds., Tectonic evolution of Asia: Cambridge, Cambridge University Press, p. 281–299.

Zhu, X., 1989, Remarks on the Chinese Mesozoic–Cenozoic sedimentary basins, *in* Zhu, X., ed., Chinese sedimentary basins: Amsterdam, Elsevier, p. 1–5.

Zou, D.B., Rao, R.B., Chen, Y.M., and Chen, L.K., 1984, Triassic turbidites in southern Bayankela Mountains, Tibetan Plateau, *in* CGQXP Editorial Committee, ed., Contribution to the Geology of the Tibet Plateau: Beijing, Geological Publishing House, p. 27–47.

Manuscript Accepted by the Society January 22, 2003

Printed in the USA

Geological Society of America
Special Paper 370
2003

Megareefs in Middle Devonian supergreenhouse climates

Paul Copper*
Department of Earth Sciences, Laurentian University, Sudbury P3E 2C6, Canada

Christopher R. Scotese*
Department of Geology, University of Texas at Arlington, Arlington, Texas 76019, USA

ABSTRACT

A newly refined reef database, modified to calculate reef tracts in relation to major tectonic plates, and with new paleogeographic maps, indicates that the largest known, and latitudinally most widespread Phanerozoic reefs developed during the Middle Paleozoic (Siluro-Devonian), with an acme in the Middle Devonian. Expanding during times of exceptional sea-level highstands and widespread epicontinental shallow seas, this 26 m.y. long acme of coral-sponge reef growth coincided with the warmest global temperatures known for the Phanerozoic, i.e., with a "supergreenhouse" climate mode well above Holocene interglacial norms. During the Middle Paleozoic, reefs were particularly abundant, occupying large, continental seaboard, carbonate platforms, and vast inland epicontinental seas. Examples of such "extremes" occurred mostly on passive margin settings, and extensive flooded continental interiors, e.g., the 1700–3000 km long tracts of the Western Canada Sedimentary Basin, Canadian arctic (Innuitian platform), eastern Laurentia "Old Red Continent" (United Kingdom to Poland), eastern Russian Platform (northeast Laurentia), Ural "Fold Belt" (eastern slopes of Urals), Siberia, northwest Africa, and South China. Smaller scale reef belts between 700 and 1300 km long were constructed on isolated tectonic terranes facing Gondwana on the north (Pyrenees, Afghanistan-Pakistan), Mongolia, Kolyma-Chukot, and North China. Large basins and flooded shelf areas, and the reefs featured within them, were not persistently developed throughout the Middle Paleozoic. They especially characterized the middle Emsian through Givetian (late Early Devonian–Middle Devonian). The following Frasnian (Late Devonian) showed more restricted and confined distribution of coral-stromatoporoid reefs, and during the Famennian, coral-stromatoporoid reefs "crashed" and were replaced by calcimicrobial reefs and platforms. During the latter phases of the Frasnian/Famennian mass extinctions, such microbial reefs were confined to relatively small areas, and metazoan reefs were nearly entirely obliterated, being confined to rare stromatoporoid patch reefs or lithistid mounds. Coral reefs were completely absent during the 21 m.y. long Famennian interval, and no real recovery of "keystone" frame-building, colonial corals took place in reef settings. The Famennian coincided with repeated glaciations, sharp sea-surface cooling events, sea-level drawdowns, and concurrent, matching stable isotope excursions.

Keywords: Phanerozoic acme, Middle Devonian, coral-sponge reefs, supergreenhouse climates.

*E-mail: pcopper@nickel.laurentian.ca; cscotese@uta.edu

Copper, P., and Scotese, C.R., 2003, Megareefs in Middle Devonian supergreenhouse climates, *in* Chan, M.A., and Archer, A.W., eds., Extreme depositional environments: Mega end members in geologic time: Boulder, Colorado, Geological Society of America Special Paper 370, p. 209–230.

INTRODUCTION

The Holocene is an unusual episode in Phanerozoic Earth history with respect to reef distribution worldwide, as well as its climate fluctuations between glacial and interglacial intervals. There are three relatively large, continuous, barrier reef platforms today with prolific coral growth: (1) the Great Barrier Reef, which is about 2100 km long stretched along the east coast of Australia to New Guinea, (2) the shelf along New Caledonia, which is 750 km long, and (3) the Yucatan shelf, which is 650 km long. Only the Great Barrier Reef can match some of the much larger scale reefs of the Devonian reef optimum, and thus the Holocene cannot be considered to have typical large-scale barrier reef development for the Phanerozoic, the past 544 m.y. of Earth history. This can most likely be attributed to the fact that for most of post-Cretaceous time, Earth has been in an icehouse mode, oscillating around cool episodes defined by bipolar glaciation. Holocene reefs of the present interglacial warm episode, the past 12,000 years, are normally confined to tropical low latitudes roughly below 27°N and S on eastern seaboards (Wells, 1988; Veron, 2000). Reefs occur at higher latitudes on the eastern seaboards because they are influenced by warm currents, e.g., Bermuda 32°25′N (generated by the Gulf Stream), Gulf of Aqaba 29°30′N (Wells, 1988), Sea of Japan at Iki Island 34°N (Yamano et al., 2001), and Lord Howe Island, Australia, 31°30′S (Veron, 2000). The southernmost west Australian reefs, the Houtman Abrolhos, occur at 29°S, while on the Australian east coast, the most southerly reefs occur near Brisbane at 27°S (Veron, 2000). Reefs in the Gulf of Aqaba are sustained at relatively high latitudes because of minor riverine input and very clear waters, despite sporadic cool winter temperatures. Reefs on the east African coast range to latitudes between 26° and 27°S (Veron, 2000). Shallow water coral communities (biostromes) with reasonably dense stands of low diversity corals may occur at even higher latitudes than reefs.

On the western sides of continents, due to cold currents and nutrient upwelling, Holocene reefs are unknown at latitudes higher than 10°N and S. On the west coast of Africa there are only coral thickets, and zooxanthellate hermatypic corals are restricted to waters less than 20 m deep (Wells, 1988). As for much of the eastern coastline of Brazil affected by the Amazon drainage (where the southernmost Abrolhos reefs are located at 18°S), a secondary constraint for west African reefs is high river runoff with low salinities, and high rates of sediment influx. These relatively low latitude modern reef distribution patterns are not the norm for most of the Phanerozoic, except for the cooler Late Paleozoic, the Early Triassic, and much of the Cenozoic (but for the late Paleocene and Eocene warm episodes). Despite an improved database (Kiessling et al., 1999; Kiessling, 2001), the reef picture resolution is still relatively coarse and insufficient for producing complete global maps at the stage level or less. In other words, the fossil reef database lacks the precision of the 5 m.y. long Pliocene-Quaternary reef database that has been accumulated over the past 50 years. Several reasons for this bias are evident: (1) large carbonate platforms have disappeared via uplift and erosion or subduction (e.g., Tibet, eastern Australia); (2) little is known about the very small, but widespread, fringing reefs on ancient island arcs and atolls (one of the chief types of modern reefs so well known to Darwin); (3) largely unexplored subsurface drill core data exist in remote areas; and (4) many carbonate platforms are simply inadequately described, or little known, in a modern carbonate sedimentology sense, e.g., Mongolia, Tian Shan, North China, and northeastern Russia. Thus, at best, our database remains only preliminary. Often the information that is published is controversial, with some authors interpreting the distribution of reefs in terms of present-day environments. (e.g, James and Bourque, 1992; Copper, 2001), and others promoting ideas suggesting that (1) Phanerozoic reef distribution is either unrelated to latitudinal dispersal or to temperature (Kiessling, 2001); (2) that ancient reefs have no, or few, modern analogues, and therefore do not fit actualistic patterns (Wood, 1999); or (3) that modern Cenozoic reefs have entirely different modes of skeletonization favoring aragonite supersaturation, and thus show temperature and carbon dioxide solubility constraints, or responses, that are diametrically opposed to those of the dominant Middle Paleozoic calcite ocean mode (Kleypas et al., 1998).

Reefs are herein defined as three-dimensional, biogenic, carbonate structures raised above the surrounding seafloor, generally possessing a fringing apron of reef-derived flank deposits. Reefs are defined sedimentologically by massive bedding in the reef core, usually a skeletal, reef-builder framework (e.g., calcimicrobes, algae, sponges), but also carbonate muds, producing fluid-conducting cavities and containing carbonate cement. In this we, therefore, include the spectrum of mudmounds, following Wood (2001). The term tropical, as used in this paper, does not strictly refer to the modern region between the tropics of Cancer and Capricorn, 23°28′N and S. Purely artificial latitudinal constraints do not adequately describe the distribution of warm marine environments during greenhouse modes more typical of much of the Phanerozoic. The marine tropics are herein defined by climate and temperature constraints, i.e., as marking the limits of vigorous coral skeleton growth (and $CaCO_3$ sponge skeletons) in shallow waters where the sea surface/atmosphere interface never drops below freezing. In a supergreenhouse, Middle Paleozoic world (Berner, 1997), the tropics expanded poleward, with average global temperatures 4–14 °C above the Holocene average of 16 °C. Thus, it is not only the mean tropical sea-surface temperature that provides the constraint to reefs, but also the annual maximal and minimal temperature, much as most human agriculture is constrained by the onset of winter frost during harvest, or late frosts for seeding in spring. On land, the comparable tropical climate belt would be defined by the limit of normal, humid tropical plant growth and reproduction, e.g., palms, coconuts, mangroves, etc.

A further latitudinal constraint for modern reefs may be cyclical, indiscriminate bleaching events, such as those that accompanied the 1997–1998 El Niños (Wilkinson, 2000), though the fossil record for bleaching, and coral reef recovery therefrom, remains very uncertain. This means that tropical metazoan reefs of the Middle Paleozoic could move to as high as the 40° to 50°, and possibly 60°, latitudes. Such an extended range for reef occurrences fits well with paleogeographic maps of the Middle Devonian presented in this paper (Fig.1). Broad patterns and regionally detailed surveys of the distribution of Emsian through Givetian reefs, their thickness and broad faunal construction, are now extractable from our own database. This paper updates the distribution of Emsian through Givetian reefs in a plate tectonic framework, i.e., as Devonian carbonate platforms or island arc belts flanking major plates (see also Scotese, 2001; Copper, 2002). We have broadly defined the term *reef tracts* as large, carbonate platforms in which reefs are a common, and more or less continuous, though not an evenly distributed feature. This thereby also includes "Great Barrier Reef" tract types (GBRs), epicontinental seas dominated by large or small patch reef complexes, tectonically isolated Bahama-type platforms, or the narrower platforms of chains of fringing reefs around tectonically active island arcs, or collisional plate margins with fringing reefs.

DEVONIAN (EMSIAN-GIVETIAN) REEFS

During the beginning of Early Devonian time (Lochkovian-Pragian), reef development was generally constricted worldwide due to relative sea-level lowstands and limited accommodation space (Copper, 2002). However, during the succeeding Emsian, reef expansion was rapid. There are at least three major and divergent paleogeographic reconstructions for the Emsian through Givetian (late Early Devonian–Middle Devonian). These include: (1) a Laurentia-centric version, in which the dominant reef growth patterns encircle Laurentia, and in which collision between Laurentia and western North Africa is shown to have taken place by the Gedinnian (Golonka et al., 1994; Kiessling et al., 1999; Scotese, 2001), (2) a version centered around Russia-Siberia-Kazakhstan with major reef belts in the 5–20° latitudes north (Zonenshain et al., 1990a, 1990b; Fig. 13 *in* Kuznetsov, 2000), and (3) an eastern Gondwana, Sino-Australo-centric version that places the Chinese plates proximal to Australia (Talent et al., 2000). The lengths of these reef belts varied from ~3100 km to as little as 400 km (Table 1).

These three stages of the Devonian, beginning in the latter part of the Early Devonian (middle to late Emsian), and ending in the Givetian, are broadly agreed to mark the Phanerozoic peak of reef development on a worldwide basis, with reefs reaching much higher latitudes than the Holocene climatic optimum (Copper, 1994; Kiessling et al., 1999; Kiessling, 2001; Copper, 2002). A peak in latitudinal reef distribution and in the size of carbonate platforms and epicontinental seas with reef belts was reached during the Givetian, after which there was a sharp decline and retreat of reefs from many areas, generally accompanied by sea-level drawdown; note that the reef abundance database is artificially skewed toward the Frasnian because of extensive drill core data from western Canada (Kiessling et al., 1999). However, reef development was far from uniform during the 26 m.y. long Emsian through Givetian reef optimum. A revised absolute time scale for the Devonian (Okulitch, 1999) shows that the Emsian represented 15 m.y., the Eifelian 7 m.y., and the Givetian 4 m.y. By the late Emsian, reef and peri-reefal faunas generally became very cosmopolitan, dominated by an Old World component that may be found east to west from Australia through South China, the Urals, western Europe, and northwestern Canada. Malvinokaffric cool climates centered around Gondwana (South America and southern and central Africa), were dominated by siliciclastics and lacked limestones and

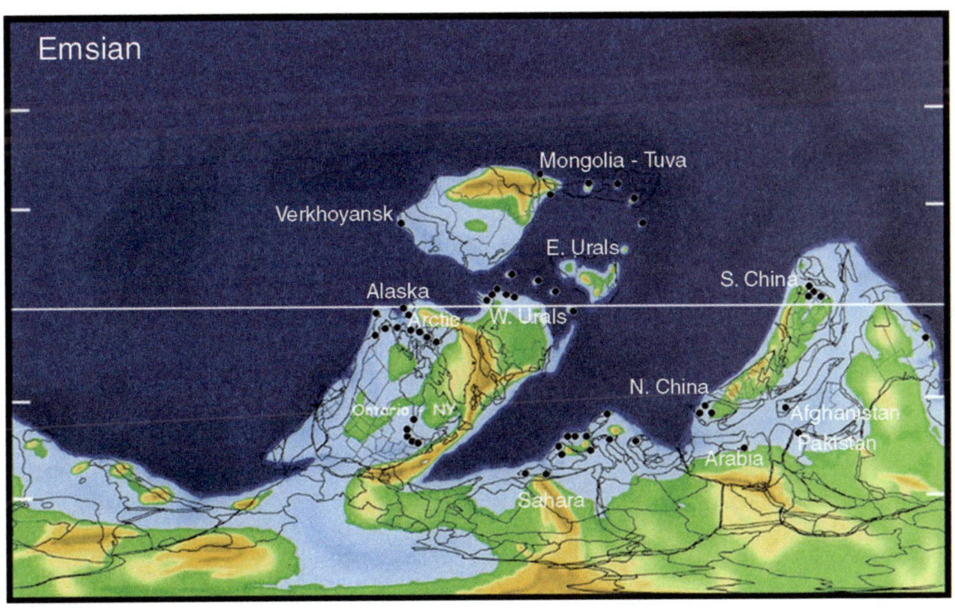

Figure 1. Plate reconstruction for the Emsian (late Early Devonian) showing initiation of Devonian reef megacycle, mostly in the middle to late Emsian. Reefs were developed in 30–35° latitudes in Mongolia-Tuva as part of active tectonic margin (see Scotese, 2001). Reefs in northern Gondwana (Morocco) grew at 45–50°S, and microplates of south European archipelago nearby, e.g., Spain, Sardinia, Carnic Alps, Montagne Noire, Barrandian high, and Armorica, at ~40–45°S. Few Emsian reefs occurred in eastern or western Laurentia, except in the Appalachian Basin and equatorial Inuitian Platform of Arctic Canada. A long belt probably developed on active eastern Australia margin, but much of this was lost by subsequent subduction of platform or erosion. Kazakhstan should have Emsian reefs (to date unreported), since reefs were developed earlier in the Lochkovian, and later in the Givetian.

TABLE 1. TOP 20 "GREAT BARRIER REEFS" OF THE EMSIAN-EIFELIAN-GIVETIAN, WITH ESTIMATED LENGTH OF THE REEF TRACT IN ROUGH ORDER OF SIZE

1.	Western Canada reef province (stable platform, western Laurentia): ~3100 km
2.	Ural fold belt (active, island arc belt): ca. 3000 km
3.	Northeast Baltica or Russian Platform (stable platform, Pechora to Volga Urals): ~2600 km
4.	Alaska-Canadian arctic (stable Innuitian platform): ~2500 km
5.	East Australia–New Guinea margin (tectonically active, ? isolated carbonate platforms): ~2000 km
6.	Near East tract (eastern Iran-Afghanistan-Pakistan): ~1900 km
7.	Southeastern Laurentia or "Old Red Continent" (Northwest Europe, United Kingdom–Poland): ca. 1800 km
8.	Mongolia-Tuva plate (southern Mongolian Gobi): ~1800 km
9.	South China plate (Vietnam-Guangxi-Hunan): ~1700 km
10.	Belarus to Caspian Basin (eastern Laurentia, stable platform, but discontinuous): ~1600 km
11.	Kazakhstan (central Kazakhstan to Junggar, Xinjiang): ~1600 km
12.	Spanish plate (Pyrenees: 800 km), or alternatively Bohemian facies tract Spain-Prague Basin, 1600 km.
13.	Siberia (southern and central): ~1500 km
14.	Saharan reef tract (Western Sahara-Morocco-Algeria): ~1200 km.
15.	Eastern Laurentia (Eifelian only: Pennsylvania-Hudson Bay): 1100 km
16.	Russian Tian Shan fold belt (Uzbekistan-Pamirs): ~1000 km.
17.	Tarim-Karakorum plate (northern flank): ~1000 km
18.	Kolyma plate: ~800 km arcuate tract
19.	North China plate: ~700 km
20.	Western Australia (Givetian, Canning Basin): ~400 km.

Note: Compare top three today: Australia–Papua New Guinea, Great Barrier Reef 2100 km, New Caledonia 750 km, Yucatan 650 km.

reefs; not even microbial mudmounds were present. The northern fringes of Gondwana, i.e., those located in the Saharan belt (Morocco-Libya), had microbial mudmounds, modest development of coral-stromatoporoid reefs and carbonate platforms, and shared some common faunas with Appalachian North America.

REEF TRACTS AND PLATE LOCATIONS

Laurentia (North American Plate)

There is broad agreement in terms of climate sensitive sediments, faunas, and reefs, that Laurentia straddled the equator during the Devonian, with most of the plate in the southern hemisphere (Fig. 1). The equator cut through Alaska and passed just north of Greenland. By early Emsian time, if not earlier, Baltica had fused to Laurentia. The Avalon-Meguma terrane, a rifted fragment broken from Gondwana, had probably collided in the Late Ordovician or Early Silurian, sealing off part of the Iapetus Ocean (Lin et al., 1994). The Appalachians were rising and continued along the eastern margin of Greenland. An Arctic mountain belt arched around the north Greenland coast westward, to the northern fringe of Laurentia.

Active volcanism, and the development of large deltas in a humid tropical setting, prevented active reef development along much of the eastern margins of Laurentia in the Early-Middle Devonian, due to the Acadian orogeny in the Appalachians. However, small, isolated Lochkovian stromatoporoid and bryozoan reefs were present in the Keyser Limestone of the Appalachian Basin as far south as Tennessee (Smosna, 1988). To the northeast Gaspé, pinnacle reefs were present in the Pridolian–early Lochkovian West Point Limestone; these were associated with volcanics, but were buried by the Pragian-Emsian (Bourque and Amyot, 1989; Bourque, 2001). During the Emsian through early Eifelian, minor patch reef growth occurred in the southern Ontario Detroit River Limestone, and New York's lower Onondaga Limestone (Fagerstrom 1983; Lindemann, 1989), located at about 40°S. Onondaga pinnacle reefs occur in the subsurface of Pennsylvania (Woodrow et al., 1988). Onondaga reefs may have been inhibited by a quasi-estuarine circulation pattern, such as that seen in the Java Sea today (Edinger et al., 2002). In northern Ontario, the late Emsian, reefal Kwataboahegan Formation was capped by evaporites at 30°S (Telford, 1988). The Pennsylvania–New York through Hudson Bay Emsian reef belt probably extended along a sea lane ~1100 km long, as carbonate xenoliths in kimberlite pipes indicate that areas presently lacking Devonian strata were covered by Devonian seas (McCracken et al., 2000). On the western side of Laurentia, including the western United States and Canada, Emsian reef development was absent, and reefs were not initiated until Middle Devonian time.

The most extensive Emsian reef belt, more than 2500 km long, was wrapped around the northern margin of Laurentia, which is presently within the Canadian arctic (Fig. 1). Though its location is uncertain, the Alexander Terrane was fringed by Pragian-Emsian reefs in an outer island forearc setting (Soja, 1988: possibly an extension of the Arctic reef belt). Emsian to early Eifelian coral-stromatoporoid reefs of the Prongs Creek and Ogilvie formations in the Yukon (Perry, 1979), and in nearby eastern Alaska (Clough and Blodgett, 1988; Soja, 1988), were part of an isolated stable carbonate platform linked to the Arctic belt to the east. Along the stable carbonate shelf of the Northwest Territories, Eifelian coral patch reefs were present (Noble and Ferguson, 1971). In Arctic Canada, reefs were even more widely developed during the Emsian than in the preceding Early Devonian (Smith, 1985). This stable platform reef belt, mostly assigned to the Emsian Stuart Bay and lower Blue Fiord formations, stretched ~1600 km westward from Ellesmere Island, where ~100-m-thick patch reefs were present (Smith, 1985). The most westerly extension of this platform was at the Prince of Wales Strait, along western Victoria and Princess Royal Islands (see Fig. 2), where Emsian-Eifelian reefs of the Blue Fiord Formation reached diameters of a kilometer or more, and are exposed as prominent reef clusters for ~170 km of outcrop (Thorsteinsson and Tozer, 1962). Emsian and early Eifelian reefs

Figure 2. Two views of typical large-scale Middle Devonian reefs, part of Arctic Innuitian "Great Barrier Reef" province, which extended 2500 km from Banks and Victoria Islands, past Melville to Ellesmere Island. Part of reef belt is on western flank of Victoria Island. A: The ~150-km-long Blue Fiord Formation tectonically undisturbed reef tract with large coral patch reefs up to 1 km in diameter on the western shore of Victoria Island (Prince Albert peninsula), ~20 km north of Hay Point, ~N72°, 50′, E117°. B: view of Princess Royal Island (between Banks and Victoria Islands) reef perspective; this flat-lying coral-stromatoporoid reef portion, looking southeast, is ~2.2 km long, <1 km wide, and <50 m thick, with reef flank facies, deeper water dark grey calcareous shales, and debris flows dipping northwest, toward Banks Island (see also Thorsteinsson and Tozer, 1962), Eifelian–early Givetian, Blue Fiord Formation, Princess Royal Island, Prince of Wales Strait, at N72°45′; E118°05′. (Photos by Paul Copper.)

to the northeast of Banks Island included those on large Bahama-type carbonate platforms of Cameron and Prince Patrick Islands (Smith, 1985). These Arctic Emsian reefs were all developed within about 10° from the equator, open to eastward-flowing, warm tropical gyres.

Following deepening and extensive siliciclastic supply from the Appalachian deltas to the east, reefs were not present in the Appalachian Basin until the early Givetian, as small tabulate coral patch reefs of the Silica Shale in Ohio (Stumm, 1969). By the late Givetian, a modest return of reefs to eastern North America is marked by small, coral-capped mudmounds of the Tully Formation in New York (Heckel, 1973), and stromatoporoid-coral patch reefs of the Traverse Group in Michigan (Kesling et al., 1974; Meyer, 1989). To the south, in Iowa and Missouri, extensive, very shallow water carbonates, frequently alternating with intertidal and evaporitic sediments, showed episodic, coral-rich biostromes, but reefs were lacking during the Eifelian-Givetian, as well as Frasnian (Witzke et al., 1988). To the northeast, reef development had ceased in the Gaspé in Quebec, Canada, before Emsian time, and no later reef developed until microbial mounds in the Carboniferous (Bourque et al., 2001).

The Middle Devonian "Great Barrier Reef" Belt of western Canada and the northwestern United States has been well documented and reviewed previously (McMillan et al., 1988; Moore, 1988, 1989). Reefs were most extensively developed in western Canada from the middle Eifelian through middle Givetian, ranging from subsurface pinnacle reefs of the Winnipegosis Formation south of the Canadian border (Edie, 1959) to the Hume Formation on the Carnwath River, Mackenzie delta (Mackenzie, 1969), for more than 3100 km. In Eifelian time, an isolated 300-km-long reef tract was also present as far south as southern Nevada, though in the Givetian, this switched to evaporites (Johnson et al., 1991). During growth of the Givetian Keg River reefs in Alberta, on the western margin of Laurentia, inter-reef and back-reef sediments were usually organic-rich and bituminous, suggesting high rates of shelf plankton productivity, and adaptation of reef organisms to episodic hypoxia from upwelling deeper waters below the oxygen minimum zone (Chow et al., 1995). In the late Givetian, this reef belt shrank to a broad platform less than 800 km long, and about 600,000 km^2 in area, from the Swan Hills area of central Alberta in the south (Fischbuch, 1968) to the Great Slave Lake (Moore, 1989). Another smaller, late Givetian isolated reefal platform of the Kee Scarp Formation, <1000 km^2 in area, exists in the Normal Wells area (Muir et al., 1984). By the Frasnian, reefs shrank and retreated largely to central and southern Alberta; their decline was possibly related to periodic regressive cycles (Moore, 1989).

From middle Eifelian to Frasnian time, reefs almost entirely disappeared in the Canadian arctic, except for a restricted early-middle Frasnian reef platform belt on northeastern Banks Island (Embry and Klovan, 1971; Embry and Klovan, 1989; Embry, 1991). This is attributable to development of a giant siliciclastic delta wedge, derived from the northeast as far as Greenland, commencing by draping over Eifelian Blue Fiord reefs, with the delta expanding in the Frasnian to middle Famennian (Figs. 3 and 4). This clastic wedge was several kilometers thick, and more than 1.5 million km^2 in area, containing coal units and abundant plant remains, locally dumping fossil logs into the isolated Frasnian reef platform of northeastern Banks Island to the distal west. Embry (1991) attributed the build-up of this mega-delta to uplift, high subsidence rates, and climate change from humid to dry savanna, but episodic late Frasnian and Famennian sea-level drawdowns and cooler, drier climates may also have been important factors.

Northwestern Europe (Eastern Laurentia: The "Old Red Continent")

On the southeastern flanks of the Old Red Continent, i.e., the areas represented today by England, northwestern France (Boulonnais), Belgium, Germany, and Poland, the Early Devonian (Emsian) was dominated by siliciclastic facies, with plant remains, fish faunas, and other evidence indicating broad coastal delta plains (Blieck et al., 1988). These were generally unsuitable for reef development until the middle to late Eifelian, though some thin coral-rich biostromal carbonates are known. Similarly, the Russian Platform and Baltic Basin, though located on the easterly tropical flank of Laurussia, were apparently unsuitable for reef growth during the Early Devonian, probably as a result of river drainage and extensive wet coastal deltaic plains, providing metahaline and mud-rich shelf waters prohibitive for reefs. Such a situation seems analogous to the easterly drainage of the Amazon River today, also in the sense that early, shrubby pteridophyte land plants were being established in such coastal areas of the Old Red Continent.

During the Eifelian, as the east and west margins of the Rheic Ocean were approaching, reef development was initiated almost the full length of the northeast-southwest–trending eastern seaboard of Laurussia, from Devon, England, through northwest France, Belgium, Germany, and Poland (~1800 km), and another 1600 km distally to the edge of the Caspian embayment. The northeast tip of this continental margin touched the equator in the Middle Devonian. Whether this was a continuous reef belt is not clear, as major gaps exist in the outcrop and possibly in the subsurface record. The shelf was narrow to the south latitudes at about 30° and widened toward the Russian Platform, flooded partly by a large, shallow, epicontinental sea. Ziegler (1982) constructed the Armorican Massif as an integral part of Laurentia, extending eastward along the plate margin toward the Saxo-Thüringian Basin and Barrandian high and separated to the south from the Montagne Noire and other plates by a long island continent, the Ligerian-Vosgian-Moldanubian cordillera (but see below where the Armorican Massif and Saxo-Thüringen belt are joined to the Southern European Archipelago). In the maps presented here (Figs. 3 and 4), Armorica is shown adjacent to the northern margin of Gondwana.

The literature of northwestern European Middle Devonian reefs is now vast: it has been reviewed by Jux (1960), Krebs (1974), Burchette (1981), and Blieck et al. (1988). The late Eifelian–Givetian represents the acme of reef development, though in some areas reefs continued well into the late Frasnian. Considerable siliciclastic influence (the "Rhenish facies") curtailed reef development in the early and middle Eifelian, and the total pile of carbonates-siliciclastics was up to 1.5 km thick. Reefal facies in England seem to have flourished best in the late Eifelian through Givetian, as it did northwestern France, in Belgium, and east of the Rhine. In southwest England, platform margin reef facies commenced in the Givetian (Scrutton, 1977). In the Boulonnais outcrops of northern France, small patch reefs mark the Givetian Blacourt Formation (Mistiaen and Poncet, 1989). The south and west margins of Armorica (Massif Armoricain) featured coral patch reefs near Brest, Nehou, and Angers during the Pragian, but reef decline was evident in the early Emsian (Cavet et al., 1967; Poncet, 1977; Plusquellec, 1980; Morzadec et al., 1988). These were, in view of their brachiopod-coral faunas, considered to be part of the eastern proto-Atlantic south European archipelago by Hladil et al. (1999), but part of the Old Red Continent as proposed by Ziegler (1982) and Franke (1999). The Belgian mixed siliciclastic-carbonate ramp appears to have had strong sea-level and cyclonic controls restraining reef development in the Eifelian, but by Givetian time, numerous stromatoporoid, coral, and microbial reefs developed on a wide platform with back reef lagoonal facies (Préat and Weiss, 1994). Severe back-stepping of reefs in the Belgian Frasnian ultimately ended in the rapid burial of reefs by muds, silts, and sands in the late Frasnian.

In the classic Eifel region west of the Rhine, which extended as a shelf to Bergisches Land east of the Rhine for about 250 km, patch reefs were developed in a mid-to-distal, possibly outer ramp to platform margin (facies type B in Faber et al., 1977; Faber, 1980), distal to land sources of siliciclastics to the northwest. These included planar to tabular stromatoporoid reefs as well as patch reefs with abundant corals, and perireefal, biostromal communities with crinoid meadows, phaceloid and branching rugose corals, and platy tabulates. In late Eifelian time (Ahbach beds), mudmounds with coral-stromatoporoid caps sporadically developed in the Eifel, e.g., in the Hillesheim and Dollendorf synclines (Malmsheimer et al., 1996; Pohler et al., 1999). Reef development rich in corals continued into the early Givetian (Loogh-Cürten Beds) and middle Givetian (Dreimühlen-Rodert Beds) in the same area, sometimes shallowing into intertidal facies (Birenheide et al., 1991). In the Prüm syncline, coral patch reefs up to 6 m thick and 30 m in diameter are present in the late Givetian Bolsdorf and Wallersheimer dolostones; the latter upper units are possibly of early Frasnian age (Jux, 1960). Reefs effectively ceased in the Eifel region during the Frasnian

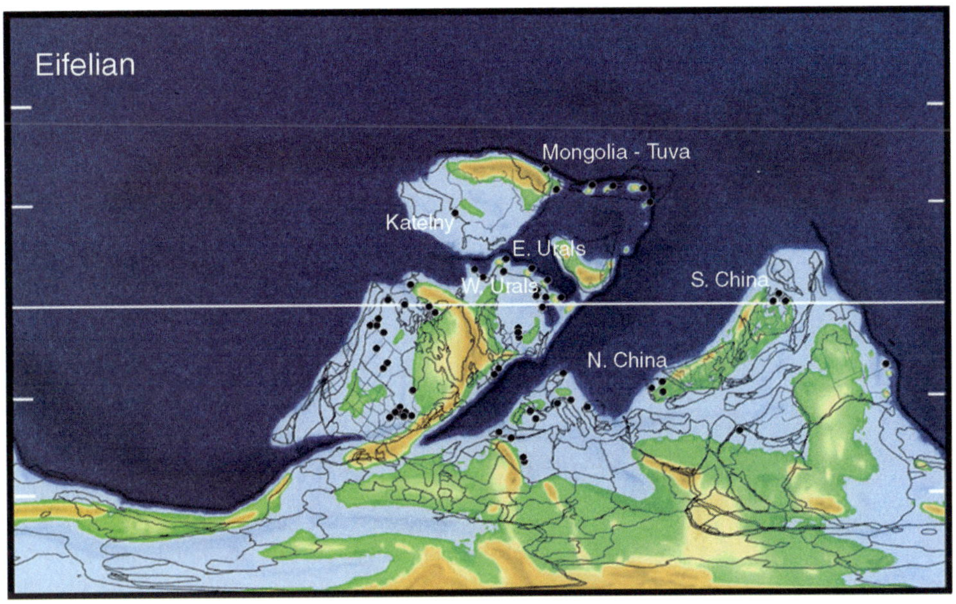

Figure 3. Plate reconstruction for Eifelian time showing a major expansion of reef belts for Middle Devonian. Reef tracts expanded considerably in western and central North America (Western Canada Sedimentary Basin), southern and east-central Laurentia (Hudson Bay and south), the east side of the Old Red Continent (i.e., platform to ramp of England through Poland), and the Russian Platform (W slope Urals) and eastern Urals. Reef growth was less extensive in areas on north margins of Gondwana (Morocco to Prague Basin). A Near Eastern reef belt stretched from Iran through Pakistan. Reefs were present on the eastern side of Kazakhstan. Reefs continued in Mongolia, Hingganling ranges of northern China, and Amur region.

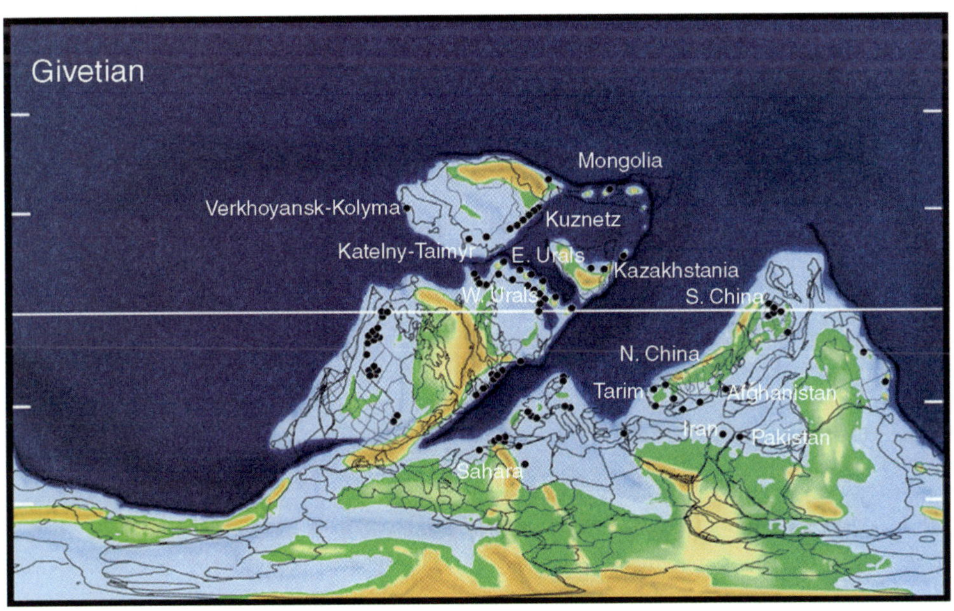

Figure 4. Maximal expression of Devonian (Givetian) saw widespread barrier reef provinces and high-latitude reef growth. Marine barrier between Laurentia and Gondwana had narrowed to a shallow seaway less than 500 km wide. At least eight or nine giant barrier reef provinces, greater in length than, or comparable to, the Holocene Great Barrier Reef of Australia were present (see Table 1): (1) northwest Gondwana (Marocco-Pyrenees); (2) England-Poland on eastern Old Red Continent; (3) western Canada; (4) central to north Urals (both on stable platform of east Baltica and the offshore, island arc Urals; (5) Siberia and northeast Russia (Kolyma-Verkhoyan); (6) Tarim-Afghanistan; and (7) south China and proximal areas of Vietnam, Laos, Cambodia, and Thailand.

and were succeeded by intertidal micritic facies, or black shales rich in goniatites.

East of the Rhine, small patch reefs of middle Eifelian age first occur in the Hobräcker Formation (Scheibe, 1965) and continued into the late Eifelian Honsel Formation (Scheibe, 1965; Buggisch and Flügel, 1992; Malmsheimer et al., 1996). Givetian reef development varied in different basins east of the Rhine (May, 1983, 1994). To the north, the stable shelf featured stromatoporoid reef facies at the deeper part of the shelf margin (May, 1997), though Machel and Hunter (1994) favored interpretation of the Massenkalk as a wave-resistant barrier reef like the Great Barrier Reef of Australia. In the basinal facies, reefs beginning in the Givetian *varcus* zone had regional tectonic controls and were commonly associated with local volcanic highs or "Schwellen" (Königshof et al., 1991; Weller, 1991; Buggisch and Flügel, 1992; Braun et al., 1994; Braun and Königshof, 1997). The Lahn syncline, part of the Rhenohercynian fold belt, could be viewed as developing in a backarc basin parallel to the Old Red Continent stable platform reef carbonates, which ranged from the Eifel west of the Rhine some 250 km through Brilon, east of the Rhine. Alternative views consider the Rhenohercynian massifs to be part of microplates flanking other Gondwana terranes to the south; see below. The Massenkalk reef facies continued into the early Frasnian, possibly with a break, but it was ultimately disrupted by deposition of the organic-rich Kellwasser limestones. For Poland, following Eifelian evaporites, an extensive Givetian reef platform with coral-rich shoals and coral banks was constructed, but reefs also developed on a reduced platform later in the Frasnian (Racki, 1988; 1992).

Russian Platform (NE Laurussia)

On the western side of the Russian platform, a Middle Devonian carbonate platform with reefs ~400 km wide occurs in the subsurface of Belarus (Makhnach et al., 1986); this was apparently a carbonate platform isolated from the Polish reefs to the west. The Pripyat-Donets Graben (aulacogen), to the southeast of Belarus, was not opened to reef development until Late Devonian time (Ioganson, 1990d). Farther on the eastern flanks of Baltica is the large, subcircular Caspian Basin, flanked by a rim of Devonian reefs of Eifelian-Givetian age (the "Precaspian syneclise" of Rusetskaya and Yaroshenko, 1990, or the Caspian Basin of others), located at the paleo-equator. This ~400,000 km^2 region represents the buried southeastern, stable flank of the Baltica plate, with drillcore data revealing coral-stromatoporoid reefs from a few meters in thickness and diameter to structures 150 m thick and over 1 km in diameter. Late Devonian (Famennian) microbial reefs, part of the north Caspian Basin, are known in drillcore from the northeast corner of the Caspian Sea in western Kazakhstan, just as they were located on a rimmed carbonate platform in the Karatau ranges to the east (Cook et al., 1994). This subsurface Devonian Caspian reefal complex on the southeast corner of Baltica ranged past Ufa for ~1000 km northward as the Volga-Ural Basin, though generally to the north, the Middle Devonian turned to siliciclastics, and the Frasnian contained barrier reefs, atolls, and isolated reef platforms separated or underlain by black Domanik limestone facies (Ulmishek, 1988; Rusetskaya and Yaroshenko, 1990). Kirikov (1988, p. 522) described the southeast Russian platform (the "Kama-Kinel trough") with "barrier-type [reefs] extending for hundreds of kilometers…and a width of no more than 15 km…dominated by algal limestone" during the Frasnian, and continuing into the Famennian. The stable shelf of northeastern Baltica in the Timan-Pechora region shows mostly dolomitized, platform-edge Frasnian reefs in the subsurface and outcrop (Grachevskii and Solomatin, 1977; Belyaeva, 1986). These are similar to and slightly separated from those of the Caspian Basin and Volga-Ural parts of the Russian Platform to the south (Ulmishek, 1988). If the reef belt from Pechora is connected to the Caspian basin, this stable carbonate platform on the northeast margins of Baltica, at ~2600 km long, would be the second longest reef tract of Russia.

Eastern Slopes of the Urals (Active, Unstable Margin, Island Arcs)

The "Uralian reef province" on the northeast margin of Laurussia or Euramerica (Figs. 3–5) is usually shown as two separate tectonic units: (1) the western slopes, part of the stable platform and eastern shelf of Baltica, from the Pechora region southward (Laurussia or Euramerica: see above paragraph), and (2) the unstable island arc complexes on the eastern slopes of the Urals (the Ural fold belt). These long, parallel carbonate belts developed Middle and Upper Devonian reefs, though many were concentrated in the Late Devonian (Sokolov, 1986). Zonenshain et al. (1990a) identified the eastern Ural reef belt oriented in a more or less east-west alignment as flanking the north side of the Baltica plate or Russian Platform in the Lower and Middle Devonian. This is the longest and most extensive Devonian reef belt of Russia, developing almost continuously, with some breaks, from the Late Silurian into the Late Devonian over more than 3000 km from the southern end of the Urals to northern Novaya Zemlya (Zadoroshnaya et al., 1990a; Kuznetsov, 2000).

Novaya Zemlya Island, the extension of the Ural fold belt, shows reef development in the early Emsian, then middle to late Eifelian through middle Givetian, with erosion at the end of the Givetian (Cherkesova, 1988; Zadoroshnaya et al., 1990a). In the Russian arctic, Taimyr and Severnaya Zemlya represent the northern flanks of the Siberian plate, while the New Siberian islands, e.g., Kotelny, were part of the Kolyma plate. On the northeastern slopes of the Urals, Givetian patch reefs were accompanied by volcanics, indicating probable island fringing reef types (Antoshkina, 1998). Vaigach Island and Pai-Khoi peninsula of the polar Urals had extensive seaward- and eastward-prograding reef massif outcrops, with reefs to 80 m thick, in the Eifelian and Givetian. Similar reef complexes stretched southward almost the entire length of the eastern slopes of the northern Urals to northern Kazakhstan for over 2000 km (Zadoroshnaya et al., 1990a). On the tectonically active east

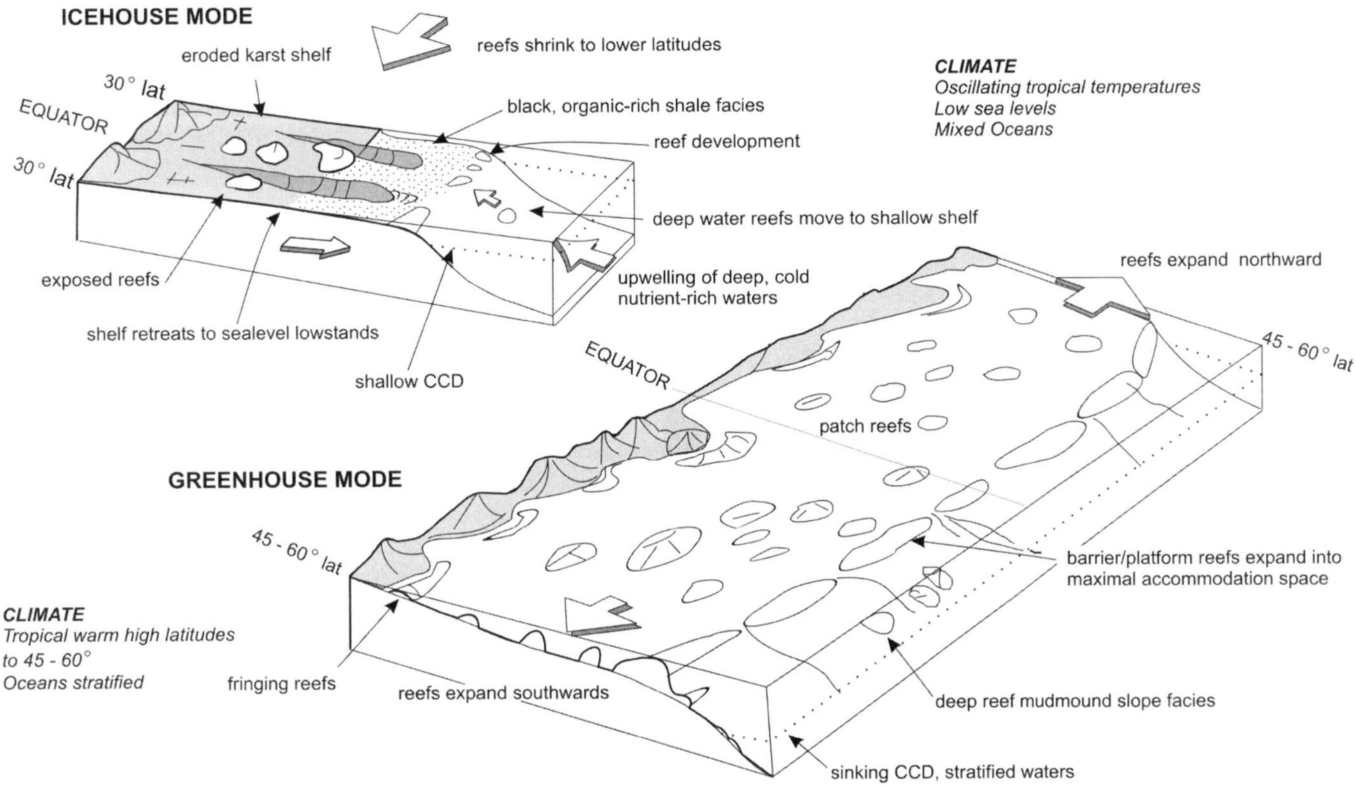

Figure 5. Schematic sketch of reefs in the Devonian calcite ocean system. During the Middle Devonian supergreenhouse (lower figure) with sealevel highstands, and stratified warm surface waters, reefs expanded into high latitudes (40°–60°) and occupied vast accommodation space on flooded continental interiors, as well as wide platforms. During the Late Devonian (Famennian: upper figure) sealevel lowstands and regressions favored higher Carbonate Compensation Depths (CCDs), and more vigorous oceanic advection, as icehouse episodes shrank reefs to lower, equatorial latitudes (corresponding to Holocene interglacial norms).

flanks of the south Urals, large and thick reef complexes were also developed from the Emsian-Eifelian through late Givetian, associated with andesitic-basaltic tuffs, followed by an erosional hiatus and further Frasnian reef growth (Stepanova et al., 1985). Zadoroshnaya et al. (1990b) noted the establishment of Famennian microbial ("algal") fringing reefs and atolls on the mobile, volcanic eastern island arc belts, showing that such reef complexes were maintained as carbonate platforms from the Middle Devonian metazoan-dominated suites.

South European Microplates (The South European Archipelago, or SEA)

Flanking the north edge of Gondwana proximal to northeast Africa were a suite of micro-island plates up to 500,000 km^2 in area, e.g., (1) Spain, (2) Sardinia, (3) possibly parts of Armorica, and extension to the Prague Basin, (4) the Montagne Noire, and (5), the Carnic Alps to Croatia. Here, these are labeled the South European Archipelago (Fig. 1). Whether these were fused into a single plate that subsequently separated into smaller units or were always separate plates forming an archipelago is unclear. Vai (1991) suggested that most of these microplates were contiguous to North Africa and that the Carnic Alps were perhaps located on the south flank of neighboring Kazakhstan. Another aspect not generally agreed upon is how close these plates were to the north margins of Gondwana in the Devonian, or whether some plates were indeed part of Laurussia, the "Old Red Continent." Franke (1999), in contrast, took the view that the Rhenohercynian belt was a passive plate margin extending from the Avalonia microplate and fused to Laurussia, and that Armorica was incorporated therewith, as well as the Bohemian massif, the Prague Basin, and the Sudetes. Hladil et al. (1999) also placed Moravia and the Prague Basin adjacent to the Old Red Continent to the north. In this paper, the South European Archipelago is shown as an archipelago along the northern margin of Gondwana, as indeed there appear to be closer faunal similarities between the Massif Armoricain, Montagne Noire, Prague Basin, and the Carnic Alps, and distinctions from the Old Red faunas. If these microplates are considered as a discontinuously distributed, "Bohemian facies" reef tract from Spain through the Montagne Noire and Prague Basin (Golonka et al., 1994), it would have been ~1600 km in length.

However, Hladil et al. (1999) also indicated in their palinspastic reconstruction that the Moravian karst area was part of a large, tectonically dismembered, Rhenish-type (or Bohemian-type) basin along the southern margin of Laurussia from the Emsian through Carboniferous. Considered as a unit, or as proximal units, these plates roughly straddled 30° latitude in a position due north of the Sahara, i.e., between Morocco and Libya. Baltica had already collided with the eastern margins of Laurentia in the Late Silurian or Early Devonian, closing the Rheic Ocean to form a single, large "Old Red Continent," also known as "Laurussia." There was progressive closure of the Rheic Ocean that separated Laurussia from the South European Archipelago from the Emsian through Givetian. Initially, this oceanic separation was on the order of 1500 km, but by the end of the Givetian, some millions of years later, the Rheic Ocean had shrunk to a sea lane <300–500 km wide (Fig. 4). This must have deflected warm currents back into a large ocean surrounded by the horseshoe made by Laurussia and central to eastern Gondwana, perhaps stimulating the dramatic losses in the late Givetian, Old World benthic coral, stromatoporoid and brachiopod faunas.

During the Emsian, reefs and coral biostromes were patchily, but vigorously developed along relatively short platforms and margins of northern Spain (Cantabria), and the westerly Pyrenees (Fernandez-Martinez et al., 1994). Carbonates flanked the northern side of the Spanish plate, from the Oviedo area to the Mediterranean Pyrenees, for ~800 km. Massive, possibly reefal carbonate facies of Emsian age occur in the Basque Pyrenees and in the subsurface, according to Blieck et al. (1988). Though reefs were apparently missing in the Eifelian of Spain, they were present in the central Pyrenees, and the whole region was reefal in the Givetian, from east to west. In the Montagne Noire of France, deepwater, aphotic, nonrigid sponge mudmounds with abundant stromatactis structures, in slope facies, were the main late Emsian reef development (Bourrouilh and Bourque, 1995; Bourrouilh et al, 1997; Flajs et al., 1996a), though they disappeared thereafter in the Devonian.

In Sardinia, the more than 60-m-thick mudmounds in the Pragian Mason Porcus Formation contain tabulate and colonial rugose corals (Gnoli et al., 1981, 1988, 1990). Emsian-age reefs with Uralian affinities, especially the *Karpinskia* giant brachiopod fauna, were present in western Slovenia (Krstic et al., 1988) and the Carnic Alps (Vai, 1998). Vai (1991) reconstructed a deep-water Bohemian facies and shallow water, isolated, Bahama-type platforms with reefs in Uralian facies for the Devonian. Reefs appeared here in the Lochkovian; larger reefs up to 350-m-thick identify the Pragian and extended maximally during the Eifelian, culminating in the Givetian. Vai (1998) placed the Carnic Alps–Slovenia platform as an extension of the Uralian platform on the east margin of the Russian Platform, a position fully justified on the basis of faunal similarities of the reefs. During the Frasnian, smaller Carnic reefs were present, and Late Devonian extensional tectonics made these platforms founder and drown (Vai, 1998). An alternative explanation might be that due to the closure of the gap between Gondwana and Laurussia, and the separation of the Rheic from the Uralian Ocean, as shown by Vai (1998), reefs were eliminated via cold waters upwelling from the southeast, as warm waters were diverted on the Uralian side. This concept is supported by increased radiolarites spreading into shallow waters in the latest Devonian and Carboniferous (Vai, 1998). From the Frasnian to Famennian, carbonate production on the Carnic platforms dropped from 150 m to <25 m thickness, a loss of more than 90% (Ferrari and Vai, 1966). On the Austrian side of the Carnic Alps, Schönlaub (1998) suggested a depocenter for reefs in the Kellerwand and Hohe Warte area, which produced 1100 m of shallow water carbonates with patch reefs in the Lochkovian through Pragian, and a massive platform reef peak in the Givetian-Frasnian.

In the Prague Basin, there were both mudmounds and small coral-stromatoporoid patch reefs (the Koneprusy reefal limestone of mostly Pragian age; Chlupac, 1988), but the Prague area was already 10° closer to the equator than the Carnic Alps. Emsian through early Eifelian mudmounds and bedded limestones with stromatactis fabrics in the Koneprusy area were described by Flajs et al. (1996b); this facies persisted for several million years. On the Bohemian Massif, Moravian reef buildups covered roughly 6670 km^2 during the Eifelian and Givetian, continuing into the Frasnian (Hladil, 1986, 1988). Reef growth appears to have been temporarily disrupted during the late Givetian during one of several megacycles (Hladil, 1994). Reef decline began in the late Frasnian, and reefs disappeared by the Famennian. Hladil et al. (1999) identified considerable collision activity and crustal shortening in the Emsian of the Saxo-Thüringen through Moravian areas, some of which were detrimental to reef growth. In the former Yugoslavia, massive reef limestones are known from Emsian and Middle Devonian rocks astride the Bosnian sill in the Dinaric Alps (Vranica Mountain: Krstic et al., 1988). Some of these are preserved as slumped reef breccia carrying a Uralian fauna.

Northwestern Gondwana (Saharan Reef Tract)

The largest continent of the Middle Paleozoic was Gondwana, consisting of Africa, South America, Arabia, India, Antarctica, and Australia (Figs. 1, 3–4). Most of the marine shelves of Gondwana, as it centered around the south pole, represented cold climate shelf areas, replete with siliciclastic sediments and a Malvinokaffric, cold to cool temperate invertebrate fauna. Carbonate platforms and reefs were absent except on the northern fringes reaching the 45° to 30°S latitudes around the belt from Mauritania-Morocco into Algeria. Reefs are still unknown to the east in Tunisia and Libya. The reefs of the western "Spanish" Sahara near Zemmour (Saharawi People's Republic) mark the southern end, and reefs extended, probably discontinuously, to Morocco and Algeria for some 1200 km. The northern shelf to ramp tract was up to 200 km wide along the north margin of Gondwana at the limits of coral reef growth in late Emsian time. It appears to have consisted of an inner shelf, separated from an outer shelf by a land area assigned to the western Meseta (Gendrot et al., 1969; Gendrot, 1973; Benbouziane et al. 1993). El Hassani and El Kamel (2000) suggest the primary controls of patch reef, atoll, and barrier reef development in the Moroccan Meseta were tectonic highs. Elloy (1973), Benbouziane et al. (1993), Cattaneo et al. (1993), Hilali et al. (1998), and El Hassani and El Kamel (2000) noted the growth of shallow water coral-stromatoporoid reefs on the Moroccan shelf Meseta continued from the late Emsian, and may have reached a peak during the Givetian. The area was emergent by the middle Frasnian.

The relatively thinly developed, shallow water carbonates and small patch reefs of the Meseta and Maider Basin differ from deeper water Kess-Kess mudmounds in the Emsian of the Tafilalt area to the southeast, with flank deposits rich in nautiloids and ammonoids. Montenat et al. (1996) proposed that the Kess Kess mounds may have been tectonically controlled, but others suggest these were the result of microbial activity around submarine hydrothermal vents (Belka, 1998), and thus their development was unrelated to that of normal shelf reefs or slope mudmounds affected by climatic events. Moroccan (Anti-Atlas and Meseta) and West Saharan reefs of Emsian age gave way to relatively small tracts of coral-stromatoporoid patch reefs, shelf atolls, and small barrier banks in the Eifelian and early Givetian. This suggests relatively warm surface waters in the Middle Devonian, despite high latitudes (Figs. 3–4). However, reefs completely disappeared in northwestern Gondwana during the Frasnian and Famennian. Pedder (1999) indicated that the geographic affinities of the early Givetian rugose coral fauna of Morocco, at a time when reefs still occurred in the Maider Basin, were with faunas from Spain, the Pyrenees, and Moravia, situated on island plates to the north, and that no more than 17% of the Moroccan fauna had taxa in common with eastern Laurentia (i.e., the Appalachians).

North Central Gondwana (Near East: Iran, Afghanistan, Pakistan)

On the north-central flanks of Gondwana, sponge-coral reefs appear to have been absent. In the Arabian peninsula, part of the African-Arabian plate, small microbial ("stromatolitic") mudmounds of late Emsian age, up to 10 m in diameter and <5 m high occur; these are associated with bryozoans and rare solitary rugosans (Al-Laboun and Walthall, 1988), representing a relatively cool water, or stressed, reef spectrum. The Arabian microbial reefs were located at about the same high latitude of 45°S as the Saharan reef belt (above); no other reefs are reported to date from this area. To the north of this region, in Turkey and the Caucasus (Georgia, Armenia), reefs appear to have been undeveloped in the Emsian mixed siliciclastic-carbonate facies through the Middle Devonian, though coral and stromatoporoid faunas are known (Mamedov and Rzhonsnitskaya, 1985). These areas were possibly an extension of the Mediterranean terrane blocks flanking the northern margins of Gondwana.

To the north of the Gondwana supercontinent, adjacent small plates of central Asia, such as those from southern Afghanistan and Pakistan, were episodically marked by well developed reefs during the late Early and Middle Devonian (Stauffer, 1968; Gaetani, 1968). For example, Emsian reef limestone was present along a volcanic belt in central Afghanistan, according to Wolfart and Wittekindt (1980). This reef belt was said to extend into northern Pakistan. Mistiaen (1995) noted that in Afghanistan, small, <10-m-high stromatoporoid, tabulate coral, microbial reefs also began in the Emsian, corroborating that reefs there in the 30°S latitude were developed on a series of microplates, or a large, elongated plate situated close to the northern flanks of Gondwana. Whether Emsian reefs were present in the neighboring areas of Iran (Gaetani, 1968; Dastanpour, 1996) and Pakistan (Stauffer, 1968), dominated by siliciclastics and evaporites at this time, is unclear, though the regions have a similar geologic history. Assuming Middle Devonian reefs were developed from central Iran through southern Afghanistan and into neighboring Pakistan, this would have been a tract ~1800 km long (Table 1). There are extensive hiatuses in the Early Devonian and Eifelian of Iran (Dastanpour, 1996) and Afghanistan. Brookfield and Gupta (1988) reviewed the Devonian of the Pakistan through Himalayan regions, though reefs were unidentified, and the dating of many units remains in question. During the late Middle Devonian and Frasnian, reefal carbonates were expanded into eastern Iran and Pakistan (Wolfart and Wittekindt, 1980). Jux (1969) and Mistiaen (1985) have also indicated that reef growth continued in this region during the Frasnian.

Tian Shan Fold Belt (Uzbekistan, Tadzhikistan, Kirgizstan, Northern Afghanistan)

The Tian Shan Fold Belt (Russia), including areas of Uzbekistan, northern Afghanistan, and Tadzhikistan, contains widespread Emsian through Middle Devonian reefs, as well as some Frasnian reef complexes (Zadoroshnaya et al., 1990b; Dronov and Natalin, 1990). These areas appear to represent smaller, mobile "island continent" terranes, proximal to the northeastern and central margins of Gondwana. In southern Uzbekistan, thick coral-stromatoporoid-calcimicrobial reef platform facies, with back reef lagoons, appear to be more prominent in the Middle Devonian Khodzhakurgan Formation than in Early Devonian strata dominated by volcanics and siliciclastics (Kim et al., 1984). In southern Fergana (Tadzhikistan), the upper part of the atoll-like Kiziltash reefal mudmound, 5 km long and ~500 m thick, is of Eifelian age (Dronov, 1993). To the southeast of Tadzhikistan, in the central Pamir Mountain ranges, reefs of Emsian through Givetian age were also developed on a miogeosynclinal platform up to 1.5 km thick (Ioganson, 1990b). The tectonic relationship of the Pamir Reef belt to the Tarim-Karakorum plate is disputed (see below). The Uzbekistan-Pamirs reef tract, starting at the Nuratau ranges in the west, shares much in common with Middle Devonian reefs of central Tadzhikistan and the Tadzhik Pamirs, and northern Afghanistan. This Pamirs tract was just less than 1000 km long, if it ended in the Pamir ranges, and longer if it joined the Chinese areas along the borders (see Karakorum plate below).

Kazakhstan

Kuznetsov (2000) outlined a Lower and Middle Devonian reef belt ~1000 km long along the eastern, tectonically active side of Kazakhstan, influenced by calc-alkaline volcanism. Uplifted land areas were located to the west. Lochkovian reefs, located in the 8–15° latitudes north (Zonenshain et al., 1990a) are kilometers in diameter and up to 120 m thick: they do not appear to have been continued into the Emsian and Eifelian, however, though they reappeared in the Givetian. A similar outcrop belt is located in the Tarbagatai Mountains of western Junggar in Xinjiang, China, with Kazakhstan faunal affinities. In the northern and central Kazakhstan Dzungar area, reefs were renewed in a Givetian carbonate pile up to 2 km thick, but reefs were missing in the Frasnian (Zadoroshnaya and Nikitin, 1990). Famennian through Tournaisian calcimicrobial "Tubiphytes" patch reefs, atolls, and mudmounds were developed on a carbonate platform up to 150 km wide and 1000 km long on the northwestern side of Kazakhstan (Karatau ranges), but not the eastern flanks (Pavlov et al., 1988; Cook et al., 1994). In the "Junggar" region of northwestern Xinjiang in China, a rich early (Emsian) and Middle Devonian coral fauna indicates affinities with Kazakhstan (Deng, 1999), but reefs are undescribed thus far. If Middle Devonian reefs were developed from the eastern side (Kenzhebasai River, west of Balkhash Lake) of the extensive carbonate platform in Kazakhstan to proximal Xinjiang province of China, this tract would be ~1600 km long.

Karakorum-Tarim Plate

Today this plate extends from the Tarim Basin in western China through the Karakorum ranges of North Pakistan (and possibly the Pamir ranges of Tadzhikistan). An ophiolite belt separates the Tarim Block from the Kazakhstan and Dzungar plates today, indicating the probable initial separation and subsequent collision of these plates took place after the Devonian. It has been suggested that the Karakorum Block and Tarim Block may have been separated (Talent et al., 1986); here it is taken as a single unit, following current Russian usage (Belenitskaya and Zadoroshnaya, 1990). What its relationship is between the Tien Shan ranges of Tadzhikistan and the neighboring areas of China is unclear. The southern Pamir fold belt contains both stable miogeoclinal carbonate platform and unstable margins with allochthonous debris flows. Rugose, tabulate coral and stromatoporoid reefs indicate normal tropical carbonate shelf conditions, with the sequence up to 1.5 km thick on the Russian side (Ioganson, 1990b). The southern Chinese Tian Shan Range is shown as an accretionary collisional fold belt on the southern side of Kazakhstan by Yolkin et al. (2000). This suggests that the area may have been constructed of smaller plates accreted to Kazakhstan or the Dzungar at a later date. For the Karakorum-Tarim plate, Kuznetsov (2000) reconstructed a very elongate continent up to 5000 km long, with a ~1000-km-long reef belt mostly between the 5° and 10°N latitudes and in the northern Pamir to Tarim portions. Thus far, little information on reefs is available from the Tarim Basin in China (Zhou and Chen, 1992; Liao et al., 1992) and the Tian Shan and Karakorum Ranges (Wen, 1998). Where the Tian Shan fold belt, presently located southeast of Kazakhstan and neighboring China, was situated in the Devonian seems unclear, or disputed. Sun and Chen (1998) cited Emsian reefal limestones from the Hongshandaban Formation of the Karakorum-Kunlun Mountains, and regarded their brachiopod faunas as similar the Old World faunas of the Urals and South China.

Siberia

On the southwestern margins of Siberia (Salair, Kuznetsk, Minusinsk Basins), Paleozoic reefs reached their maximal development in the Devonian (Yolkin et al., 2000). There is some debate whether Siberia and Mongolia and the northeastern plates were close to each other, or distal, or where they were located latitudinally. In the Salair Range, coral-stromatoporoid reefs were present in the middle Lochkovian and Pragian, and discontinuously from the Emsian through Givetian, with a gap in the late Givetian and early Frasnian (Yolkin et al., 1997). These Emsian-Frasnian reefs were part of a major belt stretching north-south in west central Siberia, about 1500 km from the middle Enisei River to the Gornoi Altai (Ioganson, 1990c), with peak development in the Middle Devonian primarily in stable platform settings. Ivanova et al. (1964) plotted Eifelian coral-stromatoporoid reefs for the Minusinsk and Kuznetsk basins, but identified volcanism and regression in the Givetian. On the far western flanks of the Siberian platform, i.e., the eastern slopes of the Urals and the subsurface of the western Siberian lowlands, there was a tectonically active collisional zone in the Devonian (Belenitskaya and Zadoroshnaya, 1990). Fringing microbial reefs flanked areas with active flysch development in the Devonian, along the present day eastern slopes of the Urals (Chuvashov and Shuiskii, 1988). It is difficult to interpret such reefs as being either part of the Russian platform or western Siberia. In north central Siberia, the Taimyr peninsula, a platform carbonate with reefs was present in the Givetian, associated with back reef *Stringocephalus* (Cherkesova, 1988).

In southeast Siberia, close to the Amur River border (Amur-Okhotsk fold belt), massive microbial-coral and bryozoan reefs of Eifelian age range from 250 to 800 m thick and extend discontinuously for nearly 400 km (Belyaeva and Ioganson, 1990). These apparently reflect a passive margin and broad shelf setting. Whether these may be tectonically part of the Mongolia or North China plates or whether they pertain to the southeastern margin of Siberia is uncertain. Another remote Middle Devonian reef area several hundred kilometers long is indicated in the Khabarovsk region (Kuznetsov, 2000).

Verkhoyan-Kolyma Plate (Northeast Russia)

Devonian reefs were well developed along the western sides of the Kolyma block of northeastern Russia. Lower Devonian reefs of the Tas-Khayakhtakh Mountains are exposed in a belt about 3–10 km wide and 150 km long, with reefs to 200 m thick, and from 0.5 to 3 km long (Ioganson and Baganov, 1990). Reef frame-builders were calcimicrobes, algae, and tabulate corals, with growth ranging from the middle Lochkovian, peaking in the Pragian and ranging to the Emsian. A break in the Eifelian, with more siliciclastic input followed by a disconformity, saw renewed vigorous algal-tabulate-amphiporid reef growth in the Givetian with reefs extending ~100 km in length and up to 500 m thick into the Frasnian (Vishnetskii and Baganov, 1986). These Lochkovian-Pragian reef belts were further spread ~200 km into the upper Selenyakh River reaches to the north. Givetian and Frasnian reefs were also present northwest of the Kolyma River, in a sequence up to 500 m thick on the stable platform of the Ulakhan-Tasskoi area (Ioganson and Bazanov, 1990). In the southwestern Kolyma River zone, the Alazei foldbelt, decameter-thick "biohermal massifs" are dominated by corals, stromatoporoids, and algae of Early Devonian age associated with dolomites and evaporites up to 1 km thick (Ioganson, 1990a). Dolomitized reef complexes are also present in rocks of Givetian and Frasnian age, but not as thick as those preceding. If these Kolyma Reef belts are part of a large platform, this may have stretched more than 800 km. Devonian reefs appear to be generally absent, however, in the Chukot block to the northeast, though they are present in the Middle Silurian and Early Carboniferous. An exception may be along the Omolon River zone, where an active margin with siliciclastics, andesites, and tuffs featured 40-m-thick patch reefs with algae, corals, and stromatoporoids of Middle Devonian age, but no Early or Late Devonian reefs (Ioganson, 1990a). The northern margin of the Kolyma plate featured the development of early Eifelian reefs on offshore Kotelny Island in the Arctic Ocean (New Siberian Islands; Cherkesova, 1988). How much their faunas match those of the Canadian Arctic Blue Fiord reef faunas is not yet clear, but both areas lack Frasnian carbonates, except for Banks Island.

Mongolia (Tuva Plate)

Various reconstructions have placed Mongolia, parts of the Russian Altai, and Inner Mongolia (the Tuva plate) in high north latitudes of 45–60° during the Devonian, as a lateral extension of an inverted Siberian plate (Golonka et al., 1994). Yolkin et al. (2000) interpreted the Siberian Devonian continent as including Mongolia, basically sutured in the alignment they possess today. Kuznetsov (2000) showed the Altai and Salair as two separate plates in the vicinity of Mongolia. He also marked central Mongolia, with Devonian reefs, as an isolated, elliptical small continent about 1500–2000 km long, roughly halfway between north China and Kazakhstan, in the lower 20–30°N latitudes. Rong et al. (1995) argued for alignment of the Late Silurian Mongol-Okhotsk faunal province as fixed to the eastern side of Siberia, i.e., no separate plate for Mongolia. Unique *Tuvaella* brachiopod and endemic Tuva-Mongolian coral faunas mark a high degree of provinciality in the Late Silurian and Early Devonian, and these argue for separation of both parts of Mongolia from Siberia. Thus, two interpretations exist, one uniting Tuva and Siberia during the Devonian, and the other creating an isolated paleocontinent for the Tuva-Okhotsk-Mongol region, separate from Siberia and possibly extending to the Hingangling Ranges of northeastern China.

The development in Mongolia of extensive carbonate platforms, and commonly reefs, took place from the Late Ordovician (Caradoc) through Middle Devonian, ceasing in the late Givetian (Minjin et al., 2001). Sharkova (1986) plotted an 1800-km-long, ~150-km-wide reef platform, flanking the southern side of present Mongolia. Tabulate-rugose coral reefs are developed in the Orgol Formation (Lochkovian) and upper Dungnee Formation (Emsian) in Mongolia (Suetenko et al., 1977). In the Shine Jinst area of south-central Mongolia, the early Emsian Hurenboom tabulate and rugose coral "reef," or carbonate bank, is locally developed in the Chuluun Formation, extending into the Eifelian (Sharkova, 1980, 1986; Minjin, 2001). Facies indicate an active tectonic margin interrupted by volcanics to the south and a stable continent to the north (Minjin et al., 2001). Although carbonate deposition continued, reefs ceased by the Givetian, and thinly bedded black limestones and shales took over in the Famennian and Tournaisian (Alekseeva, 1993).

North China Plate

The tectonically active north margin of the North China plate saw reef development in the Emsian, Eifelian, and early Givetian in Inner Mongolia and Heilongjiang provinces. Such reefs were associated with strong siliciclastic pulses and volcanism. Tabulate-rugose coral patch reefs of Emsian age belonging to the Unur Formation occur in the Hingang Mountains (Su, 1988). The Dean Formation of Eifelian age contains coral reef limestones rich in brachiopods, and the Hoboshan Formation (early Givetian) contains microbial ("algal") reefs associated with acid volcanics (Su, 1988). Frasnian reefs are absent. In the Famennian, siliciclastics and terrestrial settings took over. This suggests a decline in reef growth from the early Givetian onward. The complete extent of the reef belt is unknown at present, but it was ~700 km long. A relationship to the Amur-Okhotsk area of Russia is possible, though not verified. North China is plotted in very high latitudes between 30° and 60°N during the Middle Devonian by Kuznetsov (2000), in low latitudes straddling the equator by Golonka et al. (1994), and adjacent to Australian Gondwana by Talent et al. (2000). The affinities of the North China plate with Mongolia-Tuva are unclear, but today the two areas are separated by an ophiolite belt. Talent et al. (2000) placed the North China plate against the north shelf margin of Australia from the Early to Late Devonian, with the South China plate to the west, both straddling the equator.

South China Plate

South China was a separate plate south of the equator during the Devonian, probably separated from northwest Australia by a relatively narrow ocean (Talent et al., 2000). The area encompasses a stable platform carbonate belt reaching eastward from Laos, Cambodia, and Vietnam, past Burma, to Yunnan, Guangxi, Guizhou, and Hunan provinces in southern China, for a passive margin more than 1700 km long (Zeng et al., 1992). A collisional, active margin with reefs has not been identified; land areas were on the northern margins of present day South China. How much of the South China plate had reef growth remains uncertain, but coral-stromatoporoid reefs appear centered in the area from Yunnan east via Guangxi, Guizhou, and Hunan provinces, for ~1000 km. Maximal area for carbonate platform development was during the Emsian, Eifelian, and Givetian (Bai et al., 1994; Yu and Shen, 1998; Shen et al., 1994). For much of the earlier Devonian, the continent was emergent, with reefs commencing in the late Pragian as small patch reefs and expanding to larger tracts during the Emsian (Yu and Wu, 1988). In the Nandan Basin of southern Guangxi province, the 830-m-thick Dachang Reef began its growth just above the top of the Emsian *perbonus* zone and continued into the late Eifelian *ensensis* Zone for a duration of ~15 m.y. (Bai et al., 1994).

Maximal reef growth in the southeastern portion of the South China plate was during the middle Givetian, especially in the *varcus* and *herrmani-cristatus* zones of the reefal Tangjiawan Formation, with a lowstand in or at the top of the *disparilis* zone (Shen et al., 1994; Wu and Shen, 1998). In South China, the Givetian-Frasnian boundary is still debated to lie either below or on top of the middle *falsiovalis* or *pristina* zone. There are multiple "nickel events," used in China to define "anoxic" episodes in the *hermanni* through *disparilis* zones, below the Givetian-Frasnian boundary, with a regressive spike in the upper *disparilis* zone, but there was no marked latest Givetian regression or reef proxy signal (Bai et al., 1994). The large Middle Devonian carbonate platform was broken up by late Givetian *varcus* time into smaller, isolated Bahama-type platforms with reefs separated by basinal Nandan-type black shales during the Late Devonian, and metazoans were replaced by calcimicrobes in Famennian reefs (Yu and Shen, 1998).

Australia-Papua New Guinea Plate

Emsian and Eifelian reefs of Australia were primarily confined to the tectonically active, unstable eastern margins of Australia, associated with volcanics and siliciclastics (Crook, 1961; Wolf, 1965). Thus, there was not an apparently large scale, continuous barrier-rimmed platform on which reefs could have been developed, as there was in western Canada, but probably a number of smaller, isolated basins (Talent et al., 1986). Many reef bodies are found as allochthonous tectonic debris flows, presumably as collapsed margins or as slumps along the eastern Australia margins for a distance of more than 2000 km, effectively representing "lost" carbonate platforms as they are only partly preserved today (Talent and Mawson, 1999; Talent et al., 2001). The Nubrigyn calcimicrobial reef blocks, occurring as allochthonous units, are of late Lochkovian through Emsian age (Wolf, 1965; Conaghan et al., 1976). Pohler and Kim (2001) outlined rugose coral framework and "organic buildup facies" in the Emsian of New South Wales. These extended as Eifel-Givetian stromatoporoid reefs close to 20 m thick within the Moore Creek Formation of New South Wales (Pohler, 1998), part of a volcanic island arc complex. In Queensland, Jell and Zhen (1994) cited small coral-stromatoporoid patch reefs, mostly of Givetian age, from the Fanning River Group. No outcrop or drillcore data is available on the extension of the east Australian active margin, e.g. the presence of reefs in Papua New Guinea and West Irian, a tectonically complex area primarily covered by tropical rainforest today but ideally located for reefs in the Devonian paleotropical equatorial belt on the northern tip of the Australian plate (Talent et al., 2000).

On the passive western margin of Australia, Middle Devonian reefs appear to have been scarcer, except for the commencement of reefal carbonates in the middle to late Givetian of the Canning Basin (Read, 1973a, 1973b; Grey, 1991; Southgate et al., 1993). These continued into the more spectacular, exposed, ~400-km-long reef platform during the early and middle Frasnian (Playford and Cockbain, 1989; George and Powell, 1997; George and Chow, 1999), with prominent shallow water stromatoporoids, corals, and shallow as well as deeper calcimicrobes within the photic zone (Paul, 1996). In the subsurface, subaerial karst has been difficult to find in the Canning Basin, and no events are specifically known yet to characterize the end Givetian regressions found elsewhere, as the Pillara Formation apparently spans the Givetian-Frasnian boundary without a break (Read, 1973a; Brownlaw et al., 1996). Emsian to Givetian reefs appear to have been absent in the Carnarvon Basin to the far west.

HOW WERE DEVONIAN MEGAREEFS PRODUCED?

Questions emerge from this global reef snapshot of the Middle Devonian. With examples from the Middle Paleozoic maximum, it is evident that megareef belts were produced in calcite oceans under global supergreenhouse climates featuring high tropical sea surface temperatures. Several factors came to play:

1. Warm sea suface temperatures stretched reefs to very high latitudes of 40–50° (or possibly higher if some reconstructions based on paleomagnetic data are correct), and may be used to constrain paleolatitudinal positions of major plates. High latitude reefs had one drawback in that, at their outer limits, there was reduced solar input during the winter months because of fewer daylight hours, though this may have been balanced by longer summer hours and rapid growth. The Middle Devonian supergreenhouse may have had parallels in the Cretaceous greenhouse model of Poulsen et al. (1999) and Johnson (1999), for which equatorward shifts of atmospheric circulation, polar warming, and high latitude increases in precipitation, climate reorganization of atmospheric low and high pressure regimes in the tropics, and a low pole-equator temperature gradient were proposed.

2. To create such megareefs required sea-level highstand systems tracts flooding wide areas of low elevation continents; i.e.,

they inhabited vast epicontinental seas, unlike limited Holocene open shelf systems (e.g., Givetian, Fig. 4).

3. Megareefs were favored on wide, open passive margins, with good open ocean access, inducing free mixing and normal salinities (Figs. 4 and 5).

4. Large Paleozoic reefs required the evolution of large, reef-building, modular, photozoans (i.e., skeleton builders utilizing photosynthetic symbionts) with rapid composite growth rates, e.g., giant colonial corals such as tabulates and rugosans, and calcareous stromatoporoid sponges, and also a wide selection of encrusting, frame-fixing calcimicrobes (for contrasting view, see Hallock, 1996). Devonian megareefs were exploited by warm temperature-adapted, rugose and tabulate coral framebuilders with calcite skeletons, but the reef niche was also occupied by nonspiculate stromatoporoid sponges with a basal aragonitic, or possibly high Mg-calcite skeleton exposed to light (probably also photozoans). Red and green calcareous algae generally played only a small role in Devonian metazoan reef building.

5. The giant and complex colonial forms of corals had tropical growth rates (40–100 mm/yr; Gao and Copper, 1997) comparable to those of modern scleractinians. This suggests that these forms were zooxanthellate, and that such symbioses assisted in achieving accelerated reef growth. Massive, multi-meter thick and wide stromatoporoid sponges had relatively slower growth rates (1–10 mm/yr), and it is unclear if such forms had dinoflagellate symbionts; i.e., were they photozoans? However, much higher CO_2 saturation states in shallow shelf areas during the Middle Paleozoic suggest that zooxanthellae probably were obligate symbionts and providers of O_2 for calcification, to maintain even modest growth in aragonitic sponges. There is, as yet, no evidence that growth banding was more prominent in high, temperate latitude reef organisms in the Devonian.

WHY DID REEFS CONTRACT WITH CLIMATE CHANGE?

The issue of climate and reef distribution in the distant past is still speculative. Kiessling (2001, p. 754) suggested that "many of the changes [in reefs] are linked to changing nutrient requirements of the prevailing reef builders and nutrient availability in the oceans…ultimately controlled by climate change," following the theme suggested by Wood (1993, 1999). It is difficult to find any direct proof of the relationship of Paleozoic reefs to nutrients or organic-rich sediments. Climate cooling during the Late Devonian (Frasnian/Famennian) mass extinctions was marked by an increase in phosphorites, and silica (radiolarite) deposition (Racki, 1999), and the concomitant loss of metazoan coral-stromatoporoid sponge reefs. It seems more logical that cooling reduced carbonate deposition rates by accelerating the death of tropically adapted, zooxanthellate reef builders (photozoans), and that the loss of metazoan reefs was not directly connected to nutrient increase. Joachimski and Buggisch (1996) confirmed positive $\delta^{13}C$ values shifting about +2‰ above background norms for the two late Frasnian Kellwasser organic-rich carbonates in the late *rhenana* and *linguiformis* conodont zones, just below the F/F boundary. They interpreted this rise as marking global flooding surfaces with relatively short-lived sea-level highs (accompanied by carbon deposition from deep, anoxic bottom waters dumped onto the shallow shelf), as mechanisms inducing cooling, i.e., that anoxia triggered cooling (sometimes called the "preservation model"). A counter explanation might be that cooling and vigorous watermass overturn triggered upwelling of nutrient-rich bottom waters, stimulating high surface plankton productivity, as well as growth of benthic algal turf on the reefs, and that high rates of carbon burial under shallowing and regressive shelf conditions produced the black shales (as suggested in modern "production models" by Pedersen and Calvert, 1990). This scenario would fit the widespread Devonian deposition of black shales under regressive coastal conditions, as seen in many late Frasnian and Famennian boundary sections (Copper, 2002). In contrast, Murphy et al. (2000) suggested a "seasonal mixing-efficient cycling" model (decoupled from water depth and temperature), to account for black shales in the Devonian Appalachian Basin.

Reef boundary conditions in the Middle Paleozoic were set largely by climatic cooling events, with the 90 m.y. long reef megacycle starting in the Silurian and terminating in the Devonian (Fig. 6). The Middle Paleozoic cycle commenced in the Early Silurian, 3–4 m.y. after the Ordovician-Silurian boundary triple glaciation, though the Silurian biota had already been initiated in the latest Ordovician stage (Hirnantian: Copper, 2001). Regressions in the Late Silurian (Pridolian) continued in the Early Devonian (Lochkovian-Pragian), when reefs retreated equatorward, became smaller in size, and endemic reef provinces were established. No glaciations are known from this time. Renewed glaciation during the Late Devonian (Famennian: Isaacson et al., 1999) marked major cooling episodes that eliminated most of the coral and stromatoporoid reef-building taxa and the metazoan reef ecosystem. Over the 21 m.y. long Famennian "recovery" period, patch reefs were populated by calcimicrobes, rare stromatoporoids, and lithistid sponges, and reefs were small-sized, or mud-mound-type structures in both shallow water and slope settings (Webb, 2001; Copper, 2002). At this time of climatic perturbation, the oceans moved to oscillating icehouse modes, the carbonate compensation depth (CCD) rose into shallower waters, as it does in cold oceans today, and oceans probably shifted largely to an aragonite mode of skeleton production (Sandberg, 1983; Webb and Sorauf, 2002), that reduced the importance of the remaining tabulate and rugose corals. Famennian surviving tabulate and rugose corals probably derived from relict refugial, deep-sea forms, persisted in the Tournaisian, but they rarely built significant reefs in such later Paleozoic shelf settings.

During the Late Paleozoic, boundary conditions for reef growth were changed by alignment of tectonically amalgamated plates that fused and shaped the supercontinent of Pangea. This horseshoe-shaped Pangea, opened eastward, blocked equatorial currents in low latitudes, producing a very warm Tethys, but creating large, cool shelf areas on the western sides of continents. Secondly, repeated glaciation-induced sea-level drawdowns

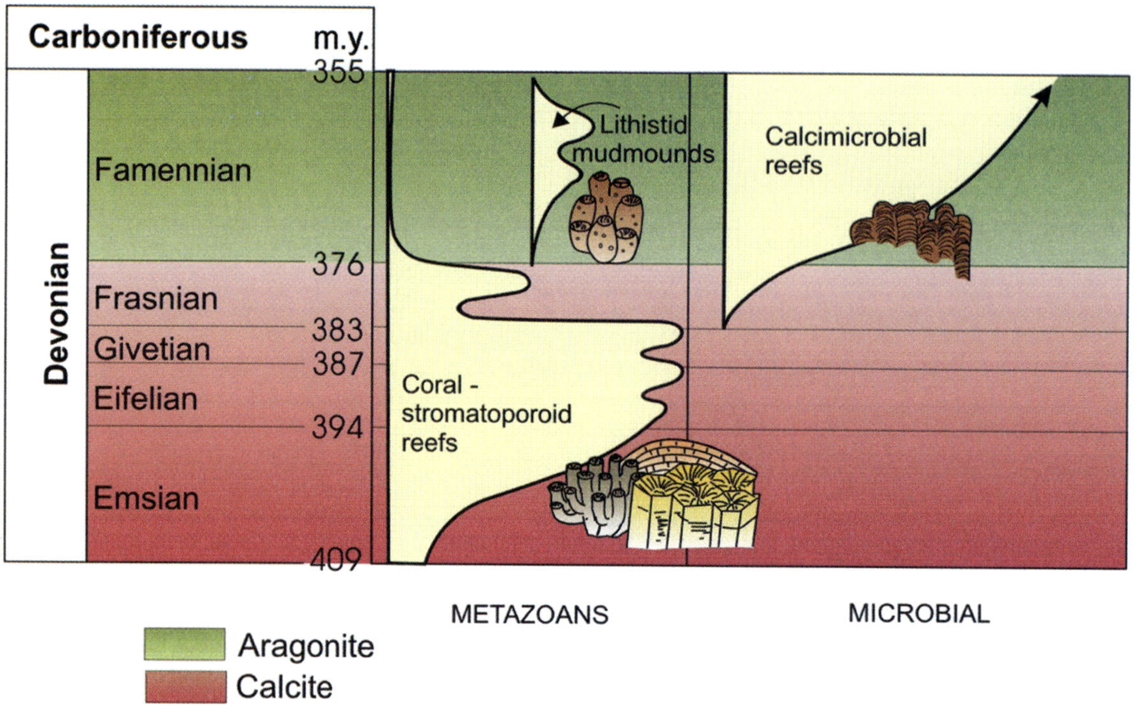

Figure 6. "Reef curve" during the Eifelian-Givetian, and reef crises during the Late Devonian as seen in expansion and subsequent loss of coral-stromatoporoid reef builders in the late Frasnian and rise of episodic development of regional mudmounds and calcimicrobial reefs in the 21 m.y. long Famennian. Full recovery of large-scale metazoan coral-sponge reefs never occurred in the switchover to aragonite oceans during the Famennian and Late Paleozoic, though small patch reefs of lithistid and stromatoporoid reefs were sporadic in the late Famennian (Strunian).

during regressive episodes eliminated reef and carbonate platform accommodation space, reduced the sizes of epicontinental seas, and shrank warm seas to lower paleolatitudes. This situation was comparable to that of the Pliocene-Pleistocene, with the closure of the Isthmus of Panama (Budd and Johnson, 1999). Thirdly, additional drawbacks in the Late Paleozoic were mountain-building cycles as a result of continental collisions: these triggered increased siliciclastic sedimentation and probably increased volcanism, drowning reef platforms in muds and silts, as in the Bay of Bengal and Kalimantan today. Fourthly, in the tropical belts, as warmer, more humid air was displaced northward into higher latitudes, heavy rainfall produced lower salinities in coastal shelf areas, probably favoring the establishment of large-scale Carboniferous Waulsortian mudmound growth, on which light- and temperature-independent heterozoan biota survived under cooler, stressed conditions (e.g., as for the Great Australian Bight today; James, 1997).

CONCLUSIONS

The Devonian lasted some 63 m.y. (Okulitch, 1999), about the same duration as the Cenozoic, with the 26 m.y. long Emsian-Givetian episode representing the peak in Phanerozoic global carbonate production and widespread reef growth. Such coral-stromatoporoid reefs expanded with high RCO_2 levels and super-greenhouse climates and crashed as O_2 levels rose (Berner, 1997). The following are some general observations:

1. During the Emsian-Givetian, reefs in low latitudes of <40° were dominated by rugose and tabulate coral-stromatoporoid frameworks and keystone taxa, commonly with a significant calcimicrobial component acting as binders and encrusters in the reef core. The Givetian marks the peak of Devonian reef distribution and carbonate platform growth, following progressive expansion of metazoan reefs from the Emsian onward. The end Givetian marks a major collapse of reefs, exceeded only by the late Frasnian.

2. Reefs in higher latitudes (40–50°) were generally mudmounds of enigmatic origin, with a sparse metazoan cover (e.g., Kess-Kess mudmounds of Morocco and Algeria, mudmounds of the Montagne Noire, and Sardinia). Mudmounds less commonly featured prominent calcimicrobial biota such as *Girvanella, Rothpletzella,* and *Wetheredella*, suggesting that many grew in deeper waters or in shallow seas with high surface plankton production, diminishing light penetration. Reefs with coral-stromatoporoid frameworks were exceptional in higher latitudes.

3. As reefs declined, higher latitude reefs vanished first during the Late Devonian, e.g., on the northern margins of Gondwana (Morocco-Libya, Montagne Noire), though large reef tracts

in some lower paleolatitudes also vanished (e.g., Frasnian of the Canadian arctic).

4. Pre-Middle Devonian (Pridolian-Lochkovian-Pragian and early Emsian) reefs were patchily developed, probably largely as a result of relative sea-level lowstands, limited carbonate platform, and epicontinental reef accommodation space. Faunal provincialism was widespread.

5. Post Middle Devonian (Frasnian) metazoan reefs showed relatively lower, albeit cosmopolitan, biodiversity of corals, stromatoporoids, and accompanying shelly biotas, related to major losses of tropical reef-dwelling subfamilies and families toward the close of Givetian time. The Frasnian featured a restricted expansion of metazoan reefs worldwide, though metazoan reefs were prominent regionally in the Canning Basin of Australia, South China, the Urals, Banks Island, and Alberta.

6. The use of the reef database to reconfigure the latitudinal position of major tectonic plates seems feasible. Our reconstructions, for example, show that the Mongolia-Tuva plate may be located considerably further south than previously indicated (Golonka et al., 1994).

7. Care should be taken in drawing broad conclusions about reef latitudinal constraints when dealing with long periods of time, e.g., over millions of years. For example, within the 63 m.y. long Devonian, sharp variations in reef development are recorded, just as for the Cenozoic, covering roughly the same time span.

ACKNOWLEDGMENTS

This Devonian reef summary is only a preliminary version. Much of the database needs to be verified and expanded, and we would be grateful to readers for pointing out omissions and corrections. Paul Copper is especially grateful to Russian colleagues Zadoroshnaya, Kim, Rzhonsnitskaya, Antoshkina, and Yolkin for contributing their knowledge and literature of Devonian reefs to the larger history of Middle Devonian carbonate platforms globally. Paul Copper is grateful to the Natural Sciences and Engineering Research Council of Canada, which provided the support needed to carry out the long-term studies and field work around the world. A more complete Devonian reef reference list is available from authors as an unpublished appendix.

REFERENCES CITED

Alekseeva, R.E., 1993, Biostratigraphy of the Devonian of Mongolia: Joint Soviet-Mongolian expedition, Moscow: Nauka, v. 44, 265 p.

Al-Laboun, A.A., and Walthall, B.H., 1988, The Devonian of the Arabian peninsula, *in* McMillan, A.F., Embry, A.F., and Glass, D.J., eds., Devonian of the world: Calgary, Canadian Society of Petroleum Geologists, v. 1, p. 557–568.

Antoshkina, A.I., 1998, Organic reefs and buildups of the Palaeozoic platform margin, Pechora Urals, Russia: Sedimentary Geology, v. 118, p. 187–211.

Bai, Shunliang, Bai, Zhiqiang, Ma, Xueping, Wang, Darui, and Sun, Yuanlin, 1994, Devonian events and biostratigraphy of south China: Beijing, Beijing University Press, 303 p.

Belenitskaya, G.A., and Zadoroshnaya, N.M., eds., 1990, Rifogennye i sulfatonosnye formatsii Fanerozoya SSSR [Phanerozoic reefal and sulfate formations of the USSR]: Moscow, Nedra, 292 p.

Belka, Z., 1998, Early Devonian Kess-kess carbonate mudmounds of the eastern Anti-Atlas (Morocco), and their relation to submarine hydrothermal venting: Journal of Sedimentary Research, v. 68, p. 368–377.

Belyaeva, G.V., 1986, Reef reservoirs of the Pechora petroleum basin, *in* Sokolov, B.S., ed., Fanerozoiskii rify i korally SSSR [Phanerozoic reefs and corals of the USSR]: Noscow, Nauka, p. 197–202.

Belyaeva, G.V., and Ioganson, A.K., 1990, Amuro-Okhotskaya skladchataya sistema [Amur-Okhotsk foldbelt system], *in* Belenitskaya, G.A., and Zadoroshnaya, N.M., eds., Rifogennye i sulfatonosnye formatsii Fanerozoya SSSR [Phanerozoic reefal and sulfate formations of the USSR]: Moscow, Nedra, Moscow, p. 85–88.

Benbouziane, A., El Hassani, A., and Fedan, B.,1993, The Silurian-Devonian carbonate formations of Oulad Abbou and Mechra Ben Abbou (Moroccan Meseta): Field Guide Excursion A3, 14th IAS Meeting Sedimentology, Marrakesh, p. 24–43.

Berner, R.A., 1997, The rise of plants and their effect on weathering and atmospheric CO_2: Science, v. 276, p. 544–545.

Birenheide, R., Coen-Aubert, M., Lütte, B.P., and Tourneur, F., 1991, Devonian coral-bearing strata of the Eifel Hills and the Ardennes: Excursion Guidebook 6th International Symposium Fossil Cnidaria and Porifera, Münster, v. B1, 113 p.

Blieck, A., Brice, D., Fesir, R., Guillot, F., Majesté-Mejoulas, C., and Meillez, F., 1988, The Devonian of France and Belgium, *in* McMillan, A.F., Embry, A.F., and Glass, D.J., eds., Devonian of the world, Canadian Society of Petroleum Geologists, Calgary, 1, p. 359–400.

Bourque, P.A., 2001, Sea level, synsedimentary tectonics, and reefs: Implications for hydrocarbon exploration in the Silurian–lowermost Devonian Gaspé Belt, Québec Appalachians: Bulletin of Canadian Petroleum Geology, v. 49, p. 217–237.

Bourque, P.A., and Amyot, G., 1989, Stromatoporoid-coral reefs of the upper West Point reef complex, Late Silurian–Gaspé Peninsula, Québec: Memoirs Canadian Society of Petroleum Geologists, v. 13, p. 251–257.

Bourque, P.A., Kirkwood, D., and Malo, M., 2001, Stratigraphy, tectono-sedimentary evolution and paleogeography of the post-Taconian-pre-Carboniferous Gaspé Belt: An overview: Bulletin of Canadian Petroleum Geology, v. 49, p. 186–201.

Bourrouilh, R., and Bourque, P.A., 1995, Marquers d'evolution de marges continental Paléozoiques: Les monticules carbonatés à stromatactis: Bulletin Société Géologique de France, v. 166, p. 711–724.

Bourrouilh, R., Bourque, P.A., Dansereau, P., Bourrouilh-leJan, F., and Weyant, M., 1997, Synsedimentary tectonics, mud-mounds and sea-level changes on a Palaeozoic carbonate platform margin: A Devonian Montagne Noire example (France): Sedimentary Geology, v. 118, p. 95–118.

Braun, R., and Königshof, P., 1997, Trockenes Fußes durch ein Riff—Stromatoporen Riffe in der Lahn Mulde: Kleine Senckenberg Reihe, v. 24, p. 77–84.

Braun, R., Oetken, S., Königshof, P., Kornder, L., and Wehrmann, A., 1994, Development and biofacies of reef-influenced carbonates (central Lahn Syncline, Rheinisches Schiefergebirge): Courier Forschungsinstitut Senckenberg, v. 169, p. 351–386.

Brookfield, M.E., and Gupta, V.J., 1988, The Devonian of northern Gondwanaland: a Himalayan viewpoint and terrane analysis, *in* McMillan, A.F., Embry, A.F. , and Glass, D.J., eds., Devonian of the world, Canadian Society of Petroleum Geologists, Calgary, v. 1, p. 579–589.

Brownlaw, R.L.S., Hocking, R.M., and Jell, J.S., 1996, High frequency sea-level fluctuations in the Pillara Limestone, Guppy Hills, Lennard Shelf, northwestern Australia: Historical Biology, v. 11, p. 187–212.

Budd, A.F., and Johnson, K.G., 1999, Origination preceding extinction during late Cenozoic turnover of Caribbean reefs: Paleobiology, v. 25, p. 188–200.

Buggisch, W., and Flügel, E., 1992, Mittel-bis oberdevonische Karbonate auf Blatt Weilburg (Rheinisches Schiefergebirge) und Randgebieten: Initialstaden der Riffentwicklung auf Vulkanschwellen: Geologisches Jahrbuch Hessen, v. 120, p. 77–97.

Burchette, T.P., 1981, European Devonian reefs: A review of current concepts and models: Society of Economic Paleontologists and Mineralogists Special Publication 30, p. 85–142.

Cattaneo, G.A., Tahiri, A., Zahraoui, M., and Vachard, D., 1993, La sédimentation récifale du Givétien dans la Meseta maroccaine nord-occidentale: Comptes Rendus Académie Sciences, Paris, v. 317, p. 73–80.

Cavet, P., Lardeux, H., and Pillet, J., 1967, Notice stratigraphique sur les formations Paléozoïques des synclinoria de Saint-Julien-de-Vouvantes et de Redon-Ancenis: Mémoires Bureau Recherches Géologiques Minières, v. 33, p. 301–320.

Cherkesova, S.V., 1988, Lower and Middle Devonian marine deposits of the Soviet arctic and the correlation with eastern Canada, *in* McMillan, N.J., Embry, A.F., and Glass, D.J., eds., Devonian of the world: Calgary, Canadian Society of Petroleum Geologists, v. 3, p. 669–679.

Chlupac, I., 1988, The Devonian of Czechoslovakia and its significance, *in* McMillan, N.J., Embry, A.F., and Glass, D.J., eds., Devonian of the world: Calgary, Canadian Society of Petroleum Geologists, v. 1, p. 481–497

Chow, N., Wendte, J., and Stasiuk, L.D., 1995, Productivity versus preservation controls on two organic-rich carbonate facies in the Devonian of Alberta: Sedimentological and organic petrological evidence: Bulletin of Canadian Petroleum Geology, v. 43, p. 433–460.

Chuvashov, B.I., and Shuiskii, V.P., 1988, Stratigraficheskie i fatsialnye kompleksy izvestkovikh vodoroslei Paleozoya Urala [Stratigraphic and facies complexes of algal limestones in the Paleozoic of the Urals], *in* Izvestkovye vodorosli i stromatoliti [Algal and stromatolitic limestones]: Novosibirsk, Institut Geologii i Geofizikii, p. 98–124.

Clough, J.G., and Blodgett, R.B., 1988, An Early Devonian reef complex in the Ogilvie Formation of east-central Alaska, *in* McMillan, A.F., Embry, A.F., and Glass, D.J., eds., Devonian of the world: Calgary, Canadian Society of Petroleum Geologists, v. 2, p. 517–523.

Conaghan, P.I., Mountjoy, E., Edgecombe, D.R., Talent, J.A., and Owen, D.E., 1976, Nubrigyn algal reefs (Devonian) eastern Australia: Allochthonous blocks and megabreccias: Geological Society of America Bulletin, v. 87, p. 515–530.

Cook, H.E., Zhemchuzhnikov, V.G., Buvtyshkin, V.M., Golub, L.Ya., Gatovskii, Yu.A., and Zorin, A.E., 1994, Devonian and Carboniferous passive-margin carbonate platform of southern Kazakhstan: Summary of depositional and stratigraphic models to assist in the exploration and production of coeval giant carbonate platform oil and gas fields in the north Caspian Basin, western Kazakhstan: Memoirs Canadian Society of Petroleum Geologists, v. 17, p. 363–381.

Copper, P., 1994, Ancient reef ecosystem expansion and collapse: Coral Reefs, v. 13, p. 3–11.

Copper, P., 2001, Evolution, radiation and extinctions in Proterozoic to Mid-Paleozoic reefs, *in* Stanley, G.D., ed., The history and sedimentology of ancient reef systems: New York, Kluwer Topics in Geobiology, v. 17, p. 89–119.

Copper, P., 2002, Silurian and Devonian reefs: 80 million years of global greenhouse between two ice ages: SEPM (Society for Sedimentary Geology) Special Publication 72, p. 181–238.

Crook, K.A.W., 1961, Stratigraphy of the Tamworth group (Lower and Middle Devonian), Tamworth-Nundle District, NSW: Journal and Proceedings of the Royal Society of New South Wales, v. 94, p. 173–188.

Dastanpour, M., 1996, The Devonian System in Iran: A review: Geological Magazine, v. 133, p. 159–170.

Deng, Zhanqiu, 1999, Some Paleozoic tabulate corals from northern Xinjiang, *in* Palaeozoic fossils of northern Xinjiang, China: Nanjing, Nanjing Institute of Geology and Paleontology, p. 187–269.

Dronov, A.V., 1993, Middle Paleozoic Waulsortian-type mudmounds in southern Fergana (southern Tien-Shan, Commonwealth of Independent States): The shallow water atoll model: Facies, v. 28, p. 169–180.

Dronov, A.V., and Natalin, N.A., 1990, Yuzhnyi Tyan-Shan [Southern Tyan-Shan], *in* Belenitskaya, G.A., and Zadoroshnaya, N.M., eds., Rifogennye i sulfatonosnye formatsii Fanerozoya SSSR [Phanerozoic reefal and sulfate formations of the USSR]: Moscow, Nedra, p. 56–65.

Edie, R.W., 1959, Middle Devonian sedimentation and oil possibilities, central Saskatchewan: Bulletin American Association of Petroleum Geologists, v. 43, p. 1026–1057

Edinger, E.N., Copper, P., Risk, M.J., and Atmojo, W., 2002, Oceanography and reefs of Recent and Paleozoic tropical epeiric seas: Facies, v. 47, p. 127–150.

El Hassani, A., and El Kamel, F., 2000, Tectonic control of Devonian reef building in Mechra ben Abbou area (northern Rehamna, Morocco): Travaux de l'Institute Scientifique, Série Géologique, Géographique, Physique, v. 20, p. 25–30.

Elloy, R., 1973, Réflexions sur quelques environnements récifaux du Paléozoique: Bulletin Centre de Recherches, Pau, v. 6, p. 1–105.

Embry, A.F., 1991, Middle-Upper Devonian clastic wedge of the Arctic islands, *in* Trettin, H.P., ed., Geology of the Innuitian orogen: Geology of Canada: Ottawa, Geological Survey of Canada, v. 3, p. 263–279.

Embry, A.F., and Klovan, J.E., 1971, A Late Devonian reef tract on northeastern Banks Island: Bulletin Canadian Petroleum Geology, v. 19, p. 730–781.

Embry, A.F., and Klovan, J.E., 1989, Mercy Bay patch reefs, Banks Island, Canadian arctic archipelago: Memoirs Canadian Society of Petroleum Geologists, v. 13, p. 429–430.

Faber, P., 1980, Fazies-Gliederung und Entwicklung im Mitteldevon der Eifel (Rheinisches Schiefergebirge): Mitteilungen Mainzer Geowisschenschaften, v. 8, p. 83–149.

Faber, P., Vogel, K., and Winter, J., 1977, Beziehungen zwischen morphologischen Merkmalen der Brachiopoden und Fazies, dargestellt an Beispielen des Mitteldevons der Eifel und Sudmarokkos: Abhandlungen Neues Jahrbuch Geologie und Paläontologie, v. 154, p. 21–60.

Fagerstrom, J.A., 1983, Diversity, speciation, endemism, and extinction in Devonian reef and level-bottom communities: Coral Reefs, v. 2, p. 65–70.

Fernandez-Martinez, E., Soto, F., and Mendez-Bedia, I., 1994, An example of reef development in the Middle Devonian (Candas Fm., Givetian) in the Cantabrian Mountains (NW Spain): Courier Forschungsinstitut Senckenberg, v. 172, p. 103–110.

Ferrari, A., and Vai, G.B., 1966, Ricerche stratigrafiche e paleoecologiche al Monte Zermula: Giornale di Geologia, v. 33, p. 389–406.

Fischbuch, N.R., 1968, Stratigraphy, Devonian of the Swan Hills reef complexes of central Alberta: Bulletin of Canadian Petroleum Geology, v. 16, p. 444–556.

Flajs, G., Hüssner, H., and Vigener, M., 1996a, Stromatactis mudmounds in the upper Emsian of the Montagne Noire (France): Formation and diagenesis of stromatactis structures: Göttinger Arbeiten in Geologie, Paläontologie, v. 2, p. 345–348.

Flajs, G., Dieken, G., and Hüssner, H., 1996b, Upper Emsian and lower Eifelian stromatactis limestones of the Koneprusy area (Barrandian, Czech Republic): Sonderband Göttinger Arbeiten Geologie und Paläontologie Sonderband, v. 2, p. 349–353.

Franke, W., 1999, Tectonic and plate tectonic units at the north Gondwana margin: Evidence from the central European Variscides: Abhandlungen Geologischen Bundesanstalt zu Wien, v. 54, p. 7–13.

Gaetani, M., 1968, Devonian of eastern Iran, northern Afghanistan and northern Pakistan, *in* Oswald, D.H., ed., International Symposium on the Devonian System: Calgary, Alberta Society of Petroleum Geologists, v. 1, p. 519–528.

Gao Jiangguo, and Copper, P., 1997, Growth rates of Middle Paleozoic corals and sponges: Early Silurian of eastern Canada: Proceedings 8th International Coral Reef Symposium, Panama: Balboa, Panama, Smithsonian Tropical Research Institute, v. 2, p. 1651–1656.

Gendrot, C., 1973, Environnements du Dévonien récifale au Maroc: Notes de Service Géologique du Maroc, v. 254, p. 55–86.

Gendrot, C., Kergomard, J., and Rabate, J., 1969, Étude des formations récifales du Dévonien de la Méséta occidentale: Bureau des Recherches Minérales, Rabat, 66 p.

George, A.D., and Chow, N., 1999, Palaeokarst development in a lower Frasnian (Devonian) platform succession, Canning Basin, northwestern Australia: Australian Journal of Earth Sciences, v. 46, p. 905–913.

George, A.D., and Powell, M., 1997, Lithofacies and sequence development on an Upper Devonian reef complex of the Canning Basin, Western Australia: Journal of Sedimentary Research, v. A67, p. 935–944.

Gnoli, M., Jaanusson, V., Leone, F., and Serpagli, E., 1981, A Lower Devonian stromatactis-bearing carbonate mound from southern Sardinia: Neues Jahrbuch für Geologie, Paläontologie, Monatshefte, v. 1981, p. 339–345.

Gnoli, M., Leone, F., Olivieri, R., and Serpagli, E., 1988, The Mason Porcus section as reference section for uppermost Silurian–Lower Devonian in SW Sardinia: Bolletin Societa di Paleontologica Italiana, v. 27, p. 323–334.

Gnoli, M., Kriz, J., Leone, F., Olivieri, R., Serpagli, E., and Storch P., 1990, Lithostratigraphic units and biostratigraphy of the Silurian and Early Devonian of southwest Sardinia: Bolletin Societa di Paleontologica Italiana, v. 29, p. 11–23.

Golonka, J., Ross, M.I., and Scotese, C.R., 1994, Phanerozoic paleogeographic and paleoclimatic modeling maps, *in* Embry, A.F., Beauchamp, B., and Glass, D.J., eds.., Pangea: Global environment and resources: Memoirs Canadian Society of Petroleum Geologists, v. 17, p. 1–47.

Grachevskii, M.M., and Solomatin, A.V., 1977, Upper Devonian potentially productive barrier reefs of the Timan-Pechora province [Verkhnie Devonskie potentsialnyi plodorodnyi barer rifov Timan-Pechorskya provintsiya]: Doklady Akademiya Nauk, SSSR, v. 232, p. 875–878.

Grey, K., 1991, A mid-Givetian miospore age for the onset of reef development on the Lennard Shelf, Canning Basin, Western Australia: Review of Palaeobotany and Palynology, v. 68, p. 3–48.

Hallock, P., 1996, Reefs and reef limestones in Earth history, *in* Birkeland, C., ed., Life and death of coral reefs: New York, Chapman and Hall, p. 13–42.

Heckel, P.B., 1973, Nature, origin and significance of the Tully Limestone: Boulder, Colorado, Geological Society of America Special Paper 138, 244 p.

Hilali, A., Lachkem, H., and Boulvain, F., 1998, Comparaison des Kess-Kess de Hmar Lakhdad (Emsian, Maroc) et des monticules micritiques de l'anticlinorium de Philippeville (Frasnian, Belgique): Geologica Belgica, v. 1, p. 17–31.

Hladil, J., 1986, Trends in development and cyclic pattern of Middle and Upper Devonian buildups: Facies, v. 15, p. 1–34.

Hladil, J., 1988, Structure and microfacies of Middle and Upper Devonian carbonate buildups in Moravia, Czechoslovakia., *in* McMillan, N.J., Embry, A.F., and Glass, D.J., eds., Devonian of the world: Calgary, Canadian Society of Petroleum Geologists, v. 2, p. 607–618.

Hladil, J., 1994, Moravian Middle and Late Devonian buildups: Evolution in time and space with respect to Laurussia Shelf: Courier Forschungsinstitut Senckenberg, v. 172, p. 111–125.

Hladil, J., Melichar, R., Otava, J., Galle, A., Krs, M., Man, O., Pruner, P., Cejchan, P., and Orel, P., 1999, The Devonian in the easternmost Variscides, Moravia: A holistic analysis directed towards comprehension of the original context: Abhandlungen Geologischen Bundesanstalt zu Wien, v. 54, p. 27–47.

Ioganson, A.K., 1990a, Alazeiskaya skladchataya sistem [Alazei fold belt system], *in* Belenitskaya, G.A., and Zadoroshnaya, N.M., eds., Rifogennye i sulfatonosnye formatsii Fanerozoya SSSR [Phanerozoic reefal and sulfate formations of the USSR]: Moscow, Nedra, p. 123–125.

Ioganson, A.K., 1990b, Pamirskaya skladchataya oblast [Pamir fold belt massif], *in* Belenitskaya, G.A., and Zadoroshnaya, N.M., eds., Rifogennye i sulfatonosnye formatsii Fanerozoya SSSR [Phanerozoic reefal and sulfate formations of the USSR]: Moscow, Nedra, p. 106–114.

Ioganson, A.K., 1990c, Sibirskaya platforma [Siberian platform], *in* Belenitskaya, G.A., and Zadoroshnaya, N.M., eds., Rifogennye i sulfatonosnye formatsii Fanerozoya SSSR [Phanerozoic reefal and sulfate formations of the USSR]: Moscow, Nedra, p. 131–139.

Ioganson, A.K., 1990d, Pripyatsko-Donetskii avlakogen [Pripyat-Donets aulacogen], *in* Belenitskaya, G.A., and Zadoroshnaya, N.M., eds., Rifogennye i sulfatonosnye formatsii Fanerozoya SSSR [Phanerozoic reefal and sulfate formations of the USSR]: Moscow, Nedra, p. 145–147.

Ioganson, A.K., and Bazanov, S.U., 1990, Omulevskaya zona [Omulev zone], *in* Belenitskaya, G.A., and Zadoroshnaya, N.M., eds., Rifogennye i sulfatonosnye formatsii Fanerozoya SSSR [Phanerozoic reefal and sulfate formations of the USSR], Ministerstvo Geologii SSSR: Moscow, Nedra, p. 114–122.

Isaacson, P.E., Hladil, J., Shen, J.W., Kalvoda, J., and Grader, G., 1999, Late Devonian (Famennian) glaciation in South America and marine offlap on other continents: Abhandlungen Geologischen Bundesanstalt zu Wien, v. 54, p. 239–257.

Ivanova, E.A., Belskaya, T.N., and Chudinova, I.I., 1964, Usloviya obitaniya morskoi fauny Silura i Devona Kuznetskogo, Minusinskogo i Tuvinskogo basseinov [Distribiution of marine faunas in the Silurian and Devonian of the Kuznetsk, Minusinsk and Tuva basins]: Trudy Paleontologicheskogo Instituta, v. 102, p. 1–226.

James, N.P., 1997, The cool-water carbonate depositional realm: SEPM (Society for Sedimentary Geology) Special Publication, v. 56, p. 1–20.

James, N.P., and Bourque, P.A., 1992, Reefs and mounds, *in* Walker, R.G., and James, N.P., eds., Facies models—Response to sea level change: Waterloo, Geological Association of Canada, p. 323–347.

Jell, J. S., and Zhen, Y.Y., 1994, Middle Devonian rugose coral biostratigraphy of the Fanning River Group, north Queensland, Australia: Courier Forschungsinstitut Senckenberg, v. 172, p. 1–6.

Joachimski, M.M., and Buggisch, W., 1996, The Upper Devonian reef crisis–Insights from the carbon isotope record: Sonderband Göttinger Arbeiten Geologie und Paläontologie, v. 2, p. 365–370.

Johnson, C.C., 1999, Evolution of Cretaceous surface current circulation patterns, Caribbean and Gulf of Mexico: Boulder, Colorado, Geological Society of America Special Paper 332, p. 329–343.

Johnson, J.G., Sandberg, C.A., and Poole, F.G., 1991, Devonian lithofacies of western United States, *in* Cooper, J.D., and Stevens, C.H., eds., Paleozoic paleogeography of the western United States: Bakersfield, Pacific Section SEPM (Society for Sedimentary Geology), v. 67, p. 83–105.

Jux, U., 1960, Die devonischen Riffe im Rheinischen Schiefergebirge: Abhandlungen Neues Jahrbuch für Geologie und Paläontologie, v. 110, p. 186–256.

Jux, U., 1969, Frasne-Riffbildner im Devon der Daste Nawar (Afghanistan): Neues Jahrbuch für Geologie und Paläontologie, Monatshefte, v. 1969, p. 680–690.

Kesling, R.V., Segall, R.T., and Sorensen, H.O., 1974, Devonian strata of Emmet and Charlevoix counties, Michigan: University of Michigan Papers in Paleontology, v. 7, p. 1–46.

Kiessling, W., 2001, Paleoclimatic significance of Phanerozoic reefs: Geology, v. 29, p. 751–754.

Kiessling, W., Flügel, E., and Golonka, J., 1999, Paleoreef maps: Evaluation of a comprehensive data base on Phanerozoic reefs: Bulletin American Association of Petroleum Geologists, v. 83, p. 1552–1587.

Kim, A.I., Elkin, E.A., Erina, M.V., Korsakov, V.S., and Tsoy, R.V., 1984, The Middle Paleozoic of southern Tien Shan: Excursion Guidebook 100: Moscow, International Geological Congress, p. 128–156.

Kirikov, V.P., 1988, Devonian period evolution of the Russian Platform, *in* McMillan, N.J., Embry, A.F., and Glass, D.J., eds., Devonian of the world: Calgary, Canadian Society of Petroleum Geologists, v. 1, p. 513–525.

Kleypas, J.A., Buddemeier, R.W., Archer, D., Gattuso, J.P., Langdon, C., and Opdyke, B.N., 1999, Geochemical consequences of increased carbon dioxide on coral reefs: Science, v. 284, p. 118–120.

Königshof, P., Gewehr, B., Kornder, L., Wehrmann, A., Braun, R., and Zankl, H., 1991, Stromatoporen-Morphotypen aus einem zentralen Riffbereich (Mitteldevon) in der südwestlichen Lahnmulde: Geologica et Palaeontologica, v. 25, p. 19–35.

Krebs, W., 1974, Devonian carbonate complexes of central Europe: Society of Economic Paleontologists and Mineralogists Special Publication 18, p. 155–208.

Krstic, B., Grubic, A., Ramovs, A., and Filipovic, I., 1988, The Devonian of Yugoslavia, *in* McMillan, N.J., Embry, A.F., and Glass, D.J., eds., Devonian of the world: Calgary, Canadian Society Petroleum Geologists, v. 1, p. 499–506.

Kuznetsov, V.G., 2000, Paleozoiskoe rifoobrazovanie na territorii rossii i smezhnykh stran [Paleozoic reef formations on the territory of Russia and neighbouring areas]: Moskva, Rossiiskaya Akademiya Estestvennykh Nauk, 227 p.

Liao Weihoa, Xia Fengsheng, Zhu Huaicheng, and Zhang Jin, 1992, Devonian of Tarim and adjacent areas, *in* Zhou Zhiyi and Chen Peiji, eds., Biostratigraphy and geological evolution of Tarim: Beijing, Science Press, p. 173–201.

Lindemann, R.H., 1989, The Leroy bioherm, Onondaga Limestone (Middle Devonian), western New York: Memoirs Canadian Society of Petroleum Geologists, v. 13, p. 487–491.

Lin, S., van Staal, C., and Dubé, B., 1994, Promontory-promontory collision in the Canadian Appalachians: Geology, v. 22, p. 897–900.

Machel, H.G., and Hunter, I. G., 1994, Facies models for Middle to Late Devonian shallow marine carbonates, with comparisons to modern reefs: Facies, v. 30, p. 155–176.

Mackenzie, W.S., 1969, Stratigraphy of the Devonian Southesk-Cairn carbonate complex and associated strata, eastern Jasper National Park: Bulletin Geological Survey of Canada, v. 184, p. 1–63.

Makhnach, A.S. Urev, I.I., Kruchek, S.A., and Moskvich, V.A., 1986, Rifogennye obrazovaniya Paleozoya i Verkhnego Proterozoya Belorussi [Paleozoic and Proterozoic reef formations of Belarus], *in* Sokolov, B.S., ed., Fanerozoiskie rify i korally SSSR [Phanerozoic reefs and corals of the USSR]: Moscow, Nauka, p. 191–197.

Malmsheimer, K.W., Flajs, G., and Koch-Früchtl, U., 1996, Middle Devonian initial reef-facies from the Rhenish Schiefergebirge (Sauerland and Eifel), western Germany: Sonderband Göttinger Arbeiten Geologie und Paläontologie, v. 2, p. 371–375.

Mamedov, A.B., and Rzhonsnitskaya, M.A., 1985, Devonian of the south Transcaucasus zonal subdivision, boundaries of series and stages, correlation: Courier Forschungsinstitut Senckenberg, v. 75, p. 135–156.

May, A., 1983, Ein Korallenriff im Oberen Mittel-Devon von Werdohl (Sauerland): Dortmünder Beiträge Landeskunde, Naturwissenschaftliche Mitteilungen, v. 17, p. 35–46.

May, A., 1994, Paleoecology and paleobiogeography of corals and other reef builders from the Middle Devonian of the Sauerland (Germany): Courier Forschungs-Institut Senckenberg, v. 172, p. 147–159.

May, A., 1997, Sind die devonischen Riffe des Sauerlandes heutigen Korallenriffen vergleichbar?: Dortmünder Beiträge zu Landeskunde Naturwissenschaftliche Mitteilungen, v. 31, p. 127–135.

McCracken, A.D., Armstrong, D.K., and Bolton, T.E., 2000, Conodonts and corals in kimberlite xenoliths confirm a Devonian seaway in central Ontario and Quebec: Canadian Journal of Earth Sciences, v. 37, p. 1651–1663.

McMillan, N.J., Embry, A.F., and Glass, D.J., eds., 1988, Devonian of the world (3 volumes): Calgary, Canadian Society of Petroleum Geologists, v. 1, p. 1–674, v. 3, p. 1–714.

Meyer, F.O., 1989, Stromatoporoid-coral patch reefs of Givetian age, Michigan: Canadian Society of Petroleum Geologists Memoir, v. 13, p. 492–496.

Minjin, C., Sersmaa, G., Ariunchimeg, Ya., Gereltsetseg, L., Undarya, J., Tumenbayar, B., and Bolortsetseg, M., 2001, Mongolian field guide: Ulaanbaatar, Part 2: Ulanbaatar, Mongolian Academy of Science, p. 12–127.

Mistiaen, B., 1985, Phénomènes récifaux dans le Dévonien d'Afghanistan (montagnes centrales): Publications Société Géologique du Nord, v. 11, p. 1–381.

Mistiaen, B., and Poncet, J., 1989, Stromatolithes, serpulides et *Trypanopora* (Vers?) associés dans des petites biohermes Givetiens du Boulonnais (France): Palaeogeography, Palaeoclimatology, Palaeoecology, v. 41, p. 125–138.

Montenat, C., Baider, L., Barrier, P., Hilali, A., Lachkem, H., and Meming, J., 1996, Contrôle tectonique de l'édification biosédimentaires dévoniens du Hamer Lakhdad d'Erfoud: Comptes Rendus de l'Académie des Sciences, Paris, v. 323, p. 297–304.

Moore, P.F., 1988, Devonian geohistory of the western interior of Canada, *in* McMillan, N.J., Embry, A.F., and Glass, D.J., eds., Devonian of the world: Calgary, Canadian Society of Petroleum Geologists, v. 1, p. 67–83.

Moore, P.F., 1989, Devonian reefs in Canada and some adjacent areas: Memoir Canadian Society Petroleum Geologists, v. 13, p. 367–390.

Morzadec, P., Paris, F., Plusquellec, Y., Racheboeuf, P., and Weyant, M., 1988, Devonian stratigraphy and paleogeography of the Armorican Massif (western France), *in* McMillan, N.J., Embry, A.F., and Glass, D.J., eds., Devonian of the world: Calgary, Canadian Society of Petroleum Geologists, v. 1, p. 401–420.

Muir, I., Wong, P., and Wendte, J., 1984, Devonian Hare Indian-Ramparts (Kee Scarp) evolution, Mackenzie Mountains and subsurface Norman Wells, NWT, *in* Eliuk, L., ed., Carbonates in subsurface and outcrop, 1984 Core Conference: Calgary, Canadian Society of Petroleum Geologists, p. 82–101.

Murphy, A.E., Sageman, B.B., Hollander, D.J., Lyons, T.W., and Brett, C.E., 2000, Black shale deposition and faunal overturn in the Devonian Appalachian Basin: Clastic starvation, seasonal water-column mixing, and efficient biolimiting nutrient recycling: Paleooceanography, v. 15, p. 280–291.

Noble, J.P.A., and Ferguson, R.D., 1971, Facies and faunal relations at edge of early Mid-Devonian carbonate shelf, south Nahanni River area, Northwest Territories: Bulletin of Canadian Petroleum Geology, v. 29, p. 540–560.

Okulitch, A.V., 1999, Geological time chart: Geological Survey of Canada Open File 3040, published in Geolog, v. 29, Part 1.

Paul, J., 1996, Stromatolites of the Canning Basin (Upper Devonian, Western Australia): Sonderband Göttinger Arbeiten Geologie und Paläontologie, v. 2, p. 377–383.

Pavlov, N.D., Salov, Yu.A., Gononenkov, G.N., Akopov, Yu.I., Tostykh, A.A., and Denisyuk, R.S., 1988, Geological-geophysical model of the Tengiz paleoatoll based on seismography: Interrnational Geology Review, v. 30, p. 1057–1069.

Pedder, A.E.H., 1999, Paleogeographic implications of a Devonian (Givetian, lower *varcus* Subzone) rugose coral fauna from the Ma'der Basin (Morocco): Abhandlungen Geologischen Bundesanstalt zu Wien, v. 54, p. 385–434.

Pedersen, T.F., and Calvert, S.E., 1990, Anoxia vs. productivity: what controls the formation of organic carbon-rich sediments and sedimentary rocks?: American Association of Petroleum Geologists Bulletin, v. 74, p. 454–466.

Perry, D.G., 1979, Late Early Devonian reef associated brachiopods of the Prongs Creek Formation, northern Yukon: Journal of Paleontology, v. 53, p. 1094–1111.

Playford, P.E., and Cockbain, A.E., 1989, Devonian reef complexes, Canning Basin, western Australia: A review: Memoirs Association of Australasian Palaeontologists, v. 8, p. 401–412.

Plusquellec, Y., 1980, Les schistes et calcaires de l'Armorique: Mémoires Société Géologique et Mineralogique de Bretagne, v. 29, p. 1–6.

Pohler, S.M.L., 1998, Devonian carbonate buildup facies in an intra-oceanic island arc (Tamworth Belt, New South Wales, Australia): Facies, v. 39, p. 1–34.

Pohler, S.M.L., Brühl, D., and Meistermann, B., 1999, Struves Mudmound am Weinberg—Carbonate buildup facies, Hillesheimer Mulde, Eifel: Senckenbergiana Lethaea, v. 79, p. 13–29.

Pohler, S.M.L., and Kim, A.I. 2001, Biofacies and paleogeography of tabulate corals from recurrent limestone facies (Emsian and Middle Devonian) of the Tamworth Group (Australia), *in* Weiss, R.H., ed., Contributions to geology and palaeontology of Gondwana in honour of Helmut Wopfner: Köln, p. 383–407.

Poncet, J., 1977, Les microfaciès: Mémoires Société Géologique et Mineralogique de Bretagne, v. 23, p. 91–94.

Poulsen, C.J., Barron, E.J., Johnson, C.C., and Fawcett, P., 1999, Links between major climatic factors and regional oceanic circulation in the mid-Cretaceous: Boulder, Colorado, Geological Society of America Special Paper 332, p. 73–89.

Préat, A., and Weiss, D., 1994, Variations du niveau marin dans le Dévonien carbonaté de Belgique: Approches sédimentologiques et séquentielle: Bulletin Société Géologique de France, v. 165, p. 469–483.

Racki, G., 1988, Middle to Upper Devonian boundary beds of the Holy Cross Mountains, central Poland: Introduction to ecostratigraphy, *in* McMillan, N.J., Embry, A.F., and Glass, D.J., eds., Devonian of the world: Calgary, Canadian Society of Petroleum Geologists, v. 3, p. 119–130.

Racki, G., 1992, Evolution of the bank to reef complex in the Devonian of the Holy Cross Mountains: Acta Palaeontologica Polonica, v. 37, p. 87–182.

Racki, G., 1999, Silica-secreting biota and mass extinctions: Survival patterns and processes: Palaeogeography, Palaeoclimatology, Palaeoecology, v. 154, p. 107–132.

Read, J.F., 1973a, Carbonate cycles, Pillara formation (Devonian), Canning Basin, Western Australia: Bulletin of Canadian Petroleum Geology, v. 21, p. 38–51.

Read, J.F., 1973b, Paleo-environments and paleogeography, Pillara Formation (Devonian), Western Australia: Bulletin Canadian Petroleum Geology, v. 21, p. 344–394.

Rong, Jiayu, Boucot, A.J., Su Yangzheng, and Strusz, D.L., 1995, Biogeographical analysis of Late Silurian brachiopod faunas, chiefly from Asia and Australia: Lethaia, v. 28, p. 39–60.

Rusetskaya, N.N., and Yaroshenko, A.V., 1990, Prikaspiiskaya Sinekliza [Pre-Caspian synclinorium], *in* Belenitskaya, G.A., and Zadoroshnaya, N.M., eds., Rifogennye i sulfatonosnye formatsii Fanerozoya SSSR, Ministerstvo Geologii SSSR: Moscow, Nedra, p. 148–156.

Sandberg, P., 1983, An oscillating trend in Phanerozoic non-skeletal carbonate mineralogy: Nature, v. 305, p. 19–22.

Scheibe, H.J., 1965, Untersuchungen im Unter- und Mittel Devon der westlichen Gummersbacher Mulde (Bergisches Land, Rheinisches Schiefergebirge): Fortschritte Geologie Rheinland und Westfalen, v. 9, p. 469–484.

Schönlaub, H.P., 1998, Review of the Paleozoic paleogeography of the Southern Alps–The perspective from the Austrian side: Giornale di Geologia, Special Issue, v. 60, p. 59–68.

Scotese, C. R., 2001, Atlas of Earth History, Volume 1: Arlington, Texas, Paleogeography, PALEOMAP Project, 52 p.

Scrutton, C.T., 1977, Reef facies in the Devonian of eastern south Devon: Mémoires Bureau des Recherches Géologiques et Minéralogiques, v. 89, p. 124–135.

Sharkova, T. T., 1980, Rifogennye postroiki rannego Devona Yuzhnoi Mongolii [Reefal beds of the early Devonian of southern Mongolia], *in* Sokolov, B.S., ed., Fanerozoiskie rify i korally SSSR [Phanerozoic reefs and corals of the USSR]: Moscow, Nedra, p. 92–98.

Sharkova, T.T., 1986, Korallovye assotsiatsi Devonskikh rifov yuzhnoi Mongolii [Coral associations of Devonian reefs in Mongolia], *in* Sokolov, B.S., ed., Fanerozoiskie rify i korally SSSR [Phanerozoic reefs and corals of the USSR]: Moscow, Nedra, p. 188–190.

Shen, J.W., Yu, C.M., Yin, B.A., and Zhang, S.L., 1994, Sequence stratigraphy of Devonian carbonate platform and reef complexes in Guilin, Guangxi: Journal of Stratigraphy, v. 18, p. 161–167.

Smith, G.P., 1985, The distribution and significance of Lower Devonian carbonate buildups in time and space: Bulletin of Canadian Petroleum Geology, v. 33, p. 471–478.

Smosna, R., 1988, Paleogeographic reconstruction of the Lower Devonian Helderberg group in the Appalachian Basin, *in* McMillan, N.J., Embry, A.F., and Glass, D.J., eds., Devonian of the world: Calgary, Canadian Society of Petroleum Geologists, v. 1, p. 265–275.

Soja, C.M., 1988, Lower Devonian platform carbonates from Kasaan island, southeastern Alaska, Alexander Terrane: Canadian Journal of Earth Sciences, v. 25, p. 639–656.

Sokolov, B.S., editor, 1986. Fanerozoiskie rify i korally SSSR [Phanerozoic reefs and corals of the USSR]: Moscow, Nauka, 232 p.

Southgate, P.N., Kennard, J.M., Jackson, M.J., O'Brien, P.E., and Sexton, M.J., 1993, Reciprocal lowstand clastic and highstand carbonate sedimentation, subsurface Devonian reef complex, Canning Basin, Western Australia: Memoir American Association of Petroleum Geologists, v. 57, p. 157–180.

Stauffer, K.W., 1968, Devonian of India and Pakistan, *in* Oswald, D.H., ed., International Symposium on the Devonian System: Calgary, Alberta Society of Petroleum Geologists, v. 1, p. 545–556.

Stepanova, G.A., Khalymbadzha, V.G., Chernysheva, N.G., Petrova, L.G., and Postoyalko, M.V., 1985, Boundaries of stages of the Upper Devonian on the south Urals (the eastern slope): Courier Forschungsinstitut Senckenberg, v. 75, p. 123–134.

Stumm, E.C., 1969, Devonian bioherms of the Michigan basin: Contributions Museum of Paleontology Michigan, v. 22, p. 241–247.

Su, Yanzheng, 1988. Devonian paleogeography of northeastern China., *in* McMillan, N.J., Embry, A.F., and Glass, D.J., eds., Devonian of the world: Calgary, Canadian Society of Petroleum Geologists, v. 1, p. 607–618.

Suetenko, O.D., Sharkova, T.T., and Ulitina, L.M., 1977, Paleozoic stratigraphy and fauna of the eastern spurs of the Gobi Altai (Mandal-Ovoo Massif): Transactions Joint Soviet-Mongolian Expedition: Moscow, Akademiya Nauk, v. 5, p. 32–48.

Sun Dongli, and Chen Xiuqin, 1998, Devonian brachiopods from Karakorum-Kunlun region, *in* Wen Shixuan, ed., Palaeontology of the Karakorum-Kunlun mountains: Beijing, Science Press, p. 163–214.

Talent, J.A., Gratsianova, R.T., and Yolkin, E.A., 1986, Protethys: Fact or phantom? Palaeobiogeography in relation to the crustal mosaic for the Asia-Australia hemisphere in Devonian–Early Carboniferous times, International Symposium on Shallow Tethys, Wagga-Wagga, Australia: Rotterdam, Balkema Press, p. 87–111.

Talent, J.A., and Mawson, R., 1999, Northeastern Molong Arch and adjacent Hill End Trough (eastern Australia): Mid-Palaeozoic conodont data and implications: Abhandlungen Geologischen Bundesanstalt zu Wien, v. 54, p. 49–105.

Talent, J.A., Mawson, R. , Aitchison, J.C., Becker, R.T., Bell, K.N., Bradshaw, M.A., Burrow, C.J., Cook, A.G., Dargan, G.M., Douglas, J.G., Edgecombe, G.D., Feist, M., Jones, P.J., Long, J.A., Phillips-Ross, J.R., Picket, J.W., Playford, G., Rickards, R.B., Webby, B.D., Winchester-Seeto, T., Wright, A.J., Young, G.C., and Zhen, Y.Y., 2000, Devonian palaoebiogeography of Australia and adjoining regions: Memoirs Association of Australasian Palaeontologists, v. 23, p. 167–257.

Talent, J.A., Mawson, R., Brock, G.A., Furey-Greig, T., Molloy, P.D., and Simpson, A.J., 2001, Tectonostratigraphic implications of "lost" Cambrian-Devonian carbonate platforms of eastern Australia. Contributions to Siberian IGCP 410/421 joint meeting, UNESCO IGCP, *in* Yolkin, E., Talent, J.A., and Webby, B., eds.: Novosibirsk, Institut Geologii Nefti i Gaza, p. 106.

Telford, P. G., 1988, Devonian stratigraphy of the Moose River Basin, James Bay Lowland, Ontario, *in* McMillan, N.J., Embry, A.F., and Glass, D.J., eds., Devonian of the world: Calgary, Canadian Society of Petroleum Geologists, v. 1, p. 123–132.

Thorsteinsson, R., and Tozer, E.T., 1962, Banks, Victoria and Stefansson islands, Arctic archipelago: Memoirs Geological Survey of Canada, v. 330, p. 1–83.

Ulmishek, G.F., 1988, Upper Devonian-Tournaisian facies and oil resources of the Russian craton's eastern margin, *in* McMillan N.J., Embry, A.F., and Glass, D.J., eds., Devonian of the world: Calgary, Canadian Society of Petroleum Geologists, v.1, p. 527–549.

Vai, G.B., 1991, Palaeozoic strike-slip rift pulses and palaeogeography in the circum-Mediterranean Tethyan realm: Palaeogeography, Palaeoclimatology, Palaeoecology, v. 87, p. 223–252.

Vai, G.B., 1998, Field trip through the Southern Alps: an introduction with geologic settings, palaeogeography and Palaeozoic stratigraphy: Giornale di Geologia, Special Issue, v. 60, p. 1–38.

Veron, J.E.B., 2000, Corals of the world: Townsville, Australian Institute of Marine Science, 463 p.

Vishnetskii, A.N., and Baganov, S.U., 1986, Organogennye postroiki yugo-zapadnogo sklona khrebtaTas-Khayakhtakh [Organic buildups on the south-western slopes of the Tas-.Khayakhtakh mountains]: Yakutsk, Mineragеniya, tektonika, i stratigrafiya skladchatykh regionov Yakutii, p. 92–98.

Webb, G.E., 2001, Famennian mudmounds in the proximal fore-reef slope, Canning Basin, Western Australia: Sedimentary Geology, v. 145, p. 295–315.

Webb, G.E., and Sorauf, J., 2002, Zigzag microstructure in rugose corals: A possible indicator of relative seawater Mg/Ca ratios: Geology, v. 30, p. 415–418.

Weller, H., 1991, Facies and development of the Devonian (Givetian/Frasnian) Elbingerode reef complex in the Harz area (Germany): Facies, v. 25, p. 1–50.

Wells, S.M., ed., 1988, Coral reefs of the world: Nairobi, United Nations Environment Program, International Union for Conservation of Mature and Natural Resources, Switzerland, v. 1, Atlantic and eastern Pacific, 373 p.; v. 2, Indian Ocean, Red Sea and Gulf, 389 p.; v. 3, Central and western Pacific, 329 p.

Wen, Shixuan, ed., 1998, Palaeontology of the Karakorum-Kunlun mountains: Beijing, Science Press, 365 p.

Wilkinson, C., ed., 2000, Status of coral reefs of the world: 2000: Townsville, Australian Institute of Marine Science, 363 p.

Witzke, B.J., Bunker, B.J., and Rogers, F.S., 1988, Eifelian through lower Frasnian stratigraphy and deposition in the Iowa area, ventral mid-continent, USA, *in* McMillan, N.J., Embry, A.F., and Glass, D.J., eds., Devonian of the world: Calgary, Canadian Society of Petroleum Geologists, v. 1, p. 221–250.

Wolf, K.H, 1965, Petrogenesis and paleoenvironment of Devonian algal limestones of New South Wales: Sedimentology, v. 4, p. 113–178.

Wolfart, R., and Wittekindt, H., 1980, Devonian of Afghanistan: Berlin, Borntraeger, 500 p.

Wood, R., 1993, Nutrients, predation and the history of reef building: Palaios, v. 8, p. 526–543.

Wood, R., 1999, Reef evolution: Oxford, Oxford University Press, 413 p.

Wood, R., 2001, Are reefs and mounds really so different?: Sedimentary Geology, v. 145, p. 161–171.

Woodrow, D.L., Dennison, J.M., Ettensohn, F.R., Sevon, W.T., and Kirchgasser, W.T., 1988, Middle and Upper Devonian stratigraphy and paleogeography of the central and southern Appalachians and eastern mid-continent, USA, *in* McMillan, N.J., Embry, A.F., and Glass, D.J., eds., Devonian of the world: Calgary, Canadian Society of Petroleum Geologists, v. 1, p. 277–301.

Yamano, H., Hori, K., Yamauchi, M., Yamagawa, O., and Ohmura, A., 2001, Highest-latitude coral reef at Iki Island, Japan: Coral Reefs, v. 20, p. 9–12.

Yolkin, E.A., Gratsianova, R.T., Izokh, N.G., Yazikov, A.Yu., and Bakharev, N.K., 1997, Devonian sea-level fluctuations on the southwestern margin of the Siberian continent: Courier Forschungsinstitut Senckenberg, v. 199, p. 83–98.

Yolkin, E.A., Gratsianova, R.T., Izokh, N.G., Yazikov, A. Yu., Bakharev, N.K., Alekseeva, R.E., Erina, M.V., Kim, A.I., and Shishkina, G.R., 2000, Devonian standard boundaries within the shelf belt of the Siberian Old Continent (southern part of Siberia, Mongolia, Russian Far East) and in the south Tien Shan: Courier Forschungsinstitut Senckenberg, v. 225, p. 303–318.

Yu, Changmin, and Shen, Jianwei, 1998, Devonian reefs and reef complexes in Guilin, Guangxi, China: Nanjing, Jiangsu Science and Technology Press, 168 p.

Yu, Changmin, and Wu Yang, 1988, Middle Devonian facies patterns and reef development in South China, *in* McMillan N.J., Embry, A.F., and Glass, D.J., eds., Devonian of the world: Calgary, Canadian Society of Petroleum Geologists, v. 2, p. 649–657.

Zadoroshnaya, N.M., Antoshkina, A.I., Chuvashov, B.I., and Shuiskii, V.P., 1990a, Uralskaya skladchataya sistem [Uralian foldbelt system], *in* Belenitskaya, G.A., and Zadoroshnaya, N.M., eds., Rifogennye i sulfatonosnye formatsii Fanerozoya SSSR [Phanerozoic reefal and sulfate formations of the USSR]: Moscow, Nedra, p. 20–41.

Zadoroshnaya, N.M., Garina, N.S., Kozenyuk, V.A., Litin, V.S., and Osyanikov, V.S., 1990b, Sredniinyi Tyan-Shan [Northern Tyan-Shan], *in* Belenitskaya, G.A., and Zadoroshnaya, N.M., eds., Rifogennye i sulfatonosnye formatsii Fanerozoya SSSR [Phanerozoic reefal and sulfate formations of the USSR]: Moscow, Nedra, p. 53–56.

Zadoroshnaya, N.M., and Nikitin, I. F.,1990, Kazakhstanskaya skladchataya oblast [Kazakhstan fold belt], in Belenitskaya, G.A., and Zadoroshnaya, N.M., eds., Rifogennye i sulfatonosnye formatsii Fanerozoya SSSR [Reefal and sulfate bearing formations of the Phanerozoic of the USSR], Ministerstvo Geologii SSSR: Moscow, Nedra, p. 41–52.

Zeng, Dingqian, Liu, Bingwen, and Huang, Yunming, 1992, Reefs through geological ages in China: Beijing, China Petroleum Industry Press, 104 p.

Zhou Zhiyi, and Chen, Peiji, eds., 1992, Biostratigraphy and geological evolution of Tarim: Beijing, Science Press, 399 p.

Ziegler, P.A., 1982, Geological atlas of western and central Europe, Shell Internationale Petroleum Maatschappij, Den Haag, 130 p.

Zonenshain, L.P., Kuzmin, M.M., and Napatov, L.M., 1990a, Tektonika litosfernykh plit territorii SSSR [Tectonics of the lithospheric plates of the USSR]: Moscow, Nedra, v. 1, 328 p., v. 2, 334 p.

Zonenshain, L.P., Kuzmin, M.M., and Napatov, L.M., 1990b, Geology of the USSR: a plate-tectonic synthesis: American Geophysical Union, Geodynamics Series, v. 21, p. 368–397.

Manuscript Accepted by the Society January 22, 2003

Printed in the USA

Geological Society of America
Special Paper 370
2003

Origin and evolution of large Precambrian iron formations

Bruce M. Simonson
Geology Department, Oberlin College, Oberlin, Ohio 44074-1044, USA

ABSTRACT

Collectively, iron formations represent Earth's preeminent supracrustal repository of iron. The largest iron formations were deposited in the Late Archean and Paleoproterozoic via a unique confluence of atmospheric, hydrospheric, lithospheric, and biospheric conditions. Understanding these conditions better requires a deeper appreciation of the sedimentary features of iron formations. Many researchers refer to them collectively as banded iron formations or the acronym BIF, but banding is not always well developed. Iron formations that lack thin banding consist of sand-sized detritus and are cross-bedded, and should be referred to as granular iron formations or the acronym GIF. The mineralogical and textural heterogeneity of iron formations is also underappreciated. The iron in many iron formations resides in siderite or iron-rich silicates rather than oxides. This implies that iron formations did not all form from local releases of oxygen by photosynthetic microbes. Both the heterogeneity of iron formations and the variety of different rock types with which they are associated indicate that large iron formations are not products of a particular depositional setting, such as evaporites. They owe their existence to a combination of: (1) copious masses of dissolved iron supplied by deep-sea hydrothermal systems, (2) the appearance of large continental shelves to serve as depositional repositories, and (3) a stratified ocean with a chemistry suitable for connecting the two. The mechanisms of precipitation are still unclear, but it probably took place along regional chemoclines. Evidence of microbial involvement is increasing. The largest iron formations of all are those of the Hamersley and Transvaal Basins in western Australia and South Africa, respectively, and they may have originally formed a single huge unit.

Keywords: iron-rich rocks, banded iron formations, BIFs, granular iron formations, Precambrian, secular variations.

INTRODUCTION

Iron-rich sedimentary rocks are those containing ≥15% metallic iron by weight (James, 1966). Most workers recognize two main categories: *iron formations*, which are generally cherty, thinly laminated, and Precambrian in age, and *ironstones*, which are generally less siliceous, more aluminous, not laminated, smaller, and Phanerozoic in age (Young and Taylor, 1989). This distinction has gained wide acceptance and highlights time-related changes in iron-rich sedimentary rocks, which have important implications for the evolution of Earth's atmosphere and hydrosphere. There is general agreement that iron formations are a distinct class of sedimentary rock whose deposition was essentially restricted to early Earth history.

Iron formations are also important because they contain the vast majority of iron that will ever be mined. Iron formations gave rise to the largest and richest ore deposits via leaching of silica and oxidation of iron during the Precambrian (Morris, 1987). These ore deposits are currently being mined all over the world. In 2000, Australia produced over 160 million metric tons of iron ore worth in excess of $2.5 billion. China and Brazil each produced even more. Supposedly, there are 10 trillion tons of iron within 300 m of the land surface in just one mining district of the former Soviet Union, the Kursk Magnetic Anomaly (Alexandrov, 1973). However, ore deposits per se are not the focus of this paper; its purpose is to assess the conditions that are needed to create the biggest iron formations.

Simonson, B.M., 2003, Origin and evolution of large Precambrian iron formations, *in* Chan, M.A., and Archer, A.W., eds., Extreme depositional environments: Mega end members in geologic time: Boulder, Colorado, Geological Society of America Special Paper 370, p. 231–244.

CHARACTERISTICS OF IRON FORMATIONS

Mineralogical Composition of Iron Formations

Regardless of size, iron formations consist of a broad range of iron-bearing minerals that are generally accompanied by quartz. The quartz represents chert that has been recrystallized to varying degrees. Geologists of the fledgling U.S. Geological Survey produced a string of monographs on the "iron ranges" in the Lake Superior region, culminating in Van Hise and Leith's (1911) overview. They recognized the diverse nature of iron formations, which James (1954) later systematized into four "facies": oxide, silicate, carbonate, and sulfide (Table 1). Chert was not incorporated into this scheme because it is a near-ubiquitous component of iron formations, but its abundance helps to distinguish them from Phanerozoic ironstones. James used the term "facies" more like metamorphic petrologists than sedimentary geologists, i.e., for rocks with consistent mineral compositions rather than certain depositional structures. However, there is some correlation between mineral facies and other sedimentary features of iron formations (James, 1954; Simonson, 1985). Debate continues about which of the mineral constituents in iron formation (if any) represent original precipitates as opposed to diagenetic phases. In view of this uncertainty, it is not a good idea to infer depositional conditions from mineralogical composition without additional evidence. For example, the fact that an iron formation belongs to the oxide facies should not be used as prima facie evidence for deposition in shallow water. For excellent overviews of the chemistry and mineralogy of iron formation, see James (1954, 1966), Klein (1983), and Lepp (1987).

Sedimentary Textures of Iron Formations

Where detrital textures are not obscured by metamorphism, sedimentary features permit the subdivision of iron formations into *banded* versus *granular* varieties. Banded iron formations (BIFs) were originally laminated chemical muds (Fig. 1), whereas granular iron formations (GIFs) originated largely as well-sorted chemical sands (Fig. 2). Most of the clasts in granular iron formations were produced by intrabasinal erosion and redeposition of pre-existing banded iron formations. Banded iron formations are by far the more abundant of the two, but the use of the term for all iron formations is unfortunate because it obscures the fact that some of the large iron formations accumulated in shallow-water, high-energy environments. This fundamental dichotomy has been recognized over the years by other terms as well. For example, the "slaty" versus "cherty" iron formations of the Lake Superior region (Morey, 1983) and the pelagic versus platform iron formations of Dimroth (1986) are essentially banded iron formations and granular iron formations respectively. The acronym BIF has gained wide acceptance in recent years, and GIF should, too. Mineralogically, most granular iron formations belong to the oxide and silicate mineral facies, whereas banded iron formations are more diverse mineralogically and include an abundance of both oxide and carbonate facies (James, 1954; Simonson, 1985). However, the textural relationships described below are not restricted to specific mineralogical compositions.

Granular Iron Formations (GIFs)

Three primary textural components are readily recognizable in granular iron formations, as in most arenites: (1) a framework of clasts, (2) matrix (finer grained interstitial material), and (3) cement (authigenic minerals filling interstitial voids). Framework clasts typically consist of a mixture of iron oxides, iron silicates, and/or chert, although there are rare examples of clasts consisting of other types of iron-rich minerals. Matrix consists of the same minerals, but it is rare in granular iron formations as a whole. The crystals we see today in most, if not all, of the detrital material in granular iron formations were derived from, but not

TABLE 1. NAMES AND IDEALIZED COMPOSITIONS OF IRON-BEARING MINERALS CHARACTERISTIC OF EACH OF JAMES' (1954, 1966) FOUR MINERALOGICAL FACIES OF LOW-GRADE IRON FORMATIONS

Mineral facies	Principal iron-rich mineral(s)	Chemical formula
Oxide	hematite	Fe_2O_3
	magnetite	Fe_3O_4
Silicate	greenalite	$Fe_3Si_2O_5(OH)_4$[†]
	minnesotaite	$(Mg,Fe)_3Si_4O_{10}(OH)_2$[†]
	stilpnomelane	$(K,Na,Ca)_{0.6}(Mg,Fe^{2+},Fe^{3+})_6Si_8Al(O,OH)_{27}\cdot 2\text{-}4H_2O$
	"chlorite"	$(Fe,Al,Mg)_3(Si,Al)_2O_5(OH)_4$[§]
Carbonate	siderite	$FeCO_3$
	ankerite	$Ca(Fe,Mg,Mn)(CO_3)_2$
Sulfide	pyrite	FeS_2

Note: Formulae are from Deer et al. (1992) unless indicated otherwise.
[†]From Miyano (1987).
[§]From Klein and Bricker (1987).

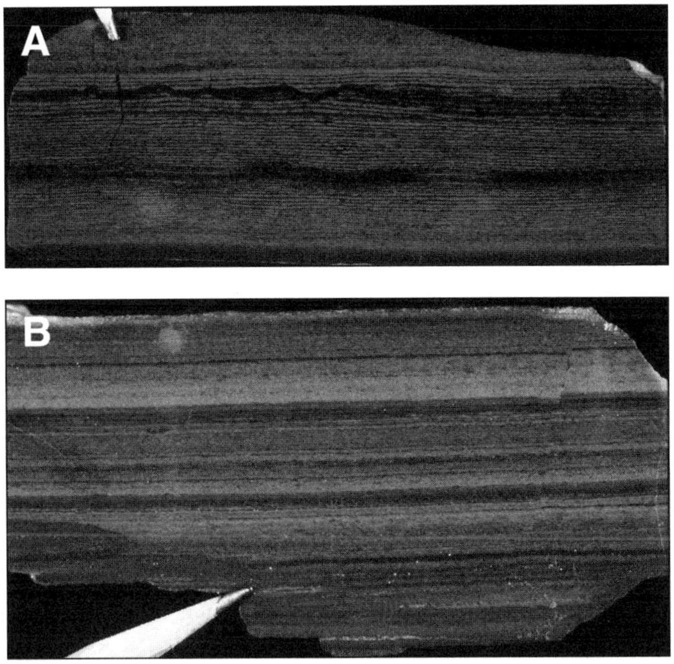

Figure 1. Banded iron formations. A: Cut face of microbanded oxide-facies BIF from Dales Gorge Member; lower three-fourths of sample has thicker layers and consists of reddish jasper (hematitic chert); upper fourth of sample (underneath pencil point) has thinner lamination and is metallic (magnetite-rich and chert-poor); sample is about 5 cm thick. B: Cut face of carbonate-facies BIF from Sokoman Iron Formation; lamination is not as rhythmic and sample is dull gray, except for a dark brown rind on exterior from siderite oxidized by surface weathering (e.g., to left of pencil point).

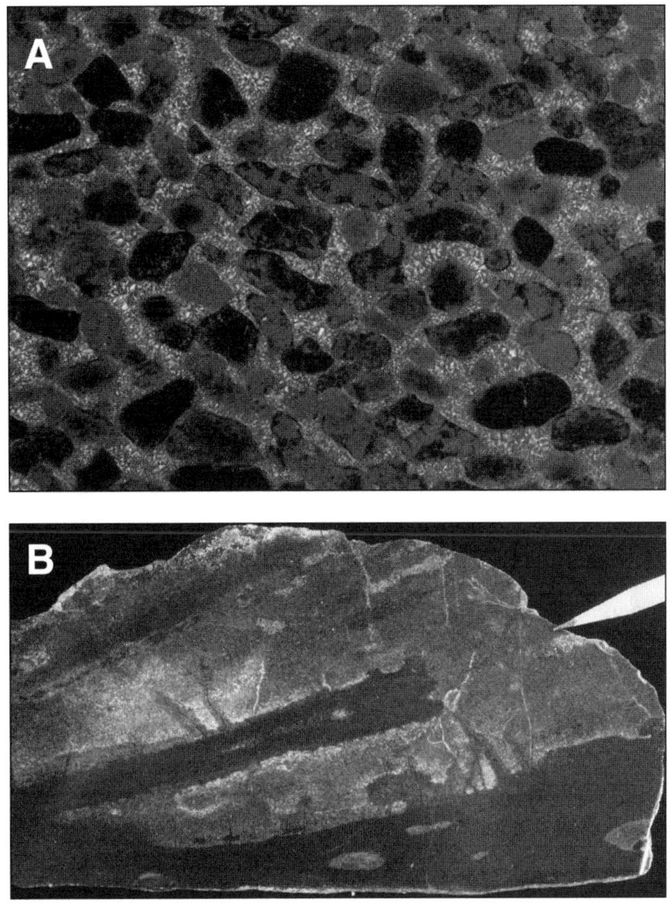

Figure 2. Granular iron formations. A: Photomicrograph of sample from the Gunflint Iron Formation (between crossed polarizers with gypsum plate inserted); sediment was originally medium to coarse sand-size "granules" that now consists of a combination of very finely crystalline hematitic chert (uniform gray) and opaque hematite (black); interstitial pores are largely filled with chalcendonic cement (speckled gray). Long dimension of field of view is about 4 mm. B: Cut face of cross bed from Sokoman Iron Formation, Howell's River area (Klein and Fink, 1976); dark areas are metallic (magnetite-rich and chert-poor) whereas white to light gray areas are chert-rich and range in color from white to red (due to disseminated hematite) to green (due to disseminated greenalite); pencil point for scale.

the same as, the crystals or other materials that were originally present. Han (1978, 1982) revealed widespread evidence of replacement in magnetite crystals by heating samples to about 300°C in a free-air circulating furnace, thereby inducing partial oxidation. Even the relict or pseudomorphic textures Han discovered are secondary rather than original depositional features. In contrast, most cements consist of iron-poor chert and/or quartz, and many show textures acquired during void filling. In addition to these three primary components, all iron formations contain various secondary or diagenetic phases. These later phases are generally more coarsely crystalline, cut across clearly detrital textures, and are not discussed further here.

The dominant clasts in granular iron formations (Fig. 2A) have long been referred to as "granules." Unmetamorphosed granules consist of finely crystalline material internally (e.g., Van Hise and Leith, 1911). They are analogous to the peloids and intraclasts of carbonate grainstones (Dimroth, 1976; Dimroth and Chauvel, 1973). Most granules range in size from fine to coarse sand and in shape from well-rounded to angular (Mengel, 1973). Some granular iron formations also contain abundant ooids, but these are much rarer than granules. Internally, ooids in granular iron formations display concentrically laminated cortices; no radial textures have been reported. Some granules and ooids in granular iron formations contain internal cracks with septarian geometries; these have been attributed to post-depositional shrinkage (Figs. 2, 3, and 9 in Simonson, 1987).

Siliceous cements showing void-filling textures are abundant in granular iron formations. The cements consist largely of drusy quartz and/or parallel-fibrous to radial-fibrous chalcedony. Several different lines of evidence indicate that these cements were emplaced very early. One is a minus-cement porosity of 40–50% in many granular iron formations (Fig. 2A), which approaches the depositional porosity of a well-sorted sand. There is also an abundance of tangential contacts, which is typical of uncompacted sand. Finally, some granular iron formations contain rare intraclasts of silica-cemented granular iron formations that were detritally reworked (Fig. 11 *in* Simonson, 1987). However, early silica cementation is not universal; many granular iron formations were heavily compacted as evidenced by tight frameworks and distorted clasts. The spatial distribution of cements in granular iron formation is typically highly irregular but clearly guided by contrasts between depositional layers (Fig. 2B).

Banded Iron Formations (BIFs)

In contrast to granular iron formations, banded iron formations originally consisted of a broad spectrum of iron-rich minerals in precipitates that were too fine-grained to reveal much via petrographic analysis. Even those banded iron formations subjected to relatively little deformation and metamorphism have been diagenetically reorganized. Nevertheless, they are still fine-grained and uniform (Fig. 1), indicating that the particles that precipitated originally must have been quite small. Banded iron formations show more diversity in iron mineralogy than granular iron formations, including substantial thicknesses of all four of James' facies. Most of these minerals are thought to have compositions close to the phases originally precipitated from basin waters, except for stilpnomelane and other aluminosilicate minerals. Aluminous minerals in either banded or granular iron formations usually reflect contamination with volcaniclastic and/or siliciclastic detritus (e.g. Pickard, 2002). Exquisite volcanic shards replaced by stilpnomelane occur in some iron formations (LaBerge, 1966a, 1996b). Finally, the abundance of silica at a given stratigraphic level can vary tremendously along bedding; this generally takes the form of what are known as chert pods (described below).

Sedimentary Structures of Iron Formations

Granular Iron Formations

Depositional structures are often obscured by diagenetically redistributed minerals, but dune-scale cross-stratification (Fig. 2B) is widespread in granular iron formations (Simonson, 1985). The few paleocurrents that have been measured show complex polymodal patterns with hints of herringbone; this is typical of shallow marine sands (Ojakangas, 1983). Flat pebble conglomerates are a minor but widespread component of granular iron formations. Most of the pebbles in these intraclastic layers are derived from silica-rich layers rather than silica-poor layers.

Siliceous stromatolites are also found in granular iron formations. Although they are quite distinctive, they are quite minor in terms of their total volume. Iron formation stromatolites vary in width from less than a centimeter to over a meter and range in morphology from columnar to domal structures. They were originally interpreted as products of sediment trapping and/or precipitation by microbial mats, but some stromatolites in granular iron formations have characteristics like those of siliceous sinters deposited in and around hot springs (Walter, 1972). Thanks to early silica cementation, these stromatolites and associated iron formations contain some of the best-preserved early Precambrian biotas in the world (Walter and Hofmann, 1983; Han and Runnegar, 1992).

Some granular iron formations also have relatively large cavities, cracks, and/or vugs filled with siliceous cements and, in some case, a bit of fine sediment (Fig. 8 *in* Simonson, 1987). These larger cracks and the small septarian-style cracks inside individual granules form a continuum and are attributed to post-depositional shrinkage. The larger cracks in granular iron formations cut indiscriminately across granules and cements at times, indicating that some of the cements also shrank. Similar cracks and vugs developed in stromatolitic cherts in granular iron formations and contain evidence of cavity-dwelling microbes (Simonson and Lanier, 1987). The presence of sediment in these cracks proves that they formed close to the sediment-water interface. However, they are not normal mudcracks formed via subaerial desiccation because they are in cemented sands (granular iron formations) rather than former muds (banded iron formations), and they do not have the requisite columnar geometries. These cracks appear to be unique to iron formations and are attributed to true syneresis, i.e., shrinkage due to the dewatering of gelatinous silica precursors (Gross 1972; Dimroth and Chauvel, 1973; Beukes, 1984).

Layers of pure granular iron formation thicker than a few meters are rare, whereas banded iron formations can continue uninterrupted by granular iron formations for up to a hundred meters stratigraphically (Simonson and Hassler, 1996; Trendall, 2002). Iron formations with a mixture of banded iron formation and granular iron formation are more abundant than pure granular iron formations, and they show bedding that is more irregular than pure banded iron formation but less massive than pure granular iron formation. The granular iron formation in mixed iron formations usually occurs as discontinuous lenses enclosed in banded iron formation. These lenses represent "starved" bedforms generated by storm waves and currents (Simonson, 1985). However, differential compaction around sediment that was preferentially cemented with silica gave rise to secondary features that look similar. In addition, some granular iron formation lenses in mixed iron formations have an oxidized, jaspery core and a more reduced outer rind. The outer rind is probably a "reaction rim" formed by incomplete equilibration between oxidized sands versus reduced muds during diagenesis.

In many large iron formations, extensive alteration and/or deformation make it difficult to assess the original proportions of banded iron formation versus granular iron formation. The large Indian iron formations in Orissa are a case in point. Although some of these iron formations display current-formed structures (Majumder and Chakraborty, 1977), indicating they must have been granular, most appear banded and are so extensively altered that depositional textures are difficult to assess (Majumder and Chakraborty, 1977). The situation is even worse in the famous Quadrilátero Ferrífero of Brazil. Banding is ubiquitous in the Cauê Itabirite, a large iron formation indeed, but it is so metamorphosed and deformed (Chemale et al., 1994) that it is impossible to say whether or not granular textures were originally present.

The Carajás formation, a large iron formation in northern Brazil, is much less deformed (Trendall et al., 1998), but little sedimentological work has been done on this unit. Perhaps it is no coincidence that most sedimentological work on large iron formations has been done in Australia, North America, and South Africa, as these appear to have the best preserved sedimentary features.

Banded Iron Formations

As the name implies, most banded iron formations have well-developed thin lamination to thin bedding with alternating iron-rich and iron-poor layers (Fig. 1). Thin lamination is the norm in fine-grained Precambrian strata, given the lack of burrowers, but the layers in banded iron formations (particularly those rich in iron oxides) are among the most striking found in sediments of any age. In some cases, exceedingly thin layers can be correlated for over 100 km (Trendall and Blockley, 1970; Ewers and Morris, 1981; McConchie, 1987), but this level of correlation has rarely been attempted, let alone achieved. Bedding can also be highly cyclic via the alternation of either iron-rich versus iron-poor layers within banded iron formation or layers of banded iron formation versus layers of fine shaly or volcaniclastic sediment (Trendall and Blockley, 1970; Trendall, 1973b; Ewers and Morris, 1981; Beukes, 1984). Trendall (1972) attempted to relate these cycles to orbital parameters, but no one has tested them for the periodicities typical of Milankovitch forcing in recent years.

The only common sedimentary structures in banded iron formations other than banding are chert pods, which are concretion-like bodies rich in silica that are typically ellipsoidal in cross-section. Individual layers can often be traced continuously through chert pods, and the chert-poor banded iron formations adjacent to the pods offer textbook examples of differential compaction (Dimroth, 1976; Beukes, 1984; Simonson, 1987). Therefore, chert pods are analogous to concretions in other types of sediment, i.e., localized pockets of early cementation. Drastic changes in the thickness of individual layers that pass through chert pods indicate that some, and perhaps most, silica-poor banded iron formations lost 90% or more of their original thickness during compaction. This indicates that the depositional porosities of banded iron formation were comparable to those of other fine-grained sediments such as argillites (70–90%; Singer and Müller, 1983) and carbonate oozes (80–95%; Cook and Egbert, 1983). Early concretions typically shield minerals from chemical alteration as well as physical compaction. A range of iron-rich minerals are preserved in chert pods, suggesting that the original sediment had a range of compositions similar to the four facies shown by present-day banded iron formations rather than any single precursor mineral.

Secular Changes in Iron Formations

Iron formations range in age from Early Archean to Neoproterozoic, but they were not formed in equal measure throughout this long time span. Banded iron formations are found among the oldest sedimentary strata on Earth, although the sedimentary origins of some of these have recently been questioned (Fedo and Whitehouse, 2002). At the other extreme, iron-rich rocks often referred to as iron formations were deposited on various continents in the Neoproterozoic. However, the Neoproterozoic units have a simple iron mineralogy dominated by hematite and are less cherty than early Precambrian iron formation (James and Trendall, 1982; Beukes and Klein, 1992). They are also much more closely associated with glaciogenic sediments (Young, 1988) and much smaller than the largest of the older iron formations, so they will receive little consideration in this paper. The largest iron formations were deposited during an interval of ca. 800 m.y. in the Late Archean to Paleoproterozoic, which ended rather abruptly on or before 1.8 Ga (Gole and Klein, 1981; Trendall, 2002). This "interval" may actually consist of two main peaks rather than a single plateau of iron formation deposition (Isley and Abbott, 1999). After 1.8 Ga, few if any iron formations were deposited until the Neoproterozoic. Although some of the details will no doubt change as research continues, there were clearly secular changes in both the size and depositional environments of iron formation, as follows.

Increase in Mass Through Time

Statistically, Early to Middle Archean iron formations tend to be smaller than those that are Late Archean to Paleoproterozoic in age. This is reflected in Gross's (1965, 1983) classification of iron formations into two major varieties, Superior-type versus Algoma-type. Gross's original formulation did not prove to be universally applicable in all respects (Trendall, 2002). However, the names will be used in this paper because they provide a handy way to distinguish iron formations associated mainly with volcanic rocks, the Algoma-type, from iron formations associated mainly with sedimentary strata, the Superior type. The main departure from Gross's original schema is that not all Superior-type iron formations contain granular iron formations (Table 2). When defined in this fashion, it turns out that all of the largest iron formations are Superior-type iron formations. Additional distinctions similar to, but not the same as, those that Gross made between Algoma-type and Superior-type iron formations are outlined as follows. These reflect secular changes in the nature of iron formation.

James and Trendall (1982) attempted a semi-quantitative analysis of variation in the size of iron formation as a function of age by placing major iron formations from five continents into four categories: small, moderate, large, and very large. Their data set indicates that the largest iron formations are all Late Archean through Paleoproterozoic in age, whereas smaller Algoma-type iron formations occur throughout the entire age range from Early Archean through Paleoproterozoic. The smaller size of the Algoma-type iron formations presumably reflects deposition in smaller basins. However, Gole and Klein (1981) correctly noted that they are typically more deformed than Superior-type iron formations and cautioned that some Algoma-type iron formations "may have been quite extensive prior to deformation and disruption."

Among the Late Archean to Paleoproterozoic iron formations, those of the Hamersley Basin of western Australia and the

TABLE 2. BASIC DATA ON LARGE IRON FORMATIONS MENTIONED IN TEXT*

Continent	Geologic province	Name of local sub-basin (where applicable)	Stratigraphic unit(s) dominated by iron formation	Type(s) present	Estimated age (in Ga)
Australia	Hamersley Basin	Not applicable	Marra Mamba Iron Formation	BIF	2.6
		Not applicable	Brockman Iron Formation, Dales Gorge Member	BIF	2.5
		Not applicable	Brockman Iron Formation, Joffre Member	BIF	2.5
		Not applicable	Weeli Wolli Iron Formation	BIF	2.4
		Not applicable	Boolgeeda Iron Formation	BIF	2.4
	Nabberu Basin	Earaheedy Basin	Frere Formation	GIF, BIF	1.8
Europe	Russian platform	Kursk Magnetic Anomaly	Various	BIF?	2.3?
India	Indian Shield	Bihar and Orissa	Various	BIF,GIF?	2.9
North America	Labrador trough	Knob Lake area	Sokoman Iron Formation	GIF, BIF	1.9
		Knob Lake area	Ruth Iron Formation	BIF	1.9
	Lake Superior region	Gogebic Range	Ironwood Iron Formation	GIF, BIF	1.9
		Gunflint Range	Gunflint Iron Formation	GIF, BIF	1.9
		Mesabi Range	Biwabik Iron Formation	GIF, BIF	1.9
		Marquette Range	Negaunee Iron Formation	BIF	1.9
South Africa	Transvaal Basin	Griqualand west	Kuruman Iron Formation	BIF	2.5
		Griqualand west	Griquatown Iron Formation	GIF, BIF	2.5
		Transvaal Basin proper	Penge Iron Formation	BIF	2.5
South America	São Fransisco Craton	Quarilátero Ferrífero	Cauê Itabirite	BIF?	2.5
	Amazon Craton	Grão Pará Group	Carajás Formation	uncertain	2.7

Note: Ages compiled from Majumder and Chakraborty (1979), Alexandrov (1973), Trendall et al. (1998), Isley and Abbott (1999), Nelson et al. (1999), Hocking et al. (2001), and Pickard (2002). BIF—banded iron formation; GIF—granular iron formation.
*See Figure 3.

Transvaal Basin of South Africa are clearly the largest. While examples of James and Trendall's (1982) "very large" iron formations are found on all five continents, only the Hamersley and Transvaal Basins each contain in excess of 10^{14} tonnes of iron. There are five major iron formations within the Hamersley succession (Trendall, 1983) versus only two in the Transvaal succession (Beukes, 1984). However, this is offset in part by the fact that the preserved area of the Transvaal Basin is roughly twice that of the Hamersley Basin. The exceptional size of the iron formations in these two basins becomes even more remarkable since they may actually be two parts of a single basin. Button (1976) summarized an impressive number of similarities in their sedimentary and economic deposits, as well as their geological evolution. Cheney (1996) formalized this hypothesis by suggesting the name "Vaalbara" for the combined landmass. Not everyone is persuaded, but subsequent studies continue to reveal more and more geological parallels between these two successions, and their geochronologies seem to grow ever closer (Nelson et al., 1999). The most recent connection is a striking similarity in the detrital zircon populations of 3.47-Ga

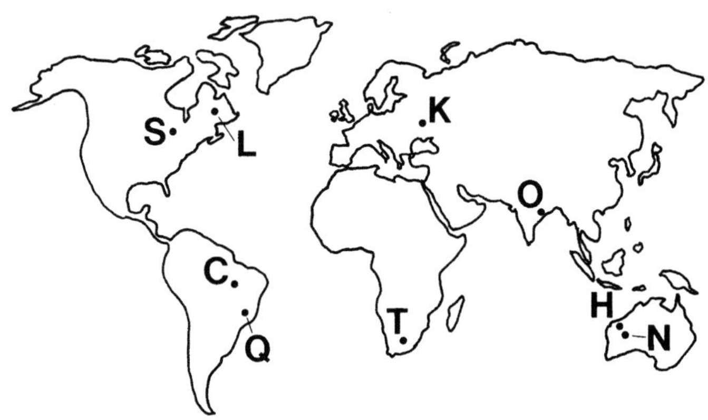

Figure 3. World map with locations of selected basins with large iron formations, indicated as follows: C—Carajás, H—Hamersley, K—Kursk Magnetic Anomaly, L—Labrador trough, N—Nabberu, O—Orissa, Q—Quadrilátero Ferrífero, S—Lake Superior, and T—Transvaal. See Table 2 for more details.

spherule layers on both cratons that appear to be the products of a single large asteroid or comet impact (Byerly et al., 2002). Individually or collectively, the Hamersley and Transvaal Basins contain a record of the largest and most sustained episode of iron sedimentation in Earth history.

Increase in Environmental Energy through Time

Changes in the sedimentary textures of iron formations signal an increase in the energy of their depositional environments through time. Algoma-type iron formations consist almost exclusively of banded iron formation; the few granular iron formations that have been reported (e.g., Manikyamba, 1999) are highly unusual. The oldest Superior-type iron formations, those of the Hamersley and Transvaal Basins, also consist mainly of banded iron formation. The fine grain size and thin, laterally persistent lamination of these banded iron formations reflect "exceptionally still and quiet" conditions (Trendall 1983), implying deposition in deep shelf and possibly slope environments well below wave base. Higher-energy conditions did occur on rare occasions in the Hamersley Basin, as indicated by a few highly restricted occurrences of granular iron formation (Simonson and Goode, 1989). In contrast, a stratigraphic unit of granular iron formation in the Transvaal Basin (Table 2) implies a sustained period of higher-energy conditions that are probably associated with a lowstand (Beukes, 1983, 1984). However, unlike most younger granular iron formations, the main granular iron formation in the Transvaal Basin belongs to the carbonate facies, i.e., it is siderite-dominated (Beukes, 1984; Beukes and Klein, 1990). This suggests that it formed in deeper, more stagnant waters than most granular iron formations.

In contrast to the older Superior-type iron formations, granular iron formations are widespread in young Superior-type iron formations, although they are still subordinate in total volume to banded iron formations. The best examples are the Superior-type iron formations in the Lake Superior area and Labrador trough of North America (Zajac, 1974; Morey 1983; Dimroth, 1986; Fralick and Barrett 1995) and the Nabberu Basin of western Australia (Hall and Goode, 1978; Goode et al., 1983; Bunting, 1986), most of which contain substantial thicknesses of granular iron formation (Table 2). These granular iron formations display a host of shallow-water features, most notably abundant cross-bedding (Fig. 2B). They also contain more limited but at times spectacular oolitic and stromatolitic layers. These characteristics clearly indicate that substantial parts of these iron formations accumulated in higher energy environments, although most were probably deposited in deeper water fairly close to wave base because they all interfinger with banded iron formations stratigraphically.

Changes in the stratigraphic units associated with iron formations provide further evidence of shallowing through time in iron formation basins. Algoma-type iron formations are generally associated with volcanic rocks that include deep-water deposits such as volcaniclastic turbidites (Dunbar and McCall, 1971; Barrett and Fralick, 1985, 1989; Shegelski, 1987). While the older Superior-type iron formations are associated largely with sedimentary rather than volcanic rocks, the associated sediments again are deeper water deposits that include turbidites and graded tuff beds (Beukes, 1983; Simonson et al., 1993; Hassler, 1993). In contrast, a number of the younger Superior-type iron formations are in conformable contact with shallow-water deposits such as tidally cross-bedded quartzarenites and stromatolitic dolomites (Hall and Goode, 1978; Ojakangas, 1983; Morey, 1983; Simonson, 1984). However, some of the young Superior-type iron formations consist of just banded iron formation and are associated with deep-water, turbidite-rich units (Larue, 1981; Simonson, 1985).

The transition from Algoma- to Superior-type iron-formations was probably gradual rather than abrupt. While the oldest Superior-type iron-formations were being deposited in the Hamersley and Transvaal Basins ca. 2.6 Ga, Algoma-type iron formations were accumulating on other continents. Moreover, some iron formations appear to be intermediate in character between Algoma-type and Superior-type iron-formations. This includes some iron formations deposited on the margins of the Kaapvaal and Zimbabwe Cratons around 3.0 Ga (Watchorn, 1980; Fedo and Eriksson, 1996) and others deposited in the Lake Superior area (Morey and Southwick, 1995). As for the increase in energy through time evident among the Superior-type iron formations, it is not clear whether this was stepwise or gradual because there are so few iron formations with well-constrained ages between about 2.4 and 2.0 Ga (Isley and Abbott, 1999).

ORIGINS OF LARGE IRON FORMATIONS

Probably because of the lack of close modern analogs, many different theories have been proposed for the origin of iron formation. Consensus has yet to be reached on the specific mechanisms whereby iron and silica were precipitated, but a broad consensus has been reached on the general setting and some of the key parameters of iron formation's deposition. Before discussing the views that currently prevail, it is perhaps simplest to outline some of the theories for the origin of iron formation that no longer seem viable.

Obsolete Hypotheses

Replaced Carbonates

As noted above, granular iron formations have textural constituents analogous to those of carbonate grainstones. The petrographic analysis of granular iron formations reached its zenith in the work of Erich Dimroth (Dimroth, 1968; Dimroth and Chauvel, 1973). He ultimately concluded that the similarities between granular iron formations and calcarenites were so striking that iron formations must have been deposited as carbonates, then replaced wholesale by iron- and silica-rich minerals during diagenesis (Dimroth, 1976). Other researchers arrived at similar conclusions (e.g., Kimberley, 1974; Lougheed, 1983; Lepp, 1987; Sommers et al., 2000), but few advocates remain for this interpretation. Arguments that militate against it include the apparent lack of any carbonate units that are half-converted to iron formations and the presence of textural features like syneresis cracks that are not found in carbonates and require a gelatinous precursor (Simonson, 1987). In the words of Kimberley (1989), "the evidence against this concept is now so overwhelming ...that diagenetic replacement is no longer ...viable."

Lacustrine/Nonmarine Deposits

Despite the fact that James (1954) marshalled a number of good arguments supporting the deposition of iron formations in marginal marine basins, it has repeatedly been suggested that iron formations were deposited in lacustrine environments completely isolated from the world ocean (e.g., Hough, 1958; Eugster and Chou, 1973; Garrels, 1987). While it is certainly possible that some smaller iron formations were deposited in lacustrine settings (Eriksson, 1983; Beukes, 1984), multiple lines of evidence indicate that the Superior-type iron formations were deposited in open marine settings, primarily on continental margins. One line of evidence is their close association with shallow marine deposits such as tidally influenced quartzarenites (Ojakangas, 1983; Simonson, 1984). In many instances, the transitions from such units to overlying iron formations coincide with transgressions (Beukes, 1983; Simonson and Hassler, 1996), which would make the connection with the world ocean even deeper. Sequence-stratigraphic analyses have also confirmed that Superior-type iron formations occur in successions typical of those deposited on Phanerozoic continental margins (Barley et al., 1992; Morey and Southwick, 1995; Krapez and Martin, 1999). Additional lines of evidence supporting a marine origin include the lack of chemical and mineralogical variability one would expect of precipitates from closed basin waters with variable chemistries (Gole and Klein, 1981; Lepp, 1987) and the sheer size and lack of internal variability of the largest iron formations (Kimberley, 1989; Simonson and Hassler, 1996).

Evaporites of the Precambrian

Iron formations have been interpreted as both marine (Trendall, 1973a) and non-marine (e.g., Eugster and Chou, 1973) evaporites. It is reasonable to expect differences in composition between evaporites formed in the Phanerzoic versus the Precambrian, particularly in light of recent documentation that marine evaporites have varied in composition within the Phanerozoic (Lowenstein et al., 2001). However, it is hard to see how the evaporation of seawater could give rise to iron- and silica-rich minerals and little else at any time in Earth history. Equally damaging to the evaporite interpretation is the total lack of any structures reflecting arid conditions in either iron formations or the strata associated with them. As noted above, shrinkage structures are present in some granular iron formations, but they are early diagenetic rather than depositional in origin and were caused by syneresis of amorphous silica precursors rather than subaerial exposure. Moreover, none of the carbonates closely associated with Superior-type iron formations contain sabkha deposits or any other evidence of aridity. They appear instead to have been deposited in deeper water, open marine settings (Dimroth, 1971; Klein and Beukes, 1989; Simonson et al., 1993). Sedimentological studies have made it clear that the clastic units associated with Superior-type iron formations were likewise deposited in open marine settings and lack evidence of arid conditions or even subaerial exposure during deposition (Ojakangas, 1983; Beukes, 1983; Simonson, 1984; Bunting, 1986).

Precipitation in Oxygen Oases

Preston Cloud was a scientist of exceptionally broad vision and one of the first to invoke non-uniformitarian differences between environmental conditions of the Precambrian and Phanerozoic to try to explain iron formations. The crux of his elaborate theory, summarized in Trendall (2002), is that dissolved ferrous iron was ubiquitous in the early oceans, so iron formations formed wherever photosynthetic microbes provided an abundance of oxygen. It was an elegant hypothesis, but precipitating iron oxides is only part of the story. Iron formations contain a variety of iron-rich phases, and minerals shielded within chert pods indicate that the precursor sediment had a range of compositions similar to those of present-day banded iron formations (Simonson, 1987). One corollary of Cloud's hypothesis would be the presence of high concentrations of iron in carbonate sediments contemporaneous with iron formations. It has since been determined that Late Archean to early Paleoproterozoic carbonates do contain somewhat more iron than Phanerozoic carbonates (Veizer et al., 1990, 1992), but it is hardly enough to fit the scenario of ubiquitous ferrous iron in the world ocean (Holland, 1984).

Biogenic Oozes

It has been suggested that iron formations represent accumulations of the skeletal remains of microorganisms. Banded iron formations and granular iron formations both contain a profusion of spheroidal microstructures that average about 30 microns in diameter and have been attributed to organic activity (LaBerge, 1973; LaBerge et al., 1987). Heaney and Veblen (1991) demonstrated that these microstructures are diagenetic on the basis of a transmission electron microscopy study. Moreover, the fossil record of silica-secreting organisms only dates back to the Lower Cambrian (Allison, 1981). Lastly, textural evidence from granular iron formations indicates that much of the silica was added as cement via interstitial precipitation (Simonson, 1987) instead of being a primary constituent added directly from the water column.

As for the iron in iron formations, certain groups of organisms including bacteria form perfect magnetite crystals. Possible examples of biogenic magnetite have been recovered from limestones in the Gunflint Iron Formation (Chang and Kirschvink, 1989). However, it is hard to envision how biogenic magnetite could accumulate in such pure concentrations over such extensive areas, then be altered to the various different minerals needed to create present-day iron formations (Table 1). Therefore, it is highly unlikely that iron formations are biogenic oozes composed of the products of biomineralization. Nevertheless, it is possible that microbes played a role in the deposition of iron formations, and quite possibly a large one (as discussed below).

Current Consensus

If the preceding theories are no longer viable, what models currently seem most reasonable? Here are some points of broad agreement that any comprehensive model for iron formations should take into account.

Hydrothermal Source of Solutes

James (1954) and most early workers believed that deep weathering on continents provided the iron needed to make iron formations. The subsequent discovery of deep-sea hydrothermal systems provided an alternative source that is more consistent with the geological characteristics of iron formations and associated deposits (Simonson, 1985). Banded iron formation deposition has been linked to hydrothermal activity with reasonably high confidence via stratigraphic context and facies relationships for some Algoma-type iron formations (e.g., Goodwin et al., 1985). Moreover, geochemical signatures of hydrothermal sources have been detected in all types of iron formations (Klein and Beukes, 1992). Geochemical indicators ranging from isotopic ratios of sulfur (Cameron, 1983) and neodymium (Jacobsen and Pimentel-Klose, 1988) to rare earth element distributions (Fryer, 1983; Derry and Jacobsen, 1990; Danielson et al., 1992) indicate that seafloor hydrothermal systems were more active in the early Precambrian (particularly in the Archean). Fryer et al. (1979) pointed out that this activity would inject "a high magnitude flux of reduced species into the Archaean Ocean" from the bottom, including large masses of dissolved ferrous iron. The fact that the younger iron formations deposited in shallower water tend to have lower concentrations of iron than the older, deeper-water iron formations is also consistent with a deep-ocean source of iron. In summary, hydrothermal sources in the deep ocean are now widely viewed as the source of the iron that is needed to make iron formations (e.g., Barley et al., 1997).

Stratified Water Column

Most researchers now believe large iron formations were deposited in basins with a stratified water column. Hypotheses that the deeper ocean waters contained uniformly higher concentrations of dissolved ferrous iron emerged in the 1970s (Holland, 1973; Drever, 1974; Degens and Stoffer, 1976) and gained in popularity in the 1980s (Button, 1982). The building of iron formations requires the transport of large masses of dissolved iron over long distances, but surface waters were too oxygenated to do so in the Late Archean to Paleoproterozoic (Trendall, 2002). Some of the best evidence of this comes from iron formations. Hematite is the dominant iron mineral in the least-altered granular iron formations, which are much likelier to reflect the oxidation state of near-surface waters than are banded iron formations. If iron were not mobile in surface waters, there would be no alternative to transporting it in mid- and/or deeper level waters under more reducing conditions. This implies that the water column is stratified. Contrasts in the trace element and isotopic compositions of iron formations and coeval iron-poor strata (Klein and Beukes, 1989; Carrigan and Cameron, 1991; Winter and Knauth, 1992) support a stratified ocean model.

Despite broad agreement that water columns in iron formation basins were stratified, consensus has yet to be reached on the character and causes of that stratification. Some workers envision a large reservoir of bottom water with relatively uniform concentrations of dissolved ferrous iron (e.g., Jacobsen and Pimentel-Klose, 1988). Cameron (1983) envisioned a different situation in which the highest concentrations of dissolved iron were at intermediate water depths owing to higher concentrations of hydrogen sulfide in deeper waters. Isley (1995) subsequently demonstrated the feasibility of connecting hydrothermal sources with shelf sinks in the early Precambrian by dispersing iron laterally at shallow to intermediate water depths. Under the conditions she posited, seafloor hydrothermal activity could supply enough dissolved iron to make even large, Superior-type iron formations. A mid-water maximum in the concentration of dissolved iron is also consistent with the stratigraphic context of many Superior-type iron formations (Simonson and Hassler, 1996).

Given this consensus, models for the deposition of large iron formations should focus on processes active along chemoclines between deeper iron-rich and shallower iron-poor water masses (e.g., Beukes and Klein, 1992). For example, iron could be precipitating via oxidation along the chemocline in a manner somewhat analogous to the formation of particulate MnO_2 in the modern Black Sea (Force and Maynard, 1991). Microbes are apt to take advantage of steep chemical gradients wherever they find them, so microbial mediation of precipitation along chemoclines is possible, if not probable. Recent isotopic work strongly suggests that microbes played a role in the precipitation and/or diagenetic reorganization of the iron in iron formations (Beard et al., 1999). Trendall (2002) marshaled additional arguments favoring the involvement of the biosphere in the deposition of iron formations. Given the variety of iron minerals found in iron formations, a variety of mechanisms are probably needed to explain all of the permutations (see review by Morris, 1993). Pinning these mechanisms down and explaining why they often happened in the cyclic patterns now seen in banded iron formations are the greatest challenges to achieving a full understanding of iron formation. Perhaps these cycles are ancient examples of microbial biofeedback!

High Primary Silica Content

The high chert content of iron formations relative to younger, iron-rich sediments is widely perceived as reflecting a higher average concentration of silica in Precambrian seawater relative to today's oceans. This reflects the fact that silica-fixing organisms did not evolve until the Phanerozoic (Maliva et al., 1989). Siever (1992) arrived at a value of 60 ppm for the average concentration of silica in late Precambrian seawater using reasonable estimates of relevant fluxes. Silica concentrations could have been even higher early in the Precambrian, given that the level of hydrothermal activity on the seafloor was presumably greater. Specific mechanisms that have been proposed for silica precipitation include biogenic inducement (LaBerge et al., 1987), slight evaporative concentration, co-precipitation with iron (Ewers, 1983), and polymerization due to electrolyte changes (Morris, 1993). These processes take place in surface waters, but there is good textural evidence that a significant fraction of the silica in iron formations precipitated interstially from pore waters close beneath the sediment/water interface (Simonson, 1987). Therefore, the high silica content of iron formations is probably a more general reflection of high silica concentrations in Precambrian seawater rather than biogenic activity or evaporative concentration in the surface waters of specific basins. Higher ambient geothermal gradients may also have increased the flux of silica into the shallow subsurface from below (Simonson, 1987).

Interpreting the high silica content of iron formation this way makes sense if iron formations were a type of background sediment that accumulated slowly. Eriksson (1983) suggested that iron formations were deposited wherever nothing else was accumulating fast enough to dilute it. This is consistent with the fact that many iron formations appear to have accumulated in sediment-starved settings (Simonson and Hassler, 1996). However, recent age dates suggest that iron formations were deposited relatively rapidly (Barley et al., 1997; Trendall, 2002; Pickard, 2002) and raise questions about this interpretation. Perhaps high ambient concentrations were increased by hydrothermal activity during episodes of iron formation deposition.

Tectonic Cause for Their Increase in Size

The increase in the average size of iron formations through time evidenced primarily by the appearance of Superior-type iron formations can be attributed in large part to a significant expansion of continental shelf and slope environments in the Late Archean. An increase in the total area of continental shelves is a corollary of a Late Archean surge in the growth of continental crust (Goodwin, 1991, Lowe, 1992, Eriksson, 1995). Continental margins typically offer larger repositories for sediment than basins in the volcanic terrains that hosted Algoma-type iron formations. An increase in the size of stable-shelf deposits in the Late Archean is not unique to iron formations. For example, the first platformal carbonates comparable in size to Phanerozoic build-ups appear in the Late Archean in the same basins that host the first large iron formations (Beukes, 1983; Grotzinger, 1994; Klein and Beukes, 1989; Simonson et al., 1993). Cratonization of shields was a highly diachronous process (Eriksson and Donaldson, 1986). This could help explain why the largest iron formations differ in age on different continents (Trendall, 2002).

The increase through time in the size of iron formations may not be entirely a direct result of lithospheric evolution. Evidence for deposition of iron formations in progressively shallower waters through time was summarized above. If iron formations were being deposited in basins with stratified water columns, this implies progressive shallowing of chemoclines through time. This could reflect secular changes in bathymetry, which ultimately depend on tectonic processes, but it could also reflect changes in the chemistry of the atmosphere and/or seawater. The increase in the abundance of granular iron formation suggests dissolved iron was being transported into shallower water and for longer distances from its hydrothermal sources through time. This runs counter to the notion that the atmosphere was becoming progressively more oxic throughout this time interval (Eriksson, 1995). As noted above, iron formations with well-constrained ages between ca. 2.4 and 2.0 Ga are scarce. Therefore, we are not in a good position to judge whether the observed changes in the sizes and depositional environments of iron formation were gradual or abrupt.

Atmospheric/Hydrospheric Cause for Their Demise

There is general consensus that the termination of iron formation deposition in the Paleoproterozoic reflects evolutionary shifts in atmospheric and hydrospheric chemistry. Changes clearly occurred in the chemistry of the world ocean by about 1.8 Ga, which radically reduced iron's mobility in deeper ocean waters. Researchers have historically attributed this to ventilation, that is, oxygenation, as many lines of evidence signal a dramatic rise in atmospheric oxygen ca. 2.2–1.9 Ga (Holland, 1994). However, deep-sea chemistry and atmospheric chemistry do not always march in lock step. The decrease in the concentration of dissolved iron in the deep oceans after 1.8 Ga could reflect an increase in the concentration of dissolved sulfide rather than dissolved oxygen (Canfield, 1998; Anbar and Knoll, 2002). This makes sense in terms of the high demand that inputs of organic carbon would place on the relatively low concentrations of dissolved oxygen one would expect in the deep ocean at times of little or no glacial activity. Either way, some change in the chemistry of the deep oceans clearly prevented the storage and/or long-distance transport of dissolved iron during the second half of Earth history on a scale comparable to the first half. This brought the deposition of large iron formations to an end.

The appearance of iron formations in the Neoproterozoic, which are similar, though not identical, to the older iron formations, probably indicates a short-lived return to conditions like those in the first half of Earth history. As with the older iron formations, the source of the iron for the Neoproterozoic iron formations appears to have been hydrothermal (Young, 1988; Breitkopf, 1988). The close association of these iron formations with glacial sediments may be a key piece of evidence in this regard (Young, 1988; Trendall, 2002). The Neoproterozoic glaciations were very extensive, so much so that Hoffman et al. (1998) believe they gave rise to a "snowball Earth." Perhaps a world ocean covered by ice could become highly stratified for the first time in over an eon and recreate some of the processes active in the Archean and Paleoproterozoic (Klein and Beukes, 1992). Whatever was responsible, conditions changed again before the start of the Phanerozoic such that significant iron formations finally disappeared from the stratigraphic record for good.

SUMMARY AND SPECULATION

The sedimentology of banded iron formations, granular iron formations, and associated iron-poor strata constrain the origin of large iron formations in critical ways. For example, it is now clear that large iron formations are not replaced carbonates, skeletal biogenic oozes, or precipitated around colonies of photosynthesizing microbes releasing oxygen. It is also clear that large iron formations accumulated in marine environments, none of which were highly evaporitic; that hydrothermal systems were the source of the iron; and that large iron formation basins had stratified water columns. It seems that the large, Superior-type iron formations owe their existence to a unique confluence of three main circumstances in the Late Archean to Paleoproterozoic: (1) the presence of large hydrothermal systems on the deep sea floor, which presumably were very active throughout the Archean; (2) a dramatic expansion in the total area of continental margins, which provided depositional repositories larger than any that existed earlier in the Archean; and (3) a stratified ocean with reduced intermediate and/or deep water masses that could transmit large fluxes of dissolved ferrous iron from deep-sea hydrothermal systems to distant depocenters. The fact that large iron formations occur in many different tectonic settings and are associated with many different rock types (Gross, 1983; Fralick and Barrett, 1995; Morey and Southwick, 1995) suggests that these conditions were met in a variety of different settings. If so, the first-order cause of large iron formations could simply be periods of unusually vigorous hydrothermal activity, coupled with sea-level highstands. At such times, precipitates formed along regional chemoclines could accumulate relatively undiluted by other types of sediment (Simonson and Hassler, 1996) and perhaps rather rapidly (Barley et al., 1997; Trendall, 2000). Isley and Abbott (1999) believe there is a statistically significant correlation between iron formations and proxies for mantle plume activity such as komatiites and flood basalts. A connection between iron formation deposition and mantle superplumes could also help explain why Superior-type iron formations do not appear to be evenly distributed in either time or space.

The existence of a hypsometry during the Late Archean to Paleoproterozoic unlike any before or since may have been a contributing factor in forming large iron formations. It is commonly assumed that continental freeboard has remained constant through geologic time, but this is not necessarily the case (Eriksson, 1999). Arndt (1999) called attention to features in greenstone belts that suggest the existence of broad, submerged continental platforms in the Late Archean to Paleoproterozoic unlike any known from later in Earth history. Widespread evidence of basaltic hydrovolcanism in large iron formation basins (Hassler and Simonson, 1989; Hassler, 1993; Altermann, 1996) provides support for extensive areas of shallow flooding. Hydrovolcanism will not occur in deep water because hydrostatic pressure prevents the runaway fuel-coolant reaction needed to power fragmentation of low-viscosity magma (Sheridan and Wohletz, 1983). Signs of the explosive felsic volcanism typical of convergent margins are present in some iron formation basins (LaBerge, 1966a, 1966b; Pickard, 2002), but they are surprisingly rare. The existence of uniquely large areas of flooded continental crust could help explain the exceptional continuity of depositional layers across the Hamersley and Transvaal Basins. Such flooding, and perhaps the small size of continents at that point in Earth's history, may also have aided in the formation of uniquely large iron formations by minimizing dilution with fine-grained siliciclastic sediment derived from continental weathering. The immense size of the Late Archean to Paleoproterozoic iron formations in these basins also indicates that they were unusually close and/or well connected to large hydrothermal sources of iron.

ACKNOWLEDGMENTS

Much of my work has been done in collaboration with Oberlin students who have greatly enhanced my understanding of iron formations. I am particularly indebted to Scott Hassler. Fieldwork was supported by grants from the National Geographic Society, the National Science Foundation, and Oberlin College, as well as logistical support from the Geological Survey of Western Australia, Hamersley Iron Pty. Ltd., the Iron Ore Company of Canada, and numerous other mining companies. N. Beukes and P. Link reviewed the first draft and made many helpful suggestions for its improvement. This overview draws heavily on Simonson (1997).

REFERENCES CITED

Alexandrov, E.A., 1973, The Precambrian banded iron-formations of the Soviet Union: Economic Geology, v. 68, p. 1035–1062.

Alison, C.W., 1981, Siliceous microfossils from the Lower Cambrian of northwest Canada: Possible source for biogenic chert: Science, v. 211, p. 53–55.

Altermann, W., 1996, Sedimentology, geochemistry and palaeogeographic implications of volcanic rocks in the Upper Archaean Campbell Group, western Kaapvaal craton, South Africa: Precambrian Research, v. 79, p. 73–100.

Anbar, A.D., and Knoll, A.H., 2002, Proterozoic ocean chemistry and evolution: A bioinorganic bridge?: Science, v. 297, p. 1137–1142.

Arndt, N., 1999, Why was flood volcanism on submerged continental platforms so common in the Precambrian?: Precambrian Research, v. 97, p. 155–164.

Barley, M.E., Blake, T.S., and Groves, D.I., 1992, The Mount Bruce megasequence set and eastern Yilgarn Craton: examples of Late Archaean to Early Proterozoic divergent and convergent craton margins and controls on mineralization: Precambrian Research, v. 58, p. 55–70.

Barley, M.E., Pickard, A.L., and Sylvester, P.J., 1997, Emplacement of a large igneous province as a possible cause of banded iron formation 2.45 billion years ago: Nature, v. 385, p. 55–58.

Barrett, T.J., and Fralick, P.W., 1985, Sediment redeposition in Archean iron formation: Examples from the Beardmore-Geraldton greenstone belt, Ontario: Journal of Sedimentary Petrology, v. 55, p. 205–212.

Barrett, T.J., and Fralick, P.W., 1989, Turbidites and iron formations, Beardmore-Geraldton, Ontario: Application of a combined ramp/fan model to Archaean clastic and chemical sedimentation: Sedimentology, v. 36, p. 221–234.

Beard, B.L., Johnson C.M., Cox, L., Sun, H., Nealson, K.H., and Aguilar, C., 1999, Iron isotope biosignatures: Science, v. 285, p. 1889–1892.

Beukes, N.J., 1983, Palaeoenvironmental setting of iron formations in the depositional basin of the Transvaal Supergroup, South Africa, *in* Trendall, A.F., and Morris, R.C., eds., Iron-formations: Facts and Problems: Amsterdam, Elsevier, p. 131–209.

Beukes, N.J., 1984, Sedimentology of the Kuruman and Griquatown iron-formations, Transvaal Supergroup, Griqualand West, South Africa: Precambrian Research, v. 24, p 47–84.

Beukes, N.J., and Klein, C., 1990, Geochemistry and sedimentology of a facies transition—from microbanded to granular iron-formation—in the Early Proterozoic Transvaal Supergroup, South Africa: Precambrian Research, v. 47, p. 99–139.

Beukes, N.J., and Klein, C., 1992, Models for iron-formation deposition, *in* Schopf, J.W., and Klein, C., eds., The Proterozoic Biosphere: a Multidisciplinary Study: Cambridge, Cambridge University Press, p. 147–151.

Breitkopf, J.H., 1988, Iron formations related to mafic volcanism and ensialic rifting in the Southern Margin Zone of the Damara Orogen, Namibia: Precambrian Research, v. 38, p. 111–130.

Bunting, J.A., 1986, Geology of the eastern part of the Nabberu Basin, western Australia: Geological Survey of Western Australia Bulletin 131, 130 p.

Button, A., 1976, Transvaal and Hamersley Basins—Review of basin development and mineral deposits: Minerals Science and Engineering, v. 8, p. 262–293.

Button, A. (rapporteur), 1982, Sedimentary iron deposits, evaporites and phosphorites—state of the art report, *in* Holland, H.D., and Schidlowski, M., eds., Mineral Deposits and the Evolution of the Biosphere: Berlin, Springer-Verlag, p. 259–273.

Byerly, G.R., Lowe, D.R., Wooden, J.L., and Xie, X., 2002, An Archean impact layer from the Pilbara and Kaapvaal cratons: Science, v. 297, p. 1325–1327.

Cameron, E.M., 1983, Genesis of Proterozoic iron-formation: Sulphur isotope evidence: Geochimica et Cosmochimica Acta, v. 47, p. 1069–1074.

Canfield, P., 1998, A new model for Proterozoic ocean chemistry: Nature, v. 396, p. 450–453.

Carrigan, W.J., and Cameron, E.M., 1991, Petrological and stable isotope studies of carbonate and sulfide minerals from the Gunflint Formation, Ontario: Evidence for the origin of Early Proterozoic iron-formation: Precambrian Research, v. 52, p. 347–380.

Chang, S.R., and Kirschvink, J.L., 1989, Magnetofossils, the magnetization of sediments, and the evolution of magnetite biomineralization: Annual Review of Earth and Planetary Sciences, v. 17, p. 169–195.

Chemale Jr., F., Rosière, C.A., and Endo, I., 1994, The tectonic evolution of the Quadrilátero Ferrífero, Minas Gerais, Brazil: Precambrian Research, v. 65, p. 25–54.

Cheney, E. S., 1996, Sequence stratigraphy and plate tectonic significance of the Transvaal succession of southern Africa and its equivalent in Western Australia: Precambrian Research, v. 79, p. 3–24.

Cook, H.E., and Egbert, R.M., 1983, Diagenesis of deep-sea carbonates, *in* Larsen, G., and Chilingar, G.V., eds., Diagenesis in sediments and sedimentary rocks, 2: Amsterdam, Elsevier, p. 213–288.

Danielson, A., Moller, P., and Dulski, P., 1992, The europium anomalies in banded iron formations and the thermal history of the oceanic crust: Chemical Geology, v. 97, p. 89–100.

Deer, W.A., Howie, R.A., and Zussman, J., 1992, An introduction to the rock-forming minerals, 2nd ed.: London, Longman Scientific and Technical, 696 p.

Degens, E.T., and Stoffer, P., 1976, Stratified waters as a key to the past: Nature, v. 263, p. 22–27.

Derry, L.A., and Jacobsen, S.B., 1990, The chemical evolution of Precambrian seawater: Evidence from REEs in banded iron formations: Geochimica et Cosmochimica Acta, v. 54, p. 2965–2977.

Dimroth, E., 1968, Sedimentary textures, diagenesis, and sedimentary environments of certain Precambrian ironstones: Neues Jahrbuch für Geologie und Paläontologie Abhandlungen, v. 130, p. 247–274.

Dimroth, E., 1971, The Attikamagen-Ferriman transition in part of the central Labrador trough: Canadian Journal of Earth Sciences, v. 8, p. 1432–1454.

Dimroth, E., 1976, Aspects of the sedimentary petrology of cherty iron formation, *in* Wolf, K.H., ed., Handbook of Strata-Bound and Stratiform Ore Deposits, v. 7: Amsterdam, Elsevier, p. 203–254.

Dimroth, E., 1986, Depositional environments and tectonic settings of the cherty iron formations of the Canadian Shield: Journal of the Geological Society of India, v. 28, p. 239–250.

Dimroth, E., and Chauvel, J.-J., 1973, Petrography of the Sokoman Iron Formation in part of the central Labrador trough: Geological Society of America Bulletin, v. 84, p. 111–134.

Drever, J.I., 1974, Geochemical model for the origin of Precambrian banded iron formations: Geological Society of America Bulletin, v. 85, p. 1099–1106.

Dunbar, C.J., and McCall, G.J.H., 1971, Archaean turbidites and banded ironstones of the Mt. Belches area (Western Australia): Sedimentary Geology, v. 5, p. 93–133.

Eriksson, K.A., 1983, Siliciclastic-hosted iron-formations in the Early Archean Barberton and Pilbara sequences: Journal of the Geological Society of Australia, v. 30, p. 473–482.

Eriksson, K.A., 1995. Crustal growth, surface processes and atmospheric evolution of the early Earth, *in* Coward, M.P., and Ries, A.C., eds., Early Precambrian Processes: Geological Society (London) Special Publication 95, p. 11–25.

Eriksson, K.A., and Donaldson, J.A., 1986, Basinal and shelf sedimentation in relation to the Archaean-Proterozoic boundary: Precambrian Research, v. 33, p. 103–121.

Eriksson, P.G., 1999, Sea level changes and the continental freeboard concept: General principles and application to the Precambrian: Precambrian Research, v. 97, p. 143–154.

Eugster, H.P., and Chou, I-M., 1973, Depositional environments of Precambrian banded iron formations: Economic Geology, v. 68, p. 1144–1168.

Ewers, W.E., 1983, Chemical factors in the deposition and diagenesis of banded iron-formation, *in* Trendall, A.F., and Morris, R.C., eds., Iron-formations: Facts and problems: Amsterdam, Elsevier, p. 491–512.

Ewers, W.E., and Morris, R.C., 1981, Studies of the Dales Gorge Member of the Brockman Iron Formation, Western Australia: Economic Geology, v. 76, p. 1929–1953.

Fedo, C.M., and Eriksson, K.A., 1996, Stratigraphic framework of the ~3.0 Ga Buhwa Greenstone Belt: A unique stable-shelf succession in the Zimbabwe Archean Craton: Precambrian Research, v. 77, p. 161–178.

Fedo, C.M., and Whitehouse, M.J., 2002, Metasomatic origin of quartz-pyroxene rock, Akilia, Greenland, and implications for Earth's earliest life: Science, v. 296, p. 1448–1452.

Force, E.R., and Maynard, J.B., 1991, Manganese: Syngenetic deposits on the margins of anoxic basins, *in* Force, E.R., Eidel, J.J., and Maynard, J.B., eds., Sedimentary and diagenetic mineral deposits: A basin analysis approach to exploration: Society of Economic Geologists Reviews in Economic Geology v. 5, p. 147–157.

Fralick, P.W., and Barrett, T.J., 1995, Depositional controls on iron formation associations in Canada, *in* Plint, A.G., ed., Sedimentary Facies Analysis: International Association of Sedimentologist Special Publication 22, p. 137–156.

Fryer, B., 1983, Rare earth elements in iron-formation, *in* Trendall, A.F., and Morris, R.C., eds., Iron-formations: Facts and problems: Amsterdam, Elsevier, p. 345–358.

Fryer, B., Fyfe, W.S., and Kerrich, R., 1979, Archaean volcanogenic oceans: Chemical Geology, v. 24, p. 25–33.

Garrels, R.A., 1987, A model for the deposition of the microbanded Precambrian iron formations: American Journal of Science, v. 287, p. 81–106.

Gole, M.J., and Klein, C., 1981, Banded iron-formations through much of Precambrian time: Journal of Geology, v. 89, p. 169–183.

Goode, A.D.T., Hall, W.D.M., and Bunting, J.A., 1983, The Nabberu Basin of western Australia, *in* Trendall, A. F., and Morris, R.C., eds., Iron-formations: Facts and problems: Amsterdam, Elsevier, p. 295–323.

Goodwin, A.M., 1991, Precambrian geology—the dynamic evolution of the continental crust: London, Academic Press, 666 p.

Goodwin, A.M., Thode, H.G., Chou, C.-L., and Karkhansis, S.N., 1985, Chemostratigraphy and origin of the Late Archean siderite-pyrite-rich Helen Iron Formation, Michipicoten belt, Canada: Canadian Journal of Earth Sciences, v. 22, p. 72–84.

Grotzinger, J.P., 1994. Trends in Precambrian carbonate sediments and their implications for understanding evolution, *in* Bengston, S., ed., Early life on Earth: Nobel Symposium 84: New York, Columbia University Press, p. 245–258.

Gross, G.A., 1965, Geology of iron deposits of Canada, v. I, General geology and evaluation of iron deposits: Geological Survey of Canada Economic Geology Report 22, v. I, 181 p.

Gross, G.A., 1972, Primary features in cherty iron formations: Sedimentary Geology, v. 2, p. 241–261.

Gross, G.A., 1983, Tectonic systems and the deposition of iron-formation: Precambrian Research, v. 20, p. 171–187.

Hall, W.D.M., and Goode, A.D.T., 1978, The Early Proterozoic Nabberu Basin and associated iron formations of western Australia: Precambrian Research, v. 7, p. 129–184.

Han, T.-M., 1978, Microstructures of magnetite as guides to its origin in some Precambrian iron-formations: Fortschritte der Mineralogie, v. 56, p. 105–142.

Han, T.-M., 1982, Iron formations of Precambrian age: Hematite-magnetite relationships in some Proterozoic iron deposits—a microscopic observation, *in* Amstutz, G.C., El Goresy, A., Frenzel, G., Kluth, C., Moh, G., Wauschkuhn, A., and Zimmermann, R.A., eds., Ore genesis—The state of the art: Berlin, Springer-Verlag, p. 451–459.

Han, T.-M., and Runnegar, B., 1992, Megascopic eukaryotic algae from the 2.1-billion-year-old Negaunee Iron-Formation, Michigan: Science, v. 257, p. 232–235.

Hassler, S.W., 1993, Depositional history of the main tuff interval of the Wittenoom Formation, Late Archean-Early Proterozoic Hamersley Group, western Australia: Precambrian Research, v. 60, p. 337–359.

Hassler, S.W., and Simonson, B.M., 1989, Deposition and alteration of volcaniclastic strata in two large, Early Proterozoic iron-formations in Canada: Canadian Journal of Earth Sciences, v. 26, p. 1574–1585.

Heaney, P.J., and Veblen, D.R., 1991, An examination of spherulitic dubiomicrofossils in Precambrian banded iron formations using the transmission electron microscope: Precambrian Research, v. 49, p. 355–372.

Hocking, R.M., Adamides, N.G., Pirajno, F., and Jones, J.A., 2001, Geology of the Earaheedy 1:100,000 sheet: Perth, Western Australia, Geological Survey of Western Australia Geological Map Series, scale 1:100,000, 1 sheet, no. 3246, 33 p.

Hoffman, P.F., Kaufman, A.J., Halverson, G.P., and Schrag, D.P., 1998, A Neoproterozoic snowball Earth: Science, v. 281, p. 1342–1346.

Holland, H.D., 1973, The oceans: A possible source of iron in iron-formations: Economic Geology, v. 68, p. 1169–1172.

Holland, H.D., 1984, The chemical evolution of the atmosphere and oceans: Princeton, New Jersey, Princeton University Press, 582 p.

Holland, H.D., 1994, Early Proterozoic atmospheric change, *in* Bengston, S., ed., Early life on Earth: Nobel Symposium 84: New York, Columbia University Press, p. 237–244.

Hough, J.L., 1958, Fresh-water environment of deposition of Precambrian banded iron formations: Journal of Sedimentary Petrology, v. 28, p. 414–430.

Isley, A.E., 1995, Hydrothermal plumes and the delivery of iron to banded iron formation: Journal of Geology, v. 103, p. 169–185.

Isley, A.E., and Abbott, D.H., 1999, Plume-related mafic volcanism and the deposition of banded iron formation: Journal of Geophysical Research, v. 104, p. 5461–15,477.

Jacobsen, S.B., and Pimentel-Klose, M.R., 1988, A Nd isotopic study of the Hamersley and Michipicoten banded iron formations: The source of REE and Fe in Archean oceans: Earth and Planetary Science Letters, v. 87, p. 856–867.

James, H.L., 1954, Sedimentary facies of iron-formation: Economic Geology, v. 49, p. 235–293.

James, H.L., 1966, Chemistry of the iron-rich sedimentary rocks: U.S. Geological Survey Professional Paper 440-W, 61 p.

James, H.L., and Trendall, A.F., 1982, Banded iron formation: Distribution in time and paleoenvironmental significance, *in* Holland, H.D., and Schidlowski, M., eds., Mineral deposits and the evolution of the biosphere: Berlin, Springer-Verlag, p. 199–217.

Kimberley, M.M., 1974, Origin of iron ore by diagenetic replacement of calcareous oolite: Nature, v. 250, p. 319–320.

Kimberley, M.M., 1989, Exhalative origins of iron formations: Ore Geology Reviews, v. 5, p. 13–145.

Klein, C., 1983, Diagenesis and metamorphism of banded iron-formations, *in* Trendall, A.F., and Morris, R.C., eds., Iron-formations: Facts and problems: Amsterdam, Elsevier, p. 417–469.

Klein, C., and Beukes, N.J., 1989, Geochemistry and sedimentology of a facies transition from limestone to iron-formation deposition in the Early Proterozoic Transvaal Supergroup, South Africa: Economic Geology, v. 84, p. 1733–1774.

Klein, C., and Beukes, N.J., 1992, Proterozoic iron formations, *in* Condie, K.C., ed., Proterozoic crustal evolution: Amsterdam, Elsevier, p. 383–418.

Klein, C., and Bricker, O.P., 1977, Some aspects of the sedimentary and diagenetic environment of Proterozoic banded iron-formation: Economic Geology, v. 72, p. 1457–1470.

Klein, C., and Fink, R.P., 1976, Petrology of the Sokoman Iron Formation in the Howells River area, at the western edge of the Labrador trough: Economic Geology, v. 71, p. 453–487.

Krapez, B., and Martin, D.McB., 1999, Sequence stratigraphy of the Palaeoproterozoic Nabberu Province of Western Australia: Australian Journal of Earth Sciences, v. 46, p. 89–103.

LaBerge, G.L., 1966a, Altered pyroclastic rocks in iron-formation in the Hamersley Range, western Australia: Economic Geology, v. 61, p. 147–161.

LaBerge, G.L., 1966b, Pyroclastic rocks in South African iron-formations: Economic Geology, v. 61, p. 572–581.

LaBerge, G.L. 1973, Possible biological origin of Precambrian iron-formations: Economic Geology, v. 68, p. 1098–1109.

LaBerge, G.L., Robbins, E.I., and Han, T.-M., 1987, A model for the biological precipitation of Precambrian iron-formations, A: Geological evidence, *in* Appel, P.W.U., and LaBerge, G.L., eds., Precambrian iron-formations: Athens, Greece, Theophrastus Publications, p. 69–96.

Larue, D.K., 1981, The Early Proterozoic pre-iron-formation Menominee Group siliciclastic sediments of the southern Lake Superior region: Evidence for sedimentation in platform and basinal settings: Journal of Sedimentary Petrology, v. 51, p. 397–414.

Lepp, H., 1987, Chemistry and origin of Precambrian iron-formations, *in* Appel, P.W.U., and LaBerge, G.L., eds., Precambrian iron-formations: Athens, Greece, Theophrastus Publications, p. 3–30.

Lougheed, M.S., 1983, Origin of Precambrian iron-formations in the Lake Superior region: Geological Society of America Bulletin, v. 94, p. 325–340.

Lowe, D.R., 1992, Major events in the geological development of the Precambrian Earth, *in* Schopf, J.W., and Klein, C., eds., 1992, The Proterozoic biosphere—A multidisciplinary study: Cambridge, Cambridge University Press, p. 67–75.

Lowenstein, T.K., Timofeeff, M.N., Brennan, S.T., Hardie, L.A., and Demicco, R.V., 2001, Oscillations in Phanerozoic seawater chemistry: Evidence from fluid inclusions: Science, v. 294, p. 1086–1088.

Majumder, T., and Chakraborty, K.L., 1977, Primary sedimentary structures in the banded iron-formation of Orissa, India: Sedimentary Geology, v. 19, p. 287–300.

Majumder, T., and Chakraborty, K.L., 1979, Petrography and petrology of the Precambrian banded iron-formation of Orissa, India, and reformation of the bands: Sedimentary Geology, v. 22, p. 243–265.

Maliva, R.G., Knoll, A.H., and Siever, R., 1989, Secular change in chert distribution: A reflection of evolving biological participation in the silica cycle: PALAIOS, v. 4, p. 519–532.

Manikyamba, C., 1999, Reworking of banded iron formation into granular iron formation in the Sandur Schist Belt, India: Possible evidence of sea level changes in an Archaean proto-ocean: Journal of the Geological Society of India, v. 53, p. 453–461.

McConchie, D., 1987, The geology and geochemistry of the Joffre and Whaleback Shale Members of the Brockman Iron Formation, western Australia, *in* Appel, P.W.U., and LaBerge, G.L., eds., Precambrian Iron-Formations: Athens, Greece, Theophrastus Publications, p. 541–597.

Mengel, J.T., 1973, Physical sedimentation in Precambrian cherty iron formations of the Lake Superior type, *in* Amstutz, G.C., and Bernard, A.J., eds., Ores in sediments: Berlin, Springer-Verlag, p. 179–193.

Miyano, T., 1987, Diagenetic to low-grade metamorphic conditions of Precambrian iron-formations, *in* Appel, P.W.U., and LaBerge, G.L., eds., Precambrian iron-formations: Athens, Greece, Theophrastus Publications, p. 155–186.

Morey, G.B., 1983, Animikie Basin, Lake Superior region, U.S.A., *in* Trendall, A.F., and Morris, R. C., eds., Iron-formations: Facts and problems: Amsterdam, Elsevier, p. 13–67.

Morey, G.B., and Southwick, D.L., 1995, Allostratigraphic relationships of Early Proterozoic iron-formations in the Lake Superior region: Economic Geology, v. 90, p. 1983–1993.

Morris, R.C., 1987, Iron ores derived by enrichment of banded iron-formation, *in* Hein, J.R., ed. Siliceous sedimentary rock-hosted ores and petroleum: New York, Van Nostrand Reinhold, p. 231–267.

Morris, R.C., 1993, Genetic modeling for banded iron-formation of the Hamersley Group, Pilbara Craton, western Australia: Precambrian Research, v. 60, p. 243–286.

Nelson, D.R., Trendall, A.F., and Altermann, W., 1999, Chronological correlations between the Pilbara and Kaapvaal cratons: Precambrian Research, v. 97, p. 165–189.

Ojakangas, R.W., 1983, Tidal deposits in the Early Proterozoic basin of the Lake Superior region—the Palms and Pokegama Formations: Evidence for subtidal shelf deposition of Superior-type banded iron-formation, *in* Medaris Jr., L.G., ed., Early Proterozoic geology of the Great Lakes region: Geological Society of America Memoir 60, p. 49–66.

Pickard, A.L., 2002, SHRIMP U-Pb zircon ages of tuffaceous mudrocks in the Brockman Iron Formation of the Hamersely Range, western Australia: Australian Journal of Earth Sciences, v. 49, p. 491–507.

Shegelski, R.J., 1987, The depositional environment of Archean iron formations, Sturgeon-Savant Greenstone Belt, Ontario, Canada, *in* Appel, P.W.U., and LaBerge, G.L., eds., Precambrian iron-formations: Athens, Greece, Theophrastus Publications, p. 329–344.

Sheridan, M.F., and Wohletz, K.H., 1983, Hydrovolcanism: Basic considerations and review: Journal of Volcanology and Geothermal Research, v. 17, p. 1–29.

Siever, R., 1992, The silica cycle in the Precambrian: Geochimica et Cosmochimica Acta, v. 56, p. 3265–3272.

Simonson, B.M., 1984, High-energy shelf deposit: Early Proterozoic Wishart Formation, northeastern Canada, *in* Tillman, R.W., and Siemers, C.T., eds., Siliciclastic shelf sediment: Society of Economic Paleontologists and Mineralogists Special Publication 34, p. 251–268.

Simonson, B.M., 1985, Sedimentological constraints on the origins of Precambrian iron-formations: Geological Society of America Bulletin, v. 96, p. 244–252.

Simonson, B.M., 1987, Early silica cementation and subsequent diagenesis in arenites from four Early Proterozoic iron formations of North America: Journal of Sedimentary Petrology, v. 57, p. 494–511.

Simonson, B.M., 1997, Early Precambrian iron-formations: SEPM (Society for Sedimentary Geology) Slide Set 7, 45 p.

Simonson, B.M., and Goode, A.D.T., 1989, First discovery of ferruginous chert arenites in the early Precambrian of western Australia: Geology, v. 17, p. 269–272.

Simonson, B.M., and Hassler, S.W., 1996, Was the deposition of large Precambrian iron formations linked to major marine transgression?: Journal of Geology, v. 104, p. 665–676.

Simonson, B.M., and Lanier, W.P., 1987, Early silica cementation and microfossil preservation in cavities in iron-formation stromatolites, Early Proterozoic of Canada, *in* Appel, P.W.U., and LaBerge, G.L., eds., Precambrian iron-formations: Athens, Greece, Theophrastus Publications, p. 187–213.

Simonson, B.M., Schubel, K.A., and Hassler, S.W., 1993, Carbonate sedimentology of the early Precambrian Hamersley Group of western Australia: Precambrian Research, v. 60, p. 287–335.

Singer, A., and Müller, G., 1983, Diagenesis in argillaceous sediments, *in* Larsen, G., and Chilingar, G.V., eds., Diagenesis in sediments and sedimentary rocks, 2: Amsterdam, Elsevier, p. 115–212.

Sommers, M.G., Awramik, S.M., and Woo, K.S., 2000, Evidence for initial calcite-aragonite composition of Lower Algal Chert Member ooids and stromatolites, Paleoproterozoic Gunflint Formation, Ontario: Canadian Journal of Earth Sciences, v. 37, p. 1229–1243.

Trendall, A.F., 1972, Revolution in Earth history: Journal of the Geological Society of Australia, v. 19, p. 287–311.

Trendall, A.F., 1973a, Iron-formations of the Hamersley Group of Western Australia: Type examples of varved Precambrian evaporites, *in* Anonymous, genesis of Precambrian iron and manganese deposits: Paris, UNESCO Earth Science Series 9, p. 257–270.

Trendall, A.F., 1973b, Varve cycles in the Weeli Wolli Formation of the Precambrian Hamersley Group, western Australia: Economic Geology, v. 68, p. 1089–1098.

Trendall, A.F., 1983, The Hamersley Basin, *in* Trendall, A. F., and Morris, R. C., eds., Iron-formations: Facts and problems: Amsterdam, Elsevier, p. 69–129.

Trendall, A.F., 2002, The significance of iron-formation in the Precambrian stratigraphic record, *in* Altermann, W., and Corcoran, P.L., eds., Precambrian sedimentary environments: A modern approach to ancient depositional systems: International Association of Sedimentologists Special Publication 33, p. 33–66.

Trendall, A.F., and Blockley, J.G., 1970, The iron formations of the Precambrian Hamersley Group, western Australia: Geological Survey of Western Australia Bulletin 119, 366 p.

Trendall, A.F., Basei, M.A.S., de Laeter, J.R., and Nelson, D.R., 1998, SHRIMP zircon U-Pb constraints on the age of the Carajás Formation, Grão Pará Group, Amazon Craton: Journal of South American Earth Sciences, v. 11, p. 265–277.

Van Hise, C.R., and Leith, C.K., 1911, Geology of the Lake Superior region: U.S. Geological Survey Monograph 52, 641 p.

Veizer, J., Clayton, R.H., Hinton, R.W., von Brunn, V., Mason, T.R., Buck, S.G., and Hoefs, J., 1990, Geochemistry of Precambrian carbonates: 3—shelf seas and non-marine environments of the Archean: Geochimica Cosmochimica Acta, v. 54, p. 2717–2729.

Veizer, J., Clayton, R.H., and Hinton, R.W., 1992, Geochemistry of Precambrian carbonates: IV—Early Paleoproterozoic (2.25 ± 0.25 Ga) seawater: Geochimica Cosmochimica Acta, v. 56, p. 875–885.

Walter, M.R., 1972, A hot spring analog for the depositional environment of Precambrian iron formations of the Lake Superior region: Economic Geology, v. 67, p. 965–972.

Walter, M.R., and Hofmann, H.J., 1983, The palaeontology and palaeoecology of Precambrian iron-formations, *in* Trendall, A.F., and Morris, R.C., eds., Iron-formations: Facts and problems: Amsterdam, Elsevier, p. 373–400.

Watchorn, M.A., 1980, Fluvial and tidal sedimentation in the 3000 Ma Mozaan Basin, South Africa: Precambrian Research, v. 13, p. 27–42.

Winter, B.L., and Knauth, L.P., 1992, Stable isotope geochemistry of cherts and carbonates from the 2.0 Ga Gunflint Iron Formation: Implications for the depositional setting, and the effects of diagenesis and metamorphism: Precambrian Research, v. 59, p. 283–313.

Young, G.M., 1988, Proterozoic plate tectonics, glaciation and iron-formations: Sedimentary Geology, v. 58, p. 127–144.

Young, T.P., and Taylor, W.E.G., 1989, Phanerozoic ironstones: Geological Society (London) Special Publication 46, 251 p.

Zajac, I.S., 1974, The stratigraphy and mineralogy of the Sokoman Formation in the Knob Lake area, Quebec and Newfoundland: Geological Survey of Canada Bulletin 220, 159 p.

Manuscript Accepted by the Society January 22, 2003

Printed in the USA

Geological Society of America
Special Paper 370
2003

Extreme paleoceanographic conditions in a Paleozoic oceanic upwelling system: Organic productivity and widespread phosphogenesis in the Permian Phosphoria Sea

Eric E. Hiatt*
Department of Geology, University of Wisconsin, 800 Algoma Boulevard, Oshkosh, Wisconsin 54901, USA

David A. Budd
Department of Geological Sciences, University of Colorado, Boulder, Colorado 80309-0399, USA

ABSTRACT

High-resolution geochemical data from phosphorites and associated lithofacies of the Permian Phosphoria Rock Complex suggest that organic matter deposition and phosphogenesis occurred in fundamentally different oceanographic conditions than those of modern oceanic upwelling systems. Unlike modern phosphorites, those of the Phosphoria accumulated in a shallow marginal sea within a semi-restricted epicontinental embayment that extended landward into proximal environments bordered by evaporative lagoons. The Phosphoria Rock Complex phosphorites formed in outer ramp (<200 m water depth), organically productive mid-ramp, and very shallow and restricted inner-ramp environments. Chemostratigraphic data (total organic carbon, sulfur, phosphate, $\delta^{13}C_{PO_4-CO_3}$, Ni, Cr, and Cd) indicate that this wide range of paleoenvironments was largely dysoxic or anoxic; euxinic conditions developed sporadically. High cadmium and nickel concentrations suggest maximum paleoproductivity (preserved total organic carbon up to 15 wt%) was associated with anoxic and euxinic conditions. Water column oxygen and trophic levels are interpreted to have been the primary controls over macrofaunal distribution in the Phosphoria, not cold-water temperatures as has been previously inferred.

These findings, augmented by recent Permian paleoclimate and ocean circulation models, suggest that an oxygen-poor, nutrient-rich intermediate water mass flowed into the Phosphoria embayment and impinged on the mid-ramp area. Seasonal coastal upwelling brought this water to the surface, where it mixed with warm waters flowing seaward from the restricted shallow lagoons in west-central Wyoming, resulting in high paleoproductivity and organic matter accumulation and oxygen depletion in the water column. Warming of the waters on the broad, shallow ramp, coupled with seasonal attenuation of the coastal upwelling system, is predicted to have led to a positive feedback between productivity and phosphogenesis through a wide range of environments. This new model and our findings illustrate that paleoceanographic setting and paleoenvironment must be taken into account to fully understand the geochemical variation seen in ancient phosphorites.

Keywords: Phosphoria, upwelling, chemostratigraphy, chemofacies, phosphorite, paleoceanography.

*Corresponding author: hiatt@uwosh.edu.

Hiatt, E.E., and Budd, D.A., 2003, Extreme paleoceanographic conditions in a Paleozoic oceanic upwelling system: Organic productivity and widespread phosphogenesis in the Permian Phosphoria Sea, *in* Chan, M.A., and Archer, A.W., eds., Extreme depositional environments: Mega end members in geologic time: Boulder, Colorado, Geological Society of America Special Paper 370, p. 245–264.

INTRODUCTION

The Phosphoria Rock Complex contains one of the largest sedimentary phosphate deposits in the geologic record (e.g., Cook and McElhinny, 1979). This series of phosphorite-siltstone-chert-carbonate-evaporite successions was deposited in the "Phosporia Sea" on the western margin of North America during the Permian (Fig. 1). The Phosphoria Rock Complex contains more than five times the total mass of phosphorus in today's oceans (McKelvey et al., 1953); an estimated 1.7×10^{12} metric tons of P_2O_5 were originally deposited (Cathcart et al., 1984). Organic carbon concentrations can exceed 15 wt% (Maughan, 1994; Hiatt, 1997), and 31×10^9 metric tons of petroleum have been generated (Claypool et al., 1978). The Phosphoria Rock Complex is truly an example of an extreme sedimentary system.

The Phosphoria Rock Complex also spans one of the most extraordinary global climate transitions in earth history (Fig. 2). The Early Permian was marked by widespread glaciation in the southern hemisphere (Pennsylvanian to Early Permian), low CO_2 levels, and probably widespread sea ice at the North Pole (Barron and Fawcett, 1995). In contrast, the Late Permian marks the onset of hot and dry climates over most of the middle and low-latitude continents, rapidly increasing CO_2 levels (two to five times modern values; Berner, 1994), widespread desert environments with vast areas of evaporite deposition (Ziegler, 1990; Zarkov, 1984), and widespread anoxic deep-water conditions in the world's oceans (Wignall and Twitchett, 1996; Knoll et al., 1995). In contrast to Early Permian glaciation, the southern continents in the Late Permian were marked by warm, high-latitude climates that brought boreal forests, coal swamps, and reptiles within 10° of the South Pole (Taylor et al., 1992; Dickins, 1984, 1996; Rees et al., 1999).

The abundance of phosphate, organic matter, and chert in the Phosphoria Rock Complex has long been cited as evidence for a coastal upwelling system in the Phosphoria Sea (e.g., McKelvey et al., 1959; Sheldon, 1989). Upwelling is also predicted by phosphate and elemental mass balance calculations (Piper and Link, 2002), atmospheric circulation models (Parrish, 1982; Kutzbach and Ziegler, 1994), and paleowind directions that suggest a net offshore movement of surface waters by Ekman transport (Sheldon et al., 1967; Parrish and Peterson, 1988). There is no doubt that upwelling-associated biological productivity led to the deposition of organic matter and phosphate in the Phosphoria Rock Complex.

Although the existence of an oceanic upwelling system in the Phosphoria Sea is considered a certainty, many unanswered questions remain concerning the oceanographic conditions that led to these extreme sedimentary deposits. As Boyd (1993, p. 183) pointed out, "A completely satisfying explanation for the origin of the phosphatic members has yet to appear." Key unanswered questions are: What was the nature (oxygen and nutrient levels, primarily) of the upwelled water? Under what oceanographic conditions did sedimentary phosphate form? And, what role did the semi-restricted nature of this embayment play in the oceanography of the Phosphoria Sea?

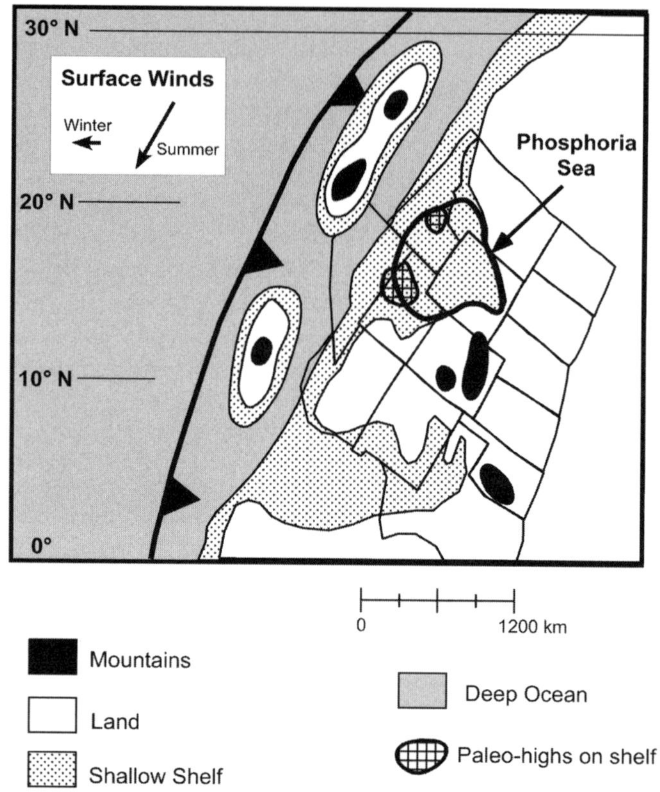

Figure 1. Position of Phosphoria Sea in context of Late Permian plate tectonic reconstruction of western North American continent and eastern Panthalassa Ocean. Position of paleoshoreline represents sea-level highstand. Paleowind vectors are based on Kutzbach and Ziegler (1994); wind strengths (inset box) are 5.0 and 1.8 m/s for summer and winter, respectively. Approximate positions of palcohighs on the Wyoming paleoshelf are from Skipp and Hall (1980) and Wardlaw (1977). Map is modified from Scotese and Langford (1995).

We believe that answers to these questions have been hindered by attempts to make the Phosphoria Rock Complex "fit" a model based on the modern Peru margin—a model that involves upwelling of cold (0–10 °C), nutrient-rich deep water onto a steep and narrow continental shelf (e.g., McKelvey et al., 1953, 1959; Sheldon, 1963, 1984; Parrish, 1982; Wardlaw and Collinson, 1984). The applicability of this interpretation to the Phosphoria Rock Complex has been bolstered by the presence of faunal elements that are interpreted as representing cold-water, "Arctic" conditions in the Phosphoria Sea (Wardlaw, 1980; Wardlaw et al., 1995).

In the last 20 years, research on the modern ocean and the Phosphoria Rock Complex has resulted in a somewhat different view of these enigmatic rocks. It now appears that the Phosphoria Sea was a relatively shallow (<200 m), semi-isolated epicontinental basin (Ketner, 1977; Scotese and Langford, 1995). Paleoclimate models suggest mean summer air temperatures over the shallow, marginal Phosphoria Sea were as high as 30–45 °C

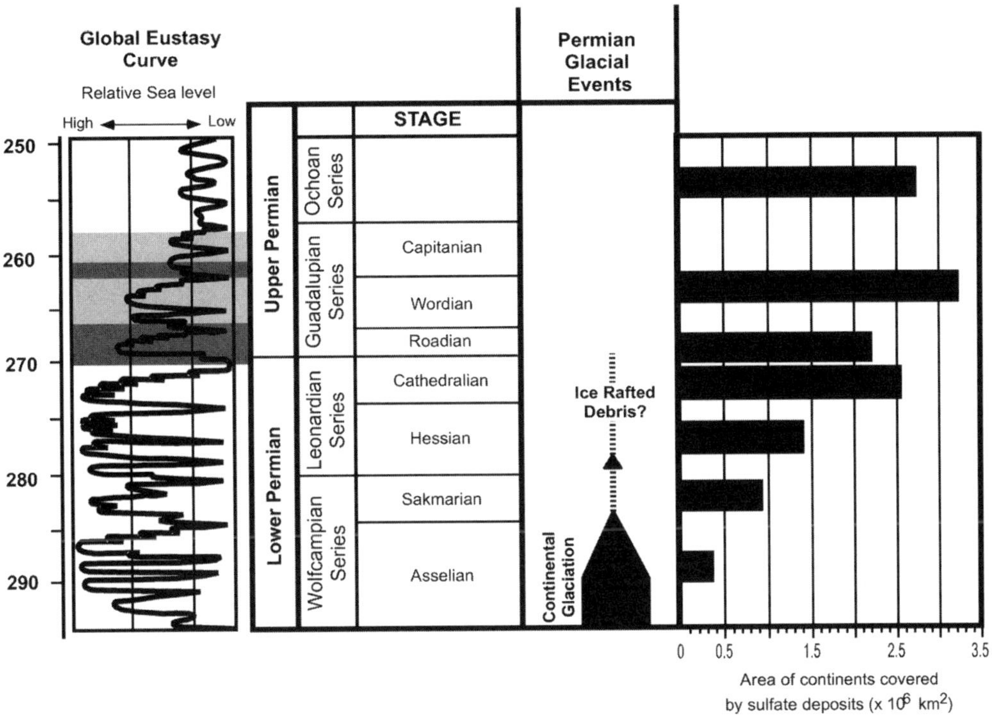

Figure 2. Permian Series and Stage names for southwestern United States and global eustasy curve; eustasy curve records sea-level variations with amplitudes of ~200 m (modified from Ross and Ross, 1994). Shaded zone on eustasy curve represents Phosphoria Rock Complex, and darkly shaded zones represent intervals of phosphate-rich Meade Peak (lower zone) and Retort (upper zone) deposition. Relative extents of Permian glacial events are shown in middle right column (width of black area provides an estimate of possible ice sheet extent; evidence for ice sheets extending to the earliest Sakmarian from González-Bonorino and Eyles, 1995, and Dickins, 1996; evidence for ice sheets in South Africa in Hessian [Artinskian] from Visser, 1996; evidence for ice-rafted debris from Frakes et al., 1992). Extent of global evaporite deposits as sulfate minerals (anhydrite and gypsum) is shown by bar graph at right (data from Zharkov, 1984).

(Kutzbach and Ziegler, 1994; Barron and Fawcett, 1995). Phosphogenesis, which is the precipitation of phosphate from seawater at or just below the sediment-water interface, occurred not only in the deeper portions of the basin, but also in shallower, inner-shelf settings (Sheldon, 1984; Peterson, 1984; Hiatt, 1997; Trappe, 1998; Stephens and Carroll, 1999; Hendrix and Byers, 2000). Paleotemperatures ranged from temperate (14–26 °C; mean of 21 °C) at sites of maximum phosphogenesis to warm (34–37 °C) across the shallow paleoshelf (Hiatt and Budd, 2001). Elevated salinities and stratification of the water column have also been proposed (Hite, 1978; Dahl et al., 1993; Stephens and Carroll, 1999). Piper and Link (2002), however, argued for temperature rather than salinity stratification. Based on the lower, phosphorite-rich portion of the Meade Peak member and mass balance calculations, Piper and Link (2002) determined that the Phosphoria Sea was "sediment-starved" and that, based on analogies with modern ocean basins, biological productivity was only moderately high. Piper (2001) used concentrations of Cd, Mo, Zn, Cu, and perhaps Ni as further evidence of elevated biological productivity in the Meade Peak member, and he coupled these with other trace elements, such as Cr, V, and U to further suggest that the Phosphoria Sea was marked by anoxic, denitrifying, but not sulfate-reducing bottom waters during deposition of this lower Meade Peak interval. Piper (2001) and Piper and Link (2002), however, did not analyze their data in a stratigraphic and regional framework. This leads to the question: How did oceanographic conditions vary regionally or temporally (stratigraphically) during the life of the upwelling system? In addition, no one has directly studied the relationship between paleoecology and chemostratigraphy of the Meade Peak to test whether the paleontologic data support either the salinity or temperature stratification model.

All of these recent findings and the fundamental unanswered questions that remain suggest it is time to reassess the nature of the original upwelling model of the Phosphoria Sea. Here, the geochemical and macrofaunal data from the Meade Peak member of the Phosphoria Formation are presented in a stratigraphic and regional framework that further clarifies the relationship between paleoproductivity, chemofacies, phosphogenesis, and oceanographic setting.

THE PHOSPHORIA ROCK COMPLEX

Deposition of the Phosphoria Rock Complex occurred in the "Phosphoria Sea" (Fig. 1), which formed in a shallow marginal foreland basin on the western margin of North America at a paleolatitude of about 20°N (Scotese and Langford, 1995). The Phosphoria Sea was partially separated from the Panthalassa Ocean by an island arc system (e.g., Scotese and Langford, 1995). The Phosphoria Rock Complex is composed of three unconformity-bounded stratigraphic sequences (Fig. 3). Phosphate [as francolite; $Ca_{10-a-b-c}Na_a,Mg_b,(PO_4)_{6-x},(CO_3)_{x-y-z}(CO_3,F)_y,(SO_4)_zF_2$, where $x = y + a + 2c$ and c = the number of Ca vacancies; Nathan, 1984] occurs throughout all three sequences, but phosphorites (beds with >10% francolite) are concentrated in the Meade Peak and Retort members of the Phosphoria Formation in the upper two sequences. In subtidal to peritidal deposits, phosphate occurs primarily as phosphatic peloids, with phosphatic ooids, intraclasts, and phosphatized skeletal grains being less common. Phosphorites are often marked by grain-supported textures, grading, and contain mechanically abraded grains, all of which suggest mechanical reworking (McKelvey et al., 1959; Trappe, 1998; Hiatt and Budd, 2001). These attributes also suggest that these beds represent relatively shallow-water deposition.

Herein, our focus is the Meade Peak member, which is the larger of the two high-productivity deposits in the Phosphoria. Meade Peak deposition was initiated in the Late Leonardian and extended through the Roadian stage of the Guadalupian (Fig. 4). Sedimentation and phosphogenesis occurred on a nearly flat ramp with an estimated shelf depositional angle of only 0.04–0.22° between the Meade Peak depocenter and the paleoshoreline (Hiatt, 1997). The Meade Peak is a seaward-thickening wedge of sediments whose depocenter was in southeastern Idaho (Maughan, 1984). Palinspastic reconstructions and regional facies patterns in the entire Phosphoria Rock Complex have been interpreted to suggest a shelf margin just to the east of that depocenter (e.g., McKelvey et al., 1959; Peterson, 1984), but no distinctive shelf-margin facies in the Meade Peak has ever been described, which is further evidence for the ramp profile. Outer- and mid-ramp deposits include thinly bedded, finely laminated fine- to medium-grained sandstone, dolomitic siltstone, and carbonate mudstone that previous workers refer to as "shales." Petrographic studies reveal, however, that shales are rare (Hiatt, 1997; Carroll et al., 1998). To the east, the Meade Peak pinches out into green siltstones and carbonates, which, in turn, grade into red beds and eventually interbedded red siltstones and evaporites in central Wyoming (Fig. 4; Maughan, 1984; Peterson, 1984).

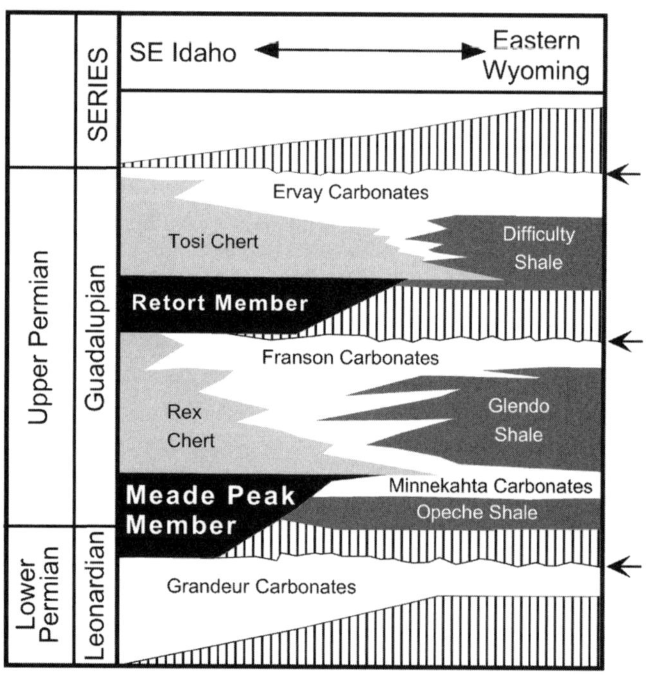

Figure 3. Stratigraphic relationships in Phosphoria Rock Complex. Vertical ruled areas denote hiatuses; arrows denote tops of unconformity-bounded sequences (modified from Maughan, 1984).

METHODS

Proxies for paleoceanographic conditions during Phosphoria Rock Complex deposition were derived from high-resolution

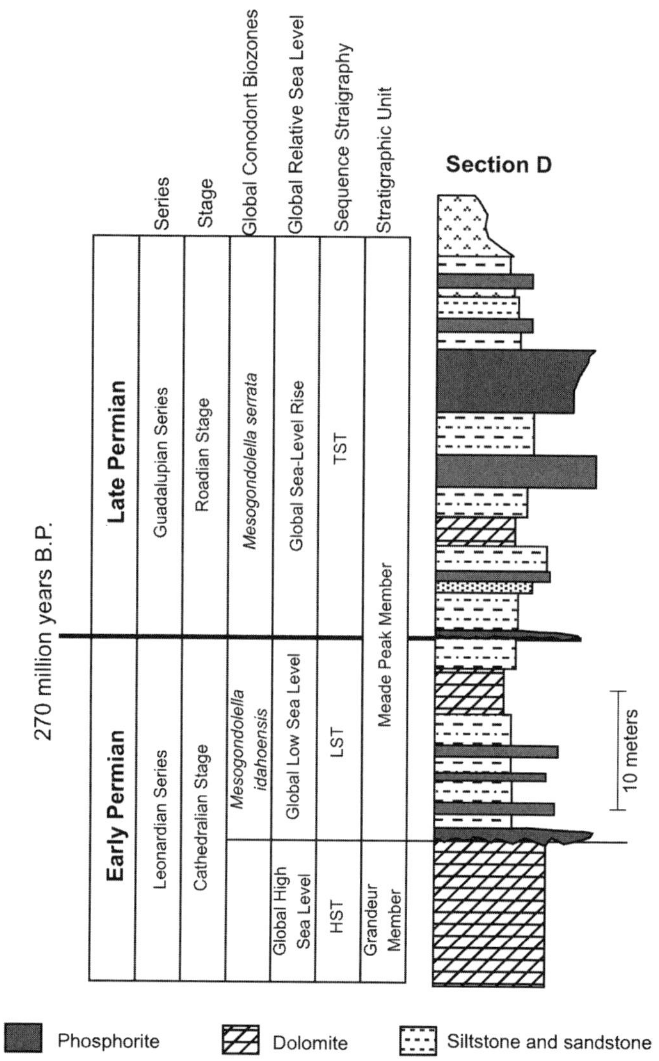

Figure 4. Generalized stratigraphic section of Meade Peak member (section D from southeastern Idaho; Fig. 5) that shows approximate position of important biostratigraphic boundary between conodonts *Mesogondolella idahoensis* and *M. serrata* (Wardlaw and Collinson, 1984), which is the accepted global boundary between Leonardian (Early Permian) and Guadalupian (Late Permian) series (Glenister et al., 1992). Series and stage names from Ross and Ross (1994). Sequence stratigraphic systems tracts from Hiatt (1997; LST—lowstand system tract, TST—transgressive system tract, HST—highstand system tract).

lithostratigraphy and chemostratigraphy. Four stratigraphic sections that form an offshore outer ramp to nearshore inner ramp transect through the Meade Peak member (Fig. 5) were utilized; they were chosen for their stratigraphic completeness and lack of visible chemical or textural alteration. Thin sections from all lithologies were examined to refine lithologic and textural classifications made during hand-sample examination. Bulk mineralogy was done by powder X-ray diffraction on a Scintag PAD-5 diffractometer. Total organic carbon (TOC) and total sulfur (TS) were determined using a Leco combustion-spectrometric device.

Phosphate grains for geochemical analyses were isolated by first disaggregating the granular phosphorite rock, sieving the resultant material into size fractions, and then passing the sand fractions through a magnetic susceptibility separator to remove glauconitic and pyritic grains. A heavy liquid separation (undiluted acetylene tetrabromide, density = 2.96 at 25 °C) was then used to remove carbonate and silicate grains. The remaining sample was washed multiple times with acetone and deionized water and dried in an oven at 80 °C for 12 hours. At this point, each size fraction was approximately 100% phosphate (francolite), as confirmed by X-ray diffraction analysis. Individual phosphatic peloids were handpicked from these concentrated and washed splits and used for the trace-element and isotopic analyses. Cadmium (Cd) concentrations were determined using X-ray fluorescence analysis; Ni and Cr concentrations were determined using instrumental neutron activation analysis. Details are given in Hiatt (1997).

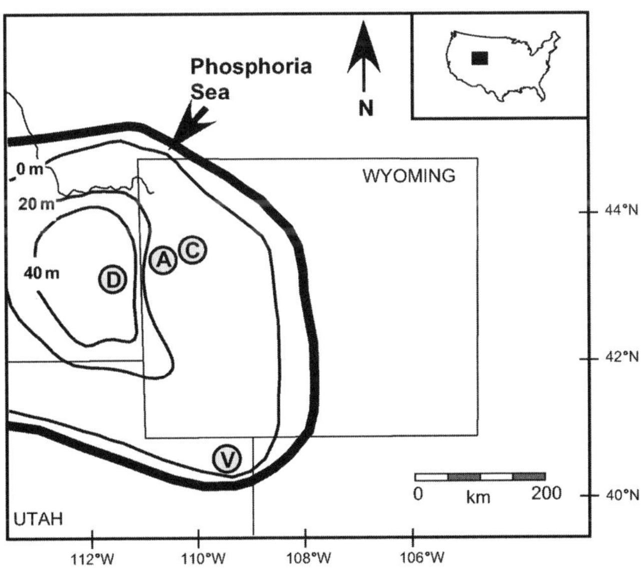

Figure 5. Map of study area showing extent of Phosphoria Sea during deposition of Meade Peak member. Isopach contours of Meade Peak member (in meters, from Maughan, 1994) and position of four stratigraphic sections utilized in this study (D—Dry Ridge; A—Astoria Hot Springs, C—Crystal Creek; V—Vernal-Brush Creek). See Hiatt (1997) for detailed locality information. Paleolatitude and base map orientation are from Scotese and Langford (1995).

Carbon isotopic values of the carbonate in the francolite crystal structure ($\delta^{13}C_{PO_4-CO_3}$) were measured in the stable isotope laboratory of the U.S. Geological Survey (USGS) in Denver. Powdered samples, averaging 100 mg in size, of isolated phosphorite peloids were reacted for 3 hours at 50 °C in individual reaction vessels with 4 ml of 100% anhydrous phosphoric acid. Evolved CO_2 was purified and isolated using two water traps maintained at –75 °C, and its isotopic composition was analyzed on a Finnigan MAT 252 mass spectrometer. Analytical precision on multiple runs (n = 35) of an internal calcite standard (CU-2) was ±0.04 for $\delta^{18}O$ and ±0.02 for $\delta^{13}C$.

All of the mineralogy, total organic carbon, total sulfur, trace element, and isotopic data generated can be found in Hiatt (1997) and is also available in tabular format from the GSA Data Repository[1]. Summary tables and figures are presented herein.

RESULTS

Meade Peak Lithofacies and Biofacies

Four general lithofacies dominate the Meade Peak member in the four stratigraphic sections analyzed. These are peloidal phosphorite packstone/grainstone, peloidal phosphorite wackestone, siltstone/sandstone, and carbonate mudstone (Fig. 6). From most seaward (section D) to most landward (section V), these four sections represent outer ramp, mid ramp, inner ramp, and nearshore paleoenvironmental settings (Fig. 7).

Sediments at the Dry Ridge (52 m thick), Astoria Hot Springs (11.6 m thick) and Crystal Creek (8.1 m thick) locations (sections D, A, and C, respectively) are dark gray to black, with phosphorites comprising about 25% of each section. All the phosphorites in these sections are planar bedded; none show any visible evidence of cross bedding. Medium- to coarse-grained phosphatic peloids predominate. Intercalated with the phosphorites at Crystal Creek and Astoria Hot Springs are thin-bedded, non-fossiliferous, organic carbon–rich silty dolomite mudstones and siltstones. Intercalated facies at the Dry Ridge locality consist of bioturbated siltstone, silty carbonate mudstone, and fine- to medium-grained sandstone with hummocky cross-stratification. In contrast, in the more landward Vernal-Brush Creek section (7 m thick), all rocks are gray to tan with decimeter-thick phosphorite beds comprising 60% of the section. None of the latter phosphorites show any visible evidence of cross bedding or fining-upward textures; all are planar bedded and consist of either well-sorted fine- to medium-grained phosphatic peloids or mixtures of peloids and small intraclasts. Intercalated lithologies are thin to very thin beds of dolomitic phosphatic wackestone and silty dolomite. Further details of the sedimentology at each section are summarized by Hiatt (1997) and Hiatt and Budd (2001).

[1]GSA Data Repository item 2003097, Meade Peak member samples and interpreted chemofacies, is available on request from Documents Secretary, GSA, P.O. Box 9140, Boulder, CO 80301-9140, USA, editing@geosociety.org, or at www.geosociety.org/pubs/ft2003.htm.

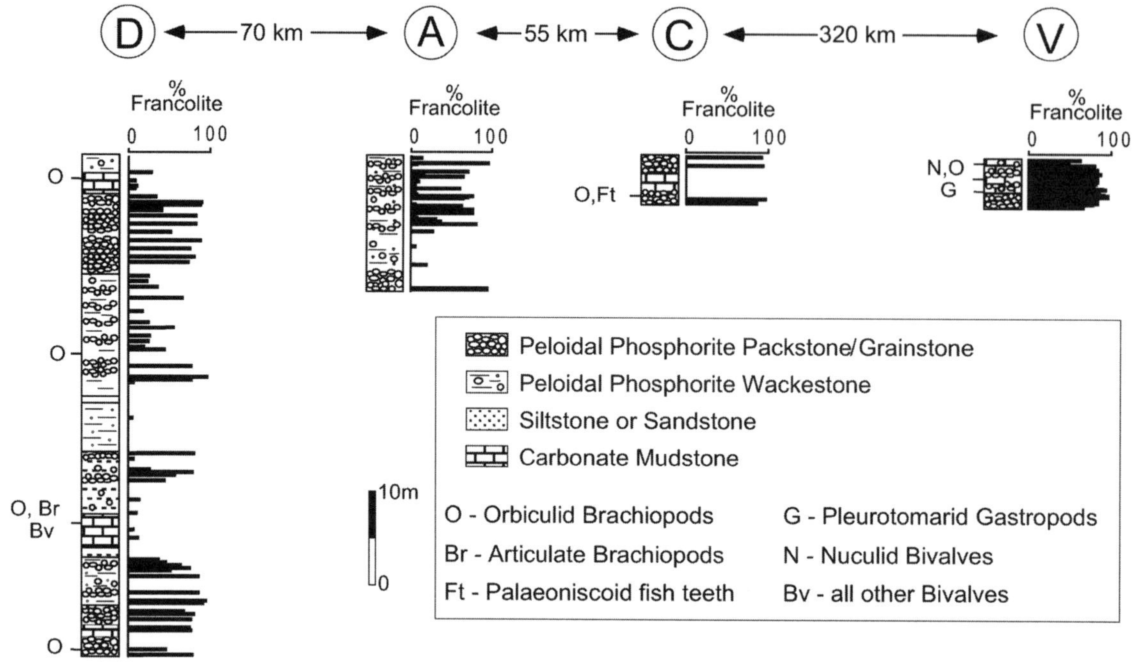

Figure 6. Stratigraphic plots of the Meade Peak member showing fauna, lithofacies, and abundance of sedimentary phosphate (as francolite).

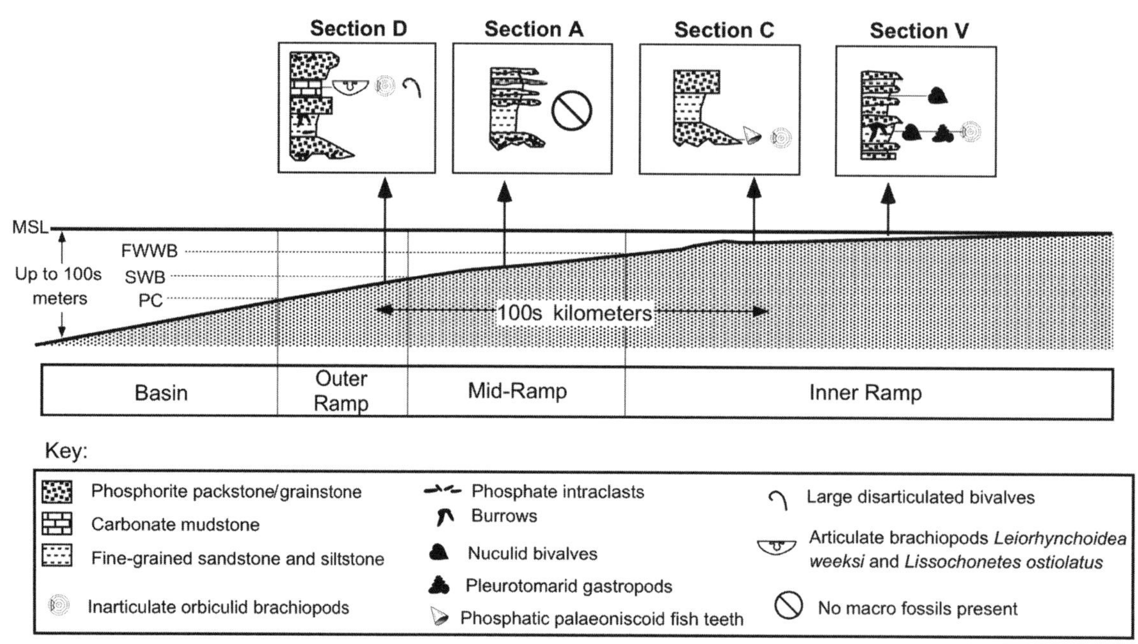

Figure 7. Diagram showing representative lithofacies associated with each Meade Peak depositional setting. See Figure 5 for section locations. Vertical exaggeration is extreme; regional slope on Wyoming paleoshelf was <0.25°. MSL—mean sea level; FWWB—fair-weather wave base; SWB—storm wave base; PC—pycnocline.

Individual phosphorite units are not randomly distributed in the Phosphoria Rock Complex. In all four sections, phosphorite beds form the grain-rich portion of individual depositional cycles (Fig. 8). Each phosphorite usually overlies a sharp basal diastem surface. About one-fifth of those surfaces are clearly erosional as they scour into the underlying bed. Exclusive of the Vernal-Brush Creek section, about half of the phosphorites exhibit a coarse sand or pebble lag (lithoclasts up to 8 cm long) and a poorly defined fining-upward texture, which suggests deposition and winnowing in the waning phase of a high-energy storm event (cf., Föllmi, 1990). The phosphorite beds grade upward into overlying finer-grained lithofacies (siltstone, sandstone, and carbonate mudstone/wackestone) that typically contain small, in situ phosphate peloids. These finer-grained units are separated from the next phosphorite by another diastem surface (Fig. 8). The beds below the upper diastem may be bioturbated, but shelly macrofauna are not common. These relationships are consistent with Föllmi's (1990) model of alternating periods of phosphogenesis, reworking, and condensation. The basic sedimentation unit is consistent across the shelf; only the scale changes, with a general decrease in diastem-to-diastem thickness from centimeters to meters at Dry Ridge to centimeters to decimeters in the other settings.

In the outer ramp Dry Ridge section, located near the depocenter of the Meade Peak (locality D), a few beds of dolomitic mudstone contain phosphatic orbiculid inarticulate brachiopods, and the articulate brachiopods *Leiorhynchoidea weeksi* and *Lissochonetes ostiolatus*, rare benthic foraminifers, and molds of disarticulated bivalves (Figs. 6 and 7). This inarticulate and articulate brachiopod assemblage characterizes dysoxic conditions in Late Paleozoic sections worldwide (Allison et al., 1995). No macrofossils were observed in the mid ramp section (Astoria Hot Springs, section A; Figs. 6 and 7). In the inner ramp section (Crystal Creek, section C; Figs. 6 and 7), the only macrofossils observed were fish teeth (order Palaeoniscoidea) and mechanically reworked orbiculid brachiopod fragments in the lowermost phosphorite packstone/grainstone lag. A faunal assemblage consisting of orbiculid inarticulate brachiopods, phosphatized nuculoid bivalves, and pleurotomarid gastropods was found in bioturbated wackestones and abraded phosphorite lags in the nearshore section (Vernal-Brush Creek, section V; Figs. 6 and 7). These mollusks are typical of nearshore, shallow-water settings in the Late Paleozoic (Stevens, 1966).

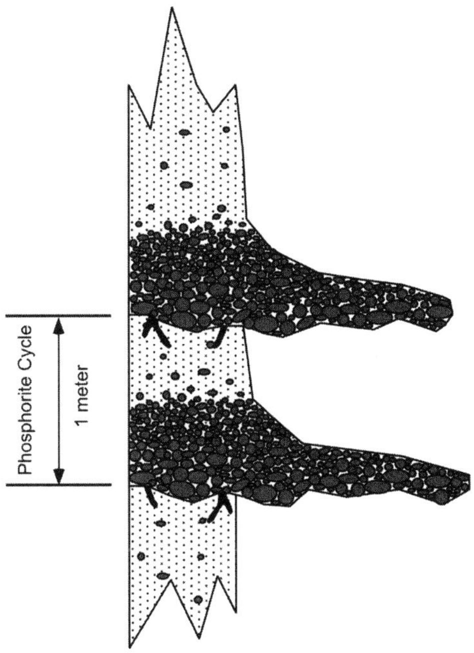

Figure 8. Generalized diagram of complete phosphorite depositional cycle in Meade Peak member of Phosphoria Rock Complex. Cycle starts with a diastem that is often erosional and bioturbated. Coarse, poorly sorted peloidal to intraclastic phosphorite is found just above diastem surface and grades into finer-grained peloidal phosphorite, followed by gradational contact with overlying organic carbon- and sulfide-rich sandstone to mudstone facies. Small, in situ phosphatic peloids are found in latter facies.

Meade Peak Chemostratigraphy

Organic Carbon, Sulfur, and Trace Elements

The stratigraphic and regional variation of phosphate (as francolite), total organic carbon, total sulfur, Cd, and carbon isotope values from phosphate peloids ($\delta^{13}C_{PO_4-CO_3}$) for the outer- to innermost-ramp transect are shown in Figure 9 and summarized in Table 1. Sedimentary phosphate is abundant throughout the Meade Peak member. The outer ramp section (locality D) is characterized by high francolite, high total organic carbon, and some high total sulfur beds. There are also two Cd-rich intervals in the outer-ramp section that also exhibit high total organic carbon and high phosphate concentrations. The concurrence of these phosphate-, TOC-, and Cd-rich intervals suggests paleoproductivity peaks with low water-column oxygen levels. Throughout the mid-ramp section (locality A), there is a similar pattern of high phosphate, high total organic carbon, high total sulfur, and high Cd concentrations. Some of the highest Cd (>300 ppm), the highest total organic carbon concentrations (>10 wt%), the highest total sulfur values, and lowest TOC:TS ratios are observed in this section. The inner ramp and nearshore sections (localities C and V) are also pervasively enriched in phosphate with thin siltstone and mudstone beds separating many of the phosphorite beds in the nearshore section (too thin to be shown in Fig. 9). These two sections, however, exhibit low total organic carbon (<1 wt%) and low total sulfur (<0.4 wt%) values, and low Cd concentrations in francolite (<5 ppm).

The number of analyses of Ni and Cr is not as large as that for Cd; thus, the distribution of these elements is not shown on Figure 9, but it is summarized in Table 1. The regional trends for Ni are the same as those affecting Cd, with highest values occurring in the outer and mid-ramp sections (D and A), which are rich in organic matter. Lower values of Ni occur in the landward sections (C and V), although Ni concentrations do not decrease to the same degree abruptly as those of Cd. In contrast, Cr values are relatively high in all sections.

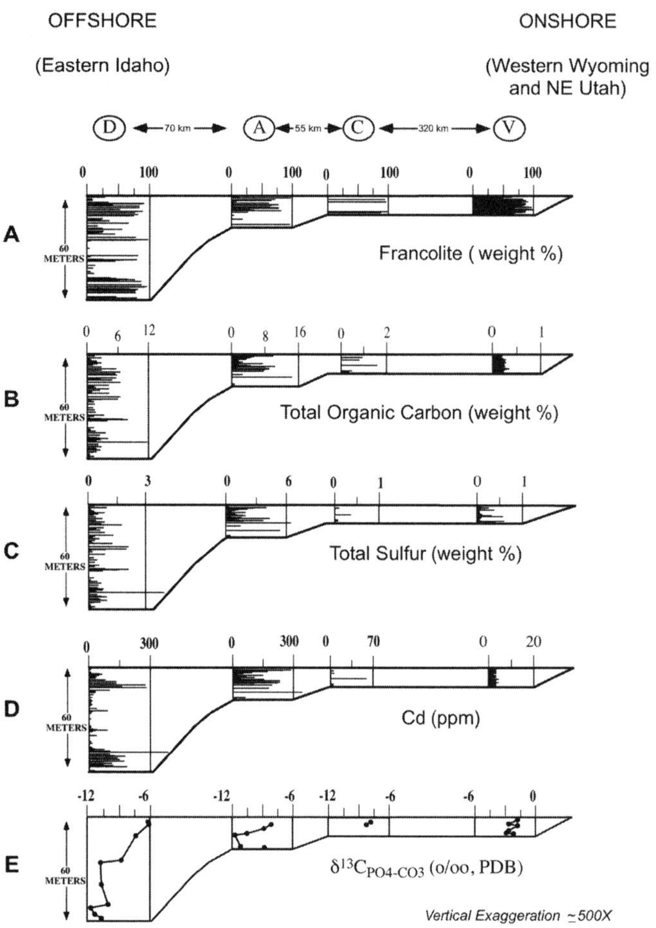

Figure 9. Regional and stratigraphic plots showing trends in (A) sedimentary phosphate (as francolite), (B) total organic carbon, (C) total sulfur, (D) Cd, and (E) $\delta^{13}C_{PO_4-CO_3}$ for Meade Peak sections in a seaward to landward transect. Francolite values are in counts per second (CPS) and are from X-ray diffraction analysis and are interpreted as semi-quantitative concentrations of francolite in rock.

Carbon Isotope Data

Table 1 and Figure 9E show the stratigraphic and regional patterns of francolite $\delta^{13}C_{PO_4-CO_3}$ values in each of the four Meade Peak sections. Francolite $\delta^{13}C_{PO_4-CO_3}$ values average <–8.0‰ Peedee belemnite in all but the nearshore section (locality V), where the mean is –2.7. The most negative individual values occur in the outer and mid ramp sections (localities D and A). There is a great deal of stratigraphic variability in these two sections as well, with more negative values (nearly –12‰) near the base and a trend toward less negative values (–8 to –6‰) stratigraphically upward (Fig. 9E).

Chemofacies

A chemofacies classification was developed based on the total organic carbon content of the rocks and their total organic carbon to total sulfur ratio. These parameters have been used to estimate paleoceanographic conditions under which fine-grained siliciclastic rocks were deposited (Berner and Raiswell, 1983; Raiswell and Berner, 1986; Allison et al., 1995) and are a result of the interplay of organic paleoproductivity, oxygen levels both in the water column and in the sediments, and in some cases, the availability of reduced iron species. This approach can present problems when used to interpret analyses of typical organic matter-bearing siliciclastics, but those shortcomings are insufficient to prevent discrimination of major geochemical facies in the Meade Peak. For example, TOC:TS ratios alone are not always able to discriminate differences in paleo-oxygen levels in organic-rich "normal marine shales" formed by typical surface productivity that have undergone significant organic burial diagenesis (Jones and Manning, 1994). However, factors such as these are not likely to generate the large, systematic, bed-by-bed differences in total organic carbon (0.1–15.7 wt%) and TOC:TS ratios (0.3–36.4) that occur in the Meade Peak (e.g., location D; Fig. 9); these are not the subtle shifts seen in "normal marine shales." Therefore, although other methods of discriminating paleoenvironmental oxygen levels are sometimes more appropriate (e.g., degree of pyritization, Raiswell et al., 1988; indicator of anoxicity, Raiswell et al., 2001), total organic carbon values and TOC:TS ratios are sufficient to discriminate between major paleoceanographic differences in the Meade Peak.

Based on published total organic carbon and total sulfur values for modern and ancient environments (Berner, 1981, 1984; Berner and Raiswell, 1983; Raiswell and Berner, 1986), and allowing for diagenesis to have lowered total organic carbon values, we used these two proxies to define three broad chemofacies (Fig. 10): dysoxic, anoxic, and euxinic. The dysoxic chemofacies was simply defined as stratigraphic units with less than 1.5 wt% total organic carbon. The anoxic chemofacies occurs in beds with greater than 1.5 wt% total organic carbon and TOC:TS ratios greater than 2.0. The euxinic chemofacies is defined by total organic carbon values greater than 1.5 wt% and TOC:TS ratios less than 2.0. Figure 10A shows the range of total organic carbon and total sulfur data for all lithologies in the phosphate- and organic carbon-rich members of the Phosphoria Formation (Meade Peak data of this study plus data from the Retort member of the Phosphoria Formation; Hiatt, 1997), as well as how these values relate to published values from modern environments and to interpreted paleoenvironments. Figure 10B shows total organic carbon and total sulfur data for just the phosphorites of the Meade Peak member and the resultant chemofacies defined herein.

The average values and ranges of bulk rock total organic carbon and total sulfur values, TOC:TS ratios; Cd, Ni, and Cr concentrations; and the $\delta^{13}C_{PO_4-CO_3}$ of phosphatic peloids in each of the three chemofacies is summarized in Table 2. Combined, these data show that the dysoxic facies is characterized by low total organic carbon (by definition), low total sulfur, high TOC:TS, the least negative $\delta^{13}C_{PO_4-CO_3}$ values, and the minimum and lowest average Cd, Ni, and Cr concentrations. Of the three, only Cr has a mean concentration above 40 ppm (mean = 395 ppm) in the phosphate peloids of the dysoxic facies.

TABLE 1. SUMMARY OF MEADE PEAK GEOCHEMICAL DATA BY STRATIGRAPHIC SECTION

	Section D (Dry Ridge)	Section A (Astoria Hot Springs)	Section C (Crystal Creek)	Section V (Vernal–Brush Creek)
Total organic carbon				
Range	0.27 to 8.50 wt	0.56 to 15.7 wt%	0.27 to 1.56 wt%	0.08 to 0.30 wt%
Average	2.63 ± 2.35	4.41 ± 3.51	0.73 ± 0.44	0.17 ± 0.05
n	79	45	6	49
Total sulfur				
Range	0.01 to 3.88 wt%	0.21 to 6.41 wt%	0.11 to 0.34 wt%	0.02 to 0.57 wt%
Average	0.48 ± 0.44	1.43 ± 1.39	0.21 ± 0.09	0.10 ± 0.13
n	79	45	6	49
Total organic carbon:total sulfur				
Range	1.8 to 36.4	0.9 to 5.6	2.0 to 7.3	0.3 to 7.3
Average	7.0 ± 5.4	3.6 ± 1.4	3.6 ± 1.9	3.4 ± 1.9
n	79	45	6	49
Cd				
Range	2 to 385 ppm	2 to 326 ppm	3 to 62 ppm	1 to 4 ppm
Average	51 ± 70 ppm	95 ± 78 ppm	13 ± 22 ppm	2.6 ± 0.7 ppm
n	79	45	6	49
Ni				
Range	40 to 540 ppm	86 to 166 ppm	40 to 115 ppm	13 to 37 ppm
Average	141 ± 169 ppm	110 ± 33 ppm	81 ± 31 ppm	21 ± 8 ppm
n	7	4	3	10
Cr				
Range	143 to 1540 ppm	177 to 1372 ppm	147 to 426 ppm	140 to 709 ppm
Average	430 ± 467 ppm	770 ± 460 ppm	282 ± 114 ppm	451 ± 191 ppm
n	7	4	3	10
$\delta^{13}C_{PO_4-CO_3}$				
Range	−11.4 to −6.0 ‰	−11.7 to −7.9 ‰	−8.1 ‰ to −7.9 ‰	−3.3 ‰ to −2.2 ‰
Average	−9.3 ± 1.9 ‰	−9.9 ± 1.3 ‰	−8.0 ± 0.1 ‰	−2.7 ± 0.3 ‰
N	11	7	2	6

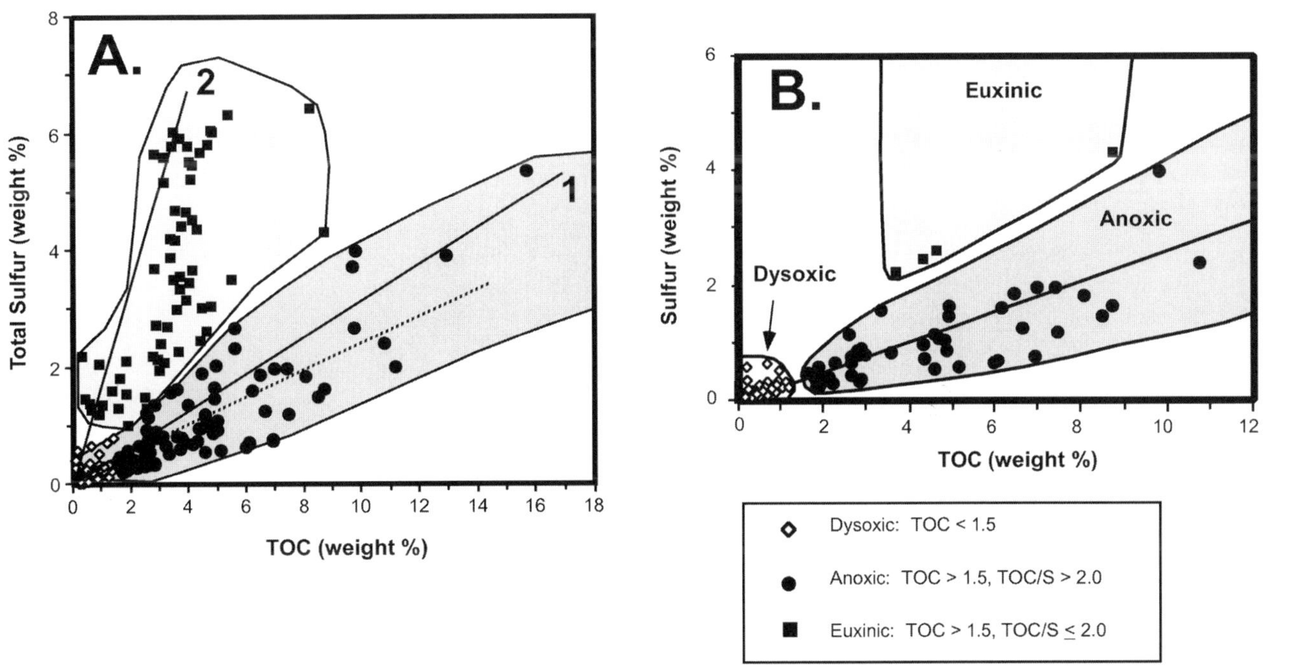

Figure 10. A: Phosphoria Rock Complex total organic carbon (TOC) and total sulfur data from all lithofacies, including non–Meade Peak rocks (additional data from Hiatt, 1997). Line 1 is trend defined by modern "normal" marine shales, and line 2 is trend for marine shales believed to have been deposited in euxinic environments (based on Berner and Raiswell, 1983). B: Total organic carbon and total sulfur data for all Meade Peak phosphorites with dysoxic, anoxic, and euxinic facies field interpretations added.

TABLE 2. SUMMARY OF MEADE PEAK GEOCHEMICAL DATA BY CHEMOFACIES

	Dysoxic chemofacies	Anoxic chemofacies	Euxinic chemofacies
Total organic carbon			
Range	0.08 to 1.46 wt%	1.56 to 15.7 wt%	1.66 to 8.75 wt%
Average	0.47 ± 0.40	4.82 ± 2.91	4.71 ± 3.12
n	93	81	5
Total Sulfur			
Range	0.01 to 0.78 wt%	0.19 to 5.35 wt%	0.99 to 6.41 wt%
Average	0.14 ± 0.05	1.03 ± 0.89	3.09 ± 1.99
n	93	81	5
Total organic carbon:total sulfur			
Range	0.3 to 36.4	2.1 to 9.6	0.9 to 1.9
Average	5.1 ± 5.4	5.2 ± 2.0	1.5 ±0.4
n	93	81	5
Cd			
Range	1 to 181 ppm	3 to 385 ppm	16 to 218 ppm
Average	13 ± 27 ppm	85 ± 82 ppm	104 ± 84 ppm
n	93	81	5
Ni			
Range	13 to 115 ppm	40 to 540 ppm	86 to 98 ppm
Rverage	39 ± 32 ppm	144 ±156 ppm	92 ± 6 ppm
n	14	8	2
Cr			
Range	140 to 709 ppm	143 to 1540 ppm	506 to 1372 ppm
Average	395 ± 192 ppm	505 ± 479 ppm	939 ± 433 ppm
n	14	8	2
$\delta^{13}C_{PO_4-CO_3}$			
Range	–10.1 to –2.2 ‰	–11.4 to –6.0 ‰	–11.7 ‰ to –7.9 ‰
Average	–6.6 ± 2.9 ‰	–9.8 ± 1.7 ‰	–9.8 ± 2.7 ‰
n	12	12	2

The anoxic chemofacies (Table 2) is characterized by high total organic carbon (by definition), moderate total sulfur, high TOC:TS (by definition), very negative $\delta^{13}C_{PO_4-CO_3}$, and high average Cd and Ni values. The euxinic chemofacies, which is the least common (n = 5), is characterized by high total organic carbon, very high total sulfur, low TOC:TS ratios (by definition), the most negative $\delta^{13}C_{PO_4-CO_3}$ values, and the highest average Cr and Cd concentrations.

Figure 11 depicts the lateral arrangement of these chemofacies and their relationship to lithofacies and macrofauna in the Meade Peak member. The sections show varying degrees of chemofacies intercalation. The one exception is the inner ramp section (locality V), which is marked exclusively by the dysoxic chemofacies. In general, the dysoxic facies dominates the nearshore Meade Peak section (locality V), whereas the anoxic chemofacies dominates both the outer ramp section and the mid ramp sections (localities D and A). The euxinic facies is found only in a few beds of the mid ramp section (Fig. 11).

DISCUSSION

Meade Peak Chemofacies and Paleoproductivity

Phosphorites are formed through chemical processes that occur independently of lithofacies. Therefore, a chemofacies approach is more indicative of the environments of phosphogenesis and a more informative paleoceanographic and paleoproductivity tool. We defined the three broad chemofacies largely on both total organic carbon values and TOC:TS ratios. The cause of high levels of organic carbon and sulfide mineral concentration and preservation has been vigorously debated. Although high concentrations of organic matter have been interpreted as simply

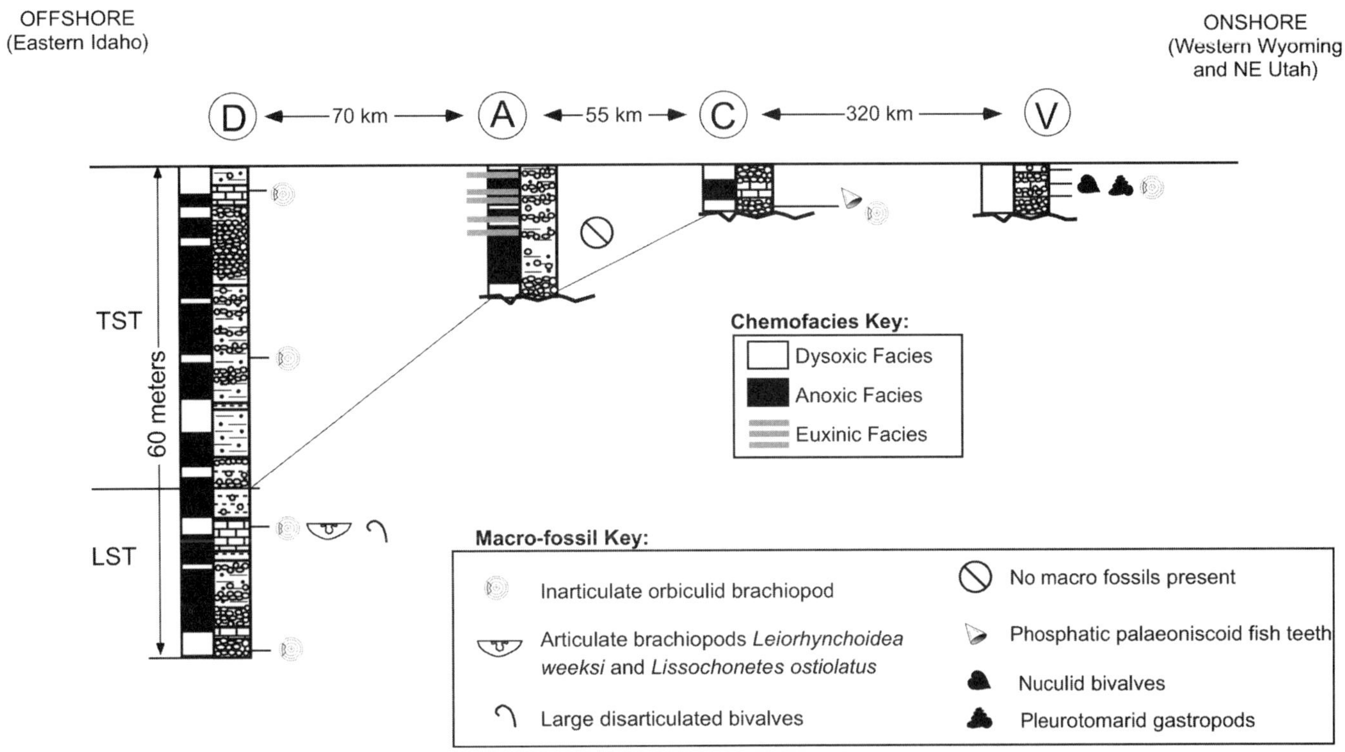

Figure 11. Meade Peak member chemofacies (left columns), lithofacies (right columns), and macrofauna (See Fig. 6 for lithofacies symbols and Fig. 5 for location information). Note that macrofossils only occur in dysoxic chemofacies. LST—lowstand systems tract; TST—transgressive systems tract.

an indicator of water column anoxia (e.g., Demaison and Moore, 1980), they are likely a complex function of biological productivity in the water column as well as sedimentation rate (e.g., Pedersen and Calvert, 1990; Canfield, 1994). Preservation of organic matter, however, is certainly enhanced under anoxic conditions (e.g., Demaison and Moore, 1980; Canfield, 1994; Ingall and Jahnke, 1997).

The important interrelationships between these factors are often lost in the debate regarding whether anoxia (e.g., Demaison and Moore, 1980) or organic productivity (e.g., Pedersen and Calvert, 1990) cause enhanced organic carbon preservation. As it falls through the water column, organic matter produced near the sea surface is broken down due to bacterial respiration (e.g., Froelich, et al., 1979; Schlesinger, 1997). Nutrients, such as phosphorus and nitrogen, are released to the water column, and oxygen is consumed in the process; if productivity is sufficient, then all water column oxygen can be consumed (e.g., Canfield, 1994). Continued breakdown of organic matter occurs after deposition leading to the release of additional phosphorus, nitrogen, and bioreactive trace elements (Froelich et al., 1979). Ingall and Jahnke (1997) pointed out that these processes are particularly important where upwelling-induced marine productivity occurs, because once nutrients are added (P, N, and sometimes Fe), increased biological productivity can quickly result in consumption of all oxygen in the water column below. Ingall and Jahnke showed that a positive feedback can develop, in which breakdown of organic matter leads to anoxic conditions that, in turn, increase phosphorus regeneration and enhance organic matter preservation. The regenerated phosphorus is released to the intrasediment porewater, where it can be fixed as sedimentary phosphate (francolite), and to the water column, where it contributes to further enhancement of productivity (Ingall and Jahnke, 1997).

Concentrations of Cd, Ni, and Cr in the phosphatic peloids augment the chemofacies approach because these elements are known to exhibit nutrient-like distributions in the modern oceans (e.g., Broecker and Peng, 1982; Calvert and Pedersen, 1993; Nathan et al., 1997; Schlesinger, 1997) and are concentrated in sediments under low-oxygen conditions (e.g., Calvert and Pedersen, 1993). Furthermore, Cd, and possibly Ni, is fixed as sulfides in sediments under sulfate reducing conditions, but Cr becomes concentrated in sediments under the less-reducing conditions of denitrification (Piper, 2001). In sediments and sedimentary rocks, these elements can be used to track past nutrient levels and paleoceanographic conditions because they are incorporated into the preserved sediments, whereas the actual nutrients (e.g., P and N) are largely recycled and remobilized in the depositional environment. These elements have been linked to biological processes and are characteristic of sediments deposited in highly productive oceanic areas today (e.g., Broecker and Peng, 1982; Schlesinger, 1997). In particular, phytoplankton concentrate Cd, and although concentrations are high, Piper et al. (2000) showed that Cd/Zn ratios in the Meade Peak member are similar to those found in modern plankton. With subsequent bacterial breakdown of organic matter, Cd can become highly concentrated in the organic residue (Gauthier et al., 1986). Ni, a micro-nutrient to many organisms, is associated with high marine productivity, and it promotes the growth of anaerobic bacteria (Cox, 1995).

The low total organic carbon and total sulfur values and low Cd, Ni, and Cr concentrations of the dysoxic Meade Peak chemofacies are all compatible with the presence of O_2 in the depositional environment and, thus, compatible with the recycling rather than burial of organic-carbon and nutrients. The low concentrations of the nutrient proxies (Cd, Ni, and Cr) suggest either low nutrient levels and/or aerobic conditions. We believe the latter to be the appropriate interpretation, because Cr exhibits a relatively high mean concentration (395 ppm). Unlike the other trace-element proxies for nutrients, Cr has multiple oxidation states and can be concentrated in sediments associated with high productivity under dysoxic conditions (Murray et al., 1983; Calvert and Pedersen, 1993). The fact that Cr concentrations in the phosphatic peloids of the dysoxic facies are not as low as the concentrations of the other trace elements suggests that nutrients were not limited, just effectively recycled by bacterial respiration of organic matter under denitrifying conditions (Piper, 2001). The oxygen present in the water, however, prevented sulfate reduction either in the water column or in the sediment that would have formed sulfide and trapped Cd and Ni, making their concentrations generally low.

In contrast, the high total organic carbon, high TOC:TS, and high average Cd and Ni values of the Meade Peak anoxic facies is compatible with an absence of O_2 in the water column and sediments. Free hydrogen sulfide in the water column must also have been absent; otherwise, total sulfur values would be higher and TOC:TS values would not be as great.

The very high concentrations of Cd in the anoxic chemofacies are probably indicative of high nutrient levels and a greater flux of organic matter to the seafloor. In the modern ocean, Cd concentrations are highest where sulfide is present in the sediment because it is incorporated into iron sulfide phases (Van Geen et al., 1994). The lack of covariance, however, between Cd and total sulfur (Fig. 9) indicates that Cd concentrations are not linked directly to pyrite content, but Cd is instead high because of extreme surface productivity and was incorporated into the francolite crystal structure under low oxygen conditions (cf., Nathan et al., 1997). The absence of O_2 suggested by this chemofacies points to ineffective recycling of organic matter and nutrients, thus the high total organic carbon values in the sediments and high trace-element concentrations in the francolite. However, the fact that such conditions persisted and dominated the outer and mid ramp settings (Fig. 11) means that the nutrient influx to these settings must have been maintained; otherwise, nutrients would have become limited, and paleoproductivity would have slowed or ceased. This, in turn, suggests that the anoxic chemofacies marks the sites of intense and persistent upwelling.

Finally, the high total organic carbon, very high total sulfur, low TOC:TS values, and high Cd and Cr concentrations of the euxinic Meade Peak chemofacies are all compatible with an absence of O_2 and the presence of free hydrogen sulfide in the water column and sediments. Total organic carbon levels are similar to those of anoxic facies (Table 1), indicating that high productivity prevailed in the water column, coupled with effective preservation. Extremely high pyrite concentrations suggest that iron was readily available and was probably sourced from terrigenous clastic sediments supplied by wind (cf. Carroll et al., 1998) or as shown for euxinic conditions in the Black Sea (Canfield et al., 1996), from breakdown of particulate organic matter in the water column. Hypereutrophic, sulfate-reducing conditions probably existed, and productivity was restricted to phytoplankton. Indeed, biomarker studies (Dahl et al., 1993; and Stephens and Carroll, 1999) indicate that phytoplankton and bacteria were abundant in the water column over the mid-ramp euxinic chemofacies. The average trace element concentrations are much higher than the dysoxic facies, and thus are compatible with high nutrient levels and reducing conditions. As with the anoxic facies, the euxinic chemofacies must also mark the site of intense and persistent upwelling.

The intercalation of anoxic and dysoxic chemofacies in the outer and inner ramp settings (Fig. 11) does suggest either temporal and/or spatial shifts in upwelling intensity away from the mid-ramp locus of upwelling and paleoproductivity. Movement in time and space of upwelling farther onto the paleoramp would generate changes in the location of maximum surface productivity and thus cause a switch from anoxic to dysoxic in the outer ramp and a concurrent switch from dysoxic to anoxic in the inner ramp. The converse, movement off the paleoramp, would generate the opposite effects in both settings. Assuming depositional rates measured in millimeters per 1000 yr or less, the thickness of the intercalated chemofacies (centimeters to meters) indicates a forcing factor with a frequency measured in 10^5 or more years. What such factors might have been is unclear to us, although fluctuations in global ocean circulation patterns are certainly feasible. A link between extreme paleoproductivity and southern-hemisphere glaciation has been suggested by prior workers (e.g., Pardee, 1917; Sheldon, 1984; Piper and Kolodny, 1987), and glacio-eustacy has also been implied in the interpretation of cyclicity in other Phosphoria Rock Complex units (e.g., Hendrix and Byers, 2000; Trappe, 2000). However, a glacio-eustatic climate and/or sea-level driver must be considered unlikely, given that recent biostratigraphic constraints establish that widespread glaciation in Gondwanaland ended millions of years before deposition of the Meade Peak phosphate and organic, carbon-rich units (Fig. 2).

Carbon Isotopic Variation

In general, the primary $\delta^{13}C_{PO_4-CO_3}$ values of phosphorites represent a largely benthic signal derived as francolite forms and recrystallizes due to microbial-mediated reactions within centimeters of the sediment-water interface (e.g., Jarvis, 1992;

Jarvis et al., 1994). Burial diagenesis can lower $\delta^{13}C_{PO_4-CO_3}$ values; the Phosphoria phosphorites are widely perceived to represent an advanced diagenetic end member (McArthur et al., 1986; Jarvis et al., 1994). Interpretation of the Meade Peak $\delta^{13}C_{PO_4-CO_3}$ data thus requires that we first evaluate the possibility of burial alteration.

The burial alteration hypothesis, as articulated by McArthur et al. (1986) and Jarvis et al. (1994) assumes that Miocene and younger phosphorites define the primary isotopic composition of francolites regardless of geologic age. Implicit in this assumption is the idea that all sedimentary francolites have formed under similar environmental conditions, which ignores secular changes in ocean-water chemistry and the diverse paleoenvironmental settings in which ancient phosphorites are known to have formed (e.g., Cook et al., 1990; Glenn et al., 1994). In fact, it is becoming clear that the Meade Peak phosphorites did not form under environmental conditions like any Neogene or Quaternary phosphogenetic environment (Dahl et al., 1993; Hiatt, 1997; Stephens and Carroll, 1999; Hiatt and Budd, 2001; Piper and Link, 2002). Thus, the diagenetically unaltered young phosphorites are not reasonable analogs for the initial $\delta^{13}C_{PO_4-CO_3}$ of the Meade Peak phosphorites.

The burial alteration hypothesis also ignores the fact that some geochemical proxies are much less likely to alter than others. Shemesh et al. (1988) concluded that $\delta^{18}O_{PO_4-CO_3}$ (isotopic value isolated from the structural carbonate site within the francolite crystal lattice) values were altered in many phosphorites, but that $\delta^{13}C_{PO_4-CO_3}$ and $\delta^{18}O_{PO_4}$ (isotopic value isolated from the structural phosphate site within the francolite crystal lattice) values were extremely resistant to diagenetic alteration. Because modern and ancient phosphorites show ranges of similar magnitude in their carbon isotopic values, the $\delta^{13}C_{PO_4-CO_3}$ values in francolite may be better preserved over geologic time than the simple burial alteration hypothesis presumes (Shemesh et al., 1988; Kolodny and Luz, 1992).

The Meade Peak $\delta^{13}C_{PO_4-CO_3}$ values reported herein do show some covariance with burial depth. The most negative values occur in the outer ramp section (Fig. 9E), which experienced the greatest burial depths (5.5 km, Hiatt and Budd, 2001). The least negative values occur in the nearshore section (Fig. 9E), which experienced the least burial depths (3 km, Hiatt and Budd, 2001). A least-square linear regression between $\delta^{13}C_{PO_4-CO_3}$ values and the maximum burial depth of the four localities yields an *R*-value of 0.69. However, this apparent covariance may just be a coincidence, as the deepest buried sections (D and A) are also the most organic-rich sections. A linear regression between $\delta^{13}C_{PO_4-CO_3}$ and total organic carbon in the same rock samples yields an R-value of 0.66, which is statistically indistinguishable from the value derived from the regression against maximum burial depth. The $\delta^{13}C_{PO_4-CO_3}$ of the Meade Peak phosphorites is thus just as likely to be a function of the total organic carbon of the rock as it is to be one of burial alteration. Of course, total organic carbon values have probably also been reduced by organic diagenesis; however, the analysis of chemofacies suggests that the main control on total organic carbon variation is changes in paleoproductivity. By analogy, we thus conclude that the primary depositional conditions across the ramp are also recorded in the $\delta^{13}C_{PO_4-CO_3}$ signal, although some diagenetic alteration cannot be completely ruled out.

As a proxy of primary depositional conditions across the Meade Peak ramp, the interpretation of the $\delta^{13}C_{PO_4-CO_3}$ values must include consideration of the chemofacies. The least negative $\delta^{13}C_{PO_4-CO_3}$ values reflect the least influence of organic ^{12}C. Such values occur in the dysoxic facies (Table 2), which is compatible with the more complete recycling of organic matter before burial below the sediment-water interface. As the francolite formed below that interface, it thus did not incorporate as much organic ^{12}C as francolite formed in the other chemofacies. The anoxic and euxinic facies would represent the opposite situation. High burial rates of organic matter due to inefficient recycling in the water column would have meant pore waters enriched in organic ^{12}C, and thus the more negative $\delta^{13}C_{PO_4-CO_3}$ values for the francolites formed in those chemofacies (Table 2). Indeed, McArthur et al. (1986) predicted that the carbon isotopic signature of francolite precipitated in dysoxic porewater should have a $\delta^{13}C_{PO_4-CO_3}$ value of –2‰ to –6‰, and those precipitated in anoxic porewater should have $\delta^{13}C_{PO_4-CO_3}$ values of –6‰ to –15‰ (PDB). These ranges are similar to those observed in the Meade Peak phosphorites (Tables 1 and 2). The lateral and vertical variations in $\delta^{13}C_{PO_4-CO_3}$ values (Fig. 9E) are thus further evidence that the process of phosphogenesis in the Meade Peak occurred across a broad spatial and temporal range of paleoceanographic conditions.

Phosphogenesis in the Phosphoria Sea

In mid and outer ramp settings where total organic carbon and total sulfur values, nutrient trace-element proxies, and phosphorite percentages are all high, phosphogenesis likely occurred much as it does in the modern ocean, albeit in much shallower waters. That is, large amounts of organic matter accumulated on the seafloor where primary productivity was extreme. Anoxic conditions within the sediments and bacterial breakdown of the organic matter (Froelich et al., 1979) released organic-bound phosphorous into the sediments (Filippelli and Delaney, 1996; Ingall and Jahnke, 1997). This process lead to phosphogenesis, while attendant anoxic conditions led to high amounts of sulfide and organic matter, and the efficient "trapping" of the nutrient-like cations in the sediments (Westerlund et al., 1986; Nathan et al., 1997). Continued influx of upwelled waters provided a source of new nutrients; thus, high productivity, high burial rates of organic carbon, and phosphogenetic processes can be long lasting.

Conditions on the inner ramp appear to have been much different. Abundant phosphorite yet low total organic carbon and total sulfur values and low nutrient-like trace-element concentrations indicate that phosphogenesis in the shallow nearshore environments must have either differed in some way from that described above, or all nearshore phosphorites were derived by long distance transport from mid-ramp settings. Although some of those nearshore phosphorites may be allochthonous, others clearly are not. In particular, the presence of small, phosphatized peloids in the intercalated carbonate muds is suggestive of in situ phosphogenesis in this setting. The phosphatization of the infaunal bivalves, many of which are still articulated and in their original burrows (Hiatt and Budd, 2001), also argues for autochthonous phosphogenesis on the inner ramp. A variation on the standard phosphogenesis model is thus needed to account for the phosphorites of the inner ramp setting.

The abundance of nearshore phosphorite formation indicates that there must have been a significant influx of phosphorous, which in turn means an influx of nutrient-rich waters to the innermost part of the ramp. Although riverine inputs have been documented for some nearshore phosphorites (e.g., Föllmi, 1996), we do not consider that a likely source for the Meade Peak phosphorites. The adjacent coastal landmass was characterized by evaporite deposits (Maughan, 1984; Peterson, 1984), and there is no published evidence for fluvial inputs to the Phosphoria Sea during Meade Peak deposition. A major source of detrital material to the sediments that make up the Meade Peak was derived by eolian influx (Carroll et al., 1998). Thus, the only viable source of nutrients was the water that upwelled farther seaward in the area of maximum upwelling (mid-ramp position) and eventually flowed into the inner ramp.

Ingall and Jahnke (1997) showed that phosphorus is more efficiently regenerated to the water column relative to organic matter under anoxic conditions. The corollary to this scenario is that where organic matter breaks down in dysoxic settings, such as the inner ramp, phosphorus is more likely to be fixed as francolite at the same time organic matter is consumed (Van Cappellen and Ingall, 1994). Therefore, phosphorus released to waters flowing from the anoxic environments in the mid-ramp could have fed productivity in the inner ramp. Oxygen supplied by air-sea exchange in these shallow-water settings, however, seems to have prevented the seawater, seafloor, and sediments from becoming anoxic, which in turn suppressed sulfide formation. Organic matter must have been efficiently recycled at and just below the sediment-water interface, resulting in low preserved total organic carbon. Over time, a net flux of phosphorous into phosphorites would occur, but because of the winnowing effect of storm-generated waves and the addition of oxygen through exchange with the atmosphere, there would be little preservation of organic carbon. Reasonably high productivity, coupled with high water temperatures on the inner ramp (Hiatt and Budd, 2001), may have resulted in a stable, possibly thermally (Piper and Link, 2002) or salinity (Dahl et al., 1993; Stephens and Carroll, 1999) stratified dysoxic water column that limited benthic dissolved oxygen levels and was supplied with enough nutrients to maintain phosphogenesis. Phosphogenesis in such a setting is distinctly different than any modern environment of phosphogenesis.

The low Cd and Ni concentrations of the inner ramp phosphorites also imply efficient recycling of these nutrient-like cations, and/or their depletion in the nearshore organic matter and the presence of oxygen in the water. Depletion is plausible if a significant percentage of the mass of these cations that was brought onto the Phosphoria ramp with upwelling waters was "trapped" in the anoxic outer- and mid-ramp sediment as sulfides. Anoxic marine basins are known to efficiently trap Cd and other metals in their sediments (Westerlund et al., 1986). Because of the superb textural preservation of all Meade Peak phosphorites, and the fact that the highest cation concentrations are found in the most deeply buried sections (localities D and A), we do not believe the low Cd, Ni, and Cr concentrations of the inner-ramp phosphorites are due to diagenetic mobilization.

Reassessment of the Paleoecology of the Phosphoria Rock Complex Biota

It has long been noted that the faunas of the Phosphoria and equivalent rocks exhibit low faunal diversity and abundance relative to age-equivalent lower-latitude sections (Yochelson, 1968; Wardlaw and Collinson, 1984; Boyd, 1993). Yochelson (1968), in the most comprehensive study of Phosphoria Rock Complex paleontology, interpreted this low diversity to be the result of low water temperatures. Wardlaw (1980) and Wardlaw et al. (1995) furthered the general acceptance of a cold-water "Arctic" fauna. The only dissenting voice to date has been that of Peterson (1980, 1984), who noted that maximum faunal diversity in the Phosphoria Rock Complex occurred in the carbonate bioherms of the Rex Chert member (Fig. 3) near the western border of Wyoming, where upwelling is predicted to have been most intense. As a result, Peterson reasoned that cool water was unlikely to have caused the low faunal diversity in the Phosphoria Rock Complex because diversity was apparently greatest where the water would have been the coldest.

We concur with Peterson (1980, 1984) that cool water was not the major biolimiting agent in the Meade Peak member and probably in the entire Phosphoria Rock Complex. Instead, the data described herein shows that macrofossils in the Meade Peak member occur almost exclusively in beds of the dysoxic chemofacies (Fig. 11). Bioturbation, described in detail by Hendrix and Byers (2000), is probably also limited to beds that would fall into our dysoxic chemofacies category. The anoxic chemofacies rarely contains macrofossils, and when present, the macrofossils always show signs of mechanical reworking. The euxinic chemofacies never contains any macrofauna or bioturbation. Where the latter two chemofacies prevailed, seafloor oxygen levels must have been so low as to exclude even the low diversity dysaerobic communities that characterize portions of the dysoxic facies.

Elevated salinity levels and possible salinity stratification caused by brines flowing westward from evaporative basins to the east of the Phosphoria Sea (Hite, 1978, Dahl et al., 1993; Stephens and Carroll, 1999) could also be a factor in explaining the faunal distribution, but Piper and Link (2002) determined that the Phosphoria Sea was probably temperature-stratified, not salinity-stratified. The absence of macrofauna in the highly productive mid-ramp section (Fig. 7A), where upwelling and influx of normal salinity ocean water from the west would have been greatest, further suggests that salinity was not the major biolimiting agent. Our data clearly suggest that the low faunal diversity and abundance of the Meade Peak member is related first and foremost to low oxygen levels.

Although we believe oxygen levels to be the foremost biolimiting factor in the Meade Peak, temperature probably did play a role. However, that role was primarily related to warm temperatures, not cold. This is evidenced by a comparison of the macrofauna in the dysoxic outer ramp facies (section D) with the macrofauna in the most landward sections (section V). Hiatt and Budd (2001) showed the outer ramp section was the site of cool, but not cold, paleotemperatures. As noted previously, the macrofauna in that section is composed of small chonetid and leiorhynchid articulate brachiopods, a few small bivalves, and orbiculid inarticulate brachiopods. This is the only Meade Peak member assemblage that closely approximates a "normal," albeit dysoxic, Late Paleozoic marine fauna (Allison et al., 1995). In contrast, the faunal assemblage of the inner ramp and nearshore sections consists only of orbiculid inarticulate brachiopods, nuculoid bivalves, and pleurotomarid gastropods (Figs. 6 and 13), which is indicative of restricted shallow water, dysaerobic conditions of variable salinity (Stevens, 1966; Kammer et al., 1986). Restriction, shallow water, and the broad nature of the ramp setting all resulted in mean paleotemperatures in excess of 30 °C (Hiatt and Budd, 2001), which may have raised salinities due to evaporation and limited the biota. In this setting, oxygen levels are the primary control on the fauna, but warm temperature is a secondary factor due to the lower solubility of oxygen at higher water temperatures and the greater consumption of oxygen by bacterial respiration.

The argument for oxygen as the dominant biolimiting factor in the paleoecology of the Meade Peak is also strengthened by a critical reanalysis of the arguments for an "Arctic" fauna. In particular, the cold water "Arctic" fauna interpretation, which is based on ammonites and one species each of conodont and brachipod (Wardlaw, 1980; Wardlaw et al., 1995), is weakened when the distribution of all Phosphoria Rock Complex fauna is considered in the context of recent plate tectonic reconstructions.

The conodont *Mesogondolella phosphoriensis* (= *Mesogondolella rosenkrantzi*; Wardlaw et al., 1995) is cited as one line of evidence for an "Arctic" fauna (Wardlaw, 1980). However, other reported occurrences of *M. phosphoriensis* are at paleolatitude less than or equal to 40° when plotted on the modern plate reconstruction of Scotese and Langford (1995). Specifically, the data of Bender and Stoppel (1965), Toula (1875), Sweet (1976), and Szaniawski and Malkowski (1979) indicate that this conodont is found at Permian paleolatitudes of 27°N (Greenland), 3°S (Sicily), and 40°N (Spitsbergen). This suggests that *M. phosphoriensis* may not be indicative of an "Arctic" fauna after all. Further, it has been reported only in the upper meter or so of the Meade Peak member (Wardlaw and Collinson, 1984), yet four other conodont species are found throughout the Meade Peak that are also widespread in low-latitude, Paleotethys and equatorial sections of western North American (e.g., *Mesogondolella serrata*, *M. gracilis*, *M. idahoensis*, and *Neostreptognathodus sulcoplicatus*; Yugan et al., 1994; Behnken et al., 1986; Igo, 1981; Szaniawski and Malkowski, 1979). The presence of these widespread "warm"-water conodonts in the Meade Peak member is not easily explained in the context of an "Arctic" fauna.

The case for a "cool water brachiopod fauna" in the Phosphoria Rock Complex (Wardlaw, 1980) is based on the presence of *Neospirifer striato-paradoxus,* which was originally described by Toula (1875) from Spitzbergen (Permian paleolatitude of 40°N). Specimens of *N. striato-paradoxus* were identified in the Phosphoria Rock Complex in rocks of Middle Wordian age (Wardlaw, 1980), which constrains the stratigraphic unit to either the Franson or the Rex Chert member of the Park City Formation (Fig. 4). Thus, this "Arctic" form is not found in the stratigraphic units associated with maximum paleoproductivity (the Meade Peak and Retort members) as has often been assumed (Parrish, 1982; Parrish and Peterson, 1988; Whelan, 1993; Inden and Coalson, 1996).

There are also several Phosphoria Rock Complex brachiopods that are characteristic of low-latitude North American and Paleotethys locations. These include *Kuvelousia leptosa*, which is common in the equatorial Paleotethys region and is found in the Rex Chert and Franson members of the Phosphoria Rock Complex in southwestern Montana (Wardlaw, 1977). There is also extensive overlap between the Phosphoria Rock Complex brachiopod fauna and sections near the paleo-equator of west Texas (Yochelson, 1968; Brittenham, 1973), Mexico (Wardlaw et al., 1979), and south China (Xu and Grant, 1994). As with the conodonts, the entire brachipod fauna does not present a convincing argument for cold water.

Lastly, support for the "cool-water" model was also deduced from ammonoids. Wardlaw et al. (1995; p. 36) pointed out that specimens of the ammonoid genus *Daubichites* found in the Meade Peak member have "....been reported (Spinosa and Nassichuk, 1985) as having a 'boreal' or cool-water, high-latitude distribution." However, Spinosa and Nassichuk (1985) also pointed out that *Daubichites* is a geographically widespread ammonoid and, in addition to its occurrence in the Meade Peak member, it is also found in the equatorial Paleothethys (Siberia, Australia, China) and northern Canada. More recently, an equatorial Permian ammonoid (*Demarezites furnishi*) was reported from the Meade Peak of southeastern Idaho (Spinosa and Nassichuk, 1994); the only other reported occurrences of this ammonoid are from west Texas and central Mexico (Spinosa and Nassichuk, 1994), areas that were situated within 10° of the Permian equator.

In summary, Phosphoria Rock Complex conodont, ammonite, and brachiopod faunas have some affinities to high-latitude Permian settings but do not provide a convincing and overwhelming argument for widespread and persistent "cool-water" conditions in the Phosphoria Sea. There are, in fact, just as many similarities in the fauna assemblage to warm, tropical settings. This suggests that other environmental factors served as the primary biolimiting agent. The chemofacies distributions defined herein clearly indicate that oxygen availability was more likely the cause of a sparse Phosphoria Rock Complex fauna in general and Meade Peak fauna in particular.

NEW PALEOCEANOGRAPHIC MODEL FOR THE PHOSPHORIA UPWELLING SYSTEM

Information from Recent Climate Models

Findings from climate models for the Permian are relevant to any reinterpretation of phosphogenesis and paleoceanography in the Phosphoria Rock Complex. Kutzbach and Ziegler (1994) produced a high-resolution model that included inland seas, large lakes, and marine embayments like the Phosphoria Sea. Their model indicates that the atmosphere over the Phosphoria Sea would have had a mean annual surface air temperature of 30–35 °C with summer surface air temperatures rising as high as 45 °C. These temperatures are in agreement with paleotemperatures of phosphogenesis in the inner ramp setting determined by Hiatt and Budd (2001). East-to-west eolian transport is also predicted from modeled wind directions, which is compatible with an eolian source for the Meade Peak siltstones and sandstones (Carroll et al., 1998). The modeled temperature and rainfall results also agree with geological evidence that indicates an extremely hot desert surrounding the Phosphoria Sea (Sheldon et al., 1967; Ziegler, 1990). Kutzbach and Ziegler's (1994) model thus seems to be geologically reasonable.

Kutzbach and Ziegler's (1994) climate model also shows a drastic wintertime weakening of the atmospheric circulation system and thus attendant reduction in coastal upwelling in the Phosphoria Sea (Fig. 1). The possibility that upwelling was seasonal is particularly important. Surface waters during summer upwelling and phosphogenesis would have been subject to hot (35–45 °C) air temperatures and would have become very warm in the shallow inner ramp. In the winter, the cessation of coastal upwelling would have meant that the waters in the Phosphoria Sea became restricted, possibly stratified, and extreme warming could have occurred in nearshore settings.

The New Paleoceanographic Model

New research and our improved understanding of the modern ocean indicate that the modern Peru-margin analog traditionally used to explain the Phosphoria upwelling system needs to be reassessed. That model cannot account for the variety of paleoenvironments that led to phosphogenesis or the relationship between paleoproductivity and macrofauna distribution. A new paleoceanographic model for the Phosphoria upwelling system is necessary.

The new paleoceanographic model that we propose (Fig. 12) takes into account the shallow and broad ramp setting, warm paleotemperatures (Hiatt and Budd, 2001), seasonal upwelling predicted by climate models (Kutzbach and Ziegler, 1994), severe oxygen depletion, high nutrient levels, and the new geochemical data presented herein. In this interpretation, oxygen-depleted, nutrient-rich water impinged on the Wyoming paleoramp at the mid-ramp position, where anoxia and euxinic conditions prevailed (present-day western Wyoming). Given the semi-restricted, shallow, marginal nature of the Phosphoria Sea, it is unlikely that an open-ocean deep-water mass could have been accessed by the wind-driven upwelling system. The more probable source of this nutrient-rich, oxygen-depleted water was a northward-flowing intermediate water mass (Jewell, 1995). Indeed, nutrient-rich, oxygen-poor water is near the surface over large areas between 20° N and 20° S today (Levitus, 1982), and a much more extreme oxygen-depleted, warm, nutrient-rich intermediate water mass is predicted for the west coast of equatorial Pangea during the Permian (Jewell, 1995; Hotinski et al., 2001). This intermediate layer probably extended into the Phosphoria Sea and was brought to the surface during the summertime, when wind patterns produced coastal upwelling by Ekman transport (Fig. 12). Upwelling may have ceased in winter, which allowed the waters, especially in the inner-ramp setting, to warm and approximate the overlying air temperatures.

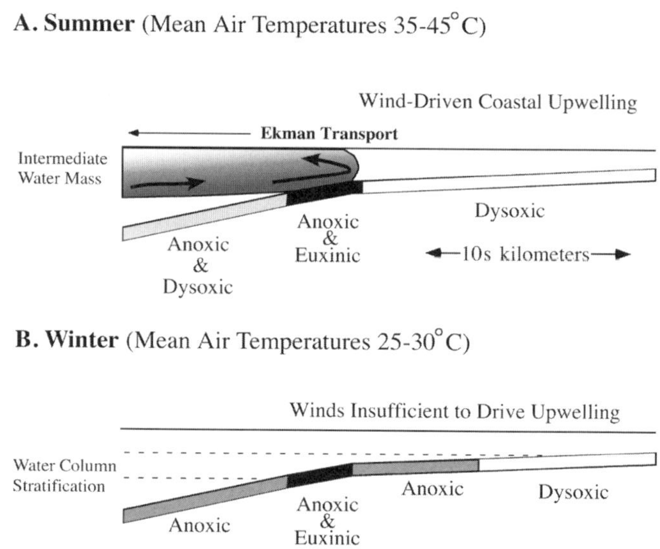

Figure 12. Conceptual model showing distribution of chemofacies, upwelling, and seasonal variation within Phosphoria Sea during Meade Peak deposition. A: Summertime case in which an oxygen-depleted, nutrient-rich water mass enters Phosphoria Sea and is driven to surface and seaward by coastal upwelling. High biological productivity in mid-ramp caused high organic particulate flux to seafloor, where bacterial respiration consumed all dissolved oxygen. Regeneration of phosphorus provided a supply of nutrients to inner ramp setting, where phosphogenesis occurred under dysoxic conditions. B: Wintertime case in which wind strength decreased and direction changed (Kutzbach and Ziegler, 1994) such that it was not able to maintain significant upwelling. Possible thermal stagnation may have occurred, water temperature of inner to mid-ramp settings would have risen, and continued bacterial respiration of organic matter in water column and below sediment-water interface would have consumed oxygen and resulted in expansion of anoxic and euxinic conditions.

The water flowing into the Phosphoria embayment probably had little or no dissolved oxygen to begin with, but as it warmed, its oxygen-carrying capacity remained extremely low, which prevented significant oxygen uptake from the atmosphere. This, combined with bacterial respiration of organic matter in the water column, favored development of widespread anoxia and even euxinic conditions. Even in nearshore environments, oxygen uptake would have been limited, and the water never became more than dysaerobic. As a result, macrofauna were suppressed due to low oxygen stress while phosphogenesis occurred in environments that ranged from dysoxic shallow inner-ramp settings to predominately anoxic mid- and outer-ramp settings. Salinity variation between the inner ramp and outer ramp probably influenced which organisms lived in which of those two settings, but the complete absence of macrofossils in the mid ramp suggests that elevated salinity did not play a major role in controlling the presence/absence of macro organisms in any setting. Paleoproductivity indicators suggest that maximum upwelling occurred in the mid-ramp position, and thus there was an ample influx of ocean water that would have prevented the elevation of salinity at this location.

This new model and our findings have great significance for the interpretation of other ancient phosphorites and suggest that paleoceanographic setting and paleoenvironment must be taken into account to fully understand the geochemical variation seen in ancient phosphorites.

CONCLUSIONS

The late Early to Late Permian is an important transitional time in earth history; the Phosphoria Rock Complex spans this change in global oceanographic, climatic, and biotic regimes. The widespread continental glaciation that marked the southern continents during the Early Permian had ended long before the phosphorites of the Phosphoria Rock Complex began forming. Extreme oceanographic conditions marked the Phosphoria Sea, and sedimentary phosphate and organic carbon were deposited across a broad range of paleoceanographic and paleoenvironmental settings, a range of phosphogenetic environments not seen today. Specifically:

1. Phosphogenesis in the Phosphoria Rock Complex occurred in depositional environments ranging from "basinal" outer ramp settings in <200 m water depth to very shallow, restricted inner ramp environments.

2. The organic carbon, sulfur, trace element, and $\delta^{13}C_{PO_4-CO_3}$ signatures vary systematically with position on the shelf, reflecting diverse paleoceanographic conditions across the paleoshelf and phosphogenesis in dysoxic to euxinic conditions.

3. Phosphogenesis in anoxic settings is strongly related to paleoproductivity indicators and occurred much as it does in the modern ocean, albeit in much shallower waters. Large amounts of organic matter accumulated on the seafloor due to the anoxic conditions and breakdown of the organic-matter-released, organic-bound phosphorous that in turn led to phosphogenesis.

4. Abundant phosphorite, yet low total organic carbon and total sulfur values and low, nutrient-like, trace-element concentrations indicate that phosphogenesis in nearshore, shallow-water, dysoxic settings was dramatically different. Oxygen supplied by air-sea exchange in these shallow-water settings maintained low oxygen levels and promoted the recycling of organic matter from the sediments. Yet, some of the liberated phosphorous produced phosphorites. Reasonably high productivity coupled with high water temperatures resulted in a stable dysoxic water column that was supplied with just enough nutrients to maintain phosphogenesis.

5. Cations that serve as paleoproductivity indicators are in low concentrations in the nearshore dysoxic settings due to either their efficient recycling or their "trapping" in the anoxic outer- and mid-ramp sediment.

6. Based on integration of lithofacies, biofacies, and chemofacies it is clear that the traditional cold- and deep-water upwelling model for the Meade Peak is not tenable. A more realistic paleoceanographic model is one in which a northward-flowing, oxygen-depleted, nutrient-rich intermediate water mass impinged on the seafloor at the mid-ramp position where anoxic and euxinic conditions prevailed. This shallow intermediate layer extended into the Phosphoria Sea and was brought to the surface during the summertime, when wind patterns produced coastal upwelling by Ekman transport. The water probably had very little dissolved oxygen to begin with, and as it warmed in the shallow waters of the Phosphoria Sea, its oxygen-carrying capacity remained extremely low. As a result, the upwelled water mass never became more than dysaerobic in nearshore environments, yet it was still capable of driving phosphogenesis in shallow-water settings.

7. Macrofauna in the Meade Peak member are only found in dysoxic facies, and their absence elsewhere (anoxic and euxinic chemofacies) is related first and foremost to low oxygen levels. Macrofauna also exhibit lateral variations that can be related to warm temperatures on the paleo ramp, but these variations are secondary to the control exerted by oxygen levels. All prior workers who use the occurrence, distribution, and nature of the Phosphoria Rock Complex fauna to argue for a simple cold-water upwelling model failed to recognize this point.

ACKNOWLEDGMENTS

We thank the donors of the Petroleum Research Fund, administered by the American Chemical Society, for financial support. Additional support was provided to EEH by Texaco, the Colorado Scientific Society, Exxon, and an American Association of Petroleum Geologists student grant. We thank Noel James, Josef Werne, Peir Pufahl, and Laura Gates for thoughtful reviews that improved the final version of this manuscript. Robert Rye kindly provided access to the U.S. Geological Survey stable isotope laboratory, Fred Luizer (University of Colorado) performed the X-ray fluorescence analyses and Halina Szymczyk (Texaco) provided the total organic carbon and total sulfur analyses.

REFERENCES CITED

Allison, P.A., Wignall, P.B., and Brett, C.E., 1995, Palaeo-oxygenation: Effects and recognition, *in* Bosence, D.W.J., and Allison, P.A., eds., Marine palaeoenvironmental analysis from fossils: Geological Society [London] Special Publication 83, p. 97–112.

Barron, E.J., and Fawcett, P.J., 1995, The climate of Pangea: A review of climate model simulations of the Permian, *in* Scholle, P.A., Peryt, T.M., and Ulmer-Scholle, D.S., eds., The Permian of northern Pangea, Volume 1: Paleogeography, paleoclimates, stratigraphy: Berlin, Springer-Verlag, p. 37–52.

Behnken, F.H., Wardlaw, B.R., and Stout, L.N., 1986, Conodont biostratigraphy of the Permian Meade Peak Phosphatic Shale member, Phosphoria Formation, southeastern Idaho: Laramie, Wyoming, University of Wyoming, Contributions to Geology, v. 24, p. 169–190.

Bender, V.H., and Stoppel, D., 1965, Perm-Conodonten: Geologisches Jahrbuch, v. 82, p. 331–364.

Berner, R.A., 1981, A new geochemical classification of sedimentary environments: Journal of Sedimentary Petrology, v. 51, p. 359–365.

Berner, R.A., 1984, Sedimentary pyrite formation: An update: Geochimica et Cosmochimica Acta, v. 48, p. 605–615.

Berner, R.A., 1994, GEOCARB II; a revised model of atmospheric CO_2 over Phanerozoic time: American Journal of Science, v. 294, p. 56–91.

Berner, R.A., and Raiswell, R., 1983, Burial of organic carbon and pyrite sulfur in sediments over Phanerozoic time: A new theory: Geochimica et Cosmochimica Acta, v. 47, p. 855–862.

Boyd, D.W., 1993, Paleozoic history of Wyoming, *in* Steidtmann, J.R., and Roberts, S.M., eds., Geology of Wyoming, Volume 1: Geological Society of Wyoming Memoir 5, p. 164–187.

Brittenham, M.D., 1973, Permian Phosphoria bioherms and related facies, southeastern Idaho [Master's thesis]: Missoula, University of Montana, 213 p.

Broecker, W.S., and Peng, T.-H., 1982, Tracers in the sea: Palisades, New York, Eldigio Press, Lamont-Doherty Geological Observatory, 690 p.

Calvert, S.E., and Pedersen, T.F., 1993, Geochemistry of Recent oxic and anoxic marine sediments: Implications for the geological record: Marine Geology, v. 113, p. 67–88.

Canfield, D.E., 1994, Factors influencing organic carbon preservation in marine sediments: Chemical Geology, v. 114, p. 315–329.

Canfield, D.E., Lyons, T.W., and Raiswell, R., 1996, A model for iron deposition to euxinic Black Sea sediments: American Journal of Science, v. 296, p. 818–834.

Carroll, A.R., Stephens, N.P., Hendrix, M.S., and Glenn, C.R., 1998, Eolian-derived siltstone in the Upper Permian Phosphoria Formation: Implications for marine upwelling: Geology, v. 26, p. 1023–1026.

Cathcart, J.B., Sheldon, R.P., and Gulbrandsen, R.A., 1984, Phosphate-rock resources of the United States: U.S. Geological Survey Circular 888, 48 p.

Claypool, G.E., Love, A.H., and Maughan, E.K., 1978, Organic geochemistry, incipient metamorphism, and oil generation in black shale members of Phosphoria Formation, western interior United States: America Association of Petroleum Geologists Bulletin, v. 62, p. 98–120.

Cook, P.J., and McElhinny, M.W., 1979, A reevaluation of the spatial and temporal distribution of sedimentary phosphate deposits in the light of plate tectonics: Economic Geology, v. 74, p. 315–330.

Cook, P.J., Shergold, J.H., Burnett, W.C., and Riggs, S.R., 1990, Phosphorite research: A historical overview, *in* Notholt, A.J.G., and Jarvis, I., eds., Phosphorite research and development: Geological Society [London] Special Publication 52, p. 1–22.

Cox, P.A., 1995, The elements on Earth: Oxford, Oxford University Press, 287 p.

Dahl, J., Moldowan, J.M., and Sundaraman, P., 1993, Relationship of biomarker distribution to depositional environment: Phosphoria Formation, Montana, U.S.A.: Organic Geochemistry, v. 20, p. 1001–1017.

Demaison, G.J., and Moore, G.T., 1980, Anoxic environments and oil source bed genesis: American Association of Petroleum Geologists Bulletin, v. 64, p. 1179–1209.

Dickins, J.M., 1984, Late Palaeozoic glaciation: B.M.R. Journal of Australian Geology and Geophysics, v. 9, p. 163–169.

Dickins, J.M., 1996, Problems of a Late Palaeozoic glaciation in Australia and subsequent climate in the Permian: Palaeogeography, Palaeoclimatology, Palaeoecology, v. 125, p. 185–197.

Filippelli, G.M., and Delaney, M.L., 1996, Phosphorus geochemistry of equatorial Pacific sediments: Geochimica et Cosmochimica Acta, v. 60, p. 1479–1495.

Föllmi, K.B., 1990, Condensation and phosphogenesis: Example of the Helvetic mid-Createaceous (northern Tethyan margin), *in* Notholt, A.J.G., and Jarvis, I., eds., Phosphorite research and development: Geological Society [London] Special Publication 52, p. 237–252.

Föllmi, K.B., 1996, The phosphorus cycle, phosphogenesis and marine phosphate-rich deposits: Earth Science Reviews, v. 40, p. 55–124

Frakes, L.A., Francis, J.E., and Syktus, J.I., 1992, Climate modes of the Phanerozoic: Cambridge, Cambridge University Press, 274 p.

Froelich, P.N., Klinkhammer, G.P., Bender, M.L., Luedtke, N.A., Heath, G.R., Cullen, D., Dauphin, P., Hammond, D., Hartman, B., and Maynard, V., 1979, Early oxidation of organic matter in pelagic sediments of the eastern equatorial Atlantic—Suboxic diagenesis: Geochimica et Cosmochimica Acta, v. 43, p. 1075–1090.

Gauthier, M.J., Clément, R.L., Flatau, G.N., and Amiard, J.-C., 1986, Accumulation du cadmium par les bactéries marines à Gram négatif selon leur sensibilité au métal et leur type respiratoire: Oceanologica Acta, v. 9, p. 333–337.

Glenister, B.F., Boyd, D.W., Furnish, W.M., Grant, R.E., Harris, M.T., Kozur, H., Lambert, L.L., Nassichuk, W.M., Newell, N.D., Pray, L.C., Spinosa, C., Wardlaw, B.R., Wilde, G.L., and Yancey, T.E., 1992, The Guadalupian: Proposed international standard for a Middle Permian Series: International Geology Review, v. 34, p. 857–888.

Glenn, C.R., Föllmi, K.B., Riggs, S.R., Baturin, G.N., Grimm, K.A., Trappe, J., Abed, A.M., Galli-Olivier, C., Garrison, R.E., Ilyin, A.V., Jehl, C., Rohrlich, V., Sadaqah, R.M.Y., Schidlowski, M., Sheldon, R.P., and Siegmund, H., 1994, Phosphorus and phosphorites: Sedimentology and environments of formation: Eclogae Geologicae Helvetiae, v. 87, p. 747–788.

González-Bonorino, G., and Eyles, N., 1995, Inverse relation between ice extent and the Late Paleozoic glacial record of Gondwana: Geology, v. 23, p. 1015–1018.

Hendrix, M.S., and Byers, C.W., 2000, Stratigraphy and sedimentology of Permian strata, Unita Mountains, Utah: Allostratigraphic controls on the accumulation of economic phosphate, *in* Glenn, C.R., Lucas, J., and Lucas, J., eds., Marine authigenesis: From global to microbial: SEPM (Society for Sedimentary Geology) Special Publication 66, p. 349–367.

Hiatt, E.E., 1997, A paleoceanographic model for oceanic upwelling in a Late Paleozoic epicontinental sea: A chemostratigraphic analysis of the Permian Phosphoria Formation [Ph.D. thesis]: Boulder, University of Colorado, 294 p.

Hiatt, E.E., and Budd, D.A., 2001, Sedimentary phosphate formation in warm shallow waters: New insights into the paleoceanography of the Permian Phosphoria Sea from analysis of phosphate oxygen isotopes: Sedimentary Geology, v. 145, p. 119–133.

Hite, R.J., 1978, Possible genetic relationships between evaporites, phosphorites, and iron-rich sediments: The Mountain Geologist, v. 14, p. 97–107.

Hotinski, R.M., Bice, K.L., Kump, L.R., Najjar, R.G., and Arthur, M.A., 2001, Ocean stagnation and end-Permian anoxia: Geology, v. 29, p. 7–10.

Igo, H., 1981, Permian conodont biostratigraphy of Japan: Palaeontological Society of Japan Special Paper 24, p. 1–50.

Inden, R.F., and Coalson, E.B., 1996, Phosphoria Formation (Permian) cycles in the Bighorn Basin, Wyoming, with emphasis on the Ervay member, *in* Longman, M.W., and Sonnenfeld, M.D., eds., Paleozoic systems of the Rocky Mountain region: Denver, The Rocky Mountain Section of SEPM (Society for Sedimentary Geology), p. 379–404.

Ingall, E., and Jahnke, R., 1997, Influence of water-column anoxia on the elemental fractionation of carbon and phosphorus during sediment diagenesis: Marine Geology, v. 139, p. 219–229.

Jarvis, I., 1992, Sedimentology, geochemistry and origin of phosphatic chalks: The Upper Cretaceous deposits of NW Europe: Sedimentology, v. 39, p. 55–97.

Jarvis, I., Burnett, W.C., Nathan, Y., Almbaydin, F.S.M., Attia, A.K.M., Castro, L.N., Flicoteaux, R., Hilmy, M.E., Husain, V., Qutawnah, A.A., Serjani, A., and Zanin, Y.N., 1994, Phosphorite geochemistry: State-of-the-art and environmental concerns: Eclogae Geologicae Helvetiae, v. 87, p. 643–700.

Jewell, P.W., 1995, Geologic consequences of globe-encircling equatorial currents: Geology, v. 23, p. 117–120.

Jones, B., and Manning, D.A.C., 1994, Comparison of geochemical indices used for the interpretation of palaeoredox conditions in ancient mudstones: Chemical Geology, v. 111, p. 111–129.

Kammer, T.W., Brett, C.E., Boardman, D.R.I., and Mapes, R.H., 1986, Ecologic stability of the dysaerobic biofacies during the Late Paleozoic: Lethaia, v. 19, p. 109–121.

Ketner, K.B., 1977, Late Paleozoic orogeny and sedimentation, southern California, Nevada, Idaho, and Montana, *in* Stewart, J.H., Stevens, C.H., and Fritsche, A.E., eds., Paleozoic paleogeography of the western United States, Pacific Coast Paleogeography Symposium 1: Los Angeles, The Pacific Section of Society of Economic Paleontologists and Mineralogists, p. 363–369.

Kolodny, Y., and Luz, B., 1992, Isotope signatures in phosphate deposits: Formation and diagenetic history, *in* Clauer, N., and Chaudhuri, S., eds., Isotopic signatures and sedimentary records: Berlin, Springer-Verlag, p. 69–121.

Knoll, A.H., Bambach, R.K., Canfield, D.E., and Grotzinger, J.P., 1995, Comparative earth history and Late Permian mass extinction: Science, v. 273, p. 452–457.

Kutzbach, J.E., and Ziegler, A.M., 1994, Simulation of Late Permian climate and biomes with an atmosphere-ocean model: Comparisons with observations, *in* Allen, J.R.L., Hoskins, B.J., Sellwood, B.W., Spicer, R.A., and Valdes, P.J., eds., Palaeoclimates and their modeling: London, Chapman and Hall, p. 119–132.

Levitus, S., 1982, Climatological atlas of the world ocean: National Oceanic and Atmospheric Administration Professional Paper 13, p. 173.

Maughan, E.K., 1984, Geological setting and some geochemistry of petroleum source rocks in the Permian Phosphoria Formation, *in* Woodward, J., Meissner, F.F., and Clayton, J.L., eds., Hydrocarbon source rocks of the Greater Rocky Mountain region: Denver, Rocky Mountain Association of Geologists, p. 281–294.

Maughan, E.K., 1994, Phosphoria Formation (Permian) and its resource significance in the Western Interior, U.S.A, *in* Embry, A.F., Beauchamp, B., and Glass, D.J., eds., Pangea: Global environments and resources: Calgary, Canadian Society of Petroleum Geologists Memoir 17, p. 479–495.

McArthur, J.M., Benmore, R.A., Coleman, M.L., Soldi, C., Yeh, H.-W., and O'Brien, G.W., 1986, Stable isotope characterisation of francolite formation: Earth and Planetary Science Letters, v. 77, p. 20–34.

McKelvey, V.E., Cheney, T.M., Cressman, E.R., Sheldon, R.P., Swanson, R.W., and Williams, J.S., 1959, The Phosphoria, Park City, and Shedhorn formations in the western phosphate field: U.S. Geological Survey Professional Paper 313-A, 47 p.

McKelvey, V.E., Swanson, R.W., and Sheldon, R.P., 1953, The Permian phosphorite deposits of western United States, *in* Saint Guilhem, M.R., ed., Origine des gisements de phosphates de chaux: 19th International Geological Congress (1952), Comptes rendus, sec. 11, no. 11, Algiers, p. 45–64.

Murray, J.W., Spell, B., and Paul, B., 1983, The contrasting geochemistry of manganese and chromium in the eastern tropical Pacific Ocean, *in* Wong, C.S., Boyle, E., Bruland, K.W., Burton, J.D., and Goldberg, E.D., eds., Trace metals in sea water: New York, Plenum Press, p. 643–669.

Nathan, Y., 1984, The mineralogy and geochemistry of phosphorites, *in* Nriagu, J.O., and Moore, P.B., eds., Phosphate minerals: Heidelberg, Springer-Verlag, p. 275–291.

Nathan, Y., Soudry, D., Levy, Y., Shitrit, D., and Dorfman, E., 1997, Geochemistry of cadmium in the Negev phosphorites: Chemical Geology, v. 142, p. 87–107.

Pardee, J.T., 1917, The Garrison and Philipsburg phosphate fields, Montana: U.S. Geological Survey Bulletin 640, p. 195–228.

Parrish, J.T., 1982, Upwelling and petroleum source beds, with reference to the Paleozoic: American Association of Petroleum Geologists Bulletin, v. 66, p. 750–774.

Parrish, J.T., and Peterson, F., 1988, Wind directions predicted from global circulation models and wind directions determined from eolian sandstones of the western United States—A comparison: Sedimentary Geology, v. 56, p. 261–282.

Pedersen, T.F., and Calvert, S.E., 1990, Anoxia vs. productivity: What controls the formation of organic-carbon-rich sediments and sedimentary rocks?: American Association of Petroleum Geologists Bulletin, v. 74, p. 454–466.

Peterson, J.A., 1980, Depositional history and petroleum geology of the Permian Phosphoria, Park City, and Shedhorn formations, Wyoming and southeastern Idaho: U.S. Geological Survey Open File Report 80–667, p. 42.

Peterson, J.A., 1984, Permian stratigraphy, sedimentary facies, and petroleum geology, Wyoming and adjacent area, *in* Goolsby, J., and Morton, D., eds., The Permian and Pennsylvanian geology of Wyoming: Wyoming Geological Association, Guidebook 35, p. 25–64.

Piper, D.Z., 2001, Marine chemistry of the Permian Phosphoria Formation and basin, southeast Idaho: Economic Geology, v. 96, p. 599–620.

Piper, D.Z., and Kolodny, Y., 1987, The stable isotopic composition of a phosphorite deposit: $\delta^{13}C$, $\delta^{34}S$, and $\delta^{18}O$: Deep-Sea Research, v. 34, p. 897–911.

Piper, D.Z., and Link, P.K., 2002, An upwelling model for the Phosphoria sea: A Permian, ocean-margin sea in the northwest United States: American Association of Petroleum Geologists Bulletin, v. 86, p. 1217–1235.

Piper, D.Z., Skorupa, J.P., Presser, T.S., Hardy, M.A., Hamilton, S.J., Huebner, M., and Gulbrandsen, R.A., 2000, The Phosphoria Formation at the Hot Springs Mine in Southeast Idaho: A source of selenium and other trace elements to surface water, ground water, vegetation, and biota: U.S. Geological Survey Open-File Report 00–050, 73 p.

Raiswell, R., and Berner, R.A., 1986, Pyrite and organic matter in Phanerozoic normal shales: Geochimica et Cosmochimica Acta, v. 50, p. 1967–1976.

Raiswell, R., Buckley, F., Berner, R.A., and Anderson, T.F., 1988, Degree of pyritization of iron as a paleoenvironmental indicator of bottom-water oxygenation: Journal of Sedimentary Petrology, v. 58, p. 812–819.

Raiswell, R., Newton, R., and Wignall, P.B., 2001, An indicator of water-column anoxia: Resolution of biofacies variation in the Kimmeridge Clay (Upper Jurassic, U.K.): Journal of Sedimentary Research, v. 71, p. 286–294.

Rees, P.M., Gibbs, M.T., Ziegler, A.M., Kutzbach, J.E., Behling, P.J., 1999, Permian climates: Evaluating model predictions using global paleobotanical data: Geology, v. 27, p. 891–894.

Ross, C.A., and Ross, J.R.P., 1994, Permian sequence stratigraphy and fossil zonation, *in* Embry, A.F., Beauchamp, B., and Glass, D.J., eds., Pangea: Global environments and resources: Canadian Society of Petroleum Geologists Memoir 17, p. 219–231.

Schlesinger, W.H., 1997, Biogeochemistry: San Diego, Academic Press, 588 p.

Scotese, C.R., and Langford, R.P., 1995, Pangea and the paleogeography of the Permian, *in* Scholle, P.A., Peryt, T.M., and Ulmer-Scholle, D.S., eds., The Permian of northern Pangea. Volume 1: Paleogeography, paleoclimates, stratigraphy: Berlin, Springer-Verlag, p. 3–19.

Sheldon, R.P., 1963, Physical stratigraphy and mineral resources of Permian rocks in western Wyoming: U.S. Geological Survey Professional Paper 313-B, 273 p.

Sheldon, R.P., 1984, Polar glacial control on sedimentation of Permian phosphorites of the Rocky Mountains, USA: Proceedings of the 27th International Geological Congress, Moscow: Utrecht, The Netherlands, VNU Science Press, v. 15, p. 223–243.

Sheldon, R.P., 1989, Phosphorite deposits of the Phosphoria Formation, Western United States, *in* Notholt, A.J.G., Sheldon, R.P., and Davidson, D.F., eds., Phosphate deposits of the world: Volume 2, phosphate rock resources: Cambridge, Cambridge University Press, p. 53–61.

Sheldon, R.P., Maughan, E.K., and Cressman, E.R., 1967, Sedimentation of rocks of Leonard (Permian) age in Wyoming and adjacent states, *in* Hale, L.A., ed., Anatomy of the western phosphate field: Salt Lake City, Utah, 15th Annual Field Conference, Intermountain Association of Geologists, p. 1–13.

Skipp, B., and Hall, W.E., 1980, Upper Paleozoic paleotectonics and paleogeography of Idaho, *in* Fouch, T.D., and Magathan, E.R., eds., Paleozoic paleogeography of the west-central United States: Rocky Mountain Paleogeography Symposium 1: Denver, Rocky Mountain Section of Society of Economic Paleontologists and Mineralogists, p. 387–419.

Shemesh, A., Kolodny, Y., and Luz, B., 1988, Isotopic geochemistry of oxygen and carbon in phosphate and carbonate of phosphorite francolite: Geochimica et Cosmochimica Acta, v. 52, p. 2565–2572.

Spinosa, C., and Nassichuk, W.W., 1985, The Permian ammonoid *Uraloceras* in North America and its global significance: Geological Society of America Abstracts with Programs, v. 17, p. 724.

Spinosa, C., and Nassichuk, W.W., 1994, The Permian ammonoid *Demarezites ruzhencev* from the Phosphoria Formation, Idaho: Journal of Paleontology, v. 68, p. 1036–1040.

Stephens, N.P, and Carroll, A.R., 1999, Salinity stratification in the Permian Phosphoria Sea: A proposed paleoceanographic model, Geology, v. 27, p. 899–902.

Stevens, C.H., 1966, Paleoecologic implications of Early Permian fossil communities in eastern Nevada and western Utah: Geological Society of America Bulletin, v. 77, p. 1121–1130.

Sweet, W.C., 1976, Conodonts from the Permian-Triassic boundary beds at Kap Stosch, East Greenland, *in* Teichert, C., and Kummel, B., eds., Permian-Triassic boundary in the Kap Stosch area, East Greenland: Meddelelser Om Gronland, v. 197, no. 5, p. 51–54.

Szaniawski, H., and Malkowski, K., 1979, Conodonts from the Kapp Starostin Formation (Permian) of Spitsbergen: Acta Palaeontologica Polonica, v. 24, p. 231–264.

Taylor, E.L., Taylor, T.N., and Cúneo, N.R., 1992, The present is not the key to the past: A polar forest from the Permian of Antarctica: Science, v. 257, p. 1675–1677.

Toula, F., 1875, Permo-Carbon-Fossilien von der Westkuste von Spitzbergen: Neues Jahrbuch fur Mineralogie, Geologie und Palaontologie, p. 225–264.

Trappe, J., 1998, Phanerozoic phosphorite depositional systems: A dynamic model for a sedimentary resource system: Berlin, Springer-Verlag, Lecture Notes in Earth Sciences 76, 316 p.

Trappe, J., 2000, Pangea: Extravagant sedimentary resource formation during supercontinent configuration, an overview: Palaeogeography, Palaeoclimatology, Palaeoecology: v. 161, p. 35–48.

Van Cappellen, P., and Ingall, E.D., 1994, Benthic phosphorus regeneration, net primary production, and ocean anoxia: A model of the coupled marine biogeochemical cycles of carbon and phosphorus: Paleoceanography, v. 9, p. 677–692.

Van Geen, A., McCorkle, D.C., Klinkhammer, G.P., 1995, Sensitivity of the phosphate-cadmium-carbon isotope relation in the ocean to cadmium removal by suboxic sediments: Paleoceanography, v. 10, p. 159–169.

Visser, J.N.J., 1996, Post-glacial Permian stratigraphy and geography of southern and central Africa: boundary conditions for climatic modeling: Palaeogeography, Palaeoclimatology, Palaeoecology, v. 118, p. 213–243.

Wardlaw, B.R., 1977, The biostratigraphy and paleoecology of the Gerster Limestone (Upper Permian) in Nevada and Utah: U.S. Geological Survey Open-File Report 77–470, p. 68.

Wardlaw, B.R., 1980, Middle-Late Permian paleogeography of Idaho, Montana, Nevada, Utah, and Wyoming, *in* Fouch, T.D., and Magathan, E.R., eds., Paleozoic paleogeography of the west-central United States: Rocky Mountain Paleogeography Symposium 1: Denver, Rocky Mountain Section of Society of Economic Paleontologists and Mineralogists, p. 353–361.

Wardlaw, B.R., 1995, Permian conodonts, *in* Scholle, P.A., Peryt, T.M., and Ulmer-Scholle, D.S., eds., The Permian of northern Pangea, Volume 1: Paleogeography, paleoclimates, stratigraphy: Berlin, Springer-Verlag, p. 186–195.

Wardlaw, B.R., and Collinson, J.W., 1984, Conodont paleoecology of the Permian Phosphoria Formation and related rocks of Wyoming and adjacent areas, *in* Clark, D.L., ed., Conodont biofacies and provincialism: Geological Society of America Special Paper 196, p. 263–281.

Wardlaw, B.R., Furnish, W.M., and Nestell, M.K., 1979, Geology and paleontology of the Permian beds near Las Delicias, Coahuila, Mexico: Geological Society of America Bulletin, v. 90, p. 111–116.

Wardlaw, B.R., Snyder, W.S., Spinosa, C., and Gallegos, D.M., 1995, Permian of the western United States, *in* Scholle, P.A., Peryt, T.M., and Ulmer-Scholle, D.S., eds., The Permian of northern Pangea, Volume 2: Sedimentary basins and economic resources: Berlin, Springer-Verlag, p. 23-40.

Westerlund, S.F.G., Anderson, L.G., Hall, P.O. J., Iverfeldt, A., Rutgers, L., Michiel, M., and Sundby, B., 1986, Benthic fluxes of cadmium, copper, nickel, zinc and lead in the coastal environment: Geochimica et Cosmochimica Acta, v. 50, p. 1289–1296.

Wignall, P.B., and Twitchett, R.J., 1996, Oceanic anoxia and the end Permian mass extinction: Science, v. 272, p. 1155–1158.

Yochelson, E.L., 1968, Biostratigraphy of the Phosphoria, Park City, and Shedhorn formations: U.S. Geological Survey Professional Paper 313-D, p. 571–660.

Yugan, J., Jing, Z., and Qinghua, S., 1994, Two phases of the end-Permian mass extinction, *in* Embry, A.F., Beauchamp, B., and Glass, D.J., eds., Pangea: Global environments and resources: Canadian Society of Petroleum Geologists Memoir 17, p. 813–822.

Xu, G., and Grant, R.E., 1994, Brachiopods near the Permian-Triassic boundary in South China: Smithsonian Contributions to Paleobiology, No. 76, p. 68.

Zharkov, M.A., 1984, History of Paleozoic salt accumulation: Berlin, Springer-Verlag, 308 p.

Ziegler, A.M., 1990, Phytogeographic patterns and continental configurations during the Permian period, *in* McKerrow, W.S., and Scotese, C.R., eds., Palaeozoic palaeogeography and biogeography: Geological Society [London] Memoir 12, p. 363–379.

Manuscript Accepted by the Society January 22, 2003

Printed in the USA

Index

A

Acadian orogeny: 203, 212
accelerator mass spectrometry: 35
accommodation, potential: 76, 79
accumulation space: 45
Acre Basin: 11
actualism: 2
Adriatic: 203
aeolian systems. *See* eolian systems
Afghanistan: 209, 219
Africa. *See also* specific countries
climate of: 48, 211
glaciation of: 10–12, 14, 47
intraplate margins in: 197
iron formations in: 231, 235–237
lacustrine deposits in: 91, 92, 102
lakes in: 76
location of: 11
loess deposits in: 55
reef formation in: 209–212, 218
rifting in Jurassic: 159
Sahara Desert. *see* Sahara Desert
Sahel region of: 64
sand sea activity in: 47–48
wind erosion of: 64
African plate
in Arabian Peninsula: 219
India, separation from: 159
Laurentia, collision with: 211
SEA, relationship to: 217
Agneteir Sand Sea: 47–48
Ahbach beds: 215
Akchar Sand Sea: 47–48
Alakol Basin: 96
Alang Beach: 159
Alaska
dunes in: 64
dust storms: 59
eolian sand in: 64
extensional zones in: 165
location of, during Devonian: 212
loess deposits in
eolian sand, relationship to: 64
geochemical analysis of: 54
location of: 55
magnetic susceptibility testing: 69
particle sizes in: 69
sources of: 67
outburst floods in: 38
paleosol studies in: 67, 69
reef formation in: 212
silt in rivers: 59
tides in: 151, 152, 165
Alazei fold belt: 221
Alberta: 214, 225
Aleutian Basin: 180, 181
Aleutian Islands: 165, 181
Alexander Terrane: 212
algae: 223
Algeria: 49, 218, 224
Algoma-type iron formation: 235, 237
alluvial-fan deposits: 12, 16, 105–125
Alps (European): 59, 203
Altai plate: 221
Altaids: 92, 195
altitude, lake size factor: 78
Alum Cave limestone: 134
Amazon Canyon: 182
Amazon Fan
geological setting of: 182–185
levee formation in: 188
sediment supply to: 175, 188
Amazon River
deltaic fan complex. see Amazon Fan
reef formation in: 210, 214
Amazonas Basin: 11
ammonoids
in Gondwana deposits: 219
in Phosphoria Rock Complex: 259
shale correlation from: 138
Amur River: 220
Amur-Okhotsk fold belt: 220
Anacanthotriletes spinosus: 135
Anchorage: 165
Andaman Sea: 194
Andes Mountains: 162, 188
andesite: 217, 221
Angers: 214
Angola Abyssal Plain: 185
Anna Shale
description of: 136
marine assemblages in: 132, 144
anoxia
climate changes from: 223
in deep ocean: 246
definition of: 252
organic matter preservation: 255
in Phosphoria Rock Complex: 245, 252–261
Antarctica
glaciation of: 10–17
glacigenic deposits in: 10, 14
lakes in, fluctuations in: 87
location of, during Paleozoic: 11
outburst flood evidence in: 28
quartzarenite deposition in: 12
redbed formation in: 12
reef formation in: 218
Anti-Atlas reefs: 219
Anvil Canyon: 106, 123
Anvil Rock Sandstone: 132
Apennines: 197
Appalachian Basin
coal bed formation in: 127, 128, 136–139, 142–143
correlations to: 135, 139
cyclothem formation in: 7, 130
dating of deposits: 142–143
depositional environment: 145
fauna in, during Emsian: 212
formation of: 212
reef formation in: 212–213, 219
shale in: 223
tectonic processes in: 145–146
Appalachian-Ouachita system
during Carboniferous/Permian: 191, 192
collision suture belts in: 193, 195
dating of deposits: 197, 198
during Jurassic: 49–50
remnant ocean basins in: 197–199
submarine fan deposition in: 198
Apulia margin: 203
Arabia: 218
Arabian Peninsula
glacigenic deposits in: 10, 14
location of, in Devonian: 219
reef formation in: 219
Arabian plate: 219
Arabian Sea: 155, 195, 197
aragonite: 210, 223
Aral Sea: 87
Archean
hydrothermal vent activity during: 239
iron deposition during: 231, 235
Archeosigillaria Zone: 14
Arctic Ocean
location of, in Devonian: 212, 213
reef formation in: 212–213, 221
sea ice in: 246
tides in: 165
Ardmore Limestone: 139
arenite: 57, 233
Argentina
eolian sand in: 64
glaciation of: 16
loess deposits in: 55, 64
tides in: 162
argillites: 235
Aridosols: 139, 144
Arizona: 7
Arkansas: 191, 197
Arkoma Basin: 136
Armenia: 219
Armorican Massif: 214, 217
Artinskian: 5–10, 14–16
Asan: 168
Asbian: 7
Ascension Canyon: 185
ash: 140, 145
Ashe Inlet: 159
Asia. *See also* specific countries in
loess deposits in: 55, 64
remnant ocean basins in: 191–197, 199–203
uplifted collisional orogen in: 188, 191–197, 201
Asian plate
convergent boundary, with Pacific plate: 168
interocean arc collision
with Luzon: 197
with Taiwan: 195
oceanic convergence, with Bengal Fan: 194
subduction zone
with Indian plate: 195, 196
with Philippine plate: 168
Asselian: 14–16
asteroid impact: 236–237
Astoria Canyon: 188
Astoria Hot Springs: 249–251
Atlantic Ocean. *See also* North Atlantic
basin development: 155
dust transport across: 64
lacustrine deposits in: 92, 100
sand transport into, from Africa: 48, 49
south margin rift basins: 100
Straits of Magellan: 151, 152, 162

Atlantic Ocean. (*continued*)
- submarine canyons in
  - Amazon Canyon: 175, 181, 182–185
  - Laurentian Channel: 186
  - Zaire Canyon: 185
- turbidite formation: 49

Atlas Mountains: 48–50
atmosphere
- carbon depletion: 92, 102–103
- carbon dioxide depletion: 37
- circulation, modeling of: 246
- loess deposition from: 54
- oxygenation of: 240

Atoka Formation: 138, 198
Australia
- alluvial-fan deposits in: 12
- ammonoids in deposits: 259
- basin development: 12
- carbonate platform preservation: 210
- coal bed formation in: 128
- dating of deposits: 12, 14
- dust storms in: 64
- fluvial deposits in: 12
- glaciation of: 10–17
- iron formations in: 231, 235–237
- lacustrine deposits in: 12, 79
- location of
  - during Devonian: 221–222
  - during Permian: 11
- loess deposits in: 55, 64
- in Paleotethys: 259
- parna deposits in: 55, 64
- reef formation in
  - during Devonian: 12, 211, 218, 222
  - during Holocene: 210, 224
- remnant ocean basins in: 203
- silt sources in: 64
- tectonic processes in: 12–14, 197
- tides in: 151, 152, 168

avalanche: 105
Avalonia microplate: 217
Avalon-Meguma terrane: 212
Avranche: 162
Azefal Sand Sea: 47
Aztec Sandstone: 49

B

backarc basins
- extensional zones in: 165, 168
- glacigenic deposits in: 11–12
- lacustrine deposits in: 93
- reef formation around: 216

Baffin Island: 35, 159
Bahamas: 211, 213
Bahia Thetis: 162
Baker Coal: 139
Balkhash Lake: 220
Baltic Basin: 214
Baltic Shield: 165
Baltica: 212, 216, 218
Banco Direccion: 162
banded iron formations: 231–241
Banks Island: 212–214, 221, 225
Baranof Fan: 185
Baranof Island: 185
Barbados: 64
Barents shelf: 28
Barrandian high: 214
Barremian: 92
basalts: 108, 241
Bashkirian: 14
basin development
- in Archean: 240
- during Chañic Orogeny: 12
- from continental collision: 191
- drainage analysis, modeling of: 152
- eolian sand accumulation, impact on: 46–47
- glacigenic deposits in: 10–16
- iron deposition during: 239
- lacustrine deposits in: 80–81
- for lakes
  - controls on: 82–84, 87
  - size of: 78
  - strange attractors: 85
- material distribution in: 2
- modeling of: 192
- shape factors: 87
- strata deformation: 12

Baudin Island: 168
Bay of Bengal: 195–197, 204, 224. *See also* Bengal Fan
Bay of Fundy: 151–155, 162
Bay of Mont-Saint-Michel: 162
bays, tides in: 172
Beishan: 96
Belarus: 216
Belgium: 214
Beloye More: 165
Bengal Fan
- correlations to: 198, 202
- dating of deposits: 196
- deformation of: 196
- levee formation in: 188
- sediment supply to
  - from continental collision: 195
  - from delta-front canyons: 175, 188
  - evolution of: 191–193, 196–197
- size of: 178, 180–181, 196
- tectonic processes in: 194

Bergisches Land: 215
Bering Canyon: 175, 181–182, 186–188
Bering Glacier: 38
Bering Sea: 175, 178–182, 186–187
Bermuda: 210
Bhavnagar: 159
Big Island: 159
biostromal communities: 210, 215
Bismarck volcanic arc: 197
bitumen: 98
bivalves
- in Eastern Interior Basin: 132
- in Phosphoria Rock Complex: 251, 259
- in South Africa: 12

Black Sea
- elemental precipitation in: 239
- euxinix conditions in: 256
- sapropel deposition in: 102
- tectonic processes in: 197

Blackjack Creek Limestone: 138
Blacourt Formation: 214
Block Coal: 139
Blue Fiord Formation: 212, 214, 221
Bogdashan Range
- evolution of: 95–98
- geological setting of: 92–93
- lacustrine deposits in: 93–95
- organic carbon deposits in: 98
- petrology of: 98–102
- stratigraphy of: 93–95

bogs: 131, 142
Bohemian Massif: 217, 218
Bolivia: 11–12
Bolsdorf dolostones: 215
Bolsovian: 138
Bombay: 159
Bonaparte Basin: 12, 16
Bosporus: 102
Botrychiopsis Zone: 14
Boulonnais outcrops: 214
Bowen Basin: 128
brachiopods
- in China: 221
- for dating of deposits: 12–16
- in Eastern Interior Basin: 132
- eustatic amplitude marker: 8
- in Karakoram-Kunlun Mountains: 220
- in Laurentia: 214
- in *Levipustula* Zone: 14
- in Mongolia: 221
- oxygenation indicator: 251
- in Phosphoria Rock Complex: 251, 259
- in SEA: 218
- in South Africa: 12

Brahmaputra River
- delta system: 203
- deltaic fan complex. *see* Bengal Fan
- sediment transportation in: 196

Brazil
- eolian system in: 43
- glacigenic deposits in: 11–12
- iron formations in: 231, 234
- lacustrine deposits in: 79, 102
- marine shale deposits in: 12
- reef formation in: 210

Brereton Limestone
- in coal bed cyclothem: 130, 136, 144
- marine assemblages in: 132
- sulfur level factor: 134

Brest: 214
Briar Hill Coal: 143
Brigantian: 7
Brisbane: 210
Bristol Channel: 151, 152, 155
Broome: 168
Brown Shale: 79
bryozoans
- for dating of deposits: 12, 14
- in Eastern Interior Basin: 132
- in *Levipustula* Zone: 14
- in mudmounds: 219
- in reefs: 212
- in South Africa: 12

Bucomazi Formation: 100
Burma: 222
Burntcoat Head: 155
Button Islands: 159

C

Calabria: 203
Calais: 165
calcarenites: 237
calcimicrobes in reefs: 209, 220, 223, 224
calcite, depletion of, in soil: 65
Caledonide: 192
California
  alluvial-fan deposits in
    Anvil Canyon: 123
    Cucomungo Canyon: 105–125
    Owens Valley: 123
    Smith Mountains: 123
  eolian sediment deposits in
    Aztec Sandstone: 49
    Cucomungo Canyon: 122
  remnant ocean basins in: 203
California Borderland: 186, 187
Calingasta-Uspallata Basin: 12, 16
Cambay: 159
Cambodia: 222
Cambrian
  rifting during: 197
  silica-secreting organisms in: 238
Cameron Island: 213
Canada. *See also* North America; specific locations in
  ammonoids in deposits: 259
  Bay of Fundy tides: 151, 152–155
  drumlin formation in: 26–27
  glaciogenic deposits in: 26–27, 55
  loess deposits in: 55
  morainal deposits in: 28
  outburst flood evidence in: 25–29, 35, 37–39
  peat mires in: 131, 141–142
  reef formation in
    during Carboniferous: 213
    during Devonian: 211–214, 222
    Western Sedimentary Basin: 209
  silt in rivers: 59
  Ungava Bay tides: 151, 152, 159
Canadian Shield: 37
Cancale: 162
Cangfanggou Group: 95
Canning Basin: 12, 222, 225
Cantabria: 218. *See also* Spain
Cantwell Shale: 79
Cape Abramov: 165
Cape Astronomicheski: 168
Cape Kanin: 165
Capitanian: 91
Capitol Reef: 51
Carajás Formation: 234
carbon dioxide
  depletion of, by meltwater: 37
  levels, in Permian: 246
  reef response to: 210
  saturation states in corals: 223
carbon, organic, burial of
  climate impact on: 91
  in lakes: 91, 102–103
  in Lucaogou Formation: 98, 101–102
  in Permian strata: 102
  in phosphate deposits: 246
  in Phosphoria Rock Complex: 251
  in shallow/regressive shelf conditions: 223
carbonate
  in alluvial-fans: 115
  in cyclothems: 7, 130
  depletion of, in soil: 65
  deposits
    in Africa: 48
    in China: 201
    in Eureka Valley area: 108
    in Labrador Sea cores: 35–38
    in North Atlantic Mid-Ocean Channel: 25, 35–38
    in Wyoming: 248
  in Devonian successions: 12
  formation of: 37
  in iron formations: 232
  in lacustrine deposits: 94
  leaching in soils: 68
  in loess: 54, 58
  production peak: 224
  reef tracts: 211
  replacement, in iron formations: 237–238
  transportation of: 37
carbonate compensation depth: 223
carbonate grainstones: 233, 237
carbonate ooze: 235
Carbondale Formation: 128–146
Carboniferous
  brachiopods in strata from: 14
  bryozoans in strata from: 14
  Chañic Orogeny during: 12
  clams in strata from: 14
  coal, compaction ratio for: 131
  continental reconstruction for: 11
  crinoids in strata from: 14
  cyclothem formation during: 5–8, 18
  fluvial channels in: 136
  flysch deposits during: 191, 197
  glaciation during: 5–18
  inland seas, tidal analysis of: 152
  Levipustula Zone during: 14
  palynomorphs in strata from: 16
  pelagic limestone deposits from: 201
  radiolarites distribution in: 218
  reef formation during: 213, 221, 224
  remnant ocean basins from: 198, 202
  rhythmites from: 162
  sea level changes during
    from eustasy: 7–8
    from glacioeustasy: 18
    from tides: 152
  strata, analysis of
    in China: 92–94
    in North America: 198
    in South America: 14
    in Titicaca Basin: 12
  tectonic processes: 97
Caribbean: 203
Carlsburg Ridge: 159
Carnarvon Basin: 222
Carnic Alps: 217, 218
Carnwath River: 214
Carpathian Mountains: 203
Caspian Basin: 102, 216
Caspian embayment: 214
Caspian Sea
  fluctuations in: 87
  mixis of: 78
  origin of: 80
  plate tectonics of: 197
  reef formation around: 216
  water chemistry of: 78
catchment: 105, 106
Caucasus Mountains: 219
Cauê Itabirite: 234
Cayeux: 165
Cenozoic
  basin development during: 192, 201
  glaciation during: 187
  reef formation during: 210
  sediment deposition during: 159
  strata, analysis of: 92
  strike-slip faulting: 99
  tectonic processes in: 197
  time scale for: 224
Central America: 197
cephalopods: 12
Chaco loess: 55
Chad: 79
chalcedony: 233
Chaloneria: 139
Changhsingian: 91–103
Chañic Orogeny: 12
Cheju-do Island: 168
chert pods: 234, 235
Chiang Jiang River: 168
Chignecto Bay: 155
Chile: 162
China
  ammonoids in deposits: 259
  arc magmatism in: 93
  biota in deposits: 259
  carbonate platforms in: 210
  climate of
    in Jurassic: 201
    modern: 66, 67
  dust storms in: 66
  eolian sand deposition in: 53, 64
  glaciation of: 60
  iron formations in: 231
  lacustrine deposits in
    dating of deposits: 75
    geological setting of: 92
    Hami Basin: 91–103
    Hongyanchi Formation: 80
    Jianghan Basin: 79
    Jingjingzagao Formation: 80
    Junggar Basin. *see* Junggar Basin
    Lucagao Formation: 80
    Qinshankou: 79, 81
    Quantou Formations: 81
    Shahejie Formation: 79
    Songliao Basin: 81
    Turpan Basin: 91–103
  location of, in Devonian: 221
  loess deposits in
    cycles in: 66–67
    with eolian sand deposits: 53
    geochemical analysis of: 58
    location of: 55

loess deposits (*continued*)
magnetic susceptibility testing: 68–69
modeling of: 67
particle sizes in: 55–57
sources of: 60–61, 66–67
ocean fan systems in: 191, 193, 195
paleosol studies in: 68–69
in Paleotethys: 259
petroleum source rock in: 91
reconstruction of, Carboniferous - Triassic: 202
reef formation in: 209, 211, 220–222, 225
remnant ocean basin development in: 191
Songpan-Ganzi Complex in
collision suture belts in: 193
evolution of: 191, 195, 199–203
size of: 199
tectonic processes in: 193, 199–203
TOC analysis of strata from: 91
Yellow Sea tides: 151, 152, 168–172
Chinle Formation: 50
chlorine: 134
chonetids: 259
Chott Rhasa Basin: 49
Chugach Mountains: 59
Chugwater Formation: 50
Chukot block: 209, 221
Chuluun Formation: 221
clams: 14
clastics: 12, 130, 139
clay
in alluvial-fan deposits: 113, 117, 120, 123
in cyclothem: 130
formation of: 123
in Labrador Sea cores: 35–38
in loess: 53, 54, 57
climate. *See also* precipitation
changes to
from anoxia: 223
from carbon dioxide depletion: 37
from carbonate production: 37
from collisional orogens: 195
for erosion: 44, 195
from Heinrich events: 37
from loess: 54
from outburst floods: 26
in Permian: 246
reef response to: 223–224
deposition factor: 2, 3
for eolian systems: 44–47
indications of, in isotope levels: 8
lake size factor: 77, 87
Malvinokaffric: 211, 218
Milankovitch cycles in: 47
modeling of: 260–261
peat accumulation factor: 144, 145
precipitation-evaporation ratio: 75
proxy for, latitude: 77
record of, in paleosols: 65–66
reef response to: 210, 223
sediment transportation factor: 87
supergreenhouse: 209, 210, 222
tropical: 210
coal
in cyclothems: 7, 130, 143–144
in flood-plain deposits: 16
lithotype changes in: 142
from molasse deposits: 203
to peat compaction ratio: 127, 130–131
in polar deposits: 246
studies of: 127–146
coesite: 200
Colchester Coal
composition of: 133
correlations to: 135, 138–139
depositional environment: 127, 133, 139–140, 143–146
durains in: 140
eustatic controls on: 143–144
paleoclimate of mire: 144
paleolocation of mire: 142
palynology of: 134, 145
size of: 127–129, 133, 140, 142
spore assemblage zones in: 138
Colorado
alluvial-fan deposits in: 123
eolian sand in: 64, 65
glaciation of: 61
lacustrine deposits in: 79, 80, 100–101
loess deposits in
dating of deposits: 65
eolian sand, relationship to: 64, 65
particle sizes in: 56
sources of: 55, 61, 64, 65
Colorado Plateau: 43, 49
Colorado River (USA): 178
Columbia River (USA): 188
comet impact: 236–237
comminution: 60
Congo Canyon: 185
Congo Fan: 185
conodonts
in *linguiformis* zone: 223
in Phosphoria Rock Complex: 259
in *rhenana* zone: 223
in shale: 144
studies of, for correlation
in North America: 135, 139
in Tibet: 14
continental margins
intraplate: 192
iron formation deposition in: 238
slopes of: 152
subduction zone collision: 192
submarine canyons in: 178
tidal amplification factor: 151, 152
continental seas. *See* inland seas
continental shelf
glaciation of
off Canada: 186
off North America: 28
growth, in Archean: 240, 241
iron formation size factor: 240
modeling of: 246
outburst flood evidence in: 28
reefs on, study of: 210–225
subduction zone impact on: 168
submarine canyons in
characteristics of: 176
off Africa: 185
off Alaska: 175, 178–182, 186–187
tidal amplification factor: 151–152, 172
off Australia: 168
off France: 159
off India: 159
off South America: 162
off UK: 159
convergent margin basins
glacigenic deposits in: 12–14
lake formation in: 79
remnant ocean, development of: 192
tides in: 168
Cook Inlet: 151, 152, 165
Coppermine area: 26–27
coppice dunes: 122. *See also* dunes
corals: 132, 209–225. *See also* reefs
Cordilleran Belt: 51
Cordilleran Ice Sheet: 59
Coriolis effect: 29, 162
Corsica: 203
Crassispora: 134
Crassispora kosankei: 135
cratonic basins
eolian sand preservation in: 47
glacigenic deposits in: 10–14
peatland development in: 127
Sahara Desert. *see* Sahara Desert
Cretaceous
basin development during: 203
Brazilian eolian system from: 43
chalk deposits from: 165
climate during: 222
fracture zones from: 159
lacustrine deposits from
in Atlantic Ocean rift basins: 100
Doba Basin: 79
Doseo Basin: 79
Lagoa Feia Formation: 79
Qinshankou: 79, 81
Quantou Formations: 81
Songliao Basin: 81
south Atlantic: 75
Termit Graben: 79
mudstone deposition during: 64
sandstone deposition during: 64
crinoids
for dating of deposits: 14
in Eastern Interior Basin: 132
in *Levipustula Zone*: 14
in reefs: 215
Croatia: 217
Croweburg Coal: 138–139, 144
Crystal Creek: 249–251
Cucomungo Canyon
alluvial-fan deposits: 105, 110–125
geological setting of: 108
hydrology of: 108–110
origin of: 122–125
sedimentology of: 110–122
size of: 105, 106, 110, 122–123
studies of: 105–125
Cumberland Sound: 38
cyclothem formation
climate controls on: 144
coal bed development in: 127, 130, 143–144
definition of: 130

deposition controls: 130, 143–145
duration of: 8, 130
eustatic amplitude link: 7–8, 18, 130, 143–144
in Gondwana: 5–8, 18
Milankovitch cycle link: 8, 130
in North America
during Desmonesian: 127, 130, 143–144
during Paleozoic: 6–8, 18
during Pennsylvanian: 144
peatland development in: 127, 130, 143–144
sequence analysis of: 130

D

Dabie Mountains: 200–202
Dachang Reef: 222
Dahanu: 159
Damodar Basin: 14
Danville Coal: 139
Daubichites: 259
Davis Coal: 143
Deadman Canyon: 123
Dean Formation: 221
Death Valley: 106, 108, 123
debris flow: 105, 117–120
Dekoven Coal: 143
delta switching: 130
deltas
cyclothemic deposition in: 130
fan channel system: 181, 192–193
formation of, in Canadian arctic: 214
geological setting of: 193
modeling of: 203
sediment supply through: 192–193
Demarezites furnishi: 259
density flows: 179
deserts
dunes in. *see* dunes
loess in: 53, 59–64
Desmoinesian
coal bed formation during: 127–146
correlations to, time period: 138
cyclothem formation during: 130
depositional environment: 139–140
peatland development in: 127–146
plant assemblages in mires during: 139
tectonic processes during: 129, 145–146
Devonian
basin development during: 203
brachiopods in strata from: 12
bryozoans in strata from: 12
Chañic Orogeny during: 12
climate during: 209, 210, 222
cyclothem formation during: 18
fauna, in strata, in China: 201
glaciations during: 6, 10, 223
glacigenic deposits during: 11–12
glacioeustasy in: 18
lacustrine deposits from: 79
mollusks in strata from: 12
palynomorphs in strata from: 12
radiolarites distribution in: 218
reconstruction of: 211
reef formation during: 209–225
sediment deposition during: 12
time scale for: 211, 224
diamictites: 10–14
diamond: 200
Diana Bay: 159
Diaphorodendron: 134
Diaphorodendron scleroticum: 139
Dinaric Alps: 218
disparilis zone: 222
Doba Basin: 79
Dockam Formation: 50
Dollendorf syncline: 215
dolomite
depletion of, in soil: 65
in iron formations: 237
in lacustrine deposits: 94
in loess: 58
reefs, association with: 221
in shallow sea basin: 249
Domanik Limestone: 216
domed mires. *See* ombrogenous mires
Doseo Basin: 79
Douglas Group: 162
Dreimühlen-Rodert Beds: 215
drumlins, formation of: 26–29
Dry Ridge: 249–251
Dryas: 47, 48
Du Quoin Monocline: 132, 146
Dunbarella: 132, 144
dunes. *See also* yardangs
in alluvial-fan deposits: 122
construction of: 44–45
in eolian systems: 43–51
grain-size analysis: 44
loess, relationship to: 53, 64
Navajo Sandstone: 49–51
pedogenesis in: 47
preservation of: 46–47
in Sahara Desert: 47–49
types of: 45, 46
Dungnee Formation: 221
dust. *See also* loess
climate changes from: 54
fluxes: 61–64
storms
in China: 66–67
in the Sahara Desert: 64
Dykersburg shale: 134
dysoxia
definition of: 252
indicator of, brachiopods: 251
in Phosphoria Rock Complex: 245, 251, 252–261
Dzungar plate: 220

E

East China Sea: 168
East Irish Sea Basin: 155
Easter Island, tidal ranges: 152
Eastern Interior Basin
coal bed formation in: 128–146
correlations to: 135–139
cyclothem formation in: 7, 130, 143–144
depositional environment: 143–146
hydrology of: 145
paleoclimate of: 144
paleolocation of: 142, 144
paleosols in: 139, 143–144
palynology of: 134–135, 144
tectonic processes in: 145–146
topography of: 144–145
Ecca Group: 128
Egypt: 64, 197
Eifel region: 215
Eifelian: 211–225
Eighty Mile Beach: 168
Ekman transport: 246, 260
El Niño: 210–211
elevation hypsometry: 81
Ellesmere Island: 212
Ellsworth Mountains: 16
Emsian: 209–225
Energy shale: 134
English Channel: 151, 152, 159–165
Enisei River: 220
ensensis zone: 222
Eocene
lacustrine deposits from
Green River Formation: 79, 80, 100–101
Rundle Formation: 79
reef formation during: 210
subduction zones during: 196
submarine fan deposition during: 196
eolian abrasion: 53, 60
eolian sediment
accumulation of: 45–46
in alluvial-fan deposits: 105, 122
construction of dunes: 44–45
deposition of
in Aztec Sandstone: 49
in Cucamonga Canyon: 105, 122
in Glen Canyon Sandstone: 49
in Illinois: 66
in Navajo Sandstone: 43, 49–51
in Nugget Sandstone: 49
in Sahara Desert: 43, 47–49
in Wingate Sandstone: 50
loess, relationship to: 53, 56, 64–65
preservation of, formation of rock: 46–47
source of: 44–45, 54
transportation of: 44–47, 65
eolian systems
modeling of: 86, 260
types of: 45–46
epeiric seas. *See* inland seas
epicontinental basin: 246
epicontinental seas. *See* inland seas
erg: 108
erosion
in eolian system lifecycle: 44–47
lateral distribution of: 81
mass wasting of sediment: 179
from meltwater: 26–29, 37–39
Escanaba Trough: 188
eskers, formation of: 26
estuaries: 151, 152
Eurasian Cimmeride system: 199
Eureka Valley: 105–125
Europe. *See also* specific countries
cyclothems in: 6–8, 18
eustasy in: 6–8
glaciation of: 8, 59

Europe. (*continued*)
glacioeustasy in: 18
loess deposits in
cycles in: 66, 68
eolian sand, relationship to: 64
location of: 55
outwash deposits in: 26
reef formation in: 211, 214–216
remnant ocean basins in: 203
Eurydesma transgression: 16
euxinix conditions: 245, 252–261
evaporite
deposits
in Africa: 48
in Permian: 246
in Wyoming: 248, 258
in iron formations: 238
Everglades: 142
Excello Shale: 138
exfoliation domes: 123

F

Fairfield Basin: 129, 146
Falkland Islands: 10, 14, 162
falsiovalis zone: 222
Famennian
black limestone/shale formation in: 221
brachiopods in strata from: 12
bryozoans in strata from: 12
clastic/carbonate successions during: 12
climate during: 223
cyclothem formation during: 18
glaciations during: 12, 209, 223
glacigenic deposits from: 12
glacioeustasy in: 18
mega-delta formation during: 214
mollusks in strata from: 12
mudstone deposition during: 12
palynomorphs in strata from: 12
quartzarenite deposition during: 12
reef formation during: 209
in Australia: 12
in Canada: 214
in China: 221, 222
in Gondwana: 219
in Kazakstan: 220
on Russian Platform: 216
in SEA: 218
in Urals: 217
shale deposition during: 223
Fanning River Group: 222
Farmdale soil, analysis of: 65
faults, alluvial fan size factor: 106
feldspar
in loess: 53, 54, 57
petrographic analysis of: 108
Fennoscandian Ice Sheet: 59
fens: 142
Fergana: 219
ferns: 134–135, 139, 144
Fire Clay coal: 128
fish
in Devonian strata: 12
in Junggar Basin: 95
in Phosphoria Rock Complex: 251
Fish Lake Valley: 108
flood
eolian sand deposition from: 45
turbidite deposition from: 188
Florida, dust from Africa in: 64
flutings
on abyssal plain: 29–30
in eroded rocks: 37
formation of: 26, 28
fluvial systems
modeling of: 86
peat, impact on: 145
flysch. *See* also turbidite processes
chronostratigraphy of: 194
deposits
in China: 191, 195, 199–203
in India: 196
modeling of: 203
in North America: 191, 197–199
in Siberia: 220
molasse, transition to: 204
paleocurrent indicators in: 198
in remnant ocean basins: 193, 198
fold-thrust belts: 193
Folsomville shale: 134
foraminifera: 251
forearc basins
glacigenic deposits in: 10–12
reef formation around: 212
foreland basins
coal bed formation in: 135
eolian sand activity in
in Africa: 47–49, 50
in North America: 50–51
glacigenic deposits in: 10–14
lithospheric loading in: 194
modeling of: 203
molasse deposition in: 196, 198
peripheral. *see* proforeland basins
phosphate deposition in: 247
river drainage in: 193
sediment transportation in: 191
subsidence of: 151
Fowey: 162
fractal geometry: 81–82
France
reef formation in: 214, 218
tides in: 151, 152, 159–165
Francis Creek Shale: 139
francolite: 247–257
Franson member: 259
Frasnian
brachiopods in strata from: 12
bryozoans in strata from: 12
clastic/carbonate successions during: 12
climate during: 223
cyclothem formation during: 18
dating of glacigenic deposits to: 12
glaciation during: 5, 6, 10
glacioeustasy in: 18
marine shale deposition during: 12
mega-delta formation during: 214
mollusks in strata from: 12
mudstone deposition during: 12
palynomorphs in strata from: 12
quartzarenite deposition during: 12
reef formation during: 12, 209–225
shale deposition during: 223
Freeport Coal, Lower: 136
Frobisher Bay: 159
frost shattering: 53, 60
Fundy Fault: 155
Furnace Creek fault: 105, 108–110, 125
fusain: 139
fusulinids: 8, 132

G

Galatia paleochannel
coal deposits: 133
depositional environment: 139–140
palynology of: 135
sulfur level, in coal from: 134
tidal facies in: 136
Gandak River: 106
Ganges River
delta system: 203
deltaic fan complex. *see* Bengal Fan
drainage from: 193
sediment transportation in: 196
Gaspé: 212, 213
gastropods, pleurotomarid: 251, 259
Gedinnian: 211
Georgia (Eurasia): 219
Geranium Harbor: 168
Germany
loess deposits in: 68
outwash deposits in: 26
reef formation in: 214–216
Ginkgophyllum Zone: 14
Girvanella: 224
Givetian: 209–225
glaciation
in Africa: 47
cyclothemic deposition from: 5–8, 18, 130
eolian sand accumulation, impact on: 45–47, 54
eustasy during
cycle periodicity: 7
vs. mass balance changes: 8–10, 17–18
peatland development from: 130, 143
in Gondwana: 5–18
Hadley Cells, impact on: 47
indications of, in isotope levels: 8
iron formation deposition during: 235
lake size factor: 78
loess, origin of: 53, 58, 66–67
in North America
outburst flood evidence in: 25–29, 35, 37–39
during Pleistocene: 25–29, 35, 37–39
subglacial deformation: 26–28
outburst flood evidence in: 28
paleoproductivity link: 256
during Paleozoic: 10–17
during Permian
in Gondwana: 5–11, 14–18
in North America: 246
sand sea activity during: 47, 48
sea level changes from
from outburst floods: 26, 28, 29
during Paleozoic: 5–10, 17–18
wind strength, impact on: 46
Glen Canyon Sandstone: 49

Gondwana
basin development in: 191, 195
climate in
during Carboniferous: 8
during Emsian: 211
Malvinokaffric: 211, 218
cyclothem formation in: 5–8, 18
glaciation of: 5–18
glacigenic deposits in: 7–8, 17–18
lacustrine deposits in: 102
location of, during Paleozoic: 11
reconstruction of, for Givetian: 218
reconstruction of, for Paleozoic: 11
reef formation in: 209, 211–212, 218–219
remnant ocean basin development: 199
rift zone in
Avalon-Meguma, separation from: 212
India, separation from: 196
SEA, relationship to: 217
suture zones in: 192, 195, 199
Gornoi Altai: 220
gradualism: 2
Granasporites: 134
Granasporites medius: 135
Grand Banks earthquake: 186
Grand Canyon: 178, 185
granite
in alluvial-fan deposits: 115, 120, 125
in Jurassic strata: 108
in Mesozoic strata: 108
in Paleozoic strata: 5, 17, 92
granular iron formations: 231–241
Granville: 159–162
gravel
in alluvial-fan deposits: 105, 110–123
in glaciogenic deposits: 28
Great Australian Bight: 224
Great Barrier Reef: 210
Great Lakes, glacial sedimentation in: 28
Great Slave Lake: 214
Green River Formation: 79, 80, 100–101
Greenbay Hollow: 65–66
Greenland
eolian sand in: 64
location of, during Devonian: 212
location of, during Permian: 259
mega-delta material from: 214
greenstone: 241
Grenville: 195
Guadalupian Sedimentation: 248
Gulf of Alaska: 165, 185
Gulf of Aqaba: 210
Gulf of Cambay: 151, 152, 155–159
Gulf of Mexico: 49, 197
Gulf of Mezan: 151, 152, 165
Gulf of St. Malo: 151, 152, 159–162
Gulf Stream: 210
Gunflint Iron Formation: 238
Gypsum Springs Formation: 51

H

Hadley Cells: 47, 50
hairpin scours: 28
Hall Point: 168
Halmahera arc: 197
Hamersley Basin: 231, 235–237, 241
Hami Basin
dating of deposition: 101
evolution of: 95–98
geological setting of: 92–93
lacustrine deposits in: 91, 92
organic carbon deposits in: 98, 101–102
petrology of: 98–102
stratigraphy of: 93–95
Hamlin Coal: 139
Hangzhou Estuary: 168
Hawaii
tidal ranges: 152
volcano collapse in: 186
Hawaiian Ridge: 186
heath: 141
Heilongjiang: 221
Heinrich events: 35, 37–39
Hell's Gate alluvial fan: 106
hematite: 235, 239
Henderson paleochannel: 133, 134, 140
Herrin Coal
composition of: 134
correlations to: 135–136
depositional environment: 127, 132, 139–140, 143–146
durains in: 140
eustatic controls on: 143–144, 145
mining of: 129
paleolocation of mire: 142
palynology of: 134, 145
size of: 127, 131, 140, 142
thickness of mire: 142
herrmani-cristatus zone: 222
Hillesheim syncline: 215
Himalayas
climate impact on erosion: 195
collision suture belts in: 196
convergent margin: 159
dating of deposits: 196
glacigenic deposits in: 12
sediment supplier: 188, 191–197
Hingang Mountains: 221
Hingangling Range: 221
Hoboshan Formation: 221
Hobräcker Formation: 216
Hohe Warte area: 218
Holocene
dust fluxes during: 61–64
reef formation in: 210, 224
sand sea activity during: 47–48
Hongshandaban Formation: 220
Hongyanchi Formation: 80, 95
Honsel Formation: 216
Hopes Advance Bay: 159
Horizon Channel: 175, 185
hornblende: 38
horseshoe vortices: 28
Houchin Creek Coal: 143
Houtman Abrolhos: 210
Hudson Bay
carbonate transport from: 38, 39
drainage from lowlands: 145
erosion in: 25, 37
peat mires near: 141–142, 144, 145
reef formation in: 212
sedimentation restrictions: 145
tides in: 159
Hudson Strait: 29, 38, 39, 159. *See also* Ungava Bay
Hueneme Fan: 186–188
Hume Formation: 214
hummocks: 29
Hungary: 79
Huon Gulf: 195, 197
Hurenboom tabulate: 221
hydrogen sulfide: 239
hydrology, subglacial: 26
hydrothermal vents
reef formation near: 219
solute source, for iron deposits: 234, 239–241
hyperpycnal flows
into Labrador Sea: 25, 35, 37–39
into Pacific Ocean: 26, 188

I

Iapetus Ocean: 203, 212
Iceland: 27–28, 37, 59
Idaho
ammonoids in deposits: 259
loess deposits in: 55
Navajo Sandstone in: 43, 49–51
sediment depocenter: 248
Iki Island: 210
Illinois. *See also* Illinois Basin
coal bed formation in: 128–146
cyclothem formation in: 130
glaciogenic deposits from: 58
loess deposits in: 55, 58, 68
paleosol studies in: 65–66
palynology of: 134–135
peatland development in: 128–146
Illinois Basin. *See also* Eastern Interior Basin
coal bed formation in: 128–146
cyclothem formation in: 7, 130, 143–144
Ouachita foreland, connection to: 140
paleolocation of: 142
peatland development in: 128–146
Inch'on: 168
India. *See also* Himalayas
glacigenic deposits in: 10, 12, 14
iron formations in: 234
proforeland basins in: 196
reef formation in: 218
rift zone in
separation from Africa: 159
separation from Gondwana: 196
separation from Madagascar: 159
river systems in
Brahmaputra River. *see* Brahmaputra River
Ganges River. *see* Ganges River
Indus River. *see* Indus River
rotation of: 197
tectonic processes in: 191–197
tides in: 151, 152, 155–159
uplifted collisional orogen in: 191–197, 201

Indian Ocean
  collision suture belts in: 193
  dust transport towards: 64
  magnetic anomalies in: 196
  subduction zone accretion in: 196
  submarine canyons in
    "Swatch": 185
    "Swatch of No Ground": 178, 180–181, 182
  submarine deltaic fans in
    Bengal Fan. *see* Bengal Fan
    Indus Fan. *see* Indus Fan
Indian plate: 159, 195–197
Indiana
  coal bed formation in: 128–146
  cyclothem formation in: 130
  loess deposits in: 55, 68
  palynology of: 134–135
  peatland development in: 128–146
Indoburman Range: 193, 196
Indo-Gangetic Plain: 196
Indonesia
  lacustrine deposits in: 79
  peat mires in: 131, 142, 144, 145
  tectonic processes in: 168
  tides in: 152, 168
Indus Fan
  dating of deposits: 196
  deformation of: 196
  geological setting of: 185
  levee formation in: 188
  sediment supply to
    from continental collision: 195
    from delta-front canyons: 175, 188
    evolution of: 191–193, 196–197
  size of: 196
Indus River
  delta system: 203
  deltaic fan complex. *see* Indus Fan
  drainage from: 193
  sediment transportation in: 196
inertinite: 134
inland seas
  development of: 152
  vs. lakes: 76
  modeling of: 151
Innuitian Platform: 209
Intertropical Convergence Zone: 144
intracratonic basins: 12
Inyo Mountains: 108, 123
Ionian Sea: 197
Iowa
  coal bed formation in: 136, 138
  glacigenic deposits in: 7
  loess deposits in: 55, 57, 68
  marine deposits in: 142
  reef formation in: 213
Iran
  reef formation in: 219
  remnant ocean basins in: 203
Irati Formation: 79, 102
Irish Sea: 155
iron formations
  aluminous minerals in: 234
  classification of: 235
  dating of deposits: 231, 235–237, 240
  definition of: 231
  deposition cycles: 235, 239
  depositional environment: 237–240
  facies in: 232
  mass of: 235–237
  microbial activity in: 234, 239
  mineralogical composition of: 232
  modeling of: 239
  organic activity in: 238
  origin of: 237–241
  petrographic analysis of: 237
  production of: 231
  sedimentary textures of: 232–234
  size of: 240
  structure of: 234–235
  studies of: 231–241
ironstones: 231
Irtysh-Zayshan area: 99
Isle of Man: 155
isotope levels
  in Carboniferous deposits: 8
  in Famennian strata: 209
  in Frasnian strata: 223
  glaciation indicator: 8
  in Paleozoic strata: 92
  Phosphoria Rock Complex: 249, 252–254, 256–257
  in reefs: 223
  temperature indicator: 8
Israel: 55, 64

J

Jackfork Formation: 198
Japan: 168
Java: 168, 204
Java Sea: 212
Jianghan Basin: 79
Jingjingzagao Formation: 80
jökulhlaups. *See* outburst floods
Joshua trees: 108
Junggar Basin
  dating of deposition: 101
  evolution of: 95–98
  geological setting of: 92–93
  lacustrine deposits in: 79, 91, 92
  mapping of, subsurface: 98
  organic carbon deposits in: 98, 101–102
  petrology of: 98–102
  reef formation in: 220
  stratigraphy of: 93–95
  total organic carbon content: 91
Jurassic
  Appalachian-Ouachita uplands in: 49–50
  basin development during: 203
  climate during, in North America: 50
  coal deposits: 203
  molasse deposition during: 203
  paleomagnetic studies: 200
  reconstruction for, North America: 50
  rifting during: 155, 159
  sand sea activity during: 49–51
  sandstone deposition during: 43
  strata, analysis of
    in China: 202
    in Cucomungo Canyon area: 108
  unconformities (J-0 thru J-2): 50, 51
  uplift during: 92–93

K

Kaapvaal Craton: 237
Kalahari Basin: 14, 16
Kalimantan: 224
Kama-Kinel Trough: 216
Kamchatka Peninsula: 168
Kamenka: 165
Kanin Peninsula: 165
Kansas
  coal bed formation in: 136, 138–139
  cyclothem formation in: 7
  glaciation of: 61
  loess deposits in: 55, 61
  palynology of: 136–138
  rhythmite deposits in: 162
Kansas Uplift: 136
Karakoram Mountains: 220
Karakorum-Tarim plate: 220
Karatau Range: 216, 220
Karoo Basin
  coal bed formation in: 128
  dating of deposits: 16
  glacigenic deposits in: 14
  lacustrine deposits in: 79, 102
Karpinskia: 218
Kasimovian: 16
Kayenta Formation: 49, 50
Kazakstan
  lacustrine deposits in: 99
  location of, in Lochkovian: 220
  plate tectonics: 220
  reef formation in: 211, 216, 220
  SEA, relationship to: 217
Kazanian: 6, 8
Kee Scarp Formation: 214
Keewatin sector: 26
Keg River: 214
Kellerwand area: 218
Kellwasser Limestone: 216, 223
Kentucky: 128–146
Kenzhebasai River: 220
kerogen: 98, 99–100
Kess-Kess mudmounds: 219, 224
Keyser Limestone: 212
Khabarovsk region: 220
Khambhat: 159
Khodzhakurgan Formation: 219
Kid Islet: 168
kimberlite pipes: 212
Kirgizstan: 219
Kittanning Coal
  correlations to: 139
  formation of: 136, 138
  lithotype changes in: 142
  paleosols beneath: 144
Kiziltash mudmound: 219
Koel Basin: 14
Koksoak River: 159

Kola Peninsula: 165
Kolyma plate: 209, 216, 220
Kolyma River: 221
komatiites: 241
Koneprusy Limestone: 218
Korea: 151, 152, 168–172
Kosi River: 106
Kotelny Island: 216, 221
Kula plate: 187
Kungurian: 6, 8, 18
Kunlun Mountains: 200, 202, 220
Kuril Trench: 168
Kursk Magnetic Anomaly: 231
Kuskokwim River: 181–182, 187
Kuvelousia leptosa: 259
Kuznetsk Basin: 220
Kwadjalen Atoll, tidal ranges: 152
Kwataboahegan Formation: 212

L

La Jolla Canyon: 175, 178, 186–188
Labardie Limestone: 136
Labrador Sea
  abyssal plain: 29–30, 35, 37–39
  cores from: 30–38
  dating of deposits: 35, 37–38
  description of: 29–30
  formation of: 159
  Heinrich layers in sediment: 35, 37, 38
  hyperpycnal flows into: 25, 35, 37–39
  location of: 29
  outburst floods into: 25, 35, 37–39
Labrador Trough: 237
Lachlan fold belt: 203
lacustrine systems. *See also* lakes
  controls on: 2, 75, 76, 81–88
  facies associations: 84–85
  iron formation deposition in: 238
  meromictic conditions in: 87
  modeling of: 81–87
  petroleum source rock in: 92
  rates of change in level: 76
  size factors: 75–88
Ladakh area: 196
Laevigatosporites: 134
Lagoa Feia Formation: 79
lagoons
  dune formation around: 48
  reef formation in: 214, 219
Lahn syncline: 216
Lake Agassiz: 28
Lake Baikal: 76
Lake Chad: 49, 87
Lake Eyre Basin: 64
Lake Malawi: 76
Lake Missoula: 188
Lake Superior region: 232, 237
Lake Tanganyika: 76
Lake Victoria: 76
lakes. *See also* lacustrine systems
  altitude of: 78
  area covered by: 76–77, 81
  bathymetry of: 81
  character of: 76, 88
  climate impact on: 77, 87
  controls on: 2, 75, 76, 81–88
  depth of: 76–77, 81
  eutrophication in: 86–87
  formation factors: 82–84, 152
  largest modern: 75
  latitude of: 77–78
  modeling of: 81–87
  modes of level response: 84–85
  origin of: 78–79
  potential accommodation rates: 76, 84, 87
  size factors: 75–88
  volume of: 76–77, 81, 84
landslide, sediment deposition in: 186
Laney Member: 101
Laos: 222
Last Chance Range
  alluvial-fan deposits in: 120, 122, 123
  geological setting of: 108
latitude: 77–78
Laurasia
  basin development in: 191, 195
  breakup of: 155
  suture zones in: 192, 195
Laurentia. *See also* North American plate
  Africa, collision with: 211
  Avalon-Meguma, collision with: 212
  Baltica, collision with: 212, 218
  climate of: 212
  coastal deltas in: 214
  location of, during Devonian: 212
  reef formation in: 209, 211–216, 219
  remnant ocean basin development: 199
  suture zones in: 199
Laurentian Fan: 175, 186, 187
Laurentide Ice Sheet
  drainage into Hudson Strait: 29
  glaciogenic deposits from
    in Canada: 26–27, 55
    in Illinois: 58
    in Montana: 28
    in USA: 28, 55
  gravel deposits surrounding: 25, 28
  loess, from outwash of: 58, 59, 66
  mapping of Keewatin sector: 26
  outburst floods from: 25, 35, 37–39
Laurussia: 214, 217–218
Le Havre: 165
Leaf Lake: 159
Learmouth: 168
Leichardt coal seam: 128
Leiorhynchoidea weeksi: 251
Leonardian deposits: 248
Lepidodendron hickii: 135
Lepidodendron mannabachense: 135
Lepidodendropsis Zone: 14
Lepidopholoios: 134, 139
Lepidopholoios hallii: 139
Leslie Cemetery paleochannel
  coal deposits: 133
  depositional environment: 139–140
  palynology of: 135
  sulfur level, in coal from: 134
Levipustula Zone: 14
Lexington Coal: 136
Libya
  fauna in, during Emsian: 212
  location of, in Devonian: 218
  loess deposits in: 55, 64
  mudmounds in: 212
  Nubian Sandstone in: 64
  reef formation in: 212, 218
  shale deposits in: 64
  siltstone deposits in: 64
Ligerian-Vosgian-Moldanubian cordillera: 214
Liguride Complex: 203
limestone
  in China: 201
  in cyclothem: 130
  in Eastern Interior Basin: 130, 132
  fauna in: 14
  in Gondwana: 211
  in Junggar Basin: 94–95
  magnetite in: 238
  in Tibet: 14
linguiformis zone: 223
Lissochonetes ostiolatus: 251
lithistid mounds: 209, 223
Little Osage Shale: 138, 144
Liverpool Cyclothem: 138
Livingstone Lake: 26, 28–29
Llanoria: 197, 198
Lochkovian
  glaciation during: 223
  reef formation during: 223, 225
    in Australia: 222
    constraints on: 211
    in Gaspé: 212
    in Kazakstan: 220
    in Laurentia: 212
    in Mongolia: 221
    in Russia: 221
    in SEA: 218
    in Siberia: 220
loess. *See also* silt
  characteristics of: 54
  composition of: 53
  dating of deposits: 59, 69
  definition of: 53, 54–55
  deposition of: 54, 68
  distribution of: 55
  eolian sand, relationship to: 53, 56, 64–65
  formation of: 54
  geochemical analysis of: 57–58, 65–66
  magnetic susceptibility testing: 68–69
  origin of: 53, 55–64, 70
  particle sizes in: 55–57, 67, 68
  vs. soil formation: 66–67
  sorting within deposits: 56
  studies of: 53–72
  transportation of: 54, 65
Loess Plateau: 55, 64
Lone Pine alluvial fan: 123
Longmen Shan fold-thrust belt: 199
Loogh-Cürten Beds: 215
Lord Howe Island: 210
Lower Freeport Coal: 136
Lucagao Formation: 80
Lucaogou Formation
  dating of deposition: 101

Lucaogou Formation (*continued*)
kerogen in: 98
lacustrine deposits in: 98, 99
organic carbon deposits in: 98, 101–102
petrology of: 98–102
stratigraphy of: 95
Luzon arc: 195, 197
lycopsids: 127, 134–135, 139–140
Lycospora: 134–135, 140

M

Mackenzie Delta: 214
Madagascar: 14, 159
maghemite: 68
magnetic susceptibility testing: 31, 68–69
magnetite
biogenic origin of: 238
in loess paleosols: 68
proxy for: 69
replacement in crystals: 233
Mahanadi Basin: 14
Mahi River: 159
Maider Basin: 219
Makran Range: 196–197
Malanzán sub-basin: 16
Malvinokaffric climates: 211, 218
Marathon-Ouachita system
dating of deposits: 198
depositional phases in: 198
evolution of: 197–199
flysch deposits in: 191, 197–198
orogenic belt in: 197
remnant ocean basins in: 192, 195
volcanic activity near: 198
during Wolfcampian: 198
Mariposa Formation: 203
Markham delta: 203
Marnes Noires Formation: 100
Mars: 29
marsh: 142
Mason Porcus Formation: 218
mass movement: 179, 186–187
mass wasting of sediment: 179
Massenkalk: 216
Mauritania: 43, 47–49, 218
Mauritanian Basin: 48
Mauritanide: 192
Mazon Creek fossil bed: 139
Meade Peak
chemostratigraphy of: 252–256
dating of deposits: 248
fossils in: 258–260
lithofacies: 249–251
studies of: 247–261
testing of samples from: 249
Mecca Quarry Shale: 133, 138, 144
Mediterranean Sea: 102, 197, 219
medullosans: 139
Meguma-Avalon terrane: 212
meltwater
erosion by: 26
flooding by
discharge rate: 28–29
duration of: 26
evidence of: 26–28, 35
glacioeustasy from: 28, 29
into Labrador Sea: 35, 37–39
magnitude of: 28–29
sediment removal by: 28–30, 35, 37–39
meromictic conditions
in the Caspian Sea: 78, 80
lacustrine systems with: 87
lake formation under: 81
Meseta reefs: 218–219
Mesogondolella gracilis: 259
Mesogondolella idahoensis: 259
Mesogondolella phosphoriensis: 259
Mesogondolella rosenkrantzi: 259
Mesogondolella serrata: 259
Mesozoic
basin development during
in Bristol Channel: 155
Songpan-Ganzi Complex: 200, 203
magmatic arc formation during: 201
rifting during
in English Channel: 162
in Gulf of Mexico: 197
siltstone deposition during: 64
strata, analysis of
in China: 99
in Eureka Valley area: 108
Mexico
biota in deposits: 259
location of, in Permian: 259
loess deposits in: 55, 61
Ouachita orogenic belt in: 197
Mezan River: 165
mica: 54
Michigan: 213
Milankovitch cycles
cyclothemic transgression/regression link: 8, 130
eustatic cycle periodicity: 7
iron formation deposition: 235
in Sahara Desert: 47
Minas Basin: 155
Mindanao-Talaud collision zone: 197
Minnesota: 91
Minusinsk Basin: 220
Miocene
climate during
in Africa: 48
in Asia: 195
continental strata from: 47
francolite standard in: 257
lacustrine deposits from: 79
sand sea activity during: 50
sediment deposition in: 48
submarine fan deposition during: 196
miospores: 12, 134–135, 139
mires
classification of: 131
coal from: 127
interpretation of deposits: 139
modern: 141–142
paludification in: 145
peat formation factors: 143–146
plant assemblages in
during Desmonesian: 139
during Pennsylvanian: 144
Mississippi Fan: 188
Mississippi River
glaciation of headwater source: 66
sand transportation by: 49
Mississippi River Arch: 136, 140
Mississippi River Embayment: 136
Mississippi River Valley
fan deposits: 28
loess deposit source: 65
outwash deposits in: 26
Mississippian
continental reconstruction for: 11
Llanoria sediment deposition during: 198
Missouri
coal bed formation in: 136–138
glacigenic deposits in: 7
loess deposits in: 55
reef formation in: 213
Missouri River: 57
Moenave Formation: 50
Mojave Desert: 61–64
molasse
dating of deposits: 203
deposits
fluvio-deltaic: 196
modeling of: 203
nonmarine coal-bearing: 198, 203
volcaniclastic: 196
flysch, transition from: 204
in foreland basins: 198
Moldanubian-Ligerian-Vosgian cordillera: 214
mollusks
for dating of deposits: 12
in glacigenic strata: 12
in Junggar Basin: 95
in Phosphoria Rock Complex: 251
Molucca Sea: 195, 197
Mongolia
carbonate platforms in: 210
lacustrine deposits in: 99
location of, in Devonian: 221
loess deposits in: 60
reef formation in: 209, 220, 221
Mongolia-Tuva plate: 221, 225
Montagne Noire: 214, 217, 218, 224
Montana
biota in deposits: 259
glacigenic deposits in: 28
Monterey Canyon: 178, 185
Monterey Fan: 185
Mont-Saint-Michel Abbey: 162
Moore Creek Formation: 222
Moorman Syncline: 129
morainal deposits
in Australia: 12
in North America: 26, 28
from outburst floods: 26
in Scandinavia: 26
in South America: 12
Moravia: 217–219
Morlaix River: 162
Morocco
clastic/carbonate successions in: 12
fauna in: 212
location of, in Devonian: 218
reef formation in: 212, 218–219, 224
Morrowan: 129, 145–146

Morzhovetz Island: 165
Moscow Basin: 7
Mu Us Desert: 64
mudflows. *See also* debris flow
  in Cucamonga Canyon area: 105, 110–123
  in Death Valley: 123
mudmounds: 210, 212–224
mudstone. *See also* oil shale
  in China: 91
  in Eastern Interior Basin: 132
  in glacigenic deposits: 12
  in Junggar Basin: 94–95
  in lacustrine deposits: 91–92
  in Phosphoria Rock Complex: 248
  in shallow sea basin: 249–251
  in South Africa: 12
Mukluk Channel: 185
Muree Supergroup: 196
Myrdalsjökull: 59
Myrick Station Limestone: 136
Mystic Coal: 136, 140

N

N. pachyderma: 35
Nabberu Basin: 237
Namibia: 55
Namurian
  brachiopods in strata from: 14
  bryozoans in strata from: 14
  clams in strata from: 14
  crinoids in strata from: 14
  cyclothem formation during: 18
  glaciation during
    in Australia: 14
    evidence of, isotope levels: 8
    in Gondwana: 5, 6, 10
    in South America: 12, 14
  glacioeustasy in: 18
  Levipustula Zone during: 14
  sea level changes during: 7
Nandan Basin: 222
Narmada River: 159
Natal Coal Field: 128
National Oceanic and Atmospheric Administration (NOAA): 152
nautiloids: 219
Navajo Sandstone: 43, 49–51
Navarin Canyon: 178, 181–182, 188
Navy Fan: 186
Nebraska
  eolian sand in: 64
  glaciation of, in Pleistocene: 61
  glacigenic deposits in: 7
  loess deposits in: 55, 61, 64
Nehou: 214
Nemaha Uplift: 135–136, 138, 144
Neocomian: 92
neodymium: 239
Neogene: 48, 108, 162
Neoproterozoic
  glaciation during: 240
  iron deposition during: 235, 240
  rifting during: 197
Neospirifer striato-paradoxus: 259
Neostreptognathodus sulcoplicatus: 259
Neotethys Basin: 196
Nerninski Point: 165
Nevada
  alluvial-fan deposits in: 123
  cyclothem formation in: 7
  eolian sediment deposits in: 49
  reef formation in: 214
Nevadan orogeny: 203
New Caledonia: 210
New England Fold Belt (Australia): 12–14
New England (USA): 28, 203
New Guinea: 210
New Jersey: 28
New Mexico: 7
New South Wales: 222
New York: 212, 213
New Zealand: 54, 55, 57, 64
nickel: 134
Nicobar Fan: 196. *See also* Bengal Fan
Niger: 79
Nigeria: 55
Nikiski: 165
Nile River: 197
Ningpo: 168
Normal Wells area: 214
North America. *See also* specific locations in
  climate in
    during Desmonesian: 127, 144
    during Jurassic: 50
    during Triassic: 50
  coal bed formation in: 127–146
  cyclothem formation in
    climate controls on: 144
    during Desmonesian: 127, 130, 143–144
    eustatic controls on: 143–144
    during Paleozoic: 6–8, 18
  eolian mantles in: 61
  eolian sand deposition in: 49–51, 53
  eustasy in: 6–8
  glaciation of
    during Carboniferous: 8
    during Pleistocene: 25–29, 35, 37–39
  glaciers: 59. *see also* Laurentide Ice Sheet
  glacioeustasy in
    during Desmonesian: 130, 143–144
    during Paleozoic: 18
    during Pleistocene: 28–29
  glaciogenic deposits in: 26–27, 28, 55, 58
  iron formations in: 237
  location of
    during Desmonesian: 127, 142
    during Devonian: 212
    during Jurassic: 50
    during Permian: 247
    during Triassic: 50
  loess deposits in
    cycles in: 66
    eolian sand, relationship to: 53, 64
    geochemical analysis of: 54, 58
    location of: 55
    particle sizes in: 55–57
    sources of: 55, 57, 61–64
  morainal deposits in: 26, 28
  outburst flood evidence in: 25–29, 35, 37–39
  peatland development in: 127–146
  phosphate deposits in: 245–261
  reconstruction of
    during Jurassic: 50
    during Paleozoic: 50
  reef formation in: 211–214
North American plate: 165, 197. *See also* Laurentia
North Atlantic
  Heinrich layers in sediment: 35, 37, 38
  submarine canyons in: 175
North Atlantic Mid-Ocean Channel
  cores from: 30–38
  Coriolis effect on: 29
  dating of deposits: 35, 37–38
  description of: 29
  hyperpycnal flows into: 25, 35, 37–39
  levee sediments: 35–39
  morphology of: 25
North Caspian Depression: 197
North China Block: 199, 221
North Sea: 159–162
Northern Patagonian Massif: 12
Northwest Territories
  outburst flood evidence in: 26–27
  reef formation in: 212
Nothorhacopteris Zone: 14
Nova Scotia: 155
Novaya Zemlya Island: 216
Nubian Sandstone: 64
Nubrigyn reef blocks: 222
Nugget Sandstone: 49
Nuratau Range: 219

O

Oak Grove Limestone: 144
Oakley Shale: 138
Ob River: 142, 145
oblique-slip deformation: 12
Ocean Drilling Program
  Amazon Fan studies: 182, 185
  Escanaba Trough studies: 188
oceanic margin basins: 47
oceans
  chemistry of, changes in: 240
  environmental controls on: 2
  sea surface temperatures: 209, 222
  stratification of water columns: 239
  submarine canyons in: 175–188
  tidal ranges: 152
Ogilvie Formation: 212
Ohio
  coal bed formation in: 134, 136, 138, 139
  reef formation in: 213
Oil Bay: 165
oil shale
  in China: 91–92
  in Green River Formation: 100
  in Lucaogou Formation: 98
Okefenokee: 142
Okhotsk region: 168, 220, 221
Oklahoma
  coal bed formation in: 136–139
  depositional environment: 139

Oklahoma (*continued*)
flysch deposits in: 191
marine deposits in: 142
Ouachita orogenic belt in: 197
sulfur, in coal deposits: 138
Old Red Continent. *See* Laurentia
Oligocene: 79
olistostromes: 201, 202
ombrogenous mires
climate for: 144
definition of: 131
plant production in: 142
Omolon River: 221
Onondaga Limestone: 212
Ontario (Canada): 212
Ontario Detroit River Limestone: 212
ooids: 233, 237, 247
Oologah Limestone: 136
ophiolite belt: 220, 221
Orbiculoidea: 132
orbiculoids: 144, 251
Ordovician
glaciation during: 223
reef formation during: 221
strata, analysis of: 92
tectonic processes during: 212
Oregon, loess deposits in: 55
Orgol Formation: 221
Orissa: 234
Ouachita Mountains
during Carboniferous/Permian: 192, 197–198
collision suture belts in: 193
during Desmonesian: 136
evolution of: 197–199
flysch deposits in: 191, 198
Illinois Basin, connection to: 140
during Jurassic: 49–50
lithology of: 197–199
origin of: 197
during Pennsylvanian: 198
remnant ocean basins in: 197–199
volcanic activity near: 198
outburst floods
duration of: 26, 188
evidence of: 26–28, 35
glacioeustasy from: 26, 28, 29
into Labrador Sea: 35, 37–39
magnitude of: 28–29
modeling of: 29
into Pacific Ocean: 26, 188
Scabland formation by: 26
sediment removal by: 28–30, 35
turbidite deposition from: 188
Owens Valley: 123
oxide: 232
Ozark Uplift: 135–136

P

Pacific Basin: 202
Pacific Ocean
dust transport towards: 64
hyperpycnal flows into: 26
remnant ocean basins in: 197
submarine canyons in: 175
Ascension Canyon: 185
Astoria Canyon: 188
Bering Canyon: 175, 181–182
Horizon Channel: 185
La Jolla Canyon: 186
Monterey Canyon: 178, 185
Navarin Canyon: 178, 181–182
Zhemchung Canyon: 175, 178–182
tides in: 151, 152, 162, 165
Pacific plate
convergent boundary: 168
subduction zone
with North American plate: 165, 187
with South American plate: 162
in Yellow Sea region: 168
Pagonzo Basin: 12, 16
Pai-Khoi peninsula: 216
Paimpol: 162
Pakistan
alluvial fan size studies: 106
Devonian strata in: 219
Himalayan detritus deposits in: 196
reef formation in: 209, 219, 220
Palaeoniscoidea: 251
Paleocene
coal bed formation during: 128
lacustrine deposits from: 79
reef formation during: 210
Paleogene: 203
Paleoproterozoic
glaciation during: 240
iron deposition during: 231, 235, 240
paleosols
characteristics of: 53
coal bed formation over: 139
formation of: 65–66
magnetic susceptibility testing: 68–69
Paleotethys: 259
Paleozoic. *See also* Carboniferous; Devonian; Permian
basin development during
in Bristol Channel: 155
Lachlan fold belt: 203
from Laurasia breakup: 155
Songpan-Ganzi Complex: 200
continental reconstruction for: 11, 50
deposits in Russia: 165
glaciation during: 5–18
reef formation during: 209, 210
sand sea activity during: 49
siltstone deposition during: 64
strata, analysis of
in China: 92
in Eureka Valley area: 108, 120
tectonic processes: 195
Palouse loess: 55
paludification: 145
palynomorphs: 12, 16
Pamir Mountain Range: 219, 220
Pampas loess: 55, 64
Panama, Isthmus of: 224
Pangea
coastal water mass: 260
cyclothem formation in: 6–8, 18
eustasy in: 6–8
glaciation of: 6–8
glacioeustasy in: 18
lacustrine deposits in: 91
reconstruction of, for Paleozoic: 11
reef formation in: 223
remnant ocean basin development: 199
Pannonian Basin: 79, 102
Panthalassa Ocean: 247
Papua New Guinea
collisional orogenic events: 193
reef formation in: 222
remnant ocean basins in: 195, 197
tectonic processes in: 197
Parachute Creek Member: 101
Paradise Coal: 132
Paraguay: 55
Paralycopodites: 134
Paralycopodites brevifolius: 139
Paraná Basin
glacigenic deposits in: 14
lacustrine deposits in: 102
marine shale deposition in: 12
Park City Formation: 259
parna: 55, 64
Parnaíba Basin: 11
peat. *See also* mires
to coal compaction ratio: 127, 130–131
coalification of: 130–131
depositional environment: 143–146
formation factors: 127, 143–146
modern: 141–142
plant assemblages in: 144
plant production in: 142
Pechora reef belt: 216
peloids: 233, 247
Pennsylvania
coal bed formation in: 138, 139
reef formation in: 212
Pennsylvanian. *See also* Carboniferous
coal bed formation during: 128, 129, 144
continental reconstruction for: 11
cyclothem formation in: 144
deposit deformation during: 198
flysch deposits during: 198
glaciation during: 246
sediment deposition in: 136
tectonic processes during: 145–146
thrusting during: 197
tree ferns in: 139
Peoria Loess: 57, 65, 68
perbonus zone: 222
Peri-Gondwanian block: 12
Permian
basin development during
Appalachian-Ouachita system: 192
in China: 91
Karoo Basin: 14, 16
Laurasian-Gondwanan system: 195
Songpan-Ganzi Complex: 201
carbon burial during: 91
coal bed formation during: 128
continental reconstruction for: 11
cyclothem formation during: 5–8, 18
fauna, in strata: 201
flysch deposits from: 191

glaciation during
in Gondwana: 5–11, 14–18
in North America: 246
glacioeustasy in: 18
lacustrine deposits from
China: 75
Hami Basin: 91–103
Hongyanchi Formation: 80
Irati Formation: 79
Jianghan Basin: 79
Jingjingzagao Formation: 80
Junggar Basin. *see* Junggar Basin
Karoo Basin: 79
Lucagao Formation: 80
Turpan Basin: 91–103
paleomagnetic studies: 200
palynomorphs in strata from: 16
peatland development in: 128
phosphate deposition during: 245–261
sea level changes during: 7–8
subsidence rates: 98
Persian Gulf: 196
Peru
glacigenic deposits in: 11–12
modeling of upwelling of coastal waters: 246
Peterborough, Ontario: 29, 37
petrology: 98
p-form erosional marks, formation of: 28
Phanerozoic, definition of: 210
Philippine plate: 168
phosphate: 246–249
phosphogenesis: 247, 257–258
Phosphoria Formation: 247
Phosphoria Rock Complex
biota in: 258–260
chemostratigraphy of: 252–256
dating of deposits: 248
depositional environment: 246–247, 259
glacioeustatic changes to: 256
isotope levels in: 252–254, 256–257
lithofacies: 249–251
mass of deposit in: 246
modeling of: 246, 251, 260–261
organic carbon deposits in: 246
phosphogenesis in: 247, 257–258
stratigraphic sections in: 247–248
studies of: 245–261
testing of samples from: 249
Phosphoria Sea
chemofacies analysis: 252–256
depth of: 246
drainage into: 258
formation of: 247
modeling of: 260–261
paleoproductivity in: 251
stratification of: 247, 258
studies of: 245–261
topography of: 248
upwelling in: 246
phosphorites: 223, 245–248, 254
photozoans: 223
phytoplankton: 255, 256
piggy-back basins: 196
Pillara Formation: 222
Pine Island: 28
pine trees: 108, 141
Piper Mountain Wilderness: 108
Pittsburgh: 128
plagioclase: 65, 108
planar mires. *See* topogenous mire
plankton: 223
platy tabulates: 215
playas
in alluvial-fan system: 108
hydrology of: 86
Pleistocene
Ascension Canyon fan deposits: 185
dust fluxes during: 61–64
glaciation during
of Africa: 47, 48
marine regression from: 48
of North America: 25–29, 35, 37–39
lacustrine deposits from: 80
loess deposits during: 61–64
mudstone deposition during: 64
Panama, closure of Isthmus of: 224
sand sea activity during: 47, 48
sea level changes during: 8
siltstone deposition during: 64
soil formation in: 65
strata, analysis of: 108
submarine canyon fan deposits during: 184
Pliensbachian: 49–51
Pliocene
continental strata from: 47
glaciation during: 47, 48
mudstone deposition during: 64
Panama, closure of Isthmus of: 224
sand sea activity during: 48, 50
siltstone deposition during: 64
strata, analysis of: 108
podzolization: 145
Poland: 209, 214, 216
Pontabault: 162
Porbandar: 159
Port Chatham: 165
Port Hedland: 168
Powder River Basin: 128
Pragian
glaciation during: 223
reef formation during: 223, 225
in Armorica: 214
in China: 222
constraints on: 211
in Gaspé: 212
in Russia: 221
in Sardinia: 218
in SEA: 218
in Siberia: 220
Prague Basin: 217, 218
Precambrian
Baltic Shield boundaries: 165
basement strata from: 47
hydrothermal vent activity during: 239
iron formations during: 231
rifting during: 136
seawater, composition of: 239–240
strata, analysis of: 92
Precaspian syneclise: 216
precipitation
changes to
from collisional orogens: 195
for erosion: 44, 195
by glaciation: 67
lake size factor: 77
record of, in paleosols: 65–66
reef growth factor: 224
soil formation factor: 66
precipitation-evaporation ratio
in dry climates: 75
lake size factor: 77
preservation model: 223
preservation space: 47
Pribilof Ridge: 179
Pridolian: 212, 223, 225
Prince of Wales Strait: 212
Prince Patrick Island: 213
Princess Coal: 138, 139
Princess Royal Island: 212
Pripyat-Donets Graben: 216
pristina zone: 222
production model: 223
proforeland basins
evolution of: 193, 198, 204
fold-thrust belts from: 199
formation of: 192, 196, 203
Prongs Creek Formation: 212
Proterozoic: 195
Protocanites Zone: 14
Prüm syncline: 215
Psaronius: 139
Pseudoreticulatispora confluens palynomorph Zone: 16
pteridophyte land plants: 214
pteridosperms: 139
pull-apart basins: 10–12
Punctatosporites: 134
Punta Arenas: 162
Pyrenees Mountains: 209, 218, 219
pyrite: 256
pyroxene: 38

Q

Qaidam: 199
Qilian: 202
Qinling Mountains: 199–203
Qinshankou: 79, 81
Quadrilátero Ferrifero: 234
Quantou Formations: 81
quartz
in glacigenic deposits: 14
in iron formations: 232–234
in Labrador Sea cores: 35–38
in loess: 53, 54, 57
quartzarenite
in Antarctica: 12
in iron formations: 237, 238
in Navajo Sandstone: 49
in South Africa: 12
quartzite: 108, 120
Quaternary
climate changes during: 54–72
dune preservation during: 49
lacustrine sediment preservation: 49
sandstone deposition during: 123
Quebec: 212, 213
Queensland: 128, 222

R

radiolarites: 218, 223
radiometric dating: 10–16
rain forests: 144
raised mires. *See* ombrogenous mires
Rajapur River: 159
Rance Estuary: 162
redbeds: 12, 248
reef tracts: 211
reefs
  accommodation space for: 223–224
  aragonite supersaturation of: 210
  bleaching events: 210–211
  carbon dioxide response by: 210
  during Carboniferous: 213, 221, 224
  definition of: 210
  during Devonian: 209–225
  disparilis zone: 222
  ensensis zone: 222
  falsiovalis zone: 222
  formation of
    in Australia: 12, 222
    in China: 220–222
    in Gondwana: 218–219
    in Kazakstan: 211, 216, 220
    in Laurentia: 211–216
    in Laurussia: 216–218
    in Papua New Guinea: 222
    in Russia: 220
    on Russian Platform: 216
    in Siberia: 220
    in Tian Shan Fold Belt: 219
    in Urals: 216–217
  fossil database: 210
  herrmani-cristatus zone: 222
  hypoxia in: 214
  isotope levels in: 223
  linguiformis zone: 223
  locations of
    in Devonian: 12, 211–225
    in Holocene: 210, 224
  modeling of: 223
  perbonus zone: 222
  plankton productivity in: 214
  preservation of: 210
  pristina zone: 222
  rhenana zone: 223
  sea level: 222–223
  studies of: 209–225
  tectonic controls on: 216, 224
  temperature response by: 210
  varcus zone: 216, 222
Reelfoot Rift: 136, 140, 143, 146
remnant ocean basins
  evolution of: 191–204
  lithosphere beneath: 194
  modeling of: 196, 203
  subsidence of: 194
reptiles: 246
Resolution Island: 159
Retort member: 247
Rex Chert member: 258, 259
Rheic Ocean: 214, 218
rhenana zone: 223
Rhenish facies: 214, 218
Rhenohercynian fold belt: 216, 217
Rhine River Valley: 26, 68, 215
rhythmite: 162
rift basins
  eolian sand preservation in: 47
  formation of
    Atlantic Ocean Basin: 155
    Bristol Channel Basin: 155
    Cambay, Gulf of: 159
  glacigenic deposits in: 10–14
  lakes in
    formation of: 79
    size factors: 80–81
  tides in: 172
Rio Gallegos: 162
Roadian deposits: 248
Roaring River: 123
Rock Eval techniques: 99
Rocky Mountains (Canada): 59
Rocky Mountains (USA): 61
Rogen moraines: 26
Rolling Downs Group: 64
Rothpletzella: 224
Rough Creek Graben: 146
Rubielos de Mora: 79
rugosans: 215, 219, 223
Rundle Formation: 79
Russia. *See also* specific locations in
  carbonate platforms in: 210
  glaciation of: 165
  Gulf of Mezan tides: 151, 152, 165
  loess deposits in: 56, 58, 69
  reef formation in: 211, 220
Russian Platform
  cyclothem formation in: 7
  reef formation in: 209, 214, 216

S

Sabarmati River: 159
sabkha: 48, 238
sag basin: 81
Sahara Desert
  eolian sand in: 43, 47–49, 50
  fauna in, during Emsian: 212
  location of, in Devonian: 218, 219
  loess deposits in: 64
  Milankovitch cycles in: 47
  mudmounds in: 212
  origin of: 47
  reef formation in: 212, 218, 219
  shale deposits in: 64
  silt production in: 64
  siltstone deposits in: 64
Saharawi People's Republic: 218
Sahel region: 64
Sakmarian
  cyclothem formation during: 18
  Eurydesma transgression during: 16
  glaciation during: 5–10, 14–16
  glacioeustasy in: 18
  lacustrine deposits from: 91–103
  sea level changes during: 7
Salair Basins: 220, 221
Salcombe: 162
Saline Range: 108
salt weathering: 53, 60, 65
San Diego Trough: 186
San Rafael Basin: 12
sand
  in alluvial-fan deposits: 110–124
  deposition of
    in Africa: 48
    in Cucamonga Canyon: 105
    by hyperpycnal flows: 26
    in Labrador Sea: 38
  in eolian systems: 43–51
  grain-size analysis: 2
  in granular iron formations: 234
  quartzolithic: 193
  in submarine fans: 188
  volume of: 49
sandstone
  arkosic: 202
  in Carboniferous strata
    in Ouachita Mountains: 198–199
    in South America: 12
    in Tibet: 14
  in cyclothem: 130
  geochemistry of: 57
  in glacigenic deposits: 14
  in Junggar Basin: 94–95
  in Jurassic strata: 43
  in Phosphoria Rock Complex: 248
  in remnant ocean basins: 193, 202
  in shallow sea basin: 248, 249–251
  volcanic: 202–203
Sangamon soil: 65–66
Sangihe arc: 197
Santa Cruz: 162
Santa Monica Basin: 186
Santamu Basin: 99
sapropel: 102
Sardinia: 203, 217, 218, 224
Savika Bay: 165
Saxo-Thüringian Basin: 214, 218
Scabland, formation of: 26
scale invariance: 81–82
Scandinavia: 26
Schwellen: 216
Scotian Arc: 162
Scotian Shelf: 28
Scotland: 203
scours, formation of: 28
sea level
  cycle periodicity: 7
  deposition factor: 2, 3
  drawdown of: 209
  erosion, impact on: 44–45
  glacioeustatic changes to
    eolian sand accumulation, impact on: 46
    from Gondwana glaciation: 5–10, 17–18
    from outburst floods: 26, 28, 29
    in submarine canyon formation: 179
  during Jurassic: 50
  outburst flood impact on: 26, 28, 29
  during Paleozoic: 7–8, 17–18
  for reef formation, maximum: 222–223
Sea of Japan: 210
Sea of Okhotsk: 151, 152, 168

seaway. *See* inland seas
Secor Coal: 139
sedges: 141
sediment
  from basement uplift: 193–194
  conservation, equation for: 45
  deposition of
    from climate changes: 87
    during continental collision: 192
  for eolian systems: 44–47
  from magmatic arcs: 193–194
  from recycled-orogenic sources: 193–194
Selenyakh River: 221
Selsey Bill: 165
Semerang: 168
Semzha River: 165
Severn Estuary: 155
Severnaya Zemlya: 216
s-form erosional marks: 26, 28
Shahejie Formation: 79
shale
  in Carboniferous strata
    in North America: 7
    in South America: 12
    in Tibet: 14
  in cyclothems: 130, 143
  depositional environment: 145
  in Devonian strata: 12, 223
  fauna in: 14
  in Frasnian strata: 216
  geochemistry of: 57
  in glacigenic deposits: 14
  marine assemblages in: 132–134
  in Mesozoic strata: 64
  in Paleozoic strata: 64, 108
  sulfur level factor: 134
Shandong Peninsula: 200, 201
sheetflood: 105, 123
Shine Jinst: 221
Shopfites colchesternsis Zone: 138
Siberia
  ammonoids in deposits: 259
  coastline geometry: 168
  flysch deposits in: 220
  loess deposits in: 69
  in Paleotethys: 259
  paludification in: 145
  peat mires in: 131, 141–145
  reef formation in: 209, 211, 220
  Sea of Okhotsk tides: 151, 152, 168
  sedimentation restrictions: 145
Siberian plate: 216
Sicily: 259
siderite: 231, 237
Sierra Nevada Mountains (USA): 108
Sierras Pampeanas: 12
Sigillaria: 134, 139
silica
  in iron formations: 239–240
  in Precambrian seawater: 239–240
Silica Shale: 213
silicate: 232
silt. *See also* loess
  in alluvial-fan deposits: 117, 120, 123, 125
  formation of: 123
  in Labrador Sea cores: 35–38
  origin of: 53
siltstone: 61, 64, 248–251
Silurian
  basin development during
    Acadian orogeny: 203
    Appalachian-Ouachita system: 192
    Laurasian-Gondwanan system: 195
  glaciation during: 223
  reef formation during: 209, 221, 223
  tectonic processes during: 212, 218
Sino-Korea: 199
sinters: 234
Siwalik Group: 196
Skeidarársandur: 27–28
Slovenia: 218
Smith Mountains: 123
Snake River Plain: 55
Sogwi-p'o: 168
Sohm Abyssal Plain: 186
soil: 65–66
soligenous mire
  definition of: 131
  peatland development from: 127
  plant production in: 142
Solimões Basin: 11, 16
Solomon Sea: 195, 197
Son Basin: 14
Songliao Basin: 81
Songpan-Ganzi Complex
  collision suture belts in: 193
  dating of deposits: 200, 202, 203
  evolution of: 191, 195, 199–203
  size of: 199
  sources of detritus: 202
Sorry Harbor: 159
source potential index: 99
South Africa
  coal bed formation in: 128
  glacigenic deposits in: 14, 16
  iron formations in: 231, 235–237
  lacustrine deposits in: 79
  mudstone deposition in: 12
  quartzarenite deposition in: 12
South America
  basin development: 12
  climate in: 211
  convergent margin on coast: 162
  eolian system in: 43
  glaciation of: 10–16
  glacigenic deposits in: 10–16
  lacustrine deposits in
    in Barremian strata: 92
    Irati Formation: 79, 102
    in Neocomian strata: 92
    during Permian: 91
  loess deposits in: 55, 64, 67
  morainal deposits in: 12
  reef formation in: 211–212, 218
  Straits of Magellan tides: 151, 152, 162
  strata deformation: 12
  tides in: 151, 152, 162, 168
South American plate: 162
South China Block: 199, 221–222
South European Archipelago (SEA): 217–218
South San Clemente Basin: 186
Soviet Union: 231. *See also* Russia; Siberia; specific locations in
Spain: 79, 217–219
Spanish plate: 218
spectrophotometry: 34
Sphagnum moss: 141
Sphenophyllum: 139
Spitsbergen: 259
Spodosols: 139, 144
Springfield Coal
  composition of: 134
  correlations to: 135, 136–138
  depositional environment: 127, 139–140, 143–146
  durains in: 140
  eustatic controls on: 143–144, 145
  mining of: 129
  paleolocation of mire: 142
  palynology of: 134–135, 145
  size of: 127, 132, 140, 142
  thickness of mire: 142
Stanley Group: 198
Stephanian
  Botrychiopsis Zone during: 14
  cyclothem formation during: 18
  dating of deposits from: 16
  Ginkgophyllum Zone during: 14
  glaciation during
    evidence of, isotope levels: 8
    in Gondwana: 5, 10, 14–16
  glacioeustasy in: 18
  Nothorhacopteris Zone during: 14
  palynomorphs in strata from: 16
stilpnomelane: 234
Storrega slide: 186
Straits of Magellan: 151, 152, 162
striations, formation of: 26–27
strike-slip fault zones
  in Furnace Creek: 105
  river drainage in: 193
  in suture zones: 194
Stringocephalus: 220
stromatolites: 234, 237
stromatoporoid reefs: 209, 212–224
Stuart Bay Formation: 212
subduction complexes: 204
Sublepidophloios: 135
submarine canyons: 175–188
submarine hydrothermal vents: 219
successor basins: 10–14
Sudetes: 217
sulfide: 232, 240
sulfur
  in coal: 133, 134
  isotopic ratios in iron formations: 239
  in Phosphoria Rock Complex: 251
Sumatra: 204
Summit Coal: 136–138
Sunda accretionary zone: 196–197
Sunda arc: 196–197
Sunda Trench: 168
Superior-type iron formation: 235, 237–241
Survant Coal: 143

suture zones
accretionary wedges in: 204
evolution of: 194
modeling of: 203–204
river drainage in: 193
swamps: 131, 142
Swan Hills: 214
"Swatch": 185
"Swatch of No Ground": 178, 180–182
Sydney Basin: 14, 16
Sylvania Mountains: 105, 108
Synchysidendron: 134
Synchysidendron resinosum: 139
syntaxes, river drainage at: 193

T

Tadzhikistan: 219, 220
Tafilalt area: 219
Taimyr: 216
Taimyr peninsula: 220
Taiwan: 193, 195
Talaud-Mindanao collision zone: 197
Tamworth Basin: 12–14
Tangjiawan Formation: 222
Tan-Lu strike-slip fault: 200
Taoshuyuan fault: 96
Tapti River: 159
Tarbagati Mountains: 220
Tarija Basin: 12, 16
Tarim Basin: 96, 199, 220
Tarim-Karakorum plate: 219
Tarlong Formation: 94–95, 98
Tas-Khayakhtakh Mountains: 220–221
Tasman Sea: 64
Tasmania: 16
Tastubian: 16
Tatarian: 6
tectonic processes. *See also* convergent margin basin
alluvial fan size factor: 106
in basin development: 12, 87, 192–194
collisional orogenic events: 192–193
cyclothemic deposition from: 130
deposition factor
in Australia: 12–14
in iron formations: 240, 241
worldwide: 2, 3
erosion, impact on: 44
hydrovolcanism: 241
lake size factor: 78–79, 87
lithospheric flexure: 98
mantle plume activity: 241
remnant ocean basin development: 191
in submarine canyon formation: 179
subsidence from: 194
tidal amplification factor: 151
Temple Cap Sandstone: 51
Tennessee: 212
Tepuel Basin: 12
Termit Graben: 79
Tertiary
basin development during: 155
loess deposition during: 69
siltstone formation during: 61
subduction zones: 168
Tesnus Formation: 198
Tethys Ocean: 223
Texas: 8, 191, 259
Thymospora: 134, 135
Thymospora pseudothiessenii Zone: 138
Tian Shan Range
carbonate platforms in: 210
fold belt in: 219, 220
loess deposits in: 60
Precambrian strata in: 92
reef formation in: 219
strike-slip faulting in: 99
Tarlong Formation in: 94–95
uplift of: 98
Tibet
carbonate platform preservation: 210
dating of deposits: 14
glacigenic deposits in: 12, 14
ocean fan systems in: 191, 193, 195
Tibetan Plateau: 191, 193, 195
tides: 151–172
Tim Mersoï Basin: 11
Timan-Pechora region: 216
Titicaca Basin: 11, 12
Toarcian: 49–51
topogenous mire
definition of: 131
peatland development from: 127
plant production in: 142
total organic carbon content: 251–257
Tournaisian
black limestone/shale formation in: 221
cyclothem formation during: 18
fauna, in strata: 14
glaciation during
evidence of, isotope levels: 8
in Gondwana: 5, 10
glacioeustasy in: 18
Protocanites Zone: 14
reef formation during: 220, 223
strata, analysis of: 12
Transantarctic Mountains: 5, 14, 17–18
transform margins: 193
Transhimalayan subduction zone: 196
transtensional deformation: 12
Transvaal Basin: 231, 235–237, 241
Traverse Group: 213
Triassic
basin development during: 200–203
climate during: 50
fluvial activity in: 50
flysch deposits during: 201–203
ocean fan systems from: 191, 195, 199–203
olistostrome deposits from: 201
pelagic carbonate deposits from: 201, 202
reef formation during: 210
rifting during: 155, 165
turbidite deposits from: 201
uplift during: 92–93
tropical: 210
Tubiphytes: 220
Tufts Abyssal Plain: 185
Tully Formation: 213
tundra: 68
Tunisia
eolian sand activity in: 49
loess deposits in: 55, 64
reef formation in: 218
turbidite processes. *See also* flysch
chronostratigraphy of: 194
glaciation marker: 16
in iron formations: 237
in Labrador Sea cores: 38
material distribution in: 2
in outburst floods: 38
in remnant ocean basins: 192, 193
in submarine canyons: 175–188
in suture zones: 194–195
volcanogenic: 92
Turkey: 219
Turpan Basin
dating of deposition: 101
evolution of: 95–98
geological setting of: 92–93
lacustrine deposits in: 91, 92
organic carbon deposits in: 98, 101–102
petrology of: 98–102
stratigraphy of: 93–95
Tuttle alluvial fan: 106, 123
Tuva plate: 221
Tuvaella: 221

U

Udacha Bay: 168
Ulakhan-Tasskoi area: 221
Ultisols: 139, 144
Ungava Bay
formation of: 159
geological setting of: 152
meltwater floods from: 39
tides in: 151, 152, 159
uniformitarianism: 2, 238
United Kingdom (UK)
Bristol Channel tides: 151, 152, 155
climate of: 155
Dee River tides: 155
English Channel tides
comparison of: 151
eastern part of: 162–165
geological setting of: 152
western part of: 159–162
eustatic amplitude in: 7
lacustrine deposits in: 79
Mersey River tides: 155
reef formation in: 209, 214
Ribble River tides: 155
shape of estuaries in: 152
Solway Firth tides: 155
tides in: 151, 152, 155
United States of America. *See also* North America; specific states
cyclothem formation in: 7
eolian system in: 43
eustatic amplitude in: 7–8
flysch deposits in: 191
glaciogenic deposits in: 28, 55
inland seas, tidal analysis of: 152
lacustrine deposits in: 79, 80, 100

morainal deposits in: 28
Navajo Sandstone in: 43, 49–51
outburst flood evidence in: 26, 28
outwash deposits in: 28
reef formation in: 212
silt in rivers: 59
Unur Formation: 221
Upper Freeport coal: 128
Ural Mountains
collisional zone: 220
cyclothem formation in: 7
flysch deposits in: 220
reef formation in: 209, 211, 216–217, 225
Uralian Ocean: 218
Utah
cyclothem formation in: 7
eolian sand deposit in: 43, 49–51
lacustrine deposits in: 79, 80, 100–101
Uzbekistan: 219

V

Vaalbara: 236
Vaigach Island: 216
vanadium: 134
varcus zone: 216, 222
Vasyugan bog complex: 142
Vatnajökull: 59
Verdigris Limestone: 138, 139
Verkhoyan plate: 220–221
Vermont coal seam: 128
Vernal-Brush Creek: 249–251
Vertisols: 139, 144
Victoria Island: 212
Victoria Land: 12, 16
Vietnam: 222
Viséan: 5–8, 14
vitrinite: 127, 133–134, 140
vitrinite reflectance: 98
Volga-Ural Basin: 216
Vosgian-Moldanubian-Ligerian cordillera: 214
Vranica Mountain: 218

W

wackestone: 249–251
Walker Lake Basin: 123
Wallersheimer dolostones: 215
Walshville paleochannel
coal deposits: 132, 136
depositional environment: 140, 146
drainage from: 136
size of: 145
sulfur level, in coal from: 134
Waltman Shale: 79
Warm Springs alluvial fan: 106
Washington
loess deposits in: 55
outburst floods in: 26, 188
Scabland formation: 26
Washingtonville Shale: 138
watershed. *See* erg
Waucoba Mountain: 108
Waulsortian mudmound growth: 224
wave velocity testing: 34
wedge-top basins: 196
West Irian: 222
West Point Limestone: 212
West Virginia: 139
Western Canada Sedimentary Basin: 209
Western Interior Basin
coal bed formation in: 127, 128, 136–139
correlations to: 135, 136–138
cyclothem formation in: 130, 143–144
depositional environment: 144
durains in: 140
paleoclimate of: 144
palynology of: 136–138
sandstone formation in: 49
topography of: 144
Westphalian
Botrychiopsis Zone during: 14
brachiopods in strata from: 14
bryozoans in strata from: 14
clams in strata from: 14
coal bed formation during: 135, 138
crinoids in strata from: 14
cyclothem formation during: 18
Ginkgophyllum Zone during: 14
glaciation during: 5, 10, 16–17
glacigenic deposits during: 12–16
glacioeustasy in: 18
Nothorhacopteris Zone during: 14
Wetheredella: 224
White River Group: 61
White Sea: 165
Whitebreast Coal: 138
Whitehill Formation: 102
Williston Basin: 128
Willow Wash: 108–110, 120
Wilton Creek: 36–37
wind
deposition agent
for dust: 66
for eolian sand: 44–47, 65
for loess: 54, 65, 69
modeling of: 50
surface water movement by: 246
transport capability of
for eolian systems: 44–47
Hadley Cells, impact on: 47
Wingate Sandstone: 50
Winnipegosis Formation: 214
Wisconsinan: 28, 39
Wolfcampian: 198
Wordian: 91, 259
wrench tectonics: 12
Wyodak-Anderson coal zone: 128
Wyoming
lacustrine deposits in: 79, 80, 100–101
phosphate deposits in: 245–261

X

xenoliths, carbonate: 212
Xinjiang: 220

Y

Yakutat Block: 185
Yakutia region: 56
Yangtze craton: 199
Yangtze River: 168
yardangs: 27, 29. *See also* dunes
Yellow Sea: 151, 152, 168–172
Yierxitu Formation: 94
Yili Basin: 99
Yucatan: 197, 210
Yugoslavia: 218
Yukon: 55, 212
Yukon River: 181–182, 187

Z

Zagros zone: 196
Zaire Canyon: 185
Zaire Fan: 185
Zaobishan Formation: 94
Zemmour: 218
Zhemchung Canyon
vs. Bengal Fan Channel: 181
vs. Bering Canyon: 181
description of: 178–180
formation of: 179, 181–182, 186
vs. Grand Canyon: 178
vs. Navarin Canyon: 181
size of: 175, 178, 186, 188
Zimbabwe Craton: 237
zinc: 134
Zion National Park: 49, 50
zircon: 236–237
Zona de Intervalo: 14
zooxanthellate hermatypic corals: 210, 223